Carotenoids

Physical, Chemical, and Biological Functions and Properties

Edited by

John T. Landrum

CRC Press
Taylor & Francis Group
Boca Raton London New York

CRC Press is an imprint of the
Taylor & Francis Group, an **informa** business

CRC Press
Taylor & Francis Group
6000 Broken Sound Parkway NW, Suite 300
Boca Raton, FL 33487-2742

Printed in the United States of America on acid-free paper
10 9 8 7 6 5 4 3 2 1

International Standard Book Number: 978-1-4200-5230-5 (Hardback)

Library of Congress Cataloging-in-Publication Data

Carotenoids : physical, chemical, and biological functions and properties / editor, John T. Landrum.
p. cm.
Includes bibliographical references and index.
ISBN 978-1-4200-5230-5 (hardcover : alk. paper)
1. Carotenoids. I. Landrum, John Thomas. II. Title.

QP671.C35C376 2010
612.4'9--dc22 2009036998

Visit the Taylor & Francis Web site at
http://www.taylorandfrancis.com

and the CRC Press Web site at
http://www.crcpress.com

This book is dedicated to my wife, Eileen,
who is always loving, encouraging, and understanding;
and to my children, James, Elizabeth, and Jeffrey.

Contents

PART III Applications of Spectroscopic Methodologies to Carotenoid Systems

PART IV Chemical Breakdown of Carotenoids In Vitro and In Vivo

PART V Antioxidant and Photoprotection Functions and Reactions Involving Singlet Oxygen and Reactive Oxygen Species

Foreword

CAROTENOIDS: A COLORFUL AND TIMELY RESEARCH FIELD

For those readers who are less familiar with this fascinating field of research, it is worth introducing a few key concepts about carotenoids. There are over 600 fully characterized, naturally occurring molecular species belonging to this class of essential pigments. Carotenoid biosynthesis occurs only in bacteria, fungi, and plants where they have established functions that include their role as antenna in the light-harvesting proteins of photosynthesis, their ability to regulate light–energy conversion in photosynthesis, their ability to protect the plant from reactive oxygen species, and coloration. If these were the only known functions/properties of carotenoids in the natural world, it would be adequate; but these molecules are also part of the diet in higher species, and in animals and humans, carotenoids assume a completely different set of important functions/actions. In humans, some carotenoids (the provitamin A carotenoids) are best known for converting enzymatically into vitamin A; diseases resulting from vitamin A deficiency remain among the most significant nutritional challenges worldwide. Carotenoids serve a number of other roles in the animal kingdom including in the coloration of plumage in birds, which has now been recognized to play a significant role in the selection of mates. In humans, the role that carotenoids play in protecting those tissues that are most heavily exposed to light (e.g., photoprotection of the skin, protection of the central retina) is perhaps most evident, while other potential roles for carotenoids in the prevention of chronic diseases are still being investigated. Because carotenoids are widely consumed and their consumption is a modifiable health behavior (via diets or supplements), health benefits for chronic disease prevention, if real, could be very significant for public health.

This book on carotenoids spans the breadth of ongoing work by researchers around the world, ranging from basic studies to advanced applied biomedical research.

As in many fields of research, new tools and techniques for measuring carotenoids in various systems are critical to support research progress. Several chapters discuss new methodologies to measure carotenoids (see Chapter 4), carotenoid metabolites/radicals (see Chapter 9), or carotenoids *in vivo* in complex biological systems, especially in the human eye (e.g., see Chapters 5 and 6). Other chapters describe the oxygenase enzymes that are essential components of carotenoid metabolism to active metabolites (see Chapter 19). The study of active metabolites includes the in-depth evaluation of carotenoid cleavage products (see Chapter 11) and carotenoid radicals (see Chapter 14) that may account for some of the biological actions observed for these unique substances.

Carotenoids are highly lipophilic; an active area of research concerns how carotenoids interact with and affect membrane systems (see Chapters 2 and 10). Also, the lipid solubility of these compounds has important implications for carotenoid intestinal absorption (see Chapter 17); models such as the Caco-2 cell model are being used to conduct detailed studies of carotenoid absorption/competition for absorption (Chapter 18). The lipid solubility of these carotenoids also leads to the aggregation of carotenoids (see Chapter 3). Carotenoids aggregate both in natural and artificial systems, with implications for carotenoid excited states (see Chapter 8). This has implications for a new indication for carotenoids, namely, serving as potential materials for harnessing solar energy.

The hydrophobicity of these compounds requires protein binding to move carotenoids through aqueous environments; an emerging area of research includes the identification of carotenoid transport proteins that determine, in part, carotenoid tissue concentrations. As carotenoids are found throughout nature, various models can be studied; for example, Chapter 24 describes carotenoid transport in silkworms, while Chapter 25 uses the monarch butterfly larvae to evaluate carotenoid accumulation/coloration.

The aforementioned chapters provide an excellent overview of carotenoid absorption/metabolism/transport, while the other chapters provide detailed analyses of either selected carotenoids or selected functions/actions. Coloration remains an important (and pleasing) function of carotenoids in nature; Chapter 23 describes various avian species that control coloration via the incorporation of carotenoids, and discusses the functions of this coloration in these birds (e.g., sexual signaling). Perhaps the most well-known function of carotenoids in nature is the critical role they play in photosynthesis (see Chapter 7). Ongoing research is revealing the details of the intra- and intermolecular mechanisms of light sensing, signal propagation, and energy dissipation, both in plants and in cyanobacteria (see Chapter 1), also known as blue-green algae, which have an elaborate membrane system that functions in photosynthesis.

Photoprotection and the potential for the prevention of diseases of the eye by carotenoids continue to be active areas of investigation. Chapter 13 comprehensively describes the rationale for a functional role of lutein/zeaxanthin in the human and primate macula, including supporting evidence from epidemiological studies that the higher consumption of these two carotenoids is associated with a lower risk of age-related macular degeneration. Newer evidence suggests that these two carotenoids are critical in maintaining retinal pigment epithelial cell health (see Chapter 15). Chapter 16 discusses the potential antioxidant role of lutein/zeaxanthin in the photoreceptor outer segment, noting that A2PE photooxidation is inhibited in the presence of these carotenoids.

Cancer is another chronic disease for which carotenoids have been evaluated for their efficacy in the prevention of disease. Chapter 22 summarizes the growing list of molecular pathways involved in cell proliferation, differentiation, and apoptosis that are thought to be modulated by various carotenoids. The carotenoid lycopene, in particular, is being studied for a potential role it may play in the prevention of prostate cancer. Chapter 21 reviews the current state of this literature, including mechanistic studies, and notes that lycopene is typically encountered along with numerous poorly characterized metabolites, complicating both the study and the interpretation of studies, even in cell systems. Chapter 20 expands upon this with a detailed discussion of the biological cleavage of lycopene into apo-lycopenoid compounds. These latter compounds may affect several key signaling pathways and molecular targets for carcinogenesis. Thus, much work is needed to better understand the potential role of lycopene/apo-lycopenoid compounds in the prevention of cancer.

In summary, the amazing breadth and depth of research in carotenoids are reasons why it draws investigators are drawn to this fascinating field of research. The research spans the continuum, from detailed studies of the roles of photoprotective carotenoids in plants to the potential application in the prevention of disease in humans. This is translational research at its best and I commend the editor, Dr. John Landrum, for assembling such an interesting and informative collection of current research.

Susan T. Mayne
Yale University School of Medicine

Editor

John T. Landrum, PhD, is a professor in the Department of Chemistry and Biochemistry at Florida International University (FIU). In addition to this, he serves as a director at the Office of Pre-Health Professions Advising for the College of Arts and Sciences. He joined the faculty of FIU in August 1980.

Dr. Landrum received his BSc in chemistry (*cum laude*) from California State University, Long Beach, California, in 1975. He completed his thesis ("The cooperative binding of oxygen by hemocyanin") and was awarded his MSc in chemistry in 1978, also from the California State University, Long Beach. In 1980, he received his PhD in chemistry from the University of Southern California (USC). He was recognized by USC for his graduate research in 1978 and was awarded the USC Graduate Research Award for Outstanding Research. His PhD dissertation ("Synthetic models toward cytochrome c oxidase") used small molecular models to investigate the structural and magnetic properties of porphyrin complexes to provide fundamental insight into the possible structures of the two copper, two iron active site of the terminal electron acceptor of the electron transport chain.

A faculty member at a young and developing university, Dr. Landrum has taught courses at all academic levels within the Department of Chemistry and Biochemistry and was honored with an Excellence in Teaching Award in 1991. He was instrumental in establishing a master of science degree program in chemistry at FIU and served as the first graduate program director (1987–1992). Dr. Landrum served as an associate dean of the FIU graduate school between 2006 and 2007. In 2008, he was invited to assume his current position as a director of the College of Arts and Sciences' Office of Pre-Health Professions Advising. After arriving at FIU, he established an active research program involving undergraduates and focused initially on the investigation of porphyrin metal complexes as models for the biological function of transition metals in natural systems. His interest in carotenoids and their functions in biological systems was triggered by a collaboration with Dr. Richard A. Bone (Department of Physics, FIU), which began in the early 1980s and led to the first definitive characterization of the human macular pigment.

His research efforts over the last 25 years have been primarily devoted to understanding the nature of the carotenoids present in the human macula, including their identity, distribution, transport, and metabolism. Over this period, he and his collaborators have shown that the macular pigment is composed of the carotenoids lutein, zeaxanthin, and meso-zeaxanthin. He has been able to demonstrate that these carotenoids have a protective function within the retina. They reduce the risk of age-related macular degeneration, which is the leading cause of vision loss among adults. Dr. Landrum's research has shown that dietary supplements of these carotenoids can increase pigmentation. His current research efforts are focused on understanding the mechanisms of biological recognition of individual carotenoids, their absorption and transport, and their role in the developing human eye.

In 2004, Dr. Landrum's research contributions in the field of chemistry were recognized by presentation of an Excellence in Research Award at FIU. Since becoming a faculty member at FIU he has been awarded 26 grants in support of his research efforts. He has directed the research of 15 graduate and over 100 undergraduate students, and has authored or coauthored 58 articles in peer-reviewed journals and books. He has become a frequent speaker and has been invited to present his research to audiences at 34 major international conferences and symposia since the early 1990s. In 2004, he served as a vice-chairman for the Gordon Research Conference on Carotenoids, and in 2007 he served as a chairman for this prestigious conference. He served as a chairman for the Macula and Nutrition Group (2000–2004), as a chairman (2008) and steering committee member of the Carotenoid Interactive Research Group (2002–2008), as a council member and treasurer for the International Carotenoid Society (2005–present), and as an associate editor for the *Archives of Biochemistry and Biophysics*. He has also served as an editor or coeditor for several special editions on the current progress in the field of carotenoid research for the journal *Archives of Biochemistry and Biophysics*.

Dr. Landrum is a member of the American Chemical Society; the Association for Research in Vision and Ophthalmology; the International Carotenoid Society (founding member); the Carotenoid Interactive Research Group; the International Research Society, Sigma Xi; the Macula and Nutrition Group (founding member); the American Society for Nutrition; and the Optometric Nutrition Society.

Contributors

Klaus Albert
Institute of Organic Chemistry
University of Tuebingen
Tuebingen, Germany

Maxime Alexandre
Department of Biophysics
Faculty of Sciences
Vrije Universiteit
Amsterdam, the Netherlands

Francesca Alvarez-Calderon
Department of Chemistry and Biochemistry
Florida International University
Miami, Florida

Paul S. Bernstein
Moran Eye Center
University of Utah
School of Medicine
Salt Lake City, Utah

Jonathan D. Blount
Centre for Ecology and Conservation
School of Biosciences
University of Exeter
Cornwall Campus, United Kingdom

Richard A. Bone
Department of Physics
Florida International University
Miami, Florida

Patrick Borel
Lipidic Nutrients and Prevention of Metabolic Diseases Unit
INRA, INSERM, Université de Aix-Marseille
Marseille, France

Claudio D. Borsarelli
Instituto de Química del Noroeste Argentino (INQUINO–CONICET)
Facultad de Agronomía y Agroindustria
Universidad Nacional de Santiago del Estero
Santiago del Estero, Argentina

Phyllis E. Bowen
Department of Kinesiology and Nutrition
University of Illinois at Chicago
Chicago, Illinois

Derick Callejas
Department of Chemistry and Biochemistry
Florida International University
Miami, Florida

Catherine Caris-Veyrat
Safety and Quality of Plant Products
INRA, Avignon University
Avignon, France

Assunta Catalano
Institute of General Pathology
Catholic University
Rome, Italy

Ruth Edge
School of Chemistry
The University of Manchester
Manchester, United Kingdom

Igor V. Ermakov
Department of Physics and Astronomy
University of Utah
Salt Lake City, Utah

Ligia Focsan
Department of Chemistry
The University of Alabama
Tuscaloosa, Alabama

Werner Gellermann
Department of Physics and Astronomy
University of Utah
Salt Lake City, Utah

Wieslaw I. Gruszecki
Department of Biophysics
Institute of Physics
Maria Curie-Sklodowska University
Lublin, Poland

Earl H. Harrison
Department of Human Nutrition
The Ohio State University
Columbus, Ohio

Karsten Holtin
Institute of Organic Chemistry
University of Tuebingen
Tuebingen, Germany

Cheryl A. Kerfeld
United States Department of Energy
Joint Genome Institute
Walnut Creek, California

and

Department of Plant and Microbial Biology
University of California
Berkeley, California

So Ra Kim
Department of Visual Optics
Seoul National University of Technology
Seoul, South Korea

Diana Kirilovsky
Commissariat à l'Energie Atomique
Institut de Biologie et Technologies de Saclay
Gif sur Yvette, France

and

Centre National de la Recherche Scientifique
Gif sur Yvette, France

Lowell D. Kispert
Department of Chemistry
The University of Alabama
Tuscaloosa, Alabama

Tatyana Konovalova
Department of Chemistry
The University of Alabama
Tuscaloosa, Alabama

John T. Landrum
Department of Chemistry and Biochemistry
Florida International University
Miami, Florida

Samuel F. Lockwood
Baselodge Group
Austin, Texas

Erin K. Marasco
Department of Biochemistry, Molecular Biology and Biophysics
University of Minnesota
Minneapolis, Minnesota

Kevin J. McGraw
School of Life Sciences
Arizona State University
Tempe, Arizona

Jonathan R. Mein
Nutrition and Cancer Biology Laboratory
Jean Mayer USDA Human Nutrition Research Center on Aging
Tufts University
Boston, Massachusetts

Adriana Z. Mercadante
Department of Food Science
Faculty of Food Engineering
University of Campinas
Campinas, Brazil

Paola Palozza
Institute of General Pathology
Catholic University
Rome, Italy

Vassilia Partali
Department of Chemistry
Norwegian University of Science and Technology
Trondheim, Norway

Tomáš Polívka
Institute of Physical Biology
University of South Bohemia
Nové Hrady, Czech Republic

and

Institute of Plant Molecular Biology
Biological Centre
Czech Academy of Sciences
České Budějovice, Czech Republic

Emmanuelle Reboul
Lipidic Nutrients and Prevention of Metabolic Diseases Unit
INRA, INSERM, Université de Aix-Marseille
Marseille, France

Małgorzata Różanowska
School of Optometry and Vision Sciences
Cardiff Vision Institute
Cardiff University
Cardiff, United Kingdom

Bartosz Różanowski
Department of Cytology and Genetics
Institute of Biology
Pedagogical University
Krakow, Poland

Alexander V. Ruban
School of Biological and Chemical Sciences
Queen Mary University of London
London, United Kingdom

Takashi Sakudoh
Division of Radiological Protection and Biology
National Institute of Infectious Diseases
Tokyo, Japan

Wolfgang Schalch
DSM Nutritional Products Ltd.
Kaiseraugst, Switzerland

Claudia Schmidt-Dannert
Department of Biochemistry, Molecular Biology and Biophysics
University of Minnesota
Minneapolis, Minnesota

Mohsen Sharifzadeh
Department of Physics and Astronomy
University of Utah
Salt Lake City, Utah

Rossella Simone
Institute of General Pathology
Catholic University
Rome, Italy

Hans-Richard Sliwka
Department of Chemistry
Norwegian University of Science and Technology
Trondheim, Norway

Janet R. Sparrow
Department of Ophthalmology
Columbia University
New York, New York

Witold K. Subczynski
Department of Biophysics
Medical College of Wisconsin
Milwaukee, Wisconsin

George Truscott
School of Physical and Geographical Sciences
Keele University
Staffordshire, United Kingdom

Kozo Tsuchida
Division of Radiological Protection and Biology
National Institute of Infectious Diseases
Tokyo, Japan

Xiang-Dong Wang
Nutrition and Cancer Biology Laboratory
Jean Mayer USDA Human Nutrition Research Center on Aging
Tufts University
Boston, Massachusetts

Justyna Widomska
Department of Plant Physiology and Biochemistry
Faculty of Biochemistry, Biophysics and Biotechnology
Jagiellonian University
Krakow, Poland

Part I

The Structural Properties, Characteristics, and Interactions of Carotenoids

1 The Orange Carotenoid Protein of Cyanobacteria

Cheryl A. Kerfeld, Maxime Alexandre, and Diana Kirilovsky

CONTENTS

1.1 INTRODUCTION

Molecular, spectroscopic, and functional genomics studies have demonstrated the remarkable similarity among the components of the photosynthetic machinery of cyanobacteria, algae, and plants. These organisms also share the need to balance the collection of energy for photosynthesis with the threat of photodestruction. Carotenoids are central to attaining this balance.

The photoprotective processes of photosynthetic organisms involving the dissipation as heat of the excess of absorbed energy in the antenna of the photosystem II are collectively known as nonphotochemical quenching (NPQ). In this mechanism, there is a decrease in the amount of energy funneled to the reaction center (RC) with a concomitant reduction in the amount of the reactive oxygen species generated. NPQ is well characterized in plants (Demmig-Adams 1990, Horton et al. 1996, Niyogi 1999, Muller et al. 2001). It relies on the same components used for light harvesting in photosynthesis. The absorption of light is accomplished by light-harvesting complexes (LHCs) that surround RCs; a RC and its LHC together form a photosystem (PS). There are two PSs in organisms that carry out oxygenic photosynthesis, PSI and PSII. In eukaryotic PSs, the RCs and LHCs are integral membrane pigment protein complexes located in the thylakoid membranes. The carotenoids in these complexes are thought to provide structural stability and act as accessory light-harvesting pigments as well as mediate photoprotection. In plants, the carotenoid-based photoprotection in PSII is triggered by acidification of the thylakoid lumen under saturating light conditions (Demmig-Adams 1990, Horton et al. 1996, Niyogi 1999, Muller et al. 2001). The drop of the lumen pH induces the interconversion of specific LHC carotenoids (Yamamoto 1979, Gilmore and Yamamoto 1993) and the protonation of a PSII subunit (PsbS), a member of the LHC superfamily (Li et al. 2000, 2004). This process also involves conformational changes in the LHCII, modifying the interaction between chlorophylls and carotenoids (Ruban et al. 1992, 2007, Pascal et al. 2005). This thermal energy dissipation is accompanied by a decrease of PSII-related fluorescence emission, known as high-energy quenching (qE), one of the NPQ processes.

In contrast to our understanding of NPQ processes in plants, until recently, relatively little was known about the mechanisms of photoprotection in cyanobacteria. Yet it is an important feature of these organisms' lifestyles. The cyanobacteria as a group differ from the eukaryotic photosynthetic organisms in their ability to thrive in a wide range of extreme habitats, many characterized by temperature extremes, high salinity, and drought conditions that exacerbate the threat of photodamage. Many cyanobacteria are known to be UV-B tolerant, perhaps through vestiges of molecular adaptations that arose during several billion years of intense UV radiation before the formation of the earth's protective ozone layer.

There is a fundamental difference between the LHCs of the cyanobacteria and those of eukaryotic photosynthetic organisms. In contrast to the integral membrane pigment (chlorophylls and carotenoid) protein LHCs of plants, the main cyanobacterial (with the exception of the prochlorophytes) light-harvesting antenna, the phycobilisome, has a very different architecture. Instead of transmembrane LHCs, the cyanobacterial phycobilisome consists of soluble phycobiliproteins and linker proteins that form a complex (core and rods) attached to the outer surface of thylakoid membranes. The phycobilisome is devoid of intrinsic carotenoids. The rod pigments (principally phycocyanin and phycoerythrin) transfer the absorbed energy to the allophycocyanin core, which contains two terminal energy acceptors, L_{CM} and $APC\alpha^B$ (MacColl 1998, Adir 2005). The energy is transferred then to the chlorophylls of the inner chlorophyll antenna and to RCII. Phycobilisomes can also transfer energy to PSI (Mullineaux 1992, Rakhimberdieva et al. 2001).

Despite their absence in phycobilisomes, carotenoids, especially the so-called secondary carotenoids such as echinenone, were presumed to play a role in cyanobacterial photoprotection. Indeed, classic biochemical approaches have led to several reports of cyanobacterial carotenoid-proteins and evidence for their photoprotective function (Kerfeld et al. 2003, Kerfeld 2004b). One of these, the water soluble orange carotenoid protein (OCP), has been structurally characterized and has recently emerged as a key player in cyanobacterial photoprotection.

The OCP was first described by David Krogmann more than 25 years ago (Holt and Krogmann 1981). Highly conserved homologs of the 34 kDa OCP are found in most cyanobacteria for which genomic data are available, as shown in Table 1.1. The genomic context of the OCP gene varies considerably, as shown in Figure 1.1. In some of the marine *Synechococcus* species there is some conservation among the putative coding sequences in the vicinity of the OCP gene; homologs of a putative β-carotene ketolase flank the OCP, followed by a homolog of a conserved hypothetical protein (*slr1964* in *Synechocystis* PCC6803), which is present and adjacent to the OCP in most cyanobacterial genomes (see Table 1.1 and Figure 1.1). This small protein (106–134 amino acids), is of unknown function. A global yeast two-hybrid analysis in *Synechocystis* PCC6803 neither links the OCP and *slr1964* gene product functionally (Sato et al. 2007) nor does this screen of protein–protein interactions offer insight into the function of the OCP. Instead, our understanding of the function of the OCP is based on molecular, genetic, and spectroscopic approaches complemented by structural biology.

1.2 RECENT STUDIES ON THE FUNCTION OF THE OCP

In contrast to the photosynthetic eukaryotes, photoprotection in cyanobacteria is not induced by the presence of a transthylakoid ΔpH or the excitation pressure on PSII. Instead, intense blue–green light (400–550 nm) induces a quenching of PSII fluorescence that is reversible in minutes even in the presence of translation inhibitors (El Bissati et al. 2000). Fluorescence spectra measurements and the study of the NPQ mechanism in phycobilisome- and PSII-mutants of the cyanobacterium *Synechocystis* PCC6803 indicate that this mechanism involves a specific decrease of the fluorescence emission of the phycobilisomes and a decrease of the energy transfer from the phycobilisomes to the RCs (Scott et al. 2006, Wilson et al. 2006). The site of the quenching appears to be the core of the phycobilisome (Scott et al. 2006, Wilson et al. 2006, Rakhimberdieva et al. 2007b).

TABLE 1.1
Occurrence of the OCP, Its Paralogs, and Co-Occurring Conserved Hypothetical Protein

Organism	*slr 1964*	OCP	OCP N-ter	OCP C-ter
Synechococcus CC9902	syncc9902_0971	syncc9902_0973		
Crocosphaera watsonii WH 8501			CWATdraft_0985	CWATdraft_5349
Lyngbya sp PCC8106	L8106_29205	L8106_29210	L8106_0668 L8106_29395 L8106_04666	 L8106_29390
Synechococcus sp BL107	BL107_14115	BL107_14105		
Synechococcus sp WH7805	WH7805_01192	WH7805_01202		
Nostoc sp PCC7120	All3148	All3149	All1123 Alr4783 All4941 All3221	All4940
Synechococcus WH7803	synwh7803_0927	synwh7803_0929		
Synechococcus WH5701	WH5701_04000	WH5701_04010 WH5701_00210 (219 a a)		
Synechococcus WH8102	SYNW1369	SYNW1367		
Anabaena varaibilis ATCC29413	Ava_3842	Ava_3843	Ava_2052	Ava_2231 Ava_2230 Ava_4694
Synechococcus CC9311	Sync_1805	Sync_1803		
Synechococcus RS9917	RS9917_00682	RS9917_00692		
Cyanothece CCY0110	CY0110_09682	CY0110_09677	CY0110_08696	CY0110_8806
Synechococcus RCC307	SynRCC307_1994	RCC307_1992		
Nostoc punctiforme PCC73102		NpR5144	NpF5133 NpR0404[a] NpF5913[a] NpR5130 NpF6243	 NpF6242[a]
Nodularia spumigena CCY9414		N9414_13085	N9414_12098 N9414_22258	N9414_22253
Gloeobacter violaceus PCC7421		glr0050 glr3935 (274)	gll0259 gll0260 (217)	gll2503
Thermosynechococcus elongates BP-1			tll1269	tll1268
Acaroychloris marina		AMI_5842		

[a] Known to be expressed by proteomic analysis.

Rakhimberdieva (Rakhimberdieva et al. 2004) showed that the action spectrum for the phycobilisome fluorescence quenching resembled the absorption spectrum of cyanobacterial carotenoids. Subsequently, it was demonstrated that the blue-light responsive carotenoid was associated with a protein that had been structurally characterized, but of unknown function—the OCP (Wilson et al. 2006). In the absence of the OCP, the NPQ induced by strong white or blue–green light in *Synechocystis* PCC6803 cells was completely inhibited and, as a consequence, the cells were more sensitive to light stress. Moreover, the action spectrum of the cyanobacterial

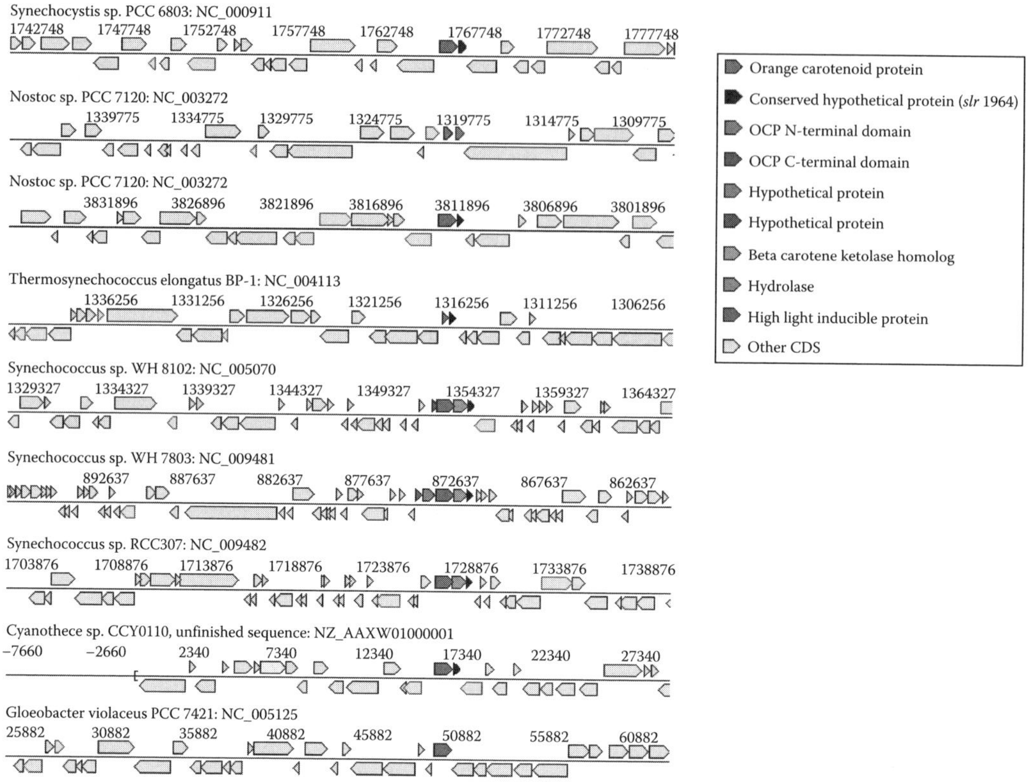

FIGURE 1.1 Representative ortholog neighborhoods for the OCP and OCP N-terminal paralogs. Arrowhead length is approximately proportional to gene length. (Adapted from Integrated Microbial Genomes, http://img.jgi.doe.gov/cgi-bin/pub/main.cgi.)

NPQ (Rakhimberdieva et al. 2004) exactly matches the absorption spectrum of the carotenoid, 3′-hydroxyechinenone (Polivka et al. 2005) in the OCP. The OCP is now known to be specifically involved in the phycobilisome-associated NPQ and not in other mechanisms affecting the levels of fluorescence such as state transitions or D1 damage (Wilson et al. 2006). Studies by immunogold labeling and electron microscopy showed that most of the OCP is present in the interthylakoid cytoplasmic region, on the phycobilisome side of the membrane, Figure 1.2 (Wilson et al. 2006). The existence of an interaction between the OCP and the phycobilisomes and thylakoids was supported by the co-isolation of the OCP with the phycobilisome-associated membrane fraction (Wilson et al. 2006, 2007).

In *Synechocystis* PCC6803 the OCP is constitutively expressed, present even in mutants lacking phycobilisomes (Wilson et al. 2007). Stress conditions (high light, salt stress, iron starvation) increases the levels of OCP transcripts and proteins (Hihara et al. 2001, Kanesaki et al. 2002, Fulda et al. 2006, Wilson et al. 2007). Under iron-starvation conditions, blue light also induces a large reversible fluorescence quenching much greater than in the presence of iron (Cadoret et al. 2004, Bailey et al. 2005, Joshua et al. 2005). It was proposed that the IsiA protein (iron-stress-induced protein), a chlorophyll-binding protein, was essential in this NPQ process. However, using *Synechocystis* PCC6803 mutants lacking IsiA, the OCP or phycobilisomes, it has been recently demonstrated that in iron-starved cells (as in iron-containing cells), the blue-light-induced fluorescence quenching is associated with the phycobilisomes and with the OCP and not with IsiA (Rakhimberdieva et al. 2007b, Wilson et al. 2007). In the ΔIsiA mutant a large reversible fluorescence quenching was always induced by blue light. Moreover, during iron starvation the increase

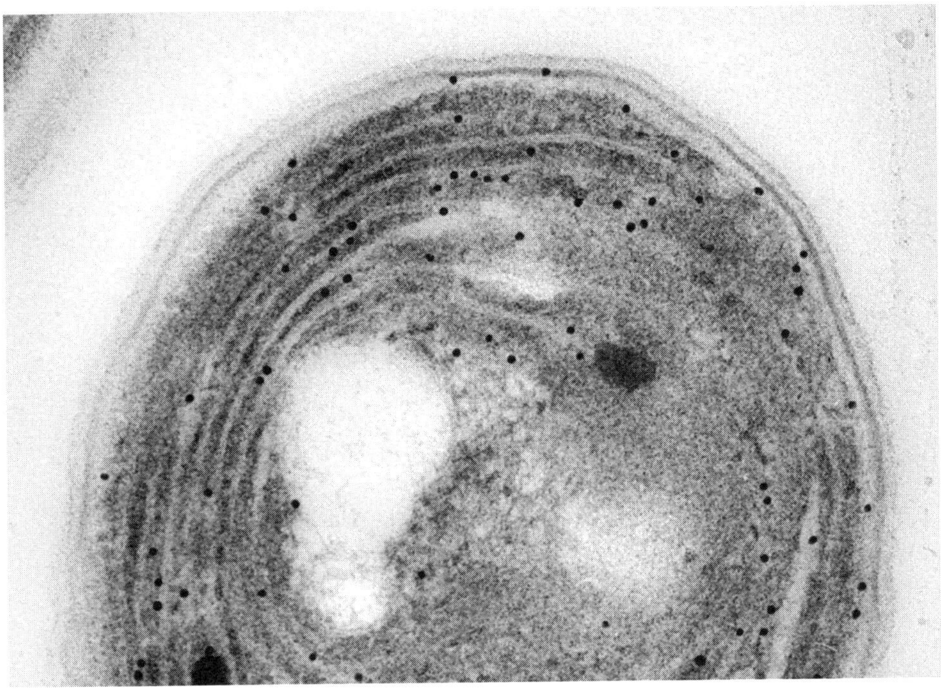

FIGURE 1.2 *In situ* localization of the OCP–green fluorescence protein (GFP) fusion protein: Immunogold labeling of a thin section of OCP–GFP transformed *Synechocystis* PCC6803; OCP–GFP cells were labeled with a polyclonal antibody against the GFP coupled to 10 nm gold particles. Bar = 0.5 μm.

in fluorescence quenching was faster in ΔIsiA cells than in WT cells. This is explained by the relationship between the quenching of fluorescence and the concentration of the OCP: In iron-starved WT *Synechocystis* PCC6803 cells, the concentration of the OCP is higher than in the presence of iron, and in iron-starved ΔIsiA cells the concentration is even higher (Wilson et al. 2007). In all cyanobacterial strains containing OCP-like genes that have been tested, the full-length OCP is present and the NPQ mechanism is induced by blue light, suggesting that this photoprotective mechanism is widespread in cyanobacteria (Boulay et al. 2008a). Additional details about this blue-light-induced NPQ mechanism are described in Karapetyan 2007, Kirilovsky 2007, Bailey and Grossman 2008.

1.3 THE OCP: PRIMARY TO QUATERNARY STRUCTURE

The crystal structure of the OCP from *Arthrospira maxima* has been solved to 2.1 Å resolution (Kerfeld et al. 2003). It is composed of two domains and the carotenoid, 3′-hydroxyechinenone, spans both. The carotenoid is almost completely buried within the protein; only 3.4% of the pigment surface is accessible to solvent (see Figure 1.3a). The OCP is a dimer in solution; the intermolecular interactions are largely mediated by hydrogen bonding among the N-terminal 30 amino acids, as shown in Figure 1.3b

The N-terminal domain of the OCP is an orthogonal alpha-helical bundle, subdivided into two four-helix bundles (Figure 1.3a and c). These subdomains are composed of discontinuous segments of the polypeptide chain (gray and white in Figure 1.3c). To date, the OCP N-terminal domain is the only known protein structure with this particular fold (Pfam 09150). The hydroxyl terminus of the 3′-hydroxyechinenone is nestled between the two bundles. The C-terminal domain (dark

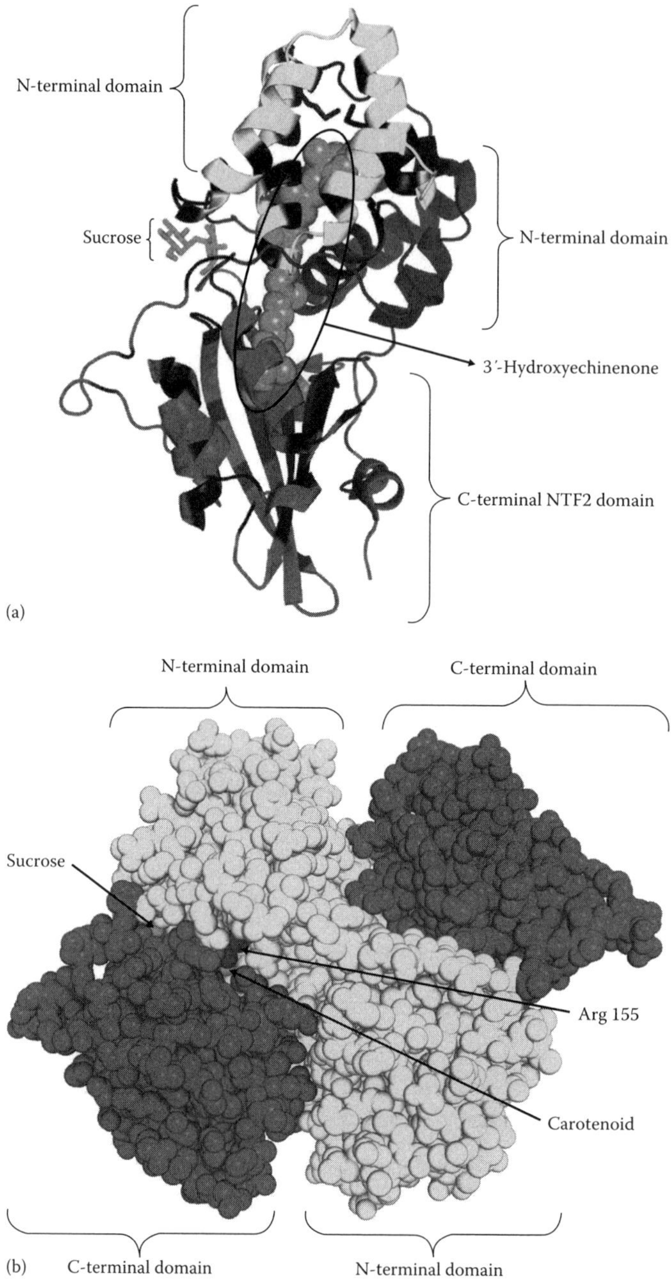

FIGURE 1.3 (See color insert following page 336.) The structure of the OCP. (a) Ribbon diagram of the *A. maxima* OCP structure. The two helical bundles making up the N-terminal domain are uppermost; the C-terminal NTF2 domain is shown in red. The 3′-hydroxyechinenone molecule is shown in space-filling representation and the sucrose molecule and the side chains of conserved Met residues are shown in sticks. Absolutely conserved amino acids are shown in black. (b) The OCP dimer is shown in space filling to emphasize cavities and protuberances. The N-terminal domain is light gray, the C-terminal domain is dark gray. The carotenoid, Arg 155, and sucrose molecule are visible for the left monomer of the dimer. The view is oriented similar to the left OCP monomer in (a). (c) Connectivity of the N-terminal domain of the OCP. (Shading as in (a); tubes correspond to alpha-helices; arrows, beta-strands; the amino acid numbers comprising each element of secondary structure are indicated). (d) Connectivity of the C-terminal domain of the OCP (Shading as in (a); tubes correspond to alpha-helices; arrows, beta-strands; the amino acids comprising each element of secondary structure are indicated). (Created using Pymol, http://www.pymol.org.)

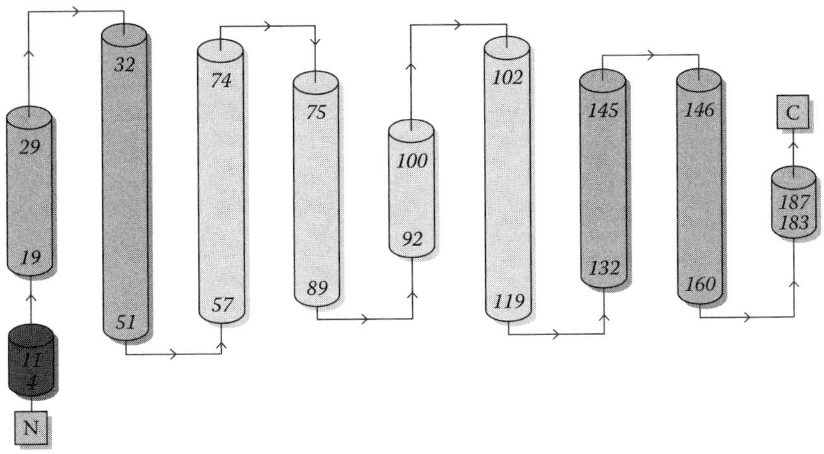

(c)

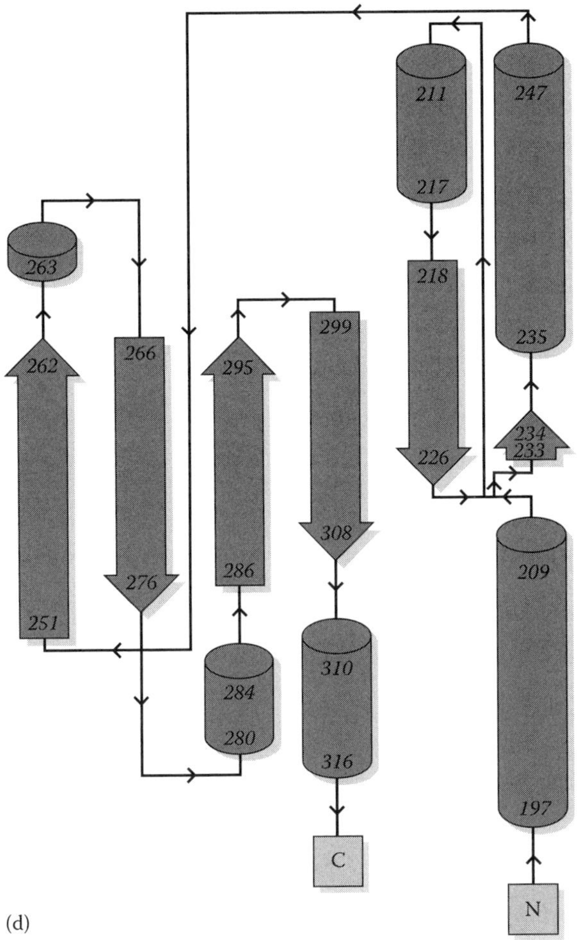

(d)

FIGURE 1.3 (continued)

gray in Figure 1.3a through d) is a member of the nuclear transport factor II (NTF2; Pfam 02136) superfamily, a group of α/β folds that form a five-stranded beta-sheet with a deep hydrophobic pocket. In addition to nuclear transport factors, other proteins containing this domain include enzymes such as the NTF2-like delta5-3-ketosteroid isomerases and other light-responsive signaling proteins, discussed below.

In *Thermosynechococcus elongatus*, the two domains of the OCP occur as separate but adjacent genes (and appear to be coordinately controlled) (Kucho et al. 2004), suggesting that in the evolutionary history of the OCP, a gene fusion occurred (Figure 1.1). Likewise, in *Crocosphaera watsonii*, there is no full-length OCP gene; single copies of the genes for the N- and C-termini are present, but they are in different parts of the chromosome. Other organisms contain, in addition to a full-length OCP gene, separate genes for the domains and/or various combinations of shorter paralogs, as shown in Table 1.1. Several cyanobacterial genomes have multiple copies of genes for the N-terminal domain and a single copy of the gene for the C-terminal domain (Table 1.1), located in disparate parts of the genome. This suggests that in some organisms, full-length OCPs may be assembled from smaller proteins. These putative modular full-length OCPs, containing a unique C-terminus combined with different N-terminal domains, is reminiscent of the modular assembly of light oxygen voltage (LOV) domain-containing proteins. Among the different kingdoms of life, LOV domain serves as an input light-sensing domain connected to very diverse functional groups (Briggs 2007). By analogy, this suggests that in the OCP, the conserved C-terminal NTF2 domain could serve as the input through which the signal is propagated to the different N-terminal modules.

In addition, in some organisms, multiple paralogs for only the N-terminal domain are scattered throughout the genome. There are several lines of evidence to suggest that these are playing a functional role: In *Nostoc punctiforme* several of the N-terminal paralogs are known to be expressed, Table 1.1 (Anderson et al. 2006). Krogmann and his colleagues (Holt and Krogmann 1981, Wu and Krogmann 1997, Knutson 1998) have isolated what appears to be a functional homolog of the N-terminal domain of the OCP. This protein appears red; the absorbance maximum is at 505 nm instead of 495 nm as in the OCP. This red carotenoid protein (RCP) from cell extracts of several cyanobacterial species including *Synechocystis* PCC6803 was assumed to be a proteolytic fragment of the OCP. A 16 kDa RCP can be generated by proteolysis *in vitro* (Kerfeld, unpublished). Based on the structure of the OCP, removal of the NTF2 domain would render the carotenoid exposed to solvent in the 16 kDa RCP; more likely, the structure of the RCP differs in conformation and/or oligomerization state from the N-terminal domain of the OCP. For example, in the 16 kDa RCP the carotenoid could be shielded by oligomerization; the 16 kDa RCP isolated from cells appeared to be a dimer (Holt and Krogmann 1981). In addition or alternatively, a substantial rearrangement of the tertiary structure may be involved. Domains composed entirely of alpha-helices are thought to be able to reorganize relatively readily (Minary and Levitt 2008). Another intriguing clue, suggestive of a conformational change, comes from the observation that exposing the OCP to low pH causes its spectrum to resemble that of the 16 kDa RCP. This low pH induced form of the RCP has a different secondary structure profile as measured by circular dichroism (Kerfeld 2004a,b).

1.4 THE STRUCTURE OF THE OCP IN THE CONTEXT OF FUNCTION

The structure of the OCP from the cyanobacterium *A. maxima* was reported in 2003 (Kerfeld et al. 2003) before its function had been established. The recent revelations about the OCP's function make a reconsideration of the structure timely. In addition, there are available structure–function data for other light responsive proteins. Blue–green light (400–550 nm), which can trigger OCP-mediated photoprotection is an important environmental signal; blue-light receptors are widespread among the prokaryotes and eukaryotes—blue-light photoreceptors such as flavin binding phototropins that contain LOV domains are known in bacteria, plants (Briggs 2006), and algae (Crosson and Moffat 2001, Takahashi et al. 2007) while photoactive yellow protein (PYP) mediates negative phototaxis in response to blue light in bacteria. LOV domains and PYP are members of the PAS (Per/Arndt/Sim) superfamily (Pfam 00989); PAS domains bind a wide range of chromophores required for the detection of sensory input signals. The PAS fold represents an important sensory domain present in all kingdoms of life. Another family of blue-light receptors is the blue-light using FAD (BLUF) domains; these domains relay light signals into a variety of outputs in bacteria.

Structural data is available for the PYP, LOV, and BLUF domains. Interestingly, these proteins and the NTF2 domain of the OCP, as shown in Figure 1.4, contain a structural core of a four- to five-stranded beta-sheet, although the connectivity, number, and disposition of the surrounding alpha-helices vary. For PYP and the LOV and BLUF domains, multiple x-ray crystal structures in combination with NMR and Fourier transform infrared (FTIR) spectroscopic data have provided details about the structural basis of light-mediated signaling. By analogy, this can be considered in the formulating hypotheses about the OCP's signal transduction mechanism.

The known structure of the OCP is a snapshot of the presumably dark-state-adapted form of the protein. From the model, it is difficult to imagine how the concealed carotenoid could interact with one of the components of the phycobilisome in order to quench the absorbed energy. However, the surface of the OCP has numerous surface cavities and clefts, as shown in Figure 1.3b, including two

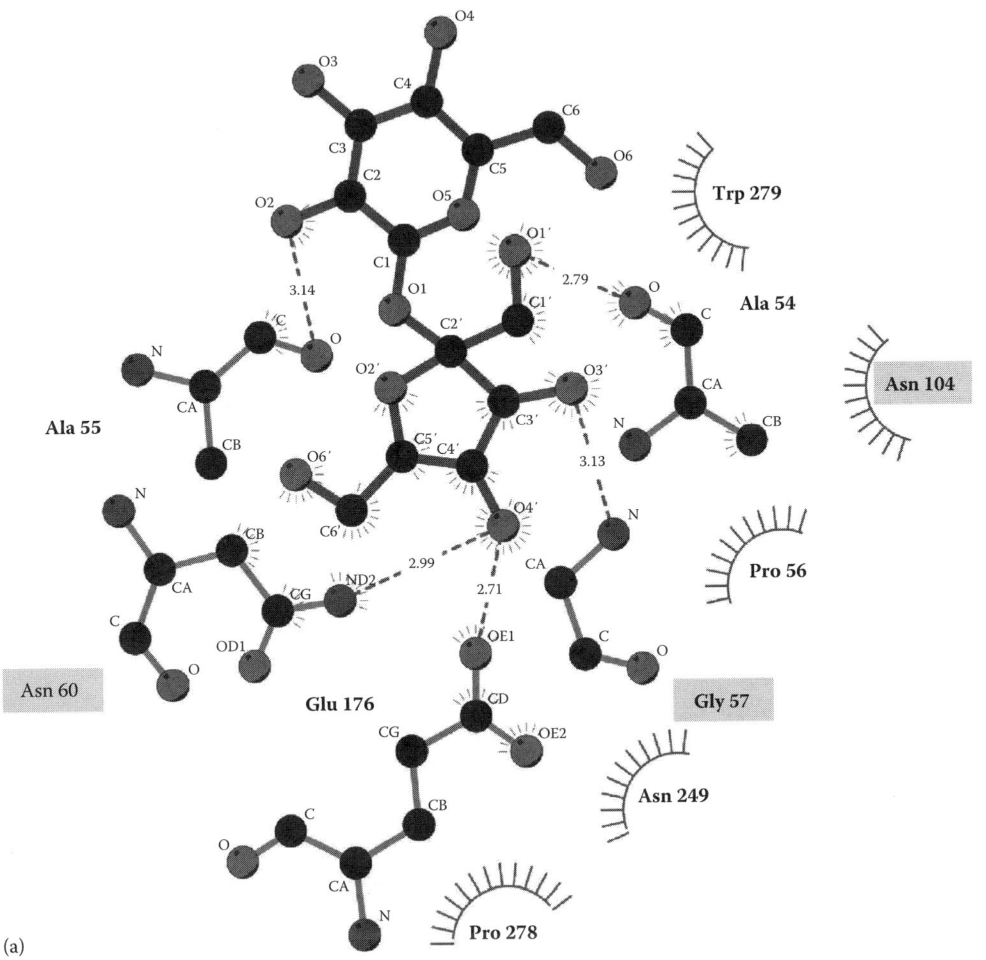

(a)

FIGURE 1.4 Ligand-binding plots showing hydrogen-bonding interactions and distances and hydrophobic contacts for (a) the sucrose molecule in the *A. maxima* OCP structure and (b) the 3′-hydroxyechinenone molecule. Residues labeled in bold are absolutely conserved in the primary structure of the OCP. (From Wallace, A.C. et al., *Protein Eng.*, 8, 127, 1995.)

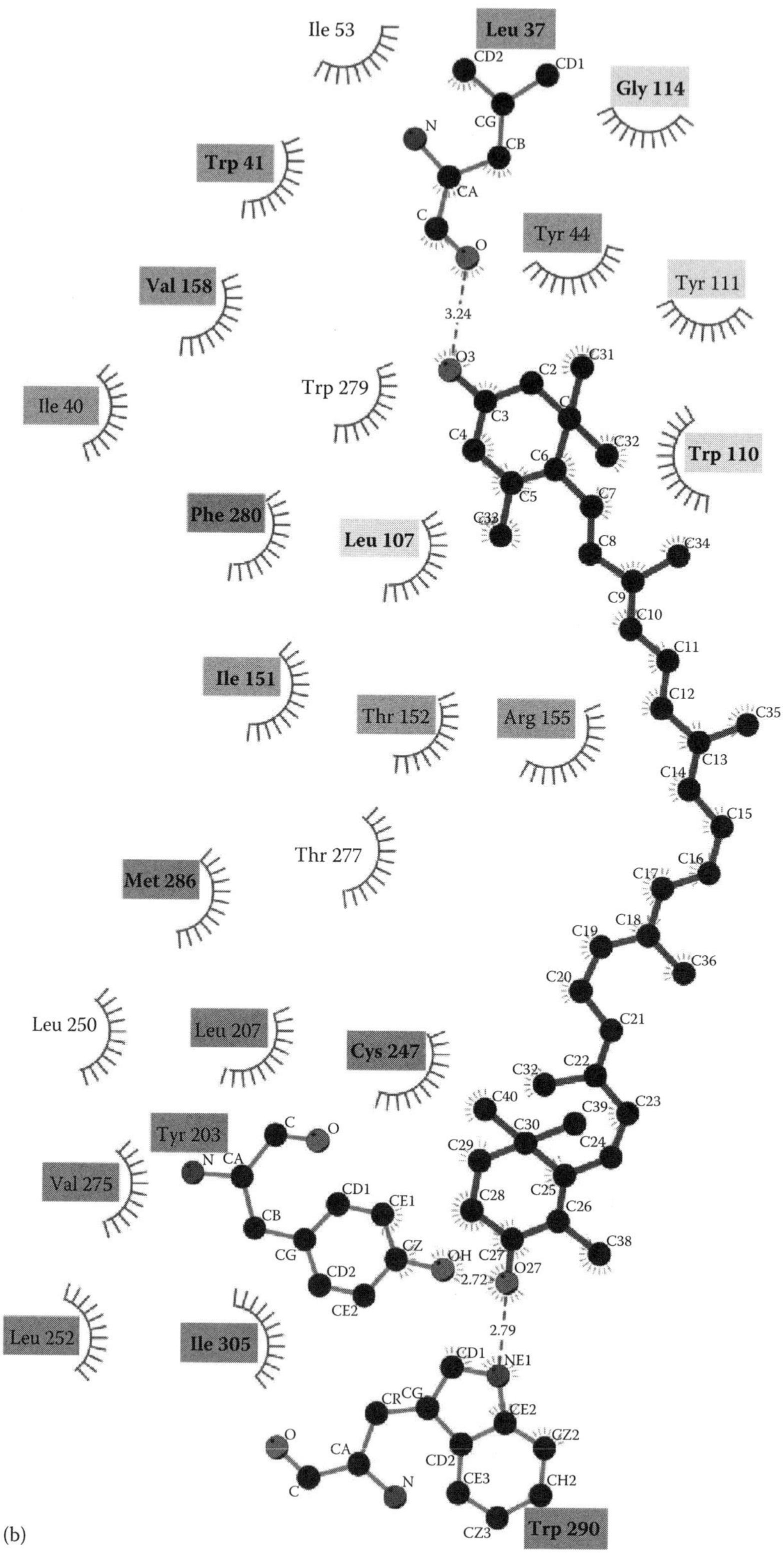

(b)

FIGURE 1.4 (continued)

that provide solvent accessibility to the carotenoid. These surface features could be the site of the interaction of the OCP with other chromophores or proteins.

Protein–protein interactions and protein conformational changes, which may unmask binding sites, alter surface shape, and induce changes in local electrostatic potential are likely essential to OCP's NPQ mechanism (Scott et al. 2006, Rakhimberdieva et al. 2007a). Glutaraldehyde and high concentrations of glycerol and sucrose completely eliminate NPQ formation in *Synechocystis* PCC6803 (Scott et al. 2006, Rakhimberdieva et al. 2007a), suggesting that this process must involve changes in the association or conformation of the proteins (phycobilisome and/or the OCP). This is of interest in the context of similar experiments on photosensors; dehydration or the addition of glycerol abolishes the large-scale and long-range protein motions of a plant LOV domain and affects the formation of the physiological signaling state (Iwata et al. 2007). These experiments also highlight the participation of internal and surface water molecules in the conformational fluctuations, which are required for large-scale and/or long-range motions of proteins.

The OCP's photoprotective function may rely on its dynamic structure in several ways. A cluster of highly conserved residues that converge at the interface of the two domains and line the pocket in which a sucrose molecule was observed in the *A. maxima* OCP structure, Figures 1.3a and 1.4a. The positioning of the sucrose molecule is reminiscent of an allosteric effector, as it is situated in a loop between the two domains of the protein. Furthermore, the binding of the sucrose molecule also involves the linker connecting the two domains of the OCP; the flexibility of this region could facilitate large changes in the disposition of the two domains with respect to each other. For example, if in the "activated" protein the interface between the two domains was opened with the linker acting as a hinge, it would increase the surface exposure of the carotenoid.

The crystals of the OCP contained two molecules in the asymmetric unit; these were refined independently including manual fitting of the carotenoid molecule into each protein chain. In both, the 3′-hydroxyequinenone adopts an *all-trans* configuration in the protein, however, with a slight bowing across its length (the average deviation from all-trans is 16°). In contrast to its conformation in solution, where both terminal rings are in the *s-cis* conformation with respect to the conjugated backbone, the terminal ring of the hECN containing the keto group is locked into an *s-trans* conformation via the hydrogen bonds to Tyr 203 and Trp 290. The absorption of blue light by the carotenoid is a potential trigger that may regulate a mechanism to modulate the protein conformation. Indeed, upon illumination with blue–green light, the OCP (which appears orange) is photoconverted to a red active form (Wilson et al. 2008). Resonance Raman spectroscopy and light-induced FTIR difference spectra demonstrated that light absorbance by the OCP induces structural changes not only in the carotenoid but also in the protein (Wilson et al. 2008). Upon illumination of the OCP, the apparent conjugation length of hECN increased by about one conjugated bond, and hECN reaches a less distorted, more planar structure. Although the hECN is still all-trans in the red form, the relatively small conformational changes of the carotenoid are sufficient to induce protein conformational changes due to the locked conformation of the carotenoid in the dark-state structure. This "activated," OCP, through interaction with the core of the phycobilisome, could elicit an alteration of the phycobilisome structure leading to the quenched state. Alternatively, the carotenoid of the OCP could directly interact with a phycobilin chromophore (most probably the terminal acceptor) and dissipate the absorbed energy. High blue-light intensities could induce changes that can lower the energy of the carotenoid S_1 state rendering possible the energy transfer from the terminal acceptor of the phycobilisome.

Those residues that are absolutely conserved (129 of 318) in the primary structure of the OCP are likely candidates for important functional roles. Many of these surround the pigment, as shown in Figures 1.3a and 1.4b. Side-chain conformations and hydrogen-bonding patterns that may involve internal water molecules are known to play critical roles in the mechanisms by which other photosensitive proteins function. Light-mediated signaling in the PYP, BLUF, and LOV domains relies on a conformational change in the protein mediated by changes in hydrogen bonding (Anderson et al. 2004, Kort et al. 2004, Jung et al. 2006). By analogy, the alteration of hydrogen-bonding patterns could be one means to propagate the light-responsive signal to the surface of the OCP. Hydrogen bonding in the OCP is extensive. There are two hydrogen bonds to the keto-oxygen of the 3′-hydroxyechinenone via invariant C-terminal residues Tyr 203 and Trp 290, as shown in Figure 1.4b. Tyr 203 is further hydrogen-bonded to the main chain atoms of Leu 207 and Thr 199; the latter residue is conserved and surface exposed. Trp 290 is hydrogen-bonded to the invariant residues Val 271 and Phe 292; these residues in the strands of the beta-sheet are also surface exposed. The surface accessibility of the hydrogen-bonded residues poises them to possibly communicate the status of the chromophore to the surface of the OCP. Similarly, at the hydroxyl terminus of the carotenoid, where it is most solvent accessible, there is a potential for forming a weak hydrogen bond to the conserved residue Leu 37 which is, in turn, hydrogen-bonded to the main chain of invariant residues Ala33 and Trp 41. These residues are also surface exposed.

Likewise, in the LOV domains of plants and fungi, light-driven structural changes in the chromophore result in a hydrogen-bond switch that causes beta-sheet motion and subsequent displacement of a small segment of alpha-helix, which is packed against the beta-sheet in the resting state (Harper et al. 2003, 2004, Nozaki et al. 2004, Halavaty and Moffat 2007). The hydrogen bond that is altered is between the flavin mononucleotide chromophore and the side chain of a conserved Gln, which belongs to the central strand of the LOV beta-sheet. An analogous mechanism is possible for the OCP via the hydrogen bond between the 3′-hydroxyechinenone carbonyl oxygen and Trp 290; Trp 290 is part of the central strand of the beta-sheet of the OCP's C-terminal domain. Light-triggered conformational changes of the 3′-hydroxyechinenone could alter the strength of this hydrogen bond. This, as in LOV domains, could influence the conformation of the central beta-sheet, affording signal propagation pathway from the carotenoid to the surface of the OCP. Furthermore, as in the LOV domains, a short alpha-helix from the N-terminus of the protein interacts with the central beta-sheet of the OCP, as shown in Figure 1.3a and c. In a mechanism analogous to the signal triggering in the LOV domain caused by the displacement of this helix (Harper et al. 2003, 2004, Halavaty and Moffat 2007) light-induced changes in the equilibrium of bound and unbound state of this N-terminal helix in the OCP could underlie the signaling/quenching switch.

The photoresponse of PYP also involves an "arginine gateway": altered hydrogen bonding to a conserved Arg displaces the side chain allowing access to the chromophore (Genick et al. 1997). The structure of a long-lived PYP_M intermediate has been determined by millisecond time-resolved crystallography (Genick et al. 1997). During the bleaching of the protein an arginine gateway opens, allowing solvent exposure and protonation of the phenolic oxygen. In the OCP, invariant Arg 155 is found at the interface of the N- and C-terminal domains, as shown in Figures 1.3b and 1.4b, occluding solvent access to the carotenoid. The alteration of the disposition of this residue in the OCP would, as in PYP, increase substantially the solvent accessibility of the 3′-hydroxyechinenone molecule.

At the time of its elucidation, one of the most intriguing features of the OCP structure was the preponderance of Met residues with their thioether groups oriented toward the carotenoid. Many of these are absolutely conserved among the primary structures of the OCP. There are several potential roles for the Met side chains in the function of the OCP. The potential for the oxidation of Met residues could confer a protective function for the carotenoid, by intercepting reactive oxygen species (via oxidation to methioinine sulfoxide and methionine sulfone) that would otherwise damage the pigment. All of the conserved Met residues make at least three hydrogen bonds to residues that are surface exposed. Of the conserved N-terminal domain Met residues (47, 61, 74, and 83), only Met 83 is buried within the protein. In contrast, Met 286, the single conserved Met in the C-terminal domain, is entirely buried. Alternatively, the Met residues may function in signal propagation, perhaps through bound water molecules. The polarizability of the sulfur atom and the distinctive geometries of Met observed in its interaction with a nucleophile and an electrophile provide structural versatility that could facilitate signaling. The structural basis of function in the BLUF domain offers an example of the role of Met residues in signaling through the protein (Jung et al. 2006). A comparison of the BLUF domain in both the dark adapted and the photoexcited, redshifted form suggests a path through the protein for signal propagation that involves a large displacement of a Met side chain in one of the terminal beta-strands of its sheet; this conveys the status of the chromophore to the surface of the protein. The associated 1 Å displacement of the Met sulfur atom is likely part of the signal relay (Jung et al. 2006).

1.5 CONCLUSIONS AND PROSPECTS

Admittedly speculative, these features of the light responsive changes in the PYP, LOV, and BLUF domains suggest some interesting hypotheses to test in the effort to define the roles of the specific amino acids in the function of the OCP. In addition, it points to the need for new structural studies on mutants as well as wild-type orthologs of the OCP and its variants to provide additional insights into the role of protein conformation and structural water molecules in the function of the OCP. To this end, we have determined the structure of *Synechocystis* PCC6803 OCP at 1.65 Å resolution (Klein et al., manuscript in preparation). The elucidation of this and other structures in conjunction with functional studies promises to reveal details of the intra- and intermolecular mechanisms of light sensing, signal propagation, and energy dissipation.

ACKNOWLEDGMENTS

We thank Michael Klein, Jay Kinney, and David Krogmann for their helpful discussions; Clémence Boulay for preparation of Table 1.1; and Edwin Kim and Jean Marc Verbavatz for assistance with the figures.

REFERENCES

Adir, N. (2005). Elucidation of the molecular structures of components of the phycobilisome: Reconstructing a giant. *Photosynth Res* **85**(1): 15–32.

Anderson, D. C., E. L. Campbell, and J. C. Meeks (2006). A soluble 3D LC/MS/MS proteome of the filamentous cyanobacterium *Nostoc punctiforme*. *J Proteome Res* **5**(11): 3096–3104.

Anderson, S., S. Crosson, and K. Moffat (2004). Short hydrogen bonds in photoactive yellow protein. *Acta Crystallogr* **60**: 1008–1016.

Bailey, S. and A. Grossman (2008). Photoprotection in cyanobacteria: Regulation of light harvesting. *Photochem Photobiol* **84**(6): 1410–1420.

Bailey, S., N. Mann, C. Robinson, and D. J. Scanlan (2005). The occurrence of rapidly reversible non-photochemical quenching of chlorophyll a fluorescence in cyanobacteria. *FEBS Lett* **579**(1): 275–280.

Boulay, C., L. Abasova, C. Six, I. Vass, and D. Kirilovsky (2008a). Occurrence and function of the orange carotenoid protein in photoprotective mechanisms in various cyanobacteria. *Biochim Biophys Acta* **1777**(10): 1344–1354.

Briggs, W. R. (2006). *Photomorphogenesis in Plants and Bacteria*. Dordrecht, the Netherlands: Springer.

Briggs, W. R. (2007). The LOV domain: A chromophore module servicing multiple photoreceptors. *J Biomed Sci* **14**: 499–504.

Cadoret, J. C., R. Demouliere, J. Lavaud et al. (2004). Dissipation of excess energy triggered by blue light in cyanobacteria with CP43′ (IsiA). *Biochim Biophys Acta* **1659**(1): 100–104.

Crosson, S. and K. Moffat (2001). Structure of a flavin-binding plant photoreceptor domain: Insights into light-mediated signal transduction. *Proc Natl Acad Sci* **98**: 2995–3000.

Demmig-Adams, B. (1990). Carotenoids and photoprotection in plants: A role for the xanthophyll zeaxanthin. *Biochim Biophys Acta* **1020**: 1–24.

El Bissati, K., E. Delphin, N. Murata, A. Etienne, and D. Kirilovsky (2000). Photosystem II fluorescence quenching in the cyanobacterium *Synechocystis* PCC 6803: Involvement of two different mechanisms. *Biochim Biophys Acta* **1457**(3): 229–242.

Fulda, S., S. Mikkat, F. Huang et al. (2006). Proteome analysis of salt stress response in the cyanobacterium *Synechocystis* sp. strain PCC 6803. *Proteomics* **6**(9): 2733–2745.

Genick, U. K., G. E. Borgstahl, K. Ng et al. (1997). Structure of a protein photocycle intermediate by millisecond time-resolved crystallography. *Science* **275**: 1471–1475.

Gilmore, A. and H. Yamamoto (1993). Linear models relating xanthophylls and lumen acidity to non-photochemical fluorescence quenching, evidence that antheraxanthin explains zeaxanthin-independent quenching. *Photosynth Res* **35**: 67–68.

Halavaty, A. S. and K. Moffat (2007). N- and C-terminal flanking regions modulate light-induced signal transduction in the LOV2 domain of the blue-light sensor phototropin 1 from *Avena sativa*. *Biochemistry* **46**: 14001–14009.

Harper, S. M., L. C. Neil and K. H. Gardner (2003). Structural basis of a phototropin light switch. *Science* **301**(5639): 1541–1544.

Harper, S. M., J. M. Christie, and K. H. Gardner (2004). Disruption of the LOV-J alpha helix interaction activates phototropin kinase activity. *Biochemistry* **43**: 16184–16192.

Hihara, Y., A. Kamei, M. Kanehisa, A. Kaplan, and M. Ikeuchi (2001). DNA microarray analysis of cyanobacterial gene expression during acclimation to high light. *Plant Cell* **13**(4): 793–806.

Holt, T. K. and D. W. Krogmann (1981). A carotenoid-protein from cyanobacteria. *Biochim Biophys Acta* **637**(3): 408–414.

Horton, P., A. V. Ruban, and R. G. Walters (1996). Regulation of light harvesting in green plants. *Annu Rev Plant Physiol Plant Mol Biol* **47**: 655–684.

Iwata, T., A. Yamamoto, S. Tokutomi, and H. Kandori (2007). Hydration and temperature similarly affect light-induced protein structural changes in the chromophoric domain of phototropin. *Biochemistry* **46**: 7016–7021.

Joshua, S., S. Bailey, N. H. Mann, and C. W. Mullineaux (2005). Involvement of phycobilisome diffusion in energy quenching in cyanobacteria. *Plant Physiol* **138**(3): 1577–1585.

Jung, A., J. Reinstein, T. Domratcheva, R. L. Shoeman, and I. Schlichting (2006). Crystal structures of the AppA BLUF domain photoreceptor provide insights into blue light-mediated signal transduction. *J Mol Biol* **362**: 717–732.

Kanesaki, Y., I. Suzuki, S. I. Allakverdiev, K. Mikami, and N. Murata (2002). Salt stress and hyperosmotic stress regulate the expression of different sets of genes in *Synechocystis* sp PCC6803. *Biochem Biophys Res Commun* **290**: 339–348.

Karapetyan, N. V. (2007). Non-photochemical quenching of fluorescence in cyanobacteria. *Biochemistry (Moscow)* **72**(10): 1127–1135.

Kerfeld, C. A. (2004a). Structure and function of the water-soluble carotenoid-binding proteins of cyanobacteria. *Photosynth Res* **81**(3): 215–225.

Kerfeld, C. A. (2004b). Water-soluble carotenoid proteins of cyanobacteria. *Arch Biochem Biophys* **430**(1): 2–9.

Kerfeld, C. A., M. R. Sawaya, V. Brahmandam et al. (2003). The crystal structure of a cyanobacterial water-soluble carotenoid binding protein. *Structure* **11**(1): 55–65.

Kirilovsky, D. (2007). Photoprotection in cyanobacteria: The orange carotenoid protein (OCP)-related non-photochemical-quenching mechanism. *Photosynth Res* **93**: 7–16.

Knutson, R. (1998). The red carotenoid protein from *Arthrospira maxima*. MS thesis, Purdue University, West Lafayette, IN.

Kort, R., K. J. Hellingwerf, and R. B. G. Ravelli (2004). Initial events in the photocycle of photoactive yellow protein. *J Biol Chem* **279**: 26417–26424.

Kucho, K.-I., Y. Tsuchiya, Y. Okumoto et al. (2004). Construction of unmodified oligonucleotide-based arrays in the thermophilic cyanobacterium *Thermosynechococcus elongatus* BP-1: Screening of the candidates for circadianly expressed genes. *Genes Genet Syst* **79**: 319–329.

Li, X.-P., O. Björkman, C. Shih et al. (2000). A pigment-binding protein essential for regulation of photosynthetic light harvesting. *Nature* **403**: 391–395.

Li, X.-P., A. M. Gilmore, S. Caffarri et al. (2004). Regulation of photosynthetic light harvesting involves intrathylakoid lumen pH sensing by the PsbS protein. *J Biol Chem* **279**(22): 22866–22874.

MacColl, R. (1998). Cyanobacterial phycobilisomes. *J Struct Biol* **124**(2–3): 311–334.

Minary, P. and M. Levitt (2008). Probing protein fold space with a simplified model. *J Mol Biol* **375**(4): 920–933.

Muller, P., X.-P. Li, and K. Niyogi (2001). Non-photochemical quenching. A response to excess light energy. *Plant Physiol* **125**: 1558–1566.

Mullineaux, C. W. (1992). Excitation energy transfer from phycobilisomes to photosystem-I in a cyanobacterium. *Biochim Biophys Acta* **1100**(3): 285–292.

Niyogi, K. K. (1999). Photoprotection revisited: Genetic and molecular approaches. *Annu Rev Plant Physiol Plant Mol Biol* **50**: 333–359.

Nozaki, D., T. Iwata, T. Ishikawa et al. (2004). Role of Gln1029 in the photoactivation processes of the LOV2 domain in *Adiantum* phytochrome3. *Biochemistry* **43**: 8373–8379.

Pascal, A. A., Z. F. Liu, K. Broess et al. (2005). Molecular basis of photoprotection and control of photosynthetic light-harvesting. *Nature* **436**(7047): 134–137.

Polívka, T., C. A. Kerfeld, T. Pascher, and V. Sundström (2005). Spectroscopic properties of the carotenoid 3′-hydroxyechinenone in the orange carotenoid protein from the cyanobacterium *Arthrospira maxima*. *Biochemistry* **44**(10): 3994–4003.

Rakhimberdieva, M. G., V. A. Boichenko, N. V. Karapetyan, and I. N. Stadnichuk (2001). Interaction of phycobilisomes with photosystem II dimers and photosystem I monomers and trimers in the cyanobacterium *Spirulina platensis*. *Biochemistry* **40**(51): 15780–15788.

Rakhimberdieva, M. G., Y. V. Bolychevtseva, I. V. Elanskaya, and N. V. Karapetyan (2007a). Protein–protein interactions in carotenoid triggered quenching of phycobilisome fluorescence in *Synechocystis* sp. PCC 6803. *FEBS Lett* **581**(13): 2429–2433.

Rakhimberdieva, M. G., I. N. Stadnichuk, I. V. Elanskaya, and N. V. Karapetyan (2004). Carotenoid-induced quenching of the phycobilisome fluorescence in photosystem II-deficient mutant of *Synechocystis* sp. *FEBS Lett* **574**(1–3): 85–88.

Rakhimberdieva, M. G., D. V. Vavilin, W. F. Vermaas, I. V. Elanskaya, and N. V. Karapetyan (2007b). Phycobilin/chlorophyll excitation equilibration upon carotenoid-induced non-photochemical fluorescence quenching in phycobilisomes of the cyanobacterium *Synechocystis* sp. PCC 6803. *Biochim Biophys Acta* **1767**(6): 757–765.

Ruban, A. V., D. Ress, P. A. A., and P. Horton (1992). Mechanism of pH-dependent dissipation of absorbed excitation energy by photosynthetic membranes. II: The relationship between LHCII aggregation and qE in isolated thylakoids. *Biochim Biophys Acta* **1102**: 39–44.

Ruban, A. V., R. Berera, C. Ilioaia et al. (2007). Identification of a mechanism of photoprotective energy dissipation in higher plants. *Nature* **450**(7169): 575–578.

Sato, S., Y. Shimoda, A. Muraki et al. (2007). A large-scale protein–protein interaction analysis in *Synechocystis* sp. PCC6803. *DNA Res* **14**: 207–216.

Scott, M., C. McCollum, S. Vasil'ev et al. (2006). Mechanism of the down regulation of photosynthesis by blue light in the Cyanobacterium *Synechocystis* sp. PCC 6803. *Biochemistry* **45**(29): 8952–8958.

Takahashi, F., D. Yamagata, M. Ishikawa et al. (2007). AUREOCHROME, a photoreceptor required for photomorphogenesis in stramenopiles. *Proc Natl Acad Sci* **104**: 19625–19630.

Wallace, A. C., R. A. Laskowski, and J. M. Thornton (1995). LIGPLOT: A program to generate schematic diagrams of protein–ligand interactions. *Protein Eng* **8**: 127–134.

Wilson, A., G. Ajlani, J. M. Verbavatz et al. (2006). A soluble carotenoid protein involved in phycobilisome-related energy dissipation in cyanobacteria. *Plant Cell* **18**(4): 992–1007.

Wilson, A., C. Boulay, A. Wilde, C. A. Kerfeld, and D. Kirilovsky (2007). Light-induced energy dissipation in iron-starved cyanobacteria: Roles of OCP and IsiA proteins. *Plant Cell* **19**(2): 656–672.

Wilson, A., C. Punginelli, A. Gall et al. (2008). A photoactive carotenoid protein acting as light intensity sensor. *Proc Natl Acad Sci* **105**(33): 12075–12080.

Wu, Y. P. and D. W. Krogmann (1997). The orange carotenoid protein of *Synechocystis* PCC 6803. *Biochim Biophys Acta* **1322**(1): 1–7.

Yamamoto, H. (1979). Biochemistry of the violaxanthin cycle in higher plants. *Pure Appl Chem* **51**: 639–648.

2 Carotenoids in Lipid Membranes

Wieslaw I. Gruszecki

CONTENTS

2.1 INTRODUCTION

Carotenoid pigments play diverse physiological functions in various environments specific for living organisms (Britton, 1995). In particular, they are associated with proteins and embedded within the lipid membranes (Britton, 1995). Some important biological functions of carotenoid pigments, such as photoprotection against the oxidative damage of biomembranes, are directly dependent on the molecular organization of carotenoids in membranes and on the effect of carotenoids on the dynamic and the structural properties of the membranes (Gruszecki and Strzalka, 2005; Krinsky et al., 2003; McNulty et al., 2007; Sujak et al., 1999; Woodall et al., 1997). In this chapter, the problem of binding of carotenoid pigments to lipid membranes, solubility within the lipid phase, the pigment orientation with respect to the membrane, and the effects on the physical properties of the lipid membranes will be overviewed and briefly discussed.

2.2 BINDING OF CAROTENOIDS TO LIPID MEMBRANES

2.2.1 LOCALIZATION

Owing to their chemical structure, carotenes as polyterpenoids are hydrophobic in nature (Britton et al., 2004). Therefore, as it might be expected, the carotenes are bound within the hydrophobic core of the lipid membranes. Polar carotenoids, with the molecules terminated on one or two sides with the oxygen-bearing substitutes, also bind to the lipid bilayer in such a way that the chromophore, constituted by the polyene backbone is embedded in the hydrophobic core of the membrane. There are several lines of evidence for such a localization of carotenoids with respect to the lipid bilayers.

Owing to the solvatochromic effect, the position of the electronic absorption maximum on energy scale depends on the dielectric properties of the medium (Andersson et al., 1991). The positions of the maxima in the absorption spectra of several carotenoid pigments incorporated into the lipid membrane systems, indicate that chromophores are embedded in the environment characterized by the polarizability term of the hydrophobic core of the membrane (Gruszecki, 1999, 2004; Gruszecki and Sielewiesiuk, 1990; Milon et al., 1986; Sujak et al., 2005). Figure 2.1 presents such a dependency plotted for violaxanthin incorporated into liposomes formed with DMPC. Detailed information concerning the segmental motion of acyl lipid chains, inferred on the basis of the EPR-spin label technique (Strzalka and Gruszecki, 1994; Subczynski et al., 1992, 1993; Wisniewska and Subczynski, 1998), NMR spectroscopy (Gabrielska and Gruszecki, 1996; Jezowska et al., 1994; Sujak et al., 2005), and FTIR spectroscopy (Sujak et al., 2005, 2007a), indicates unequivocally that the membrane-bound carotenoids modify profoundly the organization of the hydrophobic core of the lipid bilayers.

2.2.2 Orientation

Despite the fact that both the apolar and the polar carotenoids incorporated into the hydrophobic core of the membrane, the orientation of the long, bar-shaped molecules depends very much on the extent of the substitution on the polar end-group, and the ability to form hydrogen bonds within the polar headgroup zones of the membrane (Gruszecki, 1999, 2004). In general, the apolar carotenoids, such as β-carotene or lycopene, display a certain orientational freedom with respect to the membrane (see the model presented in Figure 2.2). The linear dichroism study of the orientation of β-carotene led to the conclusion that the transition dipole moment of the pigment molecule, close to the long axis of the polyene chromophore (≈15°; Shang et al., 1991), was oriented close to the plane

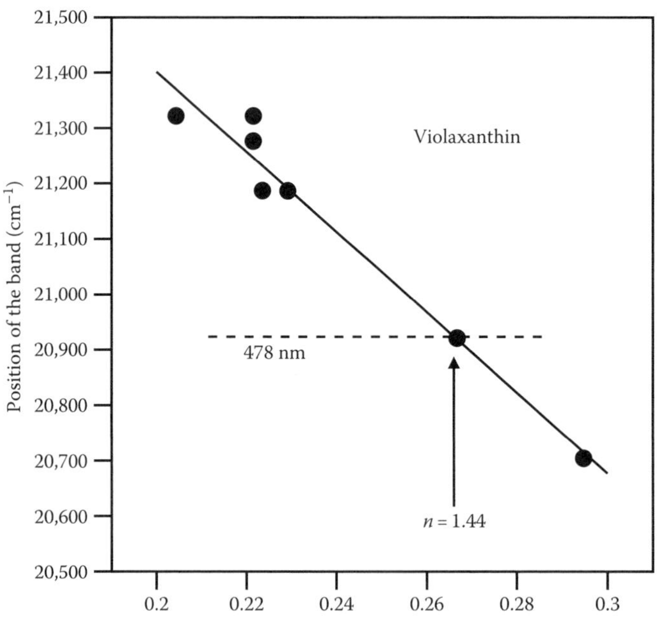

FIGURE 2.1 Energy of the 0–0 vibrational transition in the principal electronic absorption spectrum of violaxanthin ($1^1A_g^- \rightarrow 1^1B_u^+$), recorded in different organic solvents, versus the polarizability term, dependent on the refraction index of the solvent (n). The dashed line corresponds to the position of the absorption band for violaxanthin embedded into the liposomes formed with DMPC (Gruszecki and Sielewiesiuk, 1990) and the arrow corresponds to the polarizability term of the hydrophobic core of the membrane (n = 1.44).

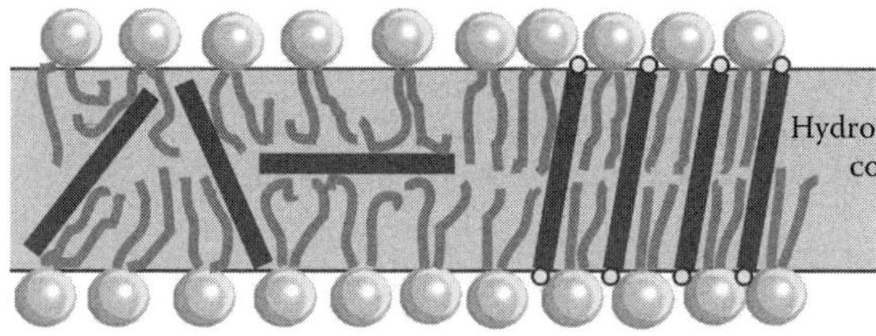

FIGURE 2.2 Model representation of organization of the lipid membrane containing apolar and polar carotenoid pigments.

of the membrane formed with DOPC (Johansson et al., 1981) or was close to the magic angle in EYPC (Gruszecki, 1999). The orientation angle of the transition dipole moment with respect to the axis normal to the plane of the membrane, is equal to the magic angle (54.7°) and can be interpreted as an indication of this particular mean orientation angle but one will arrive at the same result in the case of homogeneous distribution of the transition dipoles. The angle-resolved resonance Raman studies show that β-carotene is oriented roughly parallel to the plane of the membrane formed with DOPC but roughly perpendicular with respect to the membrane formed with SBPC (van de Ven et al., 1984). In the case of lycopene, the mean orientation angle of the transition dipole moment with respect to the normal to the plane of the membrane was determined as 74°, in the membranes formed with EYPC (Gruszecki, 1999). Such a mean angle shows that the orientation of lycopene is neither determined by the plane of the bilayer nor by the direction of the alkyl lipid chains. The x-ray analysis of the electron density profiles across the lipid membranes formed with POPC (with 0.2 mol fraction cholesterol) demonstrated that, in contrast to the polar carotenoids (in particularly astaxanthin), lycopene and β-carotene disordered the membrane bilayer (McNulty et al., 2007). In the case of the polar carotenoids, linear dichroism studies determined the orientations to be close to the axis normal to the plane of the bilayer. The polar groups bound to the end-rings of the pigments examined will tend to form hydrogen bonds with the lipid membrane headgroups and water at the membrane interface. The acute orientation angles, found in the case of polar carotenoids, indicate that the molecules adopt an orientation that allows the polar groups localized on the opposite sides to be anchored in the opposite polar membrane zones. In the case of zeaxanthin ((3*R*,3′*R*)-β,β-carotene-3,3′-diol), linear dichroism studies determined the orientation angles of the transition dipole to be 33° in EYPC (Sujak et al., 1999), 25° in DMPC (Gruszecki and Sielewiesiuk, 1990), and 9° in DGDG (Gruszecki and Sielewiesiuk, 1991). As can be seen, the orientation angles negatively correlate with the thickness of the hydrophobic core of the membrane: the greater the thickness of the membrane (ca. 2.3 nm for EYPC, 2.8 nm for DMPC, and ca. 3.0 nm in the case of DGDG) the lower the orientation angle. Such a correlation can be interpreted as a demonstration of the general rule that the orientation of polar carotenoids is determined by a matching of the distance between the opposite polar groups of the pigment and the thickness of the hydrophobic core of the membrane.

The studies of monomolecular layers formed by zeaxanthin–lipid mixtures at the air–water interface have shown that, in contrast to the pigment molecules having an all-*trans* configuration, molecules having a *cis* configuration adopt an orientation within the film such that they are anchored within the polar–apolar interface by both of the hydroxyl groups found at the 3 and 3′ positions (Milanowska et al., 2003). A similar orientation of zeaxanthin molecules having *cis* configurations can be expected in lipid bilayer systems. Interestingly, recent EPR experiments also led to the conclusion that zeaxanthin in a *cis* configuration is able to span the lipid bilayer, providing that the thickness of the hydrophobic core of the membrane does not exceed the distance between the polar groups of the pigment (Widomska and Subczynski, 2008).

Interestingly, the orientation angle of lutein ((3*R*,3′*R*,6′*R*)-β,ε-carotene-3,3′-diol), determined in different lipid membrane systems, was always larger (less acute orientation angle) relative to the normal to the membrane than in the case of zeaxanthin, despite the very similar chemical structure of these pigments. For example, the angle was greater by 30° in EYPC, by 20° in DPPC, and by 10° in DHPC (Sujak et al., 1999). Moreover, the molecular organization of two-component carotenoid–DPPC monomolecular layers was substantially different in the case of lutein and zeaxanthin (Sujak and Gruszecki, 2000). The higher molecular area values, observed in the lutein-containing lipid films, as compared to the zeaxanthin-containing monolayers, correlates with greater orientation angles when observed in the lipid bilayers. From the structural point of view, the main difference between zeaxanthin and lutein, is the position of the double bond in one terminal ring: between C5′ and C6′ in the case of zeaxanthin and between C4′ and C5′ in the case of lutein. Owing to such a difference, this particular double bond is conjugated with the double bonds system in the case of zeaxanthin and is not in the case of lutein. It seems that the relative rotational freedom of the entire ε-ring of lutein about the C6′–C7′ single bond, may be a determinant of the different orientation of the xanthophylls with respect to the lipid membranes. It has to be kept in mind that the orientation angle, determined by means of the linear dichroism technique, represents a mean value. Consequently, one can determine the orientation angle in the sample in which chromophores are all oriented in the same direction or in a very different sample, in which the chromophores are distributed in two orthogonal pools: for example, parallel and perpendicular to the plane of the membrane. Interestingly, the x-ray analysis of the electron density profiles of lipid bilayers show that the effect of lutein on the membrane is very different from the effect of zeaxanthin and resembles the effect of apolar carotenoids oriented homogenously or parallel to the plane of the membrane (McNulty et al., 2007). No differences have been observed in the orientation of lutein and zeaxanthin in the membranes formed with DMPC in the samples in which the pigments were largely in the aggregated form (Sujak et al., 2002).

As a rule, the carotenoid pigments substituted with the keto groups in the C4 and C4′ positions, demonstrate almost vertical orientation with respect to the axis normal to the plane of the membrane: 26° astaxanthin in EYPC, 29° canthaxanthin in EYPC (Gruszecki, 1999), or 20° canthaxanthin in DPPC (Sujak et al., 2005). In the latter case, the orientation of the pigment molecules was found to depend on the actual concentration in the membrane and larger orientation angles were found at canthaxanthin concentrations approaching the aggregation threshold in the lipid phase (Sujak et al., 2005). Formally, the terminal keto groups of carotenoids can only form hydrogen bonds in which they act as proton acceptors. Owing to this fact, a direct interaction of the ketocarotenoids with the ester carbonyl groups, at the border of the hydrophobic and the polar zones of the membrane, via hydrogen bonding is not possible and the pigment molecule may penetrate more deeply within the hydrophilic membrane zone. In the case of vertical orientation of the carotenoid pigment with respect to the membrane, one can expect the strongest ordering effect of the rigid, apolar backbone of the pigment with respect to the alkyl lipid chains.

2.2.3 Incorporation Rates

The binding of carotenoids within the lipid membranes has two important aspects: the incorporation rate into the lipid phase and the carotenoid–lipid miscibility or rather pigment solubility in the lipid matrix. The actual incorporation rates of carotenoids into model lipid membranes depend on several factors, such as, the kind of lipid used to form the membranes, the identity of the carotenoid to be incorporated, initial carotenoid concentration, temperature of the experiment, and to a lesser extent, the technique applied to form model lipid membranes (planar lipid bilayers, liposomes obtained by vortexing, sonication, or extrusion, etc.). For example, the presence of 5 mol% of carotenoid with respect to DPPC, during the formation of multilamellar liposomes, resulted in incorporation of only 72% of the pigment, in the case of zeaxanthin, and 52% in the case of β-carotene (Socaciu et al., 2000). A decrease in the fluidity of the liposome membranes, by addition of other lipids or cholesterol, resulted in decreases in the incorporation rate of both carotenoids (Socaciu et al., 2000). Similar differences in the incorporation rate have been observed in the lipid mixture based on SBPC and cholesterol: 80% incorporation in the case of zeaxanthin and 64% in the case of β-carotene (Grolier et al., 1992). The lower initial pigment concentration (2.5 mol% with respect to DPPC) resulted in higher incorporation rate in the case of zeaxanthin (ca. 92%) but lower in the case of β-carotene (ca. 28%) (Socaciu et al., 1999). Moreover, the similar nonlinear relationship between the initial concentration and the incorporated carotenoid fraction has been shown for zeaxanthin in the liposomes formed with EYPC (Lazrak et al., 1987). Interestingly the incorporation rate of zeaxanthin was found to be higher in the case of the liposomes formed with DMPC as compared to DPPC but the opposite effect has been observed in a longer zeaxanthin homologue, decaprenozeaxanthin (Lazrak et al., 1987). Such a finding clearly indicates the importance of the match, between the distance separating the polar groups of the carotenoid and the thickness of the hydrophobic core of the bilayer, in incorporation of polar carotenoids into the lipid membranes.

2.2.4 Solubility

The miscibility of carotenoids and lipids within the membrane represents a kind of two-dimensional solubility of pigment molecules within the lipid bilayer. The lateral diffusion of the pigment molecules incorporated can cause an aggregation. Owing to the fact that the carotenoid backbone is a polyene chain, characterized by the conjugated double bond system, the molecules readily polarize and bind to each other by van der Waals interactions. Carotenoid aggregation in the lipid phase restricts their effect with respect to the membrane. It appears that even at the low molar fractions of the pigments with respect to lipid, despite the efficient incorporation rate, carotenoids form molecular aggregates in the membranes. Pigment aggregation is associated with dipole–dipole interactions responsible for the excitonic splitting of the electronic energy levels (Kasha et al., 1965; Parkash et al., 1998). The splitting results in the hypsochromic or/and the bathochromic spectral shift(s), depend on the actual structure formed. The analysis of the shifts in the electronic absorption spectra has been applied to investigate the process of carotenoid aggregate formation, both in the water environment and in the lipid membranes (Gruszecki et al., 1999; Kolev and Kafalieva, 1986; Mendelsohn and Van Holten, 1979; Sujak et al., 2000, 2005). Figure 2.3 presents the temperature dependency of the absorption spectrum of zeaxanthin incorporated into the liposomes formed with DPPC. As can be seen, even at relatively low pigment concentration (the initial concentration used for the liposome preparation 5 mol%) the zeaxanthin absorption spectrum is different from the characteristic absorption spectrum of the monomeric form and is elevated in the short-wavelength spectral region. Lowering of the temperature results in abrupt spectral changes that accompany the $L_\alpha \rightarrow P_\beta'$ phase transition of the membranes formed with DPPC (~41°C). According to the absorption spectrum, below the transition temperature the carotenoid exists entirely in the aggregated form within the membrane, despite relatively low concentration. This is demonstrated by the hypsochromic shift of the main absorption maximum to 385 nm and by the loss of the fine vibrational substructure (Sujak et al., 2000, 2002). The miscibility threshold of the same system (zeaxanthin in DPPC liposomes) has been determined as 29 and 6 mol% in the fluid phase and the crystalline phase of the membrane, respectively, by differential scanning calorimetric experiments (Kolev and Kafalieva, 1986). The comparison of these findings with the spectral analysis shows that the pigment aggregates can have similar effects on the membrane properties to those of monomers and therefore one has to be very cautious in concluding a carotenoid miscibility on the basis of different experimental techniques. A monomolecular layer approach seems to provide a good system to study carotenoid–lipid miscibility because the analysis of molecular area in a monolayer, in terms of the additivity rule, is very sensitive to the phenomenon of perfect miscibility, poor miscibility, and phase separation (Gruszecki et al., 1999; Milanowska et al., 2003; N'soukpoe-Kossi et al., 1988; Sujak and Gruszecki 2000; Sujak et al., 2007a). The comparative monomolecular layer study of the organization of DPPC membranes containing the xanthophyll pigments, zeaxanthin and lutein, shows pronounced

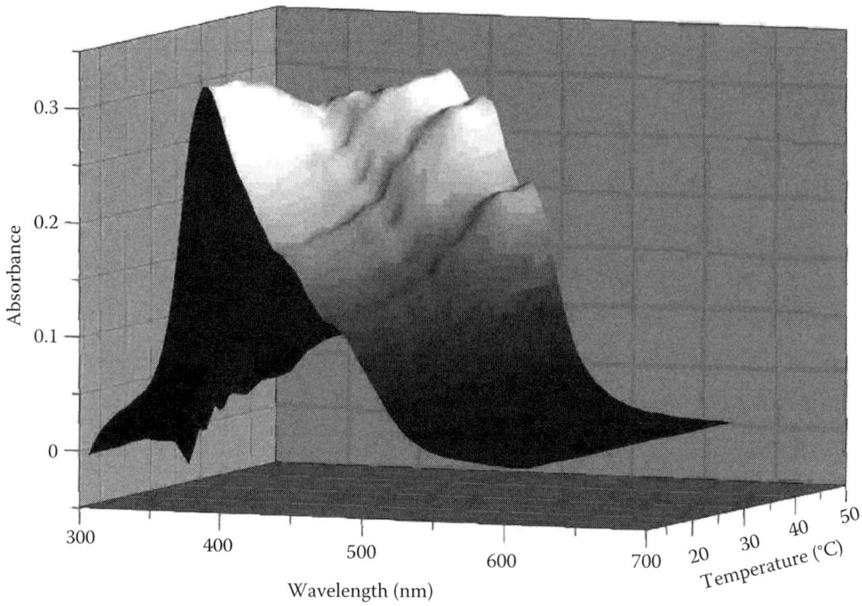

FIGURE 2.3 Temperature dependency of the absorption spectra of zeaxanthin incorporated into the liposomes formed with DPPC. The initial concentration of zeaxanthin in the medium used to prepare liposomes was 5 mol% with respect to lipid. (Based on the results presented in Sujak, A. et al., *Biochim. Biophys. Acta*, 1509, 255, 2000.)

differences expressed in much higher over-additivity of the molecular area in the lutein-containing membranes as compared to the zeaxanthin-containing membranes (Sujak and Gruszecki, 2000). Such a difference has been interpreted in terms of the different orientation of the xanthophylls in the lipid environment. The fact that the differences were not observed at the higher concentrations of carotenoids, promoting their aggregation, allowed the evaluation of the aggregation threshold concentration above which pigments remained in the form of molecular assemblies within the lipid phase. The aggregation threshold values for lutein and zeaxanthin in monomolecular layers, 30 and 20 mol% respectively, correspond to the values of 15 and 10 mol% with respect to a lipid bilayer (Sujak and Gruszecki, 2000). Below those concentrations, the pigments are distributed between the pools of monomeric and aggregated molecules. Interestingly, the aggregation threshold determined for canthaxanthin, using the same approach, was considerably lower than in the case of lutein and zeaxanthin in the monomolecular layers, equal to 2 mol%, which corresponds to 1 mol% in the case of a lipid bilayer. Such a low aggregation threshold of canthaxanthin in the membranes formed with DPPC has been confirmed in the spectroscopic studies of lipid bilayers (Sujak et al., 2005). The very strong ability of canthaxanthin to form molecular aggregates is most probably directly responsible for the formation of the crystal inclusions in the natural biomembranes of retina (Goralczyk et al., 1997, 2000).

2.3 EFFECTS OF CAROTENOIDS ON LIPID MEMBRANES

2.3.1 Model Membranes

Carotenoid molecules incorporated into the lipid membranes considerably interfere with both the structural and the dynamic membrane properties. Both effects are directly related to the chemical structure of carotenoid molecules. Importantly, it is the rigid, rod-like backbone of the carotenoids, consisting of the conjugated double-bond system of the polyene chain that appears to mediate the effect on the membrane system. The interactions of rigid carotenoid molecules with alkyl lipid chains, which undergo fast molecular motions (including the *gauche-trans* isomerization), restricts the lipid motional freedom and therefore modulates the fluidity of the lipid bilayer in the fluid phase. On the other hand, the membrane-bound carotenoids can destabilize the ordered lipid matrix in the gel phase. The actual effect of carotenoids with respect to the lipid membranes (ordering or fluidization) depends also on the chemical structure of the carotenoid that is incorporated. The latter determinant is mostly based on the presence of the polar groups that can be anchored in the polar zones of the lipid bilayer. For example, the incorporation of β-carotene and astaxanthin ((3*R*,3′*R*)-3,3′-dihydroxy-β,β-carotene-4,4′-dione) at 5 mol% into the lipid membranes composed of DOPC:cholesterol (molar ratio 1:5) result in very different effects on the membrane structure, as demonstrated by the small angle x-ray scattering (McNulty et al., 2007). In the cases involving astaxanthin, the electron density profile across the membrane, was almost unaffected. By contrast, β-carotene markedly affected the order of the hydrophobic core, especially in the central region. The effect of lycopene was even stronger than that observed in the case of β-carotene, particularly in the methyl group region of alkyl chains (McNulty et al., 2007). The differences observed directly correlate with the orientation of the carotenoid pigment with respect to the membrane, as discussed earlier. The effect of carotenoid pigments on different membrane segments can also be analyzed by means of ^{1}H-NMR spectroscopy (Gabrielska and Gruszecki, 1996; Sujak et al., 2005) or ^{13}C-NMR and ^{31}P-NMR technique, based on the natural abundance of the ^{13}C and ^{31}P isotopes (Jezowska et al., 1994). One aspect of the NMR studies is based on the analysis of the resonance lineshape that reflects the molecular dynamics of a particular group located at defined membrane zone, for example, the CH_3 groups located in the central region of the bilayer (Gabrielska and Gruszecki, 1996; Jezowska et al., 1994; Sujak et al., 2005). Figure 2.4 presents the analysis of the width of the ^{1}H-NMR band corresponding to the terminal methyl groups of the alkyl chains of the membranes modified with β-carotene and canthaxanthin (β,β-carotene-4,4′-dione). Broadening of the band,

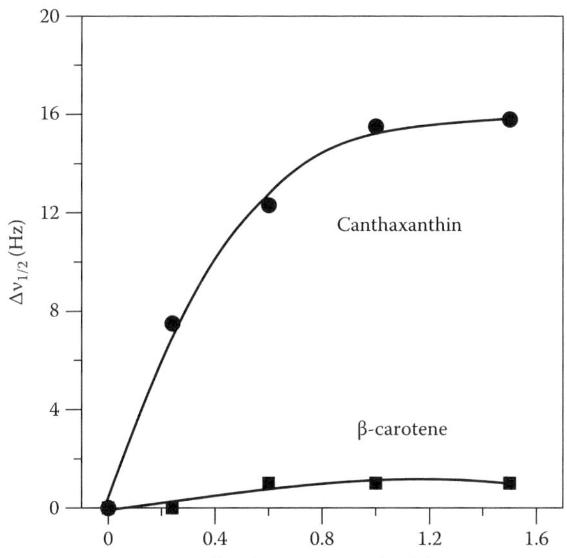

FIGURE 2.4 Carotenoid presence-induced increase in the full width at half height of the ^{1}H-NMR band corresponding to the CH_3 groups of alkyl chains of liposomes formed with EYPC and containing β-carotene and formed with DPPC and containing canthaxanthin. (Based on Gabrielska, J. and Gruszecki, W.I., *Biochim. Biophys. Acta,* 1285, 167, 1996; Sujak, A. et al., *Biochim. Biophys. Acta,* 1712, 17, 2005.)

significant in the case of the presence of canthaxanthin, reflects the restriction to the molecular motion of alkyl chains, in the center of the bilayer. The increase of the canthaxanthin concentration above 2 mol% results in a decrease of the effect (Sujak et al., 2005). Such a decrease can be interpreted in terms of the pigment aggregation and separation from the lipid phase. No significant effect has been observed in the case of apolar β-carotene in this particular membrane zone but, interestingly, the molecular motion in the polar headgroup region gains even more freedom, as concluded on the basis of the analysis of the ^{1}H-NMR band corresponding to the choline group (Gabrielska and Gruszecki, 1996). The opposite (ordering) effect with respect to the polar headgroup region has been observed by the same approach in the case of polar carotenoids, zeaxanthin (Gabrielska and Gruszecki, 1996) and canthaxanthin (Sujak et al., 2005). The ordering effect of canthaxanthin with respect into the lipid membranes has also been concluded on the basis of the analysis of the infrared absorption (FTIR) spectra (Sujak et al., 2005). The spectral band corresponding to the scissoring deformation vibrations of the CH_2 groups of alkyl lipid chains (at 1470 cm^{-1}) was shifted toward lower frequencies and became narrower as a consequence of the incorporation of canthaxanthin within the membranes formed with DPPC. Such an effect has been interpreted as a result of the ordering pigment–lipid interactions. Moreover, in the headgroup region, the spectral band corresponding to the stretching vibrations of the C–O–P–O–C group (at 1068 cm^{-1}) was considerably shifted toward lower frequencies, in the membranes modified with canthaxanthin. Such a pronounced shift is typical for hydrogen bond formation and indicates the possible molecular mechanisms of interaction of the ketocarotenoids with lipid membranes. The FTIR analysis of the spectral region corresponding to the methylene group stretching vibrations (~2850 cm^{-1}), for the two-component canthaxanthin-DPPC monolayer, reveals that the presence of the xanthophyll is associated with appearance of a separate, highly ordered membrane region, characterized by the band centered at 2839 cm^{-1} (Sujak et al., 2007a). Interestingly, all the effects observed, accompanied incorporation to the membranes of canthaxanthin at relatively low concentration (0.5 mol%) and were almost absent at higher concentrations (2–5 mol%), promoting the pigment aggregation in the lipid phase (Sujak et al. 2005, 2007a). Detailed information concerning the segmental molecular motion in the carotenoid-modified lipid membranes can be also obtained with the application of the spin label-ESR technique. These aspects are presented in detail in Chapters 9 and 10.

The overall information regarding the effect of carotenoids on the thermotropic phase behavior of lipid membranes can be obtained through the application of the differential scanning calorimetric technique (Castelli et al., 1999; Chaturvedi and Kurup, 1986; Kostecka-Gugala et al., 2003; Rengel et al., 2000; Shibata et al., 2001; Sujak et al., 2007b). In general, both the apolar and the polar carotenoid pigments incorporated into the lipid membranes decreased the cooperativity of the main $P_{\beta}' \rightarrow L_{\alpha}$ phase transition, as manifested by the broadening of the DSC thermograms, decreased the enthalpy of the transition and shifted the transition temperature toward lower values. Such effects clearly demonstrate that the carotenoid additives may be regarded as an "impurity" with respect to the well-ordered liquid-crystalline lipid phase. The local effects of carotenoid pigments incorporated into the membranes, both ordering and acting in the direction of introducing a disorder in the lipid bilayer, are transmitted to the lipid molecules in the fraction that remains in a direct contact with the carotenoids. Carotenoid presence-induced formation of the distinct phases of the membrane can be deduced from a detailed analysis of thermograms, based on the component (Gaussian) analysis (Shibata et al., 2001, 2007b). Interestingly, the thermograms of the canthaxanthin-containing membranes contain the relatively small component shifted to higher temperatures (Sujak et al., 2007b) that can correspond to the minor, highly ordered lipid phase, the existence of which was concluded on the basis of the analysis of the infrared absorption spectra (discussed earlier; Sujak et al., 2007a).

2.3.2 Natural Membranes

There are several reports concerning the modification of the physicochemical properties of biomembranes by the presence of a carotenoid within the lipid phase. Under physiological conditions, all of the xanthophyll pigments are bound to the photosynthetic pigment–protein complexes in the thylakoid membranes (Liu et al., 2004). However, under light stress conditions, a fraction of the pigments involved in the reactions of the xanthophyll cycle (Latowski et al., 2004) appears transiently within the lipid phase of the membrane. It has been shown that the appearance of the xanthophyll cycle pigment, zeaxanthin, in the thylakoid membrane is associated with a decrease in the membrane fluidity (Gruszecki and Strzalka, 1991; Havaux and Gruszecki, 1993; Havaux and Tardy, 1996). The incorporation of exogenous zeaxanthin into the isolated thylakoid membranes also decreases the fluidity of the lipid phase, as demonstrated by the spin label technique (Strzalka and Gruszecki, 1997). The same technique was applied to demonstrate the rigidifying effect of the endogenous carotenoids in the plasma membranes of *Acholeplasma laidlawii* (Huang and Haug, 1974; Rottem and Markowitz, 1979). From an evolutionary standpoint, bacterial membranes share several similarities with the chloroplast membranes. It has been proposed that in the bacterial membranes carotenoids play a similar, membrane-stabilizing role to that of sterols in the membranes of *Eukaryota* (Rohmer et al., 1979). In accordance with this hypothesis, the accumulation of the polar carotenoid pigment, zeaxanthin, has been proposed to be one of the mechanisms that operates in the cell envelope membranes of cyanobacterium *Anacystis nidulans*, to maintain the physiological membrane fluidity level (Gombos and Vigh, 1986; Gombos et al., 1987). Moreover, the enhanced carotenoid production in the membranes of *Staphylococcus aureus*, has been correlated with a decrease in the membrane fluidity (Chamberlain et al., 1991). A very interesting example of the membrane-stabilizing action of polar carotenoids seems to be the presence of the glucoside esters of zeaxanthin (called thermozeaxanthins) in the membranes of thermophilic bacteria such as *Thermus thermophilus* (Hara et al., 1999) or *Erwinia uredovora* (Nakagawa and Misawa, 1991).

ACKNOWLEDGMENTS

The author thanks Prof. J. Sielewiesiuk, Prof. K. Strzalka, Prof. J. Gabrielska, Dr. A. Sujak, Dr. J. Widomska, Dr. W. Grudzinski, Dr. M. Herec, Dr. M. Gagos, Mgr W. Wolacewicz, Mgr Z. Konarzewski, and other coworkers for years of friendly collaboration in the research on carotenoids in membranes.

ABBREVIATIONS

DGDG digalactosyl diacylglycerol
DHPC dihexadecyl phosphatidylcholine
DMPC dimyristoyl phosphatidylcholine
DOPC dioleoyl phosphatidylcholine
DPPC dipalmitoyl phosphatidylcholine
EYPC egg yolk phosphatidylcholine
POPC 1-palmitoyl 2-oleoyl-phosphatidylcholine
SBPC soya bean phosphatidylcholine

REFERENCES

Andersson, P.O., T. Gilbro, and L. Fergusson. 1991. Absorption spectral shifts of carotenoids related to medium polarizability. *Photochem. Photobiol.* 54:353–360.

Britton, G. 1995. Structure and properties of carotenoids in relation to function. *FASEB J.* 9:1551–1558.

Britton, G., S. Liaaen-Jensen, and H. Pfander. 2004. *Carotenoids Handbook*. Basel, Switzerland: Birkhauser Verlag AG.

Castelli, F., S. Caruso, and N. Giuffrida. 1999. Different effects of two structurally similar carotenoids, lutein and beta-carotene, on the thermotropic behaviour of phosphatidylcholine liposomes. Calorimetric evidence of their hindered transport through biomembranes. *Thermochim. Acta* 327:125–131.

Chamberlain, N.R., B.G. Mehrtens, Z. Xiong, F.A. Kapral, J.L. Boardman, and J.I. Rearick. 1991. Correlation of carotenoid production, decreased membrane fluidity, and resistance to oleic acid killing in *Staphylococcus aureus* 18Z. *Infect. Immun.* 59:4332–4337.

Chaturvedi, V.K. and C.K.R. Kurup. 1986. Interaction of lutein with phosphatidylcholine bilayers. *Biochim. Biophys. Acta* 860:286–292.

Gabrielska, J. and W.I. Gruszecki. 1996. Zeaxanthin (dihydroxy-beta-carotene) but not beta-carotene rigidifies lipid membranes: A 1H-NMR study of carotenoid-egg phosphatidylcholine liposomes. *Biochim. Biophys. Acta* 1285:167–174.

Gombos, Z., M. Kis, T. Pali, and L. Vigh. 1987. Nitrate starvation induces homeoviscous regulation of lipids in the cell envelope of the blue-green alga, *Anacystis nidulans*. *Eur. J. Biochem.* 165:461–465.

Gombos, Z. and L. Vigh. 1986. Primary role of the cytoplasmic membrane in thermal acclimation evidenced in nitrate-starved cells of the blue-green alga, *Anacystis nidulans*. *Plant Physiol.* 80:415–420.

Goralczyk, R., F.M. Barker, S. Buser, H. Liechti, and J. Bausch. 2000. Dose dependency of canthaxanthin crystals in monkey retina and spatial distribution of its metabolites. *Invest. Ophthalmol. Vis. Sci.* 41:1513–1522.

Goralczyk, R., S. Buser, J. Bausch, W. Bee, U. Zuhlke, and F.M. Barker. 1997. Occurrence of birefringent retinal inclusions in cynomolgus monkeys after high doses of canthaxanthin. *Invest. Ophthalmol. Vis. Sci.* 38:741–752.

Grolier, P., V. Azais-Breasco, L. Zelmire, and H. Fessi. 1992. Incorporation of carotenoids in aqueous systems: Uptake by cultured rat hepatocytes. *Biochim. Biophys. Acta* 1111:135–138.

Gruszecki, W.I. 1999. Carotenoids in membranes. In *The Photochemistry of Carotenoids*, H. A. Frank, A. J. Young, G. Britton and R. J. Cogdell (eds.). Dordrecht, the Netherlands: Kluwer Academic Publishers, pp. 363–379.

Gruszecki, W.I. 2004. Carotenoid orientation: Role in membrane stabilization. In *Carotenoids in Health and Disease*, N. I. Krinsky, S. T. Mayne, and H. Sies (eds.). New York: Marcel Dekker, pp. 151–163.

Gruszecki, W.I. and J. Sielewiesiuk. 1990. Orientation of xanthophylls in phosphatidylcholine multibilayers. *Biochim. Biophys. Acta* 1023:405–412.

Gruszecki, W.I. and J. Sielewiesiuk. 1991. Galactolipid multibilayers modified with xanthophylls: Orientational and diffractometric studies. *Biochim. Biophys. Acta* 1069:21–26.

Gruszecki, W.I. and K. Strzalka. 1991. Does the xanthophyll cycle take part in the regulation of fluidity of the thylakoid membrane. *Biochim. Biophys. Acta* 1060:310–314.

Gruszecki, W.I. and K. Strzalka. 2005. Carotenoids as modulators of lipid membrane physical properties. *Biochim. Biophys. Acta* 1740:108–115.

Gruszecki, W.I., A. Sujak, K. Strzalka, A. Radunz, and G.H. Schmid. 1999. Organisation of xanthophyll-lipid membranes studied by means of specific pigment antisera, spectrophotometry and monomolecular layer technique lutein versus zeaxanthin. *Z. Naturforsch. C* 54:517–525.

Hara, M., H. Yuan, Q. Yang, T. Hoshino, A. Yokoyama, and J. Miyake. 1999. Stabilization of liposomal membranes by thermozeaxanthins: Carotenoid-glucoside esters. *Biochim. Biophys. Acta* 1461:147–154.

Havaux, M. and W.I. Gruszecki. 1993. Heat- and light-induced chlorophyll a fluorescence changes in potato leaves containing high or low levels of the carotenoid zeaxanthin: Indications of a regulatory effect of zeaxanthin on thylakoid membrane fluidity. *Photochem. Photobiol.* 58:607–614.

Havaux, M. and F. Tardy. 1996. Temperature-dependent adjustment of the thermal stability of photosystem II in vivo: Possible involvement of xanthophyll-cycle pigments. *Planta* 193:324–333.

Huang, L. and A. Haug. 1974. Regulation of membrane lipid fluidity in *Acholeplasma laidlawii*: Effect of carotenoid pigment content. *Biochim. Biophys. Acta* 352:361–370.

Jezowska, I., A. Wolak, W.I. Gruszecki, and K. Strzalka. 1994. Effect of beta-carotene on structural and dynamic properties of model phosphatidylcholine membranes. II. A 31P-NMR and 13C-NMR study. *Biochim. Biophys. Acta* 1194:143–148.

Johansson, L.B.-A., G. Lindblom, A. Wieslander, and G. Arvidson. 1981. Orientation of b-carotene and retinal in lipid bilayers. *FEBS Lett.* 128:97–99.

Kasha, M., H.R. Rawls, and M. Ashraf El-Bayoumi. 1965. The exciton model in molecular spectroscopy. *Pure Appl. Chem.* 11:371–392.

Kolev, V.D. and D.N. Kafalieva. 1986. Miscibility of beta-carotene and zeaxanthin with dipalmitoylphosphatidylcholine in multilamellar vesicles: A calorimetric and spectroscopic study. *Photobiochem. Photobiophys.* 11:257–267.

Kostecka-Gugala, A., D. Latowski, and K. Strzalka. 2003. Thermotropic phase behaviour of alpha-dipalmitoylphosphatidylcholine multibilayers is influenced to various extents by carotenoids containing different structural features—evidence from differential scanning calorimetry. *Biochim. Biophys. Acta* 1609:193–202.

Krinsky, N.I., J.T. Landrum, and R.A. Bone. 2003. Biologic mechanisms of the protective role of lutein and zeaxanthin in the eye. *Annu. Rev. Nutr.* 23:171–201.

Latowski, D., J. Grzyb, and K. Strzalka. 2004. The xanthophyll cycle—molecular mechanism and physiological significance. *Acta Physiol. Plant.* 26:197–212.

Lazrak, T., A. Milon, G. Wolff, A.M. Albrecht, M. Miehe, G. Ourisson, and Y. Nakatani. 1987. Comparison of the effects of inserted C40- and C50-terminally dihydroxylated carotenoids on the mechanical properties of various phospholipid vesicles. *Biochim. Biophys. Acta* 903:132–141.

Liu, Z., H. Yan, K. Wang, T. Kuang, J. Zhang, L. Gui, X. An, and W. Chang. 2004. Crystal structure of spinach major light-harvesting complex at 2.72 A resolution. *Nature* 428:287–292.

McNulty, H.P., J. Byun, S.F. Lockwood, R.F. Jacob, and R.P. Mason. 2007. Differential effects of carotenoids on lipid peroxidation due to membrane interactions: X-ray diffraction analysis. *Biochim. Biophys. Acta* 1768:167–174.

Mendelsohn, R. and R.W. Van Holten. 1979. Zeaxanthin ([3R,3′R]-beta, beta-carotene-3′-diol) as a resonance Raman and visible absorption probe of membrane structure. *Biophys. J.* 27:221–235.

Milanowska, J., A. Polit, Z. Wasylewski, and W.I. Gruszecki. 2003. Interaction of isomeric forms of xanthophyll pigment zeaxanthin with dipalmitoylphosphatidylcholine studied in monomolecular layers. *J. Photochem. Photobiol. B: Biol.* 72:1–9.

Milon, A., G. Wolff, G. Ourisson, and Y. Nakatani. 1986. Organization of carotenoid-phospholipid bilayer systems. Incorporation of zeaxanthin, astaxanthin, and their C50 homologues into dimyristoylphosphatidylcholine vesicles. *Helvet. Chim. Acta* 69:12–24.

Nakagawa, M. and N. Misawa. 1991. Analysis of carotenoid glycosides produced in gram-negative bacteria by introduction of the *Erwinia uredovora* carotenoid biosynthesis genes. *Agric. Biol. Chem.* 55:2147–2148.

N'soukpoe-Kossi, Ch., J. Sielewiesiuk, R.M. Leblanc, R.A. Bone, and J.T. Landrum. 1988. Linear dichroism and orientational studies of carotenoid Langmuir-Blodgett films. *Biochim. Biophys. Acta* 940:255–265.

Parkash, J., J.H. Robblee, J. Agnew, E. Gibbs, P. Collings, R.F. Pasternack, and J.C de Paula. 1998. Depolarized resonance light scattering by porphyrin and chlorophyll a aggregates. *Biophys. J.* 74:2089–2099.

Rengel, D., A. Diez-Navajas, A. Serna-Rico, P. Veiga, A. Muga, and J.C. Milicua. 2000. Exogenously incorporated ketocarotenoids in large unilamellar vesicles. Protective activity against peroxidation. *Biochim. Biophys. Acta* 1463:179–187.

Rohmer, M., P. Bouvier, and G. Ourisson. 1979. Molecular evolution of biomembranes: structural equivalents and phylogenetic precursors of sterols. *Proc. Natl. Acad. Sci. USA* 76:847–851.

Rottem, S. and O. Markowitz. 1979. Carotenoids acts as reinforcers of the *Acholeplasma laidlawii* lipid bilayer. *J. Bacteriol.* 140:944–948.

Shang, Q., X. Dou, and B.S. Hudson. 1991. Off-axis orientation of the electronic transition moment for a linear conjugated polyene. *Nature* 352:703–705.

Shibata, A., Y. Kiba, N. Akati, K. Fukuzawa, and H. Terada. 2001. Molecular characteristics of astaxanthin and beta-carotene in the phospholipid monolayer and their distributions in the phospholipid bilayer. *Chem. Phys. Lipids* 113:11–22.

Socaciu, C., R. Jessel, and H.A. Diehl. 2000. Competitive carotenoid and cholesterol incorporation into liposomes: Effects on membrane phase transition, fluidity, polarity and anisotropy. *Chem. Phys. Lipids* 106:79–88.

Socaciu, C., C. Lausch, and H.A. Diehl. 1999. Carotenoids in DPPC vesicles: Membrane dynamics. *Spectrochim. Acta A Mol. Biomol. Spectrosc.* 55:2289–2297.

Strzalka, K. and W.I. Gruszecki. 1994. Effect of beta-carotene on structural and dynamic properties of model phosphatidylcholine membranes. I. An EPR spin label study. *Biochim. Biophys. Acta* 1194:138–142.

Strzalka, K. and W.I. Gruszecki. 1997. Modulation of thylakoid membrane fluidity by exogenously added carotenoids. *J. Biochem. Mol. Biol. Biophys.* 1:103–108.

Subczynski, W.K., E. Markowska, W.I. Gruszecki, and J. Sielewiesiuk. 1992. Effects of polar carotenoids on dimyristoylphosphatidylcholine membranes: A spin-label study. *Biochim. Biophys. Acta* 1105:97–108.

Subczynski, W.K., E. Markowska, and J. Sielewiesiuk. 1993. Spin-label studies on phosphatidylcholine-polar carotenoid membranes: Effects of alkyl-chain length and unsaturation. *Biochim. Biophys. Acta* 1150:173–181.

Sujak, A., J. Gabrielska, W. Grudzinski, R. Borc, P. Mazurek, and W.I. Gruszecki. 1999. Lutein and zeaxanthin as protectors of lipid membranes against oxidative damage: The structural aspects. *Arch. Biochem. Biophys.* 371:301–317.

Sujak, A., J. Gabrielska, J. Milanowska, P. Mazurek, K. Strzalka, and W.I. Gruszecki. 2005. Studies on canthaxanthin in lipid membranes. *Biochim. Biophys. Acta* 1712:17–28.

Sujak, A., M. Gagos, M.D. Serra, and W.I. Gruszecki. 2007a. Organization of two-component monomolecular layers formed with dipalmitoylphosphatidylcholine and the carotenoid pigment, canthaxanthin. *Mol. Membr. Biol.* 24:431–441.

Sujak, A. and W.I. Gruszecki. 2000. Organization of mixed monomolecular layers formed with the xanthophyll pigments lutein or zeaxanthin and dipalmitoylphosphatidylcholine at the argon-water interface. *J. Photochem. Photobiol. B: Biol.* 59:42–47.

Sujak, A., P. Mazurek, and W.I. Gruszecki. 2002. Xanthophyll pigments lutein and zeaxanthin in lipid multibilayers formed with dimyristoylphosphatidylcholine. *J. Photochem. Photobiol. B: Biol.* 68:39–44.

Sujak, A., W. Okulski, and W.I. Gruszecki. 2000. Organisation of xanthophyll pigments lutein and zeaxanthin in lipid membranes formed with dipalmitoylphosphatidylcholine. *Biochim. Biophys. Acta* 1509:255–263.

Sujak, A., K. Strzalka, and W.I. Gruszecki. 2007b. Thermotropic phase behaviour of lipid bilayers containing carotenoid pigment canthaxanthin: A differential scanning calorimetry study. *Chem. Phys. Lipids* 145:1–12.

van de Ven, M., M. Kattenberg, G. van Ginkel, and Y.K. Levine. 1984. Study of the orientational ordering of carotenoids in lipid bilayers by resonance-Raman spectroscopy. *Biophys. J.* 45:1203–1209.

Widomska, J. and W.K. Subczynski. 2008. Transmembrane localization of cis-isomers of zeaxanthin in the host dimyristoylphosphatidylcholine bilayer membrane. *Biochim. Biophys. Acta* 1778:10–19.

Wisniewska, A. and W.K. Subczynski. 1998. Effects of polar carotenoids on the shape of the hydrophobic barrier of phospholipid bilayers. *Biochim. Biophys. Acta* 1368:235–246.

Woodall, A.A., G. Britton, and M.J. Jackson. 1997. Carotenoids and protection of phospholipids in solution or in liposomes against oxidation by peroxyl radicals: Relationship between carotenoid structure and protective ability. *Biochim Biophys Acta* 1336:575–586.

3 Hydrophilic Carotenoids: Carotenoid Aggregates

Hans-Richard Sliwka, Vassilia Partali, and Samuel F. Lockwood

CONTENTS

3.1 INTRODUCTION

At first glance, the designation "hydrophilic carotenoid" may appear to be an oxymoron. Therefore, the phrase requires more precision: a hydrophilic carotenoid is a highly unsaturated compound, synthetic or natural, which has particular functional groups generating substantial water affinity for the compound. What then is a "carotenoid aggregate"? This term has somehow evaded accurate characterization. In the same sense that a carotenoid protein (carotenoprotein) is not formed by conjugation with carotenoid amino acids, but rather is an inclusion of a carotenoid or carotenoids within a protein macrostructure (Dreon et al. 2007), a carotenoid aggregate is not necessarily understood as an aggregate of pure carotenoids. In fact, many of the investigated carotenoid aggregates consist of carotenoids enclosed in vesicles of common surfactants (Burke et al. 2001, Chen and Djuric 2001). We will henceforth use the expression "carotenoid aggregate" in a strict manner: **carotenoid aggregates are supramolecular assemblies of carotenoid compounds in water and nothing else**. This implies that the carotenoid molecules adhere mutually in a "self-aggregating" process. Another equally justified designation perhaps would be "self-assembling." However, expressed in colloquial style, molecules self-assemble on a surface, forming two-dimensional self-assembling monolayers or Langmuir–Blodgett films (Wolf et al. 1937, Tomoaia-Cotisel and Quinn 1998, Ion et al. 2002, Liu et al. 2002, Miyahara and Kurihara 2004, Foss et al. 2006a). Self-aggregation creates three-dimensional objects or structures. Self-aggregation and self-assembly describe the more general phenomena of self-organization, which is explained within the framework of supramolecular chemistry (Wolf et al. 1937, Lehn 1988, Zana 2004). Intermolecular associations, which create aggregates, can induce properties in the resulting multimolecular structure remarkably different from those of the monomers (Jelley 1936, 1937, Scheibe 1936, 1937). The particular features of the highly dynamic and elastic aggregates are studied in the discipline of "soft matters," an emerging branch of materials science (Hamley and Castelletto 2007).

In nature, carotenoids exist as only two varieties: (1) unelaborated hydrocarbons, or (2) with functional groups, these are always attached via oxygen to the carotenoid skeleton. Carotenoids with heteroatoms other than oxygen have not yet been discovered in nature, but have been synthesized (Pfander and Leuenberger 1976, Sliwka 1999).

In nonpolar organic solvents, hydrocarbon carotenoids generally form colored monomolecular solutions, whereas the biologically relevant solvent, water, typically remains colorless when in contact with hydrocarbon carotenoids. If the water unexpectedly exhibits an orange tint (the highly unsaturated polyene chain acting as a hydrophilic component), the carotenoid concentration obtained will be extremely low. Strangely enough, the first carotenoid aggregates in water were obtained from β,β-carotene, **3.1** (von Euler et al. 1931). Its well-known hydrophobicity did not prevent other studies with β,β-carotene, **3.1**, and lycopene, **3.2**, an acyclic carotenoid hydrocarbon (Song and Moore 1974, Bystritskaya and Karpukhin 1975, Lindig and Rodgers 1981, Mortensen et al. 1997). The many natural carotenols and carotenones (zeaxanthin, **3.3**, lutein, **3.4**, violoxanthin, **3.5**, astaxanthin, **3.6**) are undoubtedly more suited for aggregation studies in water, Scheme 3.1 (Buchwald and Jencks 1968, Ke et al. 1968, Hager 1970, Salares et al. 1977, Mendelsohn and Van Holten 1979, Douillard et al. 1982, Gruszecki et al. 1990, Ruban et al. 1993, Mori et al. 1996, Auweter et al. 1999, Mori 2001, Zsila et al. 2001, Billsten et al. 2005, Köpsel et al. 2005). Yet, the hydrophilicity of these oxygenated carotenoids shown in Scheme 3.1 is still far too low for most practical applications. More specifically, these carotenoids fail to color water-based aliments. In order to overcome this problem, the two big commercial carotenoid producers (BASF, F. Hoffmann-La Roche—DSM) have

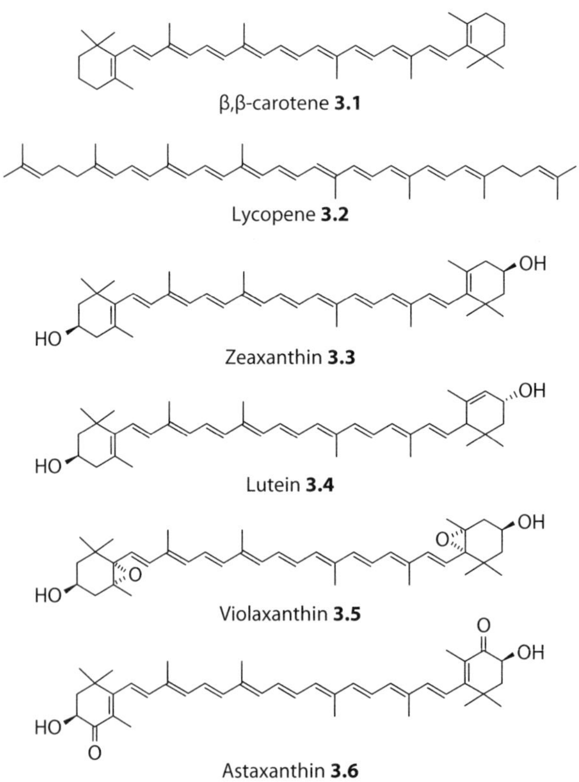

SCHEME 3.1

since 1956 developed elaborate formulation methods to color thousands of liters of soft drinks with dietary carotenoids (Bauernfeind and Howard 1956). The numerous patented formulations consist, in principle, of producing carotenoid-containing nanometer-sized particles, and subsequently coating them with a protective layer. Expressed differently, the carotenoids are incorporated into particles based on common emulsifiers (glycerides, phospholipids) or matrixes (dextrines, starches, sugars) (Inamura et al. 1989, Horn and Rieger 2001, Lockwood et al. 2003). In practical daily use these carotenoid-excipient emulsions/adducts must disperse to form sufficiently small particles to prevent the precipitation that can occur with larger entities, thus causing loss of coloration of the formulated soft drink (Borenstein et al. 1967, Runge et al. 2001). According to our definition earlier in Section 3.1, carotenoid adducts formed with emulsifiers or matrixes will not be considered as carotenoid aggregates, and will not be further mentioned in this chapter.

Since naturally occurring carotenoids typically lack strong hydrophilic functionalities (carotenoid glucosides perhaps being an exception), it is the task of the synthetic chemist to provide carotenoids with a variety of hydrophilic groups to improve the water solubility or dispersibility of this class of compounds. Attaching hydrophilic groups to hydrophobic carotenoids can impart to the resulting synthetic molecules the typical characteristics of a surfactant. Therefore, we further define **"carotenoid aggregates"** as **associations, in a water-based medium, of carotenoid surfactants**. Hydrophilic carotenoids, which aggregate unexpectedly and undesirably (Hertzberg and Liaaen-Jensen 1985, Kildahl-Andersen et al. 2007), in xenobiotic solvents (Okamoto et al. 1989) or never come into contact with water (Slama-Schwok et al. 1992), will not be mentioned in this chapter. Likewise, synthetic modifications of carotenoids that increase hydrophobicity (e.g., "from bad to worse," acetylation of a carotenoid triol to the corresponding triacetate) are also omitted from discussion in this chapter (Bikadi et al. 2002).

3.2 NATURAL HYDROPHILIC CAROTENOIDS

The overwhelming majority of the ~750 known naturally occurring carotenoids are hydrophobic (Britton et al. 2004). It is therefore a striking paradox that the most utilized carotenoid since antiquity is extremely water-soluble: crocin, **3.7**, has no saturation point in water. Crocin, **3.7**, illustrates the typical surfactant structure, the hydrophobic polyene chain being linked to two hydrophilic sugars. Crocin, **3.7**, is surface active, and the molecules associate to small oligomers at high concentration, Scheme 3.2. The surface and aggregation properties of crocin, **3.7**, have only recently been determined (Nalum Naess et al. 2006). Meanwhile, other natural sugar carotenoids have been isolated and characterized, however, the low occurrence and abundance of these "red sugar derivatives" prevents practical applications (Dembitsky 2005). Carbohydrate carotenoids have also been synthesized, unusually without the express intention to explore their properties in water (Pfander 1979). Another group of naturally occurring carotenoids—sulfates—are considerably less hydrophilic; the first characterized compound was bastaxanthin sulfate, **3.9** (Hertzberg et al. 1983). A proposed application of carotenoid sulfates as feed/flesh colorants for cultured fish requires the additional help of an organic solvent for good outcomes (Yokyoyama and Shizusato 1997). The "strange" appearance of the first recorded carotenoid sulfate visible spectrum in water was not immediately recognized as a sign of *H*-aggregation (Hertzberg and Liaaen-Jensen 1985). The aggregation of a carotenoid sulfate was later observed as a negative outcome (Oliveros et al. 1994). Norbixin, **3.8**, is the other carotenoid utilized since ancient times; it is reported to be water-soluble up to 5%, Scheme 3.2 (colorMaker, Inc.). Recent measurements could not confirm solubility; only negligible dispersibility was observed (Breukers et al. 2009).

3.3 SYNTHETIC HYDROPHILIC CAROTENOIDS

The late emergence of hydrophilic, synthetically modified carotenoids is probably the result of a too well-respected principle by traditional carotenoid chemists: "synthesize carotenoids—don't synthesize with carotenoids!" Indeed, except for some early reported functional group transformations

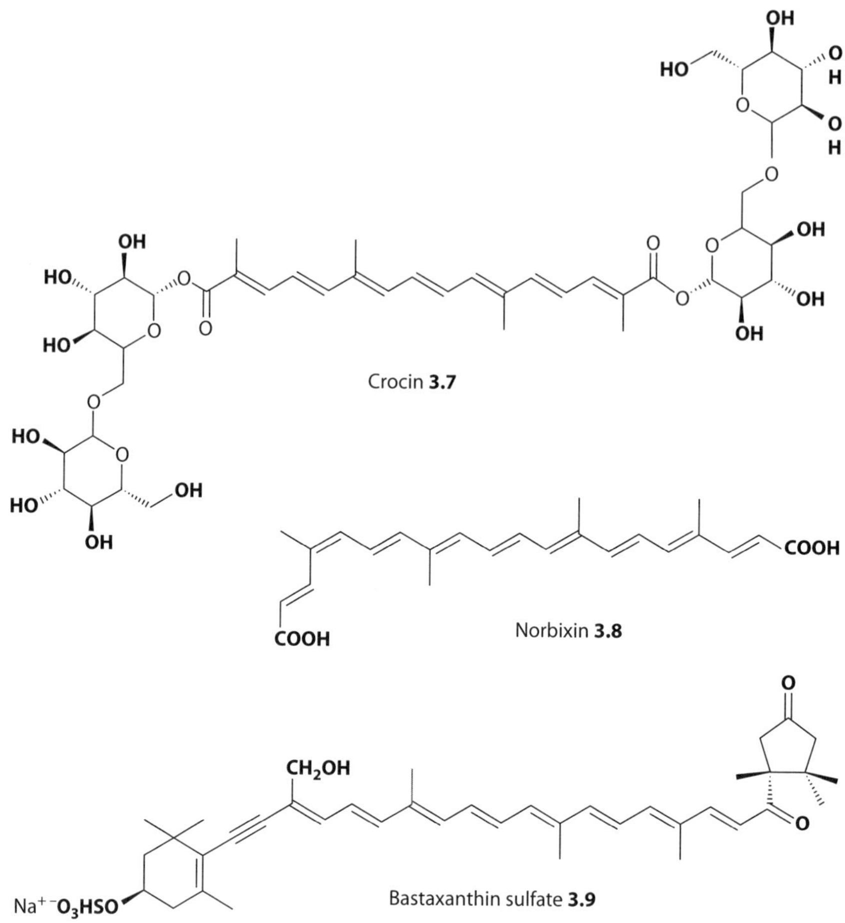

SCHEME 3.2

(Liaaen-Jensen 1971), syntheses with carotenoids were often troubled by unexpected difficulties leading to disappointing low product yields (Widmer et al. 1982). Nevertheless, the carotenoid chemist's code of conduct has been increasingly violated in recent years. Still, those neophytes who look at the synthetic schemes and think of straightforward and trouble-free organic reactions may keep in mind that even ester hydrolysis of carotenoids can become unexpectedly difficult (Larsen et al. 1998, Reddy et al. 2002). The initial topic of a PhD thesis was abandoned for the simple enough reason that it was not possible to find an appropriate method for hydrolyzing ethyl esters of long chain carotenoid diacids (Meister 2004). At times, well-established reactions do not succeed when employed with carotenoids and, occasionally, customary work-up procedures fall short of expectations; compare Sliwka and Liaaen-Jensen (1993a,b) with Kildahl-Andersen et al. (2004) and Liaaen-Jensen (1996) with Oliveros et al. (1994).

There are two approaches to synthesizing hydrophilic carotenoids: (1) appending a hydrophilic group to the carotenoid scaffold (Foss et al. 2006a) or (2) joining a carotenoid to a hydrophilic compound, Scheme 3.3 (Foss et al. 2003). Whereas the Scheme 3.3 intuitively explains the difference, these techniques cannot be clearly separated in praxis; the distinction may appear more emotional than conceptual. Both methods are habitually hampered by low yields, find their limits in the availability of functionalized carotenoids, and cause problems in the work-up procedure due to the amphiphilic character of the products.

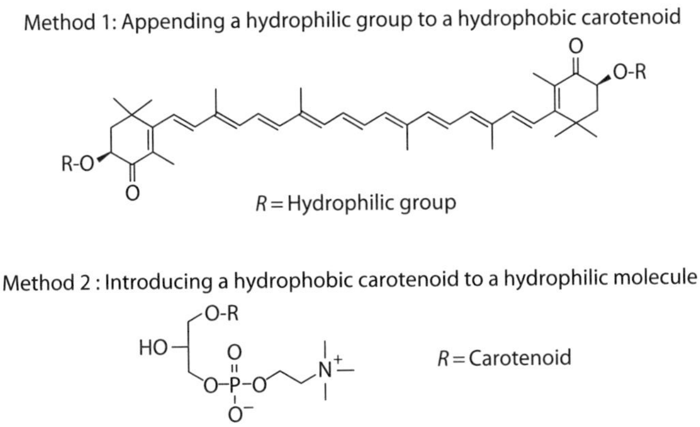

SCHEME 3.3

The first intentional synthesis of a hydrophilic carotenoid according to Method 2 of Scheme 3.3 was published in 1996 (Partali et al. 1996). Glycerol, **3.10**, was enzymatically esterified with a highly unsaturated fatty acid ester, **3.11**, to the monoglyceride, **3.12**, Scheme 3.4. Regrettably, a simple test with this unsaturated monoglyceride—adding water to the compound with subsequent shaking—did not result in "solution-coloring" properties. Therefore, a different approach was investigated. The biosynthesis of carotenoids is based on hydrophilic diphosphate intermediates, e.g., **3.13**, Scheme 3.5, and carotenoids have been previously incorporated into vesicles of phospholipids, Figure 3.1 (Milon et al. 1986, Britton 1998). It was therefore reasonable to connect carotenoids with a phosphate group, above all because hydrophobic phosphate esters had previously been synthesized (Sliwka 1997). The C30-monoglyceride, **3.12**, was therefore used as an educt for the synthesis of the zwitterionic, hydrophobic C30-lysophosphocholine, **3.15**, via aminolyse of the bromo phosphoester, **3.14**, Scheme 3.6 (Foss et al. 2003, 2005a). Enantiomeric (*R*)-**3.15** was also later prepared by direct esterification of the C30-acid, **3.11**, with glycerol phosphocholine, (*R*)-16, Scheme 3.6 (Foss et al. 2005b). The aggregates of phospholipid, **3.15**, demonstrate the competitive edge of hydrophilic carotenoids: to date, the so-called carotenoid phospholipid aggregates were heterogenic mixtures of two or more compounds, whereas the phospholipid, **3.15**, developed homogenous aggregates, which are intrinsically biocompatible, Figure 3.1 (Milon et al. 1986, Socaciu et al. 1999, Sujak et al. 2000, Shibata et al. 2001, Jemioła-Rzemińska et al. 2005). Admittedly, the synthesis of the

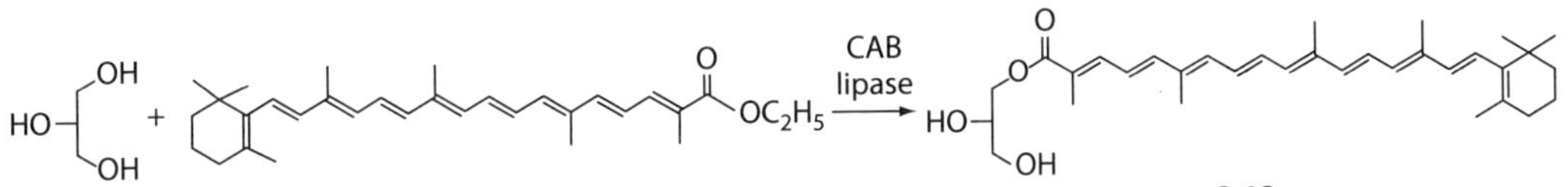

3.10 **3.11** **3.12**

SCHEME 3.4

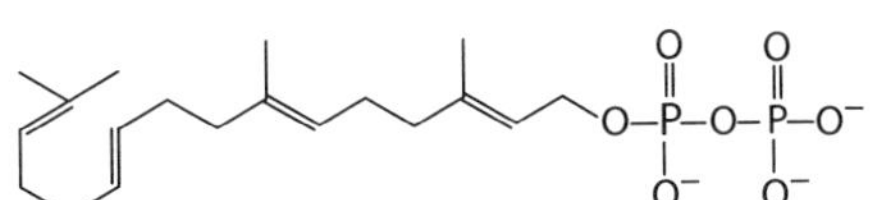

3.13

SCHEME 3.5

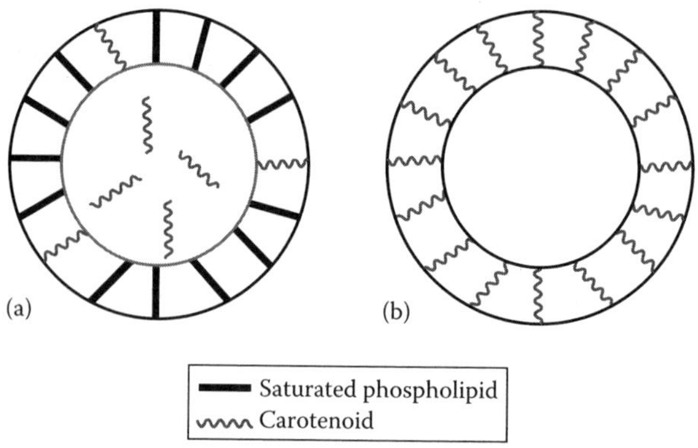

FIGURE 3.1 (a) "Carotenoid aggregate" from saturated surfactants with enclosed carotenoids. (b) Real carotenoid aggregate built of carotenoid surfactants.

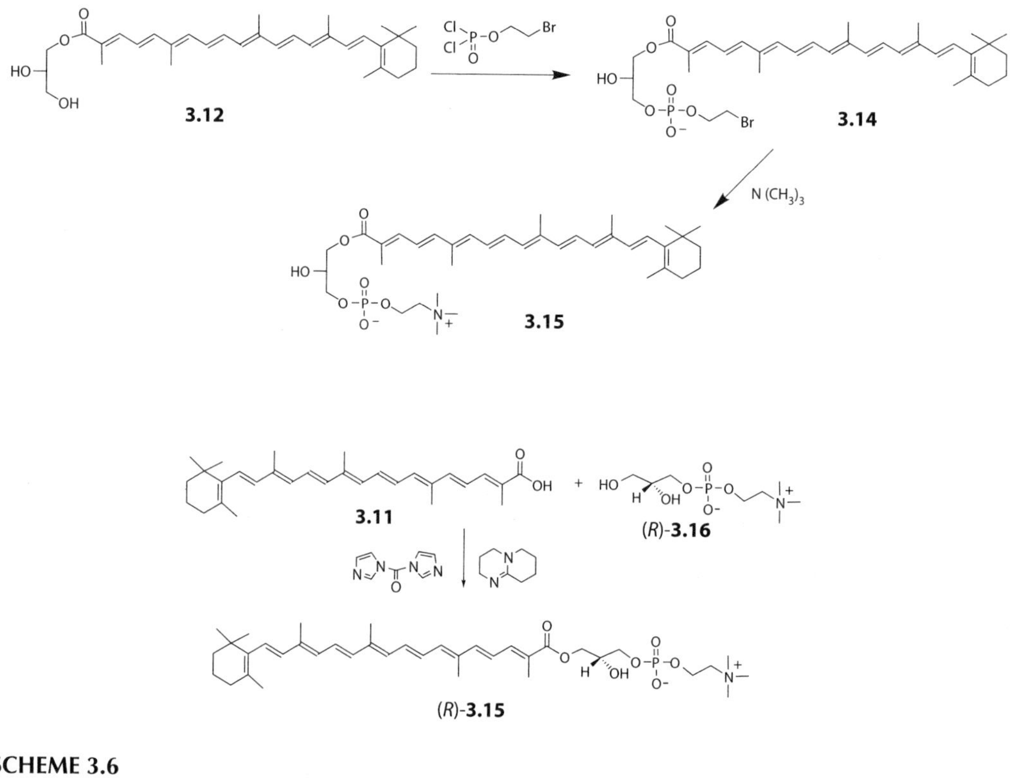

SCHEME 3.6

lyso compound, **3.15**, is tricky and, not astonishingly, previous attempts in synthesizing carotenoid phosphatidylcholines had failed (Benade 2001). Later, sodium phosphate groups were successfully introduced to lutein, **3.4**, and lycophylldiol, resulting in the phosphoesters, **3.17** and **3.18**, respectively, Scheme 3.7 (Foss et al. 2006a). Groups at Hawaii Biotech, Inc., Albany Molecular Research, Inc., and Cardax Pharmaceuticals, Inc. systematically exploited Method 1, Scheme 3.3, for the synthesis of hydrophilic carotenoids. Hydroxy carotenoids such as zeaxanthin, **3.3**, lutein, **3.4**, and especially astaxanthin, **3.6**, were systematically modified with a multitude of different hydrophilic groups, which were connected to the two hydroxyl groups in these carotenoids. This approach was

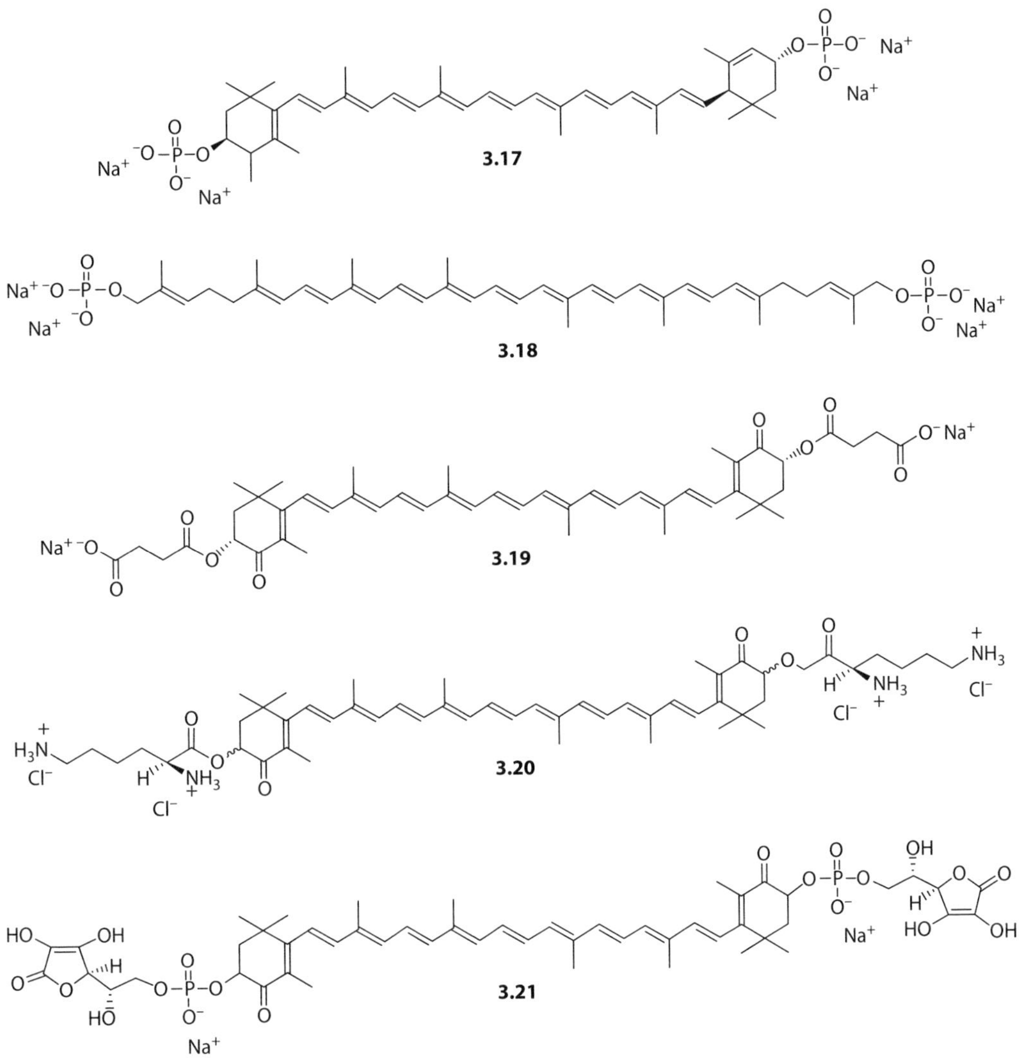

SCHEME 3.7

very successful and resulted in a remarkable number of hydrophilic carotenoids; several are shown in Schemes 3.7 through 3.9.

Cardax™ (disodium disuccinate astaxanthin, **3.19**), the first synthesized hydrophilic compound in the astaxanthin series, can today be produced in kg amounts by Cardax Pharmaceuticals, Inc., Aiea, Hawaii (Frey et al. 2004). A highly hydrophilic astaxanthin dilysine conjugate, **3.20**, although not outperforming natural crocin, **3.7**, in solubility, surpasses its natural counterpart as colorant. The lysine conjugate, **3.20**, forms deep red solutions in water and other solvents and, similar to crocin, **3.7**, aggregates only at high concentrations (Nalum Naess et al. 2006, 2007). In both compounds, **3.19** and **3.20**, the conversion of astaxanthin, **3.6**, to more soluble or dispersible compounds resulted in antioxidant activity in aqueous formulations; however, this antioxidant capacity was primarily based on the astaxanthin scaffold, and not to the conjugating moieties. Much work and resources were subsequently devoted to the combination of hydrophilic ascorbic acid, **3.22**, with several carotenoids, in particular astaxanthin, **3.6**. The many difficult initial attempts finally succeeded gratifyingly in an astaxanthin-vitamin C derivative, using a phosphate group as linker, **3.21** (Lockwood

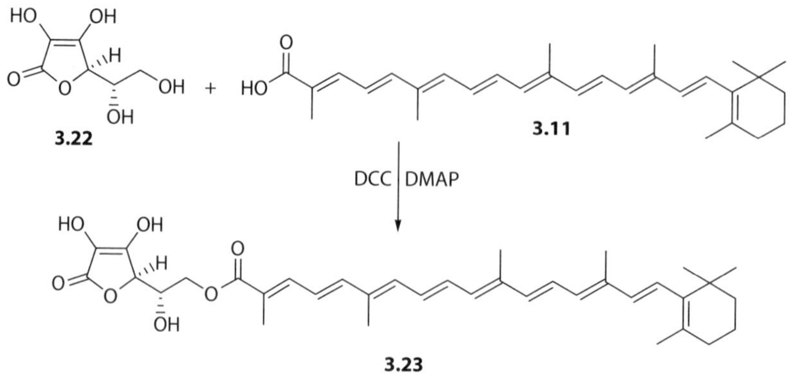

SCHEME 3.8

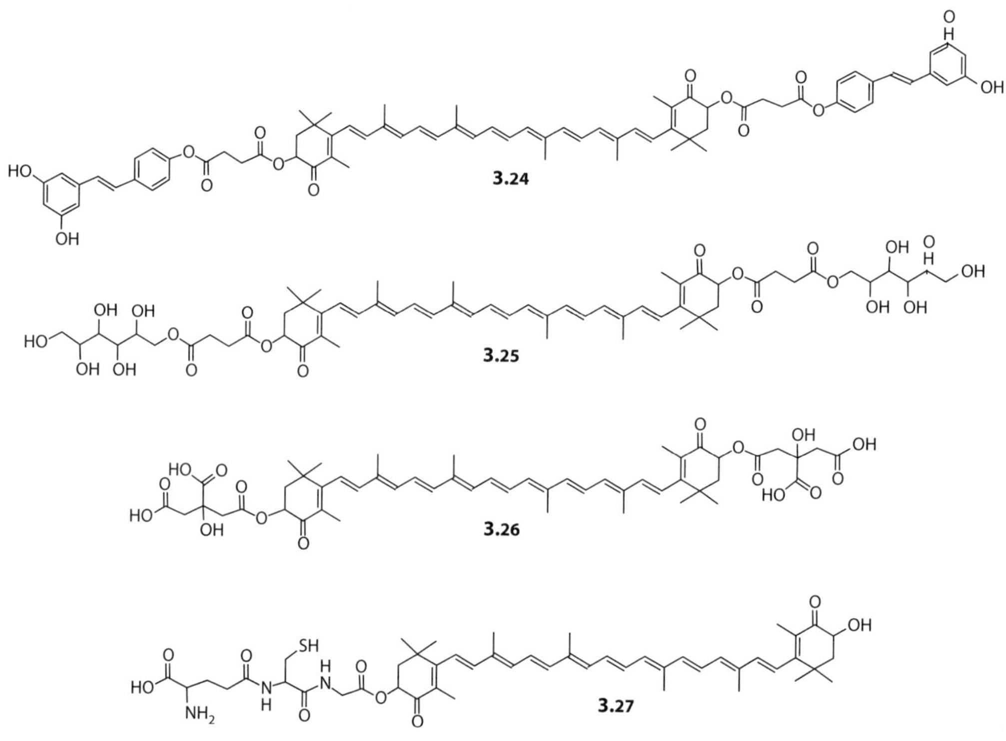

SCHEME 3.9

et al. 2005). This conjugate employed a maximum of the finesse of the synthetic chemistry inherent in Method 1 of Scheme 3.3, and resulted in a highly hydrophilic carotenoid, with orders of magnitude increased antioxidant capacity compared with astaxanthin itself. In analogy to the antioxidant food additive ascorbyl palmitate, the corresponding ascorbyl C30-carotenoate, **3.23**, was synthesized. In many instances the compound barely survived the workup procedure, and the synthesis could not always be reproduced, Scheme 3.8 (Lerfall 2002). In contrast, retinoyl-ascorbic acid is "conveniently prepared" (Yamano and Ito 1998) as well as the ascorbylester of norbixin, **3.8** (Humeau et al. 2000). The biophysical, biological, and potential human medical applications of several of these compounds are discussed later in this chapter, in both *in vitro* and *in vivo* proof-of-concept studies performed in most cases by the authors and their collaborators.

No stability problems were encountered when hydroxycarotenoids were combined with the hydrophilic antioxidant resveratrol, **3.24** (Lockwood et al. 2005), Scheme 3.9. New glyco compounds were added to the register of synthesized carotenoid sugars and carotenoid sugar alcohols, this time intentionally prepared for use in water: astaxanthin combinations with maltose, mannitol, and sorbitol, **3.25** (Lockwood et al. 2005). Further, the combination of astaxanthin-citric acid, **3.26**, and astaxanthin-glutathione, **3.27**, were obtained. The hydrophilic norcarotenoid diketones of violerythrin type (five-membered ring), **3.28**, could become interesting compounds as blue food colorants; their antioxidant ability is also very powerful (Lockwood et al. 2007), Scheme 3.10. Carotenoids with hydroxybenzene rings were first obtained in Düsseldorf; later, in Hawaii, the renieratene type carotenoid, **3.29**, was used as a parent compound for attaching hydrophilic groups (Korger 2005, Lockwood et al. 2007). Unfortunately, except for minor exceptions, the available amounts of these new hydrophilic carotenoids did not reach the necessary level for surface and aggregation studies. The phosphocholine, **3.30**, was again synthesized for self-aggregation; (Foss 2005) whereas the intention for the synthesis of the carotenoid-selenium-lipid, **3.31**, was rather for self-assembly studies, Scheme 3.11 (Foss et al. 2006b). Likewise, carotenoid phospholipids with saturated fatty acids of different chain lengths, **3.32–3.34**, and the carotenoid cholinester, **3.35**, were prepared predominantly as DNA protecting and delivery agents, although aggregation was thoroughly studied (C.L. Øpstad, Trondheim, unpublished). An intuitive approach to hydrophilic, surface active carotenoids would be the preparation of "orange soaps," alkali salts of carotenoid acids. Some potassium and sodium salts of carotenoid acids with variable chain lengths are now under investigation (e.g., potassium C30-carotenoate, **3.36**, potassium C20–C35 carotenoate, Scheme 3.11) (Foss et al. 2006b) (I.L. Alsvik, Trondheim, unpublished). Another straightforward method to introduce a hydrophilic group would be the oximation of ketocarotenoids; oximation is one of the few reactions of carotenoids with full conversion, and the oxime hydroxy group is expected to increase hydrophilicity. However, the hydrophilicity of the echinenon oxime, **3.37**, was disappointingly low, and its aggregation behavior could only be studied in acetone–water mixtures, Scheme 3.12 (Benade 2001). Improved hydrophilicity can easily be acquired when carotenoid oximes are reacted with HCl gas to oximium salts. Alas, even the oximium salt, **3.38**, was not hydrophilic enough to be used for studying surface properties in water (Willibald et al. 2009).

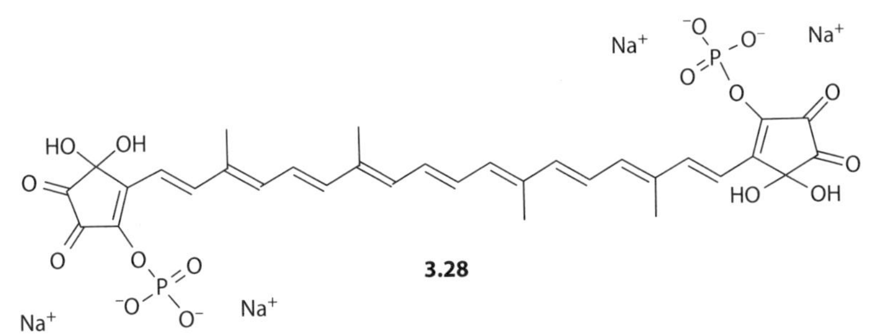

3.28

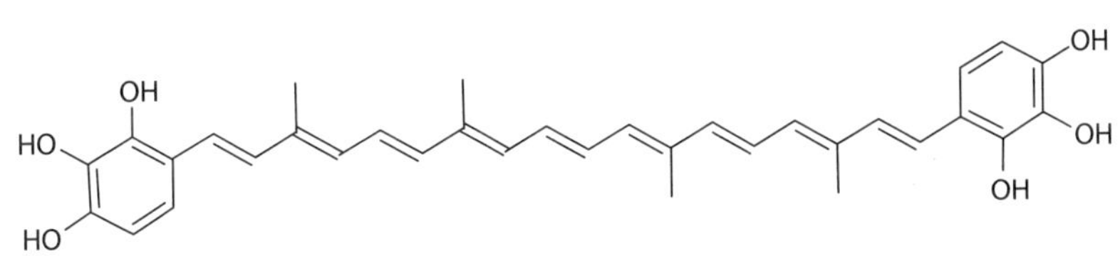

3.29

SCHEME 3.10

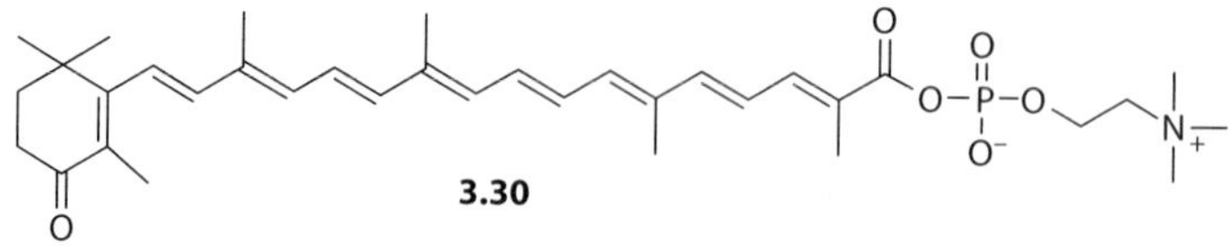

3.30

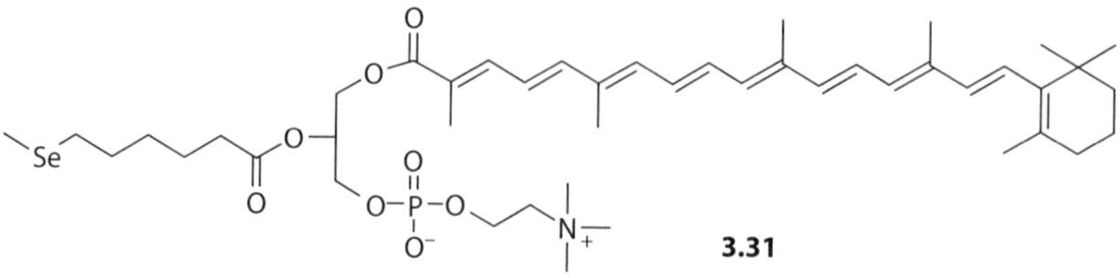

3.31

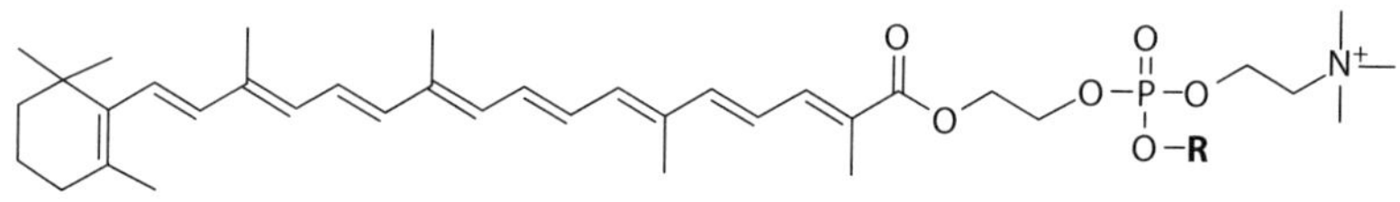

R = C2 **3.32** *R* = C6 **3.33** *R* = C12 **3.34**

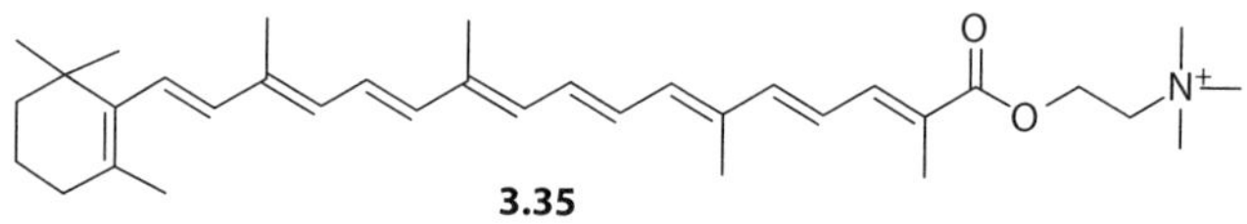

3.35

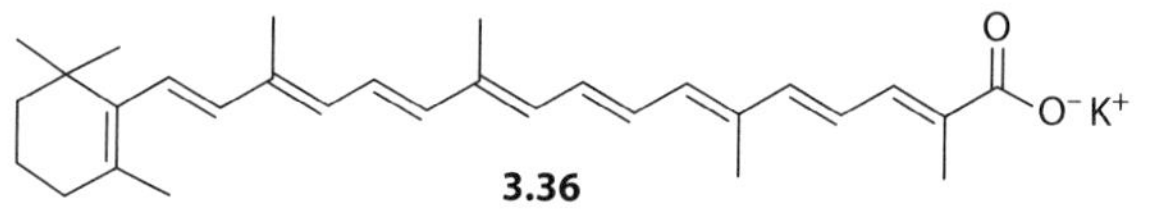

3.36

SCHEME 3.11

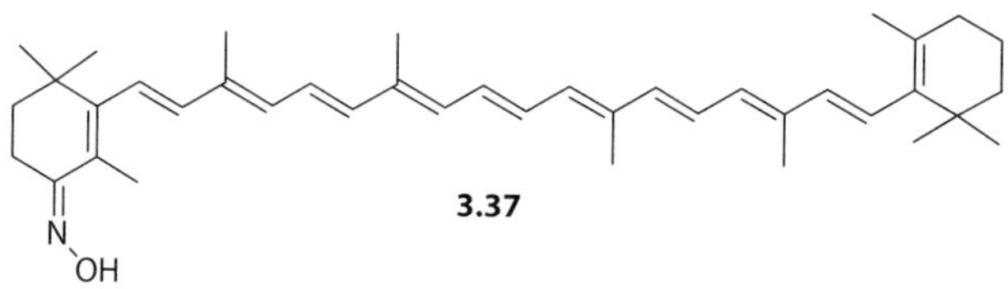

3.37

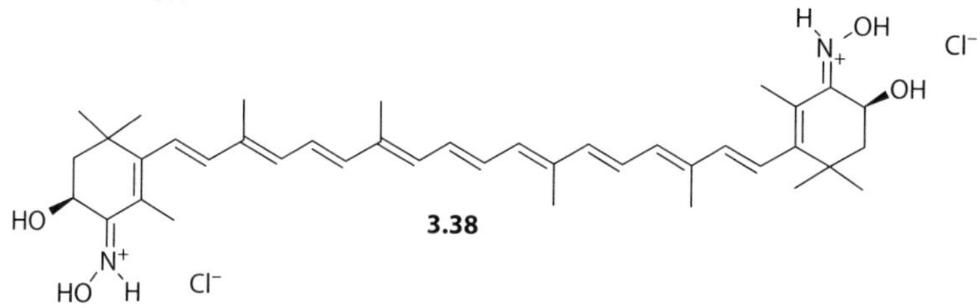

3.38

SCHEME 3.12

3.4 SURFACE PROPERTIES

Hydrophilic carotenoids behave as typical amphiphiles. The contact angle of a water drop on a dry film of the phospholipid, **3.15**, and its lifetime before spreading were signs of noticeable surfactant properties (Foss et al. 2005a). When amphiphiles are in contact with water, the molecules move to the surface; there the hydrophilic part is anchored in water, and the hydrophobic part is outstretched to the air, thus exhibiting the surface properties of a hydrocarbon solvent with decreased surface tension. When the water surface is completely occupied, the molecules are prevented from further movement to the surface. They then have to stay in the water phase where they now form energetically favored aggregates, in which the hydrophophic chains orient to the interior, the hydrophilic groups to the exterior. The concentration at which these phenomena occur is defined as critical aggregate concentration c_M corresponding to the saturated surface concentration Γ, Figure 3.2. The concentration c_M is generally determined with a tensiometer by measuring the surface tension γ in relation to the concentration *c* of the surfactant; the pendant drop method gave similar results (Foss et al. 2005a). Γ is calculated via c_M. Γ can also be measured directly by neutron reflectivity, which is, however, an elaborate, sedentary and therefore seldom used technique (Li et al. 1999). The parameters Γ and c_M allow the calculation of the molecular area a_m at the water–air interphase, as well as the equilibrium constants *k* between molecules at the surface, in the bulk and in the aggregates. A representative surface tension–concentration plot is shown in Figure 3.3. The assigned values for some surface and aggregation properties of several hydrophilic carotenoids discussed in this chapter are listed in Table 3.1. In theory, aggregation should only occur beyond c_M. Nonetheless, it was verified spectroscopically that the aggregation of the phospholipid, **3.15**, starts at exceedingly low concentrations ($c \leq 10^{-9}$ M) (Foss et al. 2005a). UV–visible (UV–VIS) spectroscopy is an obvious

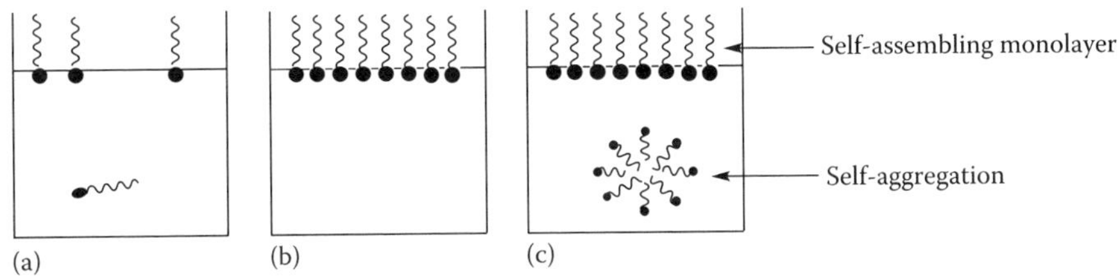

FIGURE 3.2 (a) Surface not saturated $\Gamma < \Gamma_{max}$, bulk concentration $c << c_M$; (b) surface saturated $\Gamma = \Gamma_{max}$, bulk concentration $c = 0$; (c) surface saturated $\Gamma = \Gamma_{max}$, bulk concentration $c > 0$, aggregation starts = critical aggregation concentration c_M. Surfactant molecules form (1) in a first step a self-assembling monolayer at the surface, and (2) in a second step, when the surface is saturated, the molecules self-aggregate in the bulk solution.

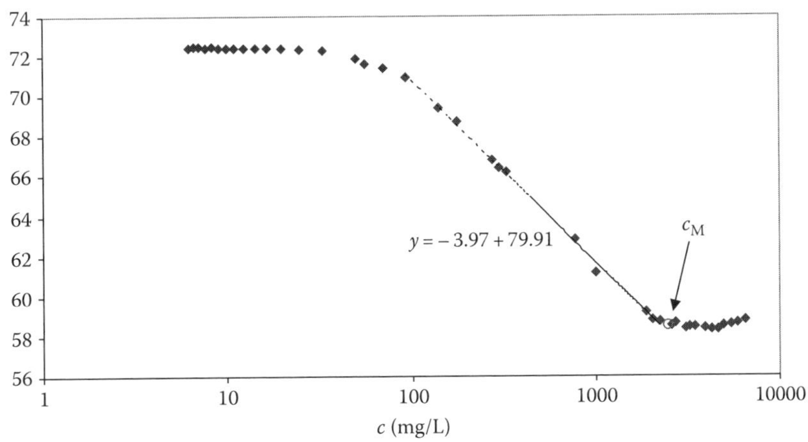

FIGURE 3.3 Surface tension γ plotted against the concentration *c* of lysine derivative **3.20**. Critical aggregate concentration (○) $c_M = 2430$ mg/L = 2.43 mM, $\gamma_{c_M} = 58$ mN/m. (Reprinted from Nalum Naess, S. et al., *Chem. Phys. Lipids*, 148, 63, 2007. With permission.)

TABLE 3.1
Surface and Aggregate Properties

	γ (mN/m)	c_M ($\times 10^{-3}$)	Γ ($\times 10^{-6}$ mol/m^2)	a_m (Å^2)	r_H (nm)
Phospholipid **3.15** (Foss et al. 2005b)	57	1.3	4.5	39	8 and 100
Crocin **3.7** (Nalum Naess et al. 2006)	52	0.82	1.4	115	150
Cardax **3.19** (Foss et al. 2005c)	60	0.45	0.7	240	1300
Lysine derivative **3.20** (Nalum Naess et al. 2007)	58.5	2.18	0.7	240	110
Phospholipid C2 **3.32** (C. L. Øpstad, Trondheim, unpublished)	47	1.66	2.4	71	290
Phospholipid C6 **3.33** (C. L. Øpstad, Trondheim, unpublished)	50	1.11	2.1	81	235
C30 acid salt **3.36** (Foss 2005)	48	1.1	2.5	66	

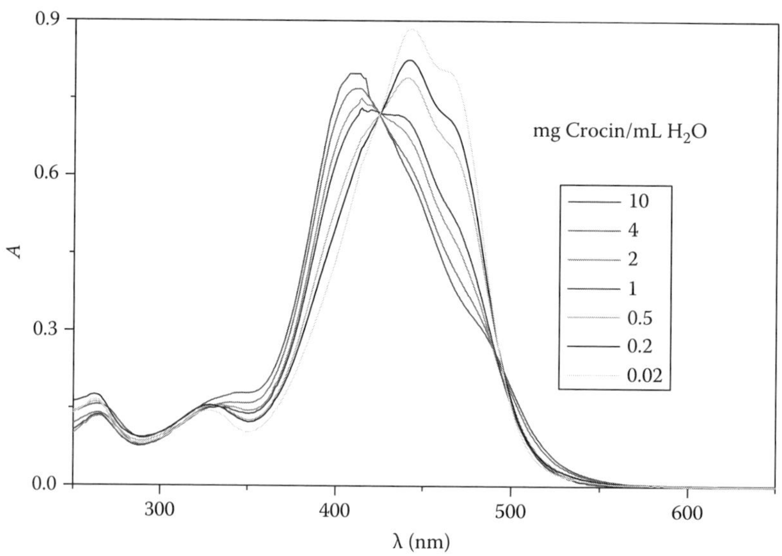

FIGURE 3.4 UV–VIS spectra of crocin **3.7** monomers (low concentration of crocin in water, $\lambda=445$ nm) and crocin aggregates (high concentration of crocin **3.7** in water, $\lambda=410$ nm). Monomer–aggregate equilibrium concentration $c=1$ mg/mL, cf. $c_M=0.8$ mg/mL from tensiometric determination. (Reprinted from Nalum Naess, S. et al., *Helv. Chim. Acta*, 89, 45, 2006. With permission.)

alternative way to determine c_M for hydrophilic carotenoids, provided the absorption maxima of the monomer and aggregate are easily distinguishable. If this is the case the concentration where aggregates and monomers are in equilibrium corresponds to c_M. The UV–VIS spectroscopically determined c_M was consistent with the c_M from tensiometric measurements, Figure 3.4 (Nalum Naess et al. 2006, 2007).

3.5 AGGREGATE STRUCTURE

"Aggregate" is a general term for molecular associations. In textbooks, aggregates are often represented as spherical structures, with good reason—since a ball or sphere is a geometrically, gravitationally, and energetically favored structure. Simple aggregates of spherical shape are micelles.

However, it is obvious that bolaamphiphiles (molecules that have hydrophilic groups at both ends of a hydrophobic hydrocarbon chain) such as crocin, **3.7**, Cardax, **3.19**, or the lysine compound, **3.20**, cannot form micelles; self-association of these molecules builds other edifices. The morphology of an aggregate can easily be predicted by determining the critical packing parameter (*cpp*), a number obtained by dividing the volume of the hydrophobic part v_L by the product of the length of the hydrophobic part l_L and the molecular area a_m, $cpp = v_L/l_L a_m$ (Israelachvilli et al. 1976). According to the calculated value, spherical, cylindrical, and bilayer structure aggregates are probable. Whereas a_m is derived from experimental values, v_L and l_L have to be calculated from molecular models. It is, however, difficult to estimate l_L, since a considerable part of the carotenoid chain is dragged into water due to the weak hydrophilicity of double bonds. The lysophospholipid, **3.15**, with its C17:8 chain (ring and methyl groups exert no significant influence on γ) corresponds to a lysophospholipid with a C10:0 or C11:0 saturated chain (Foss et al. 2005a). The ccp concept was originally developed for saturated carbon chains. (The hydrophobicity of unsaturation has no significance for the effective chain lengths of bolaamphiphiles (Foss et al. 2005c).)

The size of carotenoid aggregates have been determined by dynamic light scattering (DLS), a noninvasive method (Santos and Castanho 1996). DLS also allows distinguishing between spherical or cylindrical aggregates. The hydrodynamic radii r_H of hydrophilic carotenoids in water are given in Table 3.1. Size and molecular structure of the bolaamphiphiles crocin, **3.7**, and Cardax, **3.19**, indicate nonspherical aggregates. The aggregates of the dianionic Cardax, **3.19**—in water $r_H = 1.3\,\mu m$—slightly decreased when dispersed in physiologically relevant sodium chloride (NaCl) solutions, and then increased to $r_H = 3\,\mu m$ in 0.5 M NaCl, and to $r_H = 10\,\mu m$ in 2.0 M NaCl, Figure 3.5 (Foss et al. 2005c). The DLS-determined aggregate size of $r_H = 110\,nm$ for the lysine derivative, **3.20**, in pure water was confirmed by transmission electron microscopy (TEM) examinations, but the aggregates appeared sometimes globular, Figure 3.6, and sometimes rod-shaped. In contrast to anionic Cardax, **3.19**, the aggregates of cationic lysine derivative **3.20** did not grow or shrink in NaCl solutions (Nalum Naess et al. 2007).

The cholinester, **3.35**, formed aggregates in pure water with $r_H = 250\,nm$; after adding NaCl solutions of differing concentrations, the aggregates increased in size up to $r_H = 900\,nm$. After standing 48 h, the aggregates had returned to their initial size $r_H = 250\,nm$. When a saturated aqueous

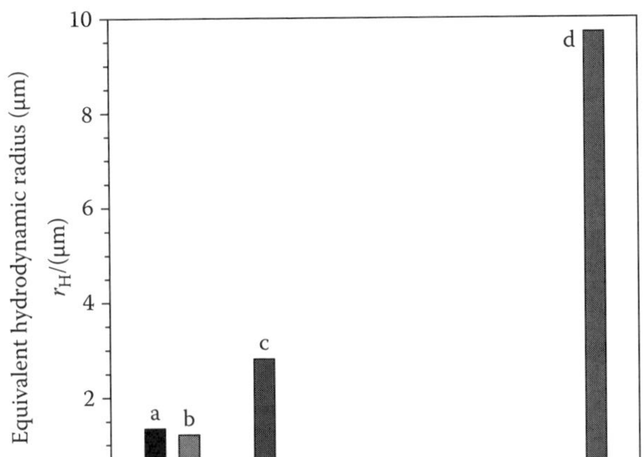

FIGURE 3.5 Cardax **3.19** forms nonspherical aggregates with an equivalent hydrodynamic radius (a) $r_H = 1.3\,mm$ (water), (b) $r_H = 1.2\,mm$ (0.155 M NaCl) (believed to be due to osmotic shrinkage), (c) $r_H = 3\,mm$ (0.5 M NaCl), and (d) $r_H = 10\,mm$ (2.0 M NaCl). (Reprinted from Foss, B.J. et al., *Chem. Phys. Lipids.*, 135, 157, 2005c. With permission.)

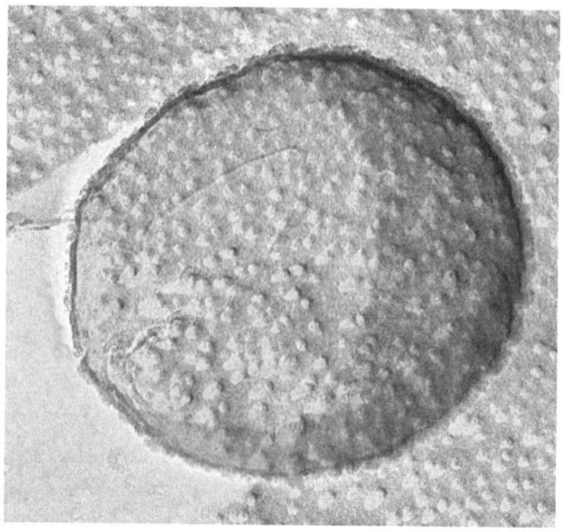

FIGURE 3.6 TEM photo of lysine derivative **3.20** showing an aggregate of $r_H = 100\,nm$. (From Nalum Naess, S. and Elgseter, A., Trondheim, unpublished.)

dispersion of **3.35** was prepared from an ethanolic stock solution, aggregates of $r_H = 1000\,nm$ were observed, which after 48 h had again contracted to aggregates of $r_H = 250\,nm$ (C.L. Øpstad, Trondheim, unpublished). The size(s) of the different aggregate dispersions converted to a common value after standing, regardless of the starting conditions.

If the size of the aggregate is known, and provided that the aggregate is a unilamellar vesicle with a geometrically defined structure (globule, ellipsoid), then the aggregation number *N* can be derived from the calculated aggregate surface area and the molecular area at the water–air interphase a_m. *N* for aggregates of the phospholipid, **3.15**, has been estimated (Foss et al. 2005a). Uncertainty about the exact morphology of the aggregate and its interior prevents a reliable determination of *N*. Whereas DLS and TEM screen the exterior of aggregates, UV–VIS spectroscopy allows the observer to evaluate the molecular arrangement inside the aggregates. A carotenoid solution in a water-miscible organic solvent absorbs at a certain λ_{max}. After adding water, a carotenoid-aggregate dispersion is formed and λ_{max} is shifted to lower or longer wavelengths; in some cases, both variations are observed. The shift in absorption is induced by weak intermolecular arrangements of the polyene chains in the aggregate, leading to a combined absorption, the exciton absorption (exciton coupling) (Davydov 1962). Exciton absorption is dependent on the type of molecular alignment: the horizontal "card-pack" orientation of the molecules forms hypsochromic-shifted *H*-aggregates, whereas the "head-to-tail" alignment of the molecules gives rise to *J*-aggregation (Horn and Rieger 2001). The exciton absorption of *H*-aggregates represents the interaction of chromophores, whose transition dipoles are oriented in a parallel alignment. For *J*-aggregates, the combined dipole transitions have to be oriented in the same direction in order to give rise to the typical bathochromic shift. *H*- and *J*-aggregates represent extreme cases. In praxis, the molecules do not aggregate exclusively in one of these arrangements, Figure 3.7. So far most of the investigated hydrophilic carotenoids prefer to arrange themselves in *H*-aggregates with minor contributions of the *J*-species; notable exceptions are the C30-aldoxime hydrochloride, **3.39**, the echinenone oxime hydrochloride, **3.40**, and canthaxanthin oxime hydrochloride, **3.41**, which form *J*-aggregates (Willibald et al. 2009), Scheme 3.13 and Table 3.2.

Aggregation is sensitive to subtle conditions during formation. Benade and Korger have carefully determined the aggregating preference for 33 carotenoids, adding acetone (ethanol) to the carotenoids in an acetone–water (ethanol–water) mixture (Benade 2001, Korger 2005). Nonetheless, their general conclusion on structure relationship for *H*- or *J*-aggregates may only be valid for

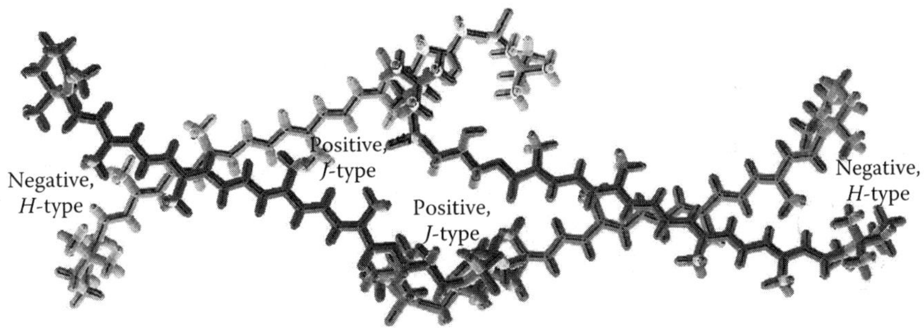

FIGURE 3.7 (See color insert following page 336.) Molecular arrangements for *H*- and *J*-aggregates. Tetrameric lysophospholipid (*R*)-**3.15** forms predominantly *H*-aggregates in addition to a small percentage of *J*-aggregates. The calculated VIS absorption of the tetramer (*R*)-**3.15** is in accordance with the experimental VIS spectra. (Reprinted from Foss, B.J. et al., *Chem. Eur. J.*, 11, 4103, 2005b. With permission.)

the specific employed experimental conditions. It may be possible that the aggregate preference for the investigated carotenoids changes when the conditions are reversed, by adding water to the carotenoid–acetone solution. When water was successively added in small increments to a methanolic solution of the astaxanthin oximium hydrochloride, **3.38**, *H*-aggregates were formed, however when the hydrochloride, **3.38**, was immediately dispersed in water, *J*-aggregates were found, Figure 3.8. Measured in the laboratory of synthesis, the phospholipid, **3.15**, gave an *H*-aggregate with λ_{max} = 380 nm. When another sample of **3.15** was later dispersed in the spectroscopy laboratory, the *H*-aggregates absorbed at 390 nm. Afterward, **3.15** was again dispersed in the laboratory of synthesis and now formed *H*-aggregates with absorption at 400 nm. Measurements with subsequently synthesized batches of **3.15** demonstrated the same alternation among the three varieties of aggregate absorption. The different aggregate dispersions were stable and did not convert to a common absorption value over time. The dependence of aggregate absorption on the method of dilution had been previously observed with a dihydroxycarotenone (Simonyi et al. 2003). Obviously, the formation and size of specific aggregates is neither exactly reproducible nor predictable. The preparation of carotenoid aggregates with predefined dimensions has to rely on other methods (E.M. Sandru, Trondheim, unpublished).

The absorption band of a monomolecular dissolved molecule expresses the energy between the molecule's ground and its excited state. In dispersions, the number of molecules associated in an aggregate can be quite high, e.g., in aggregates of palmitoylglycerophosphocholine *N* = 900 (Hayashi et al. 1994), and in heterogeneous inclusion aggregates *N* = 10,000 carotenoid molecules (Horn and Rieger 2001). Does the exciton band represent the interaction of all the many chromophores in the aggregate? Similar to a crystal, in which the crystal unit determines the properties regardless of the crystal's size, a small aggregation unit may express the properties of aggregates regardless of *N*. So far, only a couple of carotenoid aggregates have been studied with the intention to locate a simple molecule arrangement. The calculated aggregation spectra of capsorubin, **3.48**, Scheme 3.14, are considered reliable from an exciton interaction of four molecules. The absorption maxima of capsorubin tetramer, pentamer, hexamer, and heptamer are well resolved, and the octamer absorption is quite similar to that of the nonamer. The aggregate absorption for the decamer, undecamer, and dodecamer are practically identical, indicating a convergence value (Köpsel 1999), Figure 3.9. In a detailed investigation with aggregates of enantiomeric zeaxanthin, **3.3**, astaxanthin, **3.6**, capsorubin, **3.48**, and other carotenoids not only the absorption, but also the circular dichroism (CD) spectra were calculated (Köpsel 1999). It was found that the spectra for astaxanthin, **3.6**, are represented by an octamer of the *H*-aggregate type. Possible higher oligomers could not be defined, the octamer reaching the convergence value (Köpsel et al. 2005).

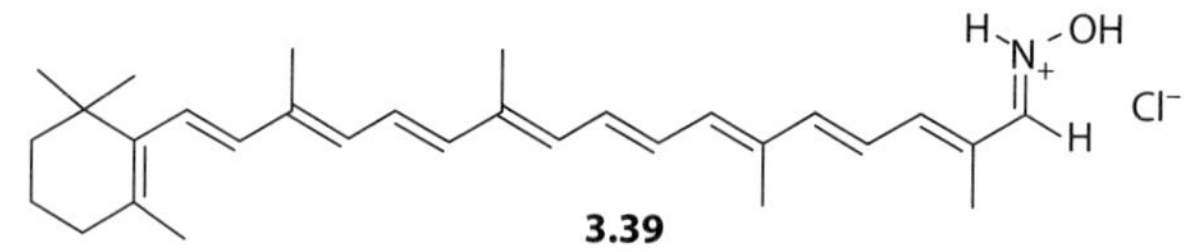

3.39

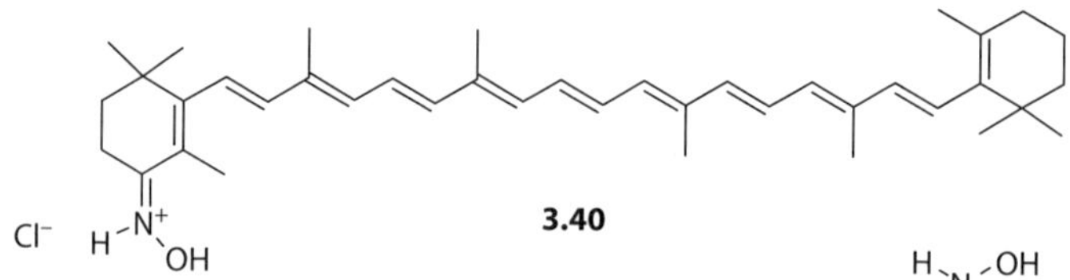

3.40

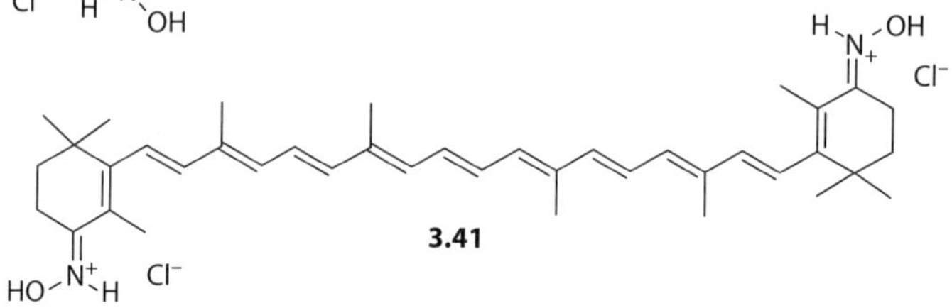

3.41

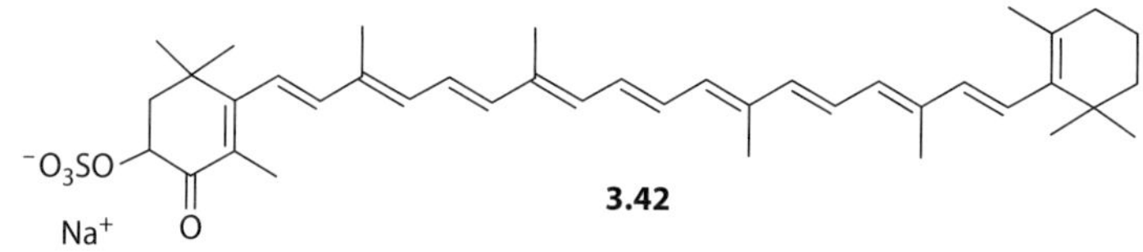

3.42

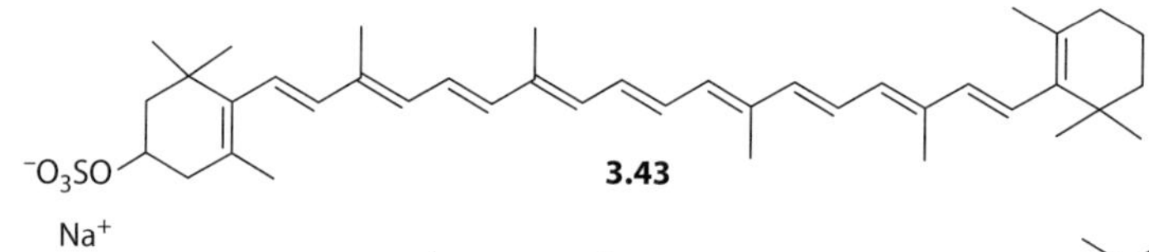

3.43

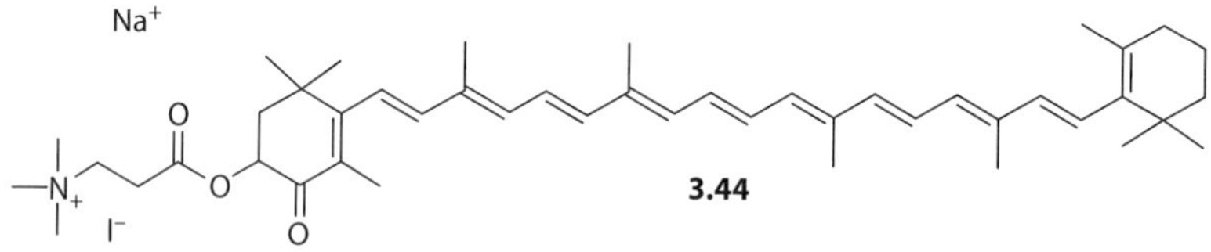

3.44

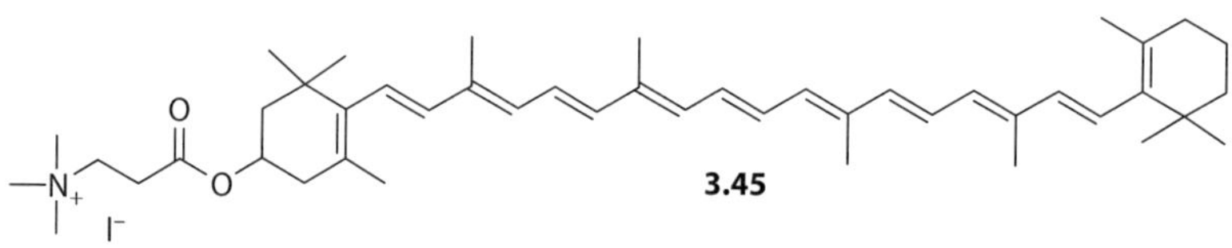

3.45

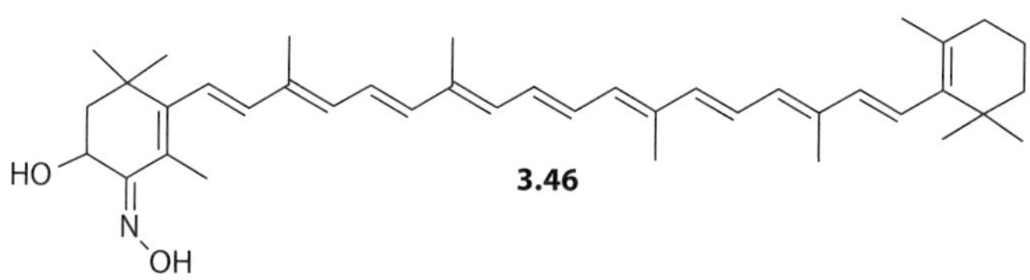

3.46

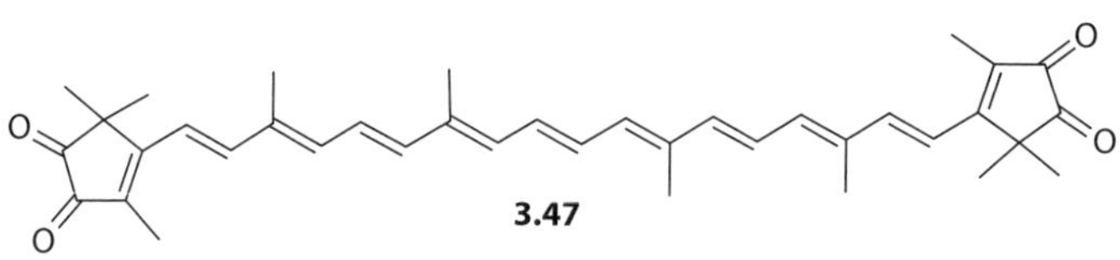

3.47

SCHEME 3.13

TABLE 3.2
Aggregation Behavior

Molecules in Scheme 3.13	Predominant Aggregate Type in H_2O or Highest H_2O Concentration
Crocin **3.7** (Nalum Naess et al. 2006)	*H*
Phospholipid **3.15** (Foss et al. 2005b)	*H*
Selenalipid **3.31** (Foss 2005)	*H*
Cardax **3.19** (Foss et al. 2005c)	*H*
Lysine derivative **3.20** (Nalum Naess et al. 2007, 77)	*H*
C30 acid salt **3.36** (Foss 2005)	*H*
Astaxanthin oxime HCl **3.38** (Willibald et al. 2009)	*J* or *H*
C30aldoxime HCl **3.39** (Willibald et al. 2009)	*J*
Echinenone oxime HCl **3.40** (Willibald et al. 2009)	*J*
Cantaxanthin oxime HCl **3.41** (Willibald et al. 2009)	*J*
Echinenone sulfate **3.42** (Benade 2001)	*H*
Cryptoxanthin sulfate **3.43** (Benade 2001)	*H*
Echinenone ammonium HCl **3.44** (Benade 2001)	*H*
Cryptoxanthin ammonium HCl **3.45** (Benade 2001)	*H*
Hydroxy echinenone oxime **3.46** (Benade 2001)	*H*
Violerythrin **3.47** (Korger 2005)	*H*

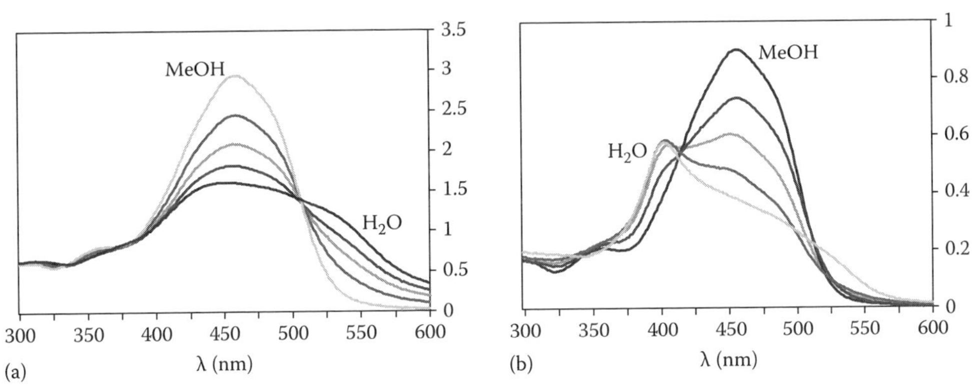

FIGURE 3.8 Aggregate disruption and formation. (a) Astaxanthin oximium hydrochloride **3.38** in water forms *J*-aggregates, which are disrupted by adding MeOH; (b) **3.38** in MeOH upon adding water forms *H*-aggregates. (From Willibald, J., *Chem. Phys. Lipids*, 161, 32, 2009. With permission.)

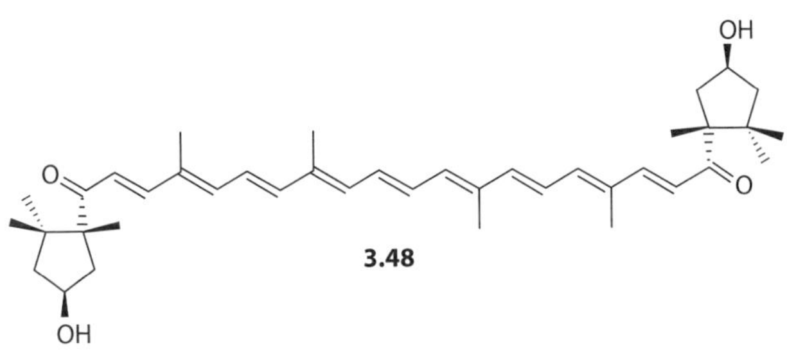

3.48

SCHEME 3.14

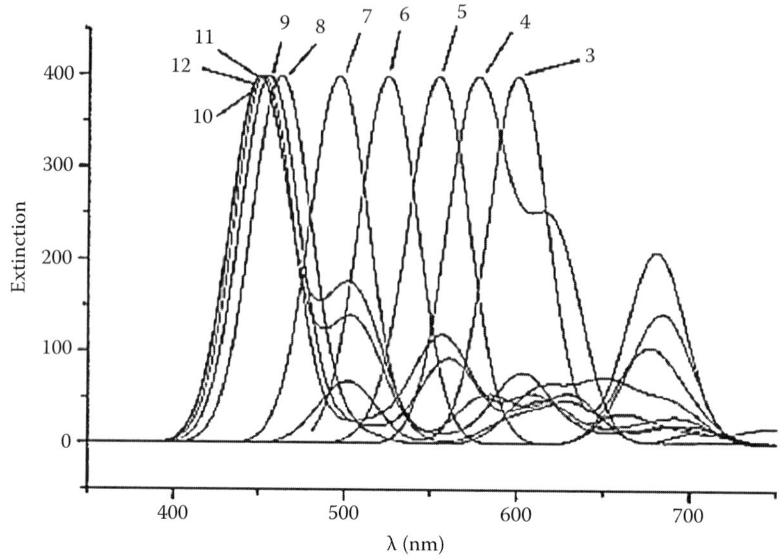

FIGURE 3.9 Number of capsorubin **(3.48)** molecules in small oligomers. Reaching a decamer, the absorbance converts to a constant value. (From Mayer, B., Düsseldorf, unpublished.)

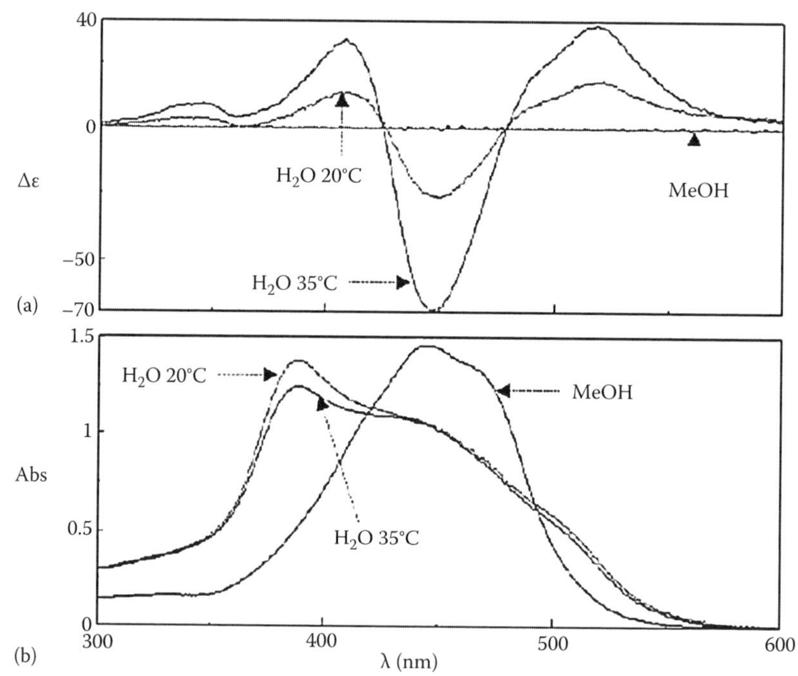

FIGURE 3.10 CD spectra (a) and absorption spectra (b) of phospholipid (*R*)-**3.15** in MeOH (no optical activity) and in water at 20°C and 35°C (strong Cotton effects). (Reprinted from Foss, B.J. et al., *Chem. Eur. J.*, 11, 4103, 2005b. With permission.)

The monomolecular solution of the phospholipid enantiomer, (*R*)-**3.15**, in MeOH is optically inactive. Surprisingly, when (*R*)-**3.15** was dispersed in water, distinct CD bands were seen, Figure 3.10. The calculation of the absorption band resulted in a tetramer with *H*-type and *J*-type

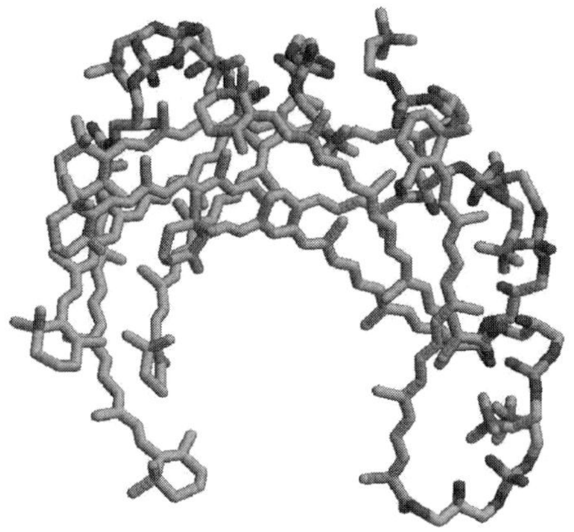

FIGURE 3.11 (See color insert following page 336.) Optically active *P*-oligomer unit, built from eight optically inactive (*R*)-**3.15** monomers. The calculated spectra of this octamer is in accordance with both the experimental VIS and CD spectra. (Reprinted from Foss, B.J. et al., *Chem. Eur. J.*, 11, 4103, 2005b. With permission.)

arrangement in accordance with the experimental visible spectrum, Figure 3.7. However, when absorption and CD spectra were calculated consecutively, it was found that the spectra originated from a helical *P*-screwed arrangement of the inactive *R*-monomers, again—accidentally—within an octamer, Figure 3.11 (Foss et al. 2005b). The irregular structure of the molecules in the octamer does not form defined *H*- or *J*-arrangements and the absorption maxima are therefore shifted to shorter as well as to longer wavelengths. It is obvious that the octamer cannot exist as an independent entity in water, since the polar and nonpolar groups are oriented in an unfavorable way. The octamers of astaxanthin, **3.6**, and the phospholipid, (*R*)-**3.15**, can be regarded as basic aggregations units, which are the lowest possible molecular associations that display the spectroscopic and chiroptical properties of the corresponding aggregates. Aggregates retain the gap between single molecules and crystals. The "basic aggregation unit" could therefore possibly be compared with elementary crystal units (Bravais). The aggregates of astaxanthin, **3.6**, and the lipid, **3.15**, may be considered as constructions built by bricks of these unit structures. The enantiomeric basic units not only form enantiomeric aggregates, they probably also form aggregates with an enantiomeric aggregate surface (Shinitzky and Haimovitz 1993). In the phospholipid, **3.15**, the asymmetric center of the monomers is located in the polar group building up the outer aggregate sphere. The enantiomeric surface may discriminate between chiral membrane-intrusion agents and could also be relevant for chiral surface reactions. The crucial tasks of enantiomeric D-sugars and L-amino acids in nature are well recognized. For lipids, the functional discrimination of enantiomers has not yet been established. Lipids with highly unsaturated carotenoid acids would be ideal compounds in elucidating the chiral requirements of the third base material in living nature. Enantiomeric aggregates of carotenoid lipids would be detectable without problems.

Whereas the morphology of a crystal determines its Bravais cell, the aggregate form may not necessarily mirror the aggregation unit. The three different absorptions of the phospholipid, **3.15**, aggregates might all originate from globules, all with an *H*-type molecule arrangement, though with different aggregation units creating the various absorption values.

In general, the UV–VIS spectra and, consequently, the CD spectra of aggregates deviate considerably from those of the monomer spectra. The (*S,S*)-astaxanthinoxime hydrochloride, **3.38**, in MeOH displays only one broad negative Cotton effect centered at 270 nm within the 215–350 nm region. When water is added the resulting aggregates display quite different Cotton effects than in MeOH, however, the signals are again similar to astaxanthin, **3.6**, Figure 3.12.

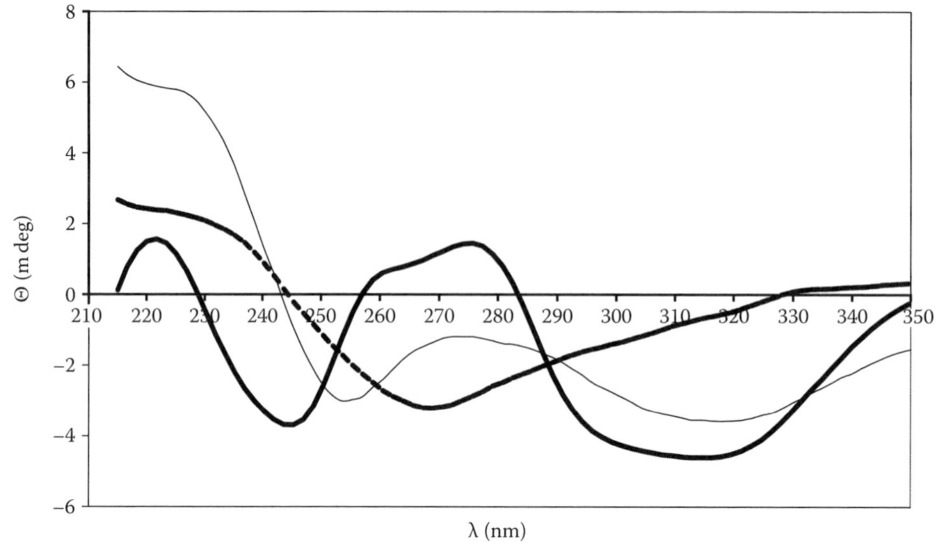

FIGURE 3.12 CD spectrum of (*S*,*S*)-astaxanthin **3.6** in MeOH (—), (*S*,*S*)-astaxanthin dioxime hydrochloride **3.38** in MeOH (- - -) and of aggregates of **3.38** in water (—). (Willibald, J. et al., *Chem. Phys. Lipids*, 161, 32, 2009. With permission.)

3.6 AGGREGATE STABILITY

Self-aggregation of carotenoids is synonymous with self-stabilization. The aggregates of the phospholipid, **3.15**, and Cardax, **3.19**, are thermostable at $T \geq 50°C$, Figure 3.13, whereas the aggregates of zeaxanthin, **3.3**, and lutein, **3.4**, are disrupted at 45°C and 55°C, respectively (Douillard et al. 1983). Carotenoid aggregates withstand much longer refluxing in MeOH/HCl than the nonaggregated monomolecular solution before bleaching occurs (Sliwka et al. 2007). Astonishingly, although the aggregate membrane is transparent, carotenoid aggregate dispersions resist light irradiation for a substantially longer time than a nonaggregated monomolecular solution (Lüddecke et al. 1999).

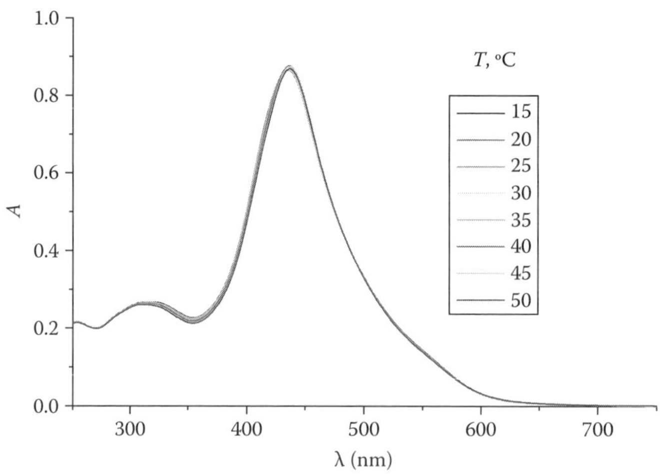

FIGURE 3.13 Thermostability of *H*-aggregates of Cardax **3.19** in water. (From Melø, T.B., Trondheim, unpublished.)

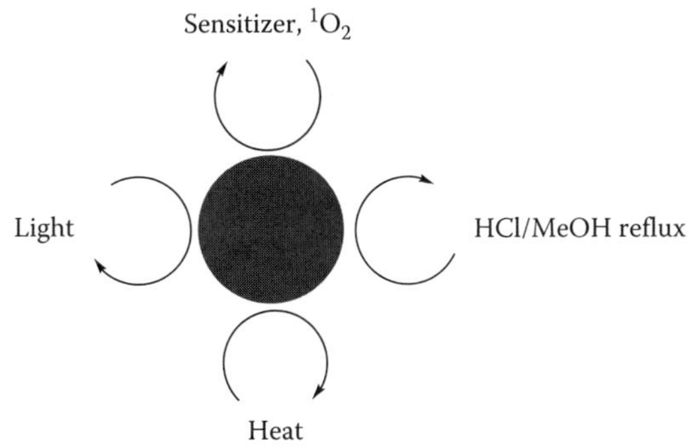

FIGURE 3.14 Stability of carotenoid aggregates.

In pure water, electron or energy transfer to carotenoid aggregates is obstructed by the membrane of outside-directed polar groups (Sliwka et al. 2007), Figure 3.14.

Water-soluble crocin, **3.7**, and the lysine derivative, **3.20**, are immediately reactive in aqueous solutions, whereas water-dispersible carotenoids only become reactive when contacting a milieu in which the aggregates are disrupted. Dispersions of carotenoid aggregates will therefore have increased shelf lives compared to monomolecular carotenoid formulations. When water is removed azeotropically or by freeze-drying from carotenoid aggregate suspensions, and the remainder is further dried at high vacuum, the residue could not always be dissolved in the solvent used for preparing the monomeric solutions. Most likely, water-containing aggregates survive the drying process, stabilize the hydrophobic membrane, and resist dissolution by organic solvents.

3.7 BIOPHYSICAL AND BIOLOGICAL ACTIVITY OF HYDROPHILIC CAROTENOIDS AND CAROTENOID AGGREGATES

As has been pointed out earlier in this chapter, the dietary consumption and historical medicinal use of carotenoids has been well documented. In the modern age, in addition to crocin, **3.7**, and norbixin, **3.8**, several carotenoids have become extremely important commercially. These include, in particular, astaxanthin, **3.6** (fish, swine, and poultry feed, and recently human nutritional supplements); lutein, **3.4**, and zeaxanthin, **3.3** (animal feed and poultry egg production, human nutritional supplements); and lycopene, **3.2** (human nutritional supplements). The inherent lipophilicity of these compounds has limited their potential applications as hydrophilic additives without significant formulation efforts; in the diet, the lipid content of the meal increases the absorption of these nutrients, however, parenteral administration to potentially effective therapeutic levels requires separate formulation that is sometimes ineffective or toxic (Lockwood et al. 2003).

Significant work began in 2002 to produce rational chemical derivatives of carotenoids that might be utilized in human medicinal applications, by a globally connected multidisciplinary group of researchers. Retrometabolic drug design was used to produce derivatives with novel characteristics to be exploited in such applications, hopefully without introducing chemical toxicity not inherent in the starting scaffold astaxanthin, **3.6**. The prototypical astaxanthin derivatives were produced at kg scale as disuccinate sodium salts (Frey et al. 2004); the trade name of the compound under development was Cardax, **3.19**. Cardax, **3.19**, underwent thorough preclinical evaluation, both as a water-dispersible radical scavenger (Cardounel et al. 2003), Table 3.3, as well as an *in vivo* oral and parenteral myocardial salvage agent, Figure 3.15 (Gross and Lockwood 2004, 2005, Lauver et al. 2005, Gross et al. 2006, Lockwood et al. 2006a).

The aggregation and surface properties of Cardax, **3.19**, in various aqueous formulations were comprehensively evaluated in 2005 (Foss et al. 2005c), as well as the potential plasma protein binding in mammalian applications with molecular modeling (Zsila et al. 2003). Cardax, **3.19**, proved

TABLE 3.3
Concentration of Hydrophilic Carotenoids in Water for Almost Complete Inhibition of Aqueous Superoxide Anion ($O_2^{\bullet -}$)

	$O_2^{\bullet -}$ Inhibition (%)	*c* in Water (mM)
Phospholipid **3.15** (Foss et al. 2006b)	94.3	10
Lutein phosphate **3.17** (Foss et al. 2006b)	91	5
Cardax **3.19** (Foss et al. 2006b)	95	3
Lysine derivative **3.20** (Lockwood et al. 2006a)	95.7	0.1

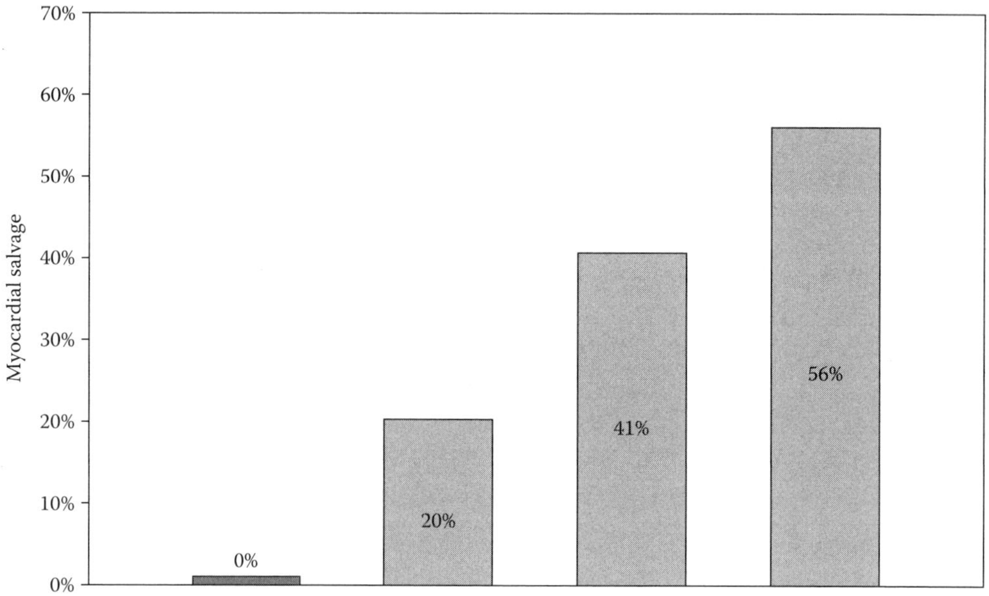

FIGURE 3.15 Mean myocardial salvage by Cardax **3.19** (0, 25, 50, 75 mg/kg) as percentage of infarct size in rats. Myocardial salvage of 56% was achieved with the highest dose 75 mg/kg. (Reprinted from Gross, G.J. and Lockwood, S.F., *Life Sci.*, 75, 215, 2004. With permission.)

to be a water-dispersible (~9 mg/mL), injectable, orally available myocardial salvage agent with distinctly favorable properties in addition to those documented for astaxanthin, **3.6**. While aggregated in solution, the disuccinate astaxanthin molecules were protected from degradation by the self-assembly, and became more biophysically active after chemical disruption (Foss et al. 2005c). The compound was active, both orally and parenterally, not only as a myocardial salvage agent; but also novel anti-inflammatory activity was documented along two important medicinal axes in addition to straight antioxidant activity: complement activation and lipoxygenase activity. These activities are detailed in the articles cited above.

Second generation astaxanthin derivatives were then pursued, with an eye on increasing solubility or dispersibility over the prototypical compound, **3.6**. The surface and aggregation properties of the highly soluble astaxanthin–lysine conjugate, **3.20**, were evaluated (Jackson et al. 2004, Zsila et al. 2004, Nalum Naess et al. 2007). The compound, **3.20**, shared many solubility properties with natural crocin, **3.7**, i.e., aggregation if at all only at very high concentrations. The lysine derivative, **3.20**, appeared to be active as a radical scavenger at low concentration immediately when solvated, Table 3.3.

The solubility of lysine derivative, **3.20**, was measured at slightly over 180 mg/mL, and molecular modeling also demonstrated potentially favorable plasma protein binding (Zsila et al. 2004).

A second highly soluble diphosphate derivate, **3.17**, was also produced (solubility ~29 mg/mL); its efficacy in an *in vitro* cancer agent was screened, and it proved to be the most active carotenoid ever tested in this system (Hix et al. 2005), and more potent than Cardax, **3.19** (Hix et al. 2004). Overall, the second-generation compounds showed increased promise over the prototypes in certain contexts, particularly those in which immediate radical scavenging by highly potent and soluble compounds are required.

Third-generation compounds were then explored. These novel conjugates combined astaxanthin, **3.6**, with other antioxidants (in particular ascorbic acid, **3.22**) with flexible linkers, at once providing covalent linkage of two powerful antioxidants in favorable stoichiometric ratios, as well as increasing the solubility/dispersibility to encouraging amounts, e.g., compound **3.21** (Lockwood et al. 2006b). These compounds underwent *in vitro* testing demonstrating these qualities (for reviews of the above chemistry and biology, see Hix et al. (2004), Foss et al. (2006a), and Lockwood et al. (2006b)). Preclinical animal testing is underway for several of these promising compounds. Hydroxy carotenoids other than astaxanthin, **3.6**, were successfully modified with retrometabolic synthesis, resulting in similar efficacy and surface and aggregation properties, e.g., lutein, **3.4** (Nadolski et al. 2006). In the case of lycopene, disymmetric lycophyll was successfully synthesized at scale and used for retrometabolic synthesis, e.g., for diphosphate, **3.18** (Jackson et al. 2005, Braun et al. 2006). These compounds should prove useful in applications in macular degeneration, cataracts, and prostate cancer, respectively. Therefore, the medicinal applications of hydrophilic carotenoids with modifiable aggregation and solubility or dispersibility properties are highly promising.

3.8 POSSIBLE ADDITIONAL COMMERCIAL AND SCIENTIFIC APPLICATION

It appears that self-aggregating and self-stabilizing hydrophilic carotenoids would be outstanding food colorants for soft drinks, health drinks, and other liquid supplements. Heretofore, carotenoids have had to pass through harsh formulation conditions and were mixed with other ingredients before they could be used in aliments (Horn and Rieger 2001). In sharp contrast, hydrophilic carotenoids can formulate themselves at room temperature with water as the only other ingredient. They appear stable to the temperatures, acidities, light exposures, and potential sensitizers typical for both canned and bottled liquid commercial preparations. However, the available carotenoid formulations, consisting of intimate physical mixtures, are considered as safe as the individual constituents (generally recognized as safe, or GRAS), whereas hydrophilic modified carotenoids are new chemical entities (NCEs), whose toxicity, physiological tolerance, and efficacy have yet to be proven, usually by the costly and elaborate tests designated by agencies such as the Food and Drug Administration (FDA) in the United States or the European Medicines Agency (EMEA). At least one hydrophilic carotenoid (Cardax, **3.19**) has undergone significant preclinical safety and efficacy testing both *in vitro* and in animals; perhaps more testing of similar compounds will begin soon. Much work has been done to elucidate the biological function of carotenoids, and such studies will inevitably come to the point where the usual biological solvent, water, is involved, perhaps alone. Data obtained with pure carotenoid aggregate preparations will then not be obscured by emulsifiers or matrixes. Aggregates with enclosed pharmacologically active compounds are currently used for drug delivery (Xu et al. 1999, O'Sullivan et al. 2004). Aggregates of hydrophilic carotenoids demonstrate a remarkable, and highly utilitarian, difference: the drug need not be enclosed in additional compounds (excipients); the drug itself becomes its own delivery system, Figure 3.1.

3.9 CONCLUSIONS

Carotenoid aggregation has now been studied for over 77 years, but it is only in the last 7 years that numerous hydrophilic carotenoids have been synthesized. It is too early to predict whether research in hydrophilic carotenoids will become an established part within "traditional" carotenoid chemistry. The production of Cardax, **3.19**, its preclinical and potential clinical testing, the possible discovery of other pharmacological effects (both beneficial and unwanted), synthesis of additional carotenoid conjugates with specific desired properties, potential chemical and biological applications of carotenolipid–DNA adducts, future procedures to obtain carotenoid aggregates of predefined size, the study of exciton interactions, and the use of enantiomeric amphiphilic carotenoids in chiral lipid research indicate at least that hydrophilic carotenoids and carotenoid aggregates will become an interesting, highly interdisciplinary research field in the years to come.

ACKNOWLEDGMENTS

We gratefully recognize the significant collaboration with the chemists in Düsseldorf and Ludwigshafen, Germany (BASF, H. Ernst), with the physicochemists in Budapest, Hungary, the physicists in Trondheim, Norway, and with the physicians and doctoral researchers in Milwaukee, WI; Columbus and Cleveland, OH; Aiea, HI; Albany, NY; Chicago, IL; Ann Arbor, MI; and Cambridge, MA (United States).

REFERENCES

Auweter H, Benade J, Bettermann H, Beutner S, Köpsel C, Lüddecke E, Martin HD, and Mayer B. 1999. Association and self-assembly of carotenoids: Structure and stability of carotenoid aggregates. In: Mosquera M, Galan M, Mendez D, eds. *Pigments in Food Technology*, Dept. Legal, Sevilla, Spain, pp. 197–201.

Bauernfeind JC and Howard R. 1956. Water-dispersible carotenoid compositions and process of making the same, Hofmann-La Roche, Nutley, U.K., US 2861891.

Benade J. 2001. Synthese von polaren und amphiphilen Carotenoiden und ihr Aggregationsverhalten in wässrigen Systemen, Dissertation, University of Düsseldorf, Düsseldorf, Germany, p. 60, http://www.ulb.uni-duesseldorf.de/diss/mathnat/2001/benade.html

Bikadi Z, Zsila F, Deli J, Mady G, and Simonyi M. 2002. The supramolecular structure of self-assembly formed by capsanthin derivatives. *Enantiomer* 7: 67–76.

Billsten HH, Sundström V, and Polivka T. 2005. Self-assembled aggregates of the carotenoid zeaxanthin: Time-resolved study of excited states. *Journal of Physical Chemistry A* 109(8): 1521–1529.

Borenstein B, Bunnell T, and Bunnell RH. 1967. Stabilizing carotenoid compositions, Hoffmann-La Roche, Nutley, U.K., US 3316101.

Braun CL, Jackson HL, Lockwood SF, and Nadolski G. 2006. Purification of synthetic all-E lycophyll (ψ,ψ-carotene-16,16′-diol). *Journal of Chromatography B—Analytical Technologies in the Biomedical and Life Sciences* 834(1–2): 208–212.

Breukers S, Øpstad CL, Sliwka HR, and Partali V. 2009. Hydrophilic carotenoids: Surface properties and aggregation behaviour of the potassium salt of the highly unsaturated diacid norbixin. *Helvetica Chimica Acta* 92(9): 1741–1747.

Britton G. 1998. Overview of carotenoid biosynthesis. In: Britton G, Liaaen-Jensen S, Pfander H, eds. *Carotenoids, Vol. 3: Biosynthesis and Metabolism*. Birkhäuser, Basel, Switzerland, p. 14, 25.

Britton G, Liaaen-Jensen S, Pfander H, Mercadante AZ, and Egeland ES. 2004. *Carotenoids, Handbook*. Birkhäuser, Basel, Switzerland.

Buchwald M and Jencks WP. 1968. Optical properties of astaxanthin solutions and aggregates. *Biochemistry* 7: 834–843.

Burke M, Edge R, Land EJ, McGarvey DJ, and Truscott TG. 2001. One-electron reduction potentials of dietary carotenoid radical cations in aqueous micellar environments. *FEBS Letters* 500(3): 132–136.

Bystritskaya EV and Karpukhin ON. 1975. Effect of the aggregate state of a medium on the quenching of singlet oxygen. *Doklady Akademii Nauk SSSR* 221: 1100–1103.

Cardounel AJ, Dumitrescu C, Zweier JL, and Lockwood SF. 2003. Direct superoxide anion scavenging by a disodium disuccinate astaxanthin derivative: Relative efficacy of individual stereoisomers versus the statistical mixture of stereoisomers by electron paramagnetic resonance imaging. *Biochemical and Biophysical Research Communications* 307(3): 704–712.

Chen G and Djuric Z. 2001. Carotenoids are degraded by free radicals but do not affect lipid peroxidation in unilamellar liposomes under different oxygen tensions. *FEBS Letters* 505(1): 151–154.

colorMaker, Inc. Anaheim, CA, http://www.colormaker.com/CM/AboutNC/annatto.asp.

Davydov A. 1962. *Theory of Molecular Excitons*. McGraw-Hill, New York.

Dembitsky VM. 2005. Astonishing diversity of natural surfactants: 3. Carotenoid glycosides and isoprenoid glycolipids. *Lipids* 40(6): 535–557.

Douillard R, Burghoffer C, and Costes C. 1982. Structure excitonique de complexes hydroéthanoliques de la lutéine et de la zéaxanthine. *Physiologie Végétale* 20: 123–136.

Douillard R, Burghoffer C, and Costes C. 1983. Conditions de formation et propriétés des complexes excitoniques formés par les xanthophylles en presence d'eau. *Physiologie Végétale* 21: 375–383.

Dreon MS, Ceolin M, and Heras H. 2007. Astaxanthin binding and structural stability of the apple snail carotenoprotein ovorubin. *Archives of Biochemistry and Biophysics* 460(1): 107–112.

Foss BJ. 2005. Synthesis and physical properties of hydrophilic carotenoid derivatives, PhD thesis, Norwegian University of Science and Technology, Trondheim, Norway.

Foss BJ, Nalum Naess S, Sliwka HR, and Partali V. 2003. Stable and highly water-dispersible, highly unsaturated carotenoid phospholipids—Surface properties and aggregate size. *Angewandte Chemie-International Edition* 42(42): 5237–5240.

Foss BJ, Sliwka HR, Partali V, Nalum Naess S, Elgsaeter A, Melø TB, and Naqvi KR. 2005a. Hydrophilic carotenoids: Surface properties and aggregation behavior of a highly unsaturated carotenoid lysophospholipid. *Chemistry and Physics of Lipids* 134(2): 85–96.

Foss BJ, Sliwka HR, Partali V, Köpsel C, Mayer B, Martin HD, Zsila F, Bikadi Z, and Simonyi M. 2005b. Optically active oligomer units in aggregates of a highly unsaturated, optically inactive carotenoid phospholipid. *Chemistry-A European Journal* 11(14): 4103–4108.

Foss BJ, Sliwka HR, Partali V, Nalum Naess S, Elgsaeter A, Melø TB, Naqvi KR, O'Malley S, and Lockwood SF. 2005c. Hydrophilic carotenoids: Surface properties and aqueous aggregation of a rigid, long-chain, highly unsaturated dianionic bolaamphiphile with a carotenoid spacer. *Chemistry and Physics of Lipids* 135(2): 157–167.

Foss BJ, Nadolski G, and Lockwood SF. 2006a. Hydrophilic carotenoid amphiphiles: Methods of synthesis and biological applications. *Mini-Reviews in Medicinal Chemistry* 6(9): 953–969.

Foss BJ, Ion A, Partali V, Sliwka HR, and Banica FG. 2006b. Electrochemical and EQCM investigation of a selenium derivatized carotenoid in the self-assembled state at a gold electrode. *Journal of Electroanalytical Chemistry* 593(1–2): 15–28.

Frey DA, Kataisto EW, Ekmanis JL, O'Malley S, and Lockwood SF. 2004. The efficient synthesis of disodium disuccinate astaxanthin (Cardax). *Organic Process Research & Development* 8(5): 796–801.

Gross GJ and Lockwood SF. 2004. Cardioprotection and myocardial salvage by a disodium disuccinate astaxanthin derivative (Cardax™). *Life Sciences* 75(2): 215–224.

Gross GJ and Lockwood SF. 2005. Acute and chronic administration of disodium disuccinate astaxanthin (Cardax™) produces marked cardioprotection in dog hearts. *Molecular and Cellular Biochemistry* 272(1–2): 221–227.

Gross GJ, Hazen SL, and Lockwood SF. 2006. Seven day oral supplementation with Cardax™ (disodium disuccinate astaxanthin) provides significant cardioprotection and reduces oxidative stress in rats. *Molecular and Cellular Biochemistry* 283(1–2): 23–30.

Gruszecki WI, Zelent B, and Leblanc RM. 1990. Fluorescence of zeaxanthin and violaxanthin in aggregated forms. *Chemical Physics Letters* 171(5–6): 563–568.

Hager A. 1970. Ausbildung von Maxima im Absorptionsspektrum von Carotenoiden im Bereich um 370 nm; Folgen für die Interpretation bestimmter Wirkungsspektren. *Planta* 91: 38–53.

Hamley IW and Castelletto V. 2007. Biological soft materials. *Angewandte Chemie-International Edition* 46(24): 4442–4455.

Hayashi H, Yamanaka T, Miyajima M, and Imae T. 1994. Aggregation numbers and shapes of lysophosphatidylcholine and lysophosphatidylethanolamine micelles. *Chemistry Letters* 23(12): 2407–2410.

Hertzberg S and Liaaen-Jensen S. 1985. Carotenoid sulfates 4. Syntheses and properties of carotenoid sulfates. *Acta Chemica Scandinavica Series B—Organic Chemistry and Biochemistry* 39(8): 629–638.

Hertzberg S, Ramdahl T, Johansen JE, and Liaaen-Jensen S. 1983. Carotenoid sulfates 2. Structural elucidation of bastaxanthin. *Acta Chemica Scandinavica* B37: 267–280.

Hix LM, Lockwood SF, and Bertram JS. 2004. Bioactive carotenoids: Potent antioxidants and regulators of gene expression. *Redox Report* 9: 181–191.

Hix LM, Frey DA, McLaws MD, Østerlie M, Lockwood SF, and Bertram JS. 2005. Inhibition of chemically-induced neoplastic transformation by a novel tetrasodium diphosphate astaxanthin derivative. *Carcinogenesis* 26(9): 1634–1641.

Horn D and Rieger J. 2001. Organic nanoparticles in the aqueous phase—Theory, experiment, and use. *Angewandte Chemie-International Edition* 40(23): 4331–4361.

Humeau C, Rovel B, and Girardin M. 2000. Enzymatic esterification of bixin by L-ascorbic acid. *Biotechnology Letters* 22(2): 165–168.

Inamura I, Isshiki M, and Araki T. 1989. Solubilization of β-carotene in water by water-soluble linear macromolecules. *Bulletin of the Chemical Society of Japan* 62(5): 1671–1673.
Ion A, Partali V, Sliwka HR, and Banica FG. 2002. Electrochemistry of a carotenoid self-assembled monolayer. *Electrochemistry Communications* 4(9): 674–678.
Israelachvilli JN, Mitchell DJ, and Ninham BW. 1976. Theory of self-assembly of hydrocarbon amphiphiles into micelles and bilayers. *Journal of the Chemical Society, Faraday Transactions 2: Molecular and Chemical Physics* 72: 1525–1568.
Jackson HL, Cardounel AJ, Zweier JL, and Lockwood SF. 2004. Synthesis, characterization, and direct aqueous superoxide anion scavenging of a highly water-dispersible astaxanthin–amino acid conjugate. *Bioorganic & Medicinal Chemistry Letters* 14(15): 3985–3991.
Jackson HL, Nadolski GT, Braun C, and Lockwood SF. 2005. Efficient total synthesis of lycophyll (ψ,ψ-carotene-16,16′-diol). *Organic Process Research & Development* 9(6): 830–836.
Jelley EE. 1936. Spectral absorption and fluorescence of dyes in the molecular state. *Nature* 138: 1009–1010.
Jelley EE. 1937. Molecular, nematic and crystal states of 1,1′-diethyl-ψ-cyanine chloride. *Nature* 139: 631–632.
Jemioła-Rzemińska M, Pasenkiewicz-Gierula M, and Strzałka K. 2005. The behaviour of beta-carotene in the phosphatidylcholine bilayer as revealed by a molecular simulation study. *Chemistry and Physics of Lipids* 135(1): 27–37.
Ke B, Green M, Vernon LP, and Garcia AF. 1968. Some optical properties of a carotenoid complex derived from *Rhodospirillum Rubrum. Biochimica et Biophysica Acta* 162(3): 467–469.
Kildahl-Andersen G, Bruas L, Lutnaes BF, and Liaaen-Jensen S. 2004. Nucleophilic reactions of charge delocalised carotenoid mono- and dications. *Organic & Biomolecular Chemistry* 2(17): 2496–2506.
Kildahl-Andersen G, Anthonsen T, and Liaaen-Jensen S. 2007. Studies on the mechanism of the Carr–Price blue color reaction. *Organic Biomolecular Chemistry* 5: 3027–3033.
Köpsel C. 1999. Struktur von Carotenoidaggregaten, Dissertation, University of Düsseldorf, Düsseldorf, Germany, pp. 113, 115, 116.
Köpsel C, Möltgen H, Schuch H, Auweter H, Kleinermanns K, Martin HD, and Bettermann H. 2005. Structure investigations on assembled astaxanthin molecules. *Journal of Molecular Structure* 750: 109–115.
Korger M. 2005. Aggregationsverhalten von Carotenoiden in Membranen unilamellarer Liposomen, Dissertation University of Düsseldorf, Düsseldorf, Germany, http://diss.ub.uni-duesseldorf.de/ebib/diss/show?dissid=1151.
Larsen E, Abendroth J, Partali V, Schulz B, Sliwka HR, and Quartey EGK. 1998. Combination of vitamin E with a carotenoid: α-Tocopherol and trolox linked to β-apo-8′-carotenoic acid. *Chemistry—A European Journal* 4(1): 113–117.
Lauver DA, Lockwood SF, and Lucchesi BR. 2005. Disodium disuccinate astaxanthin (Cardax) attenuates complement activation and reduces myocardial injury following ischemia/reperfusion. *Journal of Pharmacology and Experimental Therapeutics* 314(2): 686–692.
Lehn JM. 1988. Supramolecular chemistry—Scope and perspectives molecules, supermolecules, and molecular devices. *Angewandte Chemie—International Edition in English* 27(1): 89–112.
Lerfall J. 2002. Syntetisk kombinasjon av β-apo-8′-karotensyre, vitamin C, EPA, trolox og selenafettsyre, Master thesis, Norwegian University of Science and Technology, Trondheim, Norway.
Li ZX, Dong CC, and Thomas RK. 1999. Neutron reflectivity studies of the surface excess of Gemini surfactants at the air–water interface. *Langmuir* 15(13): 4392–4396.
Liaaen-Jensen S. 1971. Isolation, reactions. In: Isler O, ed. *Carotenoids*. Basel, Birkhäuser, pp. 73–91.
Liaaen-Jensen S. 1996. Partial synthesis of sulfates. In: Britton G, Liaaen-Jensen S, Pfander H, eds. *Carotenoids, Vol. 2 Synthesis*. Birkhäuser, Basel, Switzerland.
Lindig BA and Rodgers MAJ. 1981. Rate parameters for the quenching of singlet oxygen by water-soluble and lipid-soluble substrates in aqueous and micellar systems. *Photochemistry and Photobiology* 33(5): 627–634.
Liu DZ, Szulczewski GJ, Kispert LD, Primak A, Moore TA, Moore AL, and Gust D. 2002. A thiol-substituted carotenoid self-assembles on gold surfaces. *Journal of Physical Chemistry B* 106(11): 2933–2936.
Lockwood SF, O'Malley S, and Mosher GL. 2003. Improved aqueous solubility of crystalline astaxanthin (3,3′-dihydroxy-β,β-carotene-4,4′-dione) by Captisol (sulfobutyl ether β-cyclodextrin). *Journal of Pharmaceutical Sciences* 92(4): 922–926.
Lockwood SF, O'Malley S, Watumull DG, Hix LM, Jackson H, and Nadolski G. 2005. Preparation of carotenoid ether analogs or derivatives for the inhibition and amelioration of liver disease. US 2005026874.
Lockwood SF, Jackson HL, and Gross GJ. 2006a. Retrometabolic syntheses of astaxanthin (3,3′-dihydroxy-β,β-carotene-4,4′-dione) conjugates: A novel approach to oral and parenteral cardioprotection. *Cardiovascular & Hematological Agents* 4: 335–349.

Lockwood SF, O'Malley S, Watumull DG, Hix LM, Jackson H, and Nadolski G. 2006b. Structural carotenoid analogs for the inhibition and amelioration of disease. US 7145025.

Lockwood SF, Nadolski G, and Foss BJ. 2007. Synthesis of carotenoid analogs or derivatives with improved antioxidant characteristics, Cardax Pharmaceuticals, Hawaii. WO 2007067957.

Lüddecke E, Auweter H, and Schweikert L. 1999. Use of carotenoid aggregates as colorants, BASF, Germany. EP 930022.

Meister B. 2004. Modellverbindungen zum Studium Sialinsäure-vermittelter Erkennungsprozesse: Synthese neuer Saccharide auf Basis von Carotenoiden und Furanen. Dissertation, University of Heidelberg, Heidelberg, Germany, p. 36, http://www.ub.uni-heidelberg.de/archiv/4440.

Mendelsohn R and Van Holten RW. 1979. Zeaxanthin ([3R,3′R]-β,β-carotene-3-3′ diol) as a resonance Raman and visible absorption probe of membrane structure. *Biophysical Journal* 27: 221–235.

Milon A, Wolff G, Ourisson G, and Nakatani Y. 1986. Organization of carotenoid-phospholipid bilayer systems—Incorporation of zeaxanthin, astaxanthin, and their C-50 homologs into dimyristoylphosphatidylcholine vesicles. *Helvetica Chimica Acta* 69(1): 12–24.

Miyahara T and Kurihara K. 2004. Electroconductive Langmuir–Blodgett films containing a carotenoid amphiphile for sugar recognition. *Journal of the American Chemical Society* 126(18): 5684–5685.

Mori Y. 2001. Introductory studies on the growth and characterization of carotenoid solids: An approach to carotenoid solid engineering. *Journal of Raman Spectroscopy* 32(6–7): 543–550.

Mori Y, Yamano K, and Hashimoto H. 1996. Bistable aggregate of all-*trans*-astaxanthin in an aqueous solution. *Chemical Physics Letters* 254(1–2): 84–88.

Mortensen A, Skibsted LH, Sampson J, RiceEvans C, and Everett SA. 1997. Comparative mechanisms and rates of free radical scavenging by carotenoid antioxidants. *FEBS Letters* 418(1–2): 91–97.

Nadolski G, Cardounel AJ, Zweier JL, and Lockwood SF. 2006. The synthesis and aqueous superoxide anion scavenging of water-dispersible lutein esters. *Bioorganic & Medicinal Chemistry Letters* 16(4): 775–781.

Nalum Naess S, Elgsaeter A, Foss BJ, Li BJ, Sliwka HR, Partali V, Melø TB, and Naqvi KR. 2006. Hydrophilic carotenoids: Surface properties and aggregation of crocin as a biosurfactant. *Helvetica Chimica Acta* 89(1): 45–53.

Nalum Naess S, Sliwka HR, Partali V, Melø TB, Naqvi KR, Jackson HL, and Lockwood SF. 2007. Hydrophilic carotenoids: Surface properties and aggregation of an astaxanthin–lysine conjugate, a rigid, long-chain, highly unsaturated and highly water-soluble tetracationic bolaamphiphile. *Chemistry and Physics of Lipids* 148(2): 63–69.

Okamoto H, Hamaguchi HO, and Tasumi M. 1989. Resonance Raman studies on tetradesmethyl-β-carotene aggregates. *Journal of Raman Spectroscopy* 20(11): 751–756.

Oliveros E, Braun AM, Aminiansaghafi T, and Sliwka HR. 1994. Quenching of singlet oxygen ($^1\Delta_G$) by carotenoid derivatives—Kinetic analysis by near-infrared luminescence. *New Journal of Chemistry* 18(4): 535–539.

O'Sullivan SM, Woods JA, and O'Brien NM. 2004. Use of Tween 40 and Tween 80 to deliver a mixture of phytochemicals to human colonic adenocarcinoma cell (CaCo-2) monolayers. *British Journal of Nutrition* 91(5): 757–764.

Partali V, Kvittingen L, Sliwka HR, and Anthonsen T. 1996. Stable, highly unsaturated glycerides—Enzymatic synthesis with a carotenoic acid. *Angewandte Chemie-International Edition in English* 35(3): 329–330.

Pfander H. 1979. Synthesis of carotenoid glycosylesters and other carotenoids. *Pure and Applied Chemistry* 51(3): 565–580.

Pfander H and Leuenberger U. 1976. Chlorierte carotinoide bei der $CHCl_3$/HCl-reaktion. *Chimia* 30: 71–73.

Reddy PV, Rabago-Smith M, and Borhan B. 2002. Synthesis of all-*trans*-[10′-H-3]-8′-apo-β-carotenoic acid. *Journal of Labelled Compounds & Radiopharmaceuticals* 45(1): 79–89.

Ruban AV, Horton P, and Young AJ. 1993. Aggregation of higher-plant xanthophylls—Differences in absorption-spectra and in the dependency on solvent polarity. *Journal of Photochemistry and Photobiology B-Biology* 21(2–3): 229–234.

Runge F, Zwissler GK, End L, Schweikert L, and Horn D. 2001. Use of solubilized carotenoid for coloring food and pharmaceutical preparations, BASF, Germany. EP 848913.

Salares VR, Young NM, Carey PR, and Bernstein HJ. 1977. Excited-state (exciton) interactions in polyene aggregates—Resonance Raman and absorption spectroscopic evidence. *Journal of Raman Spectroscopy* 6(6): 282–288.

Santos NC and Castanho MARB. 1996. Teaching light scattering spectroscopy: The dimension and shape of tobacco mosaic virus. *Biophysical Journal* 71(3): 1641–1650.

Scheibe G. 1936. Über die Veränderlichkeit des Absorptionsspektrums einiger Sensibilisierungsfarbstoffe und deren Ursache. *Angewandte Chemie* 49: 563.

Scheibe G. 1937. Über die Veränderlichkeit der Absorptionsspektren in Lösungen und die Nebenvalenzen als ihre Ursache. *Angewandte Chemie* 50: 212–219.

Shibata A, Kiba Y, Akati N, Fukuzawa K, and Terada H. 2001. Molecular characteristics of astaxanthin and β-carotene in the phospholipid monolayer and their distributions in the phospholipid bilayer. *Chemistry and Physics of Lipids* 113(1–2): 11–22.

Shinitzky M and Haimovitz R. 1993. Chiral surfaces in micelles of enantiomeric *N*-palmitoyl- and *N*-stearoylserine. *Journal of the American Chemical Society* 115: 12545–12549.

Simonyi M, Bikadi Z, Zsila F, and Deli J. 2003. Supramolecular exciton chirality of carotenoid aggregates. *Chirality* 15(8): 680–698.

Slama-Schwok A, Blanchard-Desce M, and Lehn JM. 1992. Caroviologen molecular wires—Pulse-radiolysis of bis(pyridinium) polyenes. *Journal of Physical Chemistry* 96: 10559–10565.

Sliwka HR. 1997. Selenium carotenoids 3. First synthesis of optically active carotenoid phosphates. *Acta Chemica Scandinavica* 51(3): 345–347.

Sliwka HR. 1999. Conformation and circular dichroism of β,β-carotene derivatives with nitrogen-, sulfur-, and selenium-containing substituents. *Helvetica Chimica Acta* 82(2): 161–169.

Sliwka HR and Liaaen-Jensen S. 1993a. Synthetic sulfur carotenoids 2. Optically-active carotenoid thiols. *Tetrahedron-Asymmetry* 4(3): 361–368.

Sliwka HR and Liaaen-Jensen S. 1993b. Synthetic nitrogen carotenoids—Optically-active carotenoid amines. *Tetrahedron-Asymmetry* 4(11): 2377–2382.

Sliwka HR, Melø TB, Foss BJ, Abdel-Hafez SH, Partali V, Nadolski G, Jackson H, and Lockwood SE. 2007. Electron- and energy-transfer properties of hydrophilic carotenoids. *Chemistry—A European Journal* 13(16): 4458–4466.

Socaciu C, Lausch C, and Diehl HA. 1999. Carotenoids in DPPC vesicles: Membrane dynamics. *Spectrochimica Acta Part A—Molecular and Biomolecular Spectroscopy* 55(11): 2289–2297.

Song PS and Moore TA. 1974. On the photoreceptor pigment for phototropism and phototaxis: Is a carotenoid the most likely candidate? *Photochemistry and Photobiology* 19: 435–441.

Sujak A, Okulski W, and Gruszecki WI. 2000. Organisation of xanthophyll pigments lutein and zeaxanthin in lipid membranes formed with dipalmitoylphosphatidylcholine. *Biochimica et Biophysica Acta, Biomembranes* 1509: 255–263.

Tomoaia-Cotisel M and Quinn PJ. 1998. Biophysical properties of carotenoids. *Subcellular Biochemistry* 30: 219–242.

von Euler H, Hellström H, and Klussmann E. 1931. Physikalisch-chemische Beobachtungen und Messungen an Carotenoiden. *Arkiv för Mineralogi och Geologi* 10B: 1–4.

Widmer E, Lukác T, Bernhard K, and Zell R. 1982. Technische Verfahren zur Synthese von Carotenoiden und verwandten Verbindungen aus 6-Oxo-Isophoron. V. Synthese von Astacin. *Helvetica Chimica Acta* 65(3): 671–683.

Willibald J, Rennebaum S, Breukers S, Abdel Hafez SH, Patel A, Øpstad CL, Schmid R, Nalum Naess S, Sliwka HR, and Partali V. 2009. Hydrophilic carotenoids: Facile synthesis of carotenoid oxime hydrochlorides as long-chain, highly unsaturated cationic (bola)amphiphiles. *Chemistry and Physics of Lipids* 161(1): 32–37.

Wolf KL, Framm H, and Harms H. 1937. Über den Ordnungszustand der Moleküle in Flüssigkeiten. *Zeitschrift für Physikalische Chemie* B36: 237–287.

Xu XY, Wang Y, Constantinou AI, Stacewicz-Sapuntzakis M, Bowen PE, and van Breemen RB. 1999. Solubilization and stabilization of carotenoids using micelles: Delivery of lycopene to cells in culture. *Lipids* 34(10): 1031–1036.

Yamano Y and Ito M. 1998. Synthesis of the 3-O-retinoyl-L-ascorbic acid and related compounds: Characterization and reducing activity against DPPH. *Heterocycles* 47(1): 289–299.

Yokyoyama A and Shizusato Y. 1997. Carotenoid sulfate and its production, Kaiyo Biotechnology, Kenkyusho, Japan. JP 9084591.

Zana R. 2004. Micelles and vesicles. In: Atwood JL and Steed JW, eds. *Encyclopedia of Supramolecular Chemistry*. Marcel Dekker, New York, pp. 861–861.

Zsila F, Deli J, Bikadi Z, and Simonyi M. 2001. Supramolecular assemblies of carotenoids. *Chirality* 13(10): 739–744.

Zsila F, Simonyi M, and Lockwood SF. 2003. Interaction of the disodium disuccinate derivative of meso-astaxanthin with human serum albumin: From chiral complexation to self-assembly. *Bioorganic & Medicinal Chemistry Letters* 13(22): 4093–4100.

Zsila F, Fitos I, Bikadi Z, Simonyi M, Jackson HL, and Lockwood SF. 2004. In vitro plasma protein binding and aqueous aggregation behavior of astaxanthin dilysinate tetrahydrochloride. *Bioorganic & Medicinal Chemistry Letters* 14(21): 5357–5366.

Part II

Analytical Methodologies for the Measurement of Carotenoids

4 The Use of NMR Detection of LC in Carotenoid Analysis

Karsten Holtin and Klaus Albert

CONTENTS

4.1 INTRODUCTION

Bioactive compounds, such as carotenoids have strong antioxidative properties and are used as efficient radical scavengers. In some natural sources several carotenoid isomers can be found, which differ in their biochemical activities such as bioavailability or antioxidation potency. Knowing the structure and concentration of each stereoisomer is crucial for an understanding of the effectiveness of carotenoids *in vivo*.

Because carotenoids are light- and oxygen-sensitive, a closed-loop hyphenated technique such as the on-line coupling of high performance liquid chromatography (HPLC) together with nuclear magnetic resonance (NMR) spectroscopy can be used for the artifact-free structural determination of the different isomers.

4.2 EXTRACTION

The extraction of light- and air-sensitive compounds from plant material is performed with the help of the highly efficient matrix solid phase dispersion (MSPD) technique, as shown in Figure 4.1 (Barker 2000). Here, the plant material is carefully homogenized together with a C_{18} silica-based reversed-phase material with the help of a mortar and pestle. The alkyl chains of the C_{18} material serve as hydrophobic protection environment for the extracted carotenoids; the silica helps to break the plant vesicular structure. In contrast to other techniques, such as soxhlet extraction, little or no isomerization or degradation of the extracted compounds occurs. The homogenized mixture of the plant material and the sorbent material is transferred to a solid phase extraction (SPE) column with a polyethylene frit and compressed to create a compact column bed. The polar impurities are eluted first using polar solvents; the desired class of carotenoids is finally eluted and concentrated using nonpolar solvents.

4.3 SEPARATION

After performing the mild and effective extraction process the carotenoids must be separated in order to make structural assignments. Robust and reproducible separations of air and UV-sensitive

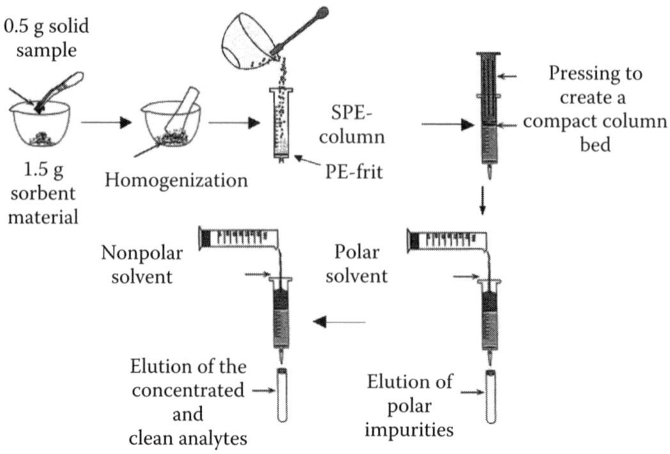

FIGURE 4.1 Extraction technique MSPD. (From Albert, K., *On-Line LC-NMR and Related Techniques*, John Wiley & Sons Ltd., 131, 2002. With permission.)

compounds such as carotenoids can be performed with the help of HPLC employing "reversed phase" stationary phases. These materials are composed of *n*-alkylsilyl ligands covalently bound via a Si–O–Si bonds to silica particles (diameter 3–5 µm, pore size 100–300 Å). Conventional reversed-phase materials have an *n*-alkyl chain length of 18 carbons. These C_{18} phases are not efficient for separating structural and stereo isomers of the many different carotenoids. Lane Sander developed a tailor-made C_{30} stationary phase where the separation of shape-constrained isomers can be achieved (Sander et al. 1994). This C_{30} phase exhibits a unique shape selectivity behavior due to a sophisticated alkyl chain organization, Figure 4.2. Here, tight clusters of alkyl chains, extended in more crystalline all-"*trans*-like" conformations alternate with more fluid clusters of alkyl chains exhibiting flexible gauche conformations (Albert et al. 1998, Raitza et al. 2000). This "slot model," outlined in Figure 4.3,

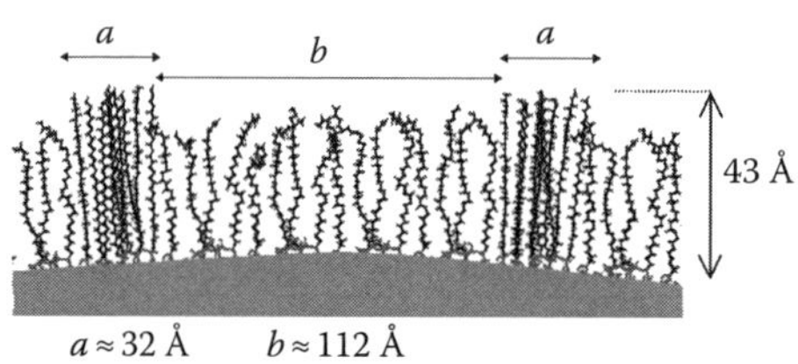

FIGURE 4.2 (See color insert following page 336.) Alkyl chain organization of a C_{30} phase. (From Raitza, M. et al., *Investigating the Surface Morphology of Triacontyl Phases with Spin-Diffusion Solid-State NMR Spectroscopy*, John Wiley & Sons Ltd., 3489, 2000. With permission.)

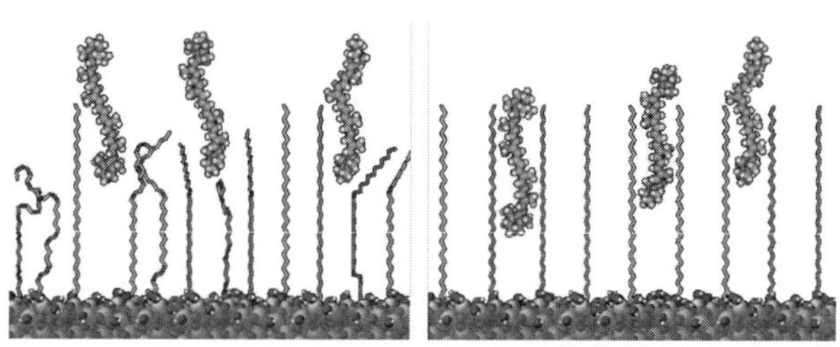

FIGURE 4.3 (See color insert following page 336.) Slot model. (From Meyer, C. et al., *Nuclear Magnetic Resonance and High Performance Liquid Chromatography Evaluation of Polymer Based Stationary Phases Immobilized on Silica*, Springer-Verlag GmbH, 686, 2005. With permission.)

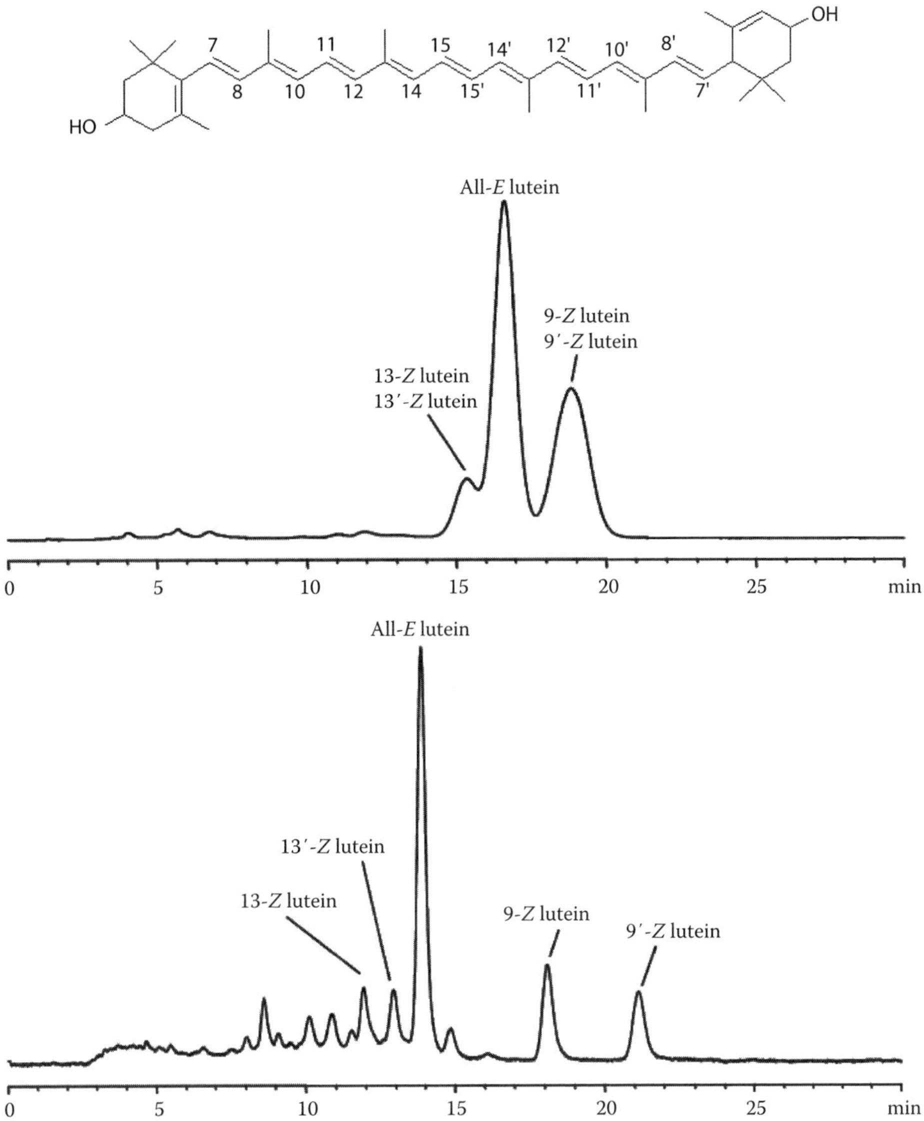

FIGURE 4.4 Separation of lutein stereoisomers, comparison between C_{18} and C_{30} phases. (From Dachtler, M. et al., *J. Chromatogr. B*, 211, 1998. With permission.)

explains the retention behavior of different isomers due to their differing abilities to penetrate the alkyl chain clusters (Albert 1988). Figure 4.4 shows a comparison of the separation of lutein derivatives performed on a C_{18} versus a C_{30} column (Wise and Sander 1985). It is obvious that the C_{18} column is unable to achieve the resolution necessary for the separation of these different compounds.

4.4 ON-LINE CAPILLARY HPLC–NMR COUPLING

HPLC–NMR analysis in a closed-circuit reveals the stereochemical information for elucidating the structures of unknown compounds (Albert 2002). In contrast to the technique of off-line separation, sample collection, and peak identification closed-circuit analysis guarantees the absence of isomerization and degradation. Very often only small amounts of sample are available after extraction.

Thus the on-line coupling of capillary HPLC with NMR is the method of choice. Unambiguous peak identification can be performed by using data obtained by HPLC–electrospray chemical ionization (ESI) MS or HPLC–atmospheric pressure chemical ionization (APCI) MS coupled together with results from on-line capillary HPLC–NMR. On-line capillary HPLC–NMR is conducted using NMR flow cells with detection volumes between 1.5 and 5.0 μL, enabling the use of deuterated solvents. With small amounts of sample, higher concentrations of analyte in the nanoliter detection cell are obtained leading to reasonable NMR acquisition times for 1D and 2D NMR spectra (Olson et al. 1995, Webb 1997).

The on-line coupling of HPLC and NMR can either be performed in the stopped-flow or in the continuous-flow mode (Krucker et al. 2004, Grynbaum et al. 2005, Putzbach et al. 2005, Albert et al. 2006, Hentschel et al. 2006, Rehbein et al. 2007). Current sensitivity levels are in the lower nanogram range for 1D ^1H NMR spectra and in the microgram range for 2D spectra.

Figure 4.5 shows the schematic design of a microcoil NMR probe. The horizontally oriented radio frequency copper coil is directly attached to the glass with an internal diameter of 100 μL. Thus an excellent filling factor (ratio of sample volume versus detection coil volume) is guaranteed. This newly designed probe with a microcoil shows significant improvements in the signal line shape and an easy magnetic field homogenization. The obtained signal-to-noise ratio of 50:1 for the anomeric proton of a 0.2 M solution of sucrose in D_2O is sufficient to perform structure elucidation of naturally occurring substances.

The instrumental setup for capillary HPLC–NMR coupling is shown in Figure 4.6. The capillary pump is connected via 50 μm capillaries between the capillary HPLC pump, the UV detector, and the NMR flow probe.

Figure 4.7 shows the structures of important carotenoids: (all-*E*) lutein, (all-*E*) zeaxanthin, (all-*E*) canthaxanthin, (all-*E*) β-carotene, and (all-*E*) lycopene. Employing a self-packed C_{30} capillary column, the carotenoids can be separated with a solvent gradient of acetone:water = 80:20 (v/v) to 99:1 (v/v) and a flow rate of 5 μL min^{-1}, as shown in Figure 4.8 (Putzbach et al. 2005). The more polar carotenoids (all-*E*) lutein, (all-*E*) zeaxanthin, and (all-*E*) canthaxanthin elute first followed by the less polar (all-*E*) β-carotene and the nonpolar (all-*E*) lycopene. Figure 4.9 shows the stopped-flow ^1H NMR spectra of these five carotenoids. The chromatographic run was stopped when the peak maximum of the compound of interest reached the NMR probe detection volume.

The spectrum of the noncentrosymmetric (all-*E*) lutein shows a multiplet (integration value four) for the protons 11/11′ (6.62 ppm) and 15/15′ (6.59 ppm). The protons 12/12′ (6.30 ppm) and

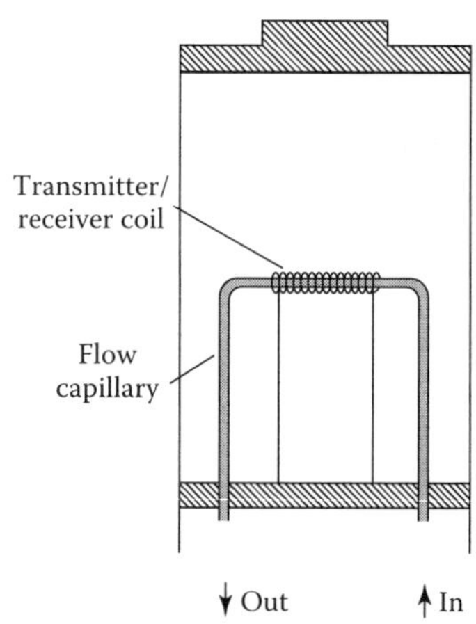

FIGURE 4.5 Schematic design of a microcoil NMR probe. (From Rehbein, J. et al., *Characterization of Bixin by LC-MS and LC-NMR*, John Wiley & Sons Ltd., 2387, 2007. With permission.)

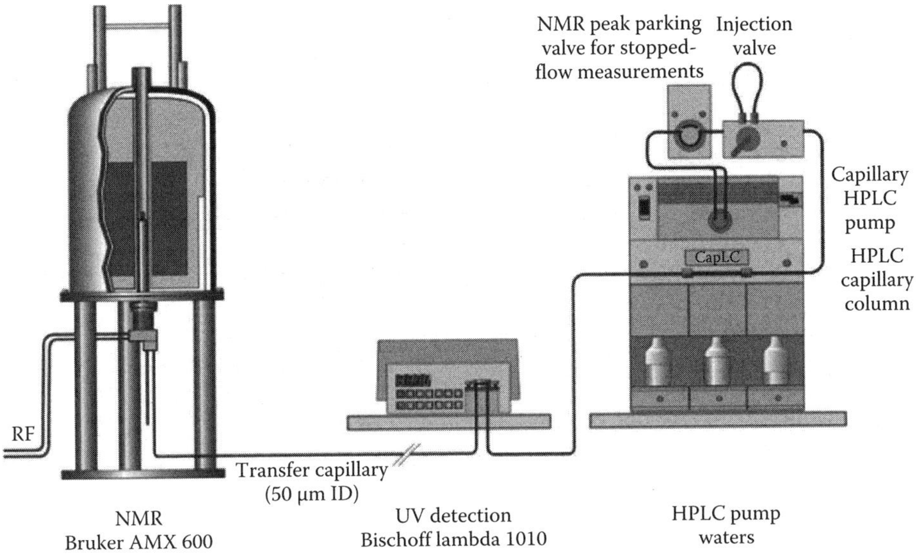

FIGURE 4.6 Instrumental setup for capillary HPLC–NMR coupling. (From Hentschel, P. et al., *J. Chromatogr. A*, 285, 2006. With permission.)

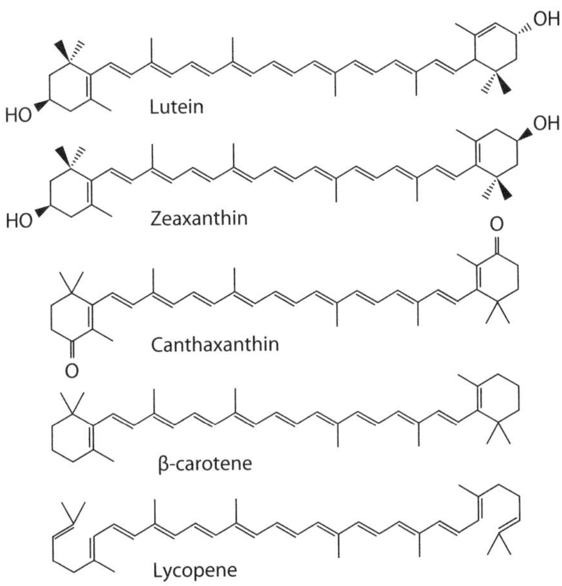

FIGURE 4.7 Structures of important carotenoids: (all-*E*) lutein, (all-*E*) zeaxanthin, (all-*E*) canthaxanthin, (all-*E*) β-carotene, and (all-*E*) lycopene.

14/14′ 6.22 ppm) are doublets (integration value two for each doublet). The signals of protons 8/8′ (6.07/6.09 ppm) together with the protons 10/10′ (6.07/6.09 ppm) and 7 (6.06 ppm) overlap to yield a multiplet (integration value of five). In comparison to proton 7 (6.06 ppm) proton 7′ is shifted to higher field (5.39 ppm) because of the shift of the double bond in the corresponding ionone ring.

The chemical shifts of the olefinic protons from the centrosymmetric (all-*E*) zeaxanthin are very similar to the chemical shifts of (all-*E*) lutein except for proton 7′. The resonances of protons 11/11′ (6.65 ppm) and of protons 15/15′ (6.62 ppm) show a multiplet with an integration value of four. The

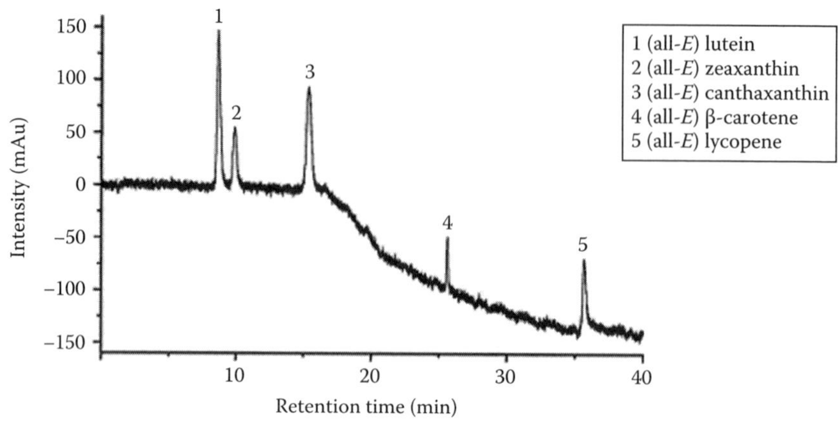

FIGURE 4.8 Capillary HPLC separation on a C_{30} column of (all-*E*) lutein, (all-*E* 7) zeaxanthin, (all-*E*) canthaxanthin, (all-*E*) β-carotene, and (all-*E*) lycopene. (From Putzbach, K. et al., *J. Pharm. Biomed. Anal.*, 910, 2005. With permission.)

chemical shifts of the protons 12/12′ (6.32 ppm) and 14/14′ (6.23 ppm) are slightly different from the chemical shifts of the corresponding protons of (all-*E*) lutein. The main difference in the ^{1}H-NMR spectra is found in the signals of protons 10/10′ (6.11 ppm), 8/8′ (6.08 ppm), and 7/7′ (6.07 ppm). The centrosymmetric structure of (all-*E*) zeaxanthin leads to one signal for protons 7 and 7′.

In comparison to the NMR spectra of (all-*E*) lutein and (all-*E*) zeaxanthin, the multiplet signal of the protons 11/11′ (6.68 ppm) and 15/15′ (6.65 ppm) of (all-*E*) canthaxanthin exhibits a slightly stronger "low-field" shift. The doublets of the protons 12/12′ and 8/8′ appear at 6.40 ppm and 6.36 ppm, respectively. A multiplet of protons 14/14′ (6.29 ppm), 10/10′ (6.27 ppm), and 7/7′ (6.25 ppm) is shifted to lower field due to the shielding effect of the carbonyl group at C-4.

The ^{1}H NMR spectrum of (all-*E*) β-carotene shows the characteristic low-field multiplet at 6.75 ppm arising from protons 11/11′ (6.76 ppm) and protons 15/15′ (6.74 ppm). Similar to the spectra of (all-*E*) lutein and (all-*E*) zeaxanthin two doublets can be seen for protons 12/12′ (6.43 ppm) and 14/14′ (6.34 ppm). Protons 7/7′ (6.24 ppm) together with protons 10/10′ (6.23 ppm) show a multiplet (integration ratio four). The doublet of protons 8/8′ is found at 6.18 ppm.

The pattern of the ^{1}H-NMR spectrum of lycopene differs from the spectra of the other carotenoids because lycopene consists of conjugated double bonds. At 6.6 ppm the multiplet of protons 11/11′ (6.63 ppm) and of proton pairs 15/15′ (6.60 ppm) resonate adjacent to the doublet of proton pair 7/7′ (6.44 ppm), the doublet of proton pair 12/12′ (6.29 ppm), the doublet of proton pair 14/14′ (6.22 ppm), the doublet of proton pairs 8/8′ (6.15 ppm), and finally the doublet of proton pair 10/10′. The resonance of proton pairs 6/6′ and 2/2′ are shifted to a higher field at 5.85 and 5.00 ppm due to their position in the conjugated system.

In all recorded spectra the $^3J_{HH}$ coupling constants between the olefinic protons are on the order of 11–12 Hz, proving the all-*E* configuration of the investigated carotenoids. Minor differences between the reported chemical shifts and literature data are due to the effect of different solvent compositions.

In addition to 1D ^{1}H-NMR spectroscopy, 2D NMR spectra recorded in the stopped-flow mode give valuable information of the homonuclear and heteronuclear scalar connectivities. Figure 4.10 shows the homonuclear correlated spectrum (^{1}H^{1}H-COSY) of (all-*E*) lycopene, proving all the assignments shown in Figure 4.9e. An inverse detected spectrum (heteronuclear single quantum coherence, HSQC) of tocopherol acetate is depicted in Figure 4.11. Here, the chemical shifts of the proton signal can be directly correlated with the chemical shifts of the adjacent carbon atoms.

In contrast to mass spectroscopy, NMR spectroscopy reveals the effect of stereoisomerization. One example is the isomerization of lutein to anhydroltutein induced by cooking (Hentschel et al.

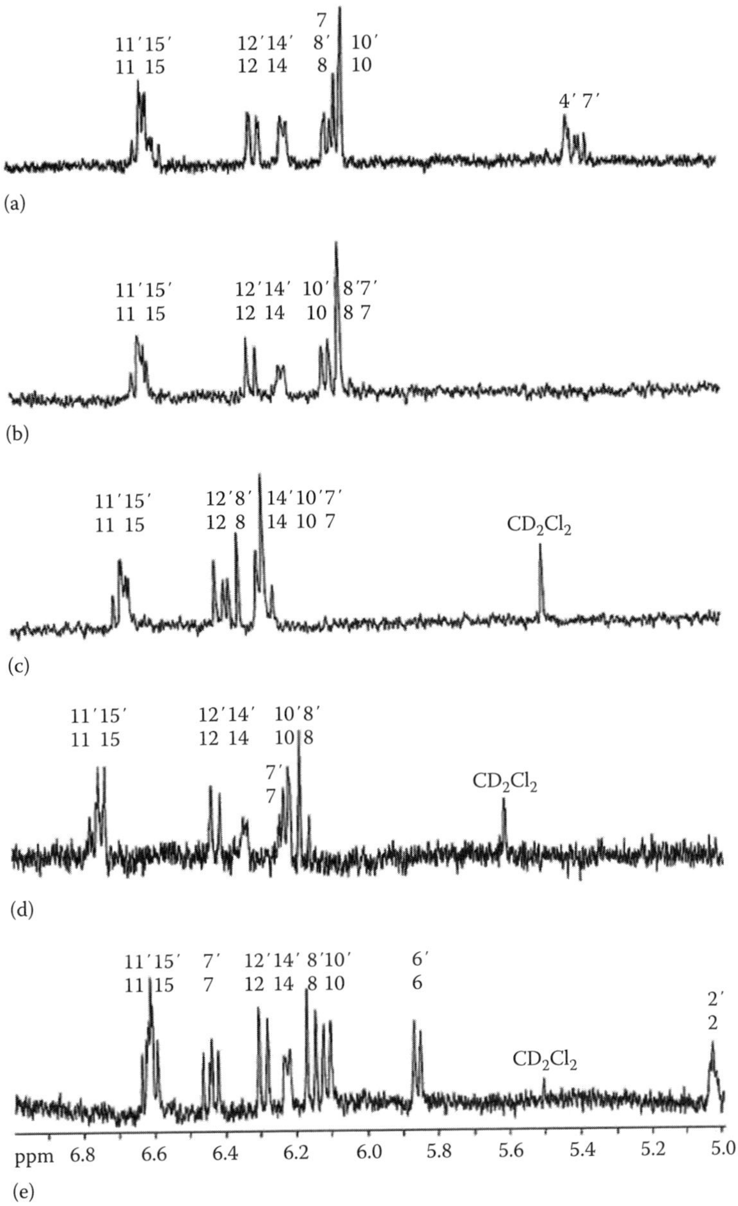

FIGURE 4.9 Stopped-flow ¹H-NMR spectra of (a) (all-*E*) lutein, (b) (all-*E*) zeaxanthin, (c) (all-*E*) canthaxanthin, (d) (all-*E*) β-carotene, and (e) (all-*E*) lycopene. (From Putzbach, K. et al., *J. Pharm. Biomed. Anal.*, 910, 2005. With permission.)

2006). Figure 4.12 shows the HPLC chromatograms of crude and cooked sorrels. In the chromatogram of cooked sorrel there is a new peak at a retention time of 34 min. With the help of a stopped-flow ¹H¹H-COSY NMR spectrum, this peak can be assigned to (all-*E*)-anhydrolutein I, Figure 4.13. A iodine-catalyzed photoisomerization leads to the formation of several stereoisomers that can be separated with the highly selective C_{30} column, Figure 4.14. As an example, Figure 4.15 shows the stopped-flow ¹H-NMR spectra of the olefinic region of (all-*E*) anhydrolutein I and 9-*Z* anhydrolutein I. The isomerization shift for proton 8 of 0.58 ppm and for proton 10 of −0.11 ppm is clearly visible. Thus the stereochemical assignment of different stereoisomers is possible.

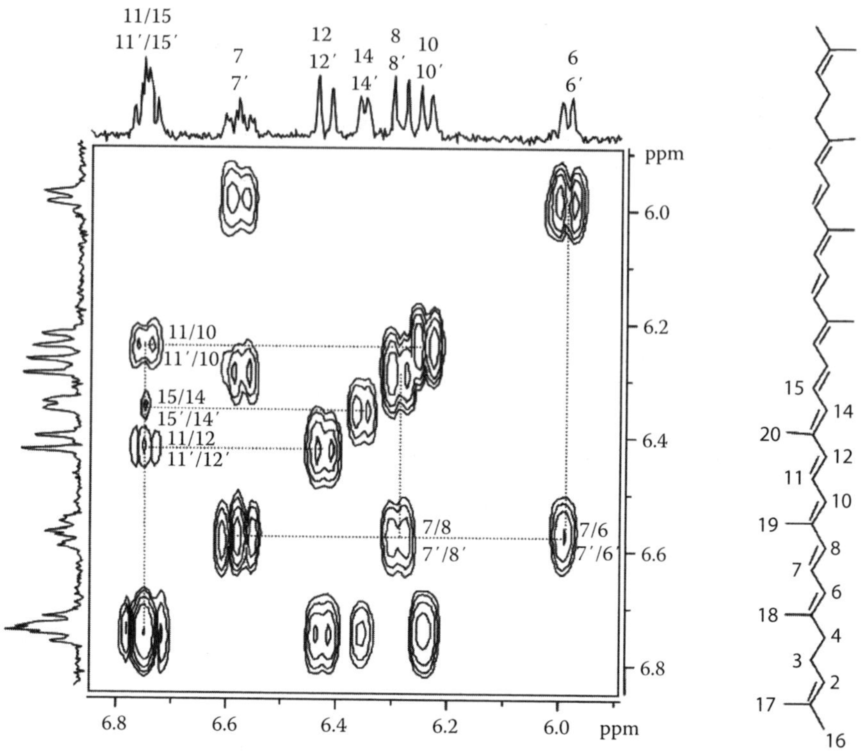

FIGURE 4.10 Stopped-flow $^1H^1H$-COSY NMR spectrum of (all-*E*) lycopene. (From Albert, K., *On-Line LC-NMR and Related Techniques*, John Wiley & Sons Ltd., 131, 2002. With permission.)

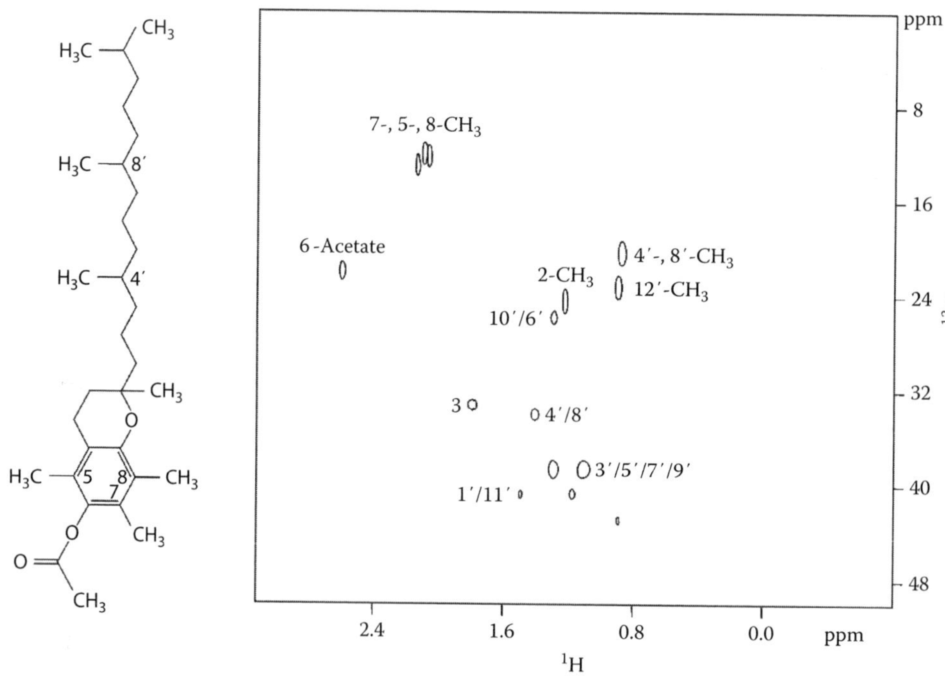

FIGURE 4.11 Stopped-flow $^1H^{13}C$-HSQC NMR spectrum of α-tocopherol acetate.

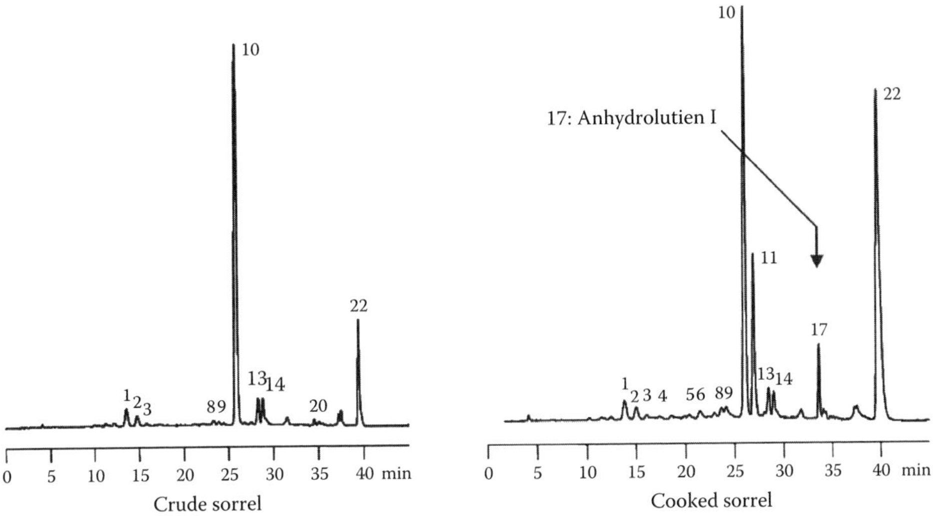

FIGURE 4.12 HPLC chromatograms of crude uncooked and cooked sorrel. 1: neochrome I, 2: neochrome II, 6: auroxanthin, 8: mutatoxanthin, 10: lutein, 11: 3-epilutein, 13: (9/9′Z)-lutein, 14: (13/13′Z)-lutein, 17: anhydrolutein I, 20: α-cryptoxanthin, and 22: β-carotene.

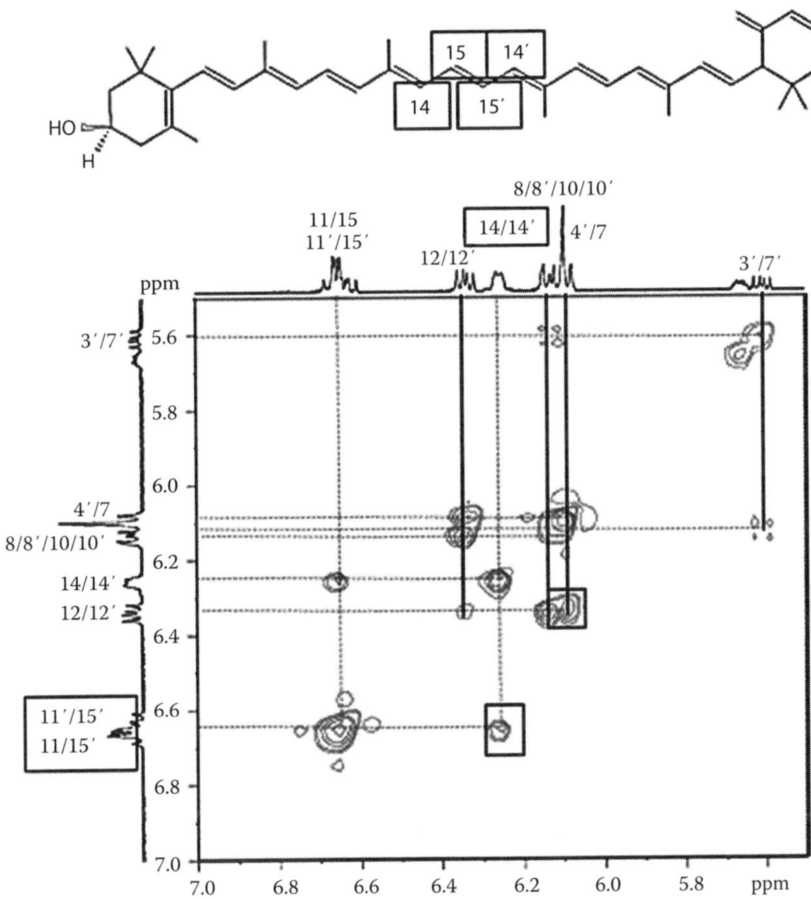

FIGURE 4.13 Stopped-flow $^1H^1H$-COSY NMR spectrum of (all-*E*) anhydrolutein I. (From Hentschel, P. et al., *J. Chromatogr. A*, 285, 2006. With permission.)

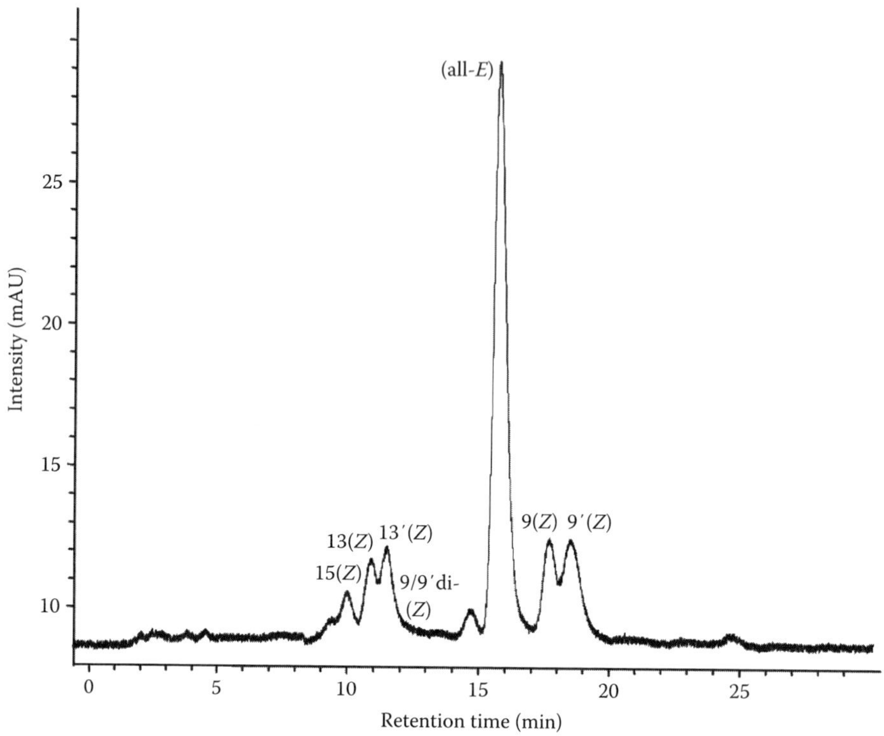

FIGURE 4.14 Capillary HPLC separation of anhydrolutein I isomers on a C_{30} phase. (From Hentschel, P. et al., *J. Chromatogr. A*, 285, 2006. With permission.)

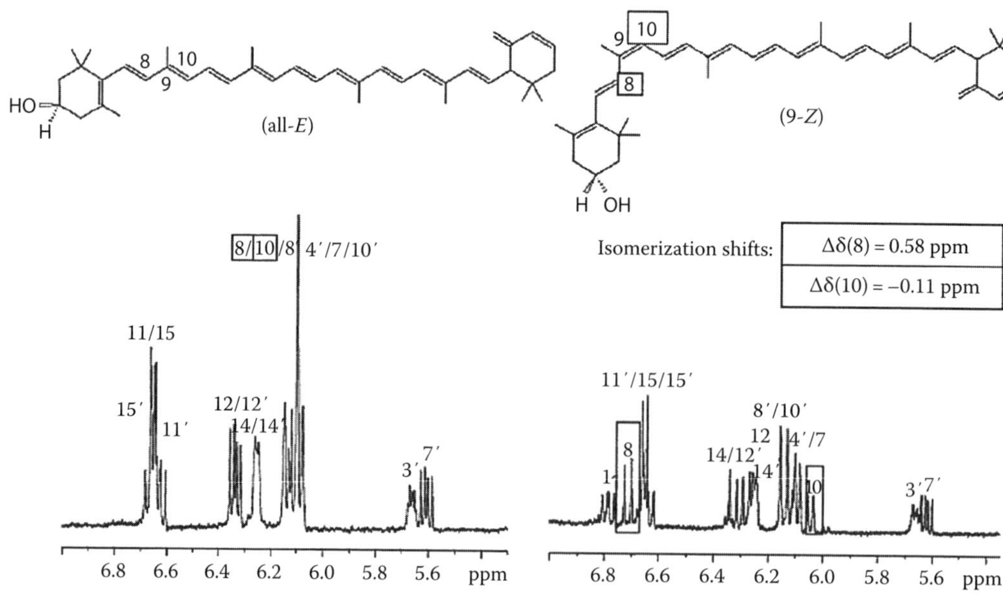

FIGURE 4.15 Stopped-flow capillary ^{1}H-NMR spectra of (all-*E*) and (9-*Z*) anhydrolutein I. (From Hentschel, P. et al., *J. Chromatogr. A*, 285, 2006. With permission.)

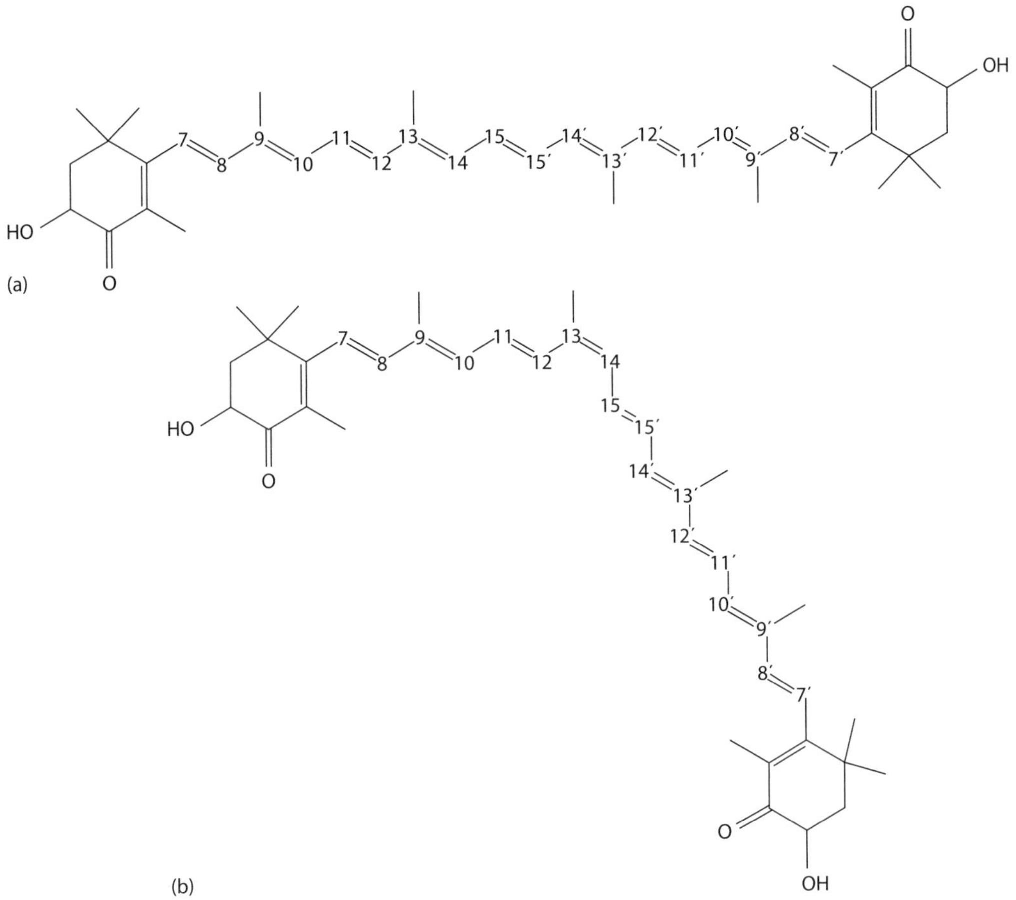

FIGURE 4.16 Structures of the stereoisomers of astaxanthin: (a) all-*E* and (b) 13-*Z*.

NMR spectroscopy is essential for the structure determination of carotenoid isomers because the ^{1}H-NMR signals of the olefinic range are characteristic for the arrangement of the isomers. The stereoisomers of astaxanthin, as shown in Figure 4.16, can be separated on a shape-selective C_{30} capillary column with methanol under isocratic conditions.

Figure 4.17a shows the ^{1}H-NMR spectrum of (all-*E*) astaxanthin recorded in the stopped-flow modus. The spectrum of the centrosymmetric (all-*E*) astaxanthin indicates an overlapped multiplet (integration value of four) for the protons 11/11′ (6.97 ppm) and the protons 15/15′ (6.77 ppm). The multiplet consists of a doublet of the protons 15/15′ and a pseudo triplet (a doublet of doublets) for the protons 11/11′. The protons 8/8′ (6.52 ppm) and 12/12′ (6.50 ppm) appear as two very close doublets with a total integration value of four. The protons 14/14′ (6.40 ppm), 10/10′ (6.38 ppm), and 7/7′ (6.36 ppm) show a multiplet generated by the three overlapping doublets with an integration value of six. The $^{3}J_{H/H}$ coupling constants of $J_{7/8}$ (16.2 Hz), $J_{10/11}$ (14.0 Hz), $J_{11/12}$ (14.0 Hz), and $J_{14/15}$ (11.8 Hz) are in a typical region of *trans* conjugated carotenoids.

Figure 4.18 displays the ^{1}H^{1}H-COSY NMR spectrum of (all-*E*) astaxanthin recorded under stopped-flow condition. The three different spin systems 7/8, (7′/8′), 10/11/12 (10′/11′/12′), and 14/15 (14′/15′) can be determined by four cross peaks (marked in Figure 4.18) between 7/8, (7′/8′), 10/11 (10′/11′), 11/12 (11′/12′), and 14/15 (14′/15′).

The (13-*Z*) isomer of astaxanthin is a noncentrosymmetric carotenoid, thus the proton shifts of both sides of the chain are not equal any longer. For example, this causes proton 15 to have a spectrum of higher order, while it exhibits a doublet in the all-*E* compound. The largest shift differences

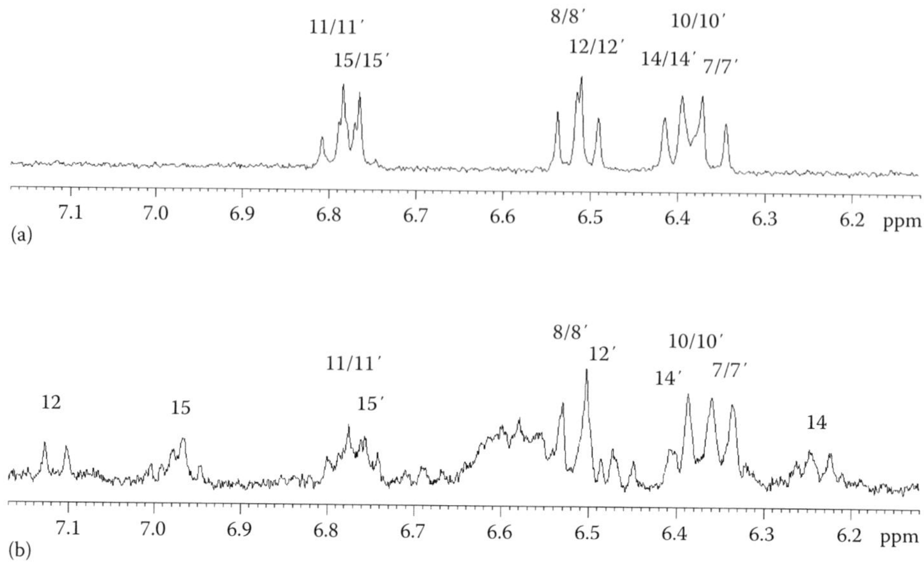

FIGURE 4.17 Stopped-flow ^{1}H-NMR spectra of (a) (all-*E*) astaxanthin and (b) (13-*Z*) astaxanthin.

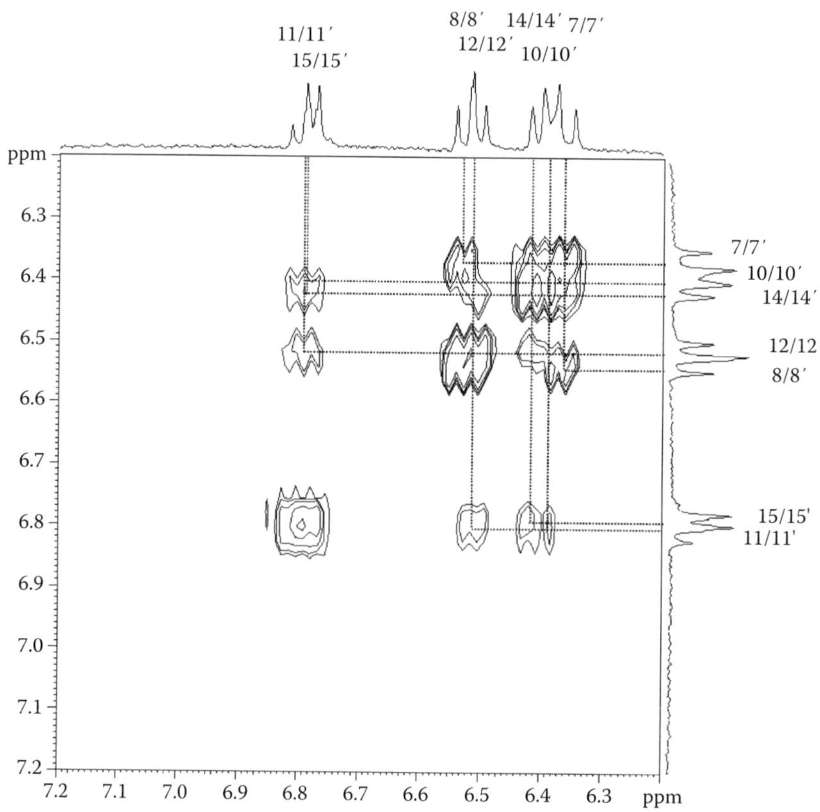

FIGURE 4.18 Stopped-flow ^{1}H^{1}H-COSY NMR spectrum of (all-*E*) astaxanthin.

TABLE 4.1
Chemical Shifts and Isomeric Shift Differences of (All-*E*) and (13-*Z*) Astaxanthin

Proton	δ (All-*E*) Astaxanthin (ppm)	δ (13-*Z*) Astaxanthin (ppm)	Δδ (ppm)
H (7)	6.36	6.36	—
H (7′)		6.36	—
H (8)	6.52	6.52	—
H (8′)		6.52	—
H (10)	6.38	6.38	—
H (10′)		6.38	—
H (11)	6.77	6.77	—
H (11′)		6.77	—
H (12)	6.50	7.12	0.62
H (12′)		6.50	—
H (14)	6.40	6.24	−0.16
H (14′)		6.40	—
H (15)	6.79	6.97	0.18
H (15′)		6.74	−0.05

($\delta\Delta = \delta_Z - \delta_{\text{all-}E}$) comparing the 13-*Z* and all-*E* spectrum are expected in the area around the 13-*Z* arrangement.

In the ^{1}H-NMR spectrum of the (13-*Z*) astaxanthin isomer, Figure 4.17b, the "convex-side" protons 14 (6.24 ppm) and 15′ (6.74 ppm) are shifted to higher field, with Δδ values of −0.16 ppm (14) and −0.05 ppm (15′). However, the "concave-side" protons 12 (7.12 ppm) and 15 (6.97 ppm) are shifted to lower field with Δδ values of 0.62 ppm (12) and 0.18 ppm (15). The protons far from the *cis* bond are unaffected from the stereochemical behavior.

The isomerization shifts conform to those that are described in literature (Englert and Vecci 1980, Englert 1995). All chemical shifts and isomeric shift differences of the olefinic region are listed in Table 4.1.

Overall, the combination of HPLC together with NMR is a very efficient tool to elucidate structure of different stereoisomers found in complex natural mixtures.

4.5 CONCLUDING REMARKS

In summary, NMR spectroscopy is an extremely versatile tool useful that enables researchers to understand the structure of natural products such as carotenoids. For a full structural assignment, the compound of interest has to be separated from coeluents. Thus, it is a prerequisite to employ tailored stationary phases with high shape selectivity for the separation in the closed-loop on-line LC–NMR system. For the NMR detection, microcoils prove to be advantageous for small quantities of sample. Overall, the closed-loop system of HPLC and NMR detection is very advantageous for the structural elucidation of air- and UV-sensitive carotenoids.

ACKNOWLEDGMENT

The authors gratefully acknowledge the help of Jan Peter Mayser in preparing the figures.

REFERENCES

Albert, K. 1988. Correlation between chromatographic and physicochemical properties of stationary phases in HPLC: C_{30} bonded reversed-phase silica. *Trends Anal. Chem.* 17:648–658.

Albert, K. 2002. *On-line LC-NMR and Related Techniques*, Chichester, U.K.: John Wiley & Sons.

Albert, K., Lacker, T., Raitza, M., Pursch, M., Egelhaaf, H.-J., and Oelkrug, D. 1998. Investigating the selectivity of triacontyl interphases. *Angew. Chem.* 110:810–812, *Angew. Chem. Int. Ed. Engl.* 37:778–780.

Albert, K., Krucker, M., Putzbach, K., and Grynbaum, M. D. 2006. LC-NMR coupling. In *HPLC Made to Measure: A Practical Handbook for Optimization*, ed. S. Kromidas, pp. 551–563. Weinheim, Germany: Wiley-VCH.

Barker, S. A. 2000. Matrix solid-phase dispersion. *J. Chromatogr. A.* 885:115–127.

Dachtler, M., Kohler, K., and Albert, K. 1998. Reversed-phase high-performance liquid chromatographic identification of lutein and zeaxanthin stereoisomers in bovine retina using a C30 bonded phase. *J. Chromatogr. B* 211–216.

Englert, G. 1995. NMR Spectroscopy: In *Carotenoids Volume 1B: Spectroscopy*, ed. G. Britton, S. Liaaen-Jensen, and H. Pfander, pp. 147–260. Basel, Switzerland: Birkhäuser.

Englert, G. and Vecci, M. 1980. *Trans/cis* isomerization of astaxanthin diacetate/isolation by HPLC and identification by 1H-NMR spectroscopy of three mono-*cis*- and six di-*cis*-isomers. *Helv. Chim. Acta* 63:1711–1717.

Grynbaum, M. D., Hentschel, P., Putzbach, K., Rehbein, J., Krucker, M., Nicholson, G., and Albert, K. 2005. Unambiguous detection of astaxanthin and astaxanthin fatty acid esters in krill (*Euphausia superba dana*). *J. Sep. Sci.* 28:1685–1693.

Hentschel, P., Grynbaum, M. D., Molnar, P., Putzbach, K., Rehbein, J., Deli, J., and Albert, K. 2006. Structure elucidation of deoxylutein-II isomers by on-line capillary high performance liquid chromatography—^{1}H nuclear magnetic resonance spectroscopy. *J. Chromatogr. A* 1112:285–292.

Krucker, M., Lienau, A., Putzbach, K., Grynbaum, M. D., Schuler, P., and Albert, K. 2004. Hyphenation of capillary HPLC to microcoil ^{1}H NMR spectroscopy for the determination of tocopherol homologues. *Anal. Chem.* 76:2623–2628.

Meyer, C., Skogsberg, U., Welsch, N., and Albert, K. 2005. *Nuclear Magnetic Resonance and High Performance Liquid Chromatography Evaluation of Polymer Based Stationary Phases Immobilized on Silica*. Springer-Verlag GmbH, p. 686.

Olson, D. L., Peck, T. L., Webb, A. G., Magin, R. L., and Sweedler, J. V. 1995. High-resolution microcoil 1H-NMR for mass-limited, nanoliter-volume samples. *Science* 270:1967–1970.

Putzbach, K., Krucker, M., Grynbaum, M. D., Hentschel, P., Webb, A. G., and Albert, K. 2005. Hyphenation of capillary high-performance liquid chromatography to microcoil magnetic resonance spectroscopy—Determination of various carotenoids in a small-sized spinach sample. *J. Pharm. Biomed. Anal.* 38:910–917.

Raitza, M., Wegmann, J., Bachmann, S., and Albert, K. 2000. Investigating the surface morphology of triacontyl phases with spin-diffusion solid-state NMR spectroscopy. *Angew. Chem.* 112:3629–3632, *Angew. Chem. Int. Ed.* 112:3486–3489.

Rehbein, J., Dietrich, B., Grynbaum, M. D., Hentschel, P., Holtin, K., Kuehnle, M., Schuler, P., Bayer, M., and Albert, K. 2007. Characterization of bixin by LC-MS and LC-NMR, *J. Sep. Sci.* 30:2382–2390.

Sander, L. C., Epler Sharpless, K., Craft, N. E., and Wise, S. A. 1994. Development of engineered stationary phases for the separation of carotenoid isomers. *Anal. Chem.* 66:1667–1674.

Webb, A. G. 1997. Radio frequency microcoils in magnetic resonance. *Prog. NMR Spec.* 31:1–42.

Wise, S. A. and Sander, L. C. 1985. Factors affecting the reversed-phase liquid chromatographic separation of polycyclic aromatic hydrocarbon isomers. *J. High Resolut. Chromatogr. Commun.* 8:248–255.

5 Quantitative Methods for the Determination of Carotenoids in the Retina

Richard A. Bone, Wolfgang Schalch, and John T. Landrum

CONTENTS

5.1 INTRODUCTION

A remarkable sequence of selective processes leads to the uptake of just two carotenoids by the primate eye. Approximately 750 naturally occurring carotenoids have been identified, some 30–50 of these are consumed as part of the human diet, and about 20 are found in the blood. Yet only the dihydroxy carotenoids, lutein and zeaxanthin undergo active uptake from the blood into various tissues in the eye. Of particular interest is the concentration of lutein and zeaxanthin in the center of the retina where they form a visible yellow spot, or "macula lutea" (Bone et al. 1985). The reason for such interest is the evidence that has been uncovered over the years for a protective function by this "macular pigment" (MP), in particular against the eye disease, age-related macular degeneration (AMD) (Schalch 2001). There are two potential modes of protection. The MP forms a blue-light-absorbing layer in the inner part of the retina and reduces the amount of toxic blue light reaching the posterior tissues that tend to become damaged in AMD patients. Additionally these carotenoids possess antioxidant activity with the ability to quench reactive oxygen species and free radicals that could otherwise lead to damage (Beatty et al. 2000b).

The recognition of the importance of MP in maintaining the health of the retina has led to the development of a number of methods for determining its concentration *in situ*. These methods, necessarily noninvasive, are routinely employed in dietary supplementation studies with lutein or zeaxanthin to monitor the uptake of the carotenoids into the retina. Every method exploits the optical properties of lutein and zeaxanthin, specifically their absorbance at visible wavelengths. The detection of a light signal, modified by the carotenoids, is accomplished either by the retinal photoreceptors themselves (psychophysical methods) or by a physical detector such as a photomultiplier, photodiode, or CCD array (physical methods). This chapter provides a review of past and current methods for quantifying MP in the living human retina. (A method employing resonance Raman spectroscopy is reviewed in Chapter 6.)

5.2 PSYCHOPHYSICAL METHODS

The majority of the MP is found in the photoreceptor axons, which spread out radially from the foveal center to an eccentricity of approximately 4° (Snodderly et al. 1984) (see Figure 13.4). This layer of mainly cone axons is known as the Henle fiber layer. The rest of the cone cell, including the photopigment-containing outer segment where phototransduction begins, is posterior to the Henle fiber layer. Therefore, any attenuation of the intensity of light reaching the outer segments as a result of the MP will be reflected in a reduced perception of luminance. In other words, a small visual stimulus imaged on the fovea will appear less bright than it otherwise would in the absence of MP. This is the basis of an entoptical phenomenon known as Maxwell's spot. In 1871, the physicist James Clerk Maxwell reported his observation that "When observing a spectrum... I noticed an elongated spot running up and down the spectrum but refusing to pass out of the blue into the other colors... The conclusion to which I have come is that the appearance is due to the yellow spot in the retina..." (Maxwell 1856). The elongation to which Maxwell referred is probably a reflection of his own elliptically, rather than circularly, shaped MP distribution that most of the methods to be described here have often revealed. In essence, the task of the psychophysical method is to determine quantitatively how dark Maxwell's spot appears to be in comparison with the surrounding, unattenuated visual field. We begin with what is probably the most well-established psychophysical method, heterochromatic flicker photometry (HFP).

5.2.1 Heterochromatic Flicker Photometry

Flicker photometry is an established, outmoded, photometric procedure for comparing the luminances provided by two lamps, for example a standard and a substandard lamp (Walsh 1953). Light from the two lamps is directed alternately onto a circular visual field, which appears to flicker if the luminances provided by the lamps are different but appears steady under the condition of equiluminance. HFP, as the name implies, involves matching the luminances of lights of different color. The procedure eliminates the difficulty of achieving a luminance match between differently colored visual fields presented side by side for direct comparison, during which the observer must attempt to ignore chromaticity differences. In HFP, the color fusion of the two lights occurs at a frequency well below that required for luminance fusion. Thus the visual field appears to be of a single hue but, depending on the frequency, appears to flicker or appears to be steady. A critical frequency must be sought so that a steady appearance is achieved only when the luminances of the two lights are equal. If the frequency is higher than this critical value, a steady appearance is observed over a range of relative luminances of the two lights; if it is lower than the critical value, the observer is unable to eliminate flicker by the adjustment of the relative luminances.

When HFP is adapted for MP measurements in a subject, the two colors are selected based on the spectral absorbance of the pigment (Bone et al. 1992). As shown in Figure 5.1, peak absorbance occurs at a wavelength of 460 nm, and beyond 530 nm absorbance is essentially zero. In a typical instrument, interference filters are used to isolate narrow wavelength bands centered on 460 nm (blue) and 540 nm (green) from an incandescent light source such as a quartz–halogen lamp. The green light provides a standard that is unaffected by MP and the blue light provides the test wavelength at which MP optical density will be determined. A means of varying the luminance of the blue light is included. A device, such as a mechanical chopper, provides a method for illuminating the visual field alternately with these two colors. In another adaptation, the colors are provided by appropriate light emitting diodes (LED) (Wooten et al. 1999). The advantage of LEDs is that the alternation between them can be achieved electronically. A disadvantage is that the bandwidth of

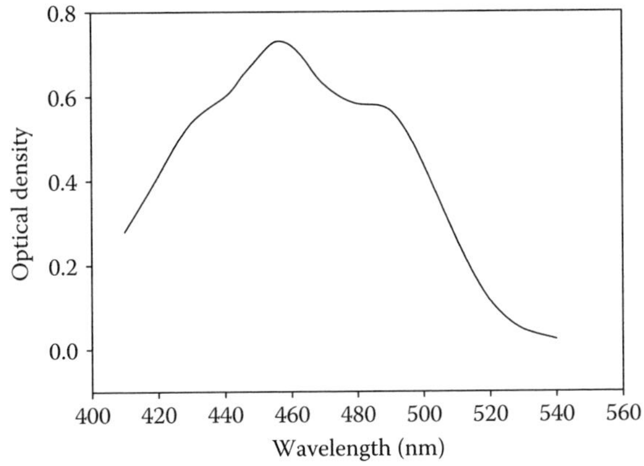

FIGURE 5.1 Absorbance spectrum (relative) of human MP.

LEDs tends to be relatively large and corrections must be applied to the data in order to be able to report the MP optical density at the test wavelength.

The instrument's visual field typically subtends an angle of about a degree at the subject's eye thus ensuring that the corresponding stimulus on the retina falls within the area of the yellow spot. While fixating on the center of the visual field, the subject adjusts the intensity of the blue light until the sensation of flicker is either eliminated or minimized (see Figure 5.2). In what has been termed "customized HFP (cHFP)," the critical frequency is customized for the individual subject (Stringham et al. 2008). Older subjects may be less sensitive to flicker and require a lower frequency compared to young subjects. The intensity setting for the blue light that eliminates flicker will, of course, depend on the optical density of the subject's MP. Subjects with a high MP optical density will require a higher intensity to compensate for attenuation by the MP compared with subjects having a low optical density. However, other factors will affect the intensity setting, such as lens yellowing that increases with age (Weale 1963) and, like MP, will attenuate the blue, but not the green

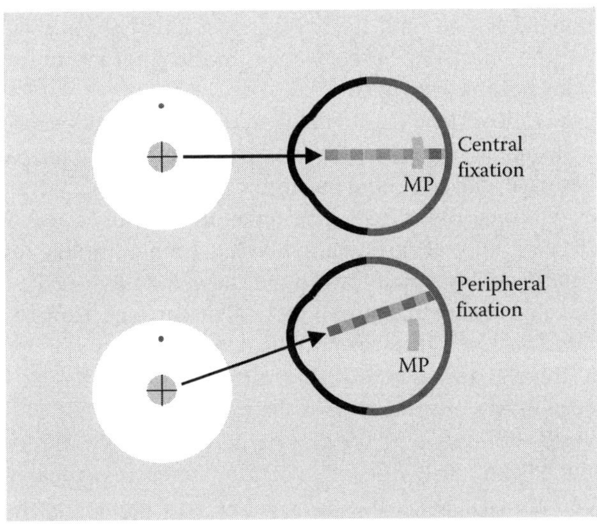

FIGURE 5.2 (See color insert following page 336.) Illustration of the method of HFP. On viewing the stimulus directly (upper), MP attenuates the blue component of the stimulus whereas with peripheral viewing (lower), no such attenuation occurs. In each case, the subject adjusts the luminance of the blue component until it matches the luminance of the green component, which is unaffected by MP.

light. Likewise differences in overall photoreceptor spectral sensitivity among otherwise identical subjects could lead to different blue-light intensity settings. Thus, a means of eliminating the effects of all but the MP is essential and the HFP procedure requires a second measurement. For this, the subject directs his or her gaze toward a fixation mark to one side of the stimulus at an eccentricity that varies among instruments from about 5° to 8°. The stimulus itself is then imaged in the parafoveal retina, an area that is assumed to have negligible amounts of MP. Once again the subject seeks to eliminate or minimize flicker, and the corresponding blue-light intensity setting reflects the degree of lens yellowing and the spectral properties of the photoreceptors, but not the MP. Because the parafoveal retina has a lower flicker threshold than the fovea, a lower frequency is used for this measurement than for the foveal test. The MP optical density at the blue test wavelength is given by the log ratio of intensity settings made by the subject:

$$\mathrm{OD} = \log_{10}(I_{\mathrm{fov}}/I_{\mathrm{parafov}}) \tag{5.1}$$

Acceptance of Equation 5.1 rests on the assumption that the only factor modifying the luminance match between the blue and green lights in the fovea compared with that in the parafovea is the attenuation of the blue light by MP in the former match. There is, however, evidence that MP optical density may not be completely negligible at the parafoveal location. It may increase with age (Berendschot and Van Norren 2005) or as a result of supplementation with lutein or zeaxanthin (Rodriguez-Carmona et al. 2006). The assumption would also be invalidated if the spectral sensitivity of the photoreceptors in the fovea differed from that in the parafovea. Specifically, if the ratio of sensitivities at the blue to green wavelengths was higher in the parafovea than in the fovea, HFP would return a value of the subject's MP optical density that was too high. Differing sensitivity ratios across the retina could, in principle, be expected if the proportions of the three cone types (long-, medium-, and short-wavelength-sensitive) and rods varied. Certainly rods and short-wavelength-sensitive cones (S-cones) are not well represented in the fovea. However there is little contribution to luminance by S-cones (Guth et al. 1980), and since S-cones have low flicker thresholds (Brindley et al. 1966), any contribution to luminance can be minimized by the use of sufficiently high flicker frequencies. Additionally, the use of stimuli with luminances well above the mesopic range will ensure that essentially only cones, and not rods, are responding. An added safeguard that has been adopted in a number of applications is the use of a blue adapting background on which the stimulus is superimposed (Hammond Jr. and Fuld 1992, Wooten et al. 1999). In principle, the background will preferentially lower the sensitivity of the S-cones to the point where their contribution to the luminance of the stimulus becomes negligible.

On the other hand, the ratio of long (L)- to medium (M)-wavelength-sensitive cones is believed to remain reasonably constant as one moves outward from the fovea to the parafovea (Wooten and Wald 1973). If this is the case, and in light of the arguments presented above, the effective spectral sensitivity in the fovea will only differ from that in the parafovea because of the presence of MP in the former region. The validity of this assumption has been put to the test by modifying HFP so that the test wavelength can be varied throughout the wavelength range of MP absorption. In this way, it has been possible to construct an MP optical density spectrum, which is in remarkably good agreement with one obtained from spectrophotometric analysis of appropriate mixtures of lutein and zeaxanthin (Bone et al. 1992). However, a recent study calls into question the assumption of a constant L-cone to M-cone ratio across the retina (Bone et al. 2007b). In this study, the test wavelength was varied not only over the absorption range of the MP, but up to a wavelength of 680 nm. Above about 580 nm, a significant, generally increasing, apparent MP optical density was observed in a number of subjects. In reality, lutein and zeaxanthin have zero optical density at these wavelengths. There is no evidence for the existence of another foveal pigment with appropriate spectral properties. One possible explanation is a higher L-cone to M-cone ratio in the parafovea compared with the fovea. However, when Wald measured spectral sensitivities in these regions by the method of absolute thresholds, he found that the log-transformed curves for his subjects were parallel above 578 nm (Wald 1945). This is consistent with the L-cone to M-cone ratio in the fovea and parafovea being the same. Additional work in this area is needed to resolve the issue.

HFP has also been used to measure the profile of MP optical density across the retina rather than a single, central optical density measurement (Hammond Jr. et al. 1997, Bone et al. 2004). In order to make such a measurement, a set of fixation marks is provided to one side of the stimulus so that it can be imaged at various distances from the center of the fovea. In addition, a number of researchers have exploited the "edge hypothesis" for the same purpose (Werner et al. 1987, Hammond Jr. et al. 1997, Beatty et al. 2000a, Hammond and Caruso-Avery 2000, Werner et al. 2000, Delori et al. 2001, Snodderly et al. 2004). This hypothesis states that for a circular stimulus, flicker sensitivity is enhanced at the edge of the stimulus. Thus, when a subject achieves a flicker null, it is because the luminances of the blue and green components of the stimulus are equalized at an eccentricity from the fovea equal to the stimulus radius. By using stimuli of different radii, one can, according to the hypothesis, obtain MP optical density measurements at several eccentricities and thereby obtain a profile. However, the validity of the edge hypothesis has been questioned (Bone et al. 2004).

5.2.2 Minimum Motion and Apparent Motion Photometry

Minimum motion photometry is a close cousin of HFP. The stimulus in this case consists of a grating of alternately colored bars that move across a circular, centrally viewed visual field. As with HFP, the two colors are selected for maximum absorption by the MP and nearly zero absorption. In Moreland's apparatus (Robson et al. 2003, Moreland 2004), wavelengths of 460 nm (blue) and 580 nm (orange) were chosen, and the stimulus superimposed on a 450 nm pedestal in order to saturate S-cones. The subject adjusts the luminance of the orange bars until the perception of motion of the bars across the field is minimized. This minimization condition occurs when the luminances of the bars are equal for the subject. Proponents of the method claim that finding this null point is easier than the corresponding task in HFP. Once again it is necessary to make a reference measurement with the stimulus imaged in an MP-free region of the retina in order to account for the factors other than MP mentioned in the previous section. For this measurement, the visual field is in the shape of an annular arc with the fixation point at the center of curvature. Profiles of the MP across the retina can be obtained using arcs of different radii.

In apparent motion photometry, colored bars appear to move across the field of view but their direction of motion reverses as the subject passes through the equiluminance condition (Anstis and Cavanagh 1983). The illusion is achieved on a CRT monitor by presenting a repetitive sequence of four square-wave gratings, each phase advanced by a quarter cycle from the previous one (West and Mellerio 2005). The first and third gratings are composed of blue and red bars; the second and fourth of light gray and dark gray bars. If the blue bars are brighter than the red, the subject will associate their luminance with that of the light gray bars of the following grating that are, for example, phase-shifted a quarter cycle to the right. When the third grating appears, its blue bars will appear at yet another quarter cycle to the right, so the subject's perception is of the pattern of bars moving continuously to the right. If, on the other hand, the red bars are brighter than the blue, their luminance will be associated with the light gray bars of the following grating that are phase-shifted to the left, and the perception is of motion to the left.

For the central measurement, the field of view is rectangular (e.g., 0.3 by 1.25°) and for other eccentricities, the field is an annular arc similar to that provided in minimum motion photometry. The use of a CRT monitor introduces the same problem as the use of LEDs in HFP, namely, the broadband nature of the screen phosphors, and a correction must be made before reporting the peak MP optical density. A system that could employ lamps and filters instead of a CRT monitor would be difficult to design because of the complexity of the visual stimulus.

5.2.3 Dichroism-Based Photometry

An orderly arrangement of carotenoid molecules in the Henle fibers is responsible for an entoptic phenomenon similar to Maxwell's spot, and provides the basis for a method of measuring MP optical density (Bone 1980, Bone and Landrum 1984, Bone et al. 1992). The entoptic phenomenon, Haidinger's brushes (Von Haidinger 1844), appears at the fixation point if a surface, uniformly illuminated with light in the 400–500 nm range, is viewed through a polarizing filter. The brushes appear as a dark, hourglass-shaped object against the background, the main axis of which is perpendicular to the plane of polarization of the light entering the eye. The dimensions of Haidinger's brushes match those of Maxwell's spot. In order to account for the brushes, it has been proposed that a fraction of the long-chain lutein and zeaxanthin molecules are aligned transversely with respect to the cylindrical membranes of the Henle fibers (Bone and Landrum 1984, Bone et al. 1992). Since the fibers are themselves arranged radially from the foveal center like the spokes on a wheel, the MP molecules would be oriented perpendicular to the "spokes." A second requirement is the dichroism of the chain-like lutein and zeaxanthin molecules, meaning that they absorb light preferentially that is polarized parallel to the chain.

Exploiting these properties, Bone et al. devised a method for quantifying the MP (Bone and Landrum 1984, Bone et al. 1992). A monochromatic stimulus (460 nm) was presented to the subject, as shown in Figure 5.3. The two triangular areas were polarized either parallel or perpendicular to the main axis of the figure whereas the surrounding field was unpolarized. Using central fixation, the subject adjusted the intensity of the triangles to match that of the surround for each of the polarization conditions. Because the colors were identical, the task was relatively easy. When the plane of polarization was perpendicular to the main axis of the figure, the triangles appeared relatively dark because they coincided with the dark sectors of Haidinger's brushes. When the plane of polarization was rotated by 90°, the triangles appeared lighter because they coincided with the light sectors of the brush pattern.

The quantity that was reported was the log ratio of intensity settings (log G) for the two polarization conditions. It was shown theoretically that this quantity was proportional to the MP optical density. The constant of proportionality included the fraction of preferentially aligned MP molecules. Additional measurements of the subjects' MP optical density by HFP indicated that this fraction was essentially the same for all subjects. If this is generally true, the method could be added to the list of other psychophysical methods for measuring MP optical density. The validation of the method was carried out by repeating the measurements at multiple wavelengths. The resulting spectrum of log G was virtually identical in shape to the optical density spectrum obtained by HFP. There is, however, an additional determination that must be made. The cornea is birefringent (Bone and Draper 2007) and, as such, transforms incident plane-polarized light into elliptically polarized light except when the former is polarized parallel to either of the principal axes of the cornea. It is

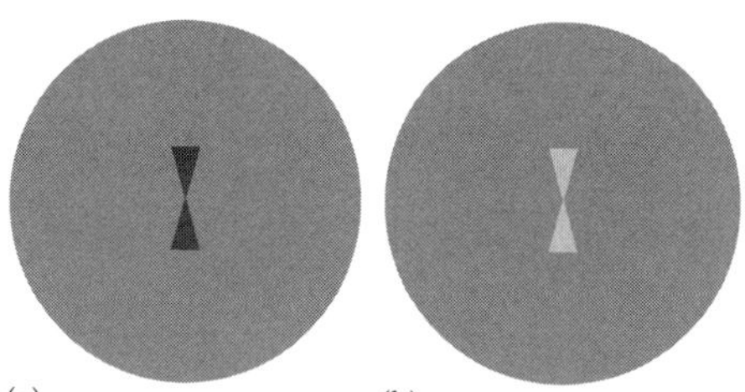

(a) (b)

FIGURE 5.3 (See color insert following page 336.) Appearance of the visual field in dichroism-based photometry. The background field is unpolarized and is of wavelength 460 nm. The triangles are also of wavelength 460 nm, but are polarized. In (a), the plane of polarization is horizontal causing the triangles to appear darker; in (b) the plane of polarization is vertical causing them to appear lighter. In each case, the subject adjusts the luminance of the triangles until they match the luminance of the unpolarized background.

necessary, therefore, to align the main axis of the stimulus (Figure 5.3) with one of these principal axis, otherwise log G will be underestimated. To determine the orientations of the principal axes, the third Purkyně–Sanson image of a plane-polarized light source was viewed through a crossed analyzer. This image, formed by reflection at the front surface of the lens, undergoes extinction when the plane of polarization of the incident light coincides with either of the principal axes. The polarizer and analyzer were rotated in tandem so that their transmitting directions remained perpendicular. As expected, two orientations of the polarizer, 90° apart, were found where extinction of the image occurred.

5.3 PHYSICAL METHODS

5.3.1 Reflectometry

The first report of the measurement of the optical density of MP by reflectometry may be attributed to Brindley and Willmer (1952). Their technique was to measure the reflectance spectra of the retina both in the fovea and the peripheral retina. The difference between these spectra, namely, lower reflectance at the shorter wavelengths, was attributed to the presence of MP in the fovea. It is likely, however, that other pigments encountered by the incident and reflected light, and also light scattering in the ocular media, will influence the MP optical density measurements. In order to minimize these influences, investigators developed models of the retina and applied curve fitting techniques to model the contribution to the reflectance spectrum from these unwanted artifacts. Initially, attempts were made to remove the effects of RPE melanin, choroidal melanin, and choroidal oxyhemoglobin (Van Norren and Tiemeijer 1986, Delori and Pflibsen 1989, Van de Kraats et al. 1996). More recently, Van de Kraats et al. (2006) developed a compact reflectance instrument for rapidly measuring MP optical density. A small white light source is imaged as a 1° spot on the subject's retina, and the reflected light is directed through an optical fiber for analysis by a spectrophotometer. The spectrum of this reflected light is assumed to be shaped by multiple chromophores in the retina (including MP), scattering and reflection by different retinal layers, and the Stiles–Crawford effect. A sophisticated model of the retina, with seven free parameters, is adjusted until its output matches the measured reflectance spectrum in the 400–800 nm range. The parameter of interest is, of course, the peak MP optical density; however, other useful parameters, such as lens optical density, are also provided by the model.

Reflectometry can also be adapted so that the spatial distribution of MP can be measured. This technique, imaging reflectometry, was pioneered by Kilbride et al. (1989) and was able to generate two-dimensional MP optical density distributions, ~7° × 7°, in the retina. Digital images of the bleached retina were captured using a modified retinal camera at a number of discrete wavelengths. These included 462 nm, close to the peak MP optical density, and 559 nm where the MP optical density is zero. By bleaching the photopigments in the cones and rods, the effects of their absorption on the remitted light, that would be expected to vary with retinal location, were minimized. Later investigators (see below), using similar techniques, obtained the MP optical density distribution by equating it to half the difference between the log-transformed and aligned 462 and 559 nm images, basing their calculations on the Brindley and Willmer (1952) model. The factor of one-half is due to the fact that the remitted light passes twice through the MP layer. On the other hand, Kilbride et al. (1989) attempted to remove the influence of melanin and hemoglobin from their MP distributions. To achieve this, they weighted the 559 nm image by the ratio of 462/559 nm extinction coefficients of the combined pigments. This procedure is valid provided the relative contributions of melanin and hemoglobin do not vary across the retina, an assumption that may not be warranted.

Closely related methods have been used in a number of studies. Chen et al. (2001) used the technique to investigate possible age-related variations in MP spatial distribution. Bour et al. (2002) used a film-based retinal camera to obtain retinal images at 480 and 540 nm, which they converted to a digital format and analyzed using the Brindley and Willmer model, that is, the difference between the log-transformed images was assumed to represent the (double) optical density distribution of MP. The majority of MP reflectometry studies have employed the scanning laser ophthalmoscope (SLO) rather than a standard retinal camera (Elsner et al. 1998, 2000, Berendschot et al. 2000, Wüstemeyer et al. 2002). While the SLO is a relatively expensive instrument, it is comparatively immune to the problem of scattered light in the eye's optical media that can degrade the images. Images of the bleached retina are typically captured at 488 and 514 nm, conveniently the wavelengths of an argon laser, and sufficiently close to the wavelengths of peak and zero absorption of MP. It is usually assumed that spatial variations in optical density of pigments in the light path other than MP may be neglected and, once again, the subtraction of the log-transformed images provides the MP spatial distribution. In an attempt to simplify the procedure and analysis as much as possible, a number of investigators have chosen to rely on a single image captured at, or close to, the peak wavelength in the MP absorption spectrum (Schweitzer et al. 2002). Such images always display a decreased intensity in the foveal part of the image and it is tempting to attribute this entirely to the MP. However, images captured in the green part of the spectrum prior to bleaching usually show a decreased foveal intensity, due to the absorption by cone photopigments, which peaks exactly where MP peaks.

A recent attempt to overcome this problem and still retain a relatively straightforward procedure has been reported by Bone et al. (2007a). Using a standard digital retinal camera in conjunction with multi-band-pass filters, it was possible to extract images of the retina at four different wavelengths from just two captured images. The retina was modeled as a sequence of four spatially varying, absorbing layers backed by a spectrally neutral reflector, the sclera. The layers consisted of MP, cone photopigments, rod photopigment, and melanin. In accordance with the model, and using published extinction spectra of the absorbing pigments, the four monochromatic images were transformed logarithmically and then combined linearly to yield optical density distribution maps of not only MP, but also cone and rod photopigments, and melanin. Because of the susceptibility of retinal cameras to intraocular light scatter that results in less than perfect images, the method may be unsuitable for older subjects for whom light scatter is more pronounced.

5.3.2 Lipofuscin Autofluorescence-Based Method

Posterior to the neural retina, and therefore to the MP, is the retinal pigmented epithelium (RPE). Lipofuscin is a fluorescent material that is sequestered in RPE cells. The fluorescence can be excited by wavelengths in the ~400–570 nm range, which includes the range of MP absorbance, and emission is in the ~520–800 nm range, which excludes MP absorbance. The principle behind the autofluorescence-based method of measuring MP is that an exciting wavelength close to the MP absorption maximum will be attenuated by the MP resulting in an MP-dependent intensity of fluorescence emission (Delori et al. 2001).

The methods can be subdivided into the one- and two-wavelength methods. In an application of the one-wavelength method, an SLO is modified so that fluorescence images of the retina can be obtained using the 488 nm line of an argon laser as the exciting wavelength (Robson et al. 2003, Trieschmann et al. 2003). A barrier filter that transmits only wavelengths above 560 nm is used in conjunction with the detector so as to exclude the excitation light. The resulting images show a decreased intensity in the foveal region due to absorption of the exciting light by the MP and, consequently, a decrease in fluorescence emission. By comparing the intensity of fluorescence at any point in this region with the intensity at, say, 6° eccentricity where MP density is assumed to be negligible, a two-dimensional MP optical density distribution at 488 nm can be generated. If needed, the distribution can be multiplied by the ratio of MP extinction coefficients, K_{460}/K_{488}, to obtain the MP optical density distribution at the peak wavelength, 460 nm.

An assumption inherent in the one-wavelength method is that the distribution of lipofuscin is uniform throughout the area of the retina being analyzed. Certainly if lipofuscin concentration peaked at the fovea, it would lead to enhanced fluorescence and an underestimate of the central MP optical density. To eliminate this problem, Delori et al. (2001) introduced the two-wavelength method. In this method, two exciting wavelengths provided by the argon laser can conveniently be used (Trieschmann et al. 2006). One wavelength (e.g., 488 nm) is attenuated by the MP; the other (514 nm) is minimally absorbed. Using the longer wavelength, the distribution of fluorescence reveals any nonuniformity in the concentration of lipofuscin and is unaffected by MP. Multiplying this distribution by the ratio of fluorescence efficiencies of lipofuscin, F_{460}/F_{514}, at the two wavelengths, we obtain the distribution of fluorescence due to excitation by the shorter wavelength as it would appear in the absence of MP. Comparing this with the actual distribution of fluorescence obtained with the shorter exciting wavelength allows one to compute the optical density of the MP. In fact, what is reported is the difference in optical density between any point in the retina and a parafoveal reference point where MP optical density is assumed to be negligible. In this calculation, the ratio of fluorescence efficiencies of lipofuscin, F_{460}/F_{514}, assumed to be the same at the two retinal locations, is eliminated from the final expression. However, the fluorescence of lipofuscin is due to the presence of more than one fluorophore (Parish et al. 1998) and, if the composition of lipofuscin changes with retinal location, it is possible that the ratio of fluorescence efficiencies is not constant across the retina. In this case, an error would occur in the calculated MP optical density.

5.3.3 Resonance Raman Spectroscopy

When carotenoids such as lutein and zeaxanthin are excited by wavelengths in the ~450–550 nm range, they exhibit particularly strong resonance Raman signals that can be used to quantify the amount of carotenoid present. The application of this technique for quantifying the macular carotenoids has been developed, thereby providing another noninvasive physical method for MP measurement. A detailed description of this method is given in Chapter 6.

ACKNOWLEDGMENTS

Support provided by NIH grants S06 GM08205 and R25 GM61347.

REFERENCES

Anstis, S. M. and P. Cavanagh (1983). A minimum motion technique for judging equiluminance. In *Colour Vision Psychophysics and Physiology*, J. D. Mollon and L. T. Sharpe (eds.). London: Academic Press, pp. 66–77.

Beatty, S. et al. (2000a). Macular pigment optical density measurement: A novel compact instrument. *Ophthalmic and Physiological Optics* **20**: 105–111.

Beatty, S. et al. (2000b). The role of oxidative stress in the pathogenesis of age-related macular degeneration. *Survey of Ophthalmology* **45**: 115–134.

Berendschot, T. T. J. M. and D. Van Norren (2005). On the age dependency of the macular pigment optical density. *Experimental Eye Research* **81**: 602–609.

Berendschot, T. T. J. M. et al. (2000). Influence of lutein supplementation on macular pigment, assessed with two objective techniques. *Investigative Ophthalmology and Visual Science* **41**: 3322–3326.

Bone, R. (1980). The role of the macular pigment in the detection of polarized light. *Vision Research* **20**: 213–220.

Bone, R. A. and G. Draper (2007). Optical anisotropy of the human cornea determined with a polarizing microscope. *Applied Optics* **46**: 8351–8357.

Bone, R. A. and J. T. Landrum (1984). Macular pigment in Henle fiber membranes a model for Haidinger's brushes. *Vision Research* **24**: 103–108.

Bone, R. A. et al. (1985). Preliminary identification of the human macular pigment. *Vision Research* **25**: 1531–1535.

Bone, R. A. et al. (1992). Optical density spectra of the macular pigment in vivo and in vitro. *Vision Research* **32**: 105–110.

Bone, R. A. et al. (2004). Macular pigment and the edge hypothesis of flicker photometry. *Vision Research* **44**: 3045–3051.

Bone, R. A. et al. (2007a). Macular pigment, photopigments and melanin: Distributions in young subjects determined by four-wavelength reflectometry. *Vision Research* **47**: 3259–3268.

Bone, R. A. et al. (2007b). Validity of macular pigment optical density measurements by heterochromatic flicker photometry. *Investigative Ophthalmology and Visual Science* **48**(ARVO E-Abstract): 2131.

Bour, L. J. et al. (2002). Fundus photography for measurement of macular pigment density distribution in children. *Investigative Ophthalmology and Visual Science* **43**: 1450–1455.

Brindley, G. S. and E. N. Willmer (1952). The reflexion of light from the macular and peripheral fundus oculi in man. *Journal of Physiology* **116**: 350–356.

Brindley, G. S. et al. (1966). The flicker fusion frequency of the blue-sensitive mechanism of colour vision. *Journal of Physiology (London)* **183**: 497–500.

Chen, S.-J. et al. (2001). The spatial distribution of macular pigment in humans. *Current Eye Research* **23**: 422–434.

Delori, F. C. and K. P. Pflibsen (1989). Spectral reflectance of the human ocular fundus. *Applied Optics* **28**: 1061–1077.

Delori, F. C. et al. (2001). Macular pigment density measured by autofluorescence spectrometry: Comparison with reflectometry and heterochromatic flicker photometry. *Journal of the Optical Society of America A* **18**: 1212–1230.

Elsner, A. E. et al. (1998). Foveal cone photopigment distribution: Small alterations associated with macular pigment distribution. *Investigative Ophthalmology and Visual Science* **39**: 2394–2404.

Elsner, A. E. et al. (2000). Scanning laser reflectometry of retinal and subretinal tissues. *Optics Express* **13**: 243–250.

Guth, S. L. et al. (1980). Vector model for normal and dichromatic color vision. *Journal of the Optical Society of America* **70**: 197–212.

Hammond, B. R. and M. Caruso-Avery (2000). Macular pigment optical density in a southwestern sample. *Investigative Ophthalmology and Visual Science* **41**: 1492–1497.

Hammond Jr., B. R. and K. Fuld (1992). Interocular differences in macular pigment density. *Investigative Ophthalmology Visual Science* **33**: 350–355.

Hammond Jr., B. R. et al. (1997). Individual variations in the spatial profile of human macular pigment. *Journal of the Optical Society of America A* **14**: 1–10.

Kilbride, P. E. et al. (1989). Human macular pigment assessed by imaging fundus reflectometry. *Vision Research* **29**: 663–674.

Maxwell, J. C. (1856). On the unequal sensibility of the foramen centrale to light of different colours. *British Association Reports* pt. **2**: 12.

Moreland, J. D. (2004). Macular pigment assessment by motion photometry. *Archives of Biochemistry and Biophysics* **430**: 143–148.

Parish, C. A. et al. (1998). Isolation and one-step preparation of A2E and iso-A2E, fluorophores from human retinal pigment epithelium. *Proceedings of the National Academy of Sciences* **95**: 2988–2995.

Robson, A. G. et al. (2003). Macular pigment density and distribution: Comparison of fundus autofluorescence with minimum motion photometry. *Vision Research* **43**(16): 1765–1775.

Rodriguez-Carmona, M. et al. (2006). The effects of supplementation with lutein and/or zeaxanthin on human macular pigment density and colour vision. *Ophthalmic and Physiological Optics* **26**: 137–147.

Schalch, W. (2001). Possible contribution of lutein and zeaxanthin, carotenoids of the macula lutea, to reducing the risk of age-related macular degeneration: A review. *HKJ Ophthalmology* **4**: 31–42.

Schweitzer, D. et al. (2002). Objektive bestimmung der optischen dichte von xanthophyll nach supplementation von lutein. *Ophthalmologe* **99**: 270–275.

Snodderly, D. M. et al. (1984). The macular pigment. II. Spatial distribution in primate retinas. *Investigative Ophthalmology and Visual Science* **25**: 674–685.

Snodderly, D. M. et al. (2004). Macular pigment measurements by heterochromatic flicker photometry in older subjects: The carotenoids and age-related eye disease study. *Investigative Ophthalmology and Visual Science* **45**: 531–538.

Stringham, J. M. et al. (2008). The utility of using customized heterochromatic flicker photometry (cHFP) to measure macular pigment in patients with age-related macular degeneration. *Experimental Eye Research* **87**: 445–453.

Trieschmann, M. et al. (2003). Macular pigment: Quantitative analysis on autofluorescence images. *Graefe's Archive for Clinical and Experimental Ophthalmology* **241**: 1006–1012.

Trieschmann, M. et al. (2006). Macular pigment optical density measurement in autofluorescence imaging: Comparison of one- and two-wavelength methods. *Graefe's Archive for Clinical and Experimental Ophthalmology* **244**: 1565–1574.

Van de Kraats, J. et al. (1996). The pathways of light measured in fundus reflectometry. *Vision Research* **36**: 2229–2247.

Van de Kraats, J. et al. (2006). Fast assessment of the central macular pigment density with natural pupil using the macular pigment reflectometer. *Journal of Biomedical Optics* **11**: 064031.

Van Norren, D. and L. F. Tiemeijer (1986). Spectral reflectance of the human eye. *Vision Research* **26**: 313–320.

Von Haidinger, W. K. (1844). Über das direkte erkennen des polarisierten lichts und der lage der polarisationsebene. *Annalen der Physik* **63**: 29–39.

Wald, G. (1945). Human vision and spectrum. *Science* **101**: 653–658.

Walsh, J. W. T. (1953). *Photometry*. London: Constable and Co. Ltd.

Weale, R. A. (1963). *The Aging Eye*. London: Lewis.

Werner, J. S. et al. (1987). Aging and the human macular pigment density. Appended with translations from the work of Max Schultz and Ewald Hering. *Vision Research* **27**: 257–68.

Werner, J. S. et al. (2000). Senescence of foveal and parafoveal cone sensitivities and their relations to macular pigment density. *Journal of the Optical Society of America A* **17**: 1918–1932.

West, P. and J. Mellerio (2005). An innovative instrument for the psychophysical measurement of macular pigment optical density using a CRT display. *International Color Vision Society Annual Meeting*, Lyons, France.

Wooten, B. R. and G. Wald (1973). Color-vision mechanisms in the peripheral retinas of normal and dichromatic observers. *Journal of General Physiology* **61**: 125–145.

Wooten, B. R. et al. (1999). A practical method for measuring macular pigment optical density. *Investigative Ophthalmology and Visual Science* **40**: 2481–2489.

Wüstemeyer, H. et al. (2002). A new instrument for the quantification of macular pigment density: First results in patients with AMD and healthy subjects. *Graefe's Archive for Clinical and Experimental Ophthalmology*. **240**: 666–671.

6 Application of Resonance Raman Spectroscopy to the Detection of Carotenoids *In Vivo*

Igor V. Ermakov, Mohsen Sharifzadeh, Paul S. Bernstein, and Werner Gellermann

CONTENTS

6.1 INTRODUCTION

Motivated by the growing importance of carotenoid antioxidants in health and disease, we investigate resonance Raman scattering, RRS, as a novel approach for the noninvasive optical detection of carotenoids in living human tissue. Raman spectroscopy is a well-known, highly molecule-specific form of vibrational spectroscopy that is commonly used to identify a vast assortment of molecular compounds through their respective, spectrally very narrow, Raman "spectral fingerprint" responses. Most frequently, off-resonance Raman techniques are used for this purpose since they avoid the strong intrinsic electronic fluorescence transitions typically encountered in complex molecules. Carotenoid molecules, however, possess a unique energy level structure and associated optical pumping cycle. While easily excited from the ground state into a higher excited state within a strong, electric dipole-allowed absorption transition, they relax quickly into a new, lower-lying excited state, from which fluorescence transitions back to the ground state are forbidden. This offers the opportunity to use the fluorescence-background-free resonant excitation of the carotenoids in their visible absorption bands, which results in a resonance enhancement of the carotenoid Raman response by about five orders of magnitude relative to non-resonant Raman scattering (Koyama 1995). It becomes possible, therefore, to explore RRS not only for the identification of carotenoids in biological tissue environments, but also, through the intensity of the RRS response, for the measurement of their tissue concentrations. The tissue environment can be expected to have only a minor effect on the molecule's vibrational energy, and thus should cause the Raman signature to be virtually identical for the isolated carotenoid molecule, the molecule in solution, or the molecule in a cell environment. However, the applicability of the method can be expected to depend heavily on potentially confounding tissue properties such as a saturation of the carotenoid Raman response at high concentrations, and the existence of other molecules with potentially interfering scattering, absorption, and/or fluorescence contributions. A crucial task therefore is the validation of the RRS detection method for the particular tissue environment. If successful, RRS could be used as a novel optical diagnostic method for the measurement of tissue carotenoid levels, potentially allowing one to measure large populations in clinical and field settings, and to track their changes occurring over time as a consequence of developing pathology and/or tissue uptake.

A tissue site that appears to be particularly interesting for the application of the Raman method is the macula lutea. It is located in the human retina and contains the highest concentration of carotenoids in the human body. Of the about ten carotenoid species found in human serum, only two carotenoids, lutein and zeaxanthin, are selectively taken up at this tissue site. Their concentrations can be as high as several 10 ng per gram of tissue, however, in the healthy human retina. Due to their strong absorption in the blue–green spectral range, the macular carotenoids, also termed macular pigment, MP, impart a yellow coloration to the macula, which contains a high density of photoreceptors, enabling high-acuity color vision. When viewed in cross section, MP is located anterior to the photoreceptor outer segments and the retinal pigment epithelium (Snodderly et al. 1984a,b) and therefore is thought to shield these vulnerable tissues from light-induced oxidative damage by blocking phototoxic short-wavelength visible light. Also, MP may directly protect the cells in this area, since lutein and zeaxanthin are efficient antioxidants and scavengers of reactive oxygen species. There is increasing evidence that MP may help mediate protection against visual loss from age-related macular degeneration, AMD (Seddon et al. 1994, Landrum and Bone 2001, Krinsky et al. 2003, Krinsky and Johnson 2005, AREDS 2007), the leading cause of irreversible blindness affecting a large portion of the elderly population. Since the MP compounds are taken up through the diet, there is a chance that early age screening of MP concentrations to identify individuals with low levels of MP, accompanied with dietary interventions such as nutritional supplementation, will help prevent or delay the onset of the disease.

MP concentrations in the healthy human retina are usually assumed to be highest in the very center of the macula, the foveola, and to drop off rapidly with increasing eccentricity, especially when using low-spatial resolution techniques such as heterochromatic flicker photometry (Snodderly et al. 2004). However, recently emerging high-resolution optical imaging techniques based on lipofuscin fluorescence (autofluorescence) excitation and reflection methods have already demonstrated a much more complex pattern of MP distributions in the living human retina, such as those with depletions and ring-shaped concentration distributions (Robson et al. 2003, Trieschmann et al. 2003, Delori 2004, Berendschot and van Norren 2006). It would be important to confirm these interesting features with an imaging Raman method, which by comparison would be a more direct, carotenoid specific method, and to track the MP distributions and any potential changes occurring in them upon dietary modifications or supplementation.

Aside from the human retina, RRS spectroscopy also appears to be interesting for the detection of carotenoids in human skin. In this tissue, which constitutes the largest organ of the human body, the carotenoid species lycopene and beta-carotene are thought to play an important protective role as antioxidants, like in the protection of skin from ultraviolet and short-wavelength visible radiation. The carotenoids lutein and lycopene may also have protective functions for cardiovascular health, and lycopene may play a role in the prevention of prostate cancer. It is conceivable that skin levels of these species are correlated with corresponding levels in internal tissues. Objective measurements of carotenoid levels are also of interest in improving dietary data collected in epidemiological studies, which in turn are used in developing public health guidelines that promote healthier diets. The protective effects of diets rich in fruits and vegetables have been observed for many disease outcomes, including various cancers (Kolonel et al. 2000, Michaud et al. 2000) and cardiovascular disease (Liu et al. 2000). Since carotenoids are a good biomarker for fruit and vegetable intake,

Raman measurements of skin carotenoid levels could be used as an indirect, rapid optical method to assess fruit and vegetable consumption in large populations.

For many decades, the standard technique for measuring carotenoids has been high-pressure liquid chromatography (HPLC). This time consuming and expensive chemical method works well for the measurement of carotenoids in serum, but it is difficult to perform in human tissue since it requires biopsies of relatively large tissue volumes. Additionally, serum antioxidant measurements are more indicative of short-term dietary intakes of antioxidants rather than steady-state accumulations in body tissues exposed to external oxidative stress factors such as smoking and UV-light exposure.

6.2 OPTICAL PROPERTIES AND RESONANCE RAMAN SCATTERING OF CAROTENOIDS

Carotenoids are molecules that possess a long polyene chain, and their structural and optical properties are principally the result of the conjugated double bonds. Distinguishing features are the number of conjugated carbon double bonds (C=C bonds), the number of methyl side groups, and the presence and nature of attached end-groups. The molecular structure of β-carotene is shown as an example in Figure 6.1, along with a configuration coordinate diagram for the three lowest lying energy states, and an indication of all optical and nonradiative transitions connecting the states. Absorption, fluorescence, and Raman transitions occur on a very short time scale ($t \leq 10^{-15}$ s) and obey the Frank–Condon principle; the nuclear positions of the constituent atoms of the molecule remain unchanged during the time interval of the transition. On the coordinate diagram of Figure 6.1 the electronic transitions are shown as vertical lines reflecting fixed configuration coordinates. A characteristic strong, electric-dipole allowed absorption transition occurs between the molecule's delocalized π-orbital from the 1^1A_g singlet ground state (S_0) to the 1^1B_u singlet excited state (S_2), giving rise to a broad absorption band (~100 nm width) in the blue–green region of the visible spectrum, peaking at ~460 nm. Clearly resolved vibronic substructures spaced at ~1400 cm^{-1} are also present, as illustrated in Figure 6.2a. Following excitation of the 1^1B_u state, the carotenoid molecule relaxes very rapidly, within ~200–250 fs (Shreve et al. 1991), via nonradiative transitions, to the lower-lying 2^1A_g excited state (S_1), from which electronic emission to the ground state is spin-forbidden. As a consequence, fluorescence resulting from a transition from the 1^1B_u (S_2) state to and 2^1A_g (S_0) ground state is very weak for carotenoids (the emission quantum yield, ϕ_c, is typically 10^{-5}–10^{-4}. This allows one to detect the RRS response of the molecular vibrations virtually free of potentially masking fluorescence signals.

For tetrahydrofuran solutions of β-carotene, zeaxanthin, lycopene, lutein, and phytofluene, we obtain the RRS spectra displayed in Figure 6.2b. All of these carotenoids reveal strong, clearly resolved Raman signals that are comparable or even stronger than the intrinsic fluorescence background, with three prominent Raman stokes lines appearing at ~1525 cm^{-1} (C=C stretch), 1159 cm^{-1} (C-C stretch), and 1008 cm^{-1} (C-CH_3 rocking motions) (Koyama 1995). In the shorter chain phytofluene molecule, only the C=C stretch appears, and it is shifted significantly to higher frequencies (by ~40 cm^{-1}) due to the shorter conjugation length in this molecule.

Raman scattering is a linear spectroscopy, in principle, meaning that the Raman scattering intensity, I_S, scales linearly with the intensity of the incident light, I_L, provided the scattering compound can be considered as optically thin. At fixed incident light intensity I_L, the Raman response scales with the population density of the scatterers, $N(E_i)$ according to

$$I_s = N(E_i) \times \sigma_R \times I_L \tag{6.1}$$

Here, σ_R is the Raman cross section, a constant whose magnitude depends on the excitation and collection geometry. In optically thick media, as in a geometrically thin but optically dense tissue, a deviation from the linear Raman response of I_s versus concentration N is to be expected. This can

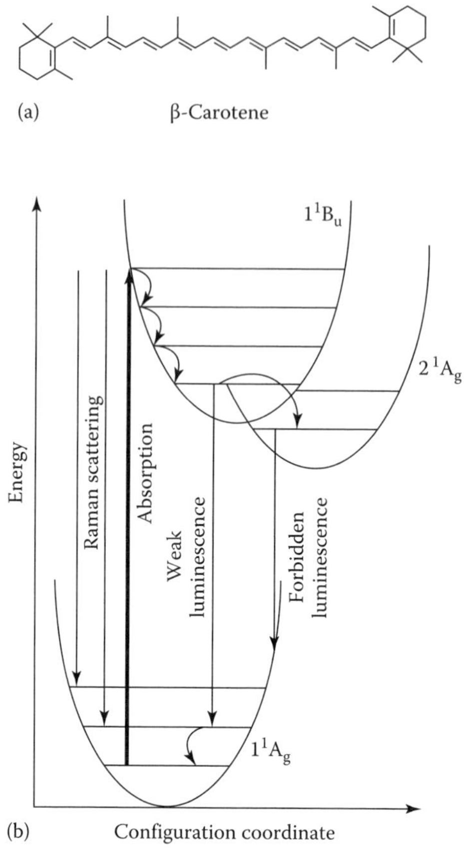

FIGURE 6.1 (a) The molecular structure of β-carotene, which consists of a linear, conjugated, carbon backbone with alternating carbon single (C–C) and double bonds (C=C), two ionone end-groups, and four methyl side groups. The structure is very similar to all other carotenoids of interest in this chapter. Resonance Raman spectroscopy detects the vibrational stretch frequencies of the carbon bonds as well as the rocking motion of the attached methyl side groups; (b) the configuration coordinate diagram for the three lowest lying energy levels of carotenoids, with indication of optical and nonradiative transitions between all levels. The configuration coordinate represents the displacement of a normal coordinate of the molecule's atoms in their equilibrium positions. Absorption transitions, from S_0 to S_2 (1^1A_g to 1^1B_u), are electric-dipole allowed while luminescence transitions are very weak due to the existence of a low-lying excited singlet state (S_1 or 2^1A_g) that has the same multiplicity as the ground state (S_0). The absence of any strong luminescence in carotenoids allows one to detect the relatively weak resonance Raman responses of the molecule without an otherwise overwhelming intrinsic luminescence background.

occur, for example, due to the self-absorption of the Stokes Raman signal by the strong electronic absorption, or due to insufficient light penetration. In these cases, a nonlinear calibration between RRS response and molecule concentration may be required using suitable tissue phantoms.

6.3 SPATIALLY INTEGRATED RESONANCE RAMAN MEASUREMENTS OF MACULAR PIGMENT

The macular region of the retina is optically relatively easily accessible. The excitation and the Raman light must traverse the cornea, lens, and vitreous, sketched in Figure 6.3a, all of which are generally of sufficient clarity for optical measurements. Correction factors can be expected to be required

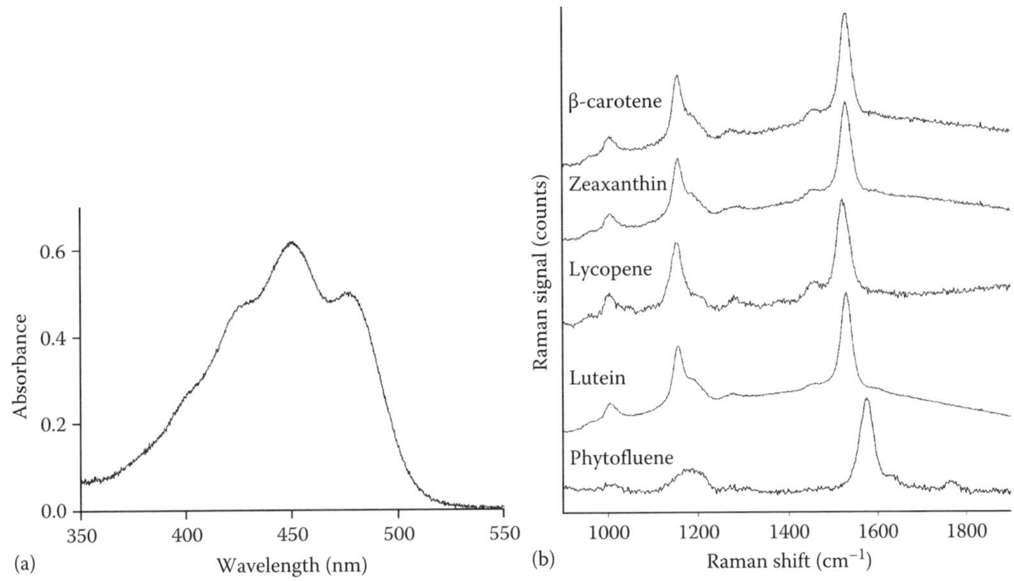

FIGURE 6.2 (a) Absorption spectrum of a β-carotene solution corresponding to the molecule's $1^1A_g \rightarrow 1^1B_u$ transition, showing the characteristic broad absorption with vibronic substructure in the blue–green spectral range; (b) resonance Raman spectra of β-carotene, zeaxanthin, lycopene, lutein, and phytofluene solutions, all displaying three characteristic sharp spectral Raman lines, originating, respectively, from the rocking motion of the methyl components (C–CH_3), the stretch vibration of the carbon–carbon single bonds (C–C), and the stretch vibration of the carbon–carbon double bonds (C=C). In all carotenoids, these peaks appear at 1008, 1159, and 1525 cm^{-1}, respectively. The exception is phytofluene, in which the C=C stretch frequency is shifted by ~40 cm^{-1} to higher frequencies due to the shorter conjugation length of the backbone. (From Ermakov, I.V. et al., *J. Biomed. Opt.*, 10(6), 064028-1, 2005b. With permission.)

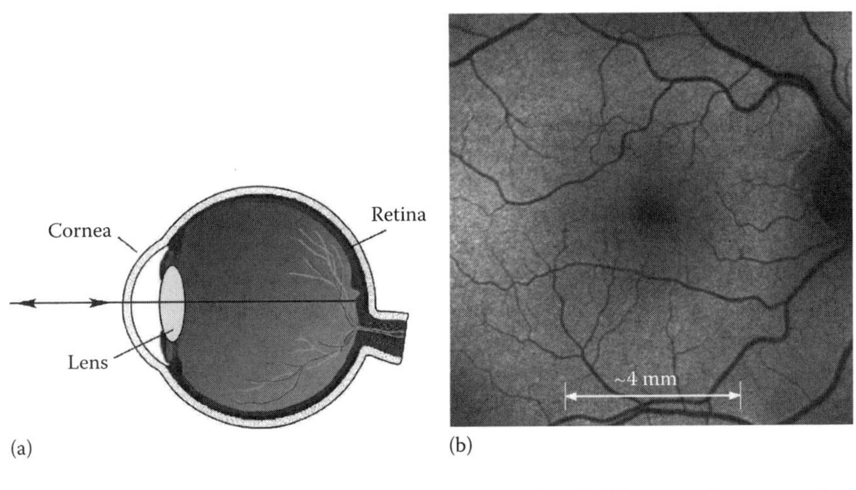

FIGURE 6.3 (a) Cross section of human eye with indication of optical beam paths propagating back and forth to the macular region of the retina; (b) autofluorescence photograph of healthy human retina, showing the macular region in the center with dark shading. Part of the optic nerve head can be seen as a dark spot at center right.

only in cases of substantial cataracts. The macula is essentially free of blood vessels, and when containing a healthy concentration of lutein and zeaxanthin pigments, appears as a gray-shaded area in black-and-white autofluorescence images of the retina, as can be seen from autofluorescence images recorded with blue excitation light, such as the one shown in Figure 6.3b. *In vivo* RRS spectroscopy of the macula can take further advantage of favorable anatomical features of the tissue structures encountered in the excitation and light scattering pathways. A cross section of the retinal tissue layers in the macular region, shown in Figure 6.4, helps to illustrate the concept. First, the major site of macular carotenoid deposition is the Henle fiber layer, which has a thickness of only about 100 microns, and to a lesser extent the plexiform layer (both layers are shown in Figure 6.4 together with the outer nuclear layer as a single layer, HPN). Considering that the optical density of MP in the peak of the absorption band is typically smaller than 1, as determined from direct absorption measurements of MP in excised eyecups, these tissue properties provide essentially an optically thin film with minimal self-absorption for both the excitation and Raman scattered light if properly excited in the long-wavelength shoulder of the absorption. Second, since Raman scattering uses only the backscattered, single-path Raman response from the lutein- and zeaxanthin-containing MP layers, and since these layers are located anteriorly in the optical pathway through the retina, absorption and fluorescence effects originating from other chromophores, such as rhodopsin in the photoreceptor layer, PhR, and melanin and lipofuscin in the retinal pigment epithelial layer, RPE, respectively, can be ignored or subtracted from the Raman spectra.

Our initial "proof of principle" studies of ocular carotenoid RRS employed a laboratory-grade high-resolution Raman spectrometer and flat mounted human cadaver retinas and eyecups. We were able to record characteristic carotenoid RRS spectra from these tissues with a spatial resolution of approximately 100 microns, and we were able to confirm linearity of the response by extracting and analyzing tissue carotenoids by HPLC, after completion of the Raman measurements (Bernstein et al. 1998). For *in vivo* experiments and clinical use, we developed Raman instruments with lower spectral resolution but highly improved light throughput (Ermakov et al. 2001b, Gellermann et al. 2002a). A current version that is combined with a fundus camera to permit independent operator targeting of the subject's macula (Ermakov et al. 2004a) is shown in Figure 6.5a. The instrument's

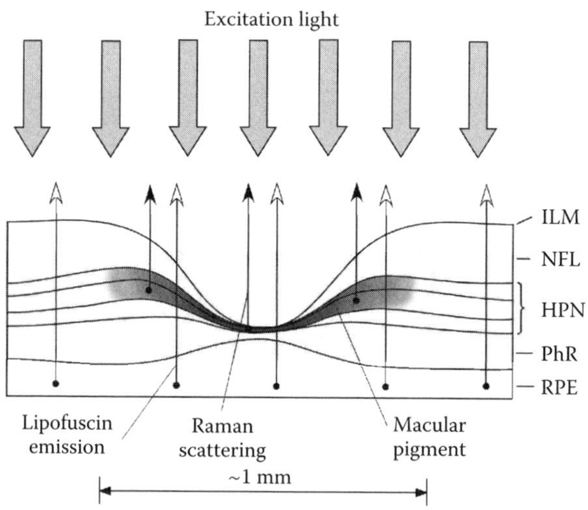

FIGURE 6.4 Schematics of retinal layers participating in light absorption, transmission, and scattering of excitation and emission light. ILM: inner limiting membrane; NFL: nerve fiber layer; HPN: Henle fiber, plexiform, and nuclear layers; PhR: photoreceptor layer; and RPE: retinal pigment epithelium. In Raman scattering, the scattering response (dark arrows) originates from MP, which is located anteriorly to the photoreceptor layer. The influence of deeper fundus layers is largely avoided since fluorescence contributions, such as those from lipofuscin in the RPE (light arrows), are spectrally broad and can be subtracted.

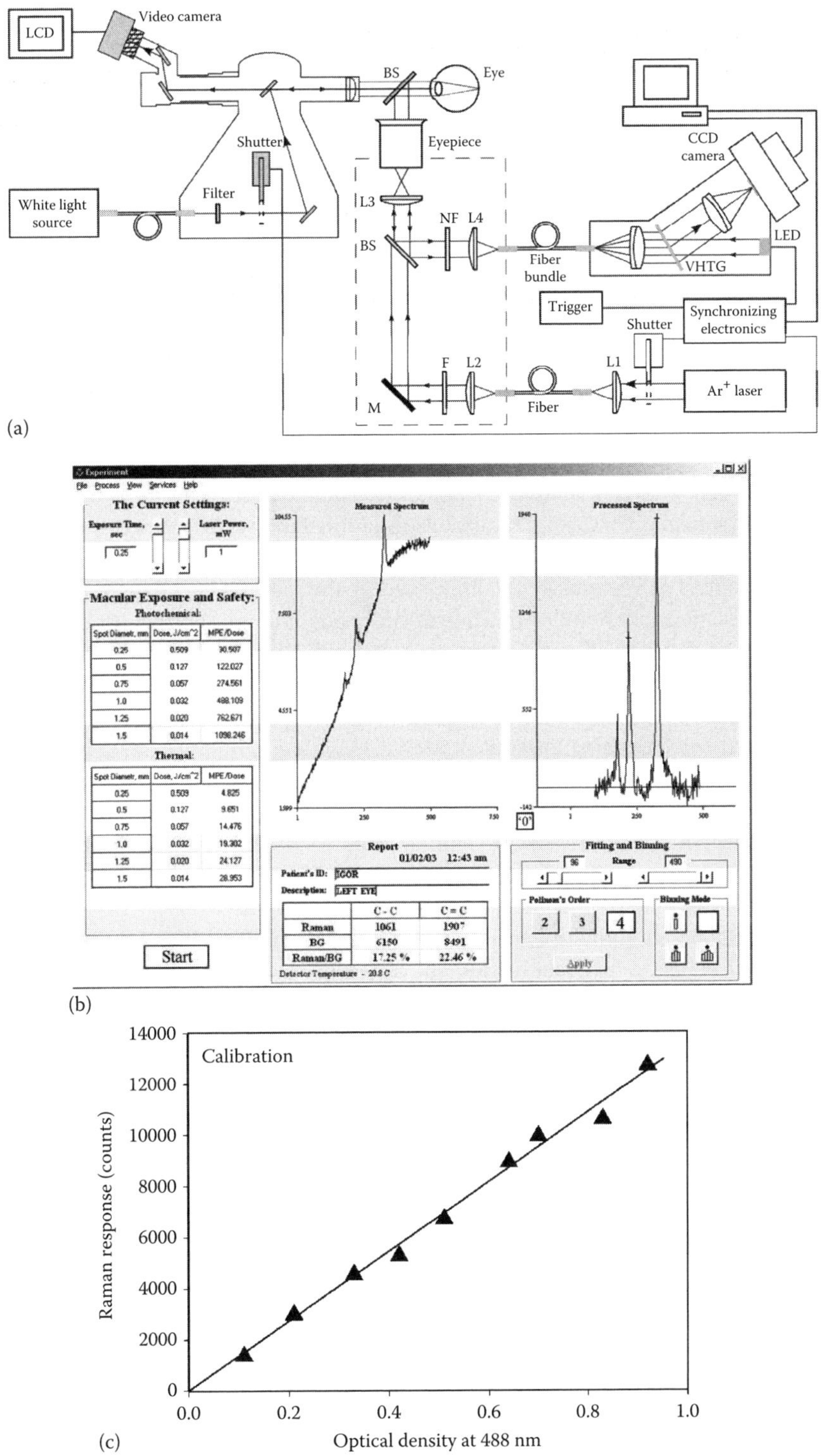

FIGURE 6.5 (a) Schematics of fundus-camera-interfaced RRS instrument for measurement of integral MP concentrations in human clinical studies; (b) computer monitor display showing raw Raman spectrum obtained after single measurement (left panel) and processed, scaled spectrum obtained after subtraction of fluorescence background (right panel); and (c) calibration curve for RRS response of tissue phantom for nine lutein and zeaxanthin concentrations. (From Ermakov, I.V. et al., *J. Biomed. Opt.*, 9(1), 139, 2004. With permission.)

Raman module, containing a 488 nm laser excitation source, a spectrograph, and a CCD array detector, is optically connected with the fundus camera using a beam splitter that is mounted between the front-end optics of the fundus camera and the eye of the subject. Once alignment is established, an approximately 1 mm diameter, 1.0 mW, light excitation disk is projected onto the subject's macula for 0.25 s through the pharmacologically dilated pupil, and the backscattered light is routed to the Raman module for detection. Retinal light exposure levels of the instrument are in compliance with ANSI safety regulations since ocular exposure levels are a factor of 19 below the thermal limit, and a factor of 480 below the photochemical limit for retinal injury (Ermakov et al. 2004a).

Typical RRS spectra, measured from the macula of a healthy human volunteer through a dilated pupil are displayed, in near real time, on the instrument's computer monitor, as shown in Figure 6.5b. The left panel shows the raw spectrum obtained from a single measurement, and clearly reveals the three characteristic carotenoid Raman signals, which are superimposed on a steep, spectrally broad fluorescence background. The background is caused partially by the weak intrinsic fluorescence of lutein and zeaxanthin, and partially by the short-wavelength emission tail of lipofuscin, which is present in the retinal pigment epithelial layer, and is excited by the portion of the excitation light that is transmitted through the MP-containing Henle fiber and plexiform layers. The ratio between the intensities of the carotenoid C=C Raman response and the fluorescence background is high enough (~0.25) that it is easily possible to quantify the amplitudes of the C=C peak after digital background subtraction. This step is automatically accomplished by the instrument's data processing software, which approximates the background with a fourth-order polynomial, subtracts the background from the raw spectrum, and displays the final result as a processed, scaled spectrum in the right panel of the computer monitor, shown in Figure 6.5b. MP carotenoid RRS spectra measured for the living human macula were indistinguishable from corresponding spectra of pure lutein or zeaxanthin solutions, measured with the same instrument.

While the fundus-camera-interfaced Raman instrument is well suited for measurements of elderly subjects, subjects with macular pathologies, and research animals, we found that simplified instrument versions can be used for healthy human subjects provided they have good visual acuity and are able to self align on a fixation target prior to a Raman measurement. An example for a particularly simple self-alignment instrument is a version in which the CCD/spectrograph combination is replaced with a single photomultiplier/filter combination (Ermakov et al. 2005a).

In order to cross-calibrate different instrument versions, we constructed a simple tissue phantom consisting of a lens and a thin, 1 mm path length, cuvette placed in the focal plane of the lens, and measured the RRS response for preset lutein and zeaxanthin solutions with optical densities in the range 0.1–1.0, a range that at the higher end exceeds typically encountered physiological concentration levels. An example of a calibration curve for a particular instrument version is shown in Figure 6.5c. It demonstrates a linear RRS response up to a relatively high optical density of 0.8. This calibration method can also be used to correlate the RRS response of a subject's MP with its corresponding optical density value.

An example for RRS clinical measurements of a relatively young subgroup (33 eyes), ranging in age from 21 to 29 years, is shown in Figure 6.6a. A striking observation is the fact, that the RRS measured MP levels can vary dramatically between individuals (up to ~10-fold difference). Since the ocular transmission properties in this age group can be assumed to be very similar, the differences must be attributed to variation in MP levels. Subjects with extremely low carotenoid levels may be at higher risk of developing macular degeneration later in life.

When measuring a large population of normal subjects, none of whom were consuming supplements containing substantial amounts of lutein or zeaxanthin, we found a striking decline of average macular carotenoid levels with age (Bernstein et al. 2002, Gellermann et al. 2002a), as shown in Figure 6.6b. Part of this decline can be explained by the "yellowing" of the crystalline lens with age and by any other optical losses existing in the anterior optical media, such as the vitreous. These losses would attenuate part of the illuminating and backscattered light. Regarding lens effects, however, we found consistently low MP levels even in patients who had previously had cataract surgery with the implantation of optically clear prosthetic intraocular lenses (pseudophakia).

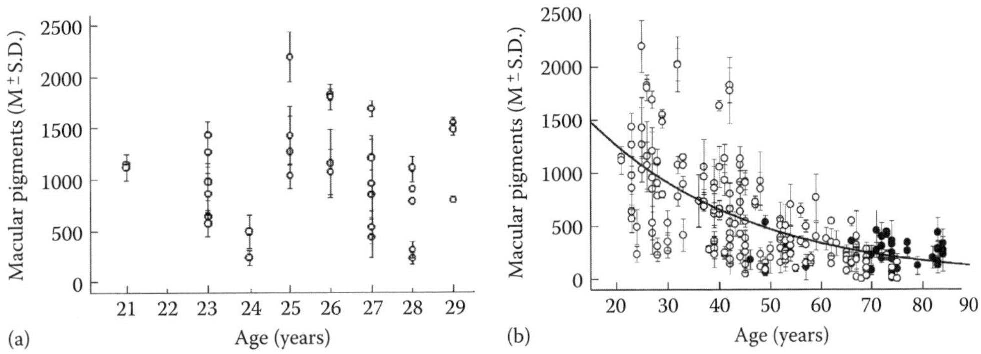

FIGURE 6.6 (a) RRS MP measurements of 33 normal eyes for a young group of subjects ranging in age from 21 to 29 years. Note the large (up to ~10-fold) variation of RRS levels that can exist between individuals. Since the ocular transmission properties in this age group can be assumed to be very similar, the variations can be assigned to differing MP levels. Subjects with low MP levels may be at higher risk of developing macular degeneration later in life. (From Ermakov, I.V. et al., *J. Biomed. Opt.*, 10(6), 064028-1, 2005b. With permission.) (b) RRS measurements of 212 normal eyes as a function of subject age, revealing a statistically significant decrease of MP concentration with age. Solid circles represent subjects with clear prosthetic intraocular lenses. Data are not corrected for decrease of ocular transmission with age (see text). (From Gellermann, W. et al., *J. Opt. Soc. Am. A*, 19: 1172, 2002. With permission.)

Also, we have noted that patients with unilateral cataracts after trauma or retinal detachment repair typically have very similar RRS carotenoid levels in the normal and in the pseudophakic eye. Thus, we have concluded that there is a decline of macular carotenoids that reaches a low steady state just at the time when the incidence and prevalence of AMD begins to rise dramatically. While this age effect has been noticed sometimes also in other studies using clinical populations and different MP detection methods (Sharifzadeh et al. 2006, Nolan et al. 2007), several groups have reported constant, age-independent MP levels. Examples include reflectance-based population studies in which respective average MP optical densities of 0.23 (Delori et al. 2001), 0.33 (Berendschot et al. 2002), and 0.48 (Berendschot and Van Norren 2004) were determined.

6.4 SPATIALLY RESOLVED RESONANCE RAMAN IMAGING OF MACULAR PIGMENT

MP distributions are often assumed to have strict rotational symmetry, high central pigment levels, and a monotonous decline with increasing eccentricity. However, initial resonance Raman imaging, RRI, results obtained with excised human eyecups demonstrated intriguing deviations, clearly revealing the existence of strong significant rotational asymmetries, distribution patterns with central depletions, patterns with widely differing widths between samples, and patterns with fragmented concentration levels (Gellermann et al. 2002b).

In order to confirm these new distribution features in the living human retina, we developed the Raman method for *in vivo* imaging applications (Sharifzadeh et al. 2008). The experimental setup for this purpose is shown in Figure 6.7. Once the subject achieves head alignment with the help of a red fixation target, blue light from a solid state 488 nm laser is projected onto the macula as a ~3.5 mm diameter excitation disk, and two images are recorded with a CCD camera. In the first image, "Raman plus fluorescence image," the light returned from the retina under 488 nm excitation is filtered to transmit only 528 nm light, which is the spectral position, λ_R, of the resonance Raman response of the 1525 cm^{-1} carbon–carbon double bond stretch frequency of the MP carotenoids. Each pixel of this image contains the Raman response of MP as well as the fluorescence components overlapping the Raman response at this wavelength. In the second image, "fluorescence image," the

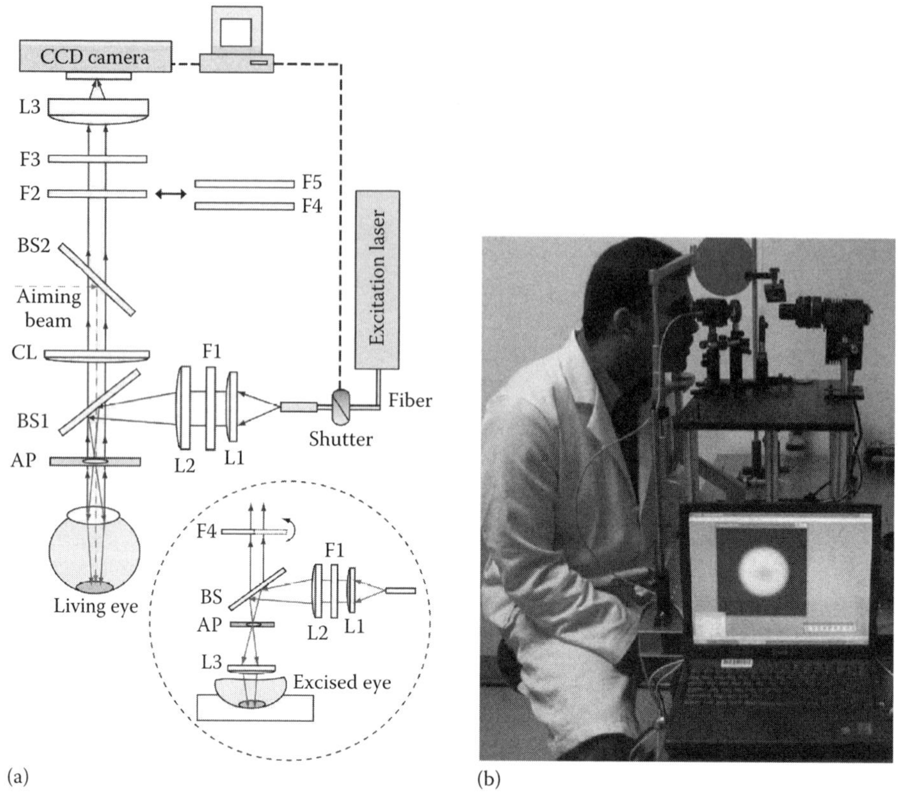

FIGURE 6.7 (a) Schematics of experimental setup used for *in vivo* resonance Raman imaging, RRI, of MP distributions. Light from a blue laser source is projected onto the macula as a ~3.5 mm diameter excitation disk. The backscattered light is collimated by the lens of the eye and imaged with a two-dimensional CCD camera array detector. Two sets of filters are used sequentially to selectively image light at the C=C Raman wavelength (Raman image) and at a slightly longer wavelength (offset image). The two images are digitally subtracted and displayed as topographic or three-dimensional pseudocolor images of the spatial MP concentrations. L1–3: lenses; F1: laser line filter; BS: dichroic beam splitters; F2: tunable filter; and F3: band pass filter. Inset shows modifications for use with excised tissue. (b) Photograph of subject measured with instrument. RRI images are recorded with 0.2 s exposure time for dilated or non-dilated pupils.

light returned from the retina is filtered to only transmit fluorescence components slightly above the Raman wavelength, at λ_{offset}.

The contribution of the broad fluorescence at the Raman wavelength λ_R is approximately the same as at the slightly offset longer wavelength position λ_{offset}. It can be shown, Equation 6.2, that the Raman component $I_R(\lambda_R)$ for each image pixel is approximately

$$I_R(\lambda_R) \approx T_{OM}^{-1}(\lambda_{exc}) \cdot T_{OM}^{-1}(\lambda_R)\left(I_{Det}(\lambda_R)/T_R - I_{Det}(\lambda_{offset})/T_{offset}\right) \tag{6.2}$$

where

$I_{Det}(\lambda_R)$ and $I_{Det}(\lambda_{offset})$ are the detector intensities
T_R and T_{offset} are the filter transmissions at the respective wavelengths
T_{OM} is the unknown transmission of the ocular media

The RRI image of an MP distribution can thus be derived with a digital image subtraction routine, where the intensities obtained for each pixel of the two images are divided by the appropriate filter transmission coefficient.

For RRI imaging of MP distributions in human subjects we recruited 17 healthy volunteers from an eye clinic. Laser power levels at the cornea were 4 mW during a measurement; exposure times were 100 ms for fluorescence measurements, and 300 ms for resonance Raman imaging. The laser light exposures caused after-images that typically disappeared within a few minutes. During this time, the setup switched from Raman to fluorescence imaging mode. At a retinal spot size of 3.5 mm diameter, the photo-thermal light exposure is a factor 16 below the limit set by the ANSI standard (Sharifzadeh et al. 2008).

When evaluating the MP distributions of all subjects, distinctly different categories are apparent, as can be seen from representative distributions displayed in Figure 6.8. These feature relatively wide spatial MP distributions with a high central level, ring-like MP distributions surrounding a central MP peak, or fragmented distributions. Corresponding intensity line plots along the nasal–temporal (solid line) and inferior–superior meridians (dotted line), also shown in Figure 6.8, further highlight the significant inter-subject variations in MP levels, symmetries, and spatial extent. The spatial resolution obtainable with the instrument is approximately sub-50 microns, as can be concluded from the size of small blood vessels discernable in the gray-scale images.

Similar to the case of integrated Raman MP detection, we validated the Raman imaging method with excised human eyecups. We imaged 11 excised human donor eyecups and compared RRI derived MP levels with HPLC derived levels (Sharifzadeh et al. 2008). Two-dimensional and three-dimensional pseudocolor Raman images are shown for two representative eyecups in Figure 6.9a, with the first one featuring a distribution with a relatively strong central peak with a small depression, and the second one a strongly elongated asymmetrical distribution with high central levels and relatively smooth decline toward increasing eccentricities. In Figure 6.9e, we plotted the integrated Raman intensities obtained from the MP RRI images of all eyecups, and compared these optically derived intensities with HPLC derived MP concentration levels. The result shows a high correlation between optical and biochemical methods ($R=0.92$; $p=0.0001$).

To further test the RRI imaging method, we compared it with a recently developed, nonmydriatic version of the lipofuscin fluorescence imaging (autofluorescence imaging) method (Sharifzadeh et al. 2006). Autofluorescence imaging, AFI, is a less specific detection method since it detects the light emitted from a compound other than MP, and thus derives the concentration of MP only indirectly. The method has to take into account light traversal through deeper retinal layers, has to carefully eliminate image contrast diminishing fluorescence and scattering from the optical media such as the lens (via confocal detection techniques, filtering, etc.), has to bleach the photoreceptors, and has to use a location in the peripheral retina as a reference point. The peripheral reference could potentially lead to an underestimation of the MP density, especially in individuals regularly consuming high-dose lutein supplements, which can cause substantial increases in even peripheral carotenoid levels (Bhosale et al. 2007). AFI has an advantage, however, since the peripheral reference location allows one to eliminate, in first order, any potentially confounding attenuation arising from the anterior optical media.

In Figure 6.10, we summarize the main results of a comparison of MP distributions and concentrations obtained with RRI and AFI method for an identical subgroup of subjects. Figure 6.10a and b compare RRI and AFI obtained for one of the subjects. Compared to the RRI image, the AFI image is nearly identical, with the exception of a smoother appearance of the distribution. This is due to the derivation of the MP density map as the logarithm of a ratio between perifoveal and foveal fluorescence intensities, which tends to slightly compress the "dynamic range" of the density map amplitudes and smoothen out the resulting MP distribution. For the whole subgroup of 17 subjects, we integrated the MP levels of images obtained with both methods for each individual over the whole macula region, and plotted the results in Figure 6.10c. Using a best fit that is not forced through zero, we obtained a high correlation coefficient of $R=0.89$ between both methods. Forcing the fit through zero, the correlation coefficient dropped slightly to $R=0.80$. The high correlation is remarkable in view of the completely different optical beam paths and derivation methods used to calculate MP densities in both methods.

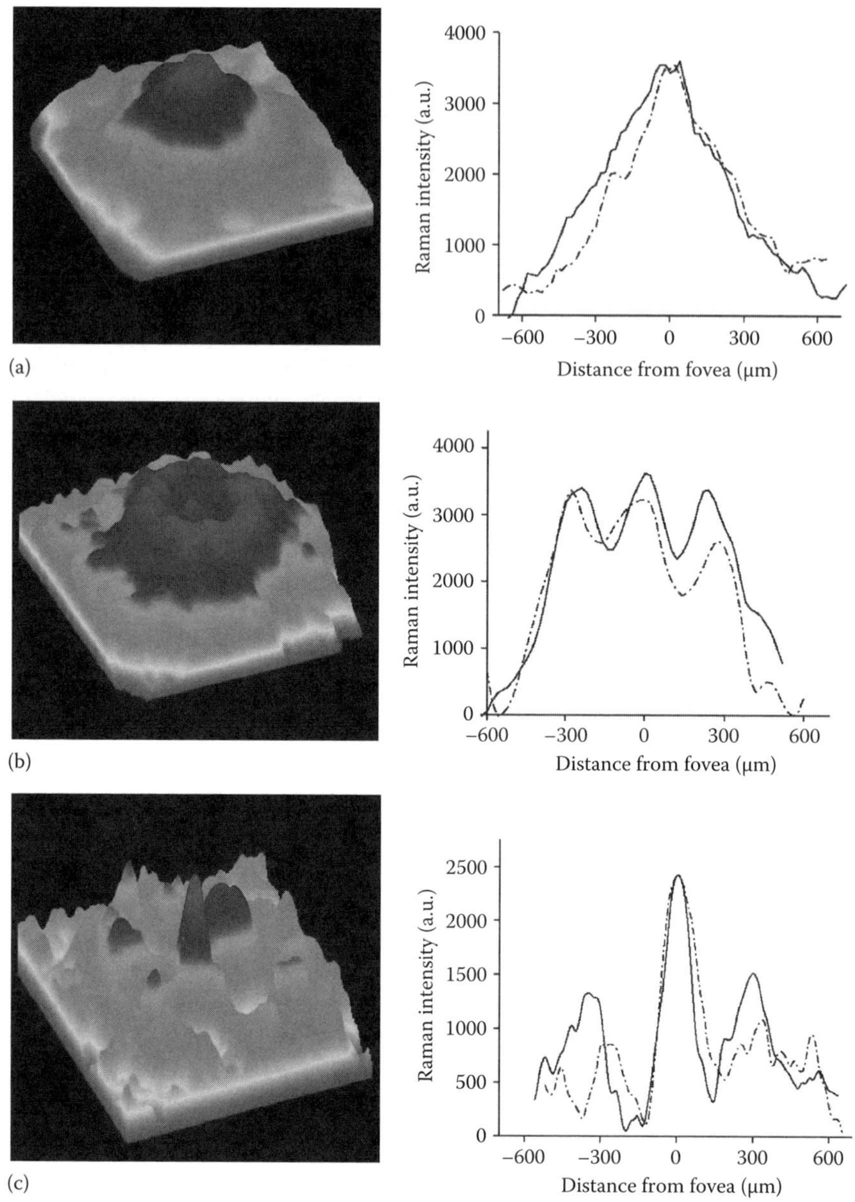

FIGURE 6.8 (See color insert following page 336.) Pseudocolor scaled, three-dimensional MP RRI images of three volunteer subjects, along with related line plot profiles, derived for each distribution along nasal–temporal (solid line) and inferior–superior meridians (dashed line), both running through the center of the macula. Distribution (a), which is representative for most healthy subjects, features a nearly rotationally symmetric MP distribution with monotonous decrease of concentration levels from the center to the periphery. Distribution (b) features a small central peak with a strong, surrounding, ring-like component. Varying in relative strength of central and ring components, this "ring-like" pattern is encountered in about 30% of the population. Distribution (c) is an example for a fragmented distribution with narrow central peak and broken-up ring structure, measured in a subject with mild form of dry macular degeneration. All images are color coded with the same intensity scale (not shown).

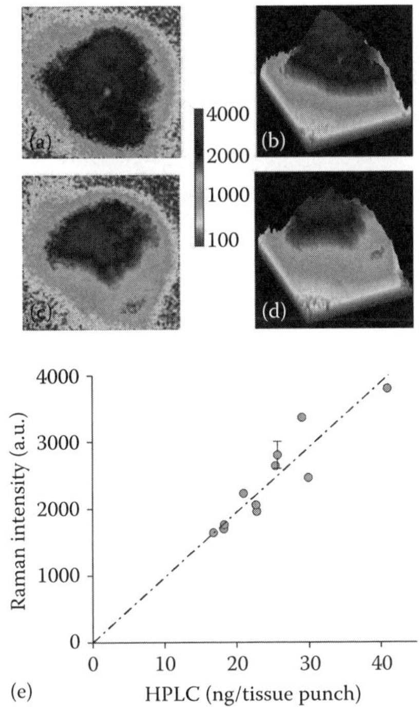

FIGURE 6.9 (a)–(d) Gray-scaled RRI images for 2 out of 11 donor eyecups, imaged to establish a correlation between Raman and HPLC derived carotenoid levels. The gray scale bar indicates the coding of the Raman intensities. (e) Plot of integrated Raman intensities versus carotenoid content derived via subsequent HPLC analysis. A high correlation exists between both methods (R=0.92). (From Sharifzadeh, M. et al., *J. Opt. Soc. Am. A*, 25, 947, 2008. With permission.)

In this context, it is interesting to also compare the Raman and AFI methods regarding age effects of MP levels. As shown in Figure 6.6b, RRS measurements indicated a decline of MP levels with age, even though the method is absolute and currently does not permit the correction of individual data for media transmissions. The AFI method, in comparison, is not influenced by the attenuation of the ocular media, since it references the MP levels to a location in the peripheral retina, and since the attenuation of the ocular media cancels out in first order. We measured AFI images of 70 healthy volunteer subjects, all very similar in demographics as compared to the subject population of Figure 6.6b, and obtained the result shown in Figure 6.10d for individual peak MP levels. Clearly, a decrease of MP levels is seen also with the AFI method (Sharifzadeh et al. 2006). The correlation of the decline of MP with age as measured by AFI is less than observed by RRS. This result may be explained by the compensation for medial opacities in the AFI method. The decline of MP with age as measured by AFI remains statistically significant (R=−0.47; $p < 0.0001$).

6.5 RESONANCE RAMAN DETECTION OF CAROTENOIDS IN SKIN

Levels of carotenoids are much lower in the skin relative to the macula of the human eye, but higher light excitation intensities and longer acquisition times can be used in Raman detection approaches to compensate for this drawback. Since the bulk of the skin carotenoids are in the superficial layers of the dermis, and since the concentrations are relatively low, the thin-film Raman equation given above, Equation 6.1, should still be a good approximation.

A cross section of excised human skin, histologically stained, is shown in Figure 6.11. It shows a layer structure of the tissue and the increased homogeneity in the bloodless *stratum corneum* layer, where the cell nuclei are absent, and where the potentially confounding melanin concentrations are

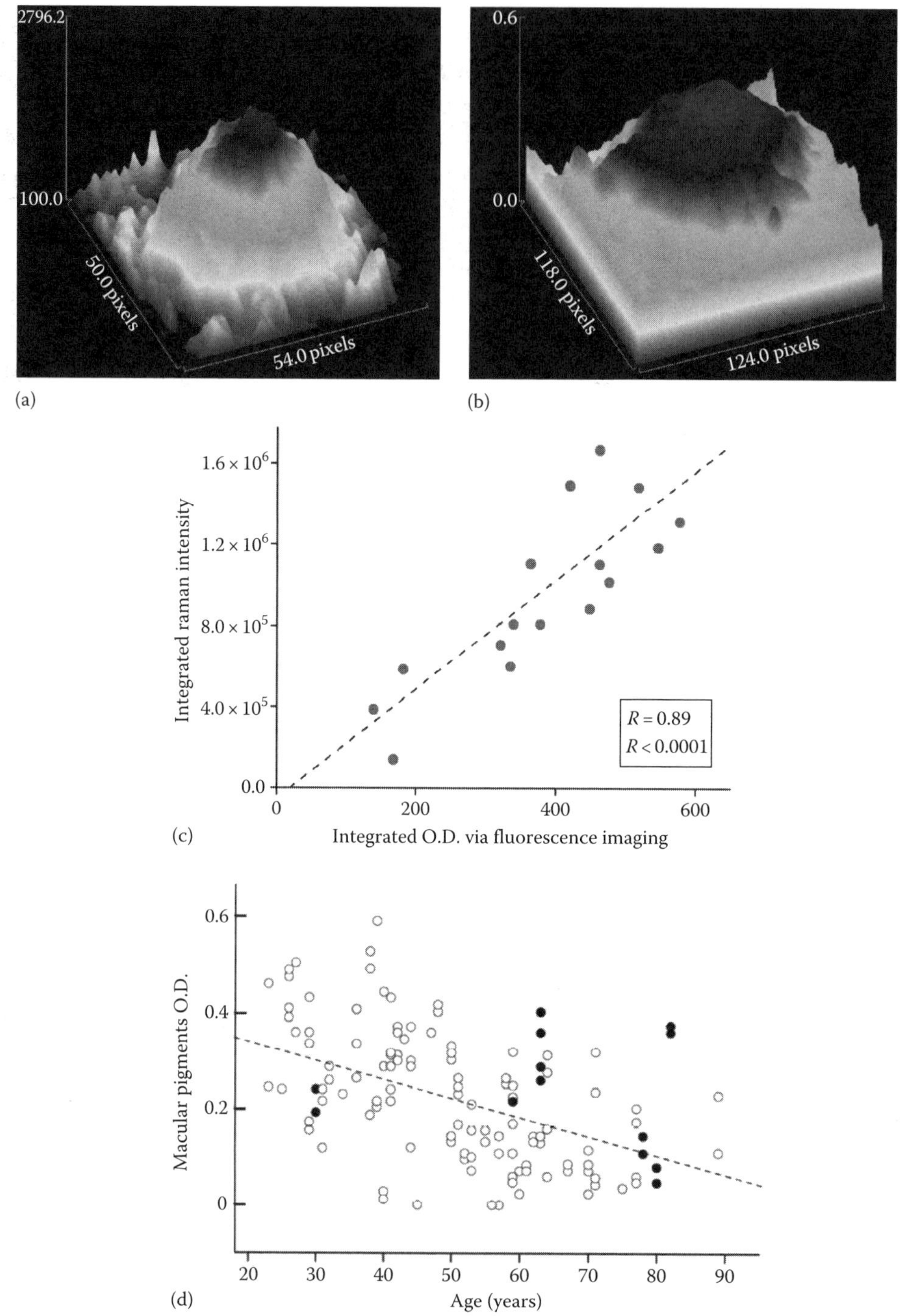

FIGURE 6.10 RRI images of MP distributions obtained for the same subject with (a) RRI and (b) lipofucsin fluorescence-based imaging. (c) Comparison of integrated MP densities obtained for 17 subjects with both imaging methods. Vertical scale shows integrated MP densities derived from RRI images by integrating intensities over the whole macular region; horizontal scale shows corresponding densities derived via fluorescence imaging. A high correlation coefficient of $R=0.89$ is obtained for both methods. (d) Age dependence of MP levels, measured with a lipofucsin fluorescence-based method ($R=-0.47$, $p < 0.0001$). (From Sharifzadeh, M. et al., *J. Opt. Soc. Am. A*, 25, 947, 2008. With permission; Sharifzadeh, M. et al., *J. Opt. Soc. Am. A*, 23, 2373, 2006. With permission.)

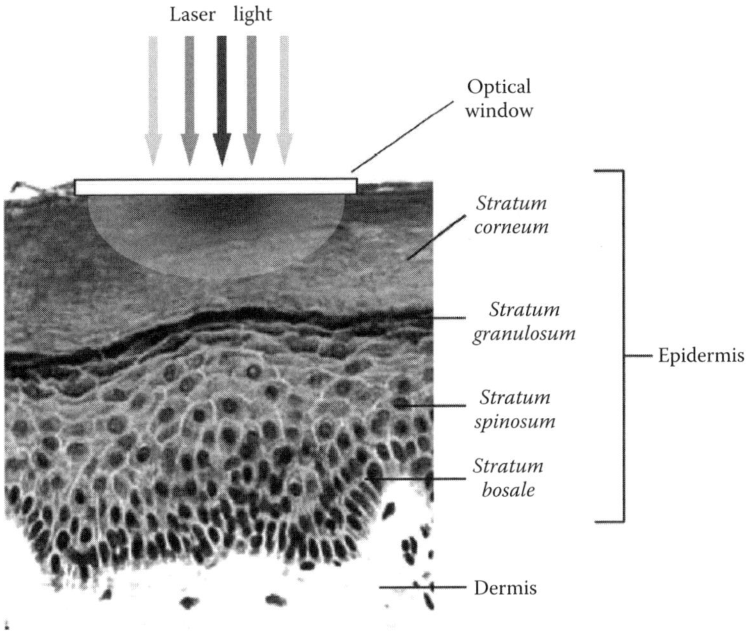

FIGURE 6.11 Layer structure of human skin as seen in a microscope after staining, showing the morphology of dermis, basal layer, *stratum spinosum*, *stratum granulosum*, and *stratum corneum*. Cells of the *stratum corneum* have no nucleus (these lack the dark staining spots), and form a relatively homogeneous optical medium, well suited for Raman measurements. For visible wavelengths, the excitation light has a penetration depth of about 400 μm, and stays within the 0.7–2 mm thick *stratum corneum*, as indicated.

minimal as well. The penetration depth of visible light into the *stratum corneum* is approximately 400 microns and therefore is confined to this outermost layer, as sketched in Figure 6.11 for a hemispherical beam penetration into the tissue. Using skin tissue sites with thick *stratum corneum* layer in RRS measurements, such as the palm of the hand or the sole of the foot, one therefore realizes measuring conditions of a fairly homogeneous uniform tissue layer with well-defined absorption and scattering conditions.

A field-usable instrument configuration that recently evolved out of the development of RRS for *in vivo* skin carotenoid measurements (Gellermann et al. 2001) is shown in Figure 6.12a. It is based on a miniaturized, fiber-based, and computer-interfaced spectrograph with high light throughput (Ermakov et al. 2001a). For an RRS skin carotenoid measurement, the palm of the hand is held against the window of the probe head module and the tissue exposed for about 10 s with 488 nm laser light at laser intensities of ~10 mW in a 2 mm diameter spot. Carotenoid RRS responses are detected with a CCD array integrated into the spectrograph. Typical skin carotenoid RRS spectra measured *in vivo* are shown in Figure 6.12b. The raw spectrum shown at the top of the panel (trace 1) was obtained directly after laser exposure and reveals a broad, featureless, strong "autofluorescence" background of skin, with three superimposed Raman peaks characteristic for the carotenoid molecules at 1008, 1159, and 1524 cm^{-1}. Even though the intensity of the skin fluorescence background is about 100 times higher than the carotenoid signals, it is possible to measure the skin carotenoid RRS responses with high accuracy by using a detector with high dynamic range. Approximation of the fluorescence background with a higher order polynomial and subsequent subtraction from the raw spectrum yields an isolated Raman spectrum of the skin carotenoids (trace 2) that is virtually undistinguishable from a solution of pure β-carotene, shown for comparison (trace 3). The skin carotenoid RRS response originates from contributions of all skin carotenoid species

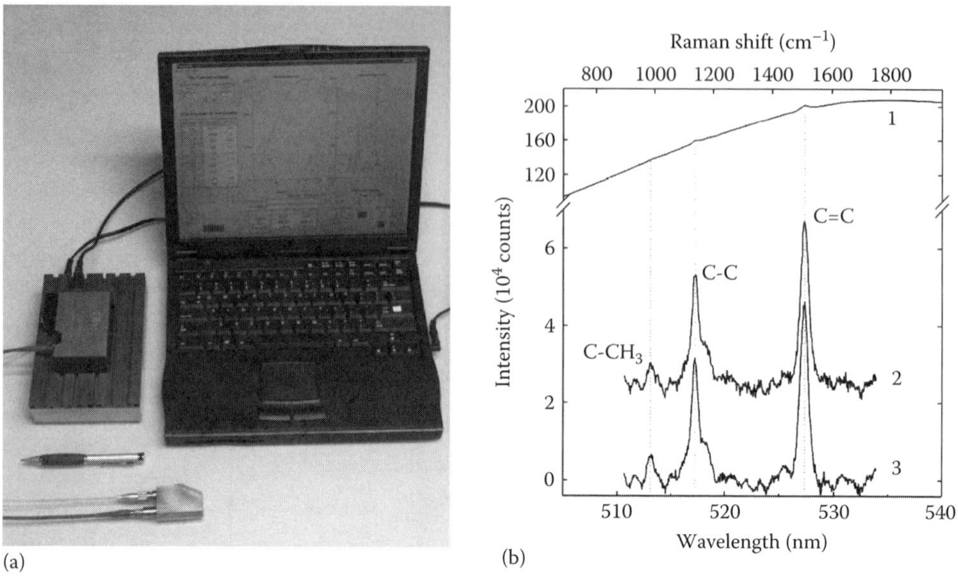

FIGURE 6.12 (a) Image of clinic- and field-usable, computer-interfaced, skin carotenoid RRS instrument, showing solid state laser, spectrograph, and light delivery/collection module. (b) Typical skin carotenoid Raman spectra measured *in vivo*. Spectrum (1) is obtained directly after exposure, and reveals a strong, spectrally broad, skin autofluorescence background with superimposed weak, but recognizable Raman peaks characteristic for carotenoids. Spectrum (2) is obtained after fitting the fluorescence background with a fourth-order polynomial, subtraction from (1), and scaling of the spectrum. Spectrum (2) is indistinguishable from a spectrum of a β-carotene solution, shown as (3) for comparison.

absorbing in the visible spectral range. Since all individual C=C stretch positions and bandwidths are indistinguishable at the instrument's spectral resolution, our RRS approach allows us to use the absolute peak height of the C=C signal at 1524 cm^{-1} as a measure for the overall carotenoid concentration in human skin.

Experiments with varying light excitation intensities showed that the skin carotenoid RRS response is stable up to the highest intensities permissible for skin applications (Ermakov et al. 2001a). To check the repeatability of the Raman measurements, we compared the RRS measurements of skin with the measurements of a tissue phantom consisting of (a) a mixture of glycerol and fine aluminum oxide powder to simulate scattering, (b) β-carotene, and (c) an organic dye (coumarin 540) that simulates the skin autofluorescence background. While the repeatability for the phantom was excellent, with a standard deviation below 1% for 10 consecutive measurements, the repeatabilities in living human tissue were significantly lower, with standard deviations ranging between 0.5% and 14% depending on the subject. To further investigate the origin of this effect, we measured the spatial distribution of a skin tissue sample with a Raman imaging instrument. The result, shown in Figure 6.13 clearly reveals that the skin carotenoid concentration varies significantly on a microscopic scale. Excitation spot sizes that are too small should be avoided due to these variations. The relatively large, 2 mm diameter beam spot size used in our skin Raman measurements appear to be an adequate solution to this effect, since it effectively integrates over these microscopic spatial concentration changes.

To validate the skin carotenoid RRS detection approach, we initially carried out an indirect validation experiment that compared HPLC derived carotenoid levels of fasting serum with RRS derived carotenoid levels for inner palm tissue sites. Measuring a large group of 104 healthy male and female human volunteers, we obtained a significant correlation ($p < 0.001$) with a correlation coefficient of 0.78 (Smidt et al. 2004). Recently, we carried out a direct validation study, in which we compared *in vivo* RRS carotenoid skin responses with HPLC-derived results, using the thick

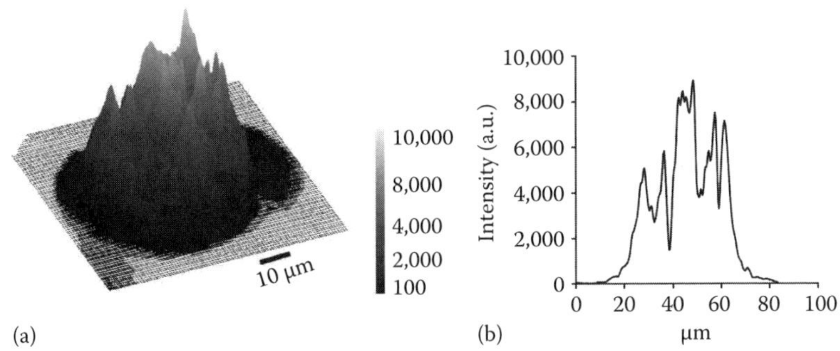

FIGURE 6.13 Gray-scale microscopic RRI image of an excised palm tissue sample (a) and intensity plot (b) along a line running through the middle of the distribution. Results show large spatial variation of the concentration of carotenoids within the skin on a microscopic scale.

stratum corneum layer of heel skin tissue sites. Following RRS measurements of the sites in eight volunteer subjects, the subjects scraped off thin skin slivers of 10–50 mg weight around the optically measured area with a razor blade for subsequent HPLC analysis. In Figure 6.14a, the comparison of RRS skin carotenoid responses is shown for all subjects with corresponding HPLC-derived

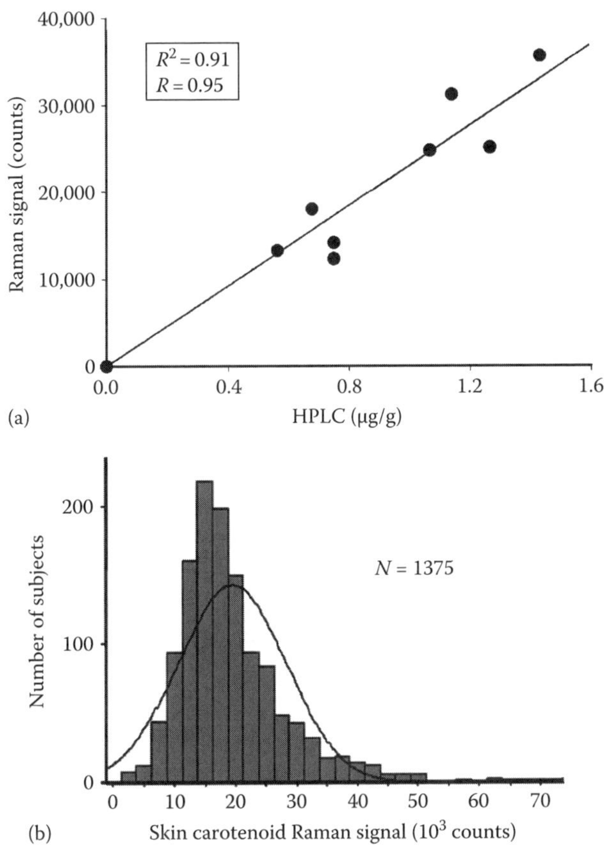

FIGURE 6.14 (a) Plot of carotenoid levels, shown as solid disks, for eight samples of human tissue measured with RRS technique *in vivo*, and subsequently, after tissue excision, with HPLC methods. The solid line is the resulting linear regression crossing the origin, and reveals a correlation coefficient R equal to 0.95. (b) Histogram of skin carotenoid RRS response measured in the palm of 1375 subjects, showing wide distribution of skin carotenoid levels in a large population.

carotenoid content. The latter is a sum of individual concentrations determined for each excised sample for the main skin carotenoids lutein, zeaxanthin, *cis*-lutein/zeaxanthin, α-cryptoxanthin, β-cryptoxanthin, *trans*-lycopene, *cis*-lycopene, α-carotene, *trans*-β-carotene, *cis*-β-carotene, and canthaxanthin. Using a regression line fit that passes through the origin we obtained a near-perfect correlation between the Raman and HPLC data, as evidenced by a correlation coefficient R as high as 0.95 (Ermakov and Gellermann, unpublished results). The results show excellent linearity of RRS derived carotenoid levels over a wide range of physiological skin carotenoid concentrations and provide a direct validation of the skin carotenoid RRS detection approach.

As a side aspect, the HPLC–Raman correlation results allow us to calibrate the RRS instruments in terms of carotenoid concentration. According to the regression analysis, the cumulative skin carotenoid content c, measured in μg per g of skin tissue, is linked to the height of the C=C RRS skin carotenoid intensity, I, via c [μg/g]$=4.3\times10^{-5}=I$ [photon counts]. Integrating the RRS spectra with the instrument's data acquiring software therefore allows us to display skin carotenoid content directly in concentration units, i.e., in μg carotenoid content per g of tissue.

Measurements of large populations with the Raman device reveal a bell-shaped distribution of carotenoid levels, as shown in Figure 6.14b for a group of 1375 healthy volunteer subjects that could be screened with the RRS method within a period of a few weeks (Smidt et al. 2004, Ermakov et al. 2005b). Analysis of the data confirmed a pronounced positive relationship between self-reported fruit and vegetable intake (a source of carotenoids) and skin Raman response. Furthermore, the study showed that people with habitual high sunlight exposure have significantly lower skin carotenoid levels than people with little sunlight exposure, independent of their carotenoid intake or dietary habits, and that smokers had dramatically lower levels of skin carotenoids as compared to nonsmokers (Ermakov et al. 2005b). Importantly, it also showed that RRS detection can track the increase of skin carotenoid levels occurring in subjects with low skin carotenoid levels within a relatively short time frame of weeks as a result of dietary supplementation with carotenoid-containing multivitamins.

Based on these capabilities, the RRS detection method has already found commercial application in the nutritional supplement industry (BioPhotonic Scanner™, Pharmanex LLC, Provo, and Utah), which has placed thousands of portable instruments with their customers for rapid optical measurements of dermal carotenoid levels, and which has further developed the instrumentation for rugged field use (Bergeson et al. 2008).

Regarding other medical applications, the method had found initial interest in dermatology, where a tentative correlation was demonstrated between certain types of cancerous lesions and depleted carotenoid levels (Hata et al. 2000). Quantitative RRS measurements in these tissues, however, which extend to layers beyond the *stratum corneum*, are more complicated due to additional chromophores, and need to be further refined for future studies. In the field of epidemiology, the RRS method has recently been applied to subjects with increased bitter taste sensitivities. Measuring the *stratum corneum* layer of palm tissue, an inverse relationship was observed between taste sensitivity and fruit and vegetable uptake (Scarmo et al., unpublished), a finding that may be helpful to promote healthy behavioral patterns of dietary change in large populations. In neonatology, skin carotenoid RRS measurements investigating correlations of carotenoid levels with retinopathy of prematurely born infants are in progress (Chan et al. 2006). Measuring the sole of the foot, it could be shown that retinopathy is influenced by the carotenoids in human milk-fed infants, and that it appears likely that carotenoids are important nutrients in decreasing the severity of the disease.

6.6 SELECTIVE RESONANCE RAMAN DETECTION OF CAROTENES AND LYCOPENE IN HUMAN SKIN

In all previous RRS measurements of dermal carotenoids we measured the total concentration of all long-chain carotenoid species since the method only detects the chain's carbon double-bond vibration, which is identical in all species. Lycopene has an increased conjugation length compared to the other carotenoids in skin and therefore features a small (~10nm) but distinguishable red shift of the absorption. This shift can be explored to measure skin lycopene levels independently of the other carotenoid concentrations (Ermakov et al. 2004b). For pure solutions of lycopene and β-carotene, the resonance Raman response has approximately the same strengths under 488nm excitation. Under excitation with 514.5nm, however, the response is about six times higher for lycopene. Taking this effect into account in a simple two-carotenoid model, it is possible to derive skin lycopene concentrations separately by measuring two RRS responses, one for 488nm excitation, and one for 514.5nm excitation (Ermakov et al. 2004b). For the ratio of the two concentrations, N_B/N_L, where N_B is the concentration of all carotenoids other than lycopene and N_L is the lycopene concentration, one obtains Equation 6.3

$$\frac{N_B}{N_L} = \frac{\sigma_L^{488} - r\sigma_L^{514}}{r\sigma_B^{514} - \sigma_B^{488}} \tag{6.3}$$

where

$r = I^{488}/I^{514}$ is the ratio of the RRS responses for blue and green excitation, respectively

σ_j^i is the respective Raman cross sections of the two carotenoid species

The RRS instrument for the selective detection of dermal lycopene levels is shown in Figure 6.15. The instrument uses a single spectrograph to detect C=C Raman responses resulting from 488 and 514nm excitation with a fixed grating position. A small air-cooled, multiline argon laser generates excitation light at both wavelengths with comparable intensities. Two shutters are synchronized such that the skin is either unexposed, exposed with 488nm light, or with 514nm light. The optical probe module contains an additional "green" excitation channel, and the detection channels each contain a separate filter to suppress scattered excitation light. A measurement starts by exposing the skin site with 488nm, while recording the RRS carotenoid C=C response. Subsequently, the electronics closes the shutter, reads out the Raman data, reactivates the CCD, and the whole process is repeated for 514nm green excitation. Finally, the software calculates and separately displays the ratio of the carotenoids and the skin lycopene levels, as shown in Figure 6.15b.

For seven volunteer subjects measured with the dual-wavelength RRS instrument, we obtained the skin carotenoid RRS results shown in Figure 6.16, where the individual lycopene and carotene levels are indicated together with the lycopene/carotene ratio for each subject. Interestingly, there is a strong, almost threefold variation in carotene to lycopene ratio in the measured subjects, ranging from 0.54 to 1.55. This means that substantially different carotenoid compositions can exist in human skin, with some subjects exhibiting almost twice the concentration of lycopene compared to carotene, and other subjects showing the opposite effect. This behavior could reflect different dietary patterns regarding the intake of lycopene or lycopene-containing vegetables, or it could point toward differing abilities between subjects to accumulate these carotenoids in the skin.

6.7 CONCLUSIONS

In ocular applications, Raman spectroscopy can quickly and objectively assess composite lutein and zeaxanthin concentrations of macular pigment using spatially averaged, integral measurements or images that quantify and map the complete MP distribution with high spatial resolution. Importantly, both variants can be validated with HPLC methods in excised human eyecups and in animal models.

Both integral and spatially resolved MP Raman methods use the backscattered, single-path Raman response from lutein and zeaxanthin in the MP-containing retinal layer, and largely avoid light traversal through the deeper retinal layers. Since they do not rely on any reflection of light at the sclera, the overlapping fluorescence signals from the ocular media can be subtracted from the overall light response. Importantly, the Raman methods make no assumptions other than approximating

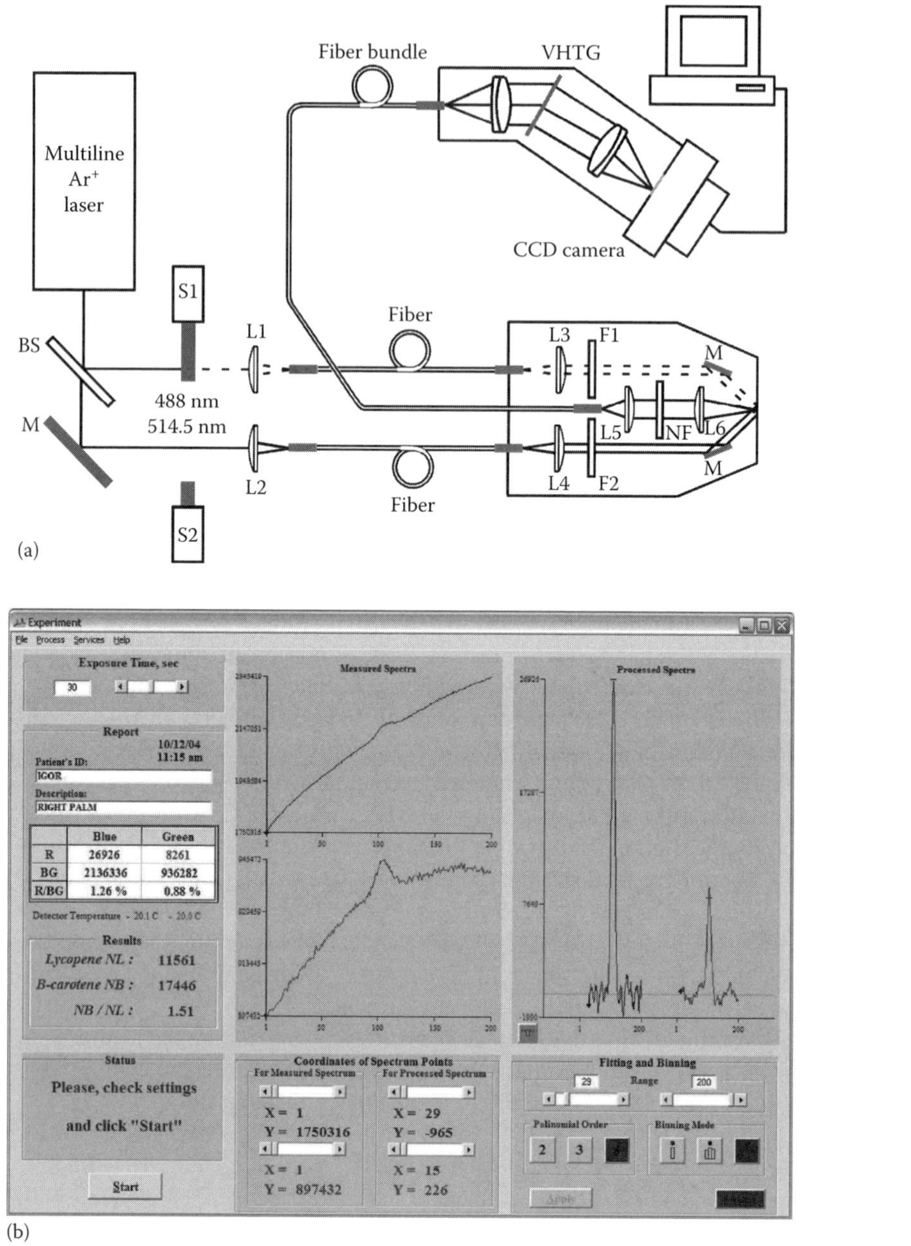

FIGURE 6.15 (a) Schematics of RRS instrument developed for selective *in vivo* measurements of lycopene and β-carotene in human skin. The optical probe features two excitation channels for blue and green light supplied by a two-color argon laser. (b) Computer monitor display of instrument. After each measurement, the software interface displays the raw and processed Raman spectrum obtained with the respective excitation wavelength. Using both spectra, it also calculates separately the concentration for lycopene and the concentration for the remaining β-carotene-like carotenoids present in the measured tissue site. (From Ermakov, I.V. et al., *SPIE Proc.*, 5686, 131, 2005. With permission.)

the spectrally broad background fluorescence with the fluorescence response at a wavelength that is slightly offset from the MP Raman response. MP Raman measurements are a measure of absolute MP concentration levels since the method does not use a reference point in the peripheral retina. Attenuation effects caused by the optical media are therefore fully effective, and have to be avoided

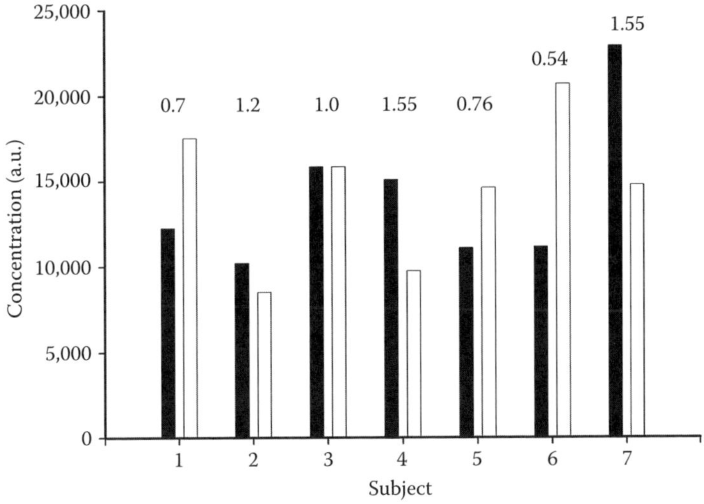

FIGURE 6.16 Bar graph of β-carotene and lycopene skin levels measured with selective RRS for seven subjects. White bars represent β-carotene levels, black bars the lycopene levels. Note strong intersubject variability of β-carotene to lycopene concentration ratios, indicated above the bar graphs.

or minimized, particularly when comparing MP levels between subjects. Optical losses from the lens can be neglected in longitudinal studies, provided these are carried out over a time span in which lens absorptions can be considered to remain constant (1–2 years), or in any studies involving subjects with lens implants. Therefore, the Raman method would be well suited, for example, in important nutritional supplementation trials, studies in which significant increases in individual MP levels have been demonstrated to be achievable in a time span of 12 months (Richer et al. 2004).

Raman imaging reveals the existence of spatially complex MP distribution patterns throughout the subject population. The distributions vary strongly regarding widths, axial and rotational asymmetries, locally depleted areas, and integrated concentration levels. RRI-derived results agree in all aspects with results obtained for the same healthy, un-supplemented population with the completely different method of lipofuscin fluorescence imaging, and therefore provide independent evidence for a more complicated nature of MP distributions in human subjects than previously thought.

In dermal applications, the Raman method can rapidly assess dermal carotenoid content in large populations. Measurements are limited to tissue sites with a thick *stratum corneum*. In this case, the probed tissue is thicker than the penetration depth of the excitation light, thus avoiding the absorption of hemoglobin. Furthermore, the *stratum corneum* tissue is free of melanin. A correlation of our Raman-derived carotenoid data with HPLC-derived serum levels again confirms the validity of the carotenoid Raman detection technique in the physiologically relevant concentration range under these measuring conditions. Any tissue opacities are of course less problematic in longitudinal studies involving the same subjects, for example, in studies designed to investigate changes of MP or dermal carotenoid levels upon dietary changes or influences of external stress.

We believe that carotenoid RRS detection has exciting application potential. In the nutritional supplement industry it is already being used as an objective, portable device for the monitoring of the effect of carotenoid-containing supplements on skin tissue carotenoid levels. In ophthalmology, it may become a fast screening method for MP levels in the general population; in epidemiology, it may serve as a noninvasive novel biomarker for fruit and vegetable intake, replacing costly plasma carotenoid measurements with inexpensive and rapid skin Raman measurements; in neonatology it may serve as a noninvasive method to assess carotenoid levels in prematurely born infants to investigate their correlation with oxidative stress related degenerative diseases. Lastly, due to its capability of selectively detecting lycopene, the technology may be useful to investigate a specific role for lycopene in the prevention of prostate cancer and other diseases.

ACKNOWLEDGMENTS

This work was supported in parts by grants from the State of Utah (Biomedical Optics Center of Excellence grant), by Spectrotek L.C., the National Eye Institute (EY 11600), and the Research to Prevent Blindness Foundation (New York).

REFERENCES

Age-Related Eye Disease Study Research Group (2007), The relationship of dietary carotenoid and vitamin A, E, and C intake with age-related macular degeneration in a case-control study, AREDS Report No. 22, *Arch. Ophthalmol.* **125**: 1225–1232.

Berendschot TTJM, Willemse-Assink JJM, Bastiaanse M, de Jong PTVM, and van Norren D (2002), Macular pigment and melanin in age-related maculopathy in a general population, *Invest. Ophthalmol. Visual Sci.* **43**: 1928–1932.

Berendschot TTJM and van Norren D (2004), Objective determination of the macular pigment optical density using fundus reflectance spectroscopy, *Arch. Biochem. Biophys.* **430**: 149–155.

Berendschot TTJM and van Norren D (2006), Macular pigment shows ringlike structures, *Invest. Ophthalmol. Visual Sci.* **47**: 709–714.

Bergeson SD, Peatross JB, Eyring NY, Fralick JF, Stevenson DN, and Ferguson SB (2008), Resonance Raman measurements of carotenoids using light emitting diodes, *J. Biomed. Opt.* **13**: 044026-1–044026-6.

Bernstein PS, Yoshida MD, Katz NB, McClane RW, and Gellermann W (1998), Raman detection of macular carotenoid pigments in intact human retina, *Invest. Ophthal. Vis. Sci.* **39**: 2003–2011.

Bernstein PS, Zhao DY, Wintch SW, Ermakov IV, and Gellermann W (2002), Resonance Raman measurement of macular carotenoids in normal subjects and in age-related macular degeneration patients, *Ophthalmology* **109**: 1780–1787.

Bhosale P, Zhao DY, and Bernstein PS (2007), HPLC measurement of ocular carotenoid levels in human donor eyes in the lutein supplementation era, *Invest. Ophthalmol. Visual Sci.* **48**: 543–549.

Chan GM, Rau C, Gellermann W, and Ermakova M (2006), Retinopathy of prematurity and carotenoids in human milk fed infants, Abstract, Meeting of the *American Academy of Pediatrics*, Washington, D.C.

Delori FC, Goger DG, Hammond BR, Snodderly DM, and Burns SA (2001), Macular Pigment density measured by autofluorescence spectrometry: Comparison with reflectometry and heterochromatic flicker photometry, *J. Opt. Soc. Am. A* **18**: 1212–1230.

Delori FC (2004), Autofluorescence method to measure macular pigment optical densities: Fluorometry and autofluorescence imaging, *Arch. Biochem. Biophys.* **430**: 156–162.

Ermakov IV et al. (2005), Two-wavelength Raman detector for noninvasive measurements of carotenes and lycopene in human skin, *SPIE Proc.*, 5686, 131–141.

Ermakov IV, Ermakova MR, Bernstein PS, and Gellermann W (2004a), Macular pigment Raman detector for clinical applications, *J. Biomed. Opt.* **9**: 139–148.

Ermakov IV, Ermakova MR, and Gellermann W (2005a), Simple Raman instrument for *in vivo* detection of macular pigments, *Appl. Spectrosc.* **59**: 861–867.

Ermakov IV, Ermakova MR, Gellermann W, and Lademann J (2004b), Non-invasive selective detection of lycopene and beta-carotene in human skin using Raman spectroscopy, *J. Biomed. Opt.* **9**: 332–338.

Ermakov IV, Ermakova MR, McClane RW, and Gellermann W (2001a), Resonance Raman detection of carotenoid antioxidants in living human skin, *Opt. Lett.* **26**: 1179–1181.

Ermakov IV and Gellermann W, unpublished results.

Ermakov IV, McClane RW, Gellermann W, and Bernstein PS (2001b), Resonant Raman detection of macular pigment levels in the living human retina, *Opt. Lett.* **26**: 202–204.

Ermakov IV, Sharifzadeh M, Ermakova MR, and Gellermann W (2005b), Resonance Raman detection of carotenoid antioxidants in living human tissue, *J. Biomed. Opt.* **10**(6): 064028-1–064028-18.

Gellermann W, Ermakov IV, Ermakova MR, McClane RW, Zhao DY, and Bernstein PS (2002a), *In vivo* resonant Raman measurement of macular carotenoid pigments in the young and the aging human retina, *J. Opt. Soc. Am. A* **19**: 1172–1186.

Gellermann W, Ermakov IV, McClane RW, and Bernstein PS (2002b), Raman imaging of human macular pigments, *Opt. Lett.* **27**: 833–835.

Gellermann W, McClane RW, Katz NB, and Bernstein PS (2001), Method and apparatus for non-invasive measurement of carotenoids and related chemical substances in biological tissue, US Patent # 6,205, 354 B1.

Hata TR, Scholz TA, Ermakov IV, McClane RW, Khachik F, Gellermann W, and Pershing LK (2000), Non-invasive Raman spectroscopic detection of carotenoids in human skin, *J. Invest. Dermatol.* **115**: 441–448.

Kolonel LN, Hankin JH, Whittemore AS et al. (2000), Vegetables, fruits, legumes, and prostate cancer: A multiethnic case-control study, *Cancer Epidemiol. Biomarkers Prev.* **9**: 795–804.

Koyama Y (1995), Resonance Raman spectroscopy, in *Carotenoids*, Vol 1B, *Spectroscopy*, G. Britton, S. Liaaen-Jensen, and H. Pfander, Eds., pp. 135–146, Birkhäuser, Basel, Switzerland.

Krinsky NI and Johnson EJ (2005), Carotenoid actions and their relation to health and disease, *Mol. Aspects Med.* **26**: 459–516.

Krinsky NI, Landrum JT, and Bone RA (2003), Biologic mechanisms of the protective role of lutein and zeaxanthin in the eye, *Ann. Rev. Nutr.* **23**: 171–201.

Landrum JT and Bone RA (2001), Lutein, zeaxanthin, and the macular pigment, *Arch. Biochem. Biophys.* **385**: 28–40.

Liu S, Manson JE, Lee IM et al. (2000), Fruit and vegetable intake and risk of cardiovascular disease: The Women's Health Study, *Am. J. Clin. Nutr.* **72**: 922–928.

Michaud DS, Feskanich DD, Rimm EB et al. (2000), Intake of specific carotenoids and risk of lung cancer in 2 prospective US cohorts, *Am. J. Clin. Nutr.* **92**: 990–997.

Nolan JM, Stack J, O'Donovan O, Loane E, and Beatty S (2007), Risk factors for age-related maculopathy are associated with a relative lack of macular pigment, *Exp. Eye Res.* **84**: 61–74.

Richer S, Stiles W, Statkute L, Pulido J, Frankowski J, Rudy D, Pei K, Tsipursky M, and Nyland J (2004), Double-masked, placebo-controlled, randomized trial of lutein and antioxidant supplementation in the intervention of atrophic age-related macular degeneration: The Veterans LAST study (lutein antioxidant supplementation trial), *Optometry* **75**: 216–30.

Robson AG, Moreland JD, Pauleikoff D, Morrissey T, Holder GE, Fitzke FW, Bird AD, and van Kuijk FJGMD (2003), Macular pigment density and distribution: comparison of fundus autofluorescence with minimum motion photometry, *Vision. Res.* **43**: 1765–1775.

Scarmo SN, Cartmel B, Gellermann W, Ermakov IV, Leffell DJ, Lin H, and Mayne ST (2009), Perceived bitter taste and fruit and vegetable intake measured by self-report and an objective indicator, unpublished.

Seddon JM, Ajani UA, Sperduto RD et al. (1994), Dietary carotenoids, vitamins A, C, and E, and advanced age-related macular degeneration, *J. Am. Med. Assoc.* **272**: 1413–1420.

Sharifzadeh M, Bernstein PS, and Gellermann W (2006), Non-mydriatic fluorescence-based quantitative imaging of human macular pigment distributions, *J. Opt. Soc. Am. A* **23**: 2373–2387.

Sharifzadeh M, Zhao DY, Bernstein PS, and Gellermann W (2008), Resonance Raman Imaging of macular pigment distributions in the human retina, *J. Opt. Soc. Am. A* **25**: 947–957.

Shreve AP, Trautman JK, Owens TG, and Albrecht AC (1991), Determination of the S2 lifetime of β-carotene, *Chem. Phys. Lett.* **178**: 89.

Smidt CR, Gellermann W, and Zidichouski JA (2004), Non-invasive Raman spectroscopy measurement of human carotenoid status, *Fed. Am. Soc. Exp. Biol. J.* **18**: A 480.

Snodderly DM, Auran JD, and Delori FC (1984a), The macular pigment. II. Spatial distribution in primate retinas, *Invest. Ophthalmol. Visual Sci.* **25**: 674–85.

Snodderly DM, Brown PK, Delori FC, and Auran JC (1984b), The macular pigment. I. Absorbance spectra, localization, and discrimination from other yellow pigments in primate retinas, *Invest. Ophthalmol. Visual Sci.* **25**: 660–73.

Snodderly DM, Mares JA, Wooten BR, Oxton L, Gruber M, and Ficek T (2004), Macular pigment measurement by heterochromatic flicker photometry in older subjects: The carotenoids and age-related eye disease study, *Invest. Ophthalmol. Visual Sci.* **45**: 531–538.

Trieschmann M, Spittal G, Lommartzsch A, van Kuijk E, Fitzke F, Bird AC, and Pauleikoff D (2003), Macular pigment: Quantitative analysis on autofluorescence images, *Graefe's Arch. Clin. Exp. Ophthalmol.* **241**: 1006–1012.

Part III

Applications of Spectroscopic Methodologies to Carotenoid Systems

7 Identification of Carotenoids in Photosynthetic Proteins: Xanthophylls of the Light Harvesting Antenna

Alexander V. Ruban

CONTENTS

7.1 INTRODUCTION TO XANTHOPHYLLS: OCCURRENCE AND MOLECULAR STRUCTURE

Carotenoids are one of the most abundant groups of pigments found in nature. Every year more than 100 million tonnes of them are being synthesized in the biosphere. Nearly 600 molecular species of carotenoids are currently identified (Del Campo et al., 2007). As powerful antioxidants, vitamin precursors, natural colorants, and odorants they became a serious global market commodity accounting for almost 1 billion dollars of the yearly trade (BCC research, 2007).

Carotenoids can be defined as lipid soluble methylated polyene derivatives or nonsaturated terpenoids. Varying numbers of conjugated carbon double bonds in carotenoids affect their delocalized excited state π-electron energy, and therefore define the color. The most abundant group of carotenoids, xanthophylls, contains oxygen atoms in their structure. The presence of polar groups makes xanthophylls less hydrophobic. The possession of hydrophobic and hydrophilic properties by a long carbon chain molecule is typical for detergents and quinones. Indeed, some xanthophylls, such as rhodopin glucoside of purple bacteria, can be classified as detergents. Rhodopin glucoside of LH2 complex possesses β-D-glucose group just like β-D-glucoside detergents and a long hydrophobic carbon tail, which differs from the one of detergents by the presence of methyl groups as in terpenes and conjugated double bonds. Molecules of the majority of xanthophylls are more symmetric than those of detergents and quinones. Xanthophylls possess two cyclic polar groups, one at each end of the molecule. This feature increases the coordination of the molecule in the membrane and determines interaction patterns with protein membrane-spanning helixes, as will be shown later in this chapter. Xanthophylls bound to proteins can play important, yet currently not well-understood, structural functions, similar to those of membrane lipids and beyond.

Fucoxanthin, lutein, neoxanthin, violaxanthin, and zeaxanthin are the most common xanthophylls on our planet. They are found in the photosynthetic machinery of algae (fucoxanthin) and higher plants (Figure 7.1). Interestingly, lutein and zeaxanthin have also been found in the retina of humans and some primates (Khachik et al., 1997; Landrum and Bone, 2001). It is likely that these carotenoids possess some universal photophysical properties essential for both photosynthesis and vision (Britton, 1995).

Fucoxanthin is the most oxygenated of these five xanthophylls. It contains six oxygen atoms, which make the molecule highly polar. Along with neoxanthin, it possesses a normally highly reactive allene group found rarely in carotenoids. Neoxanthin is found to be almost exclusively in the 9-*cis* conformation in nature. Violaxanthin, in contrast, is a very symmetric molecule, containing two epoxy oxygen atoms on the end-ring groups. Lutein is less oxygenated than violaxanthin and is asymmetric. It possesses two different types of end groups, β- and ε-rings, which differ by the position of the double bond within the ring. Zeaxanthin, an isomer of lutein, is symmetrical. Zeaxanthin possess two β-ring end groups. The reversible deepoxidation of violaxanthin into zeaxanthin occurs in the photosynthetic membrane, and is dependent on the light environment (Sapozhnikov et al., 1957; Yamamoto, 1962). As a result, an intermediate xanthophyll, antheraxanthin, which carries only one epoxy group, is transiently formed.

The variations in the end group structure and conformation are determined by the carotenoid biosynthesis enzymes. These structural features are likely to determine localization as well as functions of these xanthophylls *in vivo* (Hashimoto et al., 2001; Young et al., 2002).

7.2 ANALYTICAL APPROACHES TO IDENTIFICATION AND QUANTIFICATION OF XANTHOPHYLLS: PRINCIPLES AND CHALLENGES

The most commonly used method for the identification of carotenoids is high-pressure liquid chromatography (HPLC) combined with the UV-Vis absorption detection. The introduction of diode array detection enabled parallel collection of pigment spectra, which greatly aids the quantification and localization of unknown compounds. Coupling HPLC with the mass-spectrometer significantly

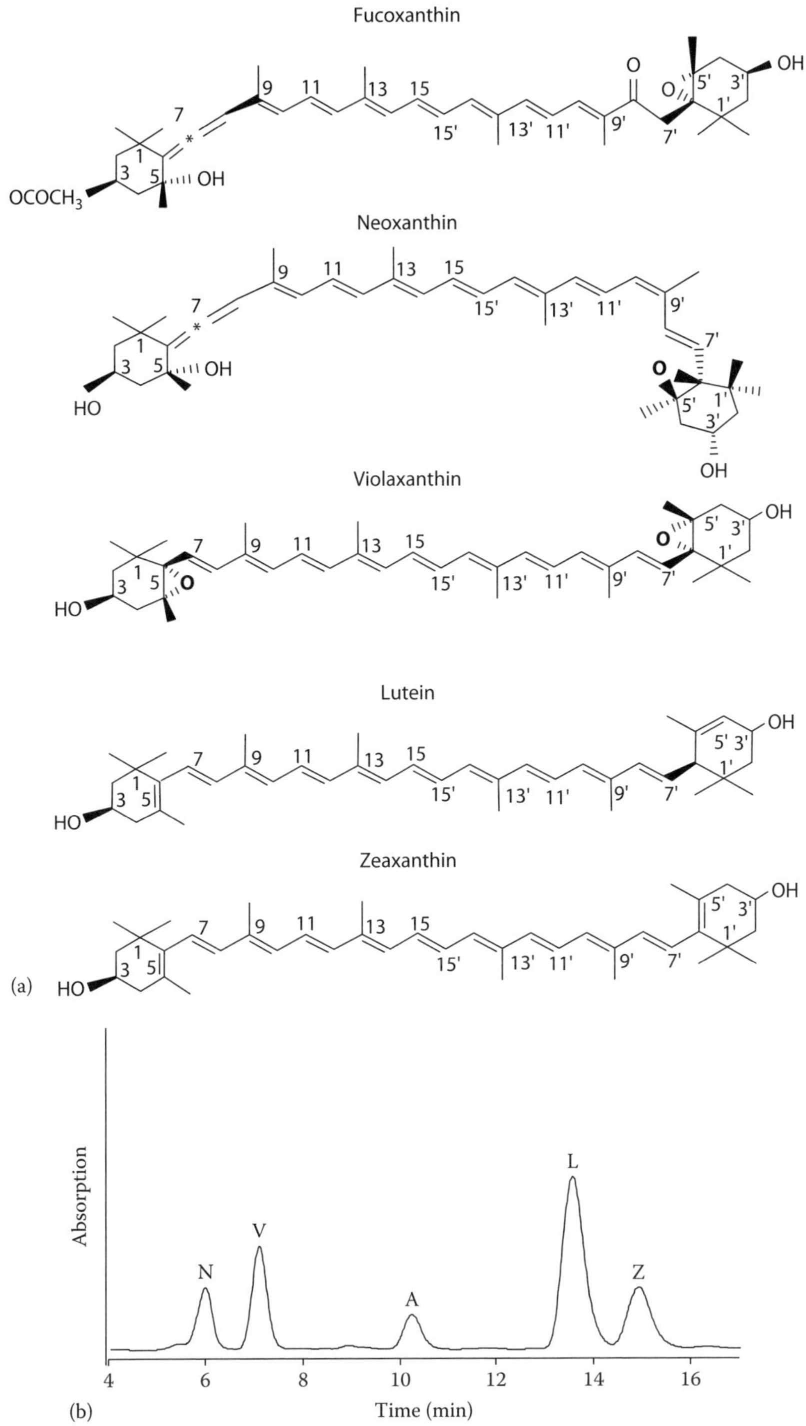

FIGURE 7.1 (a) Structures of the five most common xanthophylls. (b) HPLC separation profile of the photosynthetic membrane xanthophylls: N, neoxanthin; V, violaxanthin; A, antheraxanthin; L, lutein; and Z, zeaxanthin.

enhances the system sensitivity and analysis of structural isomers (Su et al., 2002). The optical spectroscopic analysis of carotenoids is based on the fact that the 0-0 energy of the first optically allowed transition is inversely correlated to the number of conjugated carbon double bonds in the delocalized π-electrons (Kuhn, 1949). Therefore, in theory, violaxanthin, lutein, and zeaxanthin, which have 9, 10, and 11 conjugated bonds should have all different 0-0 maxima positions. Neoxanthin has the same number of these bonds as violaxanthin, 9. However, the *cis*-conformation increases the energy of excited state leading to a slightly blueshifted 0-0 transition. In addition to this shift, a *cis*-band emerges at around 310–330 nm, which can also be used for distinguishing different isomers of the same carotenoid (Tsukida et al., 1982; Koyama et al., 1983).

Since the analytical approaches described above require the extraction of pigments from the living tissues and the membranes and protein complexes with organic solvents, the elimination of all structural and spectral features typical of the *in vivo* carotenoid state is lost. In addition, pigment degradation during sample storage and extraction conditions can frequently take place (Su et al., 2002; Feltl et al., 2005). It is also likely that the causes of some existing analytical discrepancies can be found in the method of using standards and their extinction coefficients. The hydrophobicity of carotenoid molecules and the strong environmental dependency of the excited state energy (due to high molecular polarizability) and oscillator strength could be the causes for significant variations in pigment quantification using UV-Vis detection. For example, in order to accurately separate and quantify photosynthetic membrane xanthophylls, chlorophylls, and β-carotene, a three-solvent system had to be employed (Snyder et al., 2004). All xanthophylls were separated using the polar solvent acetonitrile mixed with a fraction of methanol, whereas, in order to run β-carotene somewhat more nonpolar solvent mixture hexane/ethyl acetate was required. Figure 7.1 displays a typical HPLC profile of all higher plant xanthophylls. The more oxygenated and polar xanthophylls such as neoxanthin and violaxanthin elute much faster than the less polar lutein and zeaxanthin. In spite of the identical molecular mass, the latter two have slightly different mobility because of configuration differences in the end-group orientation leading to the differences in the molecular polarity.

Solvents with different polarities and refractive indexes significantly affect carotenoid optical properties. Because the refractive index is proportional to the ability of a solvent molecule to interact with the electric field of the solute, it can dramatically affect the excited state energy and hence the absorption maxima positions (Bayliss, 1950). Figure 7.2a shows three absorption spectra of the same xanthophyll, lutein, dissolved in isopropanol, pyridine, and carbon disulfide. The solvent refractive indexes in this case were 1.38, 1.42, and 1.63 for the three mentioned solvents, respectively.

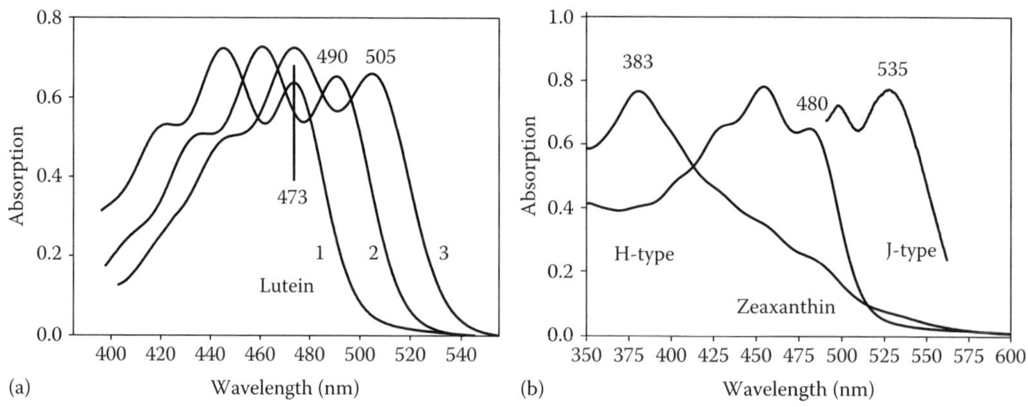

FIGURE 7.2 (a) Absorption spectra of lutein dissolved in isopropanol (1), pyridine (2), and carbon disulfide (3). (b) Absorption spectra of zeaxanthin (in ethanol) and zeaxanthin H- and J-type aggregates.

Another spectral development can take place if the solvent mixture is not able to maintain pigments in solute state. In this case, the formation of dimers and higher aggregates of all plant xanthophylls is very common (Takagi et al., 1983; Ruban et al., 1993a). Ethanol–water mixture provided us with a good system, which could not only yield xanthophyll aggregates but also test hydrophobicity of these pigments using the solvent ratio at which aggregation takes place (Ruban et al., 1993a; Horton and Ruban, 1994). Figure 7.2b displays three types of zeaxanthin absorption spectra: the pigment in solution, H- and J-type aggregates. Here, the variation in the observed spectral maxima reaches more than 150 nm (~6000 cm^{-1})—a very significant difference, indeed.

It is therefore important to bear in mind the dependency of the carotenoid spectrum upon properties of the environment for *in vivo* analysis, which is based on the application of optical spectroscopies. This approach is often the only way to study the composition, structure, and biological functions of carotenoids. Spectral sensitivity of xanthophylls to the medium could be a property to use for gaining vital information on their binding sites and dynamics. The next sections will provide a brief introduction to the structure of the environment with which photosynthetic xanthophylls interact—light harvesting antenna complexes (LHC).

7.3 LOCALIZATION AND FUNCTIONS OF XANTHOPHYLLS IN LIGHT HARVESTING ANTENNA OF PLANTS

7.3.1 The Need for Photosynthetic Antenna

The photosynthetic antenna is an assembly of pigments that is not directly involved in the charge separation process but, by the collection of light quanta and efficient energy transfer, enhances the reaction center cross section by more than two orders of magnitude. The antenna is crucial in the low-light conditions since it enhances the excitation rate of the reaction center close to its turnover rate—a requirement for maximum energy conversion efficiency (Clayton, 1980). The protein is an essential part of the antenna. It binds and orients pigments in order to optimize light energy interception and transfer. The antenna protein also tunes the excited state energies in order to provide directionality for the energy flow and enhances the absorption of light across a larger wavelength range. Without the antenna, photosynthetic organisms, particularly aquatic ones, would starve.

7.3.2 Structure of the Photosystem II Antenna: Xanthophylls in LHCII Structure

In higher plants, the photosynthetic machinery is almost exclusively localized in the thylakoid membrane of chloroplasts (Figure 7.3). Thylakoids tend to form stacks of these membranes called grana, which generally carry photosystem II (PSII) with the light harvesting antenna. PSII is organized as a dimer containing two sets of reaction center proteins, D1 and D2 with their inner-antenna complexes CP43 and CP47. The major part of the antenna is formed by a number of monomeric (minor LHCII complexes) and trimeric (major LHCII complex or LHCII) pigment–protein complexes (for review, see Dekker and Boekema (2005)). The latter can often form large oligomeric structures, which contain several interacting trimers. The integrity of the PSII complex is ensured by various noncovalent interactions between its multiple subunits.

The structure of the major trimeric LHCII complex has been recently obtained at 2.72 Å (Figure 7.3) (Liu et al., 2004). It was revealed that each 25 kDa protein monomer contains three transmembrane and three amphiphilic α-helixes. In addition, each monomer binds 14 chlorophyll (8 Chl *a* and 6 Chl *b*) and 4 xanthophyll molecules: 1 neoxanthin, 2 luteins, and 1 violaxanthin. The first three xanthophylls are situated close to the integral helixes and are tightly bound to some amino acids by hydrogen bonds to hydroxyl oxygen atoms and van der Waals interactions to chlorophylls, and hydrophobic amino acids such as tryptophan and phenylalanine.

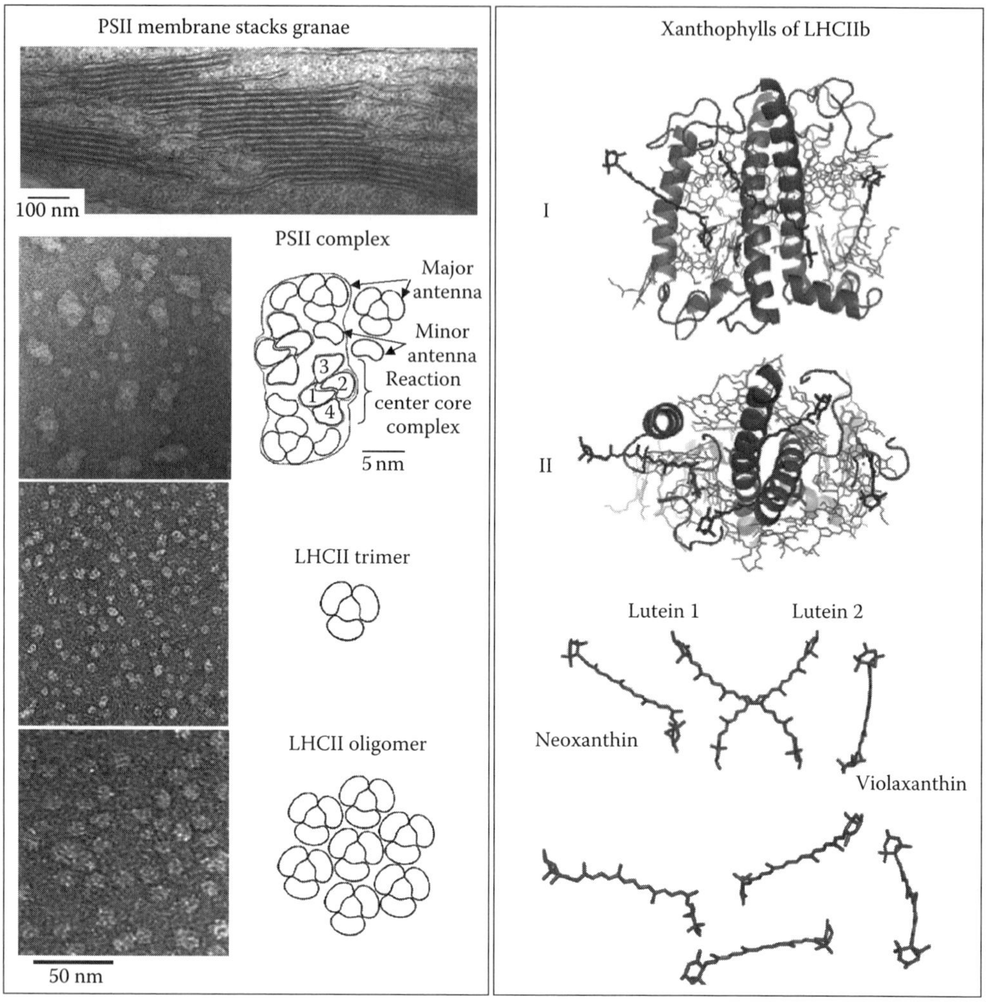

FIGURE 7.3 Structure of PSII membranes, macrocomplexes and LHCII antenna. Left from the top: electron microscopy of grana stacks, PSII macrocomplexes, LHCII trimers, and LHCII oligomers. Right from the top: Atomic structure of LHCII monomer (I and II are side and top views). Bottom part displays LHCII xanthophylls.

7.3.3 Functions of Xanthophylls in the Antenna: A Structural Perspective

Xanthophyll functions in the LHCII antenna are believed to increase the spectral cross section, complimenting chlorophyll absorption: photoprotection of chlorophyll against excess excitation energy, assembly and stability of the complex, and participation in the conformational dynamics of LHCII. The mechanisms behind these processes are still poorly understood, a major obstacle being the lack of detailed structural and spectral information available *in vivo*. There are important issues concerning xanthophylls in the LHCII antenna that remain unanswered. What is the purpose of having variations in a number of conjugated double bonds? What is the reason for the presence of the three types of xanthophylls in LHCII structure? How do differences in polarity and head group orientation determine xanthophyll binding sites. Where is zeaxanthin bound? How do the lumen-localized deepoxidase reach the violaxanthin epoxy group situated closer to the stromal side of the membrane? These are only a few questions of many, which remain to be answered in order to understand the role of xanthophylls in antenna function.

7.3.4 The Need for Identification of Xanthophylls *In Vivo*

Understanding the role of the xanthophylls in the antenna starts with the identification of their electronic excited state energy—a property that provides fingerprint information about each type of molecule and its environment. This is not always a trivial task, taking into consideration that the structure of the protein has only recently been solved. One should emphasize that the biochemical and spectroscopic information played a very crucial role in helping to make important structural assignments in the crystal structure. For example, the LHCII structure at 2.72 Å resolution contains violaxanthin, a xanthophyll whose presence in LHCII had been under debate for some years (Bassi et al., 1993; Ruban et al., 1999; Verhoeven et al., 1999). Other examples are spectral identification of neoxanthin *in vivo* as 9-*cis* (Ruban et al., 2001) and the discovery of hydrogen bonding to carbonyl groups of some chlorophyll molecules (Ruban et al., 1995; Pascal et al., 2005).

Identifying electronic and vibrational properties of xanthophylls should provide not only structural information. Gaining information about excited state energy levels would help to design and interpret kinetic experiments, which probe molecular interactions and the energetic relationship between the xanthophylls and chlorophylls.

7.4 PRINCIPLES OF IDENTIFICATION OF XANTHOPHYLLS *IN VIVO*

The identification of xanthophylls *in vivo* is a complex task and should be approached gradually with the increasing complexity of the sample. In the case of the antenna xanthophylls, the simplest sample is the isolated LHCII complex. Even here four xanthophylls are present, each having at least three major absorption transitions, 0-0, 0-1, and 0-2 (Figure 7.4). Heterogeneity in the xanthophyll environment and overlap with the chlorophyll absorption add additional complexity to the identification task. No single spectroscopic method seems suitable to resolve the overlapping spectra. However, the combination of two spectroscopic techniques, low-temperature absorption and resonance Raman spectroscopy, has proved to be fruitful (Ruban et al., 2001; Robert et al., 2004).

The Raman scattering originates from the inelastic interaction of the electromagnetic field of light with matter, resulting in an alteration of the frequency of scattered radiation. The extent of this alteration (Raman shift) depends upon the energy of molecular vibrational modes that are coupled to the electronic transition in resonance (Garey, 1982). The number and frequency of these modes depend on the type of molecule, symmetry conditions (electron–vibrational coupling), conformation and, in some cases, the electronic excited state energy. Carotenoids are effective combinational (Raman) scatterers exhibiting strong resonance enhancement (Merlin, 1985; Robert, 1999). Therefore in the mixture of spectrally different xanthophylls present in LHCII, it seemed to be feasible to achieve a selective enhancement of the Raman scattering by exciting a single absorption band belonging to only one xanthophyll species. Moreover, if every xanthophyll of LHCII possesses a specific resonance Raman fingerprint, it should be possible to identify the origin of the probed transition. Figure 7.4 shows absorption spectra of all isolated xanthophylls of LHCII. They each have different 0-0 transition energies, and therefore can *a priori* be suitable for resonance-selective experiments. Moreover, the resonance Raman spectra seem to reveal specific features for all xanthophylls (Figure 7.4). Four main regions can be seen in the xanthophyll Raman spectrum (labeled with a Greek letter ν). The first and the highest frequency region corresponds to C=C stretching vibrations. The second and the most complex region is most influenced by C–C stretching modes coupled to C–H in-plane bending/wagging or C–CH_3 stretching vibrations. The third group in the xanthophyll Raman spectrum reflects CH_3 in-plane rocking vibrations. The fourth and the smallest, seemingly featureless region corresponds to weakly coupled C–H out-of-plane bending modes. The ν_4 feature would normally be Raman-forbidden for a fully planar configuration of xanthophyll molecule. However, as appears later, these modes could become very strong in molecules, which adopt distorted configuration due to interactions with the environment.

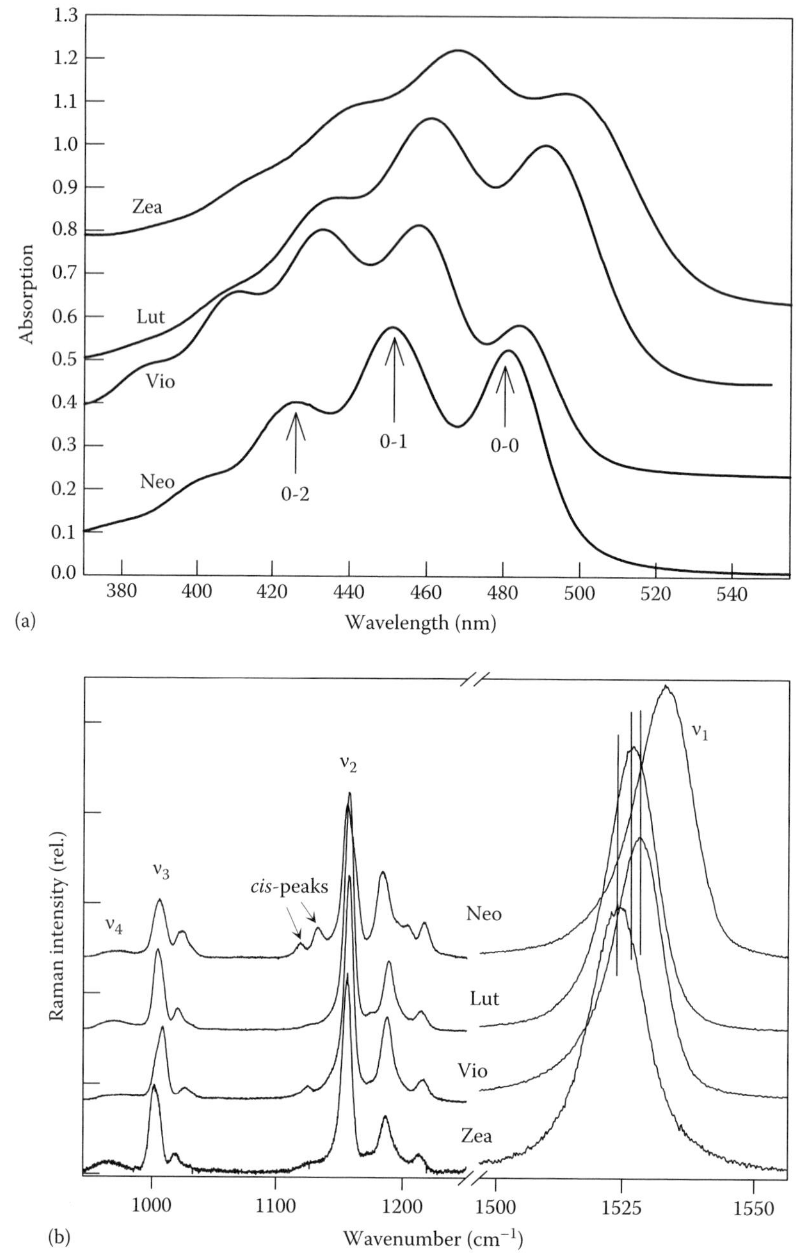

FIGURE 7.4 Absorption (a) and resonance Raman (b) spectra of the four major xanthophylls of LHCII antenna: zeaxanthin (Zea), lutein (Lut), violaxanthin (Vio), and neoxanthin (Neo).

Resonance Raman spectra of all four LHCII xanthophylls reveal differences in the ν_1 frequencies, which normally depends upon the conjugation number (Heyde et al., 1971; Rimai et al., 1973). In addition, the neoxanthin transition is further upshifted reflecting the *cis*-conformation. The ν_1 region of this xanthophyll possesses additional bands at 1120, 1132, and 1203 cm^{-1} characteristic for the 9-*cis* configuration (Hu et al., 1997). The ν_3 band frequency also differs in these xanthophylls. Finally, ν_4 is small and featureless in all isolated pigments.

Taking into consideration that antenna xanthophylls not only possess original absorption but also resonance Raman spectra, and the fact that the Raman signal is virtually free from vibrational spectroscopy artifacts (water, sample condition, etc.), it seemed of obvious advantage to apply the described combination of spectroscopies for the identification of these pigments.

7.5 IDENTIFICATION OF XANTHOPHYLLS ASSOCIATED WITH THE TRANSMEMBRANE HELIXES OF LHCII ANTENNA COMPLEX: NEOXANTHIN AND LUTEIN

Neoxanthin and the two lutein molecules have close associations with three transmembrane helixes, A, B, and C, forming three chlorophyll–xanthophyll–protein domains (Figure 7.5). Considering the structure of LHCII complex in terms of domains is useful for understanding how the antenna system works, and the functions of the different xanthophylls. Biochemical evidence suggests that these xanthophylls have a much stronger affinity of binding to LHCII in comparison to violaxanthin

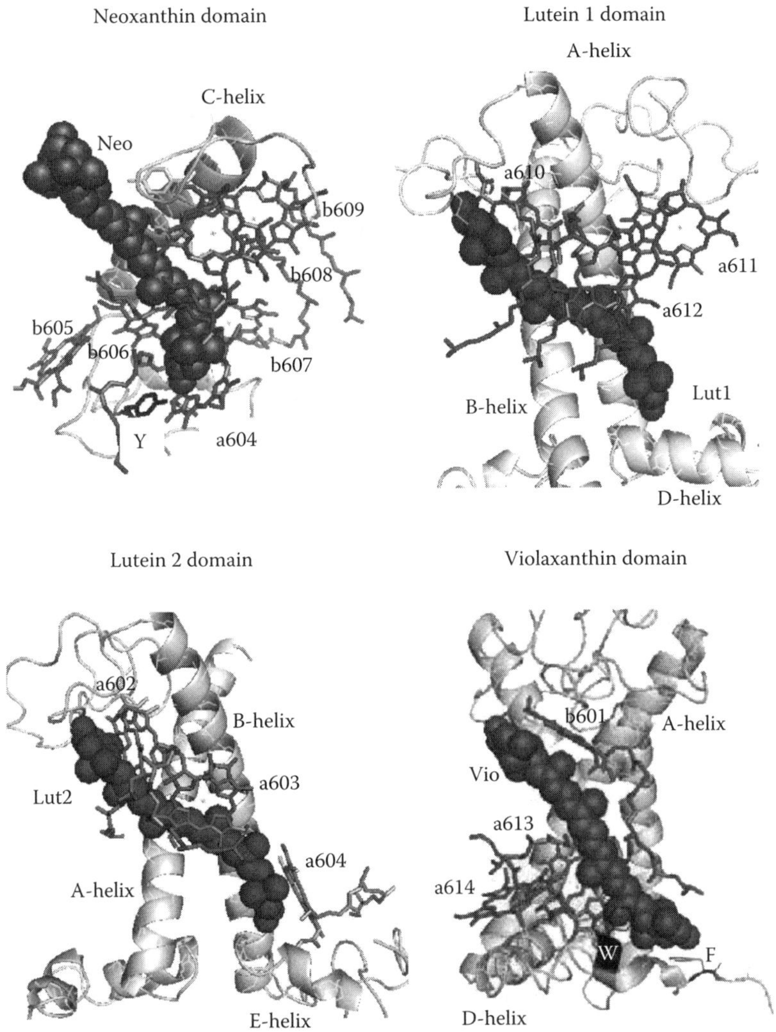

FIGURE 7.5 Structural domains of LHCII xanthophylls. Aromatic amino acids tyrosine in the neoxanthin domain and tryptophan and phenylalanine in the violaxanthin domain are labeled as Y, W, and F, respectively.

and zeaxanthin (see following paragraphs about the xanthophyll cycle carotenoids) (Ruban et al., 1999). It was relatively easy to prepare LHCII lacking the violaxanthin and zeaxanthin molecules using detergents and several steps of purification. This approach has allowed simplifying the task of spectroscopic identification of the remaining neoxanthin and lutein molecules.

7.5.1 Identification of Neoxanthin: The 9-*cis* Requirement for a Xanthophyll in the C-Helix Domain

Figure 7.6a (bottom panel) displays a low-temperature absorption spectrum of LHCII trimer in a Soret region with the second derivative, revealing some spectral details of a fine structure. Six distinct bands revealed by the derivative analysis could belong to neoxanthin and lutein. The strongest transition at 476 nm belongs, at least partially, to the absorption of chlorophyll *b* cluster of 6 pigments. The top part of the figure shows the ν_1 frequency dependence upon the excitation wavelength. Eight excitation lines have been used here to induce the resonance Raman scattering. In fact, this spectrum is a ν_1 frequency resonance Raman excitation spectrum. Normally, for isolated pigments (dashed lines on the Figure 7.6a, top panel), ν_1 is weakly dependent upon the excitation wavelength. For LHCII trimer, however, a strong wavelength dependency is revealed. The highest frequency of ν_1 was obtained for 488.0 and 457.9 nm excitations—wavelengths close to the two bands at 485 and 457 nm bands in the fine structure of absorption spectrum. Since neoxanthin has the highest ν_1 frequency of all the LHCII xanthophylls, the measurements led to the conclusion that these maxima belong to 0-0 and 0-1 transitions of neoxanthin. Moreover, the near 28 nm spacing between them is in good agreement with that observed for xanthophylls and measured *in vitro* and *in vivo* (for review see Christensen [1999]).

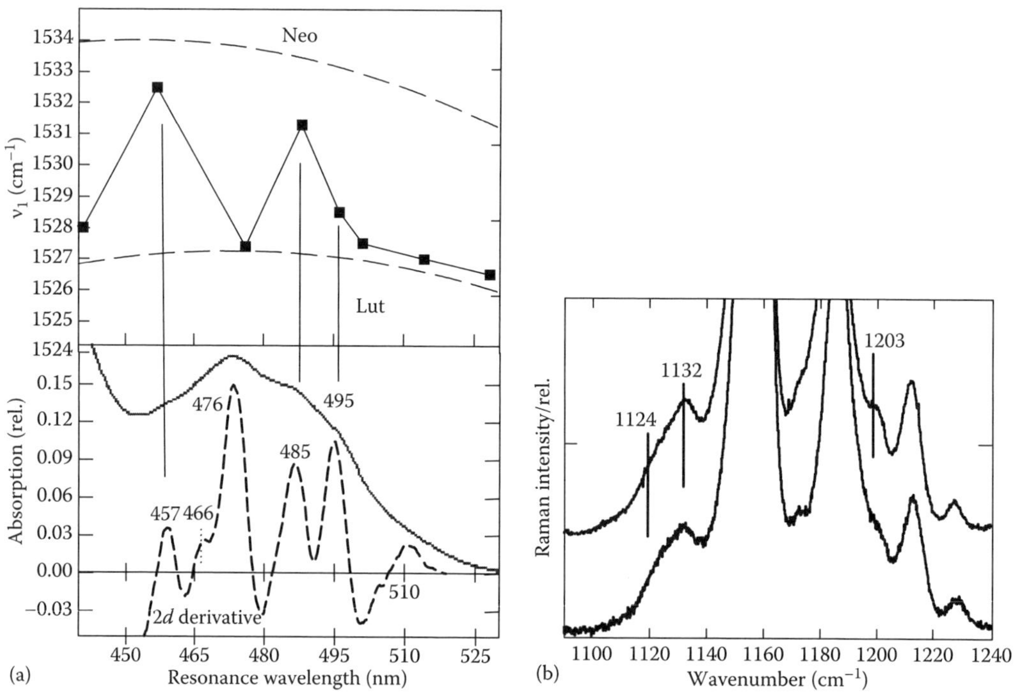

FIGURE 7.6 (a) ν_1 position dependency upon the resonance wavelength (top) and 77K absorption spectrum with the second derivative (bottom) of LHCII trimers. (b) Resonance Raman spectra in the ν_2 region with indicated 9-*cis* band positions for LHCII from spinach (top trace) and *Cuscuta reflexa* (bottom trace).

Resonance Raman spectroscopy has also revealed the 9-*cis* fingerprint features at 1124, 1132, and 1203 cm^{-1} of neoxanthin in LHCII (Ruban et al., 2001; Snyder et al., 2004). The Raman excitation profile is very similar to that of ν_1 with the maximum peaking at 488 nm (Ruban et al., 2000, 2001). In addition, the ν_3 band position for this excitation is centered at 1006 cm^{-1}—that is also characteristic for neoxanthin, while this band for lutein is positioned at 1003 cm^{-1}. The reason why neoxanthin is in the 9-*cis* conformation remains an enigma in spite of availability of the LHCII structure. The pigment is highly exposed to the environment, protruding away from the interior of the complex (Figure 7.3). Only hydrogen bond from Tyrosine 112 and tight association with a cluster of chlorophylls, in particularly Chl *a*604 and Chl *b*606, ensures the relatively strong binding affinity of neoxanthin in LHCII (Ruban et al., 1999).

In the parasitic plant *Cuscuta reflexa* where the neoxanthin biosynthesis pathway is absent (Bungard et al., 1999), LHCII is found to carry an unusually large fraction of tightly bound violaxanthin molecules (Snyder et al., 2005). The application of resonance Raman in combination with the low-temperature absorption spectroscopy reveals that violaxanthin in *C. reflexa* is in 9-*cis* conformation. 9-*cis* bands at 1124, 1132, and 1203 cm^{-1} were all present in the spectrum of LHCII from *C. reflexa* (Figure 7.6b). The band at 485 nm is also present in the absorption spectrum. This suggests that the 9-*cis* violaxanthin is in the same environment as 9-*cis* neoxanthin in LHCII. This fact along with a strong affinity of binding of 9-*cis* violaxanthin allowed us to propose that violaxanthin is bound to the C-helix domain in *C. reflexa*'s LHCII and that the 9-*cis* structure is therefore an important feature required of the xanthophyll bound into this domain.

7.5.2 Discovery of the Two Optically Different Luteins in LHCII

The excitation of the resonance Raman scattering in LHCII trimers into the two longer wavelength bands at 495 and 510 nm (using the Argon ion laser lines at 496.5, 501.7, and 514.5 nm) produces resonance Raman spectra having the ν_1 position close to that expected of lutein, i.e., ~1527 cm^{-1}. For the excitation wavelengths around 466 and 476 nm bands (the Argon laser line at 476.5 nm), the ν_1 frequency is also found to be near to that of lutein. Therefore, it is concluded that 510, 495, 466, and at least part of 476 nm band correspond to the absorption of light by lutein molecules. Since the wavelength difference between the first two long-wavelength transitions is only 15 nm, it is highly unlikely that they originate from the same pigment, since the wavelength gap between 0-0 and 0-1 transitions is almost twice larger (see above). Therefore, it was suggested that the two luteins of LHCII have different absorption spectra (Ruban et al., 2000). For the 495 nm absorbing lutein, the suitable 0-1 transition should correspond to the 466 nm band (Figure 7.6a). For the 510 nm or long-wavelength lutein, the 0-1 should be located somewhere on the slope of 476 nm band, most likely at around 482 nm. Since the 510 nm band is almost 50% broader than the 495 nm band (Ruban et al., 2001; Palacios et al., 2003), the second derivative spectrum is expected to be of reduced amplitude and poorer resolution. Early studies using fast pump-probe absorption spectroscopy have indicated that the pigment absorbing at 510 nm is closely associated with the short-wavelength chlorophyll *a* molecules (Peterman et al., 1997; Gradinaru et al., 1998). The monomerization of the LHCII trimer led to a complete disappearance of this 510 nm band (Ruban et al., 2000) and parallel enhancement and broadening of the 495 nm transition implying the shift of the 510 nm band down to the 495 nm region. These observations allowed the assignment of the 510 nm transition to lutein 2 (Lut 621 in the 2004 structure nomenclature; Liu et al., 2004). The β-ring of this xanthophyll is involved in "sandwiching" chlorophyll *a*604 with neoxanthin (compare lumenal sides of the neoxanthin and lutein 2 domains on Figure 7.5). Lutein 2 is also facing some pigments situated on neighboring monomers in the inner site of the trimer. Figure 7.7 shows the lutein 2 ε-ring is in the van der Waals contact with Chl *a*603 of the neighboring monomer. All polar oxygen groups of this chlorophyll are positioned closely near the ring. This electronic perturbation can be a very strong effect in the easily polarizable xanthophyll molecules, and may be the major cause of the 15 nm (more than 600 cm^{-1}) redshift. This explanation would be consistent with a blueshift of the 510 nm band upon monomerization of the trimer.

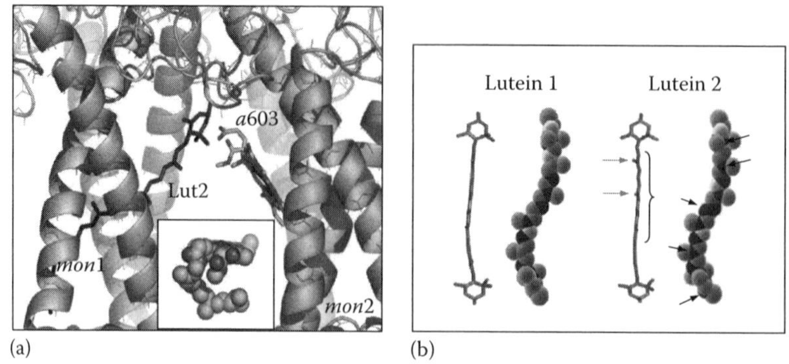

FIGURE 7.7 (a) Structure of the LHCII trimer showing lutein 2 from the monomer 1 (*mon1*) interacting with the chlorophyll *a*603 from the neighboring monomer (*mon2*). Inset displays the lutein 2–exposed side of the chlorophyll *a*603. (b) Comparison of the structures of two LHCII luteins. Arrows and black balls indicate the atoms with bonds in lutein 2, which are the most affected by distortion in addition to those of lutein 1.

7.5.3 Identification of the Chlorophyll Excitation Quencher in Aggregated LHCII

The identification studies described in Sections 7.5.1 and 7.5.2 recently played a crucial role in the search for the excitation energy quencher in LHCII. In plants, the photosynthetic apparatus responds to a harmful excess of sunlight by changing the conformation of the PSII light harvesting antenna leading to a decrease in the amount of the excitation energy funneled into the reaction center (Horton et al., 1996). This down regulation is achieved by creating a new, energy-dissipative channel in the antenna, which becomes competitive with the transfer of energy toward the PSII reaction center. This channel can be easily monitored by measurements of the chlorophyll fluorescence from antenna. A parameter known as the nonphotochemical chlorophyll *a* fluorescence quenching (NPQ), as opposed to the fluorescence quenching caused by reaction centers and called photochemical quenching (qP), can be simply derived from the measurements.

The nature of NPQ-associated alterations in LHCII, as well as the physical mechanism of quenching, has been a key focus of photosynthesis research for a number of decades. In the early 1990s, the group of Horton and coworkers put forward the LHCII aggregation model to explain the mechanism of NPQ (Horton et al., 1991, 2005). According to this model acidification of LHCII amino acid residues, resulting from the establishment of the transmembrane proton gradient, leads to the induction of a conformational change in this complex and promotion of protein–protein interactions (aggregation). Indeed, isolated aggregated LHCII has been shown to possess a very low fluorescence yield and a short excited state lifetime in comparison to the trimeric or monomeric complex (Mullineaux et al., 1992; Ruban and Horton, 1992).

Recently, pump-probe femtosecond transient absorption spectroscopy has been employed in order to search for a possible cause of the decrease in the excited state lifetime (Ruban et al., 2007). The use of diode array detection allowed us to record the spectral evolution of changes following energy equilibrium, transfer, and dissipation in LHCII. It was found that in the aggregated complex the dramatic reduction in the chlorophyll excited state lifetime is caused by a new energy transfer path to one of the xanthophylls absorbing in 490–495 nm region (Ruban et al., 2007). Since the lutein 1 absorption is found to be consistent with 495 nm, as described above, this finding implied that this xanthophyll is likely to be the quencher of the chlorophyll *a* excited states in aggregated LHCII. Lutein 1 is located near the three chlorophyll *a* molecules, Chl *a*610, *a*611, and *a*612 (Figure 7.5), which together form the terminal emitter cluster, possessing the highest exciton density in LHCII (van Grondelle and Novoderezhkin, 2006). Therefore, Lutein 1 is ideally situated for the quenching of excitation normally localized on this cluster of pigments for a period of more than 4 ns.

It remains unclear how lutein 1 becomes engaged as a quencher in aggregated LHCII. Recently, structural studies performed on crystals of this complex in various states of quenching led to the suggestion that this xanthophyll molecule can move toward the terminal emitter pigments due to structural alterations in the D-helix and its environment (Yan et al., 2007). Such a movement would alter not only the interpigment distances, but also the mutual orientation and conformation of the pigments in lutein 1 locus—these factors could be important in creating an efficient and reversible energy trap.

7.6 DISTINGUISHING CONFIGURATIONAL VARIATIONS IN XANTHOPHYLLS

The resonance Raman spectra are very rich in information. They carry not only a fingerprint of a type of carotenoid and its conformation, but also the information about molecular distortion. Even though the geometric changes are relatively small, resonance Raman can be very useful for the identification and the probing properties of the xanthophyll binding loci.

7.6.1 Lutein 2 Twisting Configuration in Trimeric LHCII

The 510 nm absorbing species in LHCII has previously revealed a strong negative CD band, implying that the molecule could be in a deformed configuration resulting from an interaction from its environment. In addition, Stark measurements showed that this species possesses a very large dipole moment of more than 14 D (Palacios et al., 2003). This is also an indicator of very specific surroundings exerting a polarization effect. Resonance Raman was yet another approach to explore the configuration of this molecule (Ruban et al., 2000). Measurements of the resonance Raman spectra excited at 501.7 as well as 514.5 nm revealed four clearly defined and pronounced bands in the ν_4 region (Ruban et al., 2000, 2001). They are fingerprints of a twisted carotenoid configuration, which can be completely abolished by the monomerization of trimers (Figure 7.8).

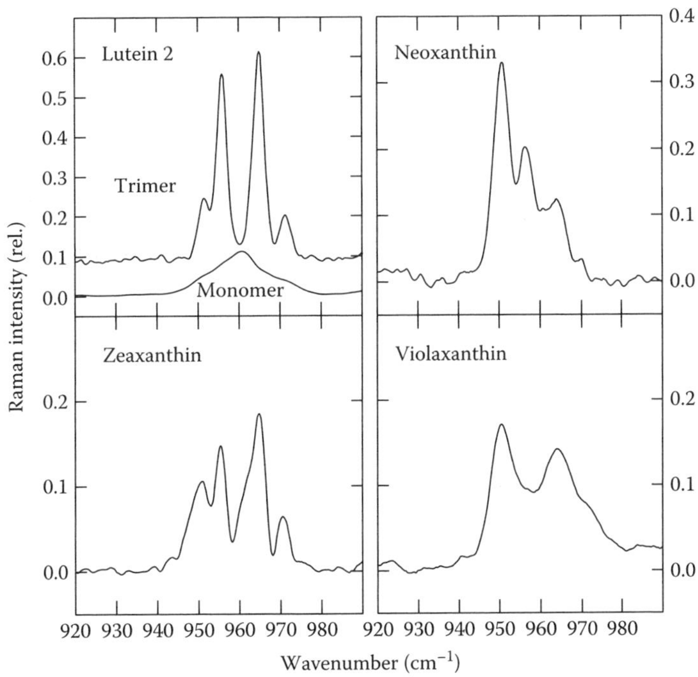

FIGURE 7.8 ν_4 resonance Raman spectra of all four LHCII xanthophylls.

The same bands were resolved in the resonance Raman spectra for the PSII membranes (Ruban et al., unpublished). Therefore, this method, for example, can be used to assess whether the LHCII trimers are intact *in vivo* at various physiological conditions.

Various spectroscopic approaches applied to the 510 nm transition indicate an unusual environment for the redshifted lutein (Figures 7.5 and 7.7a). Interaction with the Chl *a*603 could force lutein 2 molecule to adopt a twisted configuration. In addition, strong interaction with a number of aromatic residues, in particular tryptophan and phenylalanine, which possess relatively large surface areas, could further promote this distortion. It is reasonable to assume that the energy required to produce this distortion comes from the forces involved in the stabilization of LHCII trimers.

Recently, a detailed structural analysis of the luteins in the LHCII has provided further evidence to support our proposal that lutein 2 is in a twisted configuration (Yan et al., 2007). This xanthophyll appears to be more distorted along the carbon backbone than lutein 1 (Figure 7.7b). The fact that the distortion can be seen at a 2.72 Å resolution suggests its relatively large magnitude. The calculation of the local energy strain profiles reveals that lutein 2 contains more atoms with bonds affected by distortion than lutein 1 (Figure 7.7b).

7.6.2 Neoxanthin Distortion upon Aggregation and Crystallization of LHCII and *In Vivo*

Early resonance Raman experiments on trimeric and aggregated LHCII revealed small but specific differences in their spectra (Ruban et al., 1995). It has been noticed that the ν_4 region is the only one, which is affected by aggregation. A major band at 950 cm^{-1} accompanied by a group of minor upshifted bands appears in the spectrum of aggregated LHCII (Figure 7.8). The difference between the spectra of aggregates and the trimers reveals that the shape of the aggregation-associated change is different from the twisting fingerprint of lutein 2, described above. There, the 950 cm^{-1} band is strongly dominated by 955 and 965 cm^{-1} peaks (Figure 7.8). The variation in the excitation resonance Raman profiles of the amplitude of the 950 cm^{-1} band have revealed remarkable similarity to the ν_1 frequency dependence for neoxanthin in trimeric LHCII (Ruban et al., 2000). This was a clear indication that the enhancement of the 950 cm^{-1} transition originates from the twisted configuration of neoxanthin.

The amplitude of the neoxanthin distortion band is always in good correlation with the extent of the chlorophyll *a* fluorescence quenching in aggregates of LHCII. Figure 7.9 shows a series of resonance Raman spectra of the ν_4 region for neoxanthin of aggregated LHCII with differing extents of fluorescence quenching, calculated as the nonradiative constant, k_D. The amplitude of the 950 cm^{-1} band increases with the enhancement in the amount of nonradiative energy dissipation. This relationship is highly nonlinear (Ruban et al., 2007). This is likely due to a gradual increase in the interacting pigment domain size resulting from the aggregation process, which also causes the change in effectiveness of the quencher (Barzda et al., 2001).

Neoxanthin seems to be bound only by the *cis*-end within the LHCII complex, the opposite allene end of the molecule protruding into the environment (Figure 7.3). Therefore, in aggregates, neoxanthin may be easily affected by the close proximity of neighboring proteins exerting distortion forces. On the other hand, if the intrinsic conformational change takes place within the LHCII monomer it may cause a strain upon the thoroughly embedded 9-*cis* side. A critical experiment to test which of these two possibilities takes place was designed. LHCII crystals used for structural analysis were subjected to the rigorous photophysical investigation. The 77K fluorescence spectra and fluorescence lifetime analysis reveal that the complex possesses characteristics of the quenched antenna state F700 band and a decreased lifetime (Pascal et al., 2005). Therefore, it was concluded that the known LHCII structure must correspond to the dissipative or photoprotective antenna state. Remarkably, the crystals possessed a very pronounced neoxanthin twisting Raman fingerprint.

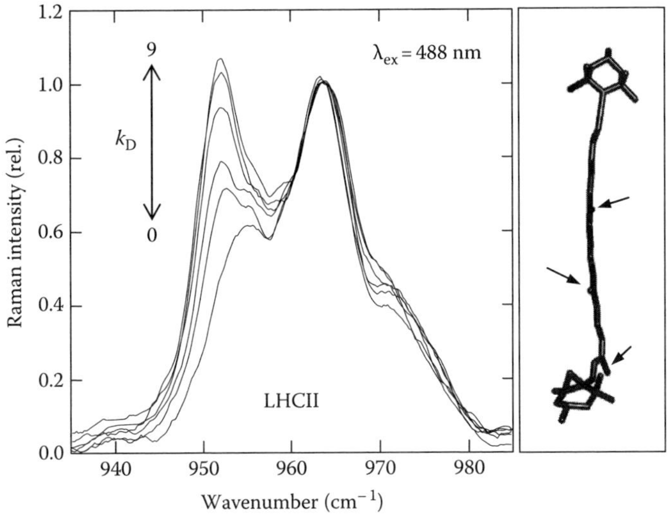

FIGURE 7.9 ν_4 resonance Raman spectra for neoxanthin in LHCII in different quenching states. The variation in the extent of quenching is illustrated by the arrow indicating variation of the nonradiative constant from 0 in trimers to 9 in highly aggregated complexes. Structure of neoxanthin is displayed on the right with arrows pointing toward the most distorted areas in the backbone of the molecule.

A close analysis of the trimers order in the crystal revealed that the exposed part of neoxanthin molecule is completely free from interactions with any protein or pigment components (Pascal et al., 2005). In addition, an examination of the neoxanthin configuration, taken from the structure of LHCII, points toward strong distortion of the *cis*-end of the molecule (Figure 7.9). This fact suggests that the twist most likely occurs within the protein interior, implying that some movement in the LHCII monomer must take place during the transition into dissipative state. Apparently, this movement affects not only lutein 1, as previously discussed, but also neoxanthin.

It has been important to determine if the neoxanthin distortion signature could be detected during the nonphotochemical quenching *in vivo*. Resonance Raman measurements on leaves and chloroplasts of various *Arabidopsis* mutants have revealed a small increase in the 950 cm^{-1} region. The relationship between the amplitude of this transition and the amount of NPQ suggests that the LHCII aggregation may be the sole cause of the protective chlorophyll fluorescence quenching *in vivo* (Ruban et al., 2007).

7.7 IDENTIFICATION OF PERIPHERAL XANTHOPHYLLS: THE XANTHOPHYLL CYCLE

The fourth binding site in LHCII structure is occupied by violaxanthin—the most polar xanthophyll of the xanthophyll cycle (Figures 7.3 and 7.5). The question of whether zeaxanthin formed upon the deepoxidation of violaxanthin is bound differently or remains in the structure at all is a controversial subject (Ruban et al., 1999; Verhoeven et al., 1999; Morosinotto et al., 2002). It is likely that the low affinity of violaxanthin/zeaxanthin binding to LHCII in the presence of detergent is responsible for these discrepancies. The fact that LHCII trimers can be prepared with one zeaxanthin per monomer using gentle solubilization procedures suggests that this xanthophyll must be a normal structural component of the antenna complex (Ruban et al., 1999; Johnson et al., 2007). The manner by which the lumen-associated deepoxidase accesses the stroma-facing epoxy group of violaxanthin also remains controversial.

7.7.1 Principles of Identification of the Xanthophyll Cycle Carotenoids

The presence of violaxanthin brings additional complexity to the absorption spectrum of both thylakoid membranes and just LHCII. Therefore, instead of trying to use ambiguous deconvolution approaches, we have developed an identification approach based on differential spectral analysis. Figure 7.10a displays the absorption spectra of thylakoid membranes measured before and after the conversion of nearly 80% of violaxanthin into zeaxanthin. The ability to record spectra at liquid helium temperature ensured the highest possible resolution of spectral structure. The deepoxidized-minus-epoxidized difference spectrum is similar to a three-maxima/minima component spectrum of zeaxanthin-minus-violaxanthin in solvents. The absorption spectrum of zeaxanthin is redshifted relatively to that of violaxanthin. This is due principally to the presence of 11 conjugated double bonds in zeaxanthin relative to only nine in violaxanthin. As is evident from the difference spectrum, the 0-0 maxima positions of zeaxanthin and violaxanthin are localized around 510 and 488 nm, respectively (Figure 7.10a). Therefore, the resonance Raman signals from these xanthophylls can be separated by selective excitation using the argon lines at 514.5 and 488.0 nm. The measurements of the Raman excitation profiles of the ν_1 band intensity before and after deepoxidation confirm the choice of these lines (see the inset in the bottom panel of Figure 7.10b). The largest difference between the two spectra was observed for the regions characteristic of violaxanthin and zeaxanthin absorption.

In order to obtain nearly absolute purity of the spectra of these xanthophylls, it was necessary to calculate the difference Raman spectra. Therefore, for zeaxanthin, two spectra of samples, one containing violaxanthin and the other enriched in zeaxanthin, were measured at 514.5 nm excitation. After their normalization using chlorophyll *a* bands at 1354 or 1389 cm^{-1}, a deepoxidized-minus-epoxidized difference spectrum has for the first time been calculated to produce a pure resonance Raman spectrum of zeaxanthin *in vivo* (Figure 7.10b). A similar procedure was used for the calculation of the pure spectrum for violaxanthin. The only difference is that the 488.0 nm excitation wavelength and epoxidized-minus-deepoxidized order of spectra have been applied in the calculation. The spectra produced using this approach have remarkable similarity to the spectra of xanthophyll cycle carotenoids in pure solvents (Ruban et al., 2001). The ν_1 peaks of violaxanthin and zeaxanthin spectra are 7 cm^{-1} apart and in correspondence to the maxima of this band for isolated zeaxanthin and violaxanthin, respectively. The ν_3 band for zeaxanthin is positioned at 1003 cm^{-1}, while the one for violaxanthin is upshifted toward 1006 cm^{-1}.

7.7.2 Fingerprints of Interaction of the Peripheral Xanthophylls with Antenna Proteins

The ν_4 region in the resonance Raman spectra for violaxanthin and zeaxanthin *in vivo* reveals a significant enhancement with the appearance of a number of bands (Figure 7.10b). The expanded ν_4 regions in these spectra are shown in the Figure 7.8. The spectrum for zeaxanthin is richer in structure than that for violaxanthin. Zeaxanthin has a Raman spectrum with six bands and clearly defined shoulders. The Raman spectrum of ν_4 for violaxanthin shows only two bands, at 950 and 965 nm, with a few minor shoulders. The pronounced ν_4 structure is indicative of the xanthophyll distortion in the binding pocket (Figure 7.5). Although the binding site of zeaxanthin in the LHCII has not been revealed, the complexity of the Raman spectra of membranes as well as isolated antenna complexes show that this xanthophyll is in association with LHCII (Ruban et al., 1999, 2002a; Johnson et al., 2007). The solubilization of PSII membranes with detergents and the use of somewhat higher detergent concentrations in the LHCII incubation medium cause a decrease in the ν_4 amplitude and disappearance of its structural features. Under these conditions, zeaxanthin becomes largely dissociated from the antenna and it is found to migrate in the free pigment band on the sucrose gradient (Ruban et al., 1999). Taken together these data indicate that the *in vivo* molecular conformation of xanthophyll cycle carotenoids relies upon the oligomeric organization

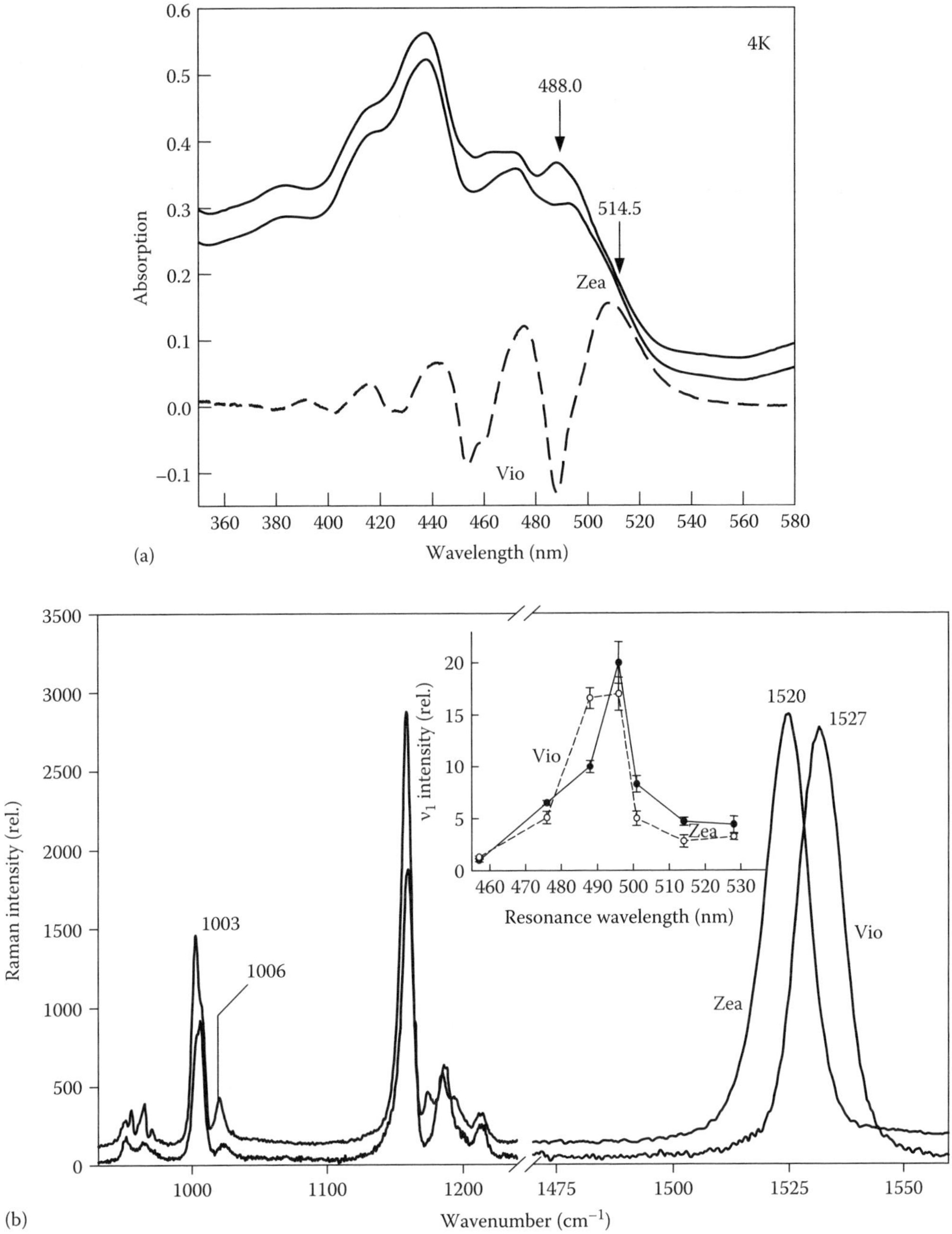

FIGURE 7.10 (a) 4K absorption spectra of thylakoid membranes containing violaxanthin (upper curve) and enriched in zeaxanthin after 80% of deepoxidation of violaxanthin (lower curve) and the deepoxidized-minus-epoxidized difference spectrum (dashed line). Zea and Vio indicate 0-0 absorption maxima of zeaxanthin and violaxanthin on the absorption difference spectrum. Arrows indicate spectral positions of the laser lines used to obtain resonance Raman spectra. (b) Calculated resonance Raman spectra of *in vivo* violaxanthin (bottom curve) and zeaxanthin (top curve). Inset: ν_1 Raman intensity dependence upon the resonance wavelength for thylakoid membranes before (Vio) and after (Zea) violaxanthin deepoxidation.

of the antenna (Ruban et al., 2002a). It is interesting to note, the ν_4 region for xanthophyll cycle carotenoids bound to the minor antenna complexes, CP26 and CP29, reveals little structure despite the fact that they remain bound in these complexes under higher detergent conditions (Ruban et al., 2002a). Therefore it is feasible to assume that the ν_4 fingerprint reflects binding of the xanthophyll within a specific site—any dislocation from this site can cause structural relaxation of the molecule without necessarily inducing its detachment from the protein.

7.8 IDENTIFICATION OF ACTIVATED ZEAXANTHIN IN THE PHOTOPROTECTIVE STATE OF ANTENNA

NPQ described in the Section 7.5.3 is always accompanied by the appearance of a small absorption change at around 535 nm (Deamer et al., 1967; Heber, 1969). The amplitude of this band correlates linearly with the nonradiative energy dissipation parameter, k_D, which was described earlier (Ruban et al., 1993). The 535 nm band has been frequently attributed to selective light scattering, which appears upon the establishment of the proton gradient across the thylakoid membrane (Heber, 1969). The maximum position of this band has now been found to depend upon the presence of zeaxanthin (Noctor et al., 1993). The absorption peaks near 535 nm in the membranes or leaves with zeaxanthin, and is blueshifted toward 525 nm without zeaxanthin. In addition, the amplitude of the 535 nm band was discovered to be in a good correlation with the amount of zeaxanthin (Ruban et al., 1993b).

These observations implicating the role of zeaxanthin in the formation of the 535 nm band have prompted us to test the nature of this absorption feature using the resonance Raman excitation near its maximum (argon line at 528.7 nm). Figure 7.11a presents the ν_1 resonance Raman spectral

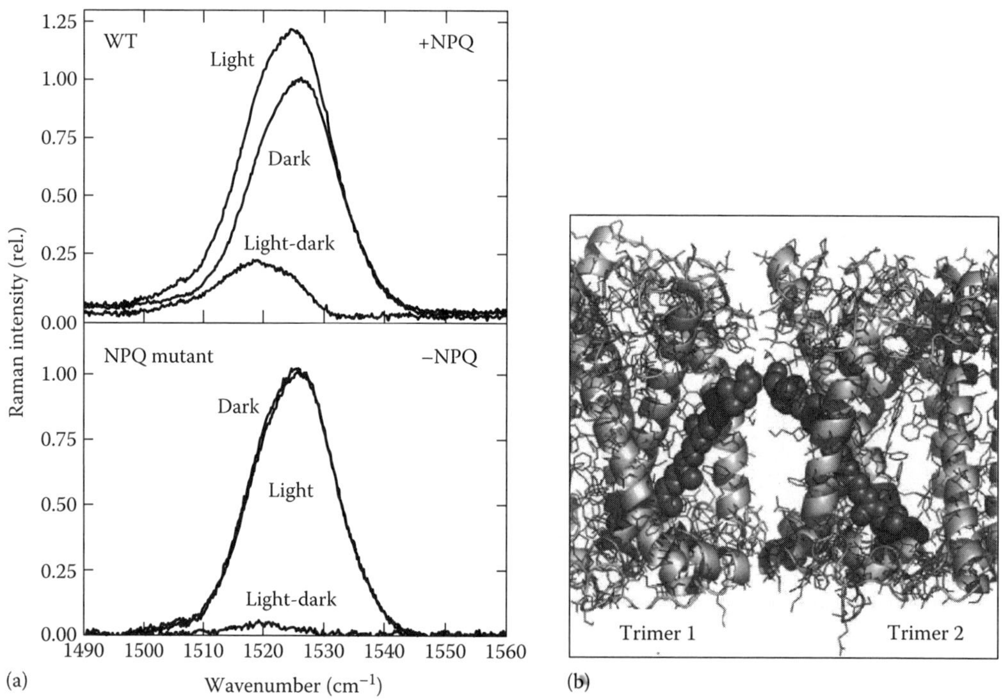

FIGURE 7.11 Identification of redshifted zeaxanthin associated with nonphotochemical chlorophyll fluorescence quenching. (a) Resonance Raman spectra in the ν_1 region for the wild type (+NPQ) and NPQ4 mutant (−NPQ) chloroplasts. Light: 15 min illumination with 1000 $\mu M \cdot m^{-2} \cdot s^{-1}$ light. Dark: 10 min recovery after the illumination. (b) Structure of two interacting LHCII trimers displaying a possible interaction giving rise to the formation of the J-type xanthophyll aggregate.

region for *Arabidopsis* plants kept in a high-light environment to induce the maximum formation of zeaxanthin and NPQ (*light*). The control was plants placed in the dark for 10 min after illumination (*dark*). The corresponding resonance Raman spectra for these two states displays a clear difference. The Raman spectrum of leaves from the light environment and therefore with both, zeaxanthin and NPQ, revealed a relative increase in the intensity and a small upshift of the 535 nm band in comparison to the spectrum measured on leaves possessing zeaxanthin but no NPQ (Figure 7.11). The light-minus-dark difference spectrum shows a ν_1 maximum blueshifted toward 1520 cm^{-1}, which is near the fingerprint frequency of zeaxanthin (Figure 7.4). Remarkably, the difference spectra of the NPQ4 mutant, which lacks the large part of NPQ, is almost nonexistent, (Figure 7.11a). These observations produced the first evidence that the 535 nm band belongs to zeaxanthin. Estimations based on the comparison of absorption and resonance Raman changes associated with NPQ have allowed us to conclude that only about two zeaxanthin molecules per PSII are involved in the formation of 535 nm band (Ruban et al., 2002b). Such a strong redshift of the absorption spectrum is explained by the formation of J-type dimers of zeaxanthin, which have a 0-0 band in 530–535 nm region (Figure 7.2). In the NPQ-associated resonance Raman spectrum, the ν_1 amplitude becomes negative for excitation wavelength below 500 nm (Ruban et al., 2002b). This observation suggests that 535 nm zeaxanthin band has been formed from some short-wavelength forms of this pigment, absorbing at 500–510 nm.

Several models can be suggested to explain the mechanism of zeaxanthin dimer formation in NPQ. One is based on the assumption that after deepoxidation, zeaxanthin remains bound to the same domain as violaxanthin. The aggregation of LHCII could force interactions between stroma-facing zeaxanthin molecules situated on the two interacting trimers (Figure 7.11b) producing J-type associates with 535 nm absorption. The latter serve as a good indicator for the conformational antenna alterations leading to NPQ. As to whether the J-type aggregate can play a direct role in the chlorophyll fluorescence quenching remains to be investigated. The fact that the 535 nm absorbing zeaxanthin displays a typical resonance Raman spectrum for nonradical all-*trans* carotenoid (Ruban et al., 2002b) suggests that this xanthophyll cannot be involved in the radical-type quenching proposed for NPQ by Holt and coauthors (Holt at al., 2005).

The other model explaining the origin of 535 nm absorbing zeaxanthin involves PSII subunit S, PsbS protein, which controls the dynamic range of NPQ by sensing the proton gradient and organizing the PSII antenna (Horton and Ruban, 2000; Li et al., 2000; Kiss et al., 2007). Isolated PsbS was found to bind zeaxanthin and shift its 0-0 maximum toward 523–536 nm region (Aspinal et al., 2002). The ν_4 in the Raman spectrum of PsbS-bound zeaxanthin possesses a similar structure to that of 535 nm absorbing zeaxanthin identified in NPQ. Circular dichroism measurements revealed the formation of a J-type dimer. The absorption of aromatic residues of the protein, mainly phenylalanine, was also strongly redshifted (Aspinal et al., 2002). This confirms the binding of zeaxanthin to PsbS. Nevertheless, the question of whether or not the zeaxanthin binds to PsbS *in vivo* during NPQ still remains controversial (Bonente et al., 2007). Alternatively, it is possible that the 535 nm signal arises from a heterogenic interaction between a PsbS-bound zeaxanthin and a LHCII-bound zeaxanthin.

7.9 MOLECULAR ORIGINS OF THE RESONANCE RAMAN TWISTING MODES OF ANTENNA XANTHOPHYLLS

The ν_4 region enhancement and structure in the resonance Raman spectra of xanthophylls reviewed in this chapter shows that it can be used for the analysis of carotenoid–protein interactions. Figure 7.8 summarizes the spectra for all four major types of LHCII xanthophylls. Lutein 2 possesses the most intense and well-resolved ν_4 bands. The spectrum for zeaxanthin is very similar to that of lutein with a slightly more complex structure. This similarity correlates with the structural similarity between these pigments. It is likely that they are both similarly distorted. The richer structure of zeaxanthin spectrum may be explained by the presence of the two flexible β-end rings with electronically conjugated π-electrons as opposed to only one in lutein (Hashimoto et al., 2001; Young et al., 2002). The rotation of these groups can increase the probability of C–H bending modes coupling to the π-electron states.

The ν_4 structure of neoxanthin is dominated by the 950 cm^{-1} band, and is similar to that of the violaxanthin. This similarity can arise from restriction on end-group rotation in these xanthophylls. Since the 950 cm^{-1} transition is present in all four types of xanthophylls it is reasonable to propose that it originates from the C–H groups situated closer to the center of the molecule, where these groups have an identical atomic environment. Indeed, a normal coordinate analysis of the β-carotene structure has shown that the band around 950 cm^{-1} belongs to the out-of-plane C–H wagging vibrations at C15=C15′ atoms positioned in the middle of the molecule (Saito and Tasumi, 1983) (for the nomenclature of atoms see Figure 7.1). The 955 and 965 cm^{-1} bands have been assigned to C7=C8 and/or C11=C12 groups. The possibility of a compositional effect on ν_4 structure, like the presence of allene and 9-*cis* conformation in violaxanthin, has been excluded by the measurements of FTIR spectra of dry xanthophylls. It was found that these compositional and structural differences do not affect the total number of C–H transitions if taken together with the complimentary Raman modes. Therefore, the differences in the ν_4 spectra among the studied xanthophylls can be reasonably explained by the differences in the rigidity of their molecular structure. It is likely that the restriction of ring mobility by the epoxy groups of violaxanthin and neoxanthin makes the C7=C8 environment more rigid. This could reduce their distortion in the binding pockets. Therefore, the characteristic out-of-plane C–H wagging modes cannot be well coupled to the π-electron motion and does not appear in the Raman spectrum.

The flexibility of xanthophyll structure, in particular that of the end-group rotation, has been recently suggested to have a strong effect on the electronic excited state energies (Drew, 2006). The change from all-s-*cis* to s-*trans* conformation of the end-group of zeaxanthin leads to a decrease in calculated S_1 energy from 15,080 to 14,152 cm^{-1}, corresponding to ~46 nm redshift below 700 nm absorption. This energy change would make this xanthophyll an efficient chlorophyll *a* excitation quencher. Therefore, the forces causing the twisting of a xanthophyll molecule in the binding locus could provide a simple switch for creating an efficient quencher in antenna during high-light exposure.

7.10 CONCLUDING REMARKS

7.10.1 Summary

Xanthophylls bound to photosynthetic light harvesting proteins can be analyzed by a combination of absorption and resonance Raman spectroscopies. This is a nondestructive and insightful methodology, which provides information on the identity of a pigment and features of its binding including conformation, configuration, and the energy of electronic excited states. This approach was successfully applied to identify the electronic excited state energies of neoxanthin and lutein bound to the LHCII complex. Neoxanthin was found to be in 9-*cis* configuration, which is apparently due to a strict condition of the binding locus. Two molecules of lutein in the trimeric complex have been found to be spectrally different, one absorbing at 495 and the other at 510 nm. The long-wavelength lutein possesses an unusually strong dipole moment, a negative CD signal, and a well-structured resonance Raman ν_4 band. These properties indicate that the molecule is in a twisted configuration. This distortion is likely to be a result of interaction with chlorophyll *a*603, localized on the neighboring monomer of the same LHCII trimer. Structural analysis confirmed a greater distortion of lutein 2 molecule as compared to lutein 1 in the trimer. The redshifted lutein increases the spectral cross section of LHCII—a good example of spectral tuning of the PSII light harvesting capacity by protein oligomerization.

The assignment of the lutein absorbing at 495 nm as lutein 1 has helped with the identification of an excitation energy quencher in LHCII, when the complex is in aggregated form.

Xanthophyll distortion revealed by Raman spectroscopy may not only be an important binding fingerprint but may also be a key functional feature, associated with the altered photophysical behavior and lead to enhancement of excitation energy exchange with chlorophyll.

The fourth xanthophyll of LHCII, violaxanthin and its deepoxidation product, zeaxanthin, were identified by the calculation of resonance Raman difference spectra of membranes enriched in either of these xanthophylls. Raman difference spectra of violaxanthin and zeaxanthin *in vivo* has been calculated providing important information on their binding character. Both of these pigments have been found to be associated with LHCII antenna in agreement with the biochemical evidence. This association causes distortion of these carotenoids. A long-wavelength form of zeaxanthin absorbing at 535 nm was discovered. This pigment form was identified as all-*trans* revealing a strong distortion along the carbon backbone. This minor form of the pigment appears transiently during the NPQ, and its intensity correlates well with the nonradiative decay constant. However, it is unlikely to be a direct excitation trap as previously suggested (Holt et al., 2005) since its spectrum lacks the characteristic features of a cation radical.

7.10.2 Future Directions

The analysis of carotenoid identity, conformation, and binding *in vivo* should allow further progress to be made in understanding of the functions of these pigments in the photosynthetic machinery. One of the obvious steps toward improvement could be the use of continuously tuneable laser systems in order to obtain more detailed resonance Raman excitation profiles (Sashima et al., 2000). This technique will be suitable for the investigation of *in vivo* systems with more complex carotenoid composition. In addition, this method may be applied for the determination of the energy of forbidden S_1 or 2^1Ag transition. This is an important parameter, since it allows an assessment of the energy transfer relationship between the carotenoids and chlorophylls within the antenna complex.

The use of selective isotope replacement of carbon and hydrogen atoms in the structure of xanthophylls in combination with LHCII reconstitution should greatly aid the assignment of multiple ν_4 twisting bands. This assignment would help localize the areas of distortion within the carotenoid molecule and understand the possible causes of this distortion.

Absorption and Raman analysis of LHCII complexes from xanthophyll biosynthesis mutants and plants containing unusual carotenoids (e.g., lactucoxanthin and lutein-epoxide) should also be interesting, since the role of these pigments and their binding properties are unknown. Understanding the specificity of binding can help to understand the reasons for xanthophyll variety in photosynthetic antennae and aid in the discovery of yet unknown functions for these molecules.

Finally, it would be interesting to extend the described spectroscopic approaches to the investigation of xanthophylls bound to antennae of other photosynthetic organisms, including various algae. Xanthophylls such as fucoxanthin, diadinoxanthin, diatoxanthin, and peridinin will be fascinating pigments to study.

REFERENCES

Aspinall-O'Dea, M., Wentworth, M., Pascal, A., Robert, B., Ruban, A.V., and Horton, P. 2002. The PsbS subunit of photosystem II binds zeaxanthin and activates it for non-photochemical fluorescence quenching. *Proc. Natl. Acad. Sci. USA* 99: 16331–16335.

Barzda, V., Gulbinas, V., Kananavicius, R., Cervinskas, V., van Amerongen, H., van Grondelle, R., and Valkunas, L. 2001. Singlet-singlet annihilation kinetics in aggregates and trimers of LHCII. *Biophys. J.* 80: 2409–2421.

Bassi, R., Pineau, B., Dainese, P., and Marquardt, J. 1993. Carotenoid-binding proteins of photosystem II. *Eur. J. Biochem.* 212: 297–303.

Bayliss, N.S. 1950. The effect of the electrostatic polarization of the solvent on the electronic absorption spectra in solution. *J. Chem. Phys.* 18: 292–296.

BCC research. 2007. The global market on carotenoids. http://www.bccresearch.com.

Bonente, G., Howes, B.D., Caffarri, S., Smulevich, G., and Bassi, R. 2008. Interactions between the photosystem II subunit PsbS and xanthophylls studied *in vivo* and *in vitro*. *J. Biol. Chem.*, Mar., 283: 8434–8445; doi:10.1074/jbc.M708291200.

Britton, G. 1995. Structure and properties of carotenoids in relation to function. *FASEB J.* 9: 1551–1558.

Bungard, R.A., Ruban, A.V., Hibberd, J.M., Press, M.C., Scholes, J.D., and Horton, P. 1999. Novel carotenoid composition and a new type of xanthophyll cycle in plants. *Proc. Natl. Acad. Sci USA* 96: 1135–1139.

Christensen, R. 1999. The electronic states of carotenoids, in *The Photochemistry of Carotenoids*, eds. H.A. Frank, A.J. Young, and G. Britton, pp. 137–159. Dordrecht, the Netherlands: Kluwer Academic Press.

Clayton, R. 1980. *Photosynthesis: Physical Mechanisms and Chemical Patterns*. Cambridge, U.K.: Cambridge University Press.

Deamer, D.W., Crofts, A.R., and Packer, L. 1967. Mechanisms of light induced structural changes in chloroplasts. *Biochim. Biophys. Acta* 131: 81–86.

Dekker, J.P. and Boekema, E.J. 2005. Supramolecular organization of thylakoid membrane proteins in green plants. *Biochim. Biophys. Acta* 1706: 12–39.

Del Campo, García-González, M., and Guerrero, M.G. 2007. Outdoor cultivation of microalgae for carotenoid production: current state and perspectives. *Appl. Microbiol. Biotechnol.* 74: 1163–1174.

Drew, A. 2006. Influence of geometry relaxation on the energies of the S_1 and S_2 states of violaxanthin, zeaxanthin and lutein. *J. Phys. Chem. A* 110: 4592–4599.

Feltl, L., Pacáková, V., Štulík, K., and Volka, K. 2005. Reliability of carotenoid analyses: A review. *Curr. Anal. Chem.* 2005: 93–102.

Garey, P.R. 1982. *Biochemical Applications of Raman and Resonance Raman Spectroscopies*. London: Academic Press Inc. Ltd.

Gradinaru, C.C., Stokkum, I.H.M., van Grondelle, R., and van Amerongen, H. 1998. Ultrafast absorption changes of the LHC-II carotenoids upon selective excitation of the chlorophylls. In *Photosynthesis: Mechanisms and Effects*, ed. G. Garab. Dordrecht, the Netherlands: Kluwer Academic Publishers.

Hashimoto, H., Yoda, T., Kobayashi, T., and Young, A.J. 2002. Molecular structures of carotenoids as predicted by MNDO-AM1 molecular orbital calculations. *J. Mol. Struct.* 604: 125–146.

Heber, U. 1969. Conformational changes in chloroplasts induced by illumination of leaves *in vivo*. *Biochim. Biophys. Acta* 180: 302–319.

Heyde, M.E., Gill, D., Kilponen, R.G., and Rimai, L. 1971. Raman spectra of Schiff bases of retinal (models of visual photoreceptors *J. Am. Chem. Soc.* 93: 6776–6780.

Holt, N.E., Zigmantas, D., Valkunas, L., Li, X.-P., Niyogi, K.K., and Fleming, G.R. 2005. Carotenoid cation formation and the regulation of photosynthetic light harvesting. *Science* 307: 433–436.

Horton P., Ruban, A.V., Rees D., A. Pascal, Noctor, G.D., and Young, A.J. 1991. Control of the light-harvesting function of chloroplast membranes by the proton concentration in the thylakoid lumen: aggregation states of the LHCII complex and the role of zeaxanthin. *FEBS Lett.* 292: 1–4.

Horton, P. and Ruban, A.V. 1994. The role of light-harvesting complex II in energy quenching. In *Photoinhibition of Photosynthesis*, eds. N.R. Baker and J.R. Bowyer, pp. 11–128. Oxford: BIOS Scientific Publishers Ltd.

Horton, P., Ruban, A.V., and Walters, R.G. 1996. Regulation of light harvesting in green plants. *Annu. Rev. Plant Physiol. Plant Mol. Biol.* 47: 655–684.

Horton, P., Ruban, A.V., and Wentworth, M. 2000. Allosteric regulation of the light harvesting system of photosystem II. *Phil. Trans. R. Soc. Lond. B* 355: 1361–1370.

Horton, P., Wentworth, M., and Ruban, A. 2005. Control of the light harvesting function of chloroplast membranes: The LHCII-aggregation model for non-photochemical quenching II. *FEBS Lett.* 579: 4201–4206.

Hu, Y., Heshimoto, H., Moie, G., Hengartner, U., and Koyama, Y. 1997. Unique properties of the 11-*cis* and 11,11'-di-*cis* isomers of β-carotene as revealed by electronic absorption, resonance Raman and 1H and 13C NMR spectroscopy and by HPLC analysis of their thermal isomerization. *J. Chem. Soc. Perkin Trans.* 2: 2699–2710.

Johnson, M.P., Havaux, M., Triantaphylidès, C., Ksas, B., Pascal, A.A., Robert, B., Davison, P.A., Ruban, A.V., and Horton, P. 2007. Elevated zeaxanthin bound to oligomeric LHCII enhances the resistance of arabidopsis to photo-oxidative stress by a lipid protective, anti-oxidant mechanism. *J. Biol. Chem.* 282: 22605–22618.

Khachik, F., Bernstein, P., and Garland, D. 1997. Identification of lutein and zeaxanthin oxidation products in human and monkey retinas. *Invest. Ophtalmol.* 38: 1802–1811.

Kiss, A., Crouchman, S., Ruban, A.V., and Horton, P. 2008. The PsbS protein controls the organisation of the photosystem II antenna in higher plant thylakoid membranes. *J. Biol. Chem.*, 283: 3972–3978.

Koyama, Y., Kito, M., Takii, T., Saili, K., Tsukida, K., and Yamashita, J. 1983. Configuration of the carotenoid in the reaction centers of photosynthetic bacteria. 2. Comparison of the resonance Raman lines of the reaction centers with those of the 14 different *cis–trans* isomers of β-carotene. *Photobiochem. Photobiophys.* 5: 139–150.

Kuhn, H. 1949. A quantum-mechanical theory of light absorption of organic dyes and similar compounds. *J. Chem. Phys.* 17: 1198–1212.

Landrum, J.T. and Bone, R.A. 2001. Lutein, zeaxanthin, and the macular pigment. *Arch. Biochem. Biophys.* 385: 28–40.

Li, X.-P., Bjorkman, O., Shih, C., Grossman, A.R., Rosenquist, M., Jansson, S., and Niyogi, K.K. 2000. A pigment-binding protein essential for regulation of photosynthetic light harvesting. *Nature* 403: 391–395.

Liu, Z.F., Yan, H.C., Wang, K.B., Kuang, T.Y., Zhang, J.P., Gui, L.L., An, X.M., and Chang, W.R. 2004. Crystal structure of spinach major light-harvesting complex at 2.72 Å resolution. *Nature* 428: 287–292.

Merlin, J.C. 1985. Resonance Raman spectroscopy of carotenoids and carotenoid-containing systems. *Pure Appl. Chem.* 57: 785–792.

Morosinotto, T., Baronio, R., and Bassi, R. 2002. Dynamics of chromophore binding to Lhc proteins *in vivo* and *in vitro* during the operation of the xanthophyll cycle. *J. Biol. Chem.* 277: 36913–36920.

Mullineaux, C.W., Pascal, A.A., Horton, P. and Holzwarth, A.R. 1992. Excitation energy quenching in aggregates of the LHCII chlorophyll-protein complex: A time-resolved fluorescence study. *Biochim. Biophys. Acta* 1141: 23–28.

Noctor, G., Ruban, A.V., and Horton, P. 1993. Interactions between the effects of potentiators and antagonists of ΔpH-dependent thermal dissipation of excitation energy in spinach thylakoids. *Biochim. Biophys. Acta* 1183: 339–344.

Palacios, M.A., Frese, R.N., Gradinaru, C.G., Premvardhan, L., Horton, P., Ruban, A.V., van Grondelle, R., and van Amerongen, H. 2003. Stark effect spectroscopy of the different oligomerisation states of light-harvesting complex II. *Biochim. Biophys. Acta* 1605: 83–95.

Pascal, A.A., Liu, Z., Broess, K., van Oort, B., van Amerongen, H., Wang, C., Horton, P., Robert, B., Chang, W., and Ruban, A. 2005. Molecular basis of photoprotection and control of photosynthetic light-harvesting. *Nature* 436: 134–137.

Peterman, E.J.G., Gradinaru, C.C., Calkoen, F., Borst, J.C., van Grondelle, R., and van Amerongen, H. 1997. The xanthophylls in light-harvesting complex II of higher plants: Light harvesting and triplet quenching. *Biochemistry* 36: 12208–12215.

Rimai, L., Heyde, M.E., and Gill, D. 1973. Vibrational spectra of some carotenoids and related linear polyenes. A Raman spectroscopic study. *J. Am. Chem. Soc.* 95: 4493–4501.

Robert, B. 1999. The electronic structure, stereochemistry and resonance Raman spectroscopy of carotenoids. In *The photochemistry of carotenoids*, eds. H.A. Frank, A.J. Young, G. Britton, and R.J. Cogdell, pp. 189–201. Dordrecht, the Netherlands: Kluwer Academic Publishers.

Robert, B., Horton, P., Pascal, A., and Ruban, A.V. 2004. Insights into the molecular dynamics of plant light-harvesting proteins *in vivo. Trends in Plant Science* 9: 385–390.

Ruban, A.V. and Horton, P. 1992. Mechanism of ΔpH-dependent dissipation of absorbed excitation energy by photosynthetic membranes. I Spectroscopic analysis of isolated light harvesting complexes. *Biochim. Biophys. Acta* 1102: 30–38.

Ruban, A.V., Horton, P., and Young, A.J. 1993a. Aggregation of higher plant xanthophylls: Differences in absorption spectra and in the dependency on solvent polarity. *J. Photochem. Photobiol.* 21: 229–234.

Ruban, A.V., Young, A., and Horton, P. 1993b. Induction of nonphotochemical energy dissipation and absorbance changes in leaves; evidence for changes in the state of the light harvesting system of photosystem II *in vivo. Plant Physiol.* 102: 741–750.

Ruban, A.V., Robert, B., and Horton, P. 1995. Resonance Raman spectroscopy of photosystem II light-harvesting complex of green plants. A comparison of trimeric and aggregated states. *Biochemistry* 34: 2333–2337.

Ruban, A.V., Lee, P.J., Wentworth, M., Young, A.J., and Horton, P. 1999. Determination of the stoichiometry and strength of binding of xanthophylls to the photosystem II light harvesting complexes. *J. Bio.l Chem.* 274: 10458–10465.

Ruban, A.V., Pascal, A., and Robert, B. 2000. Xanthophylls of the major photosynthetic light-harvesting complex of plants: Identification, conformation and dynamics. *FEBS Lett.* 477: 181–185.

Ruban, A.V., Pascal, A.A., Robert, B., and Horton, P. 2001. Configuration and dynamics of carotenoids in light-harvesting antennae of the thylakoid membrane. *J. Biol. Chem.* 276: 24862–24870.

Ruban, A.V., Pascal, A.A., Lee, P.J., Robert, B., and Horton, P. 2002a. Molecular configuration of xanthophyll cycle carotenoids in photosystem II antenna complexes. *J. Biol. Chem* 277: 42937–42942.

Ruban, A.V., Pascal, A.A., Robert, B., and Horton, P. 2002b. Activation of zeaxanthin is an obligatory event in the regulation of photosynthetic light harvesting. *J. Biol. Chem.* 277: 7785–7789.

Ruban, A.V., Berera, R., Ilioaia, C., van Stokkum, I.H.M., Kennis, J.T.M., Pascal, A.A., van Amerongen, H., Robert, B., Horton, P., and van Grondelle, R. 2007. Identification of a mechanism of photoprotective energy dissipation in higher plants. *Nature* 450: 575–578.

Saito, S. and Tasumi, M. 1983. Normal-coordinate analysis of β-carotene isomers and assignments of the Raman and infrared bands. *J. Raman Spectrosc.* 14: 310–321.

Sapozhnikov, D.I., Kransovskaya, T.A., and Maevskaya, A.N. 1957. Change in the interrelationship of the basic carotenoids of the plastids of green leaves under the action of light. *Dokl. Acad. Nauk USSR* 113: 465–467.

Sashima, T., Koyama, Y., Yamada, T., and Hashimoto, H. 2000. The $1Bu^+$, $1Bu^-$, and $2Ag^-$ energies of crystalline lycopene, β-carotene, and mini-9-β-carotene as determined by resonance-Raman excitation profiles: Dependence of the $1Bu^-$ state energy on the conjugation length. *J. Phys. Chem. B* 104: 5011–5019.

Snyder, A.M., Clark, B.M., Robert, B., Ruban, A.V., and Bungard, R.A. 2004. Carotenoid specificity of light-harvesting complex II binding sites: Occurrence of 9-*cis* violaxanthin in the neoxanthin-binding site in the parasitic angiosperm *cuscuta reflexa*. *J. Biol. Chem.* 279: 5162–5168.

Su, Q., Rowley, K.G., and Balazs, N.D.H. 2002. Carotenoids: Separation methods applicable to biological samples. *J. Chromatogr. B* 781: 393–418.

Takagi, S., Takeda, K., and Shiroishi, M. 1982. Aggregation, configuration and particle size of lutein dispersed by sodium dodecyl sulfate in various salt concentrations. *Agric. Biol. Chem.Tokyo* 46: 2217–2222.

Tsukida, K., Saiki, K., Takii, T., and Koyama, Y. 1982. Separation and determination of *cis/trans*-β-carotenes by high-performance liquid chromatography. *J. Chromatogr.* 245: 359–364.

van Grondelle, R. and Novoderezhkin, V.I. 2006. Energy transfer in photosynthesis: Experimental insights and quantitative models. *Phys. Chem. Chem. Phys.* 8: 793–807.

Verhoeven, A.S., Adams, III, W.W., Demmig-Adams, B., Croce, R., and Bassi, R. 1999. Xanthophyll cycle pigment localization and dynamics during exposure to low temperatures and light stress in low and high light-acclimated in vinca major. *Plant Physiol* 120: 1–11.

Yamamoto, H.Y., Nakayama, T.O.M., and Chichester, C.O. 1962. Studies on the light and dark interconversions of leaf xanthophylls. *Arch. Biochem. Biophys.* 97: 168–73.

Yan, H., Zhang, P., Wang, C., Liu, Z., and Chang, W. 2007. Two lutein molecules in LHCII have different conformations and functions: Insights into the molecular mechanism of thermal dissipation in plants. *Biochem. Biophys. Res. Commun.* 355: 457–463.

Young, A.J., Phillip, D.M., and Hashimoto, H. 2002. Ring-to-chain conformation may be a determining factor in the ability of xanthophylls to bind to the bulk light-harvesting complex of plants. *J. Mol. Struct.* 642: 137–145.

8 Effects of Self-Assembled Aggregation on Excited States

Tomáš Polívka

CONTENTS

8.1 INTRODUCTION

The central structural feature of all carotenoids, a linear conjugated chain, makes carotenoids highly hydrophobic molecules. Since pioneering work carried out on carotenoids more than 40 years ago (Buchwald and Jencks 1968), it has been known that this hydrophobicity promotes the formation of carotenoid aggregates when dissolved in hydrated solvents and that aggregation is characterized by dramatic changes in absorption spectra (Ruban et al. 1993, Gruszecki 1999, Simonyi et al. 2003). A number of studies carried out since the observation of astaxanthin aggregation (Buchwald and Jencks 1968) demonstrates that two types of carotenoid aggregates can be distinguished according to their absorption spectra. The first type is termed an H-aggregate and is characterized by a large blueshift of the absorption spectrum. The H-aggregate consists of molecules whose conjugated chains are oriented parallel to each other and are closely packed (the card-pack arrangement). The second aggregation type, the J-aggregate, is characterized by a redshift of the absorption spectrum, and results from a head-to-tail organization of conjugated chains (Simonyi et al. 2003).

Numerous studies of carotenoid aggregates have focused on the molecular organization of the aggregates (Simonyi et al. 2003), but little is known about aggregation-induced effects on carotenoid excited states. Classical exciton theory can qualitatively explain the aggregation-induced shifts of absorption bands (Section 8.3.1), but a detailed understanding of the parameters governing the

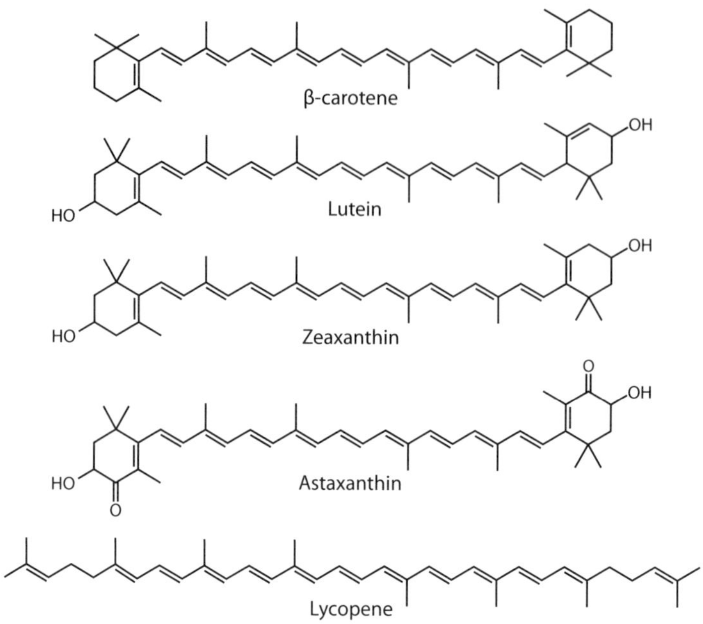

FIGURE 8.1 Molecular structures of carotenoids often used for studies of carotenoid aggregates.

aggregation (e.g., whether J- or H-aggregates are formed) and their relation to the carotenoid structure is still lacking. This is partly because aggregation studies were limited to only a few carotenoids (see Figure 8.1 for the most studied examples). Consequently, the absorption spectrum of a carotenoid aggregate cannot be reliably predicted on the basis of input parameters (carotenoid structure, solvent, water/solvent ratio, concentration, temperature, etc.). Moreover, because the majority of studies carried out so far have used steady-state (absorption and/or circular dichroism [CD]) spectroscopies (see Simonyi et al. (2003) for review), very little is known about excited-state dynamics of aggregates. Excited-state properties of carotenoids differ markedly from those of other organic dyes (Section 8.2), and a precise knowledge of excited-state properties has proven to be crucial for understanding light-driven actions of monomeric carotenoids (Polívka and Sundström 2004). Similarly, to identify the functions of carotenoid aggregates, characterizing the properties of their excited states is a crucial task. The number of natural and artificial systems, in which carotenoid aggregates have been found, is increasing and it is thus of high importance to reveal aggregation-induced effects on excited states to recognize the specific functions of carotenoid aggregates in these systems.

Apart from self-assembled aggregation in hydrated solvents, carotenoids tend to form H-aggregates when present in lipid bilayers in various biological systems, in which long-range organization of carotenoid molecules is thought to control the physical and dynamic properties of lipid membranes (Gruszecki 1999). Although the key function of carotenoids in membranes is likely protection from lipid peroxidation (Schindler and Lichtenthaler 1996), they are also found in light-sensitive environments such as the human macula (Bhosale et al. 2004). Moreover, the involvement of carotenoid aggregates in plant photoprotection has been debated for many years (Ruban et al. 1993). For example, J-aggregates of the carotenoid zeaxanthin have been suggested to be involved in chlorophyll quenching either in micelles (Avital et al. 2006) or associated with proteins (Aspinall-O'Dea et al. 2002). Carotenoid aggregates have also been identified in flower petals, where J-aggregates are almost exclusively formed. Polarization effects caused by a large-scale organization of aggregates were proposed to be an important factor in recognition of a flower by insects (Zsila et al. 2001a).

No less important are studies of the aggregation-induced effects in artificial systems with the objective of harvesting solar radiation. One of the potential applications of carotenoids is their use in solar cells based on a dye–semiconductor interface. It has been shown that carotenoid–TiO_2-based solar cells may achieve reasonable efficiency (Gao et al. 2000, Xiang et al. 2005, Wang et al. 2006). Recent studies of electron-transfer pathways between a carotenoid and TiO_2 have revealed some specific features of the carotenoid–TiO_2 interface, such as an electron recombination forming a carotenoid triplet state (Pan et al. 2002). In artificial systems, this pathway could play a role similar to its regulation function in natural systems, making the carotenoids potentially interesting materials for solar cells, especially in combination with other sensitizers. Other promising approaches are the use of carotenoids as light-harvesting chromophores, for example, carotenoid-based artificial antennas (Kodis et al. 2004, Polívka et al. 2007) or even as molecular wires (Ramachandran et al. 2003). However, to design a functional device, carotenoids are mostly deposited on surfaces where H-aggregates are often formed (Sereno et al. 1996, Gao et al. 2000, Pan et al. 2004). It is known that the aggregation of sensitizers on surfaces of semiconductor nanoparticles markedly affects the efficiency and pathways of energy and electron-transfer processes (Grätzel and Moser 2001). Since attachment of both monomeric and aggregated carotenoids have been reported (Sereno et al. 1996, Gao et al. 2000, Pan et al. 2002, 2004, Xiang et al. 2005, Wang et al. 2006), studies of excited states of aggregates are essential for the future optimization of the efficiency of potential carotenoid-based artificial photosystems.

8.2 EXCITED STATES OF MONOMERIC CAROTENOIDS

Knowledge of the excited-state properties of monomeric carotenoids in solution is a necessary prerequisite to understanding aggregation-induced effects on excited states. The key feature of carotenoids is an atypical order of energy levels, making the transition to the lowest energy state optically forbidden. Because the conjugated chain of a carotenoid molecule has a C_{2h} point symmetry, allowed transitions occur between the ground state S_0 (that is of $1A_g^-$ symmetry in the C_{2h} point group notation) and states having B_u^+ symmetry (Polívka and Sundström 2004). However, due to strong electron correlation, the second excited state with $1A_g^-$ symmetry has a lower energy than the lowest B_u^+ state, making the transition between S_0 ($1A_g^-$) and S_1 ($2A_g^-$) states forbidden for one-photon processes. The strong absorption in the spectral region 400–550 nm characteristic of carotenoids is thus due to S_0–S_2 transition, because the S_2 state is the lowest having the B_u^+ symmetry. This energy level scheme generates excited-state dynamics that differ from those known for most organic dyes. Studies of the dynamics of carotenoid excited states have established the following scheme, see Figure 8.2: after excitation of a carotenoid to its S_2 state, a fast relaxation to the S_1 state occurs on a timescale of 50–300 fs. The carotenoid subsequently relaxes to its ground state S_0 due to vibronic coupling between the S_0 and S_1 states on the timescale of 1–300 ps (Polívka and Sundström 2004). There is a clear correlation between the S_1 lifetime and the conjugation length of carotenoids: as the conjugation length increases, the energy gap between the S_0 and S_1 state becomes smaller, thereby making the S_1 lifetime shorter. While the lifetime dependence on the conjugation length is straightforward for the S_1 state, the observed dependence of the S_2 lifetime on the conjugation length does not follow that expected from the energy gap law (Kosumi et al. 2006). This observation, together with other experimental results, has led to the proposition that other dark excited states exist between the S_1 and S_2 states, Figure 8.2. Properties of these states and their roles in excited-state dynamics are frequently debated, but so far no clear consensus about their origins, energies, or lifetimes has been reached (Koyama et al. 2004, Polívka and Sundström 2004). For studies of carotenoid aggregates, two of these additional states could be of potential interest. First, the S^* state that has recently been associated with a twisted S_1 state (Niedzwiedzki et al. 2007) will clearly be affected by aggregation, because packing of molecules in an aggregate would prevent the predicted twisting that promotes the S^* state population. Second, the intramolecular charge transfer (ICT) state that is typical for carotenoids having a conjugated carbonyl group (Frank et al. 2000, Zigmantas et al. 2004) should play a role in aggregates of carbonyl carotenoids. Thus, searching for aggregation-induced effects

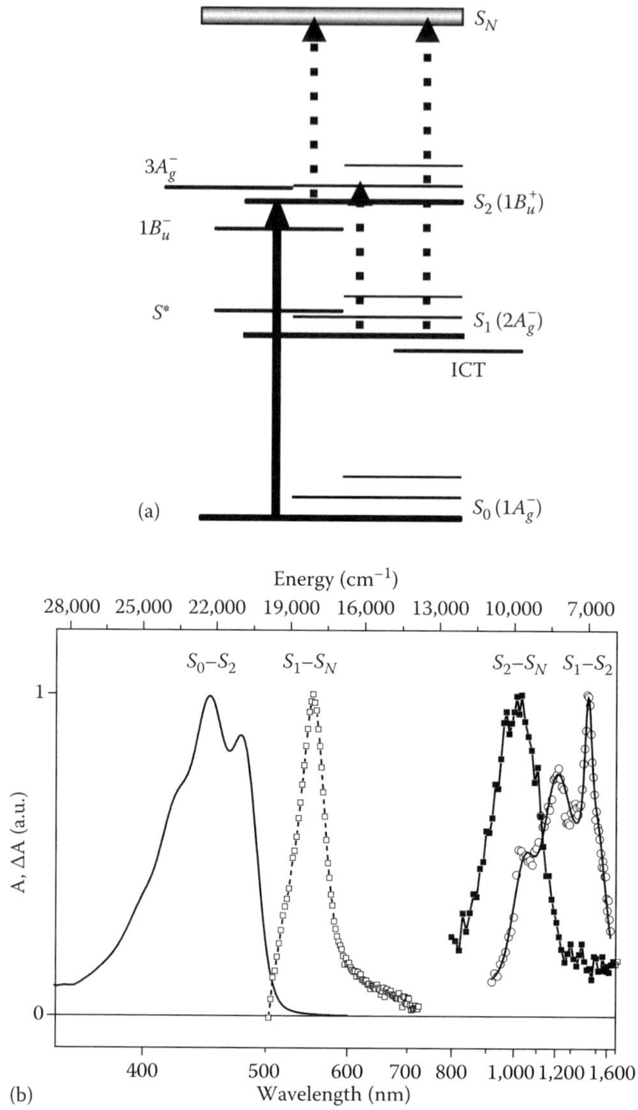

FIGURE 8.2 (a) Simplified energy-level scheme of a carotenoid molecule. The solid arrow represents the absorbing S_0–S_2 transition, the dotted arrows are transitions corresponding to transient signals occurring after excitation. The S_N state in this scheme represents only a symbolic final state for S_1–S_N and S_2–S_N transitions. In reality, the final states of these transitions must be of different symmetry and therefore the S_N state in the scheme consists actually of two different states. (b) Spectral bands corresponding to various transitions for monomeric carotenoids.

on these states may provide valuable information about the properties of these, so far, poorly described states.

The aggregation-induced effects on carotenoid excited states discussed in this chapter are limited to the S_1 and S_2 states. Although knowledge about the S_2 energy is readily obtained from the absorption spectrum, energy of the S_1 state had not been directly measured until the end of the last century when a few different approaches, resonance Raman spectroscopy (Sashima et al. 1999), detection of weak S_1 fluorescence (Fujii et al. 1998), two-photon absorption (Krueger et al. 1999), and time-resolved S_1–S_2 absorption (Polívka et al. 1999), emerged. These methods provided valuable information about the S_1 energies of some carotenoids and proved the conjecture that the S_1 energy is, due to the negligible dipole moment of the S_1 state, essentially independent of solvent. The lifetimes of the S_1 and S_2 states can be obtained from femtosecond time-resolved spectroscopy, usually from decays of the well-defined excited-state absorption (ESA) bands shown in Figure 8.2. The S_1 lifetime can be determined from the decay of either the S_1–S_N transition peaking in the 500–650 nm range for most carotenoids or the S_1–S_2 transition occurring in the near-infrared region. The profile of the S_1–S_2 transition also allows the determination of the S_1 energy (Polívka et al. 1999). The S_2 lifetime is more complicated to determine. Because it is always shorter than 300 fs, the S_2–S_N transition may overlap with other ESA bands (Figure 8.2). However, except for carotenoids with a very long conjugation and very short S_1 lifetime, the time evolution of the ESA bands usually enables extraction of "pure" S_2–S_N decay. Alternatively, a reliable method for determining the S_2 lifetime is time-resolved detection of up-converted S_2 fluorescence that represents essentially a background-free method with sub-100 fs time resolution (Macpherson and Gillbro 1998).

8.3 EXCITED STATES OF CAROTENOID AGGREGATES

While excited-state properties of monomeric carotenoids in organic solvents have been the subject of numerous experimental and theoretical studies (Polívka and Sundström 2004), considerably less is known about excited states of carotenoid aggregates. Most of the knowledge gathered so far stems from studies of aggregation-induced spectral shifts of absorption bands of carotenoid aggregates that are explained in terms of excitonic interaction between the molecules in the aggregate.

8.3.1 Excitonic Interaction: Origin of the Spectral Shifts

8.3.1.1 Intermolecular Interaction

The aggregation-induced changes of absorption spectra result from intermolecular interactions between closely spaced carotenoid molecules. For two molecules whose transition dipole moment vectors, $\boldsymbol{\mu}$, are located at places characterized by position vectors $\boldsymbol{r}_1$ and $\boldsymbol{r}_2$, with the relative position vector defined as $\boldsymbol{R}=\boldsymbol{r}_1-\boldsymbol{r}_2$, Figure 8.3, the interaction energy is expressed as (van Amerongen et al. 2000)

$$V(R)=\frac{1}{4\pi\varepsilon_0}\left[\frac{q_1q_2}{R}+\frac{q_1(\boldsymbol{\mu}_2\cdot\hat{\boldsymbol{R}})-q_2(\boldsymbol{\mu}_1\cdot\hat{\boldsymbol{R}})}{R^2}+\frac{\boldsymbol{\mu}_1\cdot\boldsymbol{\mu}_2-3(\boldsymbol{\mu}_1\cdot\hat{\boldsymbol{R}})(\boldsymbol{\mu}_2\cdot\hat{\boldsymbol{R}})}{R^3}+\cdots\right] \tag{8.1}$$

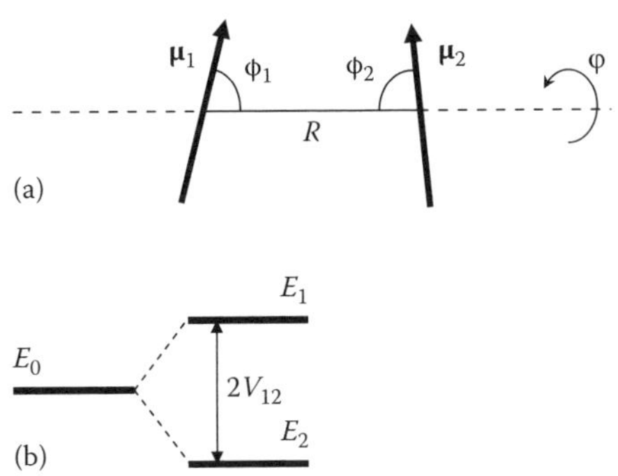

FIGURE 8.3 (a) Definition of angles between the two interacting transition dipole moments, $\boldsymbol{\mu}_1$ and $\boldsymbol{\mu}_2$, separated by the distance R. (b) Davydov splitting resulting from interaction of a pair of molecules having excited state energy E_0 and positive V_{12}.

where the hat over the vector sign indicates a unit vector. The first two terms are nonzero only for charged molecules with a total electric charge, q_i. Thus, for most cases involving carotenoid aggregates, the third term, the dipole–dipole interaction, is the first nonzero term in the expansion. For many cases, the higher order terms are significantly smaller than the dipole–dipole interaction (but see Section 8.3.1.3). Thus, in the first-order approximation, the intermolecular interaction can be well approximated by the dipole–dipole term and the interaction energy expressed as

$$V_{12} = \frac{1}{4\pi\varepsilon_0}\left[\frac{\mu_1 \cdot \mu_2 - 3(\mu_1 \cdot \hat{\boldsymbol{R}})(\mu_2 \cdot \hat{\boldsymbol{R}})}{R^3}\right] \tag{8.2}$$

For the carotenoid aggregates, we always assume aggregation of the same molecules. In this case, $\mu_1 = \mu_2$, and Equation 8.2 can be further simplified to (Scholes 2003)

$$V_{12} = \frac{1}{4\pi\varepsilon_0}\frac{\kappa\mu^2}{R^3} \tag{8.3}$$

where κ is an orientation factor defined as

$$\kappa = \hat{\mu}_1 \cdot \hat{\mu}_2 - 3(\hat{\mu}_1 \cdot \hat{\boldsymbol{R}})(\hat{\mu}_2 \cdot \hat{\boldsymbol{R}}) \tag{8.4}$$

that can be expressed in terms of angles between the transition dipoles defined in Figure 8.3 as

$$\kappa = 2\cos\phi_1 \cos\phi_2 + \sin\phi_1 \sin\phi_2 \cos\varphi \tag{8.5}$$

Knowledge of the interaction energy, V_{12}, enables the calculation of the shift of the excited-state energy of the interacting molecules in respect to their monomeric energy, E_0. In the simplest case of a pair of interacting molecules, the dimer will have two excited states denoted E_1 and E_2, whose energies are

$$E_{1,2} = E_0 \pm V_{12} \tag{8.6}$$

The energy difference $|E_1 - E_2| = 2\ V_{12}$ is known as Davydov or exciton splitting, Figure 8.3. The shift of energy levels gives rise to new bands in the absorption spectrum denoted as the upper and lower Davydov (exciton) components. These components are the H- and J-bands observed in absorption spectra of molecular aggregates.

8.3.1.2 Intensity of the Exciton Bands

The intermolecular interaction described above provides information about the magnitude of spectral shifts, but it does not explain why the absorption spectra of molecular aggregates usually have either an H- or J-band. The square of transition dipole moment (in Debye^2 units) is usually termed the dipole strength and is related to the intensity of the absorption band as (van Amerongen et al. 2000)

$$\mu^2 = 9.18 \times 10^{-3} \int \frac{\varepsilon(\omega)}{\omega} d\omega \tag{8.7}$$

where ε (ω) is the extinction coefficient in M^{-1} cm^{-1} units. In the simplest case when two identical interacting molecules have their dipoles in the same plane ($\varphi = 0°$), it is possible to show that the upper and lower exciton components have dipole strengths:

$$\mu_{1,2}^2 = \mu_0^2(1 \pm \cos\theta) \tag{8.8}$$

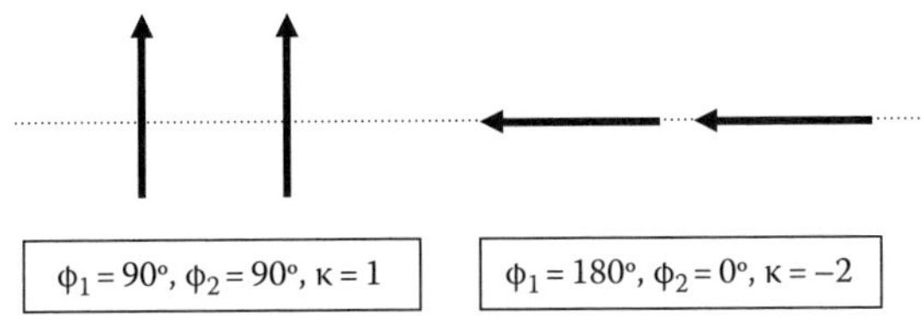

FIGURE 8.4 Orientation factor for the card-pack and head-to-tail dimers.

where θ is the angle between the two interacting dipoles (van Amerongen et al. 2000). The specific cases of the card-pack and head-to-tail aggregates are shown in Figure 8.4. Although $\theta=0°$ for both arrangements, the analysis of the orientation factor, κ, gives different values of V_{12} for the two mutual orientations of the transition dipoles. Thus, although in both cases it is the E_1 exciton component that gains the dipole strength according to Equation 8.8, for the card-pack aggregates, E_1 is the upper exciton component (V_{12} positive), whereas for the head-to-tail aggregates, the E_1 level is the lower exciton component (V_{12} negative), explaining the difference in absorption spectra of H- and J-aggregates shown in Figure 8.5.

8.3.1.3 Limitations

The dipole–dipole approximation as described above is valid only under certain conditions that must be carefully considered when applying it to carotenoid aggregates. First, the approximation is reasonable only when the distance R between the interacting molecules is larger than the size of the charge distributions of individual molecules. This represents a significant problem in interpreting spectra of carotenoid aggregates, because distances between carotenoid molecules in the aggregate are usually less than 10 Å (Zsila et al. 2001b, Billsten et al. 2005), whereas the length of the conjugated backbone, which limits the distribution of π-electrons, for most carotenoids, exceeds this value. Consequently, although the dipole–dipole approximation is a useful tool to explain the aggregation-induced changes in absorption spectra qualitatively, to obtain quantitative agreement

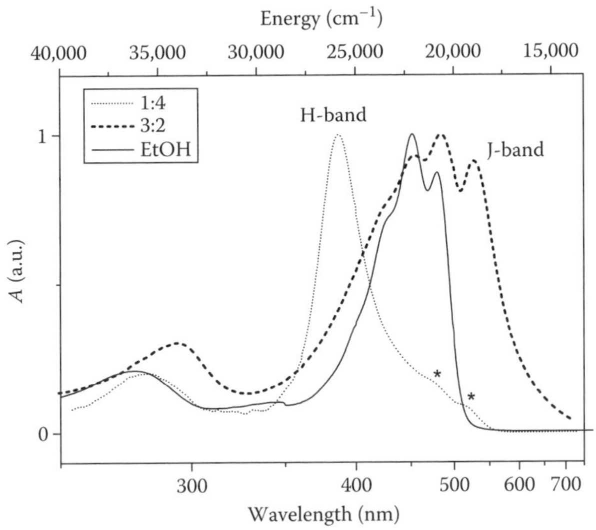

FIGURE 8.5 Absorption spectra of zeaxanthin: dissolved in pure ethanol (solid line), in ethanol/water mixture with 1:4 ratio (dotted line), and in 3:2 ethanol/water mixture. Minor bands of the H-aggregate are denoted by *.

it is often necessary to go beyond the dipole–dipole approximation and the calculation of the full Coulomb coupling is required (van Amerongen et al. 2000, Scholes 2003). To overcome the problem with the dimensions of these molecules often being larger than their intermolecular separation, it is necessary to use more sophisticated approaches that have been developed for calculations of couplings between pigments in photosynthetic systems (Krueger et al. 1998, Madjet et al. 2006). Very recent application of these advanced approaches to calculate excitonic couplings in lutein aggregates showed that the spectral features previously ascribed to J-aggregates may be also explained in terms of weakly coupled H-aggregates (Spano 2009).

A further limitation exists because Equation 8.3 is correct only in a vacuum. For molecules in a polarizable medium characterized by the dielectric constant, ε_r, the effective transient dipole moment is

$$\mu_{\text{eff}} = \frac{\varepsilon_r + 2}{3}\mu \tag{8.9}$$

and the coulombic interaction is diminished by a factor of $1/\varepsilon_r$. Consequently, Equation 8.3 in a polarizable medium has the form (Pullerits et al. 1997)

$$V_{12} = \frac{1}{\varepsilon_r}\left(\frac{\varepsilon_r + 2}{3}\right)^2 \frac{1}{4\pi\varepsilon_0}\frac{\kappa\mu^2}{R^3} \tag{8.10}$$

Another correction arises from the fact that carotenoid aggregates consist of many molecules. For an aggregate of N molecules, there are N possible aggregate excited states, with one excitation present in the aggregate. Due to the intermolecular interaction, these states are not energetic eigenstates of the aggregate and the true eigenstates, the excitons, have to be found by the diagonalization of the corresponding Hamiltonian. In the simplest case of a molecular homodimer, we obtain Equation 8.6. For a linear aggregate consisting of N molecules, assuming that the nonnearest neighbor interaction can be neglected, this procedure leads to N exciton states with energies

$$E_k = E_0 + 2V\cos\frac{\pi k}{N+1} \quad (k = 1, 2, \ldots, N) \tag{8.11}$$

where

V is the nearest-neighbor interaction
E_0 is the transition energy of monomer

The analysis of the transition dipoles (Knoester 1993, van Amerongen et al. 2000) shows that almost all of the dipole strength is collected in the E_1 state, which is, depending on the sign of the interaction term V, either the lowest (J-aggregates) or highest (H-aggregates) exciton state. Consequently, for a very large N, $(\cos(\pi/N+1) \approx 1)$, the energy of the allowed exciton state can be approximated as $E_1 = E_0 + 2V$ and the aggregation-induced shift of the transition energy is thus twice that of the dimer, Equation 8.6. This approximation, used to estimate the intermolecular distance from absorption spectra of carotenoid aggregates (Zsila et al. 2001b), has provided results in a good agreement with scanning tunneling microscopy (STM) images (Köpsel et al. 2005).

8.3.2 Absorption Spectra of Carotenoid Aggregates

Many studies of various carotenoids in hydrated solvents demonstrated a significant effect of the aggregation on the spectroscopic properties of the S_2 state. Upon aggregation, the S_0–S_2 transition undergoes a large spectral shift whose magnitude and direction depend on many conditions. The key properties facilitating formation of either the blueshifted (H-aggregate) or the redshifted (J-aggregate) absorption spectrum are the solvent/water ratio and carotenoid structure (Ruban et al. 1993, Simonyi et al. 2003, Billsten et al. 2005). Other factors, such as the initial concentration of the carotenoid in the organic solvent (Zsila et al. 2001c, Billsten et al. 2005), the pH of the water added to the carotenoid solution (Billsten et al. 2005), and the specific solvent or the temperature (Mori et al. 1996), may also tune the resulting spectral shift, because they affect the organization of molecules within the aggregate.

The dependence of the aggregation-induced shifts of the S_2 state on experimental conditions can be demonstrated for zeaxanthin. This carotenoid forms aggregates easily and also exhibits an ambivalent behavior in forming aggregates; depending on conditions either H- or J-aggregates are produced (Billsten et al. 2005, Avital et al. 2006). The formation of H-aggregates is signaled by a narrow absorption band peaking around 390 nm, Figure 8.5. This band corresponds to the upper excitonic component that gains oscillator strength due to the card-pack organization of the aggregates. In contrast, J-aggregates generate a redshifted band, because the lower excitonic component is the one with appreciable transition dipole moment. The J-band is thus due to the head-to-tail aggregates and its position varies between 510 and 540 nm, Figures 8.5 and 8.6b, and Table 8.1.

The key parameters determining whether H- or J-aggregates of zeaxanthin will be formed are the solvent/water ratio and the initial concentration of zeaxanthin. The results of two different initial concentrations of zeaxanthin are shown in Figure 8.6. For the 4×10^{-5} M ethanol solution of zeaxanthin, addition of 40% water leads to immediate suppression of the characteristic absorbance between 400 and 500 nm accompanied by the formation of a new band at 380 nm typical for H-aggregates. Since vibrational bands of the zeaxanthin S_2 state are still visible, the 3:2 ethanol/water ratio forms a system in which monomeric zeaxanthin coexists with H-aggregates. Further increase of the water content stabilizes H-aggregates; the vibrational structure disappears and the H-band dominates the absorption spectrum. It is worth noting that upon changing the ethanol/water ratio from 3:2 to 1:4 the H-band narrows and shifts from 380 to 390 nm. These effects are attributed to the stabilization

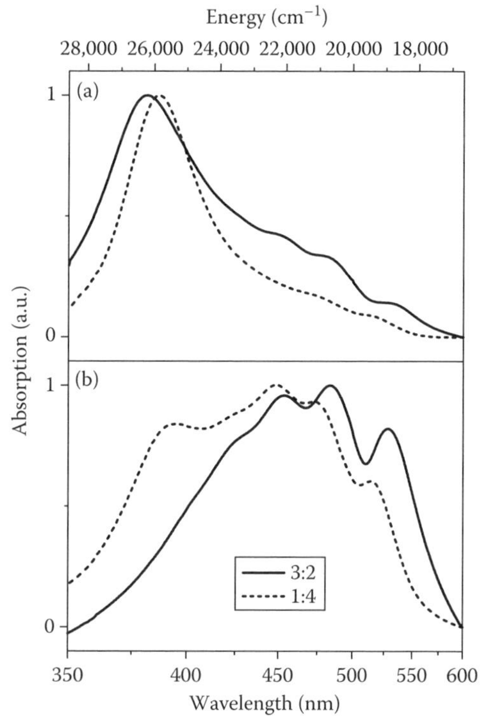

FIGURE 8.6 Absorption spectra of zeaxanthin in hydrated ethanol with two ethanol/water ratios (3:2 and 1:4) prepared from initial concentration of zeaxanthin of 4×10^{-5} M (a) and 10^{-4} M (b).

Table 8.1
Absorption Maxima of Some Carotenoid Aggregates[a]

Carotenoid	Solvent	λ_{max} (M)	λ_{max} (H)	λ_{max} (J)	Reference
ACOA[b]	Ethanol	440	390 (1:4)		T. Polívka (unpublished)
ACOA	TiO_2 film	426			Gao et al. (2000)
ACOA	TiO_2 film	400			Pan et al. (2004)
Astaxanthin	Acetone	478	450 (1:9)	562 (3:7)	Köpsel et al. (2005)
Astaxanthin	Acetone	478	403 (1:9)	560 (1:9)[c]	Mori et al. (1996)
Astaxanthin	Ethanol	476	410 (1:9)[d]		Buchwald et al. (1968)
Antheraxanthin	Ethanol	447	375 (1:3)		Ruban et al. (1993)
β-Carotene	Acetone	455	420 (1:3)	**515 (1:3)**	Zsila et al. (2001d)
β-Carotene	TX-100 micelles	505			Avital et al. (2006)
Capsanthol	Ethanol	447	385 (1:3)[e]	504 (1:3)[f]	Zsila et al. (2001e)
Lutein	Acetone	449	370 (4:15)		Zsila et al. (2001b)
Lutein	Ethanol	445	370 (~1:1)		Ruban et al. (1993)
Lutein	Thin film	380			Zsila et al. (2001b)
Lutein	Lipid bilayer	370			Sujak et al. (2002)
Lycopene	THF	480	354 (1:5)		Wang et al. (2005)
Lycopene	Ethanol	480	380 (1:4)	**580 (1:4)**	Ray and Mishra (1997)
Lycopene	LB film	348	**563**		Ray and Mishra (1997)
Spirilloxanthin	Acetone/MetOH	496	374 (1:2)		Agalidis et al. (1999)
Violaxanthin	Ethanol	440	390 (~1:1)	**500 (~1:1)**	Ruban et al. (1993)
Violaxanthin	TX-100 micelles	390	**515**		Avital et al. (2006)
Zeaxanthin	Acetone	390 (3:7)	517 (1:1)		Avital et al. (2006)
Zeaxanthin	Methanol	450	387 (1:4)	530 (3:2)	Billsten et al. (2005)
Zeaxanthin	Ethanol	450	380 (~1:1)		Ruban et al. (1993)
Zeaxanthin	TX-100 micelles	380[g]	520		Avital et al. (2006)

[a] λ_{max} refers to absorption maximum of monomers (M), H-aggregates (H), and J-aggregates (J); values in bold indicate that a J-aggregate is present in the sample together with an H-aggregate; solvent:water ratio is shown in parentheses.
[b] 8′-apo-β-carotenoic acid.
[c] At a higher temperature after several hours.
[d] In presence of 3 M sodium perchlorate.
[e] Position of the band varies with concentration (382–394 nm).
[f] J-aggregate was formed for 6′S capsanthol while H-aggregate for 6′R capsanthol.
[g] Formed from a J-aggregate after 2 h.

of H-aggregates caused by the increased water content. Since 3:2 ethanol/water ratio is close to the limit of H-aggregate formation, a large distribution of aggregate sizes is likely present in the sample. The narrowing of the H-band is caused by the delocalization of excitons in the aggregate. For individual carotenoid molecules, the spectral width of the absorption band is determined by disorder in the transition energy. However, upon aggregation, excitons are delocalized over several molecules; this results in an averaging over the energetic disorder of the individual molecules, thereby decreasing the width of the spectral band (exchange narrowing) (Ohta et al. 2001). Thus, a narrower H-band reflects an increase in the average number of molecules in the H-aggregate.

Increasing the initial concentration of zeaxanthin to 10^{-4} M, Figure 8.6b, produces a different dependence on the ethanol/water ratio. Under these initial conditions, adding water to a final ethanol/water ratio of 3:2 leads to a distinctly different absorption spectrum than that observed at lower initial concentration. The vibrational structure of the S_2 state is preserved and a new absorption band characteristic of J-aggregates appears at 530 nm. When the water content was increased further (ethanol/water ratio of 1:4), the magnitude of the J-band decreased and its position shifted to 515 nm. The H-band at 390 nm grows-in, indicating the formation of H-aggregates. Essentially the same behavior is observed when acetone is used as the primary solvent. At a 1:1 acetone/water ratio only J-aggregates are present. The J-band peaks at 517 nm, confirming that J-aggregates may be formed only in a narrow range of water concentrations (Avital et al. 2006).

Regardless of the initial concentration of zeaxanthin, when ethanol is used as the primary solvent, a water content larger than 50% always promotes the formation of H-aggregates (Ruban et al. 1993, Billsten et al. 2005). Thus, it seems that the proper choice of solvent may shift the concentration window in which the J-aggregates are formed.

Change in the initial concentration of the carotenoid also affects the spectral position of the H-band. Experiments carried out by Zsila et al. (2001c) showed that the H-band of capsanthol in a 1:3 ethanol/water mixture shifted from 394 to 382 nm when the carotenoid concentration was increased from 2.5×10^{-7} to 1.25×10^{-5} M. Since the magnitude of the blueshift reflects intermolecular interaction within the aggregate (see Section 8.3.1), this result suggests that a higher initial concentration induces tighter packing of carotenoids. Interestingly, no concentration effect on the J-band of capsanthol was observed (Zsila et al. 2001c).

8.3.2.1 Effect of Carotenoid Structure

The formation of carotenoid aggregates has been observed for many carotenoids. Effects of aggregation on absorption spectra, summarized in Table 8.1, provide basic information about the relationship of the carotenoid structure and its ability to form aggregates. It is obvious that nearly all carotenoids can form H-aggregates. The largest blueshift, and consequently the strongest intermolecular interaction, was observed for linear carotenoids without terminal rings, lycopene (Wang et al. 2005) and spirilloxanthin (Agalidis et al. 1999). This is likely due to the absence of any functional groups that promote the close alignment of the conjugated chains in the card-packed H-aggregate. On the other hand, H-aggregates of lycopene and spirilloxanthin are less stable than those of polar carotenoids having hydroxyl groups. The lower stability of H-aggregates of linear carotenoids without functional groups is manifested by the fact that the spectral position of the H-band can be significantly changed. Using ethanol instead of acetone shifts the H-band of lycopene from 354 to 380 nm, though the absorption of monomeric lycopene was not affected by the solvent change (Wang et al. 2005). The spirilloxanthin H-band was markedly affected by adding a detergent: a blueshift from 405 to 374 nm can be induced by adding lauryl dimethylamine oxide (Agalidis et al. 1999).

Polar carotenoids that have hydroxyl groups form aggregates readily suggesting that the –OH groups play a role in the formation of aggregates (Ruban et al. 1993, Simonyi et al. 2003, Billsten et al. 2005). Yet, the structure of the carotenoid is a key factor affecting the ability of aggregation. Violaxanthin, for example, has two terminal rings both possessing hydroxyl and epoxy groups, Figure 8.1. Since these bulky structures are not in conjugation with the major conjugated backbone, a possible explanation could be that their movement is less restricted, making the violaxanthin molecule rather nonplanar, thus preventing tight packing of molecules in the aggregate. As a result, the violaxanthin absorption spectrum in an ethanol/water mixture has signs of both H- and J-aggregates. This indicates that both forms coexist, but H- and J-bands are less pronounced and aggregation-induced spectral shifts are smaller than those for zeaxanthin and lutein (Ruban et al. 1993). The same effect has been observed for capsorubin and epicapsorubin, where the terminal rings are also not in conjugation with the main conjugated backbone (Simonyi et al. 2003).

A large set of results obtained in recent years for various carotenoids (see, e.g., Simonyi et al. (2003) for review) suggests that planarity of the carotenoid molecule is crucial for aggregation. This hypothesis is supported by the observation that zeaxanthin and astaxanthin, both fairly planar molecules, form aggregates more readily than other carotenoids. Moreover, zeaxanthin and astaxanthin are the only two carotenoids studied so far that can, depending on preparation conditions, form exclusively either H- or J-aggregates (Billsten et al. 2005, Köpsel et al. 2005, Avital et al. 2006). The planarity of their conjugated system allows for better packing of the molecules within the aggregate. Furthermore, other molecular forces such as π–π stacking interactions, which may contribute significantly to the attractive forces between closely packed carotenoid molecules (Wang et al. 2004), are stronger when molecules are planar. Nevertheless, the spectral position of the H-band of astaxanthin aggregates indicates weaker intermolecular interaction than in the zeaxanthin H-aggregate. The maximum of the astaxanthin H-band also exhibits a large dependence on the conditions of the experiment, Table 8.1, suggesting a lower stability of the H-aggregates of astaxanthin compared to zeaxanthin. This effect could be attributed to the presence of carbonyl groups that may interfere with tight packing of the astaxanthin molecules. It should also be noted that isomerization prevents the formation of H-aggregates; although all-*trans* zeaxanthin in 1:2 ethanol/water mixture produces H-aggregates, the absorption spectrum of 9-*cis* and 13-*cis* zeaxanthin in the same mixture exhibit characteristics of J-aggregate (Milanowska et al. 2003).

8.3.2.2 Effect of Hydrogen Bonds

The ability to form hydrogen bonds via hydroxyl groups is a decisive factor determining whether aggregation will be of H- or J-type. The role of hydrogen bonding was extensively studied by Simonyi et al. (2003) who showed that the esterification of hydroxyl groups stimulates the formation of J-aggregates. While capsanthol acetate, lutein diacetate (Bikadi et al. 2002), and zeaxanthin diacetate (Zsila et al. 2001c) form exclusively J-type aggregates, their nonesterified counterparts with hydroxyl groups form predominantly H-aggregates (Simonyi et al. 2003). A different approach to study the role of hydrogen bonding was used by Billsten et al. (2005) who varied pH of the water added to the ethanolic solution of zeaxanthin. When 40% of water at pH 4 was added to the solution, it produced the H-aggregate, while the same amount of water at pH 10 generated exclusively the J-aggregate. The pH dependence is directly related to the ability of zeaxanthin to form a hydrogen bond, because an increase in pH causes the deprotonation of the hydroxyl groups of zeaxanthin. At higher pH, zeaxanthin is not able to participate as readily in hydrogen bonding, indicating that J-aggregates are preferentially formed when hydrogen bonding is prevented. This conclusion is further supported by the fact that the nonpolar counterpart of zeaxanthin, β-carotene, having the same structure but lacking the hydroxyl groups, preferentially forms J-aggregates. The J-band at 515 nm dominates the absorption spectrum of β-carotene in acetone/water mixture, with only a hint of a blueshifted band around 420 nm (Zsila et al. 2001d).

It was also demonstrated that not only the presence of the hydroxyl groups, but also their position is an important factor in the formation of hydrogen bonds, and consequently in determining whether J- or H-aggregates will be produced. As shown by Simonyi et al. (2003), the presence of a free hydroxyl group on both sides of a carotenoid molecule is necessary for the formation of H-aggregates. However, even in this case it may eventually happen that J-aggregate is formed, as observed by comparing of aggregation properties of capsanthol stereoisomers having two hydroxyl groups either on the same or on the opposite sides of the molecular plane (Zsila et al. 2001e).

The results obtained either from the pH dependence (Billsten et al. 2005) or esterification (Simonyi et al. 2003) support the idea that the card-pack H-aggregates are stabilized via a hydrogen-bonding network. The ability of hydrogen-bond formation is thus a decisive factor determining whether J- or H-aggregates are formed. The necessity of hydrogen bonding for H-aggregate formation can be justified by the card-pack structure of the aggregates. Therefore, in the simple case of a dimer, hydrogen bonding at both sides of the carotenoid molecule helps to keep the two molecules together lying one on top of the other with their dipoles oriented almost perfectly parallel to each other.

8.3.2.3 Other Spectral Features in Absorption Spectra of Carotenoid Aggregates

Besides the main band, H-aggregates also exhibit weaker bands in the red part of the absorption spectrum (marked by * in Figure 8.5). Although in some cases the position of these bands coincides with the vibrational bands of the monomeric carotenoid and can be therefore assigned to nonaggregated carotenoid molecules, certain spectral features do not match the vibrational bands of the monomer. Moreover, even when the H-band dominates the absorption spectrum, a weak band below the low-energy edge of the monomeric absorption spectrum is often present, Figure 8.5. Even though this feature may be interpreted as the J-band, indicating that a fraction of the molecules is in the head-to-tail arrangement, fluorescence anisotropy measurements of conjugated oligomers proved that this low-energy band has a polarization nearly identical to the H-band and therefore cannot be interpreted as a J-band (Spano 2006). Very little is known about the emission properties of carotenoid aggregates. A rare measurement of the emission spectrum of an H-aggregate of lycopene organized in a Langmuir–Blodgett film (Ray and Mishra 1997) demonstrated a large redshift of the emission spectrum, but no information about polarization was provided. It is also worth mentioning that the spectral shift of the main H-band in respect to the 0–0 origin of the S_0–S_2 transition of a monomeric carotenoid can exceed 6,000 cm^{-1}, see Table 8.1, indicating that the lower, forbidden exciton state of the H-aggregate may be found below 15,000 cm^{-1}. Since the carotenoid lowest excited state, S_1, bears a negligible dipole moment and is thus barely affected by the dipole–dipole interaction, Equation 8.3, the lower exciton state of the H-aggregate may be energetically very close to the S_1 state. Therefore, the large redshift of the lycopene H-aggregate emission observed by Ray and Mishra (1997) may be interpreted as emission either from the lower exciton state of the H-aggregate or from the S_1 state. To clarify the origin of the emission band and to determine the nature of the bands in the carotenoid H-aggregate absorption spectrum, further emission data on carotenoid aggregates are clearly needed.

The absorption spectra of J-aggregates always contain bands coinciding with vibrational bands of monomeric carotenoids. Although these bands were earlier interpreted as due to the vibrational bands of the J-aggregate (Zsila et al. 2001b), studies of the excited state dynamics of the zeaxanthin J-aggregate showed that excitation of the "true" J-band at 540 nm produces distinctly different excited-state dynamics than excitation into the vibrational bands (Section 8.3.3). Since the 480 nm excitation generates excited-state dynamics similar to that of the monomeric carotenoid, the obvious assignment of the vibrational bands in the J-aggregate spectrum is that they are due to monomeric carotenoid molecules coexisting with the J-aggregate (Billsten et al. 2005).

8.3.2.4 Organization and Stability of Aggregates

A useful tool for determining the large-scale organization of carotenoid aggregates is CD spectroscopy. While monomeric carotenoids are usually nonchiral molecules, chirality is induced upon aggregation. CD spectra of carotenoid aggregates exhibit large Cotton effects that could be used to evaluate the large-scale arrangement of the aggregate, because the sign of the Cotton effect indicates the torsion angle between the neighboring molecules within the aggregate (Simonyi et al. 2003). In a set of studies, Zsila et al. demonstrated a helical arrangement of H-aggregates of certain carotenoids (Zsila et al. 2001a–e). These authors have shown that, depending on carotenoid structure, either right- or left-handed helical structures may be formed. For capsanthol, a spontaneous transition between a right-handed and a left-handed helical structure has even been observed in the time course of 2 h (Zsila et al. 2001e). When H-aggregates form on surfaces, STM reveals a neat card-pack organization. For example, astaxanthin deposited on a graphite surface exhibited an arrangement of card-packed molecules with intermolecular distances of ~6 Å (Köpsel et al. 2005). The large-scale organization of J-aggregates is less understood, but the combination of CD studies and atomic force microscopy of J-aggregates in films indicates that J-aggregates may be organized into layers of nematic crystals (Zsila et al. 2001b). J-aggregates were also observed when β-carotene was deposited on the Cu(111) surface. STM images reveal a grid of β-carotene molecules clearly organized in the head-to-tail arrangement (Baro et al. 2003), indicating that β-carotene forms predominantly J-aggregates not only in hydrated solvents, but also when deposited on surfaces.

In some cases, a transition between J- and H-aggregates has also been observed. Mori et al. (1996) showed that astaxanthin H-aggregates transform into J-aggregates in a few hours. These authors studied the transformation in the 2°C–32°C temperature range and concluded that assemblies corresponding to H- and J-aggregates are separated by an energy barrier that allows the H to J transformation above 21°C. Although the J-aggregates in the study by Mori et al. were stable in the whole temperature range, the reverse, J to H transition, was observed for zeaxanthin aggregates in TX-100 micelles (Avital et al. 2006). Immediately after the insertion of zeaxanthin into micelles from tetrahydrofuran, J-aggregates formed, but in the course of 2 h, they spontaneously turned into the card-packed H-aggregates. Stabilization of H-aggregates in TX-100 micelles was also observed for the synthetic carotenoid 7′-apo-7′,7′-dicyano-β-carotene, but the precursor was the monomer rather than the J-aggregate (He and Kispert 1999).

8.3.3 Excited-State Dynamics

Although the aggregation-induced shifts of absorption spectra have been largely investigated, very little is known about excited-state dynamics of carotenoid aggregates. Time-resolved data with sufficient time resolution are so far available only for two carotenoids, zeaxanthin (Billsten et al. 2005) and 8′-apo-β-carotenoic acid (ACOA) (T. Polívka, unpublished). Transient absorption spectra of carotenoid aggregates, compared with the corresponding monomeric carotenoids, are shown in Figure 8.7. The transient absorption spectra of aggregates are reminiscent of those recorded for monomers. They are dominated by an ESA band, which reflects the spectral profile of the S_1–S_N transition (see Section 8.2). For both zeaxanthin and ACOA, the peak position of the main ESA band of the H-aggregate is close to that of the monomeric carotenoid, but the spectral band is markedly broader. The negative band centered at 525 nm for the H-aggregate of zeaxanthin is superimposed on the high-energy wing of the ESA spectrum, giving the impression of a separate ESA band at ~500 nm. This negative feature originates from a ground state bleach of the weak 525 nm band of H-zeaxanthin.

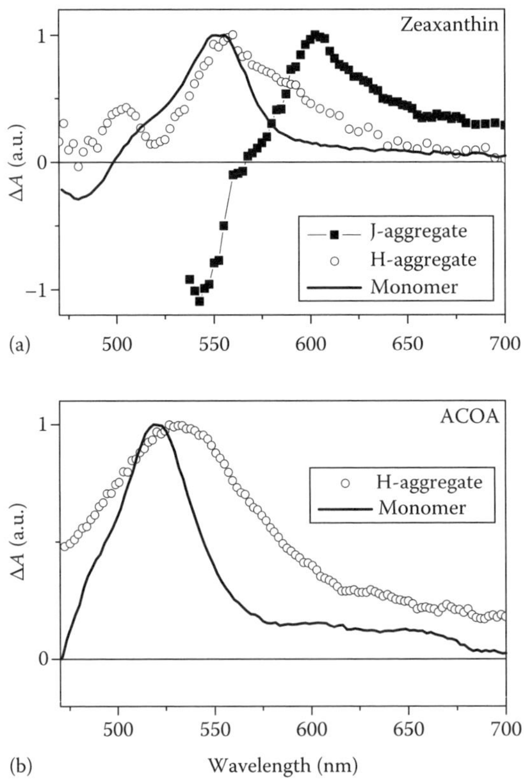

FIGURE 8.7 Transient absorption spectra recorded 3 ps following excitation for zeaxanthin (a) and ACOA (b). The spectra were measured with excitation at 400 nm (H-aggregates), 485 nm (monomers), and 525 nm (J-aggregates).

The similarity of transient absorption spectra of monomeric carotenoids and H-aggregates may be explained by the negligible dipole moment of the S_1 state, resulting in essentially no effect of aggregation on the S_1 energy, Equation 8.3. But the nearly identical energies of the S_1–S_N transitions imply that the final state also remains unaffected. This conclusion is rather surprising, because the S_N state must be of B_u^+ symmetry to have a strong signal for the S_1–S_N ESA. Consequently, the S_0–S_N transition should have an appreciable transition dipole moment and thus its energy should be affected by aggregation. However, closer inspection of the absorption spectrum of the zeaxanthin H-aggregate in Figure 8.5 indeed shows that bands below 300 nm (spectral region where the S_0–S_N transition is expected) remain unaffected by H-type aggregation, explaining the similar maxima of the S_1–S_N bands of the monomer and H-aggregate. It is worth noting, however, that intensity of the S_1–S_N ESA signal for the H-aggregate is significantly weaker than that of the monomer, indicating that although aggregation apparently does not alter the energy of the S_1–S_N transition, its transition dipole moment is weakened. The broader S_1–S_N band of H-aggregate most likely reflects the distribution of aggregate sizes.

Excited-state properties of J-aggregates have only been studied for zeaxanthin. Its ESA band redshifts to 605 nm, which, assuming that the energy of the S_1 state remains unchanged, means that the S_N state redshifts upon J-type aggregation. This is corroborated by a shift of the high-energy bands in the absorption spectrum of the J-aggregate, Figure 8.5. The S_1–S_N band of the J-aggregate also has a red tail extending beyond 700 nm indicating the distribution of aggregate sizes. The strong negative feature at ~540 nm is due to ground state bleaching of the characteristic red absorption band of J-zeaxanthin.

Kinetics recorded at the maxima of the S_1–S_N bands, monitoring dynamics of the lowest excited state, have revealed further differences, Figure 8.8. Although monoexponential decays have been observed for monomeric carotenoids (9 ps for zeaxanthin and 24 ps for ACOA), aggregates exhibit more complicated decay patterns. The S_1 decay of the H-aggregate requires at least four decay

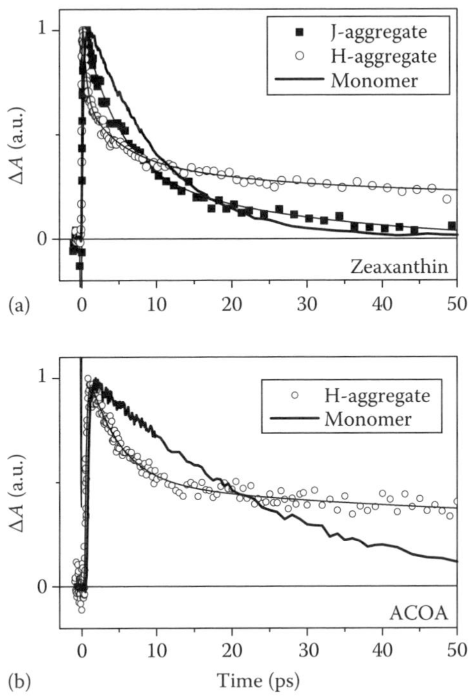

FIGURE 8.8 Kinetics at the maxima of the S_1–S_N bands of aggregated and monomeric forms of zeaxanthin (a) and ACOA (b). Probing wavelengths were 555 (zeaxanthin monomer), 560 (zeaxanthin H-aggregates), 605 (zeaxanthin J-aggregates), 520 (ACOA monomer), and 530 nm (ACOA H-aggregates).

components to obtain a satisfactory fit. For zeaxanthin, the time constant of 0.5 ps represents the major component of the decay, accompanied by two slower components of 4.5 and 20 ps (Billsten et al. 2005). Comparable behavior is observed for the excited-state properties of the H-aggregate of ACOA, where 4.2 and 38 ps components dominate the decay, but the short 0.5 ps component (pronounced in aggregated zeaxanthin) is missing. To account for the rest of the decay, a longer component (~500 ps) must be added for H-aggregates of both carotenoids.

The multiexponential decays observed for H-aggregates of both carotenoids are consistent with the annihilation dynamics that are usually present in excited-state processes of molecular aggregates (Trinkunas et al. 2001). In both zeaxanthin and ACOA, the major decay component is faster than the S_1 lifetime of monomers, suggesting a loss of the excited state population via annihilation. To confirm this conjecture, Billsten et al. (2005) have measured kinetics at the S_1–S_N ESA band maximum of H-zeaxanthin while varying the excitation intensity. The results confirmed that the amplitudes of the two fastest components increased with the increase in the excitation intensity and are therefore due to annihilation. The subpicosecond component was interpreted as due to annihilation within smaller aggregates consisting of a few molecules, in which excitation migrates only a short distance prior to the annihilation. In contrast, the ~5 ps component was assigned to a long-range annihilation occurring in larger aggregates, in which excitations must travel across a number of molecules to reach the annihilation site (Billsten et al. 2005). However, lack of the subpicosecond component in H-aggregates of ACOA indicates that the size of the aggregate is not the only factor determining the annihilation component. In ACOA, the presence of the carboxylic group at one side of the molecule likely prevents tight packing of the molecules within the aggregate. Thus, the subpicosecond component is apparently also related to the magnitude of interaction between the molecules; the nonplanar character of the ACOA molecule weakens the interaction and leads to the absence of the subpicosecond annihilation component.

The longer, 20 ps component of zeaxanthin exhibited inverse dependence on excitation intensity, that is, the amplitude increased with decreasing intensity (Billsten et al. 2005). Such dependence is expected for the intrinsic S_1 lifetime, because at lower intensities there is a lower probability of annihilation, thus a large fraction of zeaxanthin decays to the ground state with its true S_1 lifetime. Therefore, it was assigned to the S_1 lifetime in the H-aggregate. For both zeaxanthin and ACOA, the S_1 lifetime in H-aggregate is significantly longer than that of a monomer. This difference can be explained by the restrained vibrational motion of individual carotenoid molecules in the H-aggregate. It is a well-established fact that the S_1 decay is driven by vibrational coupling to the ground state via the C=C stretching mode (Nagae et al. 2000). Consequently, disturbing the vibrational motion of the conjugated backbone induces changes in the S_1 lifetime. The tight packing of carotenoid molecules in H-aggregates hinders vibrational motion of the conjugated backbone, explaining the longer S_1 lifetime in the H-aggregates. The larger difference between monomeric and aggregated zeaxanthin (9 vs. 20 ps) than in ACOA (24 vs. 38 ps) again points to a tighter packing in zeaxanthin.

Much less is known about excited-state dynamics of carotenoid J-aggregates, as only zeaxanthin J-aggregates have been studied to date. Only two decay components of ~5 and 30 ps were needed to fit the kinetics recorded at the maximum of the S_1–S_N band, Figure 8.8. Since no annihilation studies were carried out, the origin of these components is not known. It is likely that the 5 ps lifetime is due to annihilation whereas the 30 ps component corresponds to the S_1 lifetime, which is even longer than that of the H-aggregates.

It should also be noted that a change in the S_1 energy, a common reason for a change of the S_1 lifetime of monomeric carotenoids (Polívka and Sundström 2004), may also be a potential source of the longer S_1 lifetimes in aggregates. The prolongation of the observed S_1 lifetime of aggregates would require a higher S_1 energy of aggregates compared with monomers. However, due to the negligible dipole moment of the S_1 state, changes in the S_1 energy induced by aggregation will be negligible, Equation 8.3. This is also supported by comparison of the transient absorption spectra of monomers and aggregates described above. Therefore, the S_1 energy is only marginally affected by aggregation and the changes in the S_1 lifetimes are related solely to a perturbation of the vibrational coupling. This argument, however, does not provide an explanation for the long decay component (>300 ps) observed in H-aggregates of both zeaxanthin and ACOA, Figure 8.8, because a change in the vibrational coupling cannot account for the dramatic change of the S_1 lifetime. Instead, Billsten et al. (2005) showed that the spectrum of this long-lived component for zeaxanthin resembles features attributable to a triplet state. These authors proposed an enhancement of intersystem crossing induced by an H-type aggregation.

Studying the excited-state dynamics following excitation at different wavelengths has helped to assign spectral bands in the absorption spectrum of the J-aggregate, Figure 8.9. The excitation of the 530 nm band of the zeaxanthin J-aggregate results in an ESA spectrum peaking at 605 nm. This spectrum shows no resemblance to the S_1–S_N ESA spectrum of monomeric carotenoids, either in position or shape. In addition, the distinct bleaching band below 550 nm confirms that this spectrum originates from molecules forming the characteristic red band of the J-aggregates. On the contrary, the ESA bands observed following 400 and 485 nm excitations are dominated by a band at 560 nm, Figure 8.9, which is very close to that of monomeric zeaxanthin, Figure 8.7. Although H-zeaxanthin has an ESA band at the same position, kinetics shown in the inset of Figure 8.9 exclude assigning this band as originating from H-zeaxanthin; the 9 ps decay component that is present only at 560 nm after 400 and 485 nm excitation matches well the known S_1 lifetime of monomeric zeaxanthin in solution.

Thus, based on the excitation wavelength dependence, it is obvious that while excitation at 525 nm selectively excites J-aggregates, excitation at higher energies results in excited-state dynamics corresponding to carotenoid monomers in solution. This indicates that monomers contribute significantly to the absorption spectrum of the J-aggregates. The presence of a shoulder at 605 nm in the transient absorption spectrum measured after excitation at 400 and 485 nm shows that zeaxanthin J-aggregates must be excited even at these wavelengths, suggesting that the absorption spectrum of the J-aggregates extends to 400 nm (Billsten et al. 2005). These results suggest that what is

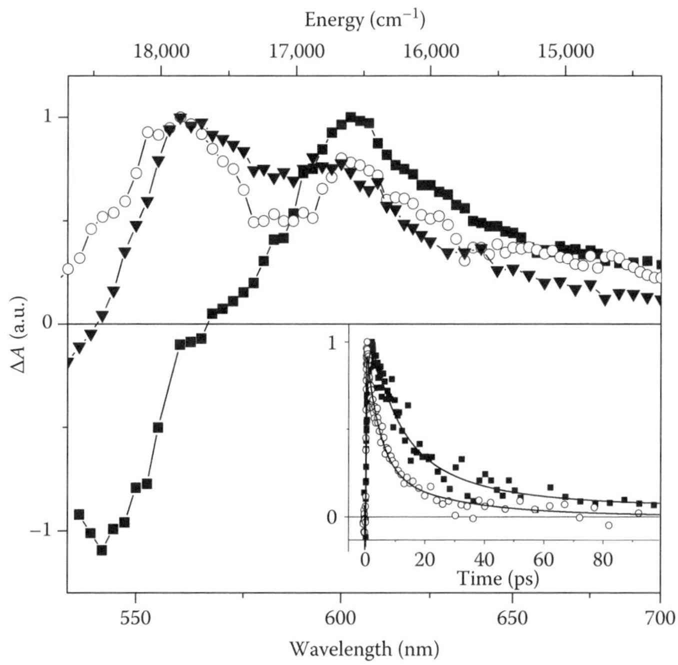

FIGURE 8.9 Transient absorption spectra of the zeaxanthin J-aggregates recorded 3 ps following excitation at 525 (full squares), 485 (open circles), and 400 nm (full triangles). All spectra are normalized to the maximum. (Inset) Kinetics of the zeaxanthin J-aggregates measured at 560 (full squares) and 605 nm (open circles) following excitation at 400 nm. Solid lines represent multiexponential fits of the data.

generally considered as the absorption spectrum of a J-aggregate of a carotenoid actually consists of contributions from both the J-aggregate and monomeric carotenoids.

8.4 SUMMARY AND OUTLOOK

The last decade has witnessed significant progress in the development of an understanding the excited-state properties of carotenoids. The majority of studies have focused on monomeric carotenoids in organic solvents, but much has also been done to improve our knowledge of carotenoid aggregates in hydrated solvents. The aggregation-induced shifts characteristic of the H- or J-aggregate spectra are now qualitatively understood, and controlled the formation of both H- and J-aggregates has been achieved for some carotenoids (Simonyi et al. 2003, Billsten et al. 2005, Avital et al. 2006). The exact relationship between the carotenoid structure and its ability to form aggregates remains incomplete. However, it is now clear that the presence of hydroxyl groups promotes the formation of H-aggregates, most likely by the stabilization of the card-pack organization involving a hydrogen-bonding network. In contrast, J-aggregates form when carotenoids lack end-ring functional groups (e.g., β-carotene), or when hydrogen-bond formation is prevented either by esterification (Simonyi et al. 2003) or pH change (Billsten et al. 2005). Consequently, J-aggregates are usually less stable and generated only in a narrow window of water/solvent ratios (Billsten et al. 2005, Avital et al. 2006). Recent microscopy studies have also revealed the organization of aggregates on surfaces, establishing that the average intermolecular distance in H-aggregates is in the 5–7 Å range (Zsila et al. 2001b, Baro et al. 2003, Köpsel et al. 2005).

Despite the considerable progress that has been achieved in the last decade, significant challenges remain. Very little is known about excited-state lifetimes and relaxation pathways in carotenoid aggregates. While this topic has been extensively studied for monomeric carotenoids, zeaxanthin is the only carotenoid whose excited-state lifetimes have been investigated in aggregated form (Billsten et al. 2005). This study has demonstrated that significant changes occur in the S_1 lifetime of zeaxanthin for both H- and J-aggregates. Additional experiments will be necessary to establish the aggregation-induced effects for other carotenoids. This is especially important for those carotenoids known to form aggregates in both natural and artificial systems. For example, some apo-β-carotenals that exhibit polarity-dependent behavior due to ICT state (Kopczynski et al. 2007) have been used as TiO_2 sensitizers in thin films where they form H-aggregates (Gao et al. 2000, Pan et al. 2004). The aggregation-induced effects on the ICT state, which may affect electron and/or energy transfer properties, remain unknown. The behavior of other dark excited states, which may be located within the S_1–S_2 gap (Koyama et al. 2004, Polívka and Sundström 2004), is completely unknown for carotenoid aggregates. Though the negligible dipole moment of these states prevents large aggregation-induced shifts, the significant change in the S_1 lifetime upon aggregation (Billsten et al. 2005) suggests that a comparable effect may also occur for other dark states. Another important issue to tackle is the effect of environment on spectroscopic properties of carotenoid aggregates. Most of studies have been carried out in hydrated solvents, but it is obvious that the spectroscopic properties of aggregates of a particular carotenoid will differ when prepared in hydrated solvent, in micelles (Avital et al. 2006), in lipid bilayers (Gruszecki 1999, Sujak et al. 2002), deposited on films (Zsila et al. 2001b), or in solid phase (Hashimoto 1999). Because aggregated carotenoids in these environments are functioning either in natural or artificial systems and are promising candidates for future applications in harnessing solar energy, both experimental and theoretical approaches will be needed to reveal details of aggregation effects on carotenoid excited states.

ACKNOWLEDGMENTS

The author thanks Tomáš Mančal for useful discussions, and Helena Billsten and Jingxi Pan for important contributions to the work surveyed here. Financial support from the Czech Ministry of Education (grants No. MSM6007665808 and AV0Z50510513) is gratefully acknowledged.

REFERENCES

Agalidis, I., T. Mattioli, and F. Reiss-Husson. 1999. Spirilloxanthin is released by detergent from *Rubrivivax gelatinosus* reaction center as an aggregate with unusual properties. *Photosynth. Res.* 62:31–42.

Aspinall-O'Dea, M., M. Wentworth, A. Pascal, B. Robert, A. Ruban, and P. Horton. 2002. *In vitro* reconstitution of the activated zeaxanthin state associated with energy dissipation in plants. *Proc. Natl. Acad. Sci. USA* 99:16331–16335.

Avital, S., V. Brumfeld, and S. Malkin. 2006. A micellar model system for the role of zeaxanthin in the non-photochemical quenching process of photosynthesis—Chlorophyll fluorescence quenching by xanthophylls. *Biochim. Biophys. Acta* 1757:798–810.

Baró, A. M., S.-W. Hla, and K. H. Rieder. 2003. LT-STM study of self-organization of β-carotene molecular layers on Cu(111). *Chem. Phys. Lett.* 369:240–247.

Bhosale, P., A. J. Larson, K. Southwick, C. D. Thulin, and P. S. Bernstein. 2004. Identification and characterization of a π-isoform of glutathione S-transferase (GSTP1) as a zeaxanthin-binding protein in the macula of the human eye. *J. Biol. Chem.* 279:49447–49454.

Bikadi, Z., F. Zsila, J. Deli, G. Mady, and M. Simonyi. 2002. The supramolecular structure of self-assembly formed by capsanthin derivatives. *Enantiomer* 7:67–76.

Billsten, H. H., V. Sundström, and T. Polívka. 2005. Self-assembled aggregates of the carotenoid zeaxanthin: Time-resolved study of excited states. *J. Phys. Chem. A* 109:1521–1529.

Buchwald, A. and W. P. Jencks. 1968. Optical properties of astaxanthin solutions and aggregates. *Biochemistry* 7:834–843.

Frank, H. A., J. A. Bautista, J. Josue, Z. Pendon, R. G. Hiller, F. P. Sharples, D. Gosztola, and M. R. Wasielewski. 2000. Effect of the solvent environment on the spectroscopic properties and dynamics of the lowest excited states of carotenoids. *J. Phys. Chem. B* 104:4569–4577.

Fujii, R., K. Onaka, M. Kuki, Y. Koyama, and Y. Watanabe. 1998. The $2A_g^-$ energies of all-*trans*-neurosporene and spheroidene as determined by fluorescence spectroscopy. *Chem. Phys. Lett.* 288:847–853.

Gao, F. G., A. J. Bard, and L. D. Kispert. 2000. Photocurrent generated on a carotenoid-sensitized TiO_2 nanocrystalline mesoporous electrode. *J. Photochem. Photobiol. A* 130:49–56.

Grätzel, M. and J.-E. Moser. 2001. In *Electron Transfer in Chemistry*, eds. V. Balzani and I. Gould, pp. 589–644. Wiley-VCH, Weinheim, Germany.

Gruszecki, W. 1999. Carotenoids in membranes. In *Photochemistry of Carotenoids*, eds. H. A. Frank, A. J. Young, G. Britton, and R. J. Cogdell, pp. 363–379. Kluwer Academic Publishers, Dordrecht, the Netherlands.

Hashimoto, H. 1999. Physical properties of carotenoids in solid state. In *Photochemistry of Carotenoids*, eds. H. A. Frank, A. J. Young, G. Britton, and R. J. Cogdell, pp. 363–379. Kluwer Academic Publishers, Dordrecht, the Netherlands.

He, Z. F. and L. D. Kispert. 1999. Electrochemical and optical study of carotenoids in TX-100 micelles: Electron transfer and a large blue shift. *J. Phys. Chem. B* 103:9038–9043.

Knoester, J. 1993. Nonlinear optical susceptibilities of disordered aggregates—A comparison of schemes to account for intermolecular interactions. *Phys. Rev.* 47:2083–2098.

Kodis, G., C. Herrero, R. Palacios, E. Marino-Ochoa, S. Gould, L. de la Garza, R. van Grondelle, D. Gust, T. A. Moore, A. L. Moore, and J. T. M. Kennis. 2004. Light harvesting and photoprotective functions of carotenoids in compact artificial photosynthetic antenna designs. *J. Phys. Chem. B* 108:414–425.

Kopczynski, M., F. Ehlers, T. Lenzer, and K. Oum. 2007. Evidence for an intramolecular charge transfer state in 12′-apo-beta-caroten-12′-al and 8′-apo-beta-caroten-8′-al: Influence of solvent polarity and temperature. *J. Phys. Chem. A* 111:5370–5381.

Köpsel, C., H. Möltgen, H. Schuch, H. Auweter, K. Kleinermanns, H.-D. Martin, and H. Bettermann. 2005. Structure investigations on assembled astaxanthin molecules. *J. Mol. Struct.* 750:109–115.

Kosumi, D., K. Yanagi, R. Fujii, H. Hashimoto, and M. Yoshizawa. 2006. Conjugation length dependence of relaxation kinetics in beta-carotene homologs probed by femtosecond Kerr-gate fluorescence spectroscopy. *Chem. Phys. Lett.* 425:66–70.

Koyama, Y., F. S. Rondonuwu, R. Fujii, and Y. Watanabe. 2004. Light-harvesting function of carotenoids in photosynthesis: The roles of the newly found $1B_u^-$ state. *Biopolymers* 74:2–18.

Krueger, B. P., G. D. Scholes, and G. R. Fleming. 1998. Calculation of couplings and energy-transfer pathways between the pigments of LH2 by the *ab initio* transition density cube method. *J. Phys. Chem. B* 102:5378–5386.

Krueger, B. P. J. Yom, P. J. Walla, and G. R. Fleming. 1999. Observation of the S_1 state of spheroidene in LH2 by two-photon fluorescence excitation. *Chem. Phys. Lett.* 310:57–64.

Macpherson, A. N and T. Gillbro. 1998. Solvent dependence of the ultrafast S_2–S_1 internal conversion rate of β-carotene. *J. Phys. Chem. A* 102:5049–5058.

Madjet, M. E., E. Abdurahman, and T. Renger. 2006. Intermolecular Coulomb couplings from *ab initio* electrostatic potentials: Application to optical transitions of strongly coupled pigments in photosynthetic antennae and reaction centers. *J. Phys. Chem. B* 110:17268–17281.

Milanowska, J., A. Polit, Z. Wasylewski, and W. I. Gruszecki. 2003. Interaction of isomeric forms of xanthophyll pigment zeaxanthin with dipalmitoylphosphatidylcholine studied in monomolecular layers. *J. Photochem. Photobiol. B* 72:1–9.

Mori, Y., K. Yamano, and H. Hashimoto. 1996. Bistable aggregate of all-*trans*-astaxanthin in an aqueous solution. *Chem. Phys. Lett.* 254:84–88.

Nagae, H., M. Kuki, J.-P. Zhang, T. Sashima, Y. Mukai, and Y. Koyama. 2000. Vibronic coupling through the in-phase, C=C stretching mode plays a major role in the $2A_g^-$ to $1A_g^-$ internal conversion of all-trans-beta-carotene. *J. Phys. Chem. A* 104:4155–4166.

Niedzwiedzki, D., J. F. Koscielecki, H. Cong, J. O. Sullivan, G. N. Gibson, R. R. Birge, and H. A. Frank. 2007. Ultrafast dynamics and excited state spectra of open-chain carotenoids at room and low temperatures. *J. Phys. Chem. B* 111:5984–5998.

Ohta, K., M. Yang, and G. R. Fleming. 2001. Ultrafast exciton dynamics of J-aggregates in room temperature solution studied by third-order nonlinear optical spectroscopy and numerical simulations based on exciton theory. *J. Chem. Phys.* 115:7609–7621.

Pan, J., G. Benkö, Y. H. Xu, T. Pascher, L. Sun, V. Sundström, and T. Polívka. 2002. Photoinduced electron transfer between a carotenoid and TiO_2 nanoparticle. *J. Am. Chem. Soc.* 124:13949–13957.

Pan, J., Y. Xu, L. Sun, V. Sundström, and T. Polívka. 2004. Carotenoid and pheophytin on semiconductor surface: Self-assembly and photoinduced electron transfer. *J. Am. Chem. Soc.* 126:3066–3067.

Polívka, T. and V. Sundström. 2004. Ultrafast dynamics of carotenoid excited states—From solution to natural and artificial systems. *Chem. Rev.* 104:2021–2071.

Polívka, T., J. L. Herek, D. Zigmantas, H.-E. Åkerlund, and V. Sundström. 1999. Direct observation of the (forbidden) S_1 state in carotenoids. *Proc. Natl. Acad. Sci. USA* 96:4914–4917.

Polívka, T., M. Pellnor, E. Melo, T. Pascher, V. Sundström, A. Osuka, and K. R. Naqvi. 2007. Polarity-tuned energy transfer efficiency in artificial light-harvesting antennae containing carbonyl carotenoids peridinin and fucoxanthin. *J. Phys. Chem. C* 110:467–476.

Pullerits, T., S. Hess, J. L. Herek, and V. Sundström. 1997. Temperature dependence of excitation transfer in LH2 of *Rhodobacter sphaeroides*. *J. Phys. Chem. B* 101:10560–10567.

Ramachandran, G. K., J. K. Tomfohr, J. Li, O. F. Sankey, X. Zarate, A. Primak, Y. Terazono, T. A., Moore, A. L. Moore, and D. Gust. 2003. Electron transport properties of a carotene molecule in a metal-(single molecule)-metal junction. *J. Phys. Chem. B* 107:6162–6169.

Ray, K. and T. N. Mishra. 1997. Photophysical properties of lycopene organized in Langmuir–Blodgett films: Formation of aggregates. *J. Photochem. Photobiol A* 107:201–205.

Ruban, A. V., P. Horton, and A. J. Young. 1993. Aggregation of higher plant xanthophylls: Differences in absorption and in the dependency on solvent polarity. *J. Photochem. Photobiol. B* 21:229–234.

Sashima, T., M. Shiba, H. Hashimoto, H. Nagae, and Y Koyama. 1999. The $2A_g^-$ energy of crystalline all-*trans*-spheroidene as determined by resonance-Raman excitation profiles. *Chem. Phys. Lett.* 290:36–42.

Schindler, C. and H. K. Lichtenthaler. 1996. Photosynthetic CO_2-assimilation, chlorophyll fluorescence and zeaxanthin accumulation in field grown maple trees in the course of a sunny and a cloudy day. *J. Plant. Physiol.* 148:399–412.

Scholes, G. D. 2003. Long-range resonance energy transfer in molecular systems. *Annu. Rev. Phys. Chem.* 54:57–87.

Sereno, L., J. J. Silber, L. Otero, M. del Valle Bohorquez, A. L. Moore, T. A. Moore, and D. Gust. 1996. Photoelectrochemistry of Langmuir–Blodgett films of carotenoid pigments on ITO electrodes. *J. Phys. Chem.* 100:814–821.

Simonyi, M., Z. Bikadi, F. Zsila, and J. Deli. 2003. Supramolecular exciton chirality of carotenoid aggregates. *Chirality* 15:680–698.

Spano, F. C. 2006. Excitons in conjugated oligomer aggregates, films and crystals. *Annu. Rev. Phys. Chem.* 57:217–243.

Spano, F. C. 2009. Analysis of the UV/Vis and CD spectral line shapes of carotenoid assemblies: Spectral signatures of chiral H-aggregates. *J. Am. Soc.* 131:4267–4278.

Sujak, A., P. Mazurek, and W. I. Gruszecki. 2002. Xanthophyll pigments lutein and zeaxanthin in lipid multibilayers formed with dimyristoylphosphatidylcholine. *J. Photochem. Photobiol. B* 68:39–44.

Trinkunas, G., J. L. Herek, T. Polívka, V. Sundström, and T. Pullerits. 2001. Exciton delocalization probed by excitation annihilation in the light-harvesting antenna LH2. *Phys. Rev. Lett.* 86:4167–4170.

van Amerongen, H., L. Valkunas, and R. van Grondelle. 2000. *Photosynthetic Excitons*. World Scientific, Singapore.

Wang, Y., L. Mao, and X. Hu. 2004. Insight into the structural role of carotenoids in the Photosystem I: A quantum chemical analysis. *Biophys. J.* 86:3097–3111.

Wang, L., Z. Du, R. Li, and D. Wu. 2005. Supramolecular aggregates of lycopene. *Dyes Pigm.* 65:15–19.

Wang, X. F., Y. Koyama, H. Nagae, Y. Yamano, M. Ito, and Y. Wada. 2006. Photocurrents of solar cells sensitized by aggregate-forming polyenes: Enhancement due to suppression of singlet–triplet annihilation by lowering of dye concentration or light intensity. *Chem. Phys. Lett.* 420:309–315.

Xiang, J., F. S. Rondonuwu, Y. Kakitani, R. Fujii, Y. Watanabe, Y. Koyama, H. Nagae, Y. Yamano, and M. Ito. 2005. Mechanisms of electron injection from retinoic acid and carotenoic acids to TiO_2 nanoparticles and charge recombination via the T_1 state as determined by subpicosecond to microsecond time-resolved absorption spectroscopy: Dependence on the conjugation length. *J. Phys. Chem. B* 109:17066–17077.

Zigmantas, D., R. G. Hiller, F. P. Sharples, H. A. Frank, V. Sundström, and T. Polívka. 2004. Effect of conjugated carbonyl group on photophysical properties of carotenoids. *Phys. Chem. Chem. Phys.* 6:3009–3016.

Zsila, F., J. Deli, and M. Simonyi. 2001a. Color and chirality: Self-assemblies in flower petals. *Planta* 213:937–942.

Zsila, F., Z. Bikadi, Z. Keresztes, J. Deli, and M. Simonyi. 2001b. Investigation of the self-organization of lutein diacetate by electronic absorption, circular dichroism spectroscopy, and atomic force microscopy. *J. Phys. Chem. B* 105:9413–9421.

Zsila, F., Z. Bikadi, J. Deli, and M. Simonyi. 2001c. Chiral detection of carotenoid assemblies. *Chirality* 13:446–453.

Zsila, F., Z. Bikadi, J. Deli, and M. Simonyi. 2001d. Configuration of a single centre determines chirality of supramolecular carotenoid self-assembly. *Tetrahedron Lett.* 42:2561–2563.

Zsila, F., J. Deli, Z. Bikadi, and M. Simonyi. 2001e. Supramolecular assemblies of carotenoids. *Chirality* 13:739–744.

9 Applications of EPR Spectroscopy to Understanding Carotenoid Radicals

Lowell D. Kispert, Ligia Focsan, and Tatyana Konovalova

CONTENTS

9.1 INTRODUCTION

Carotenoids (Car) are known antioxidants. Extensive electrochemical studies in solution have established the low oxidation potentials and demonstrated the formation in various media of carotenoid radical cations ($Car^{\bullet+}$), dications (Car^{2+}), and the loss of H^+ to form the carotenoid neutral radical ($^{\#}Car^{\bullet}$) (Gao et al. 1996, Jeevarajan et al. 1996a) according to the following equations:

$$\mathrm{Car} \overset{E_1^0}{\rightleftharpoons} \mathrm{Car}^{\bullet+} + \mathrm{e}^- \tag{9.1}$$

$$\mathrm{Car}^{\bullet+} \overset{E_2^0}{\rightleftharpoons} \mathrm{Car}^{2+} + \mathrm{e}^- \tag{9.2}$$

$$\mathrm{Car} + \mathrm{Car}^{2+} \underset{K_{\mathrm{dp}}}{\overset{K_{\mathrm{com}}}{\rightleftharpoons}} 2\mathrm{Car}^{\bullet+} \tag{9.3}$$

$$\mathrm{Car}^{2+} \overset{K_{\mathrm{dp}}}{\rightleftharpoons} {}^{\#}\mathrm{Car}^{+} + \mathrm{H}^+ \tag{9.4}$$

$$\mathrm{Car}^{\bullet+} \overset{K'_{\mathrm{dp}}}{\rightleftharpoons} {}^{\#}\mathrm{Car}^{\bullet} + \mathrm{H}^+ \tag{9.5}$$

$${}^{\#}\mathrm{Car}^{+} + \mathrm{e}^- \overset{E_3^0}{\rightleftharpoons} {}^{\#}\mathrm{Car}^{\bullet} \tag{9.6}$$

It has been demonstrated (Mairanovski et al. 1975, Park 1978, Grant et al. 1988, Chen 1991, Khaled 1992, Jeevarajan et al. 1994a–c, Jeevarajan 1995, Jeevarajan and Kispert 1996, Jeevarajan et al. 1996a, Gao et al. 1997, Deng 1999, Liu and Kispert 1999, Hapiot et al. 2001, Konovalov et al. 2002) that great care must be taken to eliminate any traces of water or oxygen during the electrochemical studies, in order to obtain reproducible results. Accurate oxidation potentials could be deduced (Hapiot et al. 2001) only if fits were made to cyclovoltammograms (CV) recorded over six orders of magnitude of sweep times. The more traditional way of recording CV (Mairanovskii et al. 1975, Park 1978, Grant et al. 1988, Chen 1991, Khaled 1992, Jeevarajan et al. 1994a–c, Jeevarajan 1995, Jeevarajan and Kispert 1996, Jeevarajan et al. 1996a, Gao et al. 1997, Deng 1999, Liu and Kispert 1999, Hapiot et al. 2001, Konovalov et al. 2002) gave oxidation potentials some 50–100 mV lower.

Radical cations and neutral radicals of carotenoids can be measured and detected by electron paramagnetic resonance spectroscopy (EPR). Such techniques have been used to detect and characterize their properties. Unfortunately, the large number of different proton hyperfine couplings (~18) results in approximately 300,000 EPR lines for symmetrical carotenoids, if all couplings were resolved, and even a greater number for asymmetrical carotenoids. There would be an even larger number of EPR lines, if it was not for the rapid rotation of the methyl groups, even at 5 K, which causes the methyl proton couplings to be averaged out, so each methyl group exhibits one set of proton couplings. The large number of proton couplings results in a single, unresolved, inhomogeneously broadened powder EPR line of 14 Gauss peak-to-peak linewidth. To resolve the hyperfine couplings, continuous wave (CW) electron-nuclear double resonance (ENDOR) measurements have been carried out (Piekara-Sady et al. 1991, 1995, Wu et al. 1991, Jeevarajan et al. 1993b). For each set of equivalent protons, only two ENDOR lines separated by the hyperfine coupling constant, A, occur instead of multiple EPR lines. The ENDOR spectrum is recorded as a function of swept radio frequency (rf) centered at the free proton frequency, while the observing magnetic field is set to the center of the EPR line. To achieve the greatest spectral resolution, ENDOR measurements should be carried out in solution where the proton dipolar anisotropy is averaged out, and at a steady-state carotenoid concentration of <1 mM. However, because the carotenoid radical cations are short-lived in chlorinated (electron acceptor) solvents (Khaled et al. 1990, Deng et al. 2000) (lifetimes on the order of 100–200 s and even shorter in other solvents), a steady-state mM concentration is not achieved for ENDOR measurement in a closed sample tube (over 30 min to 1 h needed). A possible solution to this problem is to generate the radical cations electrochemically, but not *in situ*, since the electrodes become so hot in an ENDOR cavity from the applied rf power that the solution boils destroying the radicals. Flow techniques can work if the radicals produced externally to the ENDOR cavity are stable for many minutes. Such a condition was possible for canthaxanthin. Large quantities of carotenoid dications were produced by extensive electrolysis of a concentrated carotenoid solution (Equations 9.1 and 9.2) in an EPR tube external to the cavity followed by electron transfer from the excess carotenoid to the carotenoid dication (Equation 9.3), producing the carotenoid canthaxanthin radical cation (Piekara-Sady et al. 1993) setting up a steady-state equilibrium, which existed for approximately 30 min, just long enough for a study by ENDOR.

This experiment was successful because the equilibrium constant, K, for canthaxanthin is approximately 2×10^3. By contrast, for β-carotene, K is approximately 1, favoring formation of the diamagnetic dication, so the EPR signal of β-carotene radical cation in solution was found too weak for an ENDOR measurement. Intense ENDOR spectral lines in frozen solution or powder are due to proton couplings between 0.3 and 8.3 MHz assignable to the β-methyl protons on the carbons 5, 9, and 13 as well as the α-proton couplings (Piekara-Sady et al. 1991, Jeevarajan et al. 1993b). The INDO calculated proton couplings of carotenoid radical cations were found to be overestimated (Piekara-Sady et al. 1991). As shown later by DFT calculations (Gao et al. 2006), the error was sufficient that the formation of the carotenoid neutral radical, $^{\#}Car^{\bullet}$, via the loss of a proton from the carotenoid radical cation was missed. Carotenoid neutral radicals are formed upon photoirradiation and detectable by the observation of an ENDOR line at 21–23 MHz. Since the methyl groups rapidly rotate relative to the microfrequency even at 5 K and 670 GHz (Konovalova et al. 1999), the resolved ENDOR spectra of carotenoid radicals in frozen solutions could be observed, similar to those of carotenoid radicals on Nafion film (Wu et al. 1991), silica gel (Piekara-Sady et al. 1995), and on silica-alumina (Jeevarajan et al. 1993b, Konovalova and Kispert 1998). The resolved spectra were due to couplings from the protons on the rotating methyl groups. Further studies (Jeevarajan et al. 1994a,b, Konovalova and Kispert 1998) showed the electron is transferred from the carotenoid molecules adsorbed on the activated alumina or silica-alumina to the surface Lewis acid sites since ^{27}Al couplings were detected.

In this chapter, various EPR techniques that have been used to study the carotenoid radical cations and neutral radicals will be described. These methods with references are given in Table 9.1.

9.2 SIMULTANEOUS ELECTROCHEMICAL/ELECTRON PARAMAGNETIC RESONANCE (SEEPR) TECHNIQUES

The SEEPR technique allows the simultaneous recording of the CV and the CW EPR spectrum of the radicals produced during the electron transfer reactions (Khaled et al. 1991). The SEEPR technique consists of an IBM enhanced electrolytic cell inserted in a rotating cylindrical EPR cavity. The cell is no longer sold by IBM, but a description can be found (Khaled et al. 1990, 1991). The CVs were obtained using a commercial (BAS-100) electrochemical analyzer while simultaneously recording the EPR spectra during the scan.

Comproportionation equilibrium constants for Equation 9.3 between dications and neutral molecules of carotenoids were determined from the SEEPR measurements. It was confirmed that the oxidation of the carotenoids produced π-radical cations (Equations 9.1 and 9.3), dications (Equation 9.2), cations (Equation 9.4), and neutral π-radicals (Equations 9.5 and 9.6) upon reduction of the cations. It was found that carotenoids with strong electron acceptor substituents like canthaxanthin exhibit large values of K_{com}, on the order of 10^3, while carotenoids with electron donor substituents like β-carotene exhibit K_{com}, on the order of 1. Thus, upon oxidation 96% radical cations are formed for canthaxanthin, while 99.7% dications are formed for β-carotene. This is the reason that strong EPR signals in solution are observed during the electrochemical oxidation of canthaxanthin.

TABLE 9.1
Various EPR Techniques Used to Study Radical Cations and Neutral Radicals of Carotenoids

Technique	References
SEEPR detection of the transient radical during electrochemical preparation in solution	Khaled et al. (1991), Jeevarajan et al. (1994)
TREPR measurements	Jeevarajan et al. (1993)
Photoinduced electron transfer in frozen solutions	Konovalova et al. (1997)
Carotenoid radical cations formed by chemical oxidation with I_2	Ding et al. (1988)
Carotenoid radical cations formed by chemical oxidation with $FeCl_3$	Jeevarajan et al. (1996)
EPR spin trapping methods	Polyakov et al. (2001a,b,c, 2006)
EPR studies of host–guest complexes of carotenoids	Polyakov et al. (2004, 2006)
Measuring distances between carotenoid radicals and distant metals in matrices by using ESEEM methods and pulsed EPR relaxation techniques	Gao et al. (2005)
EPR studies of radical cations on activated alumina and silica-alumina	Konovalova et al. (1999, 2001a,b), Gao et al. (2002, 2003)
Use of high-frequency/high-magnetic field techniques to resolve *g*-anisotropy	Konovalova et al. (1999)
Identify high-spin metal complexes of carotenoids	Konovalova et al. (2003)
Establish the effect of distant high-spin metals on π-radicals properties	Konovalova et al. (2004), Gao et al. (2005)
Use of CW ENDOR techniques to detect β-proton hyperfine couplings and matrix nuclei	Kevan and Kispert (1976); Goslar et al. (1994), Piekara-Sady and Kispert (1994), Kispert and Piekara-Sady (2006)
Pulsed ENDOR techniques to detect β-proton hyperfine couplings and matrix nuclei	Konovalova et al. (2001), Focsan et al. (2008), Lawrence et al. (2008)
HYSCORE techniques to detect α-proton anisotropic coupling tensors	Konovalova et al. (2001), Focsan et al. (2008)
Density functional theory (DFT) calculations to interpret the powder ENDOR and HYSCORE spectra	Gao et al. (2006), Focsan et al. (2008)
Establish the use of the *g*-tensor parameters to detect the presence of dimers	Petrenko et al. (2005)

9.3 TIME-RESOLVED EPR (TREPR)

Time-resolved X- (Jeevarajan et al. 1993a) and Q-band (Jeevarajan et al. 1996b) EPR measurements have demonstrated that upon excitation at 300 K with a 308 nm pulsed excimer laser (200 mJ/pulse, 17 ns full width at half maximum (fwhm) of a carbon tetrachloride solution of carotenoid in an EPR flat cell, an electron is transferred from the carotenoid to the solvent. It was estimated that at least 10% of the 200 mJ/pulse laser light entered the cavity, giving rise to a broad emissive high-field line and a broad low-field absorption line, 0.1–0.5 μs after the laser pulse. The *g*-values of each species were determined and a large difference in radical stability as a function of solvent was found. The EPR signals decayed below detectability after several μs. Analysis of the polarized chemically induced dynamic electron polarized (CIDEP) 35 GHz spectra showed that a solvent-separated radical ion pair $CCl_4^{\bullet-}{\bullet\bullet}CCl_4{\bullet\bullet}Car^{\bullet+}$ was formed by electron transfer from the excited singlet state of the carotenoids to the solvent (CCl_4). The low-field line was due to $CCl_4^{\bullet-}$ ($g=2.009$) and the high-field line was due to $Car^{\bullet+}$ ($g=2.002$). The TREPR study showed the photophysics and photochemistry of carotenoid in CCl_4 is given in Scheme 9.1.

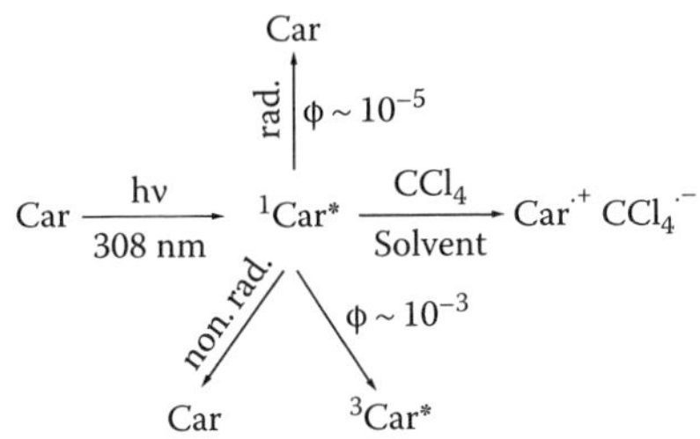

SCHEME 9.1 Photolysis and photochemistry of β-carotene. (Adapted from Jeevarajan, A.S., *J. Phys. Chem.*, 100, 669, 1996.)

9.4 PHOTOINDUCED ELECTRON TRANSFER IN FROZEN SOLUTIONS

The photoinduced electron transfer (Konovalova et al. 1997) from the excited singlet states of carotenoids to solvent molecules under 308–578 nm photolysis can also be monitored using continuous light sources. To lengthen the time available for an EPR study, carotenoid solutions of different electron acceptor solvents such CH_2Cl_2, $CHCl_3$, CCl_4, and CS_2 were frozen at 77 K. In these media, long-lived solvent-separated radical ion pairs are stable for several days and the yield is about 10 times higher in chlorinated solvents than in CS_2. Upon raising the temperature, the radicals formed were converted to dimer radical anions, $(CS_2)_2^{\bullet -}$, predissociation complexes, $(R\bullet\bullet\bullet Cl)^{\bullet -}$, and eventually to radical products. Oxygen-saturated carotenoid solutions enable one to study peroxyl radicals, $RO_2^{\bullet}$. A summary of the deduced reactions is given in Equations 9.7 through 9.22 adapted from Konovalova et al. (1997).

1. Photoexcitation of carotenoid to excited states: ($S_0 \rightarrow S_n$, $n = 2, \ldots$)

$$\text{Car} \xrightarrow{h\nu} (\text{Car}^*_{S_n}) \rightarrow \text{Car}^*_{S_2} \quad (9.7)$$

$$\text{Car}^*_{S_2} \xrightarrow{\tau_2 = 200\,\text{fs}} \text{Car}^*_{S_1} \quad (9.8)$$

$$\text{Car}^*_{S_1} \xrightarrow{\tau_1 = 10\,\text{ps}} \text{Car} \quad (9.9)$$

2. Formation of donor–acceptor complex:

$$\text{Car}^*_{S_n} + \text{Sol} \xrightarrow{kn} {}^1[\text{Car} \cdots \text{Sol}]^*, \quad n = 1, 2, \ldots \quad (9.10)$$

3. Decay of donor–acceptor complex through electron transfer to solvent molecules yielding primary radical ion pairs:

$${}^1[\text{Car} \cdots \text{Sol}]^* \rightarrow (\text{Car}^{\bullet +} \cdots \text{Sol}^{\bullet -}) \quad (9.11)$$

4. Formation of solvent-separated radical ion pairs:

$$(\text{Car}^{\bullet +} \cdots \text{Sol}^{\bullet -}) + \text{Sol} \rightarrow (\text{Car}^{\bullet +} \cdots \text{Sol} \cdots \text{Sol}^{\bullet -}) \quad (9.12)$$

a. $\text{Sol} = CS_2$

$$(\text{Car}^{\bullet +} CS_2^{\bullet -}) + CS_2 \rightarrow (\text{Car}^{\bullet +} \cdots CS_2 \cdots CS_2^{\bullet -}) \quad (9.13)$$

$$CS_2^{\bullet -} + CS_2 \rightarrow (CS_2)_2^{\bullet -} \quad (9.14)$$

b. Sol = RCl (chloroalkanes)

$$(Car^{\bullet+}\cdots RCl^{\bullet-}) + RCl \rightarrow [Car^{\bullet+}\cdots RCl\cdots(R\cdots Cl)^{\bullet-}] \quad (9.15)$$

$$(Car^{\bullet+}\cdots RCl^{\bullet-}) \rightarrow Car^{\bullet+} + R^{\bullet} + Cl^{-} \quad (9.16)$$

$$(R\cdots Cl)^{\bullet-} \rightarrow R^{\bullet} + Cl^{-} \quad (9.17)$$

$$R^{\bullet} + Car \rightarrow Car - R^{\bullet} \quad (9.18)$$

c. O_2 containing systems

$$(R\cdots Cl)^{\bullet-} + O_2 \rightarrow RO_2^{\bullet} + Cl^{-} \quad (9.19)$$

$$R^{\bullet} + O_2 \rightarrow RO_2^{\bullet} \quad (9.20)$$

$$RO_2^{\bullet} + Car \rightarrow Car^{\bullet+} + RO_2^{-} \quad (9.21)$$

$$RO_2^{\bullet} + Car \rightarrow RO_2^{\bullet} - Car \quad (9.22)$$

9.5 CHEMICALLY FORMED CAROTENOID RADICAL CATIONS

EPR techniques have also been used to detect and establish the structure of the carotenoid I_3^- complexes formed upon oxidation of carotenoids with I_2 (Ding et al. 1988). At 77 K the equilibrium is shifted so that $Car^{\bullet+}\bullet\bullet\bullet I_n^-$ forms where $n=5$, 7, or 9, and the polymeric I_n^- resides over the polyene chain in a π–π interaction giving rise to a detectable shift in the *g*-value.

The ferric ion is often used to form the carotenoid radical cation. However, care must be taken to control the concentration of the ferric ion relative to that of the carotenoid. Several existing equilibria have been studied by EPR, as well as NMR, LC-MS, and optical techniques. These studies have shown the following equilibria (Scheme 9.2) depending on the concentrations of Fe^{3+}, Fe^{2+}, and Cl^- relative to that of the neutral carotenoid and its radical cation and dication.

EPR techniques were used to show (Polyakov et al. 2001a) that one-electron transfer reactions occur between carotenoids and the quinones, 2,3-dichloro-5,6-dicyano-1,4-benzoquinone (DDQ), and tetrachlorobenzoquinone (CA). A charge-transfer complex (CTC) is formed with a *g*-values of 2.0066 and exists in equilibrium with an ion-radical pair ($Car^{\bullet+}\bullet\bullet\bullet Q^{\bullet-}$). Increasing the temperature from 77 K gave rise to a new five-line signal with $g=2.0052$ and hyperfine couplings of 0.6 G due to the DDQ radical anions. At room temperature a stable radical with $g=2.0049$ was detected, its

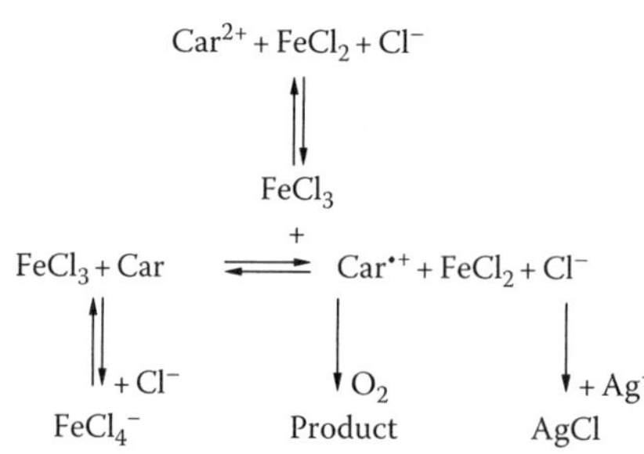

SCHEME 9.2 Equilibria that occur upon reaction of $FeCl_3$ with carotenoid in a halogenated solvent.

ENDOR spectrum exhibited chlorine and nitrogen splittings indicating a carotenoid–quinone radical adduct formation.

9.6 SPIN TRAPPING EPR METHOD

Spin trapping EPR technique and UV-Vis spectroscopy have been used (Polyakov et al. 2001b) to determine the relative rates of reaction of carotenoids with $^{\bullet}OOH$ radicals formed by the Fenton reaction in organic solvents. The Fe^{3+} species generated via the Fenton reaction

$$Fe^{2+} + H_2O_2 \rightarrow Fe^{3+} + {}^{\bullet}OH + OH^- \quad (9.23)$$

can oxidize (Polyakov et al. 2001c) the carotenoid to $Car^{\bullet+}$. At low concentrations of H_2O_2 (1 mM), the generated $^{\bullet}OH$ radical reacts with the solvent DMSO to produce $^{\bullet}CH_3$ (Figure 9.1). At increasing H_2O_2 concentration (1–10 mM), the *N-tert*-butyl-α-phenylnitrone (PBN) spin adducts of both $^{\bullet}OH$ and $^{\bullet}CH_3$ radicals appear in the EPR spectrum (Figure 9.1) (Polyakov et al. 2001b). At high concentration of H_2O_2 (500 mM), only $^{\bullet}OOH$ radicals were detected. Use of EPR along with optical absorption spectroscopy has demonstrated that the scavenging ability of carotenoid toward $^{\bullet}OOH$ increases with its oxidation potential (Figure 9.2) (Polyakov et al. 2001b). In Scheme 9.3 are listed the carotenoids for which the PBN-$^{\bullet}OOH$ adduct was formed.

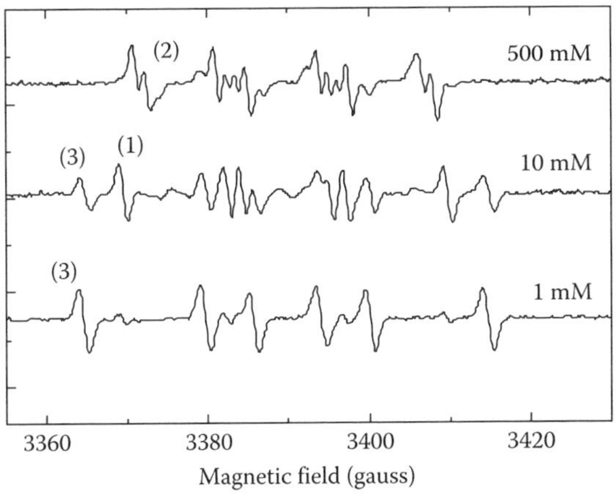

FIGURE 9.1 EPR spectra of spin adducts recorded during the Fenton reaction in DMSO at different H_2O_2 concentrations ($[FeCl_2] = 1$ mM), (1), (2), and (3) are $^{\bullet}OH$, $^{\bullet}OOH$, and $^{\bullet}CH_3$ radicals, respectively.

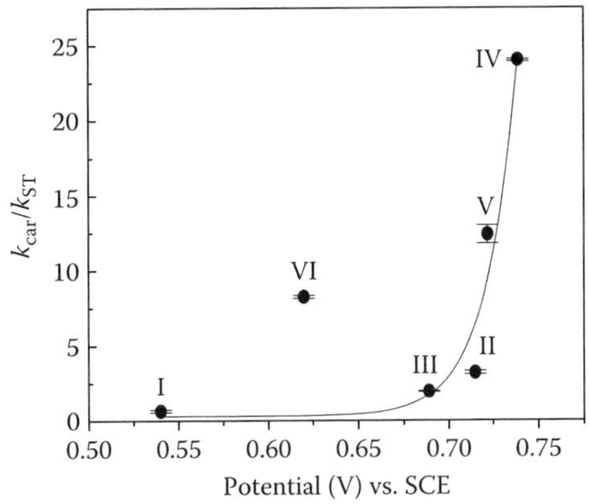

FIGURE 9.2 Dependence of scavenging ability of carotenoids in Scheme 9.3 (I–VI) with oxidation potential.

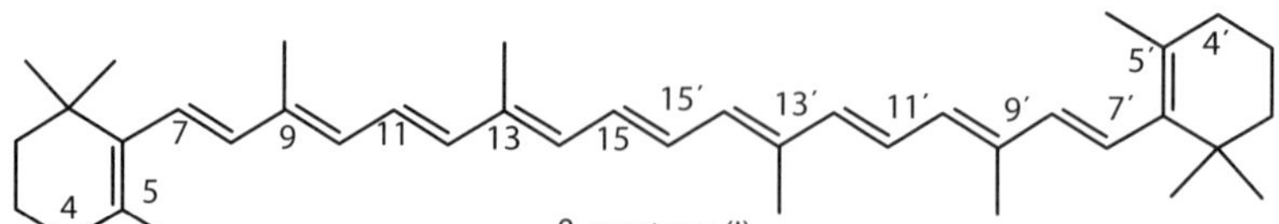

β-carotene (I)

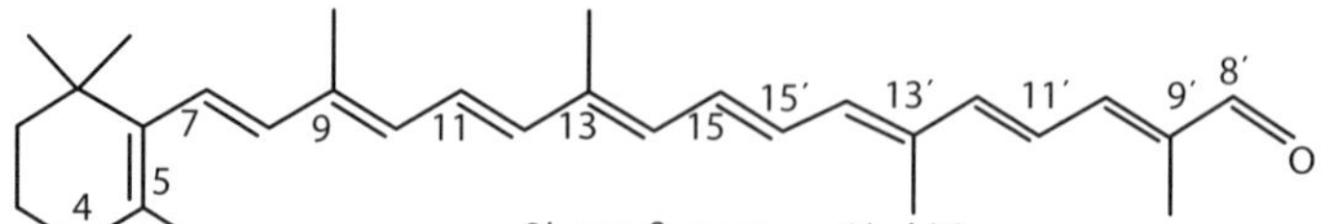

8′-apo-β-carotene-8′-al (II)

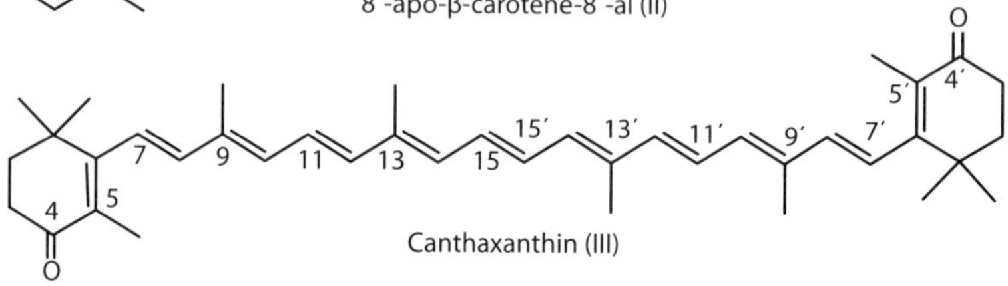

Canthaxanthin (III)

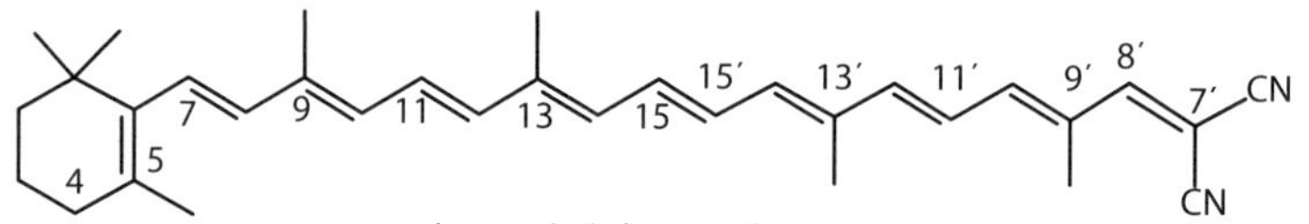

7′-apo-7′,7′-dicyano-β-carotene (IV)

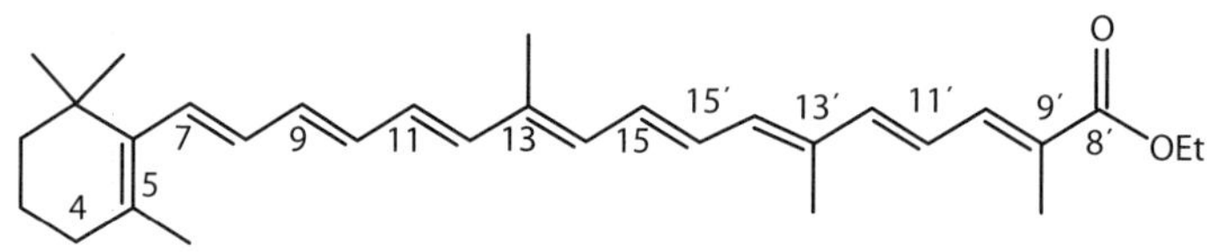

Ethyl 8′-apo-β-caroten-8′-oate (V)

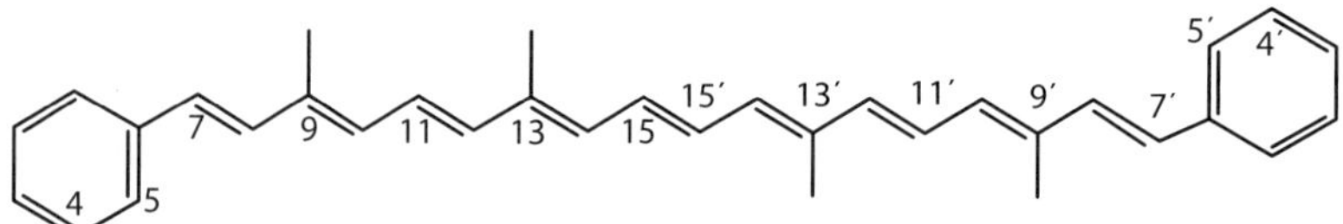

7,7′-diphenylcarotenen (VI)

SCHEME 9.3 Carotenoid structures.

It has been found (Polyakov et al. 2001c) that when carotenoids are involved in a reaction cycle with the participation of iron as Fe^{2+}, an increase of the total radical yield or a prooxidant effect will occur and will increase with decreasing carotenoid oxidation potential and its scavenging activity. The mechanism of the participating carotenoid is shown in Scheme 9.4 (Polyakov et al. 2001c).

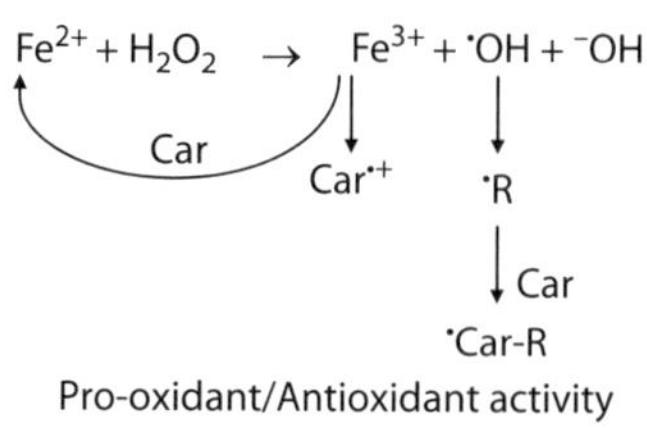

SCHEME 9.4 The formation of Car•+ is the prooxidant activity and •Car-R is the antioxidant activity (From Polyakov, N.E., *Free Rad. Biol. Med.*, 31, 398, 2001. With permission.)

9.7 SUPRAMOLECULAR COMPLEX FORMATION

The scavenging rates of canthaxathin and 7′,7′-dicyano-7′-apo-β-carotene toward •OOH after supramolecular complex formation between carotenoids and a triterpene glycoside, β-glycyrrhizic acid (GA) have been estimated using the EPR spin trapping technique. Complexation with GA results in an increase of scavenging rates by more than 10 times for both carotenoids, however complex formation had no affect on zeaxanthin. This effect was found to be due to an increase in oxidation potential upon complexation (Polyahov et al. 2006). Optical studies (Polyakov et al. 2006a,b) revealed that a 1:2 complex was formed with GA. The complex is a cyclic-like dimer of GA encapsulating a carotenoid molecule (Scheme 9.5). The stability constants are near 10^4 M^{-1}. In addition, GA forms inclusion complexes with carotenoid radical cations, which results in their stabilization.

Spin trapping methods were also used to show that when carotenoid-β-cyclodextrin 1:1 inclusion complex is formed (Polyakov et al. 2004), cyclodextrin does not prevent the reaction of carotenoids with Fe^{3+} ions but does reduce their scavenging rate toward ${}^{\bullet}OOH$ radicals. This implies that different sites of the carotenoid interact with free radicals and the Fe^{3+} ions. Presumably, the ${}^{\bullet}OOH$ radical attacks only the cyclohexene ring of the carotenoid. This indicates that the torus-shaped cyclodextrins, Scheme 9.6, protects the incorporated carotenoids from reactive oxygen species. Since cyclodextrins are widely used as carriers and stabilizers of dietary carotenoids, this demonstrates a mechanism for their safe delivery to the cell membrane before reaction with oxygen species occurs.

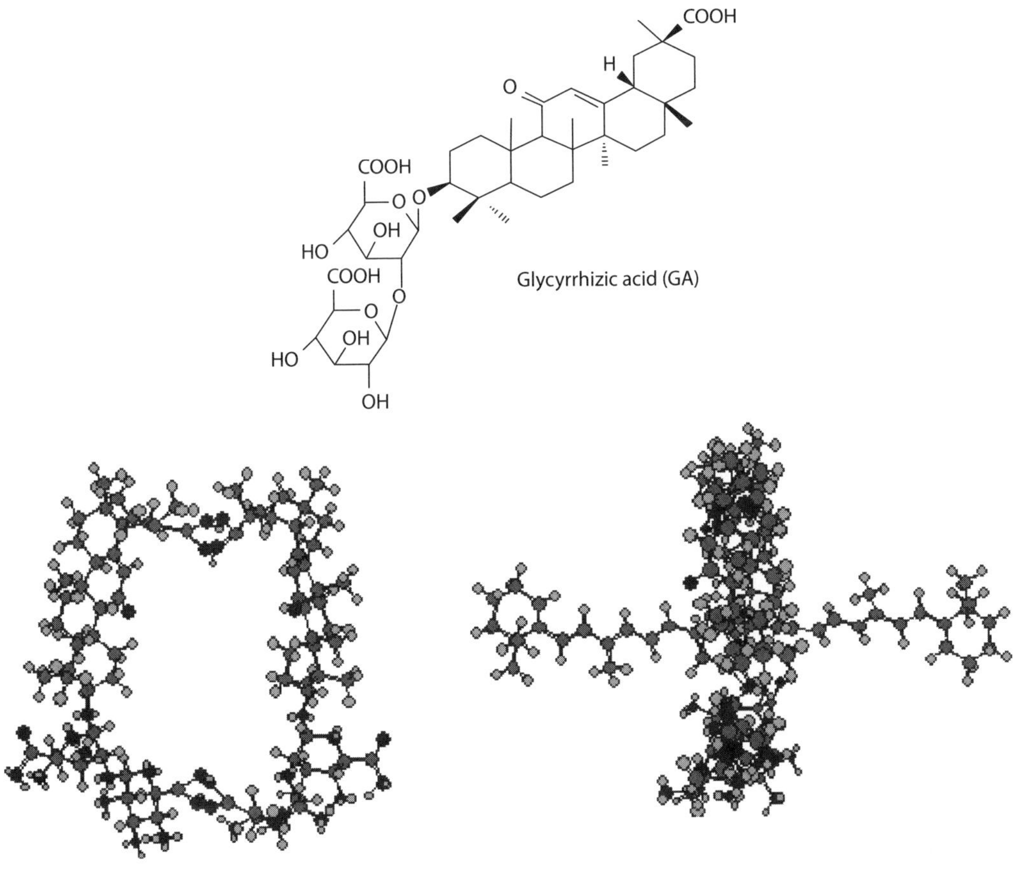

SCHEME 9.5 Suggested structures of the GA dimer and their inclusion complex with the carotenoid. Light gray: hydrogen atoms, dark gray: carbon atoms, black: oxygen atoms. (From Polyakov, N.E., *J. Phys. Chem. B*, 110, 6991, 2000. With permission.)

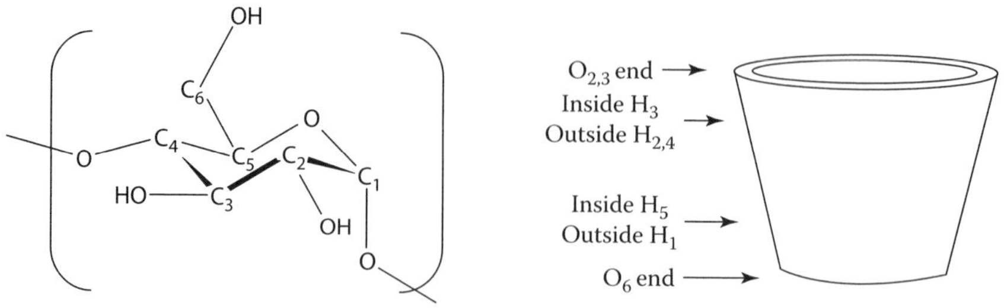

SCHEME 9.6 A macrocyclic oligosaccharide cyclodextrin forming a torus-shaped structure of cyclodextrin with rigid lipophilic cavities. (Adapted from Polyakov, N.E., *Free Rad. Biol. Med.*, 36, 872, 2004. With permission.)

9.8 CAROTENOID INTERACTION WITH SURROUNDINGS: ESEEM METHOD

Advanced EPR techniques such as CW and pulsed ENDOR, electron spin-echo envelope modulation (ESEEM), and two-dimensional (2D)-hyperfine sublevel correlation spectroscopy (HYSCORE) have been successfully used to examine complexation and electron transfer between carotenoids and the surrounding media in which the carotenoid is located.

Resolvable modulation is detected on a three-pulse echo decay spectrum of predeuterated β-carotene radical (Gao et al. 2005) as a function of delay time, *T*. The resulting modulation is known as ESEEM. Resolvable modulation will not be detected for nondeuterated β-carotene radical since the proton frequency is six times larger. The modulation signal intensity is proportional to the square root of phase sensitive detection and interfering two-pulse echoes and suppressed by phase-cycling technique (Gao et al. 2005). Analysis of the ESEEM spectrum yields the distance from the radical to the D nucleus, *a* the deuterium coupling constant, and the number of equivalent interacting nuclei (D). The details related to the analysis of the ESEEM spectrum are presented in Gao et al. 2005.

Pulsed ENDOR spectra can be recorded with a Bruker E-560 pulsed ENDOR accessory and an A-500 RF power amplifier using the Davies (π–T–$\pi/2$–τ–π–τ–echo) and Mims ($\pi/2$–τ–$\pi/2$–T–$\pi/2$–τ–echo) sequences where the additional RF π-pulse was applied during separation time, *T*. For the initial $\tau = 664$ ns for the Davies sequence of pulses, the proton couplings less than 1 MHz are not resolved, and the Davies ENDOR spectrum lacks spectral lines ±0.5 MHz around the free proton frequency. On the other hand, the Mims ENDOR will miss spectral lines due to couplings of 5, 10, and 15 MHz due to $\tau = 200$ ns. The Mims pulsed ENDOR sequence aids in detecting powder ENDOR spectra containing couplings less than 10 MHz where powder averaging may broaden and make the large couplings difficult to detect. Pulsed ENDOR hyperfine coupling simulations were carried out using the SimBud/SpecLab spectral simulation programs.

HYSCORE, is a 2D four-pulse ESEEM technique which provides correlation between nuclear frequencies originating from different electron manifolds. The sequence of four microwave pulses is $\pi/2$–τ–$\pi/2$–t_1–π– t_2–$\pi/2$–τ–echo where the echo amplitude is measured as a function of t_1 and t_2 at fixed τ. The α-proton anisotropic couplings can be detected by this technique (Konovalova et al. 2001a, Focsan et al. 2008).

9.8.1 Bonding of β-Carotene to Cu^{2+} in Cu-MCM-41

ESEEM measurements of perdeuterated all-*trans*-β-carotene imbedded in activated Cu-substituted MCM-41 molecular sieve revealed (Gao et al. 2005) that two deuterons of the carotenoid interact with the Cu^{2+} at a distance of 3.3 Å. Possible double bonds of β-carotene with one deuterium at each carbon that could interact with Cu are C7=C8, CH-C12, C15=C15, C12′ – C11′, and C8′=C7.

The narrow, resolved lines observed in the ^{1}H-ENDOR spectrum of β-carotene indicated that the methyl protons at the 5 or 5′ and the 9 or 9′ carbon positions were undergoing rapid rotation.

ENDOR lines predicted by improved DFT calculations for methyl protons at C13(13′) located at 16.5 MHz were broad and not observable due to incomplete averaging of the methyl proton couplings due to a hindering environment. Thus, the methyl groups at the C5(5′) and C9(9′) are located away from the surface of the pore and rapidly rotate, while those at C13(13′) interact with the surface. Steric hindrance by the terminal bulky trimethyl cyclohexene rings preclude attainment of the requisite distance between Cu^{2+} and the C7=C8 and C8′=C7′ bonds.

Three-pulse ESEEM spectrum of perdeuterated β-carotene imbedded in Cu-MCM-41 exhibits an echo decay with an echo modulation due to deuterons. The three-pulse ESEEM is plotted as a function of time, and curves are drawn through the maximum and minima. From ratio analysis of these curves, a best nonlinear least-squares fit determines the number of interacting deuterons, the distance (3.3±0.2 Å), and the isotopic coupling (0.06±0.2 MHz). This analysis made it possible to explain the observed reversible forward and backward electron transfer between the carotenoid and Cu^{2+} as the temperature was cycled (77–300 K).

9.9 EPR ON ACTIVATED SILICA-ALUMINA

Adsorption of carotenoids on activated silica-alumina results in their chemical oxidation and carotenoid radical formation. Tumbling of carotenoid molecules adsorbed on solid support is restricted, but the methyl groups can rotate. This rotation is the only type of dynamic processes which is evident in the CW ENDOR spectrum.

The Davies pulsed ENDOR spectrum of canthaxanthin oxidized on silica-alumina measured in the temperature range of 3.3–80 K showed no lineshape changes, which is in agreement with previous 330 GHz EPR studies of canthaxanthin radical cations (Konovalova et al. 1999). This implies very rapid rotation of the methyl groups down to 3.3 K.

Carotenoid radical formation and stabilization on silica-alumina occurs as a result of the electron transfer between carotenoid molecule and the Al^{3+} electron acceptor site. Both the three-pulse ESEEM spectrum (Figure 9.3a) and the HYSCORE spectrum (Figure 9.3b) of the canthaxanthin/$AlCl_3$ sample contain a peak at the ^{27}Al Larmor frequency (3.75 MHz). The existence of electron transfer interactions between Al^{3+} ions and carotenoids in $AlCl_3$ solution can serve as a good model for similar interactions between adsorbed carotenoids and Al^{3+} Lewis acid sites on silica-alumina.

9.10 DFT CALCULATIONS TO INTERPRET EPR SPECTRA

The Davies ENDOR spectrum of canthaxanthin radical photo-generated on silica-alumina consists of two well-resolved doublets placed around the ^1H Larmor frequency with hyperfine coupling constants (hfc) of 2.6 and 8.6 MHz and a broad low-intensity line with hfc approximately 13 MHz (Konovalova et al. 2001a). The couplings were assigned to β-protons of three different methyl groups, namely, C5(5′)–$C_βH_3$, C9(9′)–$C_βH_3$, and C13(13′)–$C_βH_3$, respectively. The original assignment of the large 13 MHz proton coupling to the C13(13′)–$C_βH_3$ group of the radical cation was based on RHF-INDO/SP calculations, the best available at the time, but shown to be in error by the more accurate DFT calculations (Gao et al. 2006, Focsan et al. 2008, Lawrence et al. 2008).

Using different DFT functionals and basis sets (Focsan et al. 2008, Lawrence et al. 2008) it was confirmed that the isotropic β-methyl proton hyperfine couplings do not exceed 9 MHz for the carotenoid radical cation, $Car^{•+}$. DFT calculations of neutral carotenoid radicals, $^{\#}Car^{•}$, formed by proton loss (indicated by #) from the radical cation, predicted isotropic β-methyl proton couplings up to 16 MHz, a fact that explained the large isotropic couplings observed by ENDOR measurements for methyl protons in UV irradiated carotenoids supported on silica gel, Nafion films, silica-alumina matrices, or incorporated in molecular sieves (Piekara-Sady et al. 1991, 1995, Wu et al.

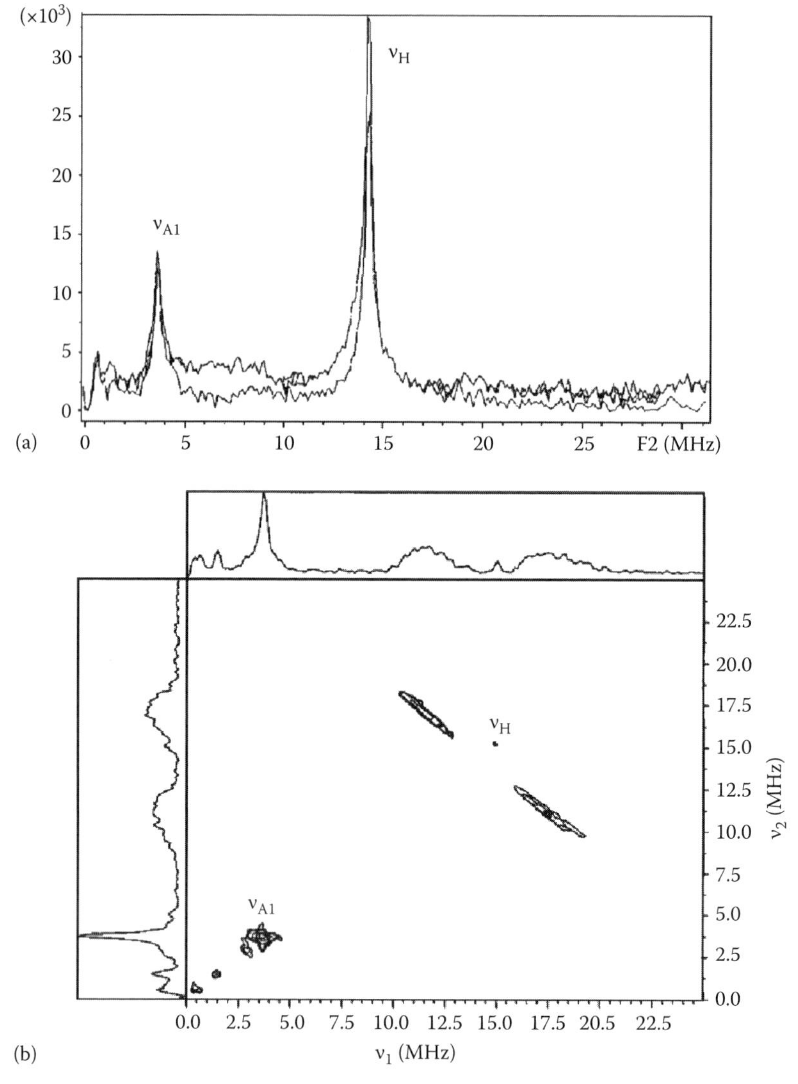

FIGURE 9.3 Spectra of the mixture of canthaxanthin (2 mM) and $AlCl_3$ (2 mM) in CH_2Cl_2 measured at 60 K at the field $\boldsymbol{B}_0=3349$ G and microwave frequency 9.3757 GHz: (a) superimposed plot of a set of three-pulse ESEEM spectra as the modulus Fourier transform and (b) HYSCORE spectrum measured with a $\tau=152$ ns. (From Konovalova, T.A., *J. Phys. Chem. B*, 105, 8361, 2001. With permission.)

1991, Konovalova and Kispert 1998, Konovalova et al. 1999, 2001a, Gao et al. 2002, 2003). It has now been shown that the carotenoid neutral radical is not formed in the absence of UV photolysis, and thus no resolvable methyl proton coupling of 13–16 MHz would have been observed.

The DFT calculations (Focsan et al. 2008) showed that the most energetically favorable neutral radical produced by deprotonation of zeaxanthin radical cation, $Zea^{\bullet+}$, is the one resulting from the loss of a proton from C4(4′)–$C_\alpha H_2$ group. In the case of violaxanthin radical cation, $Vio^{\bullet+}$, loss of a C9(9′)-methyl proton produces the most energetically favorable neutral radical. The calculations show that loss of a proton from a methyl group of position C5(5′) of the $Vio^{\bullet+}$, which contains an epoxy group at the C5(5′)–C6(6′) position, requires higher energy (~20 kcal/mol) and results in a nonconjugated neutral allyl radical. Such a radical would exhibit large couplings (Fessenden and Schuler 1963, Carrington and McLachlan 1967) which are easy to detect in the EPR spectrum, but no such peaks are observed. This difference between Zea and Vio in the location of proton loss has been shown to be critical to the quenching of chlorophyll in light harvesting center (LHCII) in the presence of excess light (Focsan et al. 2008).

The relative DFT energies ΔE for the neutral radicals formed by the loss of a proton from C4–$C_\alpha H_2$, C5–$C_\beta H_3$, C9–$C_\beta H_3$, or C13–$C_\beta H_3$ for 7′-apo-7′,7′-dicyano-β-carotene, lycopene, β-carotene, and 8′-apo-β-caroten-8′-al are all about the same (5.0±0.5, 11.0±0.5, and 12.5±0.5 kcal/mol) but higher than zeaxanthin by approximately 2.5 kcal/mol and higher than canthaxanthin by 7.0 kcal/mol (Focsan et al. 2008).

As a consequence of deprotonation, a change in the unpaired electron spin distribution of the neutral radicals produces larger methyl proton hyperfine constants (on the order of 10–16 MHz) than for the radical cation. Examples of the unpaired spin distribution of the radical cations and neutral radicals are depicted in Figure 9.4. It was observed that the unpaired spin density for the carotenoid neutral radicals increases at carbons along the chain distant from the position from where the proton was lost upon light irradiation. The #Vio•(4 or 4′) and #Vio•(5 or 5′) radicals do not show this distribution due to the presence of the epoxy group that localizes the unpaired spin density on the C4(4′) atom and C5(5′), C6(6′) atoms, respectively (see Figure 9.4). Proton loss from the C4(4′)–$C_\alpha H_2$ and C5(5′)-$C_\beta H_3$ groups of violaxanthin radical cation forms structures higher in energy than the most stable neutral radical, #Vio•(9 or 9′), and generates large couplings which are not experimentally observed in the EPR spectrum. However, loss of a proton from the methyl group at the C9(9′) or C13(13′) position exhibits similar hyperfine couplings to those of zeaxanthin neutral radicals formed by proton loss from these two positions.

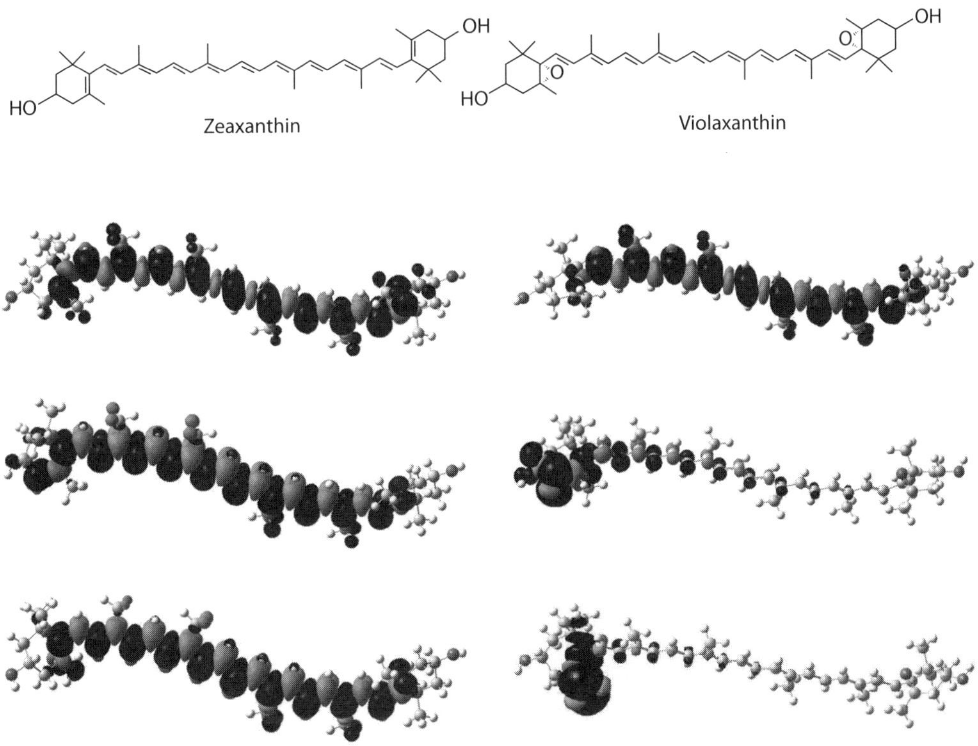

FIGURE 9.4 Unpaired spin distribution for zeaxanthin (left) and violaxanthin (right) radicals. From up to down: $Zea^{\bullet+}$, #Zea•(4), #Zea•(5) and $Vio^{\bullet+}$, #Vio•(4), #Vio•(5), respectively. The black represents excess α and the dark gray excess β unpaired spin density. (Adapted from Focsan, A.L. et al., *J. Phys. Chem. B*, 112, 1806, 2008. With permission.)

9.11 β-METHYL PROTONS FROM CW ENDOR: ADVANTAGE OF PULSED DAVIES AND MIMS ENDOR

CW ENDOR spectrum measurements carried out at 120 K (the optimum temperature for measuring resolved CW ENDOR powder spectra of carotenoid radicals) shows resolved lines from the β-methyl hfc (Piekara-Sady et al. 1991, 1995, Wu et al. 1991, Jeevarajan et al. 1993b) (see Figure 9.5). The lines above 19 MHz are due to neutral radicals according to DFT calculations (Gao et al. 2006).

Pulsed Davies and Mims ENDOR can also be used to characterize carotenoid radicals that are formed upon photo-irradiation of carotenoids on silica-alumina (Konovalova et al. 2001a, Focsan et al. 2008) or in Cu-substituted MCM-41 (Lawrence et al. 2008). Davies and Mims ENDOR complement each other because hfc are present in one but not in the other due to different pulse delay times. The pulsed ENDOR spectrum of zeaxanthin on silica-alumina was simulated using just the DFT calculated isotropic couplings for the $Zea^{\bullet+}$, $^{\#}Zea^{\bullet}$(4 or 4′), $^{\#}Zea^{\bullet}$(5 or 5′), $^{\#}Zea^{\bullet}$(9 or 9′), and $^{\#}Zea^{\bullet}$(13 or 13′) ((c) of Figure 9.6a). The best fit (b) with the experimental spectrum (a) was obtained by using both the isotropic (β-methyl proton) and anisotropic (α-proton) hyperfine couplings for the radical cation, and also for the neutral radicals which appeared to contribute to the outer peaks. Davies ENDOR spectrum of violaxanthin photo-generated on silica-alumina ((a) of Figure 9.6b) shows resolved features which can only be simulated (b) if DFT calculated isotropic and anisotropic hyperfine couplings for $Vio^{\bullet+}$, $^{\#}Vio^{\bullet}$(9 or 9′), and $^{\#}Vio^{\bullet}$(13 or 13′) are used. Hyperfine couplings of the neutral radicals $^{\#}Vio^{\bullet}$(4 or 4′) and $^{\#}Vio^{\bullet}$(5 or 5′) were not used for the simulation since these species are energetically unfavorable and the huge couplings (>10 MHz) are not observed. Similarly, if isotropic and anisotropic hyperfine couplings for only the radical cation are simulated (d), then the ENDOR peaks at positions D and E are not accounted for, showing the significant contribution of the neutral radicals to the ENDOR spectrum (Focsan et al. 2008).

Carotenoid neutral radicals are also formed under irradiation of carotenoids inside molecular sieves. Davies and Mims ENDOR spectra of lutein (Lut) radicals in Cu-MCM-41 were recorded and then compared with the simulated spectra using the isotropic and anisotropic hfcs predicted by DFT. The simulation of lutein radical cation, $Lut^{\bullet+}$, generated the Mims ENDOR spectrum in Figure 9.7a. Its features at B through E could not account for the experimental spectrum by themselves, so contribution from different neutral radicals whose features coincided with those of the experimental

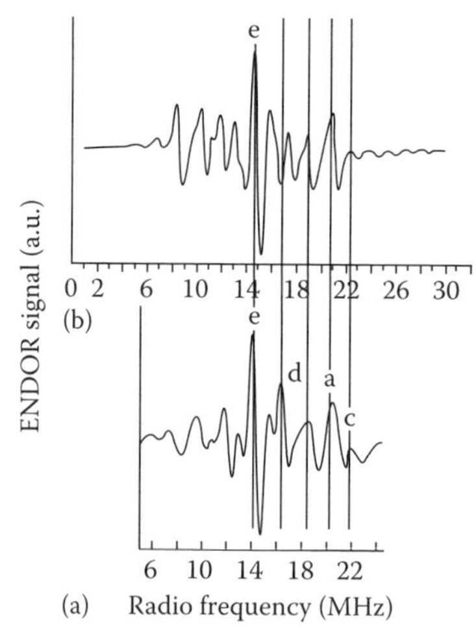

FIGURE 9.5 CW ENDOR spectrum of β-carotene radicals. (a) Experimental spectrum of Figure 9.4. (Reported in Wu, Y. et al., *Chem. Phys. Lett.*, 180, 573, 1991.) (b) Simulated ENDOR powder pattern (using linewidth of 0.6 MHz) for the sum of radical cation and neutral radicals in 5:3:1:1 ratio. (Reported in Gao, Y. et al., *J. Phys. Chem. B*, 110, 24750, 2006. With permission.)

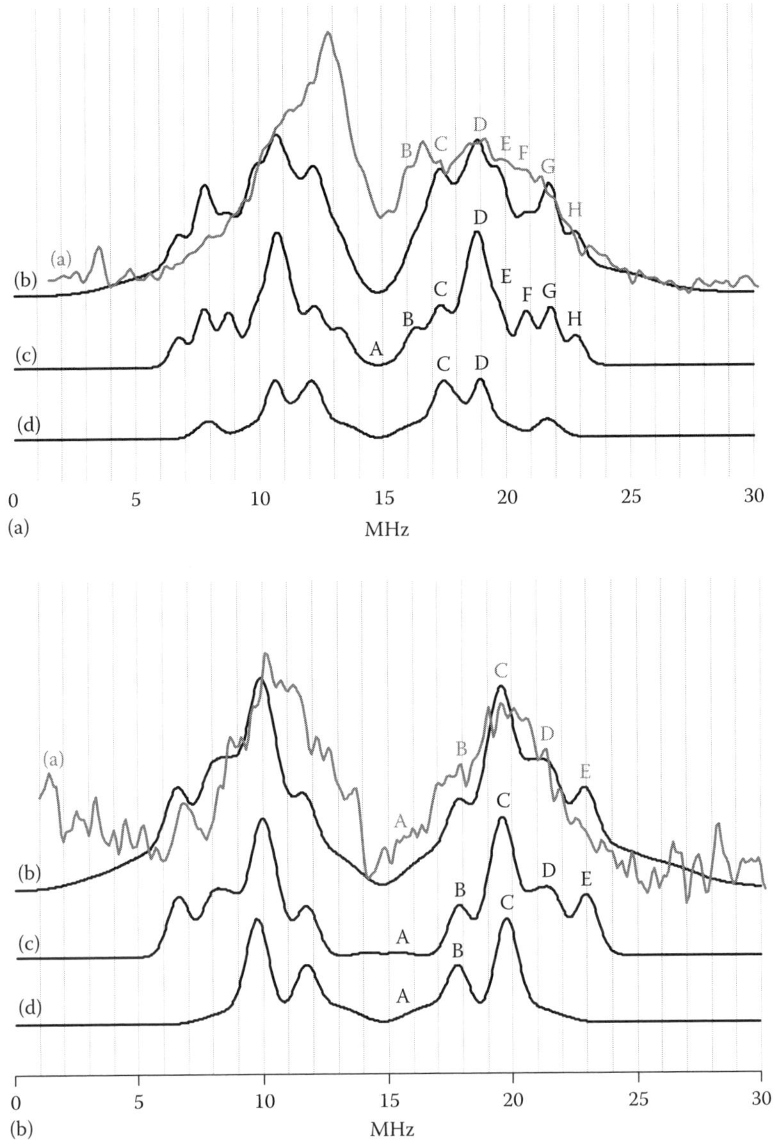

FIGURE 9.6 (a) Davies ENDOR spectrum of zeaxanthin radicals on silica-alumina: (a) Experimental parameters: T=50 K, $\boldsymbol{B}$=3460 G, ν=9.691211 GHz, τ=200 ns, MW π pulse=160 ns, RF π pulse=10 μs, and SRT=1021.02 μs, (b) simulated spectrum including both isotropic and anisotropic couplings for all five species, (c) simulated spectrum using only the isotropic coupling constants for all five species, and (d) simulated spectrum using both isotropic and anisotropic couplings of the radical cation only. (b) Davies ENDOR spectrum of violaxanthin radicals on silica-alumina: (a) Experimental parameters: T=40 K, $\boldsymbol{B}$=3460 G, ν=9.686928 GHz, τ=200 ns, MW π pulse=80 ns, RF π pulse=20 μs, and repetition time=1021.02 μs, (b) simulated spectrum including both isotropic and anisotropic couplings for all four species, (c) simulated spectrum using only the isotropic coupling constants for all four species, and (d) simulated spectrum using both isotropic and anisotropic couplings of the radical cation only. (Adapted from Focsan, A.L. et al., *J. Phys. Chem. B*, 112, 1806, 2008. With permission.)

spectrum was needed. When the spectral simulation of the neutral radical, $^{\#}$Lut$^{\bullet}$(6′), having the lowest energy was added to that of the radical cation, the features were better matching (Figure 9.7b). However, the peak at D(D′) was further improved when the radical cation Lut$^{\bullet+}$, and neutral radicals $^{\#}$Lut$^{\bullet}$(6′), $^{\#}$Lut$^{\bullet}$(4), $^{\#}$Lut$^{\bullet}$(5), $^{\#}$Lut$^{\bullet}$(9), and $^{\#}$Lut$^{\bullet}$(13) were simulated in 1:1:1:1:1:1 ratio (Figure 9.7c). It was

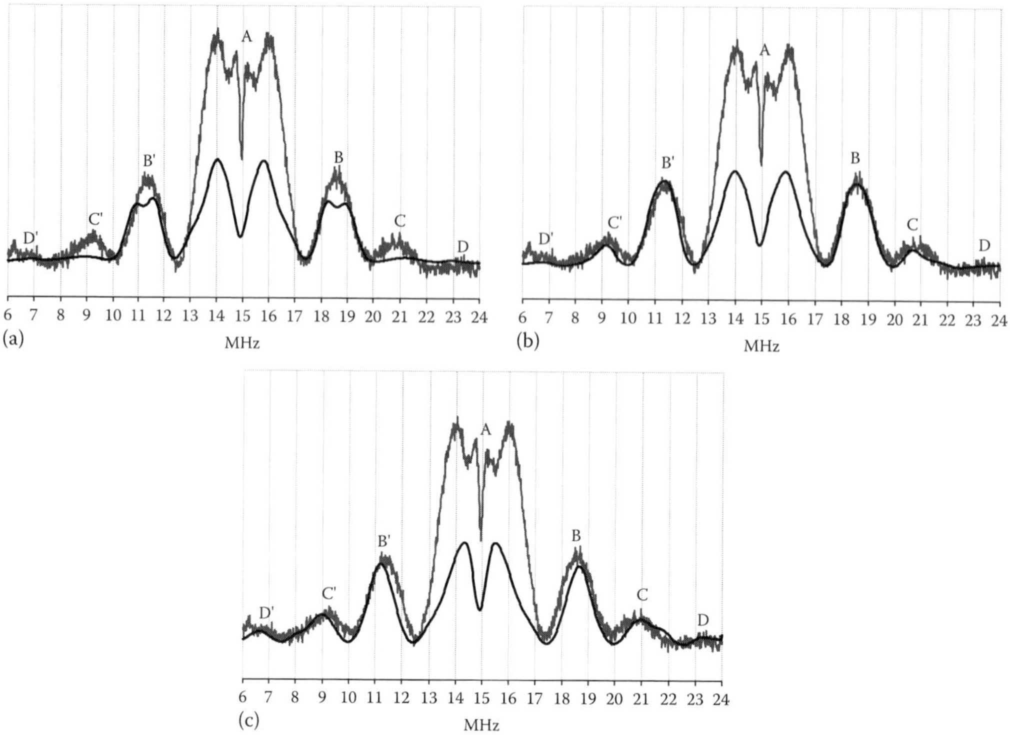

FIGURE 9.7 In gray: experimental powder Mims ENDOR spectrum of lutein radicals in Cu-MCM-41 measured at $T=15\,K$, $\nu=9.8394\,GHz$, $\boldsymbol{B}=3505.0\,G$, MW $\pi/2$ pulse$=32\,ns$, RF π pulse$=14\,\mu s$, $\tau=200\,ns$, and repetition time$=2999.82\,\mu s$. (a) black: simulated spectrum of $Lut^{\bullet+}$ using DFT proton hyperfine coupling tensors predicted by DFT; (b) black: simulated spectrum of $Lut^{\bullet+}$ and $^{\#}Lut^{\bullet}(6')$ using DFT proton hyperfine coupling tensors predicted by DFT; and (c) black: simulated spectrum of $Lut^{\bullet+}$, $^{\#}Lut^{\bullet}(6')$, $^{\#}Lut^{\bullet}(4)$, $^{\#}Lut^{\bullet}(5)$, $^{\#}Lut^{\bullet}(9)$, and $^{\#}Lut^{\bullet}(13)$ in 1:1:1:1:1:1 ratio using DFT proton hyperfine coupling tensors. (Focsan, A.L. et al., *J. Phys. Chem. B*, 112, 1806, 2008. With permission.)

determined that the hfc for the neutral radicals formed by proton loss at the primed positions of lutein do not significantly change the simulated spectrum, so they were not included in the simulation.

9.12 α-PROTONS FROM HYSCORE ANALYSIS

2D-HYSCORE was used to characterize radicals of zeaxanthin and violaxanthin photo-generated on silica-alumina and to deduce the anisotropic α-proton hyperfine coupling tensors. The couplings (MHz) were assigned based on DFT calculations. From such a comparison, the presence of the neutral radicals formed by proton loss from the radical cations was confirmed.

The hyperfine coupling tensors of carotenoids were determined from the HYSCORE analysis of the contour line-shapes of the cross-peaks (Dikanov and Bowman 1995, 1998, Dikanov et al. 2000), which provided the principal components of the tensors that appear to be rhombic. Such tensors are characteristic of planar conjugated radicals with the unpaired spin in a p_Z orbital of the carbon of the C–H group.

The cross-peak coordinates represent two frequency values, ν_α and ν_β, where $\nu_\alpha + \nu_\beta = 2\nu_I$ and ν_I is the proton frequency. When plotted in the coordinates ν^2_α and ν^2_β, the contour lineshape is transformed into a straight line segment. An extrapolation of this straight line permits the determination of the hyperfine tensors. A curve obtained by choosing some frequencies in the range will intersect the line defined by the squares of the values ν^2_α and ν^2_β in two points. The values where the curve intersects the experimental data are $(\nu_{\alpha 1}, \nu_{\beta 1})$ and $(\nu_{\alpha 2}, \nu_{\beta 2})$, where $\nu_\alpha = A_i/2 + \nu_I$ and $\nu_\beta = \nu_I - A_i/2$. This gives two values of the anisotropic coupling tensor, A_i.

HYSCORE spectra of zeaxanthin radicals photo-generated on silica-alumina were taken at two different magnetic fields $B_0 = 3450\,G$ and $B_0 = 3422\,G$, respectively. In order to combine the data from the two spectra, the field correction was applied (Dikanov and Bowman 1998). The correction consists of a set of equations that allow transformation of spectra to a common nuclear Zeeman frequency. The set of new frequencies was added to that of the former spectrum and plotted as the squares of the frequencies ν^2_α and ν^2_β. Examples of these plots can be found in Focsan et al. 2008.

9.13 *g*-ANISOTROPY: HIGH-FIELD *g*-TENSOR RESOLUTION

High-field EPR (HFEPR) spectroscopy greatly improves the resolution of the EPR signals for spectral features such as the *g*-tensor. Deviations of the *g*-value from free electron $g = 2.0023$ are due to spin-orbital interactions, which are one of the most important structural characteristics (Kevan and Bowman 1990). Using a higher frequency results in enhanced spectral resolution in accordance with the resonance equation:

$$H = \frac{h\omega}{2\pi g\beta}$$

where
- h is the Planck constant
- β is the Bohr magneton
- ω is the frequency of electromagnetic radiation

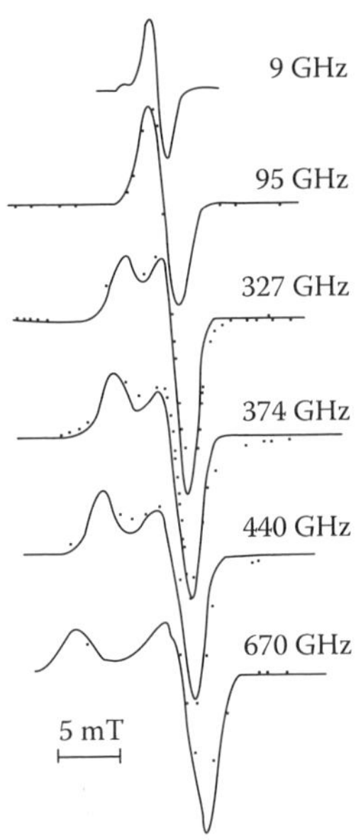

FIGURE 9.8 HF-EPR spectra of canthaxanthin radical cation adsorbed on silica-alumina: (solid line)—experimental spectra recorded at 5 K; (dotted line)—simulated spectra using *g*-tensor values $g_{zz} = 2.0032$ and $g_{xx} = g_{yy} = 2.0023$ and linewidth of 13.6 G. (From Konovalova, T.A., *J. Phys. Chem. B*, 103, 5782, 1999. With permission.)

If inhomogeneous broadening of the EPR linewidth is primarily due to unresolved hyperfine couplings (hfc), at higher frequencies the *g*-anisotropy will dominate over the hyperfine interactions, i.e., the condition $(\Delta g / g_{iso} H_o) > \Delta H_{hfc}$ must be fulfilled.

The advantage of high-frequency EPR in *g*-anisotropy resolution is provided by the spectrum of canthaxanthin radical cation adsorbed on silica-alumina (Figure 9.8). The X-band (9 GHz) EPR spectrum of a carotenoid radical cation consists of an unresolved single line with $g_{iso} = 2.0027 \pm 0.0002$, which is characteristic for organic π-radicals (Wertz and Bolton 1972). The line shape most closely resembles that of a Gaussian line, which indicates that the line is inhomogeneously broadened by unresolved proton hyperfine structure.

The 327–670 GHz EPR spectra of canthaxanthin radical cation were resolved into two principal components of the *g*-tensor (Konovalova et al. 1999). Spectral simulations indicated this to be the result of *g*-anisotropy where $g_{\text{II}} = 2.0032$ and $g_{\perp} = 2.0023$. This type of *g*-tensor is consistent with the theory for polyacene π-radical cations (Stone 1964), which states that the difference $g_{xx} - g_{yy}$ decreases with increasing chain length. When $g_{xx} - g_{yy}$ approaches zero, the *g*-tensor becomes cylindrically symmetrical with $g_{xx} = g_{yy} = g_{\perp}$ and $g_{zz} = g_{\text{II}}$. The cylindrical symmetry for the all-*trans* carotenoids is not surprising because these molecules are long straight chain polyenes. This also demonstrates that the symmetrical unresolved EPR line at 9 GHz is due to a carotenoid π-radical cation with electron density distributed throughout the whole chain of double bonds as predicted by RHF-INDO/SP molecular orbital calculations. The lack of temperature

TABLE 9.2
Comparison of *g*-Values for Various Radical Cations with Those Observed for Canthaxanthin Radical Cation

Radical Cations	g_{xx}	g_{yy}	g_{zz}	g_{iso}	Structure	Reference
$TP^{•+}/AsF_5^-$	2.0031	2.0028	2.0022	2.0027	Polymer	Robinson et al. (1985)
$TP^{•+}/PF_6^-$	2.00312	2.00230	2.00206	2.00249	Stacked array	Kispert et al. (1987)
$QP^{•+}/PF_6^-$	2.00217	2.00217	2.00310	2.00248	Stacked array	Kispert et al. (1987)
$P_{865}^{•+}$	2.00337	2.00248	2.00208	2.00264	Dimer	Klette et al. (1993)
$P_{700}^{•+}$	2.00317	2.00260	2.00226	2.0027	Dimer	Bratt et al. (1997)
Bchl*a*$^{•+}$	2.0033	2.0026	2.0022	2.0027	Monomer	Burghaus et al. (1991)
$Car^{•+}$	2.0023	2.0023	2.0032	2.0026	Symmetrical π-RC$^{•+}$	Konovalova et al. (1999)

Source: Konovalova, T.A., *J. Phys. Chem. B*, 103, 5782, 1999.

dependence of the EPR linewidths over the range of 5–80 K at 327 GHz suggests rapid rotation of methyl groups even at 5 K that averages out the proton couplings from three oriented β-protons.

Determination of *g*-tensor components from resolved 327–670 GHz EPR spectra allows differentiation between carotenoid radical cations and other C–H π-radicals which possess different symmetry. The principal components of the *g*-tensor for $Car^{•+}$ differ from those of other photosynthetic RC primary donor radical cations, which are practically identical within experimental error (Table 9.2) (Robinson et al. 1985, Kispert et al. 1987, Burghaus et al. 1991, Klette et al. 1993, Bratt et al. 1997) and exhibit large differences between g_{xx} and g_{yy} values.

9.14 HIGH-FIELD EPR MEASUREMENTS OF METAL CENTERS

The HF-EPR can also be used to good advantage to study high-spin systems, where the zero-field splitting (ZFS) term is often dominant in the spin Hamiltonian. The examples of such systems are transition metal ions like Mn(II), Ni(II), and Fe(III), which have been used for introducing active sites in mobile crystalline material (MCM-41) mesoporous materials. MCM-41 containing well-organized nanometer-sized channels has been found to be a good photoredox system where long-lived photoinduced electron transfer from bulky biomolecules such as carotenoids can occur. Although the MCM-41 framework can act as an electron acceptor, replacement of some tetrahedral Si(IV) in the MCM-41 framework by transition metal ions produces a long-lived charge separation between the carotenoid radicals and the metal electron acceptor sites.

9.14.1 Carotenoids in Ni-MCM-41

Photo-oxidation of b-carotene and canthaxanthin in mesoporous Ni(II)-containing MCM-41 molecular sieves was studied by 9–220 GHz EPR spectroscopy (Konovalova et al. 2001b). The presence of Ni(II) ions in Ni-MCM-41 was verified by 220 GHz EPR spectroscopy. Ni-containing MCM-41 samples measured at 9 GHz showed no EPR signals consistent with Ni(II) ions. The 220 GHz EPR spectrum of activated Ni-MCM-41 exhibits a broad line with g-value of 2.26 (Figure 9.9) providing direct evidence of Ni(II) incorporation into the MCM-41 framework. It has been reported (Abragam and Bleaney 1970) that Ni(II) ions in an octahedral environment give rise to very broad EPR lines with g-values of 2.10–2.33.

Irradiation of Ni-MCM-41 at 350 nm generates new paramagnetic species stable at 77 K whose spectra are superimposed. From the 110 GHz spectrum of Ni-MCM-41 (5 K), two different

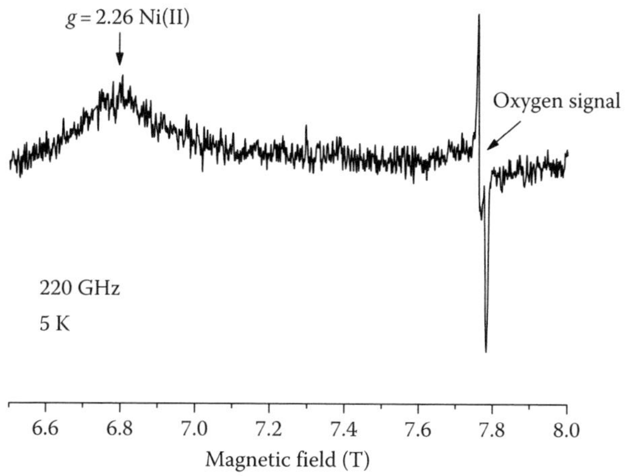

FIGURE 9.9 220 GHz EPR spectrum of Ni-MCM-41 activated at 260°C, degassed, and measured at 5 K. Modulation frequency 81 kHz, modulation amplitude 30 mV, and sweep rate 0.1 T/min. (From Konovalova, T.A., *J. Phys. Chem. B*, 105, 7549, 2001. With permission.)

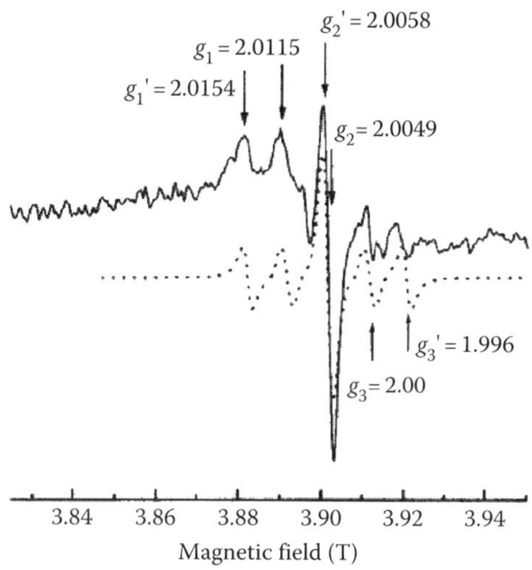

FIGURE 9.10 110 GHz EPR spectrum at 5 K of Ni-MCM-41 after 350 nm irradiation (solid line), simulated spectrum (dotted line). (From Konovalova, T.A., *J. Phys. Chem. B*, 105, 7549, 2001. With permission.)

paramagnetic species were detected (Figure 9.10). Spectral simulations determined *g*-tensors of these species. The signal with a rhombic *g*-tensor $g_1=2.0115$, $g_2=2.0049$, $g_3=2.00$ is characteristic of O_2^- species generated in MCM-41 ($g_1=2.012$, $g_2=2.003$, $g_3=2.00$) (Chang et al. 1999). We assigned the second rhombic *g*-tensor ($g_1'=2.0154$, $g_2'=2.0058$, $g_3'=1.996$) to V-centers (Figure 9.10). The so-called V-centers or trapped holes on the framework oxygens have been observed for metal-substituted MCM-41 after γ-irradiation at 77 K (Prakash et al. 1998). Similar, but less intense, signals were observed for the siliceous MCM-41.

Photo-oxidation of carotenoids in Ni-MCM-41 produces an intense EPR signal (Figure 9.11) with *g*-value 2.0027 due to the carotenoid radical; another, less intense EPR signal, with $g=2.09$ is attributed to an isolated Ni(I) species produced as a result of electron transfer from the carotenoid molecule to Ni(II). It has been reported that Ni(I) ions prepared upon reduction of Ni(II)-MCM-41 by heating in a vacuum or in dry hydrogen exhibits an EPR spectrum with $g_\perp=2.09$ and $g_\parallel=2.5$

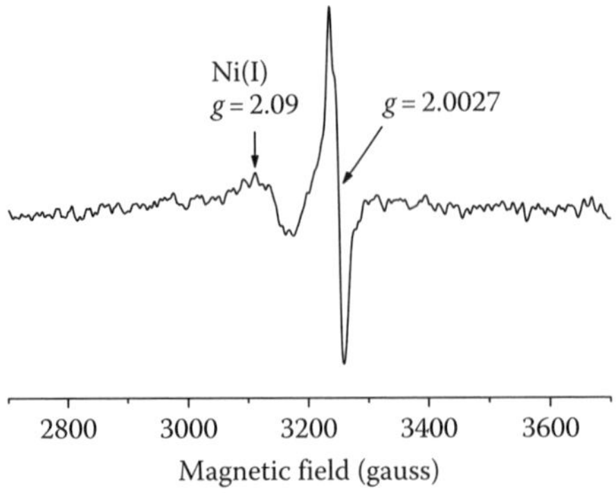

FIGURE 9.11 9 GHz EPR spectrum at 77 K of β-carotene in Ni-MCM-41 after 350 nm irradiation. (From Konovalova, T.A., *J. Phys. Chem. B*, 105, 7549, 2001. With permission.)

(Hartmann et al. 1996). The $g_{\parallel}$ component is often too weak to observe. The Ni(I) EPR signals were not detected upon 350 nm irradiation of Ni-MCM-41 samples before adsorption of carotenoids. Detected at 9 GHz, EPR signals of an isolated Ni(I) species with $g=2.09$ provide direct evidence for the reduction of Ni(II) ions by carotenoids.

9.14.2 Carotenoids in Fe-MCM-41

Multifrequency EPR spectroscopy was applied to study Fe(III)-MCM-41 mesoporous molecular sieves with incorporated carotenoids (Konovalova et al. 2003).

EPR spectroscopy is a useful technique for characterizing the iron sites in both the low-spin ($S=1/2$) and high-spin ($S=5/2$) electronic configurations. The spin Hamiltonian for high-spin iron is given by the following equation (Dowsing and Gibson 1969, Sweeney et al. 1973):

$$H_S = g\beta \boldsymbol{BS} + D\left(S_Z^2 - \frac{1}{3}\boldsymbol{S}^2\right) + \boldsymbol{E}(S_X^2 - S_Y^2) \tag{9.24}$$

In this case the g-tensor exhibits extremely small anisotropy, and the spectral characteristics are determined by the ZFS parameters D (axial) and E (rhombic). When the symmetry is axial, $D \neq 0$ and $E=0$. In the case of rhombic symmetry, $E/D=1/3$. Most high-spin d^5 systems do not belong to one of these special cases. Several different symmetries of Fe^{3+} contribute to multicomponent EPR spectra with overlapping signals. Such complex spectra arising from more than one center can be analyzed at different microwave frequencies. For high-spin Fe^{3+} in proteins, zeolites, and MCM-41 molecular sieves, the electron Zeeman interaction ($g\beta B_0 S$) is much smaller at the X-band frequency than the ZFS interaction. This makes interpretation of the 9 GHz EPR spectra difficult due to inhomogeneous broadening arising from the ZFS and overlapping signals. Use of higher microwave frequency is particularly advantageous in this case.

Studies with 9–287 GHz EPR (Konovalova et al. 2003) were carried out to characterize the Fe^{3+} sites in Fe-MCM-41 molecular sieves. Multifrequency EPR measurements were also performed to elucidate the types of iron sites which are responsible for carotenoid oxidation, their stability, and accessibility. The X-band EPR spectrum of Fe-MCM-41 activated at 260°C and recorded at 77 K consists of a strong sharp peak at $g=4.3$ with a shoulder at $g=9.0$ (Figure 9.12a). The presence of these signals originating from the middle Kramers doublet and the lowest Kramers doublet, respectively, is characteristic of high-spin Fe^{3+} when $E/D=1/3$ (Abragam and Bleaney 1970, Pilbrow 1990). The observation of a $g=4.3$ signal in zeolites and aluminophosphate molecular sieves is usually considered as evidence for the presence of framework Fe^{3+} ions (Goldfarb et al. 1994, Kosslick

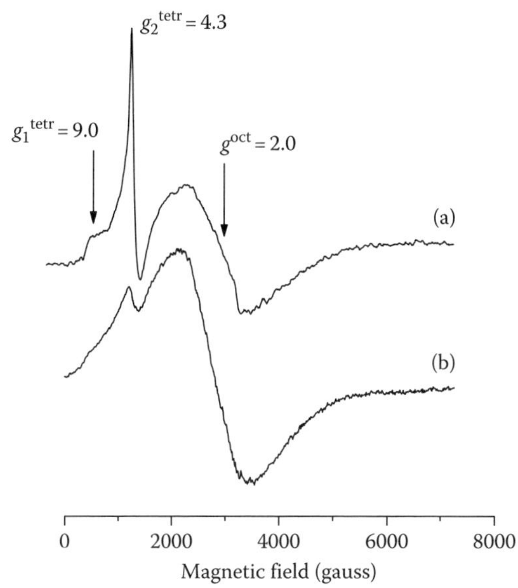

FIGURE 9.12 X-band EPR spectra of Fe(III)-MCM-41 activated at (a) 260°C and (b) 360°C, and measured at 77 K. (From Konovalova, T.A., *J. Phys. Chem. B*, 107, 1006, 2003. With permission.)

et al. 1998). The X-band spectrum of Fe-MCM-41 also exhibits a broad (~2000 G) signal with $g \approx 2$. A $g=2.0$ signal in zeolites is commonly assigned to extra-framework Fe^{3+} ions (Goldfarb et al. 1994, Kosslick et al. 1998). Figure 9.12b shows that activation of Fe-MCM-41 at higher temperature diminishes the $g=4.3$ framework iron signal, and significantly increases the extra-framework iron signal at $g=2.0$. This is consistent with the observation that tetrahedral coordination of the framework Fe^{3+} ions is not very stable (Kosslick et al. 1998).

To obtain additional information regarding the different types of Fe^{3+} sites in Fe-MCM-41 EPR measurements at higher microwave frequencies were carried out. It was found that the $g=4.3$ signal is not observed at 94.3 GHz and higher frequencies. This might be due to excessive broadening by frequency-dependent relaxation mechanisms. It is also possible that with frequency increase the electron Zeeman interaction becomes comparable to *D* resulting in inhomogeneous line broadening. In contrast, the shape of the $g=2.0$ signal is better determined at higher frequencies.

At 94–287 GHz (Konovalova et al. 2003) the $g=2.0$ line is resolved into two broad peaks and an intense narrow signal. To determine g-values and the ZFS parameters *D* and *E* for different Fe^{3+} signals, spectral simulations were performed using powder matrix diagonalization approach which is important for high-spin iron systems (Yang and Gaffney 1987, Gaffney et al. 1993). Simulations were carried out using a Gaussian lineshape and varied isotropic linewidth, *E/D* ratio and g-values. The parameters obtained at higher frequencies were used for spectral simulations at lower frequencies. Simulated parameters are given in Table 9.3. It was demonstrated (Konovalova et al. 2003) that high-frequency/high-field EPR is a promising technique to increase spectral resolution for proper assignment of different Fe^{3+} sites, which cannot be resolved by the X-band experiments. The broad unresolved EPR line at 9 GHz in the $g=2$ region is due to overlapping signals from Fe^{3+} sites with different zero-field parameters. The peak with $g=2.45$ is assigned to aggregated Fe^{3+}. The signal with $g=2.07$ can be attributed to Fe^{3+} coordinated to oxygen atoms on the surface of the pore. A narrow line with $g_x=g_y=2.003$, $g_z=1.999$, and $E/D=0.3$ was attributed to a single Fe^{3+} site.

Figure 9.13 compares X-band EPR spectra of Fe-MCM-41 before (a) and after (b) and (c) carotenoid adsorption. The sample with incorporated Car exhibits a signal with $g=2.0028 \pm 0.0002$, characteristic of carotenoid radical cation prior to irradiation (Figure 9.13b). Irradiation of the samples at 365 nm (77 K) increases the $Car^{\bullet+}$ signal intensity (Figure 9.13c). The X-band experiments (Figure

TABLE 9.3
Simulated EPR Parameters of High-Spin Fe^{3+} Sites in Fe-MCM-41

Iron Sites	g-Values	E/D	D/cm^{-1}
I. Framework Iron			
(a) Fe^{3+} in tetrahedral coordination	$g=4.3$	0.33	0.3
(b) Single Fe^{3+} site	$g_x=g_y=2.003$ $g_z=1.99$	0.3	0.013
II. Extra-Framework Iron			
(a) Iron clusters	$g=2.45$	0.033	0.6
(b) Fe^{3+} on the outer surface of the pore	$g=2.07$	0.02	0.42

Source: Konovalova, T.A., *J. Phys. Chem. B*, 107, 1006, 2003.

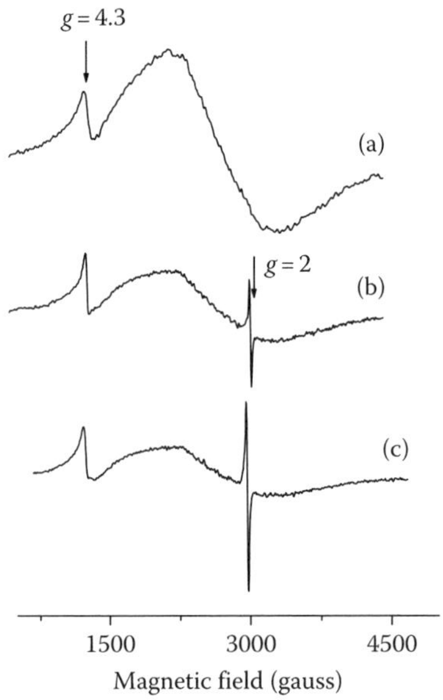

FIGURE 9.13 X-band EPR spectra of Fe(III)-MCM-41: (a) activated at 360°C and measured at 77 K, (b) after adsorption of 7′-apo-7′,7′-dicyano-β-carotene (77 K), and (c) after irradiation at 365 nm for 2 min. (From Konovalova, T.A., *J. Phys. Chem. B*, 107, 1006, 2003. With permission.)

9.13) showed that the adsorption of the carotenoid results in a decrease of the broad $g=2.0$ signal, while the intensity of the Fe^{3+} signal at $g=4.3$ does not change significantly.

The X-band measurements cannot identify which one of the iron sites can react with the carotenoid. Only the 95 GHz measurements (Figure 9.14) were able to demonstrate that adsorption of carotenoid results in a significant decrease of the $g=2.07$ signal and moderate decrease of the $g=2.45$ signal, while the intensity of the narrow line with $g_x=g_y=2.003$, $g_z=1.999$ is almost unaffected. The results show that the extra-framework Fe^{3+} ions located on the surface of the pore are primarily responsible for carotenoid oxidation. Probably, these sites are more accessible for bulky organic molecules than the framework iron within silica walls.

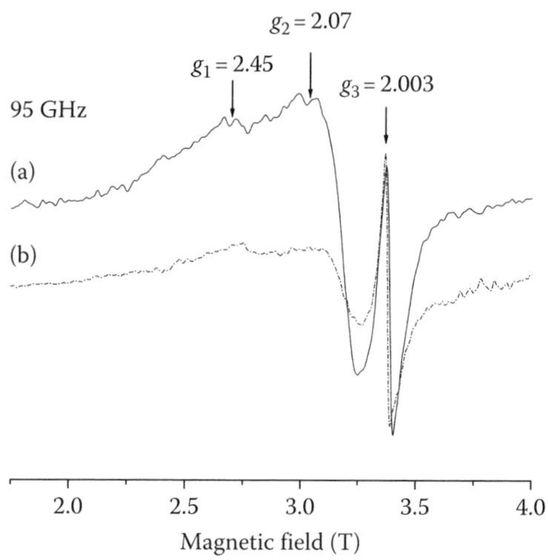

FIGURE 9.14 The 95 GHz EPR spectra of Fe-MCM-41: (a) in the absence of carotenoid and (b) after incorporation of canthaxanthin. (From Konovalova, T.A., *J. Phys. Chem. B*, 107, 1006, 2003. With permission.)

When carotenoid incorporated in Fe-MCM-41 was subjected to irradiation at 365 nm for 2 min, other paramagnetic species besides the $Car^{\bullet+}$ were detected. No such radicals were observed in the absence of carotenoids prior to or after irradiation (Konovalova et al. 2003). Spectral simulation allows determination of the carotenoid radical cation (g=2.0026) and a signal arising from species with g_1=2.015, g_2=2.006, and g_3=2.00. Signals with similar parameters have been observed in γ-irradiated siliceous, Al- and Ti-MCM-41, and attributed to Si–O$^\bullet$–Si or Al–(Ti)–O$^\bullet$–Si units of the framework, the so-called V-centers (Prakash et al. 1998). We suppose that oxidation of carotenoids in Fe-MCM-41 proceeds through electron transfer from carotenoid molecules to the electron acceptor sites (Fe^{3+} coordinated with surface oxygen atoms) producing Fe^{2+}–O$^\bullet$–Si species:

$$\text{Car} + \text{Fe}^{3+} - \text{O} - \text{Si} \rightarrow \text{Car}^{\bullet+} + \text{Fe}^{2+} - \text{O}^{\bullet} - \text{Si} \tag{9.25}$$

9.15 RELAXATION BY METALS: DISTANCE MEASUREMENTS

The long-lived charge-separation feature that is established between carotenoid radicals and the Ti-MCM-41 framework electron acceptor sites makes this system a potential photoredox system. To understand how far an electron can be transferred in this system until it is stabilized on a carotenoid forming a radical, we determined distances between a Ti^{3+} framework ion and the carotenoid radical by analyzing the enhancement in carotenoid relaxation rates caused by the metal ion. Canthaxanthin and 7′-apo-7′-(4-carboxyphenyl)-β-carotene were selected for the study because canthaxanthin was shown to exhibit no significant interaction with the Ti(IV) site in Ti-MCM-41, whereas 7′-apo-7′-(4-carboxyphenyl)-β-carotene with a terminal $-CO_2H$ group was expected to interact strongly with Ti sites. β-Ionone (BI) was chosen as a short-chain polyene system.

Carotenoids incorporated in metal-substituted MCM-41 represent systems that contain a rapidly relaxing metal ion and a slowly relaxing organic radical. For distance determination, the effect of a rapidly relaxing framework Ti^{3+} ion on spin-lattice relaxation time, T_1, and phase memory time, T_M, of a slowly relaxing carotenoid radical was measured as a function of temperature in both siliceous and Ti-substituted MCM-41. It was found that the T_M and T_1 are shorter for carotenoids embedded in Ti-MCM-41 than those in siliceous MCM-41.

The dominant effect of the rapidly relaxing metal spin on the T_M of the slowly relaxing spin is analogous to the effects of a physical motion, such as rotation of the methyl group, which averages nuclear spins to which the unpaired electron is coupled. The most dramatic effect on T_M occurs when the metal relaxation rate is of the same order of magnitude as the dipolar couplings to the slowly relaxing spin. This phenomenon is shown in Figure 9.15, which exhibits the logarithmic temperature dependence of the canthaxanthin relaxation rate, $1/T_M$, in siliceous and Ti-containing materials. As temperature increases, the metal relaxation rate increases and becomes comparable to the dipolar splittings in the vicinity of 125 K. As a result in Ti-MCM-41, both components, fast and slow, show a significant increase in $1/T_M$ around this temperature. On the other hand, siliceous samples showed little dependence on $1/T_M$. The temperature dependence of $1/T_M$ for BI in MCM-41 and in Ti-MCM-41 also showed an increase in $1/T_M$ for $BI^{\bullet+}$ in Ti-MCM-41 compared to that in MCM-41, especially near 40 K. A significant increase in $1/T_M$ for the (**1**) radical in Ti-MCM-41 compared to the MCM-41 sample indicates interaction of the carotenoid with the Ti^{3+} ion. In contrast to canthaxanthin and BI, 7′-apo-7′-(4-carboxyphenyl)-β-carotene containing the terminal carboxy group shows a monotonic increase in relaxation rate. No prominent peak in relaxation was observed.

The relaxation enhancement displayed for canthaxanthin, 7′-apo-7′-(4-carboxyphenyl)-β-carotene and BI was analyzed to provide interspin distances. The dipolar interactions and the distances can be determined according to procedures described elsewhere (Budker et al. 1995, Rakowsky et al. 1995, Eaton and Eaton 2000, Rao et al. 2000) and based on simulations of the paramagnetic metal ion contribution, W_{dd}:

$$\frac{1}{T_M} = \frac{1}{T_{M0}} + W_{dd} \tag{9.26}$$

where $1/T_M$ and $1/T_{M0}$ are the $Car^{\bullet+}$ relaxation rates in the presence and absence of the metal ion, respectively. The W_{dd} can be numerically simulated on the basis of relaxation enhancement in a two-pulse echo, $W(\tau)$, due to electron–electron interaction between the two spins. The simulation includes the experimental $1/T_M$–$1/T_{M0}$ value and T_1 of the Ti^{3+} ion. The adjustable parameter is the distance r (nm). At the proper distance, the simulations should match the experimental echo decay curves. If the relaxation rate increases significantly at a certain temperature, this procedure allows

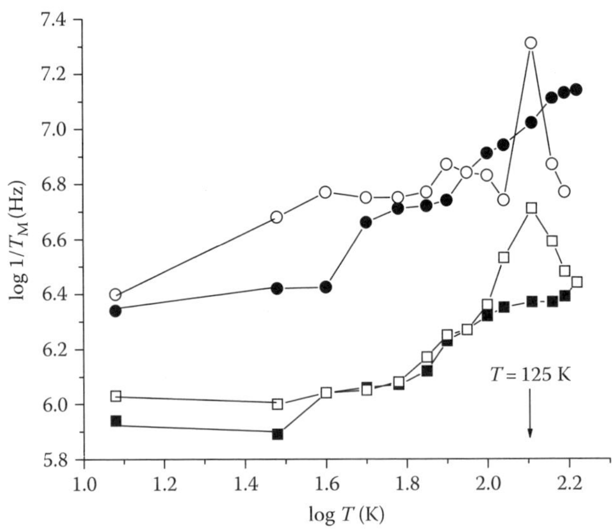

FIGURE 9.15 Temperature dependence of log $1/T_M$ (T_M [Hz]) for canthaxanthin radical. The ESEEM curves were best fitted as double exponentials: (–■–) slow component for Car/MCM-41, (–□–) slow component for Car/Ti-MCM-41, (–●–) fast component for Car/MCM-41, and (–○–) fast component for Car/Ti-MCM-41.

distance determination at this particular temperature. The distances obtained by using $1/T_M - 1/T_{M0}$ for canthaxanthin at 125 K and for BI at 40 K are found to be 13.0 and 10.5 Å, respectively.

To measure distances in the wider temperature range, this procedure was modified. Relaxation of the carotenoid occurs through several different mechanisms including the dipolar–dipolar interaction. Assuming that k_{dd} is the rate constant of the dipolar–dipolar interaction and $K=(k_1 + k_2 + k_3 + \ldots)$ is the sum of the rate constants of all other relaxation pathways, we can extract k_{dd} from the following equation:

$$e^{-k_{dd}t} = \frac{e^{-(k_{dd}+K)t}}{e^{-Kt}} \tag{9.27}$$

$W(\tau)$ was calculated from Equation 9.28 by numerical integration over the angle between the external magnetic field and the inter-nuclear axis (θ), at any given instant τ:

$$W(\tau) = \exp\left(-\frac{\tau}{T_1}\right) * \int_0^{\frac{\pi}{2}} d\theta \sin\theta \left\{ \left[\cosh(R\tau) + \frac{\sinh(R\tau)}{2RT_1} \right]^2 + \frac{D^2}{4R^2} \sinh^2(R\tau) \right\} \tag{9.28}$$

In Equation 9.28, D and R were calculated from Equations 9.29 and 9.30:

$$D = \frac{\mu_0 g_1 g_2 \beta^2}{4\pi h r^3}(1 - 3\cos^2\theta) \tag{9.29}$$

$$4R^2 = T_1^{-2} - D^2 \tag{9.30}$$

where

- T_1 is the longitudinal relaxation time of the fast relaxing Ti^{3+} ion
- D is the dipole–dipole interaction between the slow relaxing carotenoid radical and the fast relaxing Ti^{3+} ion
- r is the interspin distance
- θ is the angle between the direction of the external magnetic field and a vector connecting the two species with g-values g_1 and g_2

Prior to the integration, a change of variable was carried out by setting $x=\cos\theta$, where $\theta \in [0, \pi/2]$ and $x \in [0, 1]$, and Equation 9.28 was transformed into Equation 9.31:

$$W(\tau) = \exp\left(-\frac{\tau}{T_1}\right) * \int_0^1 dx \left\{ \left[\cosh(R\tau) + \frac{\sinh(R\tau)}{2RT_1} \right]^2 + \frac{D^2}{4R^2} \sinh^2(R\tau) \right\} \tag{9.31}$$

In Equation 9.31, D is related to x by

$$D = \frac{\mu_0 g_1 g_2 \beta^2}{4\pi h r^3}(1 - 3x^2) \tag{9.32}$$

The integration was carried out with the extended trapezoidal rule for an integral over function $f(x)$

$$I = \int_a^b f(x)dx \approx \left(\frac{f(x_0)}{2} + f(x_1) + \cdots + f(x_{n-1}) + \frac{f(x_n)}{2} \right) * \left(\frac{b-a}{n} \right) \tag{9.33}$$

The integral in Equation 9.31 converges if n is set to ~100,000 or larger. We also note that we approximated the value of sin x/x to 1 for $|x| \leq 10^{-10}$.

The simulations can be made to reproduce the initial ratio of fits in Equation 9.27 using the measured T_1 (μs) and fitting the distance r (nm), which is the only adjustable parameter. For canthaxanthin and BI the experimental fits and the integrated values showed the best match in a very narrow temperature range (±10 K) in the vicinity of the maximum enhancement in the relaxation rate. The distances obtained from the curve fits were similar to those determined from $1/T_M - 1/T_{M0}$ difference, namely, 13.0±2.0 Å for canthaxanthin and 10.0±2.0 Å for BI. It was found for canthaxanthin, which shows no prominent peak in the relaxation rate, that the distance does not depend on $1/T_M - 1/T_{M0}$. Using the ratio of curve fits, we can estimate the value of r for canthaxanthin as 9. 0±3.0 Å in TiMCM-41 in the temperature range of 110–130 K.

9.16 EFFECT OF DISTANT METALS ON *g*-TENSOR

When an organic radical is located near a high-spin metal ion, the g-tensor of the radical depends on the exchange interaction between the radical and the metal ion.

Multifrequency HF-EPR permits precise determination of the g-values of the exchange-coupled organic radical metal ion species, provides parameters for accurate simulation of the EPR spectra, and allows determination of detailed information about the radicals themselves and their environment (Gerfen et al. 1993, Un et al. 1995, Bar et al. 2001, Ivancich et al. 2001).

For instance, the Hamiltonian that describes the interacting system of an oxoferryl spin $\mathbf{S}=1$ ($\mathbf{S}^{Fe}$) with a radical spin $\mathbf{S}=½$ ($\mathbf{S}^{rad}$) is given in equation

$$\hat{H} = \beta \mathbf{S}^{rad} \cdot \mathbf{g}^{rad} \cdot \boldsymbol{B} + \beta \mathbf{S}^{Fe} \cdot \mathbf{g}^{Fe} \cdot \boldsymbol{B} + \mathbf{S}^{Fe} \cdot \mathbf{D} \cdot \mathbf{S}^{Fe} - J \cdot \mathbf{S}^{rad} \cdot \mathbf{S}^{Fe} \tag{9.34}$$

where

β is the Bohr magneton
$\boldsymbol{B}$ is the applied magnetic field
$\mathbf{g}^{rad}$ and $\mathbf{g}^{Fe}$ are the g-tensors of the radical and the iron species
$\mathbf{D}$ is the ZFS tensor for iron
J is an isotropic exchange coupling
$\mathbf{S}^{rad}$ and $\mathbf{S}^{Fe}$ are the vector spin operators

The g-tensor of the radical and the distance between the exchange-coupled radical and oxoferryl species can be obtained from spectral simulations at different frequencies. The g-values for the oxoferryl moiety and the ZFS tensor of the iron species were fixed in the simulations. The adjustable parameters in the fitting procedure were the exchange coupling, J, and the three g-values of the radical. The lower frequency EPR spectra of the radical can be well-simulated by using the parameters determined from the highest frequency spectrum. It should be emphasized that if exchange interaction (**D** and J parameters) is left out from the simulations, the lower frequency spectra cannot be well-fitted by use of the g-values obtained from the higher frequency spectrum.

9.17 DIMERS DETECTED BY *g*-TENSOR ANISOTROPY VARIATION

The g-tensor principal values of radical cations were shown to be sensitive to the presence or absence of dimer- and multimer-stacked structures (Petrenko et al. 2005). If face-to-face dimer structures occur (see Scheme 9.7), then a large change occurs in the g_{yy} component compared to the monomer structure. DFT calculations confirm this behavior and permitted an interpretation of the EPR measurements of the principal g-tensor components of the chlorophyll dimers with stacked structures like the P_{700}^{+} special dimer pair cation radical and the P_{700}^{+} special dimer pair triplet radical in photosystem I. Thus dimers that occur for radical cations can be deduced by monitoring the g_{yy} component.

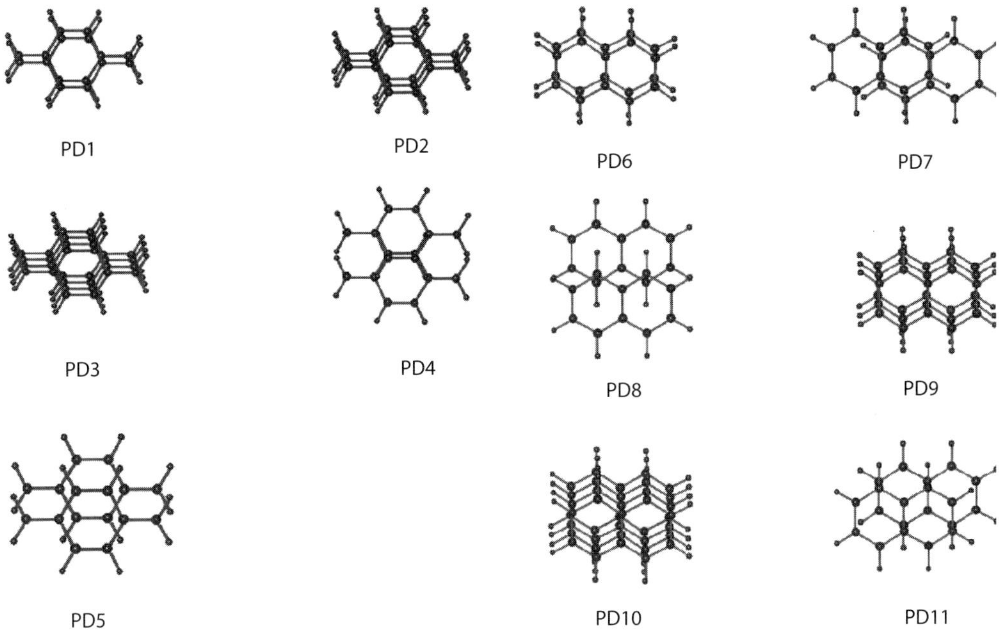

SCHEME 9.7 The geometries of Dp_2^+, Dp_3^+, and Dp_4^+ used in the g-tensor calculations (Dp is the *p*-dimethylenebenzene molecule). Face-to-face configurations PD1, PD2, and PD3 are shown for clarity. (From Petrenko, A., *Chem. Phys. Lett.*, 406, 327, 2005. With permission.)

9.18 CONCLUSIONS

Carotenoid radical intermediates generated electrochemically, chemically, and photochemically in solutions, on oxide surfaces, and in mesoporous materials have been studied by a variety of advanced EPR techniques such as pulsed EPR, ESEEM, ENDOR, HYSCORE, and a multifrequency high-field EPR combined with EPR spin trapping and DFT calculations. EPR spectroscopy is a powerful tool to characterize carotenoid radicals: to resolve *g*-anisotropy (HF-EPR), anisotropic coupling constants due to α-protons (CW, pulsed ENDOR, HYSCORE), to determine distances between carotenoid radical and electron acceptor site (ESEEM, relaxation enhancement).

ACKNOWLEDGMENTS

We thank the U.S. Department of Energy, Grant DE-FG02-86ER13465 and the National Science Foundation, Grant CHE-0079498 for support.

REFERENCES

Abragam, A. and B. Bleaney (1970). *Electron Paramagnetic Resonance of Transition Ions*. Oxford, U.K.: Clarendon Press.

Bar, G., M. Bennati et al. (2001). High-frequency (140-GHz) time domain EPR and ENDOR spectroscopy: The tyrosyl radical-diiron cofactor in ribonucleotide reductase from yeast. *J. Am. Chem. Soc.* **123**: 3569–3576.

Bratt, P. J., M. Rohrer et al. (1997). Submillimeter high-field EPR studies of the primary donor in plant photosystem I P700$^+$. *J. Phys. Chem. B* **101**: 9686–9689.

Budker, V., J.-L. Du et al. (1995). Electron-electron spin-spin interaction in spin-labeled methemoglobin. *Biophys. J.* **68**: 2531–2542.

Burghaus, O., M. Plato et al. (1991). 3 mm EPR investigation of the primary donor cation radical in single crystals of *Rhodobacter sphaeroides* R-26 reaction centers. *Chem. Phys. Lett.* **185**: 381–386.

Carrington, A. and A. D. McLachlan (1967). *Introduction to Magnetic Resonance with Applications to Chemistry and Chemical Physics*. New York: Harper and Row Publishers.

Chang, Z., Z. Zhu et al. (1999). Electron spin resonance of Ni(I) in Ni-containing MCM-41 molecular sieves. *J. Phys. Chem. B* **103**: 9442–9449.

Chen, X. (1991). An electrochemical, EPR study of carotenoids and high performance liquid chromatography separation of cis-trans isomers of canthaxantin. MS thesis, The University of Alabama, Tuscaloosa, AL.

Deng, Y. (1999). Carotenoid radical cations and dications studied by electrochemical, optical, and flow injection analysis: Lifetime, extended chain conjugation, and isomerization properties. Ph.D dissertation, The University of Alabama, Tuscaloosa, AL.

Deng, Y., G. Gao et al. (2000). Effects of polyene chain length and acceptor substituents on the stability of carotenoid radical cations. *J. Phys. Chem. B* **104**: 5651–5656.

Dikanov, S. A. and M. K. Bowman (1995). Cross-peak lineshape of two-dimensional ESEEM spectra in disordered S = 1/2, I = 1/2 spin systems. *J. Magn. Reson. A* **116**: 125–128.

Dikanov, S. A. and M. K. Bowman (1998). Determination of ligand conformation in reduced [2Fe-2S] ferredoxin from cysteine β-proton hyperfine couplings. *J. Biol. Inorg. Chem.* **3**: 18–29.

Dikanov, S. A., A. M. Tyryshkin et al. (2000). Intensity of cross-peaks in HYSCORE spectra of S = 1/2, I = 1/2 spin systems. *J. Mag. Res.* **144**: 228–242.

Ding, R., J. L. Grant et al. (1988). Carotenoid cation radicals produced by the interaction of carotenoids with iodine. *J. Phys. Chem.* **92**: 4600–4606.

Dowsing, R. D. and J. F. Gibson (1969). Electron spin resonance of high-spin d^5 systems. *J. Chem. Phys.* **50**: 294–303.

Eaton, S. S. and G. R. Eaton, Eds. (2000). Distance measurement in biological systems by EPR. *Biological Magnetic Resonance*. New York: Kluwer Academic.

Fessenden, R. W. and R. H. Schuler (1963). Electron spin resonance studies of transient alkyl radicals *J. Chem. Phys.* **39**: 2147–2195.

Focsan, A. L., M. K. Bowman et al. (2008). Pulsed EPR and DFT characterization of radicals produced by photooxidation of zeaxanthin and violaxanthin on silica-alumina. *J. Phys. Chem. B* **112**: 1806–1819.

Gaffney, B. J., D. V. Mavrophilipos et al. (1993). Access of ligands to the ferric center in lipoxygenase-1. *Biophys. J.* **64**: 773–783.

Gao, G., A. S. Jeevarajan et al. (1996). Cyclic voltammetry and spectroelectrochemical studies of cation radical and dication adsorption behavior for 7,7′-diphenyl-7,7′-diapocarotene. *J. Electroanal. Chem.* **411**: 51–56.

Gao, G., Y. Wurm et al. (1997). Electrochemical quartz crystal microbalance, voltammetry, spectroelectrochemical, and microscopic studies of adsorption behavior for (7E,7′Z)-diphenyl-7,7′-diapocarotene electrochemical oxidation product. *J. Phys. Chem. B* **101**: 2038–2045.

Gao, Y., T. A. Konovalova et al. (2002). Electron transfer of carotenoids imbedded in MCM-41 and Ti-MCM-41: EPR, ENDOR, and UV-Vis studies. *J. Phys. Chem. B* **106**: 10808–10815.

Gao, Y., T. A. Konovalova et al. (2003). Interaction of carotenoids and Cu^{2+} in Cu-MCM-41: Distance-dependent reversible electron transfer. *J. Phys. Chem. B* **107**: 2459–2465.

Gao, Y., L. D. Kispert et al. (2005). ESEEM and pulse ENDOR studies of Cu^{2+}… β-carotene interactions in Cu-MCM-41 molecular sieves. *J. Phys. Chem. B* **109**: 18289–18292.

Gao, Y., A. L. Focsan et al. (2006). Density functional theory study of the β-carotene radical cation and deprotonated radicals. *J. Phys. Chem. B* **110**: 24750–24756.

Gerfen, G. J., B. F. Bellew et al. (1993). High-frequency (139.5 GHz) EPR spectroscopy of the tyrosyl radical in *Escherichia coli* ribonucleotide reductase. *J. Am. Chem. Soc.* **115**: 6420–6421.

Goldfarb, D., M. Bernardo et al. (1994). Characterization of iron in zeolites by X-band and Q-band ESR, pulsed ESR, and UV-visible spectroscopies. *J. Am. Chem. Soc.* **116**: 6344–6353.

Goslar, J., L. Piekara-Sady et al. (1994). ENDOR data tabulations. In Poole, C. P. and Farach, H. A. (eds.), *Handbook of Electron Spin Resonance: Data Sources, Computer Technology, Relaxation and ENDOR*. New York: AIP Press.

Grant, J. L., V. J. Kramer et al. (1988). Carotenoid cation radicals: Electrochemical, optical, and EPR study. *J. Am. Chem. Soc.* **110**: 2151–2157.

Hapiot, P., L. D. Kispert et al. (2001). Single two-electron transfers vs successive one-electron transfers in polyconjugated systems illustrated by the electrochemical oxidation and reduction of carotenoids. *J. Am. Chem. Soc.* **123**: 6669–6677.

Hartmann, M., A. Püppl et al. (1996). Ethylene dimerization and butene isomerization in nickel-containing MCM-41 and AlMCM-41 mesoporous molecular sieves: An electron spin resonance and gas chromatography study. *J. Phys. Chem.* **100**: 9906–9910.

Ivancich, A., P. Dorlet et al. (2001). Multifrequency high-field EPR study of the tryptophanyl and tyrosyl radical intermediates in wild-type and the W191G mutant of cytochrome c peroxidase. *J. Am. Chem. Soc.* **123**: 5050–5058.

Jeevarajan, J. A. (1995). Electrochemical and optical studies of natural and synthetic carotenoids. PhD dissertation, The University of Alabama, Tuscaloosa, AL.

Jeevarajan, J. A. and L. D. Kispert (1996). Electrochemical oxidation of carotenoids containing donor/acceptor substituents. *J. Electroanal. Chem.* **411**: 57–66.

Jeevarajan, A. S., M. Khaled et al. (1993a). CIDEP studies of carotenoid radical cations. *Z. Phys. Chem.* **182**: 51–61.

Jeevarajan, A. S., L. D. Kispert et al. (1993b). An ENDOR study of carotenoid cation radicals on silica-alumina solid support. *Chem. Phys. Lett.* **209**: 269–274.

Jeevarajan, A. S., M. Khaled et al. (1994a). Simultaneous electrochemical and electron paramagnetic resonance studies of keto and hydroxyl carotenoids. *Chem Phys. Lett.* **225**: 340–345.

Jeevarajan, A. S., M. Khaled et al. (1994b). Simultaneous electrochemical and electron paramagnetic resonance studies of carotenoids: Effect of electron donating and accepting substituents. *J. Phys. Chem.* **98**: 7777–7781.

Jeevarajan, A. S., L. D. Kispert et al. (1994c). Spectroelectrochemistry of carotenoids in solution. *Chem. Phys. Lett.* **219**: 427–432.

Jeevarajan, J. A., A. S. Jeevarajan et al. (1996a). Electrochemical, EPR and AM1 studies of acetylenic and ethylenic carotenoids. *J. Chem. Soc. Faraday Trans.* **92**: 1757–1765.

Jeevarajan, A. S., L. D. Kispert et al. (1996b). Role of excited singlet state in the photooxidation of carotenoids: a time resolved Q-band EPR study. *J. Phys. Chem.* **100**: 669–671.

Jeevarajan, J. A., C. C. Wei et al. (1996c). Optical absorption spectra of dications of carotenoids. *J. Phys. Chem.* **100**: 5637–5641.

Kevan, L. and M. K. Bowman (1990). *Modern Pulsed and Continuous-Wave Electron Spin Resonance*. New York: Wiley.

Kevan, L. and L. D. Kispert (1976). *Electron Spin Double Resonance Spectroscopy*. New York: John Wiley and Sons.

Khaled, M. (1992). Electrochemical, transient EPR, and AM1 excited state studies of carotenoids. PhD dissertation, The University of Alabama, Tuscaloosa, AL.

Khaled, M., A. Hadjipetrou et al. (1990). Electrochemical and electron paramagnetic resonance studies of carotenoid cation radicals and dications: Effect of deuteration. *J. Phys. Chem.* **94**: 5164–5169.

Khaled, M., A. Hadjipetrou et al. (1991). Simultaneous electrochemical and electron paramagnetic resonance studies of carotenoid cation radicals and dications. *J. Phys. Chem.* **95**: 2438–2442.

Kispert, L. D. and L. Piekara-Sady (2006). *ENDOR Spectroscopy*. New York: Springer Science Publishers.

Kispert, L. D., J. Joseph et al. (1987). EPR study of arene radical cation salt crystals. *Synthetic Metals* **20**: 67–72.

Klette, R., J. T. Törring et al. (1993). Determination of the *g*-tensor of the primary donor cation radical in single crystals of Rhodobacter sphaeroides R-26 reaction centers by 3-mm high-field EPR. *J. Phys. Chem.* **97**: 2015–2020.

Konovalova, T. A. and L. D. Kispert (1998). EPR and ENDOR studies of carotenoid-solid Lewis acid interactions. *J. Chem. Soc. Faraday Trans.* **94**: 1465–1468.

Konovalova, T. A., L. D. Kispert et al. (1997). Photoinduced electron transfer between carotenoids and solvent molecules. *J. Phys. Chem. B* **101**: 7858–7862.

Konovalova, T. A., J. Krzystek et al. (1999). 95–670 GHz EPR studies of canthaxanthin radical cation stabilized on a silica-alumina surface. *J. Phys. Chem. B* **103**: 5782–5786.

Konovalova, T. A., S. A. Dikanov et al. (2001a). Detection of anisotropic hyperfine components of chemically prepared carotenoid radical cations: 1D and 2D ESEEM and pulsed ENDOR study. *J. Phys. Chem. B* **105**: 8361–8368.

Konovalova, T. A., Y. Gao et al. (2001b). Photooxidation of carotenoids in mesoporous MCM-41, Ni-MCM-41 and Al-MCM-41 molecular sieves. *J. Phys. Chem. B* **105**: 7459–7464.

Konovalov, V. V., P. Hapiot et al. (2002). *Proceedings of the 5th International Manuel M. Baizer Symposium in honor of Professor Jean Michel Savéan*. New York: Electrochemical Society.

Konovalova, T. A., Y. Gao et al. (2003). Characterization of Fe-MCM-41 molecular sieves with incorporated carotenoids. *J. Phys. Chem. B* **107**: 1006–1011.

Konovalova, T. A., L. D. Kispert et al. (2004). Multifrequency high-field electron paramagnetic resonance characterization of the peroxyl radical. Location in horse heart myoglobin oxidized by H_2O_2. *J. Phys. Chem. B* **108**: 11820–11826.

Kosslick, H., G. Lischke et al. (1998). Acidity and catalytic behavior of substituted MCM-48. *J. Catal.* **176**: 102–114.

Lawrence, J., A. L. Focsan et al. (2008). Pulsed ENDOR studies of carotenoid oxidation in Cu(II)-substituted MCM-41 molecular sieves. *J. Phys. Chem. B* **112**: 1806–1819.

Liu, D. and L. D. Kispert (1999). Electrochemical aspects of carotenoids. *Rec. Res. Dev. Electrochem.* **2**: 139–157, India: Transworld Research Network.

Mairanovski, V. G., A. A. Engovatov et al. (1975). Electron-donor and electron-acceptor properties of carotenoids: Electrochemical study of carotenes. *J. Electroanal. Chem.* **66** 123–137.

Park, S.-M. (1978). Electrochemical studies of β-carotene, all-*trans*-retinal and all-*trans*-retinol in tetrahydrofuran. *J. Electrochem. Soc.* **125**: 216–222.

Petrenko, A., K. Redding et al. (2005). The influence of the structure of the radical cation dimer pair of aromatic molecules on the principal values of a *g*-tensor: DFT predictions. *Chem. Phys. Lett.* **406**: 327–331.

Piekara-Sady, L. and L. D. Kispert (1994). *ENDOR Spectroscopy*. In Poole, C. P. and Farach, H. A. (eds.), *Handbook of Electron Spin Resonance: Data Sources, Computer Technology, Relaxation and ENDOR*. New York: AIP Press.

Piekara-Sady, L., M. M. Khaled et al. (1991). Comparison of the INDO to the RHF-INDO/SP derived EPR proton hyperfine couplings for the carotenoid cation radical: Experimental evidence. *Chem. Phys. Lett.* **186**: 143–148.

Piekara-Sady, L., S. A. Jeevarajan et al. (1993). An ENDOR study of canthaxanthin cation radical in solution. *Chem. Phys. Lett.* **207**: 173–177.

Piekara-Sady, L., A. S. Jeevarajan et al. (1995). ENDOR study of the (7,7′-dicyano)- and (7′-phenyl)-7′-apo-carotene radical cations formed by UV photolysis of carotenoids adsorbed on silica gel. *J. Chem. Soc. Faraday Trans.* **91**: 2881–2884.

Pilbrow, J. R. (1990). *Transition Ion Electron Paramagnetic Resonance*. Oxford, U.K.: Clarendon Press.

Polyakov, N. E., V. V. Konovalov et al. (2001a). One-electron transfer product of quinone addition to carotenoids: EPR and optical absorption studies. *J. Photochem. Photobiol. A* **141**: 117–126.

Polyakov, N. E., A. I. Kruppa et al. (2001b). Carotenoids as antioxidants: Spin trapping, EPR and optical studies. *Free Rad. Biol. Med.* **31**: 43–52.

Polyakov, N. E., T. V. Leshina et al. (2001c). Carotenoids as scavengers of free radicals in a Fenton reaction, antioxidants or pro-oxidants. *Free Rad. Biol. Med.* **31**: 398–404.

Polyakov, N. E., T. V. Leshina et al. (2004). Inclusion complexes of carotenoids with cyclodextrins: ^{1}H NMR, EPR and optical studies. *Free Rad. Biol. Med.* **36**: 872–880.

Polyakov, N. E., T. V. Leshina et al. (2006a). Host-guest complexes of carotenoids with β-glycyrrhizic acid. *J. Phys. Chem. B* **110**: 6991–6998.

Polyakov, N. E., T. V. Leshina et al. (2006b). Antioxidant and redox properties of supramolecular complexes of carotenoids with β-glycyrrhizic acid. *Free Rad. Biol. Med.* **40**: 1804–1809.

Prakash, A. M., H. M. Sung-Suh et al. (1998). Electron spin resonance evidence for isomorphous substitution of titanium into titanosilicate TiMCM-41 mesoporous molecular sieve. *J. Phys. Chem. B* **102**: 857–864.

Rakowsky, M. H., K. M. More et al. (1995). Time-domain electron paramagnetic resonance as a probe of electron-electron spin-spin interaction in spin-labeled low-spin iron porphyrins. *J. Am. Chem. Soc.* **117**: 2049–2057.

Rao, S. B. K., A. M. Tyryshkin et al. (2000). Inhibitory copper binding site on the spinach cytochrome *bf* complex: Implications for Q_o site catalysis. *Biochemistry* **39**: 3285–3296.

Robinson, T., L. D. Kispert et al. (1985). EPR study of acceptor doped *p*-terphenyl crystals: The oriented radical cation precursor for a conducting polymer. *J. Chem. Phys.* **82**: 1539–1542.

Stone, A. J. (1964). *g*-tensors of aromatic hydrocarbons. *Mol. Phys.* **7**: 311–316.

Sweeney, W. V., D. Coucouvanis et al. (1973). ESR of spin 5/2 systems with axial symmetry and moderately large zero-field splittings. Application of line-shape calculations to the interpretation of randomly oriented microcrystallite spectra. *J. Chem. Phys.* **59**: 369–379.

Un, S., M. Atta et al. (1995). g-values as a probe of the local protein environment: High-field EPR of tyrosyl radicals in ribonucleotide reductase and photosystem II. *J. Am. Chem. Soc.* **117**: 10713–10719.

Wertz, J. E. and J. R. Bolton (1972). *Electron Spin Resonance: Elementary Theory and Practical Applications*. New York: McGraw-Hill Book Company.

Wu, Y., L. Piekara-Sady et al. (1991). Photochemically generated carotenoid radicals on Nafion film and silica gel: An EPR and ENDOR study. *Chem. Phys. Lett.* **180**: 573–577.

Yang, A.-S. and B. J. Gaffney (1987). Determination of relative spin concentration in some high-spin ferric proteins using E/D distribution in electron paramagnetic resonance simulations. *Biophys. J.* **51**: 55–67.

10 EPR Spin Labeling in Carotenoid–Membrane Interactions

Witold K. Subczynski and Justyna Widomska

CONTENTS

10.1 INTRODUCTION

Carotenoids are synthesized by bacteria, algae, and plants where they serve as an antenna function in light-harvesting complexes and photoreactive centers (Griffiths et al. 1955, Sefirmann-Harms 1987, Koyama 1991). The highest concentration of carotenoids was reported to occur in membranes of bacteria living under extreme conditions (high or low temperatures, salinity, pH, and/or strong light) (Huang and Haug 1974, Clejan et al. 1986, Chamberlain et al. 1991, Anton et al. 2002). Carotenoids are also present at a fairly high concentration in the lipid bilayer portion of the thylakoid membrane as a free component during the violaxanthin cycle where they affect membrane fluidity (Gruszecki and Strzalka 1991, Tardy and Havaux 1997). It is hypothesized that in membranes of prokaryotes, carotenoids play a function similar to cholesterol in eukaryotes, namely, the regulation of membrane fluidity (Huang and Haug 1974, Rohmer et al. 1979, Chamberlain et al. 1991).

Carotenoids are also present in animals, including humans, where they are selectively absorbed from diet (Furr and Clark 1997). Because of their hydrophobic nature, carotenoids are located either in the lipid bilayer portion of membranes or form complexes with specific proteins, usually associated with membranes. In animals and humans, dietary carotenoids are transported in blood plasma as complexes with lipoproteins (Krinsky et al. 1958, Tso 1981) and accumulate in various organs and tissues (Parker 1989, Kaplan et al. 1990, Tanumihardjo et al. 1990, Schmitz et al. 1991, Khachik et al. 1998, Hata et al. 2000). The highest concentration of carotenoids can be found in the eye retina of primates. In the retina of the human eye, where two dipolar carotenoids, lutein and zeaxanthin, selectively accumulate from blood plasma, this concentration can reach as high as 0.1–1.0 mM (Snodderly et al. 1984, Landrum et al. 1999). It has been shown that in the retina, carotenoids are associated with lipid bilayer membranes (Sommerburg et al. 1999, Rapp et al. 2000) although, some macular carotenoids may be connected to specific membrane-bound proteins (Bernstein et al. 1997, Bhosale et al. 2004).

The membrane localization of some portion of carotenoids in bacteria, plants, and animals is commonly accepted (Havaux 1998, Gruszecki 1999). However, their function in membranes is unclear. Certainly, they protect biological systems against peroxidation and photo-damage, reacting as antioxidants with free radicals and reactive oxygen species (Krinsky 1989, Edge et al. 1997). The ability of carotenoids to quench singlet oxygen and triplet states of photoactive molecules (Fugimori and Talva 1966, Centrell et al. 2003) is especially significant. It has also been suggested in many papers that carotenoids can regulate membrane fluidity (Huang and Haug 1974, Rohmer et al. 1979, Chamberlain et al. 1991, Tardy and Havaux 1997). This is possible in bacteria and plants where the local carotenoid concentration in the lipid bilayer can reach a value of a few mole percentages. In animals, the highest carotenoid concentration can be found in the eye retina of primates, but even there, the carotenoid concentration in the lipid bilayer portion of membranes is much lower than 1 mol% (Bone and Landrum 1984). However, this concentration is high enough for effective blue-light filtration, quenching of singlet oxygen, and molecular triplet states, and effective antioxidant action. To understand the basic mechanisms of these actions it is necessary to better understand carotenoid–membrane interaction. For systems with a high carotenoid concentration, it is most significant to understand how carotenoids affect membrane physical properties, membrane structure, and membrane dynamics, as well as the lateral organization of the lipid bilayer (its domain structure). For systems with a low carotenoid concentration, it is especially important to understand how the membrane itself—membrane composition, structure, and lateral organization—affects the organization of carotenoids in the lipid bilayer, including their solubility (monomeric versus aggregated state), orientation (transmembrane versus parallel), and localization (distribution between membrane domains). Also, knowledge of the bulk membrane physical properties, which are not uniform across the lipid bilayer and can differ in different membrane domains, is significant to a better understanding of chemical reactions and physical processes that take place in the lipid bilayer membrane and involve carotenoids.

In this chapter, we explain how electron paramagnetic resonance (EPR) spin-labeling methods can be used to obtain the above-mentioned information about carotenoid–membrane interactions. We focus our presentation on how carotenoids affect membrane properties and how the membrane itself affects carotenoid organization within the lipid bilayer. We also identify membrane properties that can be easily obtained using EPR spin-labeling methods and that in our opinion are significant for chemical reactions and physical processes involving carotenoids. Using these methods, a variety of lipid spin labels were incorporated in the membrane for probing at specific depths and specific membrane domains (Figure 10.1). Application of conventional EPR as well as time-domain saturation-recovery EPR techniques are discussed and illustrated by previously published results.

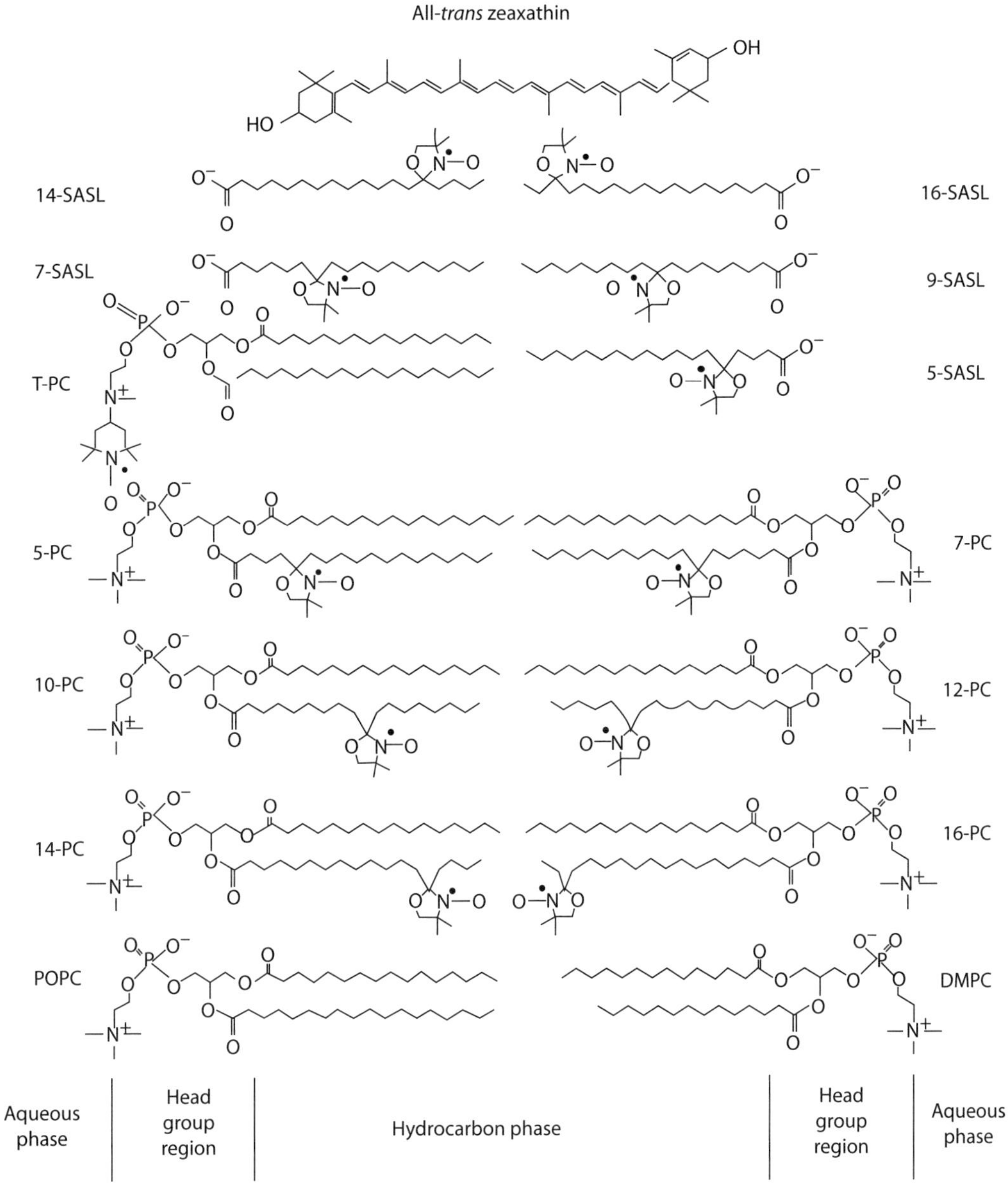

FIGURE 10.1 Chemical structures of selected spin labels 1-palmitoyl-2-(*n*-doxylsrearoyl) phosphatidylcholine (*n*-PC), tempocholine-1-palmitoyl-2-oleoylphosphatidic acid ester (T-PC), and *n*-doxylstearic acid spin label (*n*-SASL). Chemical structures of dimyristoylphosphatidylcholine (DMPC), dipalmitoylphocphatidylcholine (POPC), and zeaxanthin are included. Approximate locations of these molecules across the lipid bilayer membrane are also illustrated. However, since alkyl chains tend to have many *gauche* conformations, the chain-length projection to the membrane normal would be shorter than depicted here and the rigid structure of zeaxanthin would sink somewhat differently in the liquid–crystalline phase membranes.

10.2 HANDLING THE SAMPLE FOR EPR MEASUREMENTS

The membranes used in EPR measurements are usually multilamellar dispersions of lipids (multilamellar liposomes) containing an investigated carotenoid and 0.5–1.0 mol% of an appropriate lipid spin label (Figure 10.1). The total amount of lipids usually is 5–10 μmol per sample.

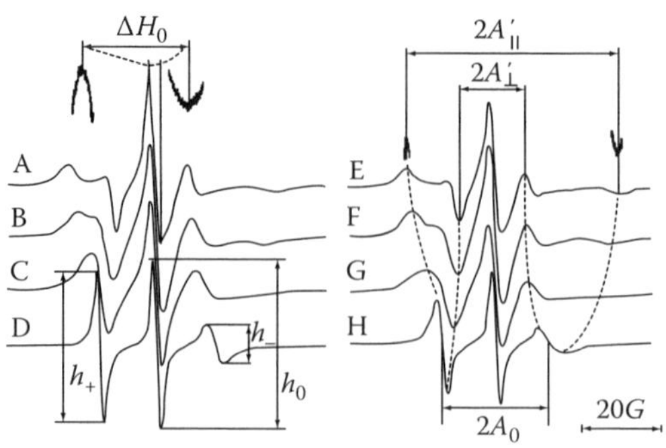

FIGURE 10.2 EPR spectra of 5-SASL (A,E), 9-SASL (B,F), 12-SASL (C,G), and 16-SASL (D,H) in DMPC membranes containing 0 (left) and 10 mol% (right) zeaxanthin recorded at 25°C. The measured values are indicated. The outer wings were also magnified by recording at 10 times higher receiver gain. Peak-to-peak central line widths were recorded with expended abscissa (magnetic field scan range by a factor of 10). (From Subczynski, W.K. et al., *Biochim. Biophys. Acta*, 1105, 97, 1992. With permission.)

It is important to add buffer to the film of the dried lipids at a temperature above the main phase-transition temperature of the investigated lipid membrane and further prepare the lipid dispersion by vortexing the sample at this temperature. The lipid dispersion is centrifuged briefly (at 16,000 g for 15 min at 4°C), and the loose pellet of multilamellar liposomes is transferred to a small-diameter gas-permeable plastic sample tube for EPR measurements. It is often desirable to concentrate the sample inside the capillary by additional centrifugation (Subczynski et al. 2005). The use of multilamellar liposomes (instead of unilamellar) and centrifugation significantly increases the signal-to-noise ratio for EPR measurements. When stearic acid spin labels (SASL) are used, a buffer with a high pH of ~9.5 has to be chosen to ensure that all SASL probe carboxyl groups are ionized in the lipid bilayer membranes (Egreet-Charlier et al. 1978, Kusumi et al. 1982a). Typical EPR spectra of 5-, 9-, 12-, and 16-SASL in fluid-phase dimyristoylphosphatidylcholine (DMPC) membranes with and without zeaxanthin are presented in Figure 10.2. All preparations and measurements with carotenoids should be performed in darkness or dim light and, when possible, under nitrogen or argon. Gas-permeable capillaries made of methylpentene polymer TPX or Teflon allow samples to be easily deoxygenated during EPR measurements: The samples are equilibrated with nitrogen gas or, if necessary, with the appropriate gas mixture (Hyde and Subczynski 1989, Subczynski et al. 2005). These gases are also used for temperature control.

Lipid spin labels are often added to biological membranes from either methanol or ethanol solutions (Tardy and Havaux 1997). This procedure is straightforward, but the final concentration of either methanol or ethanol in the membrane suspension is usually 0.2–0.7 M even when 1%–2% (v/v) of a concentrated spin label solution is added. It is recommended that biological membranes are labeled by adding the suspension to the glass test tube with the spin-label film formed on its bottom (Ligeza et al. 1998). After shaking the sample for about 30 min at room temperature, all spin-label molecules will diffuse to the membranes. This procedure is efficient for *n*-SASL but not *n*-PC spin labels.

10.3 CONVENTIONAL EPR

10.3.1 ALKYL CHAIN ORDER

In the membrane lipid alkyl chains of *n*-SASL and *n*-PC spin labels undergo rapid rotational motion about the long axis of the spin label and wobble within the confines of a cone imposed by the

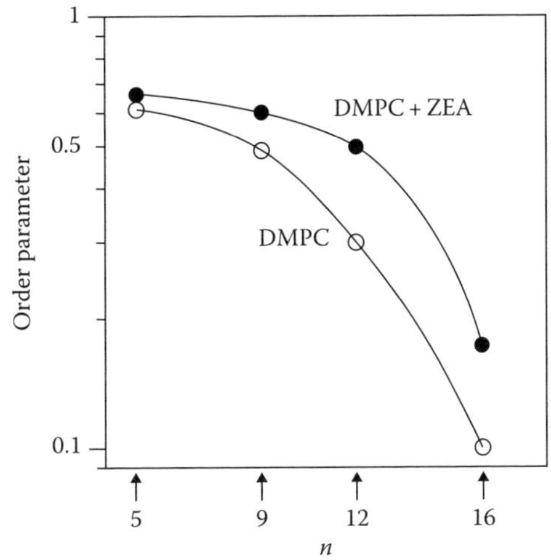

FIGURE 10.3 Profiles of the order parameter (order parameter is plotted in a log scale as a function of nitroxide position (n) along the alkyl chain of n-SASL) at 25°C in DMPC membranes with and without 10 mol% zeaxanthin. (From Subczynski, W.K. et al., *Biochim. Biophys. Acta*, 1068, 68, 1991. With permission.)

membrane environment. The anisotropic rotational motion of the spin labels gives rise to new features of the EPR spectra (shown in Figure 10.2) that can be used to calculate the order parameter, S (Marsh 1981):

$$S = 0.5407(A'_{II} - A'_{\perp})/a_0, \text{where } a_0 = (A'_{II} + 2A'_{\perp})/3 \tag{10.1}$$

Profiles of the order parameter obtained with n-SASL in DMPC membranes in the presence and absence of zeaxanthin are displayed in Figure 10.3, showing the ordering effect of this dipolar, terminally dihydroxylated carotenoid. In the case of n-SASL and n-PC spin labels, the order parameter at the nth position reflects the distribution of vectors $C_{n-1} \rightarrow C_{n+1}$ along the molecular axis. The alkyl chain order decreases gradually with an increase in depth in the membrane. It can be seen in Figure 10.3 that 10 mol% zeaxanthin significantly increases the order parameter of the hydrocarbon chains of DMPC. The increase of the order parameter is greater in the center of the bilayer (16-SASL position) than in the region near the polar headgroups (5-SASL position). However, it is suggested to compare the increase in the value of S to the decrease in temperature, which causes the same increase in the S value as incorporation of carotenoids into the bilayer. That way it is easier to compare effects of carotenoids at different positions in the membrane and in different membranes (see Subczynski et al. 1993 for more details).

10.3.2 Rotational Diffusion of Alkyl Chains

The nitroxide moiety of 16-SASL and 16-PC exhibits such a great deal of motion that the rotational correlation time can be calculated (Berliner 1978). The rotational correlation time (assuming isotropic rotational diffusion of the nitroxide fragment) can be calculated from the linear term of the line width parameter:

$$\tau_{2B} = 6.51 \times 10^{-10} \times \Delta H_0 \left(\sqrt{\frac{h_0}{h_-}} - \sqrt{\frac{h_0}{h_+}} \right) \tag{10.2}$$

and from the quadratic term

$$\tau_{2C} = 6.51\times10^{-10}\times\Delta H_0\left(\sqrt{\frac{h_0}{h_-}}+\sqrt{\frac{h_0}{h_+}}-2\right) \tag{10.3}$$

ΔH_0 is the peak-to-peak width of the central line in gauss, and h_+, h_0, h_- are heights of the low, central, and high field peaks, respectively (see Figure 10.2). When τ_{2B} and τ_{2C} are similar, it is argued that the motional model is fairly good and motion is isotropic. Figure 10.4 presents correlation times for 16-SASL in the DMPC bilayer calculated from the linear term (τ_{2B}) and the quadratic term (τ_{2C}) of the line width as a function of mole fraction of zeaxanthin. The addition of 10 mol% zeaxanthin decreases motional freedom of the 16-SASL free-radical moiety which is monitored by a large increase in correlation times. At lower temperatures (25°C and 35°C), zeaxanthin also increases the anisotropy of spin-label movement, which is manifested as a difference between τ_{2B} and τ_{2C}. However, at 45°C—well above the phase-transition temperature—calculated τ_{2B} and τ_{2C} are very similar, indicating that zeaxanthin decreases the rate of spin-label motion, but does not influence its isotropy. Additionally, from the Arrhenius display of the temperature dependence of the rotational correlation time (log τ versus reciprocal temperature), the activation energy of the rotational motion of the nitroxide moiety of 16-SASL or 16-PC can be calculated as shown in Subczynski et al. (1993).

We would like to point out that an order parameter indicates the static property of the lipid bilayer, whereas the rotational motion, the oxygen transport parameter (Section 4.1), and the chain bending (Section 4.4) characterize membrane dynamics (membrane fluidity) that report on rotational diffusion of alkyl chains, translational diffusion of oxygen molecules, and frequency of alkyl chain bending, respectively. The EPR spin-labeling approach also makes it possible to monitor another bulk property of lipid bilayer membranes, namely local membrane hydrophobicity.

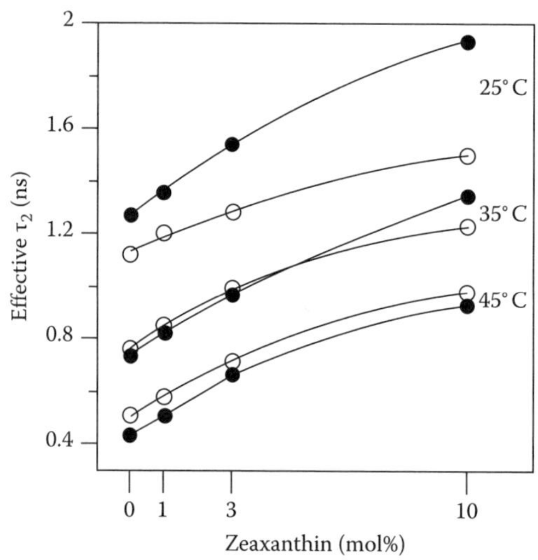

FIGURE 10.4 Effective rotational correlation time of 16-SASL in DMPC membranes plotted as a function of mole fraction of zeaxanthin at different temperatures (τ_{2B} (○) and τ_{2C} (●)). (From Subczynski, W.K. et al., *Biochim. Biophys. Acta*, 1105, 97, 1992. With permission.)

10.3.3 Hydrophobicity

The local hydrophobicity in the membrane can be monitored primarily using A_Z (Z-component of the hyperfine interaction tensor of the nitroxide spin label) as a conventional experimental observable (Griffith et al. 1974, Subczynski et al. 1994, Subczynski and Wisniewska 1998). A_Z can be obtained directly from the EPR spectra of spin labels measured for a frozen suspension of membranes (Figure 10.5a). With an increase in solvent hydrophobicity, A_Z decreases. In this type of work, a nitroxide moiety is placed at various depths in the lipid bilayer, and the hydrophobicity profiles across the membranes are obtained. Griffith et al. (1974) demonstrated that hydrophobicity in the membrane is largely determined by the extent of water penetration into the membrane, since dehydration abolishes the hydrophobicity gradient in lipid bilayer samples. Figure 10.5b shows hydrophobicity profiles across the DMPC membrane in the absence and presence of zeaxanthin. It is convenient to relate hydrophobicity as observed by A_Z at a selected depth in the membrane to hydrophobicity (or dielectric constant, ε) of bulk organic solvent, as shown in Figure 10.5c. Using this comparison, it is shown that incorporation of 10 mol% of zeaxanthin causes a considerable increase in hydrophobicity in the central region of the bilayer where hydrophobicity increases from the level of octanol ($\varepsilon = 10$) to the level of dipropylamine ($\varepsilon = 3$) (Wisniewska and Subczynski 1998). However, the presence of zeaxanthin decreases the hydrophobicity in the headgroup region.

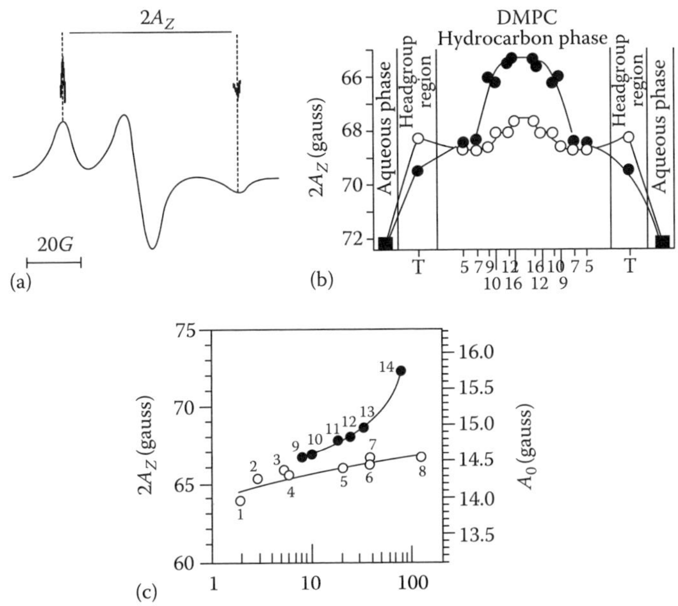

FIGURE 10.5 Ways of determining and analyzing local hydrophobicity across the lipid bilayer membranes. (a) EPR spectrum of 16-SASL in the DMPC membrane at −165°C. The measured $2A_Z$ value is indicated. The outer wings were also magnified by recording at 10 times higher receiver gain. (b) Hydrophobicity profiles ($2A_Z$) across the DMPC membrane containing 0 (○) and 10 mol% (●) zeaxanthin. Upward changes indicate increases in hydrophobicity. Approximate locations of the nitroxide moieties of spin labels are indicated by arrows. (c) $2A_Z$ and A_0 for 16-SASL plotted against the dielectric constant ε of the solvent. The solvents are numbered as follows: (1) hexane, (2) dipropylamine, (3) *N*-butylamine, (4) ethyl acetate, (5) acetone, (6) dimethylformamide, (7) acetonitrile, (8) methylpropionamide, (9) 1-decanol, (10) 1-octanol, (11) 2-propanol, (12) ethanol, (13) methanol, and (14) water. (From Wisniewska, A. et al., *Biochim. Biophys. Acta*, 1368, 235, 1998. With permission; Subczynski, W.K. et al., *Biochemistry*, 33, 7670, 1994. With permission.)

In another approach, the isotropic hyperfine coupling constant (A_0 in Figure 10.2) of 16-SASL or 16-PC can be measured for fluid-phase membranes. A decrease in the A_0 value indicates an increase in hydrophobicity at the 16-SASL position (Figure 10.5c). However, this constant reflects only the hydrophobicity of the membrane center.

Both methods of determining membrane hydrophobicity have advantages and disadvantages, which are discussed in Wisniewska et al. (2006).

10.3.4 Phase Transition

The main phase-transition temperature of the lipid bilayer membranes can be monitored by observing the amplitude of the central line of the EPR spectra of 16-SASL or 16-PC (h_0 in Figure 10.2). The decrease in signal amplitude at the phase transition can be as much as 50%. For phase-transition measurements, the temperature should be regulated by passing nitrogen gas through a coil placed in a water bath and monitored by a copper–constantan thermocouple placed in the sample just above the active volume of the cavity (Wisniewska et al. 1996). This way temperature can be regulated with accuracy better than 0.1°C. To avoid aggregation of carotenoids in the gel phase membrane, cooling experiments (fluid-to-gel phase transition) are preferred. The temperature should be lowered by the addition of a small amount of cold water to the water bath with rapid agitation, permitting a very low rate of temperature change, ~2°C/h. To avoid small cooling/heating cycles, a temperature-controlling unit should not be used.

The phase-transition temperature, T_m, and the width of transition, $\Delta T_{1/2}$, were operationally defined based on EPR data, as shown in Figure 10.6a. As a rule, in the presence of polar carotenoids the phase transition broadens and shifts to lower temperatures (Subczynski et al. 1993, Wisniewska et al. 2006). The effects on T_m are the strongest for dipolar carotenoids, significantly weaker for monopolar carotenoids, and negligible for nonpolar carotenoids. The effects decrease with the increase of membrane thickness. Additionally, the difference between dipolar and monopolar carotenoids decreases for thicker membranes (Subczynski and Wisniewska 1998, Wisniewska et al. 2006). These effects for lutein, β-cryptoxanthin, and β-carotene are illustrated in Figure 10.6b

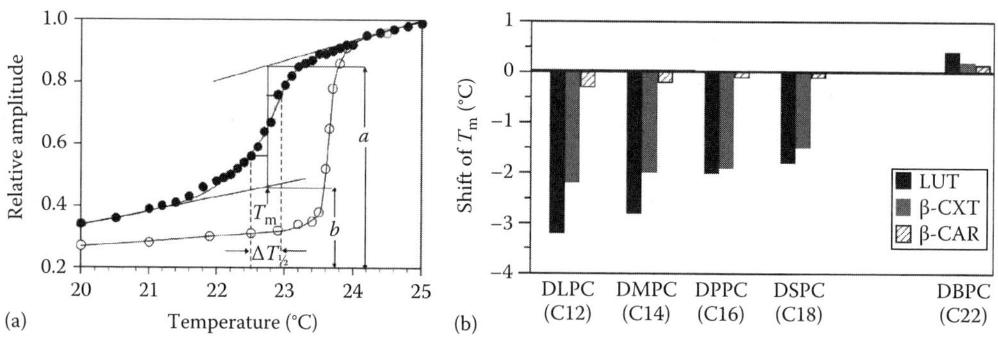

FIGURE 10.6 (a) Normalized amplitude of the central peak of the EPR spectra of 16-SASL plotted as a function of temperature (cooling experiments) in the DMPC bilayer containing 0 (○) and 10 mol% lutein (●). Definitions of T_m and $\Delta T_{1/2}$ are shown. T_m is the midpoint temperature at which the normalized EPR signal amplitude equals $(a + b)/2$, where a and b are, respectively, intensities at given temperatures in the extended linear portions of the upper and lower ends of the transition curve. As the sharpness of the transition, the width $\Delta T_{1/2}$ is employed, which is defined by two temperatures at which the EPR signal amplitude is $(a + 3b)/4$ and $(3a + b)/4$. (b) Shifts of the main phase-transition temperature, T_m, of phosphatidylcholine (PC) membranes (dilauroyl-PC (DLPC), DMPC, dipalmitoyl-PC (DPPC), distearoyl-PC (DSPC), dibehenoyl-PC (DBPC)) induced by the addition of 10 mol% carotenoid to the sample. Negative values indicate a decrease of T_m. Notice that the x-axis indicates the lipid as well as the number of carbon atoms in the alkyl chains. (From Wisniewska, A. et al., *Acta Biochim. Pol.*, 53, 475, 2006. With permission.)

where shifts of T_m induced by adding 10 mol% of these carotenoids to the samples during preparation are plotted as a function of the membrane thickness.

10.4 SATURATION-RECOVERY EPR

The saturation-recovery EPR method of measuring spin-lattice relaxation time (T_1) is a pulse technique in which recovery of the EPR signal is measured at a weak-observing microwave power after the end of the saturating microwave pulse. The time scale of this recovery is characterized by the spin-lattice relaxation time, T_1 (Eaton and Eaton 2005), which for lipid spin labels can be as long as 1–10 μs. To obtain the correct spin-lattice relaxation time, the sample should be thoroughly deoxygenated, which can be achieved by equilibrating the sample in a gas-permeable capillary with nitrogen gas, which is also used for temperature control (Hyde and Subczynski 1989, Subczynski et al. 2005). Presently, Bruker produces EPR spectrometers capable of saturation-recovery measurements at X-band.

10.4.1 Oxygen Transport Parameter

The bimolecular collision of molecular oxygen (a fast-relaxing species) and a nitroxide (a slow-relaxing species) induces spin exchange, which leads to a faster spin-lattice relaxation of the nitroxide. This effect is measured using the saturation-recovery EPR technique. An oxygen transport parameter, W(x), was introduced as a conventional quantitative measure of the collision rate between the spin label and molecular oxygen (Kusumi et al. 1982b):

$$W(x) = T_1^{-1}(\text{Air}, x) - T_1^{-1}(\text{N}_2, x) \tag{10.4}$$

T_1(Air, x) and T_1(N_2, x) are spin-lattice relaxation times of nitroxides in samples equilibrated with atmospheric air and nitrogen, respectively. Note that $W(x)$ is normalized to the sample equilibrated with the atmospheric air. $W(x)$ is proportional to the product of the local translational diffusion coefficient $D(x)$ and the local concentration $C(x)$ of oxygen at a depth x in the membrane, which is in equilibrium with the atmospheric air:

$$W(x) = AD(x)C(x), \qquad A = 8\pi p r_0 \tag{10.5}$$

where

r_0 is the interaction distance between oxygen and the nitroxide radical spin label (about 4.5 Å) (Windrem and Plachy 1980)

p is the probability that an observable event occurs when a collision does occur and is very close to 1 (Hyde and Subczynski 1984, Subczynski and Hyde 1984, Subczynski and Swartz 2005)

A is remarkably independent of the solvent viscosity, hydrophobicity, temperature, and spin-label species (Hyde and Subczynski 1984, Subczynski and Hyde 1984, Subczynski and Swartz 2005)

Figure 10.7a shows typical saturation-recovery curves for 14-PC in the DMPC bilayer containing 10 mol% 9-*cis* zeaxanthin in the presence and absence of oxygen. The recovery curves are fitted by single exponentials, and decay time constants (T_1's) are determined. To obtain the oxygen transport parameter, in principle, two saturation-recovery measurements should be performed, one for the sample equilibrated with nitrogen and the other for the sample equilibrated with air (see Equation 10.4). However, to increase accuracy, saturation-recovery measurements are carried out systematically as a function of oxygen concentration (% air) in the equilibrating gas mixture. Figure 10.7b, in which the T_1^{-1} values for 14-PC in the DMPC bilayer containing 10 mol% 9-*cis* zeaxanthin are plotted as a function of oxygen concentration (% air) in equilibrating gas mixture, shows the method of calculating the oxygen transport parameter. Experimental points show a linear dependence up to 60% air, and extrapolation to 100% air is performed to obtain the oxygen transport parameter. This process is required because accurate observation of saturation recovery becomes increasingly difficult as the oxygen partial pressure is increased due to fast relaxations. The membrane profiles of $W(x)$ (oxygen diffusion–concentration product) can be constructed on the basis of measurements with different lipid spin labels (Subczynski et al. 1989, 1991, Ashikawa et al. 1994). The effects of carotenoids on the oxygen transport parameter in different membranes were measured only at selected depths (Wisniewska and Subczynski 2006a,b, Widomska and Subczynski 2008). However, the effects of 10 mol% zeaxanthin on the profiles of the oxygen diffusion–concentration product across DMPC and egg-yolk PC (EYPC) membranes were obtained based on conventional EPR measurements of oxygen-induced line broadening of the spin-label EPR spectra (Subczynski et al. 1991). These profiles for the DMPC membranes are presented in Figure 10.8 and indicate that in the presence of 10 mol% dipolar carotenoids the oxygen diffusion–concentration product in

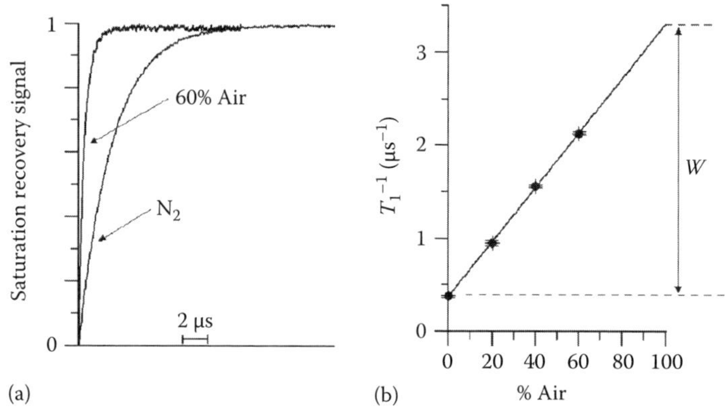

FIGURE 10.7 (a) Representative saturation-recovery signals of 14-PC in DMPC membranes containing 10 mol% 9-*cis* zeaxanthin at 35°C for samples equilibrated with nitrogen and a mixture of 60% air and 40% nitrogen. The fits to the single-exponential curves with recovery times of 2.64 μs (N_2) and 0.47 μs (60% air) were satisfactory. (b) T_1^{-1} for 14-PC in DMPC membranes containing 10 mol% 9-*cis* zeaxanthin at 35°C plotted as % air in the equilibrating gas mixture. Experimental points show a linear dependence up to 60% air, and extrapolation to 100% is performed to indicate a way of calculating the oxygen transport parameters, *W*. (From Widomska, J. et al., *Biochim. Biophys. Acta*, 1778, 10, 2008. With permission.)

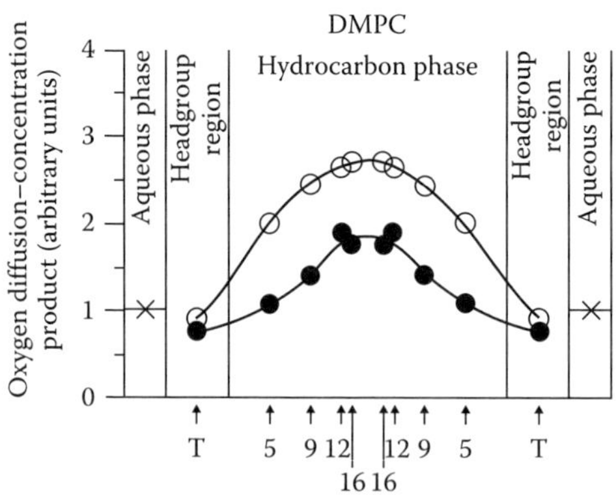

FIGURE 10.8 Profiles of the relative oxygen diffusion–concentration product across the DMPC bilayer containing 0 (○) and 10 mol% zeaxanthin (●) at 25°C. The approximate locations of nitroxide moieties of spin labels are indicated by arrows. The value of the oxygen diffusion–concentration product in water can be obtained from the oxygen diffusion coefficient and oxygen concentration in water equilibrated with air at 25°C. (From Subczynski, W.K. et al., *Biochim. Biophys. Acta*, 1068, 68, 1991. With permission.)

the hydrocarbon region of the bilayer is about 30% smaller than in the center of the pure DMPC membrane. However, zeaxanthin has little effect on the product in the polar headgroup region, which is different than the effect of cholesterol, which significantly reduces the oxygen diffusion–concentration product in and near the polar headgroup region and does not change (or even increase) it in the membrane center (Subczynski et al. 1989, 1991, Widomska et al. 2007) (see also Section 10.6). The sensitivity of the line-broadening EPR method is, however, significantly lower than the sensitivity of saturation-recovery spin-label oximetry (Subczynski and Swartz 2005).

10.4.2 Discrimination by Oxygen Transport

When located in two different membrane domains, the spin label alone most often cannot differentiate between domains and therefore gives very similar (indistinguishable) conventional EPR spectra and similar T_1 values. However, even small differences in lipid packing will affect oxygen partitioning and oxygen diffusion in these domains, which can be easily detected by observing the different T_1's from spin labels in the presence of oxygen. In membranes equilibrated with air and consisting of two lipid environments with different oxygen transport rates—the fast oxygen transport (FOT) domain and the slow oxygen transport (SLOT) domain—the saturation-recovery signal is a simple double-exponential curve with time constants of T_1^{-1}(Air, FOT) and T_1^{-1}(Air, SLOT) (Ashikawa et al. 1994, Kawasaki et al. 2001, Subczynski et al. 2007a,b).

$$W(\mathrm{FOT}) = T_1^{-1}(\mathrm{Air,FOT}) - T_1^{-1}(\mathrm{N_2,FOT}) \tag{10.6}$$

$$W(\mathrm{SLOT}) = T_1^{-1}(\mathrm{Air,SLOT}) - T_1^{-1}(\mathrm{N_2,SLOT}) \tag{10.7}$$

Here "x" from Equation 10.4 is changed to the two-membrane domain FOT and SLOT with the depth fixed (the same spin label is distributed between the FOT and SLOT domains). W(FOT) and W(SLOT) are oxygen transport parameters in each domain and represent the collision rate in samples equilibrated with air. Figure 10.9 illustrates the basis of the discrimination by oxygen transport (DOT) method, showing saturation-recovery EPR signals for 5-SASL in membranes

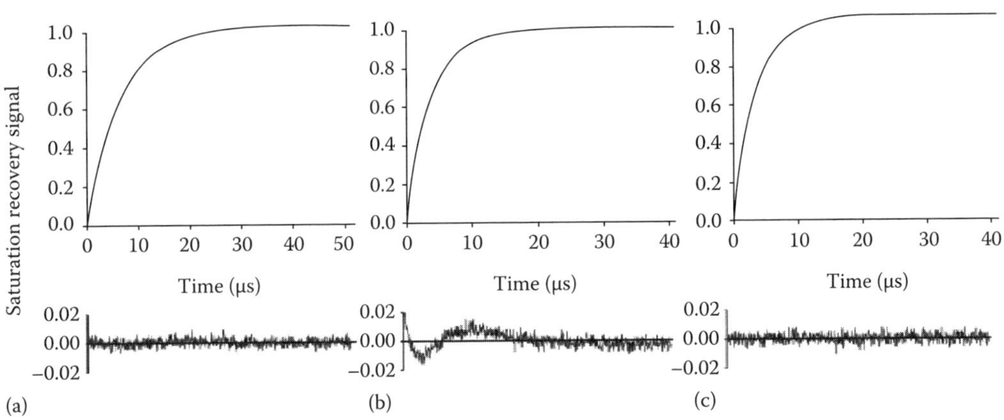

FIGURE 10.9 Typical saturation-recovery signals from 5-SASL in membranes from raft-forming mixture containing 1 mol% lutein at 20°C for samples equilibrated with (a) nitrogen and (b and c) 40% air. In the absence of oxygen, the single-exponential signal is observed with the time constant (T_1) of 6.71 μs. In the presence of oxygen, fitting the search to a (b) single exponential is unsatisfactory as shown by the residual. The fit (c), using the double-exponential mode (time constants 4.53 and 2.10 μs), is excellent. (From Wisniewska, A. and Subczynski, W.K., *Free Radic. Biol. Med.*, 40, 1820, 2006. With permission.)

made from raft-forming mixture containing 1 mol% lutein in the absence and presence of oxygen. In the absence of oxygen (Figure 10.9a), the single-exponential signal is observed. In the presence of oxygen, the single-exponential fit is unsatisfactory (Figure 10.9b), while the double-exponential fit is satisfactory (Figure 10.9c). The double saturation-recovery signal indicates the presence of two-membrane environments. In membranes made from raft-forming mixture, two domains are present: the raft domain enriched in saturated lipids and cholesterol, and the bulk domain enriched in unsaturated lipids (Dietrich et al. 2001). Lower oxygen transport parameter (W(SLOT)) results are assigned to the raft domain, while higher oxygen transport parameter (W(FOT)) results are assigned to the bulk domain (see Kawasaki et al. (2001) and Subczynski et al. (2007a) for more details). Using the DOT method, oxygen transport parameters and profiles of the oxygen transport parameter in coexisting domains can be obtained (Ashikawa et al. 1994, Kawasaki et al. 2001, Wisniewska and Subczynski 2006a,b, Subczynski et al. 2007b).

10.4.3 Ion Penetration into the Membrane

The bimolecular collision of paramagnetic metal ions or metal–ion complexes (a fast-relaxing species) with nitroxides (a slow-relaxing species) leads to spin exchange and an effective shortening of the spin-lattice relaxation time, T_1, of the spin label in proportion to the collision frequency. Thus, this collision frequency can be estimated from the saturation-recovery measurements similar to that estimated for molecular oxygen. When nitroxide moieties are located at different depths in the lipid bilayer, the collision rate should reflect the degree of penetration of metal ions into the bilayer. In analogy to the oxygen transport parameter described in Section 10.4.1 (Equation 10.5), the penetration parameter for paramagnetic iron complex, $K_3Fe(CN)_6$ was introduced as (Subczynski et al. 1994)

$$P(x) = T_1^{-1}\left(50\,\text{mM}\ K_3Fe(CN)_6, x\right) - T_1^{-1}\left(\text{no}\ K_3Fe(CN)_6, x\right) \tag{10.8}$$

This parameter is proportional to the product of the local concentration and the local translational diffusion coefficient of $Fe(CN)_6^{-3}$ at membrane depth x, where the nitroxide moiety is located. Greater $P(x)$ values indicate a greater extent of $Fe(CN)_6^{-3}$ penetration into the membrane. The ion penetration data obtained at physiological temperatures are consistent with the hydrophobicity profiles presented in Section 10.3.3, showing that hydrophobicity profiles obtained for frozen samples provide a good estimate of profiles at physiological temperatures (see Subczynski et al. (1994) and Wisniewska and Subczynski (1998) for further evidence for this statement). The membrane profiles of $P(x)$ can be constructed on the basis of measurements with different lipid spin labels (Subczynski et al. 1994). Ion penetration into the membrane can also be evaluated with the continuous-wave power-saturation method involving conventional EPR technique (Wisniewska and Subczynski 1998). In this method, $P_{1/2}$ is measured, which is the incident microwave power at which the EPR signal is half as great as it would be in the absence of saturation. Figure 10.10a shows representative power-saturation data for 12-SASL in the DMPC bilayer in the presence and absence of both lutein and $K_3Fe(CN)_6$. It is evident that the effect of $Fe(CN)_6^{-3}$ on power saturation of 12-SASL is greater in the absence of lutein. Based on saturation curves and equations derived in Wisniewska and Subczynski (1998) the $Fe(CN)_6^{-3}$ accessibility parameters for 5-, 9-, and 12-SASL in DMPC membranes were obtained in the absence and presence of 10 mol% lutein (Figure 10.10b). Penetration of $Fe(CN)_6^{-3}$ gradually decreases toward the membrane center and is significantly lowered by the presence of lutein. Penetration profiles obtained in fluid-phase membranes are consistent with the hydrophobicity profiles presented in Figure 10.5b. It should be noted that $P_{1/2}$ can be measured, reported, and duplicated in other laboratories without knowledge of the structure of the EPR spectrum and is a convenient empirical parameter.

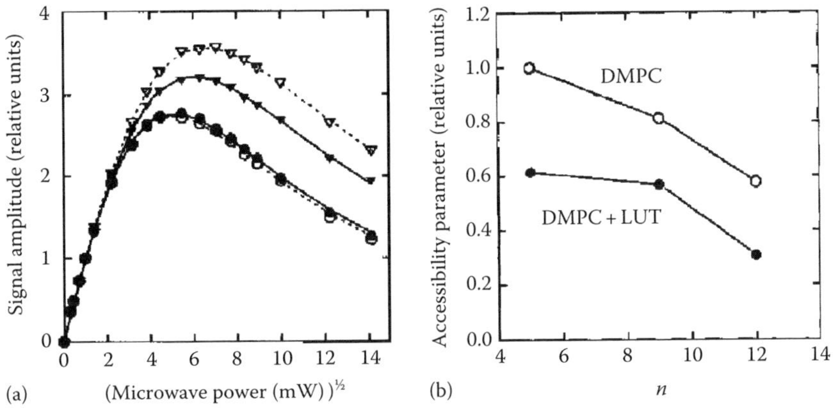

FIGURE 10.10 (a) Continuous-wave saturation data for the central line of 12-SASL in DMPC membranes at 30°C. Membranes in the absence (○, ∇) and presence (●, ▼) of 10 mol% lutein, and in the absence (○, ●) and presence (∇, ▼) of 50 mM $K_3Fe(CN)_6$ in the buffer. (b) Relative accessibility parameter obtained at 30°C in DMPC with and without 10 mol% lutein plotted as a function of the position of the nitroxide moiety of SASL in the membrane. (From Wisniewska, A. et al., *Biochim. Biophys. Acta*, 1368, 235, 1998. With permission.)

10.4.4 Alkyl Chain Bending

A pulse saturation-recovery EPR technique was used to study the effect of carotenoids on interaction of ^{14}N:^{15}N lipid spin-label pairs in fluid-phase membranes (Yin and Subczynski 1996). In the performed experiments, the ^{15}N nitroxide moiety was always attached at the C16 position of the stearic acid molecule, whereas the ^{14}N nitroxide moiety was placed at C16, C10, C7, and C5. The interaction (collision) between the ^{14}NC16:^{15}NC16 pair primarily depends on the lateral diffusion of stearic acid spin labels, whereas the interaction between pairs ^{14}NC10:^{15}NC16, ^{14}NC7:^{15}NC16, and ^{14}NC5:^{15}NC16 requires a vertical fluctuation of the nitroxide moiety at the C16 position toward the polar surface of the membrane. In Figure 10.11a, the possible collisions between nitroxide moieties are indicated. Yin and Subczynski (1996) showed that the experimental saturation-recovery curve for a given ^{14}N:^{15}N pair is a double-exponential curve with two time constants from which the collision rate constant can be calculated. Bimolecular collisions of pairs consisting of various combinations of spin labels allowed for frequency mapping of alkyl chain bending in the lipid bilayer and the observation of the effects of lutein on this type of fluidity (Figure 10.11b). These measurements confirm the occurrence of vertical fluctuations of alkyl chain ends toward the bilayer surface. The addition of lutein reduces the collision frequency for all spin-label pairs. The effect of this dipolar carotenoid is significantly different than the effect of cholesterol, which reduces collision frequency near the membrane surface and increases collision frequency at the membrane center (see also Section 10.7).

10.5 HOW CAROTENOIDS AFFECT MEMBRANE PROPERTIES (HIGH CAROTENOID CONCENTRATION)

10.5.1 Do Carotenoids Regulate Membrane Fluidity?

The hypothesis that polar carotenoids regulate membrane fluidity of prokaryotes (performing a function similar to cholesterol in eukaryotes) was postulated by Rohmer et al. (1979). Thus, the effects of polar carotenoids on membrane properties should be similar in many ways to the effects caused by cholesterol. These similarities were demonstrated using different EPR spin-labeling approaches in which the effects of dipolar, terminally dihydroxylated carotenoids such as lutein,

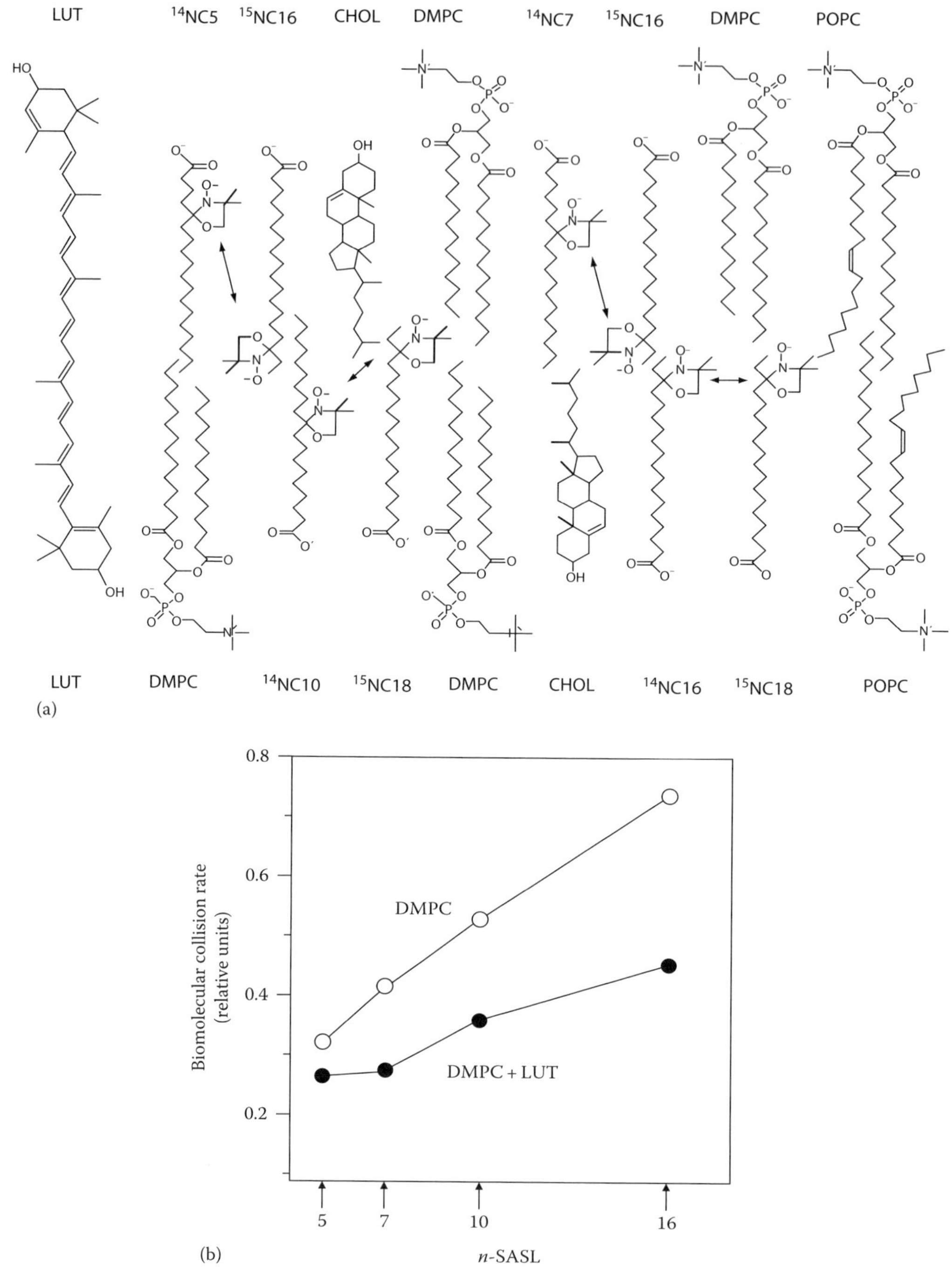

FIGURE 10.11 (a) Cross-sectional drawing of the lipid bilayer including lutein, cholesterol, and spin labels. Observed collisions between ^{14}N:^{15}N spin-label pairs are indicated. DMPC and POPC molecules are also shown. POPC represents the major component (70%) of the EYPC mixture. (b) Bimolecular collision rate for a nitroxide moiety at the C16 position of the stearic acid alkyl chain with other SASLs in the DMPC alone and the DMPC with 10 mol% lutein at 27°C. (From Yin, J.J. and Subczynski, W.K., *Biophys. J.*, 71, 832, 1996. With permission.)

zeaxanthin, and violaxanthin, on the structure and dynamics of lipid bilayer membranes were investigated (Subczynski et al. 1991, 1992, 1993, Subczynski and Wisniewska 1998, Wisniewska and Subczynski 1998, Wisniewska et al. 2006, Widomska and Subczynski 2008). It was shown that both cholesterol and dipolar carotenoids increase the order and decrease alkyl chain motion in fluid-phase membranes and disordered lipids in gel-phase membranes. Both broaden the fluid-to-gel phase transition and increase mobility of polar headgroups. As a rule, the presence of unsaturated alkyl chains moderates the effect of polar carotenoids and cholesterol (Kusumi et al. 1986, Subczynski et al. 1993). In saturated membranes, 10 mol% of polar carotenoids exert effects similar to 15–20 mol% of cholesterol. Polar carotenoids exert stronger effects on membrane properties because one molecule of polar carotenoids is located in two leaflets of the bilayer and influences both leaflets, while one molecule of cholesterol is located in one leaflet of the bilayer and influences only one leaflet. The ordering effect of cholesterol does not depend on membrane thickness (Kusumi et al. 1986), whereas the relation between the length of the polar carotenoid molecule and the thickness of the membrane is a significant factor in determining the effect of polar carotenoids on membrane properties (Subczynski et al. 1993, Wisniewska and Subczynski 1998). To manifest those effects, the rigid rod-shaped carotenoid molecule must possess two polar groups at the ends of the hydrophobic "bar." Significant differences resulting from the structure and localization of cholesterol and polar carotenoids in the membrane are discussed in Subczynski et al. (1993), Yin and Subczynski (1996), and Widomska and Subczynski (2008).

Compared with dipolar carotenoids, the effects of nonpolar carotenoids such as β-carotene on the physical properties of the membrane are negligible (Subczynski and Wisniewska 1998, Wisniewska et al. 2006). Monopolar carotenoids such as β-cryptoxanthin affect membrane properties significantly less than dipolar carotenoids (Wisniewska et al. 2006). These observations suggest that anchoring carotenoid molecules to opposite membrane surfaces with polar hydroxyl groups is important to enhance their effects on membrane properties. EPR measurements for both model and biological membranes are in agreement; they show that carotenoids rigidify the "Acholeplasma" membrane (Huang and Haug 1974). EPR measurements with lipid spin labels demonstrate that polar carotenoids, which are present transiently in the lipid bilayer portion of thylakoid membranes during the xanthophyll cycle, also regulate thylakoid membrane fluidity (Gruszecki and Strzalka 1991, Tardy and Havaux 1997).

Recently, due to increased interest in membrane raft domains, extensive attention has been paid to the cholesterol-dependent liquid-ordered phase in the membrane (Subczynski and Kusumi 2003). The pulse EPR spin-labeling DOT method detected two coexisting phases in the DMPC/cholesterol membranes: the liquid-ordered and the liquid-disordered domains above the phase-transition temperature (Subczynski et al. 2007b). However, using the same method for DMPC/lutein (zeaxanthin) membranes, only the liquid-ordered-like phase was detected above the phase-transition temperature (Widomska, Wisniewska, and Subczynski, unpublished data). No significant differences were found in the effects of lutein and zeaxanthin on the lateral organization of lipid bilayer membranes. We can conclude that lutein and zeaxanthin—macular xanthophylls that parallel cholesterol in its function as a regulator of both membrane fluidity and hydrophobicity—cannot parallel the ability of cholesterol to induce liquid-ordered–disordered phase separation.

10.5.2 Barriers of Lipid Bilayers Formed by Polar Carotenoids

Membranes of extreme halophilic (Kushwaha et al. 1975, Anwar et al. 1977, Anton et al. 2002, Lutnaes et al. 2002, Oren 2002) and thermophilic bacteria (Alfredsson et al. 1988, Yokoyama et al. 1995) contain a large concentration of polar carotenoids. Membranes of these bacteria, which live in extreme conditions, should provide a high barrier to block nonspecific permeation of polar and nonpolar molecules. Incorporation of dipolar carotenoids into these membranes at a high concentration serves this purpose well because dipolar carotenoids increase the hydrophobic barrier for polar molecules (Wisniewska and Subczynski 1998, Wisniewska et al. 2006) and increase the rigidity barrier for nonpolar molecules (Subczynski et al. 1991, 1992, 1993, Wisniewska et al. 2006). An increase in hydrophobicity should greatly increase the activation energy required for polar small molecules and ions to cross the membrane. In addition, the activation energy for translational diffusion of small molecules such as oxygen is increased by membrane rigidity. Detailed profiles obtained by EPR spin-labeling methods across lipid bilayers (order parameter (Figure 10.3), hydrophobicity (Figure 10.5b), oxygen transport parameter (Figure 10.8), and alkyl chain bending (Figure 10.11b)) illustrate the formation of these barriers in the presence of polar carotenoids.

We think that dipolar carotenoids ensure a high hydrophobic environment, which is necessary to facilitate energy transfer from light-harvesting systems to reaction centers in photosynthesis (McDermott et al. 1995). These carotenoids are present in the light-harvesting complexes of photosynthetic membranes where they act as accessory light-harvesting pigments, prevent photodynamic destruction, and stabilize the native structure of pigment–protein complexes (Conn et al. 1991, Koyama 1991, Cohen et al. 1995). It was shown that carotenoid molecules span the complex, forming a kind of "barrel" around bacteriochlorophyll molecules (McDermott et al. 1995). Formation of this "barrel" structure (McDermott et al. 1995) and the ability of carotenoids to increase local hydrophobicity (Wisniewska and Subczynski 1998) should provide a highly hydrophobic environment that will reduce the dielectric constant and facilitate the delocalization of the excited state of bacteriochlorophyll molecules over the ring of bacteriochlorophylls. The energy is then available for efficient transfer to the reaction center from any part of the ring, where it is "trapped."

10.5.3 Solubility of Carotenoids in Lipid Bilayer Membranes

Carotenoids should affect the physical properties of the membrane primarily when they are dissolved in the lipid bilayer as monomers. Undissolved carotenoid molecules that form aggregates within the lipid bilayer and/or crystals in the water phase should not affect membrane properties, or their effects should be negligible compared with the effects of monomers. It is commonly accepted that the membrane solubility of dipolar carotenoids is fairly high (Kolev and Kafalieva 1986, Milon et al. 1986, Lazrak et al. 1987, Gruszecki 1991). The high membrane solubility of lutein, zeaxanthin, and violaxanthin was also demonstrated using EPR spin-labeling methods, which allowed observation of the changes in membrane properties induced by these carotenoids. Changes in membrane properties are proportional to the amount of carotenoids added to the sample during preparation, with few signs of saturation at a concentration of 10 mol% or higher (Subczynski et al. 1992, Wisniewska and Subczynski 1998, 2006a). No discontinuity in measured properties was observed for lower carotenoid concentrations. It is significant to measure their effects on the properties in the membrane center, where these effects are very similar and independent of membrane thickness (Subczynski et al. 1993, Wisniewska and Subczynski 1998, Wisniewska et al. 2006). These EPR results and data from the literature indicate that dipolar carotenoids are miscible in lipid bilayers in the range of 0–10 mol%.

It is uncertain how much β-cryptoxanthin and β-carotene (added to the sample during preparation) can be dissolved in the lipid bilayer in the form of monomers. EPR measurements of the effects of β-cryptoxanthin on physical properties of the fluid-phase lipid bilayers indicate that the solubility of this monopolar xanthophyll strongly depends on membrane thickness, showing the threshold of solubility in DLPC membranes to be ~3 mol%, in DMPC membranes ~5–10 mol% and for thicker membranes as large as ~10 mol% (Wisniewska et al. 2006). EPR data also indicate a very low solubility of β-carotene in the lipid bilayer membranes, showing that the effects of β-carotene on membrane properties is weaker than the effect of 1 mol% of dipolar carotenoids, independent of the amount added to the sample during preparation (Subczynski and Wisniewska 1998, Wisniewska et al. 2006). These data suggest that dipolar, and possibly monopolar, carotenoids can be dissolved in the lipid bilayer as monomers with a high concentration enough to affect the physical properties of the membrane, including phase-transition temperature, membrane order, fluidity, and hydrophobicity (see also Subczynski and Wisniewska (1998) and Wisniewska et al. (2006) for more discussion).

10.6 HOW THE MEMBRANE ITSELF AFFECTS DISTRIBUTION AND LOCALIZATION OF CAROTENOIDS IN THE LIPID BILAYER (LOW CAROTENOID CONCENTRATION)

10.6.1 Accumulation of Polar Carotenoids in Unsaturated Membrane Domains

It was recently demonstrated that macular xanthophylls are substantially excluded from membrane domains enriched in saturated lipids and cholesterol (raft domains) and remain 8–14 times more concentrated in the bulk domain, which is enriched in unsaturated lipids (Wisniewska and Subczynski 2006a,b). A similar distribution was observed for β-cryptoxanthin, but not for β-carotene, which is more uniformly distributed between these domains (Figure 10.12). This distribution was demonstrated using cold Triton X-100 extraction from membranes containing 1 mol% of carotenoids. The saturation-recovery EPR DOT method was also used in these investigations showing that membrane domains are not the artifacts created by the Triton X-100 and the low temperature, but they exist in situ at physiological temperatures, indicating additionally. Results also demonstrate that macular xanthophylls, at 1 mol% do not affect the formation of these domains (Wisniewska and Subczynski 2006a,b). The location of xanthophylls in membrane domains formed from unsaturated lipids (illustrated in Figure 10.13) is ideal if they are to act as a lipid antioxidants, which is the most

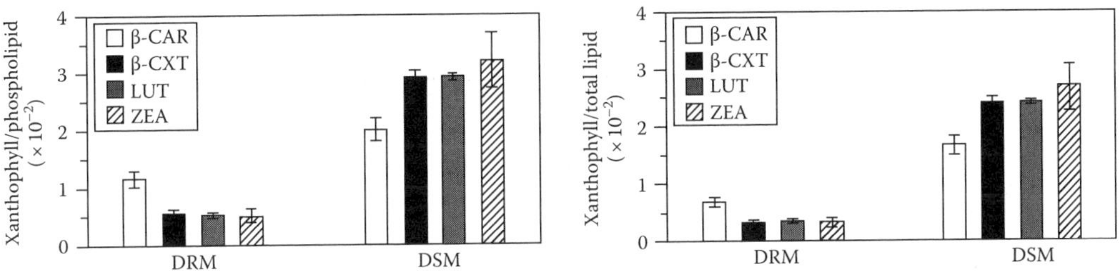

FIGURE 10.12 The mole ratio of carotenoid/phospholipid and carotenoid/total lipid (phospholipid + cholesterol) in raft domain (detergent-resistant membrane, DRM) and bulk domain (detergent-soluble membrane, DSM) isolated from membranes made of raft-forming mixture (equimolar ternary mixture of dioleoyl-PC (DOPC)/sphingomyelin/cholesterol) with 1 mol% lutein (LUT), zeaxanthin (ZEA), β-cryptoxanthin (β-CXT), or β-carotene (β-CAR).

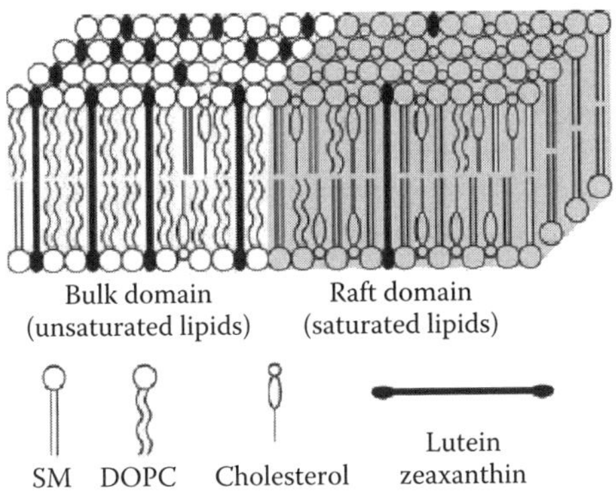

FIGURE 10.13 Schematic drawing of the distribution of xanthophyll molecules between raft domain (DRM) and bulk domain (DSM) in lipid bilayer membranes. For this illustration, the xanthophyll partition coefficient between domains is the same as obtained experimentally for raft-forming mixture. However, to better visualize the observed effect in the drawing, the number of lipid molecules was decreased and the total number of xanthophyll molecules was increased about 10 times. (From Wisniewska, A. and Subczynski, W.K., *Free Radic. Biol. Med.*, 40, 1820, 2006. With permission.)

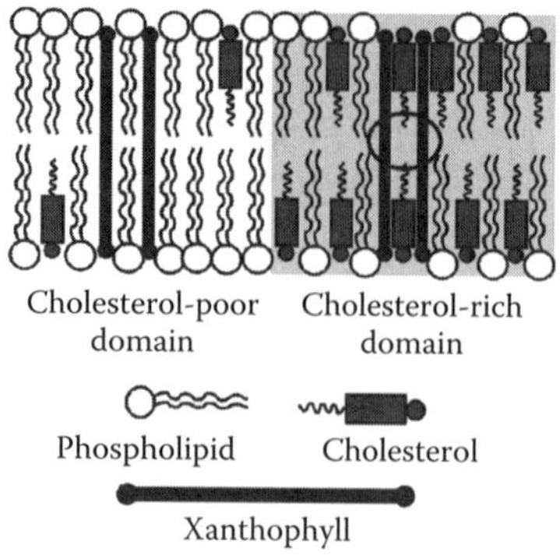

FIGURE 10.14 Schematic drawing showing the localization of xanthophyll molecules in the cholesterol-rich (raft or DRM) domain and the cholesterol-poor (bulk or DSM) domain. Unfavorable interaction with cholesterol in the cholesterol-rich domain is indicated.

accepted mechanism through which lutein and zeaxanthin protect the retina from age-related macular degeneration (Snodderly 1995, Landrum et al. 1997, Beatty et al. 1999).

The above data demonstrate that macular xanthophylls are excluded from cholesterol-rich membrane domains, which is in agreement with their poor solubility in membranes with a high cholesterol content (Socaciu et al. 2000) also shown using spin-labeled lutein and EPR spectroscopy (Wisniewska et al. 2003). This suggests that the xanthophyll–cholesterol interaction is weaker than the xanthophyll–phospholipid interaction. In the lipid bilayer, the rigid bar-like xanthophyll molecule does not conform to the cholesterol molecule which has a rigid plate-like tetracyclic ring structure and flexible isooctyl chain. When these rigid molecules are located next to each other in the lipid bilayer, a free space is created in the membrane center (Figure 10.14). Cholesterol molecules are forced to sink deeper into the bilayer, which is energetically unfavorable because it opens the access of water to the hydrophobic surface of the alkyl chains. Thus, macular xanthophylls are excluded from cholesterol-rich domains, as illustrated in Figure 10.13.

10.6.2 Transmembrane Localization of *cis*-Isomers of Zeaxanthin

Based on the molecular structure of the *cis*-isomers, localization of polar and hydrophobic parts of the molecule, and the "fit" to the membrane hydrophobic thickness, a model was proposed that placed the *cis*-isomers of zeaxanthin horizontally with respect to the plane of the membrane and with polar hydroxyl groups anchored in the same polar headgroup region (the same leaflet) of the bilayer, see review (Gruszecki 2004). However, there are no data that confirm or reject this model. Because of this the authors of this chapter have undertaken measurements, using conventional and saturation-recovery EPR spin-labeling methods, to look at the effects of *cis*-isomers of zeaxanthin on different properties of the DMPC bilayer and compare them with those caused by the all-*trans* zeaxanthin (Widomska and Subczynski 2008). All investigated properties observed in the membrane center and near the polar headgroup region were affected similarly by the 9-*cis*, 13-*cis*, and all-*trans* isomers. However, effects observed in the membrane center were different from those caused by cholesterol. The application of 14-PC, which allowed placement of the nitroxide moiety of the spin label exactly at the center of the DMPC bilayer, was a key solution for this investigation because only the measurements in the membrane center could unequivocally confirm that the transmembrane orientation of *cis*-isomers of zeaxanthin is prevalent. Obtained data suggest that *cis*-isomers, similarly to the *trans*-isomer, adopt a transmembrane orientation

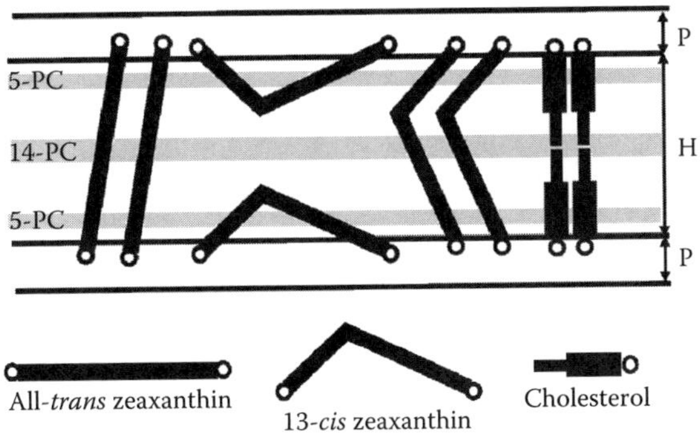

FIGURE 10.15 Schematic drawing of the localization of different isomers of zeaxanthin in the DMPC bilayer. The horizontally orientated *cis*-isomers of zeaxanthin should create more vacant pockets and increase membrane dynamics in the membrane center. Effects should be similar to those caused by cholesterol molecules. The transmembrane orientated *cis*-isomers of zeaxanthin should decrease membrane dynamics in the membrane center. Effects should be similar to those caused by all-*trans* zeaxanthin. For DMPC, the thickness of the hydrocarbon region, H, is 24.4 Å, and the polar headgroup region, P, is 5.3 Å. The distances between polar hydroxyl groups in different geometrical isomers of zeaxanthin are all-*trans*, 30.52 Å; 9-*cis*, 26.86 Å; and 13-*cis*, 24.38 Å. Hatched areas indicate regions of the membrane probed by 14-PC and 5-PC. (From Widomska, J. and Subczynski, W.K., *Biochim. Biophys. Acta*, 1778, 10, 2008. With permission.)

with the hydroxyl groups that are located in the opposite leaflets of the DMPC bilayer and that are more soluble in the lipid bilayer than those of the *trans*-isomer (*cis*-isomers do not form higher aggregates). Figure 10.15 shows possible orientations of different isomers of zeaxanthin and provides more detail.

10.7 EPR SPIN-LABELING DEMONSTRATES MEMBRANE PROPERTIES SIGNIFICANT FOR CHEMICAL REACTIONS AND PHYSICAL PROCESSES INVOLVING CAROTENOIDS

EPR spin-labeling provides a unique opportunity to obtain profiles of different membrane properties including order parameter (structural property of alkyl chains), alkyl chain bending (dynamic property of alkyl chains), oxygen and nitric oxide transport parameter (local diffusion–concentration product of these reagents in the lipid bilayer), hydrophobicity (penetration of water into the lipid bilayer), and ion penetration. These profiles can be obtained for homogenous membranes and, in some cases, in coexisting membrane domains or coexisting membrane phases without the need for their physical separation. Profiles differ in saturated and unsaturated membranes and are affected by peptides, integral membrane proteins, and, as was shown above, polar carotenoids. The most spectacular effects are observed when cholesterol is present in the lipid bilayer at a high concentration. Figure 10.16 presents different profiles across the PC and PC/Chol membranes, illustrating the extent to which the membrane properties differ. These profiles indicate that the microenvironment in which membrane-located carotenoids are immersed can change drastically with membrane composition and the depth in the membrane. It can also differ in membrane domains and phases. Thus, the microenvironmental factors presented by the lipid bilayer should be taken into account to better understand and explain chemical reactions and physical processes in which membrane-located carotenoids are involved. It should also be mentioned that appropriate profiles of membrane properties can be obtained using EPR spin-labeling measurements or found in previously published EPR data (see Subczynski et al. 2009 for more details).

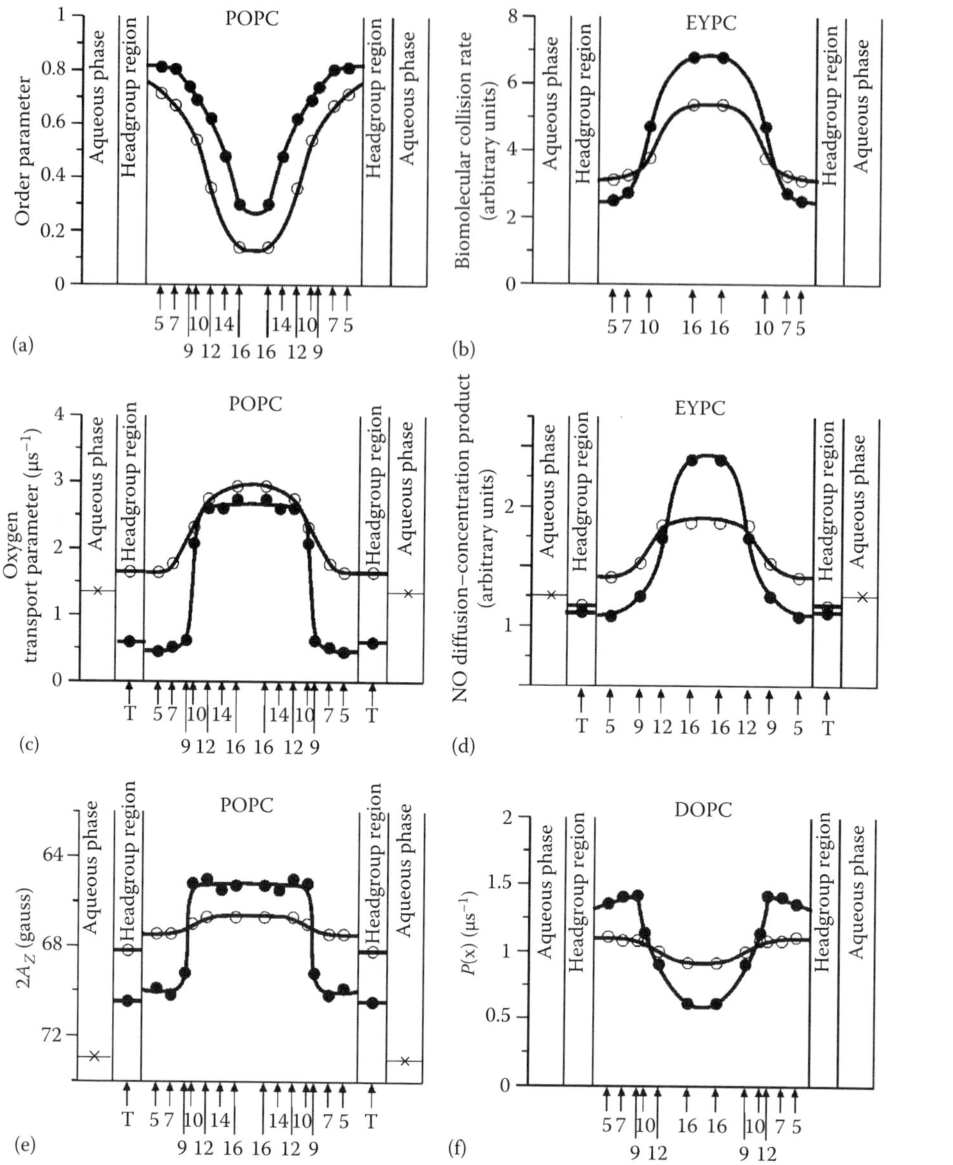

FIGURE 10.16 Profiles of different properties across PC (○) and PC/Chol (●) membranes. The approximate locations of nitroxide moieties of spin labels are indicated by arrows. (a) Order parameter in POPC membranes with and without 50 mol% cholesterol at 25°C. (b) Bimolecular collision rate for a nitroxide moiety at the C16 position of the stearic acid alkyl chain with other SASLs in EYPC membranes with and without 10 mol% lutein at 27°C. (c) Oxygen transport parameter in POPC membranes with and without 50 mol% cholesterol at 25°C. (d) Relative no diffusion–concentration product in EYPC membranes with and without 30 mol% cholesterol at 20°C. (Adapted from Subczynski, W.K. et al., *Free Radic. Res.*, 24, 343, 1996. With permission.) (e) Hydrophobicity profiles ($2A_Z$) in POPC membranes with and without 50 mol% cholesterol (results obtained at −165°C). Upward changes indicate increases in hydrophobicity. To relate hydrophobicity as observed by A_Z at a selected depth in the membrane to hydrophobicity (or ε) of bulk organic solvent, see Figure 10.5c. (f) Penetration of $Fe(CN)_6^{-3}$ ($P(x)$, defined by Equation 10.8) into the DOPC membranes with and without 30 mol% cholesterol at 25°C from the buffer containing 50 mM $K_3Fe(CN)_6$. (From Widomska, J. et al., *Biochim. Biophys. Acta*, 1768, 1454, 2007. With permission; Subczynski, W.K. et al., *Biochemistry*, 33, 7670, 1994. With permission; Subczynski, W.K. et al., *Biochemistry*, 42, 3939, 2003. With permission; Yin, J.J. and Subczynski, W.K., *Biophys. J.*, 71, 832, 1996. With permission.)

ACKNOWLEDGMENTS

This work was supported by grants EY015526, EB002052, and EB001980 of the National Institutes of Health and by the POL-POSTDOC III grant PBZ/MNiSW/07/2006/01 of the Polish Ministry of Higher Education and Science.

REFERENCES

Alfredsson, G. A., J. K. Kristjansson, S. Hjörleifsdottir, and K. O. Stetter. 1988. *Rhodothermus marinus*, gen. nov., sp. nov., a thermophilic, halophilic bacterium from submarine hot springs in Iceland. *J. Gen. Microbiol.* 134:299–306.

Anton, J., A. Oren, S. Benlloch, F. Rodriguez-Valera, R. Amann, and R. Rossello-Mora. 2002. *Salinibacter ruber* gen. nov., sp. nov., a novel, extremely halophilic member of the bacteria from saltern crystallizer ponds. *Int. J. Syst. Evol. Microbiol.* 52:485–491.

Anwar, M., T. H. Khan, J. Prebble, and P. F. Zagalsky. 1977. Membrane-bound carotenoid in *Micrococcus luteus* protects naphthoquinone from photodynamic action. *Nature* 270:538–540.

Ashikawa, I., J.-J. Yin, W. K. Subczynski, T. Kouyama, J. S. Hyde, and A. Kusumi. 1994. Molecular organization and dynamics in bacteriorhodopsin-rich reconstituted membranes: Discrimination of lipid environments by the oxygen transport parameter using a pulse ESR spin-labeling technique. *Biochemistry* 33:4947–4952.

Beatty, S., M. Boulton, D. Henson, H. H. Koh, and I. J. Murray. 1999. Macular pigment and age related macular degeneration. *Br. J. Ophthalmol.* 83:867–877.

Berliner, L. J. 1978. Spin labeling in enzymology: Spin-labeled enzymes and proteins. Rotational correlation times calculation. *Methods Enzymol.* 49:466–470.

Bernstein, P. S., N. A. Balashov, E. D. Tsong, and R. R. Rando. 1997. Retinal tubulin binds macular carotenoids. *Invest. Ophthalmol. Vis. Sci.* 38:167–175.

Bhosale, P., A. J. Larson, J. M. Frederick, K. Southwick, C. D. Thulin, and P. S. Bernstein. 2004. Identification and characterization of a Pi isoform of glutathione S-transferase (GSTP1) as a zeaxanthin-binding protein in the macula of the human eye. *J Biol. Chem.* 47:49447–49454.

Bone, R. A. and J. T. Landrum. 1984. Macular pigment in Henle fiber membranes a model for Haidinger's brushes. *Vision Res.* 24:103–108.

Bone, R. A., J. T. Landrum, and A. Cains. 1992. Optical density spectra of the macular pigment *in vivo* and *in vitro Vision Res.* 32:105–110.

Centrell, A., D. J. MacGarvey, T. G. Truscott, F. Rancan, and F. Böhm. 2003. Singlet oxygen quenching in model membrane environment. *Arch. Biochem. Biophys.* 412:47–54.

Chamberlain, N. R., B. G. Mehrtens, Z. Xiong, F. A. Kapral, J. L. Boardman, and J. I. Rearick. 1991. Correlation of carotenoid production, decreased membrane fluidity, and resistance to oleic acid killing in *Staphylococcus aureus* 18Z. *Infect. Immun.* 59:4332–4337.

Clejan, S., T. A. Krulwich, K. R. Mondrus, and D. Seto-Young. 1986. Membrane lipid composition of obligately and facultatively alkalophilic strains of *Bacillus* spp. *J. Bacteriol.* 168:334–340.

Cohen, Y., S. Yalovsky, and R. Nechushtai. 1995. Integration and assembly of photosynthetic protein complexes in chloroplast thylakoid membranes. *Biochim. Biophys. Acta* 1241:1–30.

Conn, P. F., W. Schalch, and T. G. Truscott. 1991. The singlet oxygen and carotenoid interaction. *J. Photochem. Photobiol. B. Biol.* 11:41–47.

Dietrich, C., L. A. Bagatolli, Z. N. Volovyk et al. 2001. Lipid rafts reconstitution in model membranes. *Biophys. J.* 80:1417–1428.

Eaton, S. S. and G. R. Eaton. 2005. Saturation recovery EPR. In *Biological Magnetic Resonance, Biomedical EPR–Part B: Methodology, Instrumentation, and Dynamics,* eds. S. S. Eaton, G. R. Eaton, and L. J. Berliner, Vol. 24, pp. 3–18. New York: Kluwer/Plenum.

Edge, R., D. J. McGarvey, and T. G. Truscott. 1997. The carotenoids as anti-oxidants–a review. *J. Photochem. Photobiol. B.* 41:189–200.

Egreet-Charlier, M., A. Sanson, M. Ptak, and O. Bouloussa. 1978. Ionization of fatty acids at lipid-water interface. *FEBS Lett.* 89:313–316.

Fugimori, E. and M. Talva. 1966. Light-induced electron transfer between chlorophyll and hydroquinone and the effect of oxygen and beta-carotene. *Photochem. Photobiol.* 5:877–887.

Furr, H. C. and R. M. Clark. 1997. Intenstinal absorption and tissue distribution of carotenoids. *J. Nutr. Biochem.* 8:364–377.

Griffith, O. H., P. J. Dehlinger, and S. P. Van. 1974. Shape of the hydrophobic barrier of phospholipids bilayers (Evidence for water penetration into biological membranes). *J. Membr. Biol.* 15:159–192.

Griffiths, M., W. R. Sistrom, G. Cohen-Bazire, R. Y. Stanier et al. 1955. Function of carotenoids in photosynthesis. *Nature* 176:1211–1214.

Gruszecki, W. I. 1991. Violaxanthin and zeaxanthin aggregation in lipid-water system. *Stud. Biophys.* 139:95–101.

Gruszecki, W. I. 1999. Carotenoids in membranes. In *The Photochemistry of Carotenoids*, eds. H. A. Frank, A. J. Young, and G. Britton, pp. 363–379. Dordrecht, the Netherlands: Kluwer Academic Publishers.

Gruszecki, W. I. 2004. Carotenoid orientation: Role in membrane stabilization. In *Carotenoids in Health and Disease*, eds. N. I. Krinsky, S. T. Mayne and H. Sies, 151–163. New York: Kluwer Marcel Dekker.

Gruszecki, W. I. and K. Strzalka. 1991. Does the xanthophyll cycle take part in the regulation of fluidity of the thylakoid membrane? *Biochim. Biophys. Acta* 1060:310–314.

Hata, T. R., T. A. Scholz, I. V. Ermakov et al. 2000. Non-invasive Raman spectroscopic detection of carotenoids in human skin. *J. Invest. Dermatol.* 115:441–448.

Havaux, M. 1998. Carotenoids as membrane stabilizers in chloroplasts. *Trends Plant Sci.* 3:147–151.

Huang, L. and A. Haug. 1974. Regulation of membrane lipid fluidity in *Acholeplasma laidlawii*: Effect of carotenoid pigment content. *Biochim. Biophys. Acta* 352:361–370.

Hyde, J. S. and W. K. Subczynski. 1984. Simulation of ESR spectra of the oxygen-sensitive spin-label probe CTPO. *J. Magn. Reson.* 56:125–130.

Hyde, J. S. and W. K. Subczynski. 1989. Spin-label oximetry. In *Biological Magnetic Resonance. Spin Labeling: Theory and Applications*. eds. L. J. Berliner and J. Reuben, Vol. 8, pp. 399–425. New York: Plenum.

Kaplan, L. A., J. M. Lau, and E. A. Stein. 1990. Carotenoid composition, concentrations, and relationships in various human organs. *Clin. Physiol. Biochem.* 8:1–10.

Kawasaki, K., J.-J. Yin, W. K. Subczynski, J. S. Hyde, and A. Kusumi. 2001. Pulse EPR detection of lipid exchange between protein-rich raft and bulk domains in the membrane: Methodology development and its application to studies of influenza viral membrane. *Biophys. J.* 80:738–748.

Khachik, F., F. B. Askin, and K. Lai. 1998. Distribution, bioavailability, and metabolism of carotenoids in humans. In *Phytochemicals, a New Paradigm*, eds. W. R. Bidlack, S. T. Omaye, M. S. Meskin, and D. Jahner, Vol. 5, pp. 77–96. Lancaster, PA: Technomic Publishing.

Kolev, V. D. and D. N. Kafalieva. 1986. Miscibility of beta-carotene and zeaxanthin with dipalmitoylphosphatidylcholine in multilamellar vesicles: A calorimetric and spectroscopic study. *Photobiochem. Photobiophys.* 11:257–267.

Koyama, Y. 1991. Structures and functions of carotenoids in photosynthetic systems. *J. Photochem. Photobiol. B. Biol.* 9:265–280.

Krinsky, N. I. 1989. Antioxidant functions of carotenoids. *Free Radic. Biol. Med.* 7:617–35.

Krinsky, N. I., D. G. Cronwell, and J. L. Oncley. 1958. The transport of vitamin A and carotenoids in human plasma. *Arch. Biochem. Biophys.* 73:233–246.

Kushwaha, S. C., J. K. G. Kramer, and M. Kates. 1975. Isolation and characterization of C50-carotenoid pigments and other polar isoprenoids from *Halobacterium cutirubrum*. *Biochim. Biophys. Acta.* 398:303–314.

Kusumi, A., W. K. Subczynski, and J. S. Hyde. 1982a. Effects of pH on ESR spectra of stearic acid spin labels in membranes: Probing the membrane surface. *Fed. Proc.* 41:1394.

Kusumi, A., W. K. Subczynski, and J. S. Hyde. 1982b. Oxygen transport parameter in membranes as deduced by saturation recovery measurements of spin-lattice relaxation times of spin labels. *Proc. Natl. Acad. Sci. USA* 79:1854–1858.

Kusumi, A., W. K. Subczynski, M. Pasenkiewicz-Gierula, J. S. Hyde, and H. Merkle. 1986. Spin-label studies on phosphatidylcholine-cholesterol membranes: Effects of alkyl chain length and unsaturation in the fluid phase. *Biochim. Biophys. Acta* 854:307–317.

Landrum, J. T., R. A. Bone, H. Joa, M. D. Kilburn, L. L. Moore, and K. E. Sprague. 1997. A one year study of the macular pigment: The effect of 140 days of a lutein supplement. *Exp. Eye Res.* 65:57–62.

Landrum, J. T., R. A. Bone, L. L. Moore, and C. M. Gomea. 1999. Analysis of zeaxanthin distribution within individual human retinas. *Methods Enzymol.* 229:457–467.

Lazrak, T., A. Milon, G. Wolff et al. 1987. Comparison of the effects of inserted C40- and C50-terminally dihydroxylated carotenoids on the mechanical properties of various phospholipid vasucles. *Biochim. Biophys. Acta* 903:132–141. (Published erratum appears in 1988 *Biochim. Biophys. Acta* 937:427.)

Ligeza, A., A. N. Tikhonov, J. S. Hyde, and W. K. Subczynski. 1998. Oxygen permeability of thylakoid membranes: Electron paramagnetic resonance spin labeling study. *Biochim. Biophys. Acta* 1365:453–463.

Lutnaes, B. F., A. Oren, and S. Liaaen-Jensen. 2002. New C-40 carotenoid acyl glycoside as principal carotenoid in *Salinibacter ruber*, an extremely halophilic eubacterium. *J. Nat. Products* 65:1340–1343.

Marsh, D. 1981. Electron spin resonance: Spin labels. In *Membrane Spectroscopy. Molecular Biology, Biochemistry, and Biophysics*, ed. E. Grell, Vol. 31, pp. 51–142. Berlin, Germany: Springer-Verlag.

McDermott, G., S. Prince, A. Freer et al. 1995. Crystal structure of an integral membrane light-harvesting complex from photosynthetic bacteria. *Nature* 374:517–521.

Milon, A., T. Lazrak, A. M. Albrecht et al. 1986. Osmotic swelling of unilamellar vesicles by the stopped-flow light scattering method. Influence of vesicle size, solute, temperature, cholesterol and three α, ω-dihydroxycarotenoids. *Biochim. Biophys. Acta* 859:1–9.

Oren, A. 2002. Molecular ecology of extremely halophilic archaea and bacteria. *FEMS Microbiol. Ecol.* 39:1–7.

Parker, R. S. 1989. Carotenoids in human blood and tissues. *J. Nutr.* 119:101–104.

Rapp, L. M., S. S. Maple, and J. H. Choi. 2000. Lutein and zeaxanthin concentrations in rod outer segment membranes from perifoveal and peripheral human retina. *Invest. Ophthalmol. Vis. Sci.* 41:1200–1209.

Rohmer, M., P. Bouvier, and G. Ourisson. 1979. Molecular evolution of biomembranes: Structural equivalents and phylogenetic precursors of sterols. *Proc. Natl. Acad. Sci. USA* 76:847–851.

Schmitz, H. H., C. L. Poor, R. B. Wellman, and J. W. Jr. Erdman. 1991. Concentrations of selected carotenoids and vitamin A in human liver, kidney, and lung tissue. *Am. Inst. Nutr.* 121:1613–1621.

Sefirmann-Harms, D. 1987. The light harvesting and protective function of carotenoid in photosynthetic membrane. *Physiol. Plant.* 69:501–562.

Snodderly, D. M., J. D. Aura, and F. C. Delori. 1984. The macular pigment, II spatial distribution in primate retinas. *Invest. Ophthalmol. Vis. Sci.* 25:674–685.

Snodderly, M. D. 1995. Evidence for protection against age-related macular degeneration by carotenoids and antioxidant vitamins. *Am. J. Clin. Nutr.* 62:1448S–1461S.

Socaciu, C., R. Jessel, and H. A. Diehl. 2000. Competitive carotenoid and cholesterol incorporation into liposomes: Effects on membrane phase transition, fluidity, polarity and anisotropy. *Chem. Phys. Lipids* 106:79–88.

Sommerburg, O. G., W. G. Siems, J. S. Hurst, J. W. Lewis, D. S. Kliger, and F. J. van Kuijk. 1999. Lutein and zeaxanthin are associated with photoreceptors in the human retina. *Curr. Eye Res.* 19:491–495.

Subczynski, W. K., C. C. Felix, C. S. Klug, and J. S. Hyde. 2005. Concentration by centrifugation for gas exchange EPR oximetry measurements with loop-gap resonators. *J. Magn. Reson.* 176:244–248.

Subczynski, W. K. and J. S. Hyde. 1984. Diffusion of oxygen in water and hydrocarbons using an electron spin resonance spin-label technique. *Biophys. J.* 45:743–748.

Subczynski, W. K., J. S. Hyde, and A. Kusumi. 1989. Oxygen permeability of phosphatidylcholine-cholesterol membranes. *Proc. Natl. Acad. Sci. USA* 86:4474–4478.

Subczynski, W. K., J. S. Hyde, and A. Kusumi. 1991. Effect of alkyl chain unsaturation and cholesterol intercalation on oxygen transport in membranes: A pulse ESR spin labeling study. *Biochemistry* 30:8578–8590.

Subczynski, W. K. and A. Kusumi. 2003. Dynamics of raft molecules in the cell and artificial membranes: Approaches by pulse EPR spin labeling and single molecule optical microscopy. *Biochim. Biophys. Acta* 1610:231–243.

Subczynski, W. K., E. Markowska, W. I. Gruszecki, and J. Sielewiesiuk. 1992. Effects of polar carotenoids on dimyristoylphosphatidylcholine membranes: Spin-label studies. *Biochim. Biophys. Acta* 1105:97–108.

Subczynski, W. K., E. Markowska, and J. Sielewiesiuk. 1991. Effect of polar carotenoids on the oxygen diffusion-concentration product in lipid bilayers. An EPR spin label study. *Biochim. Biophys. Acta* 1068:68–72.

Subczynski, W. K., E. Markowska, and J. Sielewiesiuk. 1993. Spin-label studies on phosphatidylcholine-polar carotenoid membranes: Effects of alkyl chain length and unsaturation. *Biochim. Biophys. Acta* 1150:173–181.

Subczynski, W. K. and H. M. Swartz. 2005. EPR oximetry in biological and model samples. In *Biological Magnetic Resonance, Biomedical EPR–Part A: Free Radicals, Metals, Medicine, and Physiology*, eds. S. S. Eaton, G. R. Eaton, and L. J. Berliner, Vol. 23, pp. 229–282. New York: Kluwer/Plenum.

Subczynski, W. K., J. Widomska, and J. B. Feix. 2009. Physical properties of lipid bilayers from EPR spin labeling and their influence on chemical reactions in a membrane environment. *Free Radic. Biol. Med.*, 46, 707–718.

Subczynski, W. K., J. Widomska, A. Wisniewska, and A. Kusumi. 2007a. Saturation-recovery electron paramagnetic resonance discrimination by oxygen transport (DOT) method for characterizing membrane domains. In *Methods in Molecular Biology: Lipid Rafts*, ed. T. J. McIntosh, Vol. 398, pp. 145–159, Totowa, NJ: Humana Press.

Subczynski, W. K. and A. Wisniewska. 1998. Effects of β-carotene on physical properties of lipid membranes–comparison with effects of polar carotenoids. *Curr. Top. Biophys.* 22:44–51.

Subczynski, W. K., A. Wisniewska, J. S. Hyde, and A. Kusumi. 2007b. Tree-dimensional dynamic structure of the liquid-ordered domain in lipid membranes as examined by pulse-EPR oxygen probing. *Biophys. J.* 92:1573–1584.

Subczynski, W. K., A. Wisniewska, J.-J. Yin, J. S. Hyde, and A. Kusumi. 1994. Hydrophobic barriers of lipid bilayer membranes formed by reduction of water penetration by alkyl chain unsaturation and cholesterol. *Biochemistry* 33:7670–7681.

Tanumihardjo, S. A., H. C. Furr, O. Amedee-Manesme, and J. A. Olson. 1990. Retinyl ester (vitamin A ester) and carotenoid composition in human liver. *Int. J. Vitam. Nutr. Res.* 60:307–313.

Tardy, T. and M. Havaux. 1997. Thylakoid membrane fluidity and thermostability during the operation of the xanthophyll cycle in higher-plant chloroplasts. *Biochim. Biophys. Acta* 1330:179–193.

Tso, P. 1981. Intestinal lipid absorption. In *Physiology of the Gastrointestinal Tract*, ed. L. R. Johanson, pp. 1867–1907, New York: Raven Press.

Widomska, J. and W. K. Subczynski. 2008. Transmembrane localization of *cis*-isomers of zeaxanthin in the host dimyristoylphosphatidylcholine bilayer membrane. *Biochim. Biophys. Acta* 1778:10–19.

Widomska, J., M. Raguz, J. Dillon, E. R. Gaillard, and W. K. Subczynski. 2007. Physical properties of the lipid bilayer membrane made of calf lens lipids: EPR spin labeling studies. *Biochim. Biophys. Acta* 1768:1454–1465.

Windrem, D. A. and W. Z. Plachy. 1980. The diffusion-solubility of oxygen in lipid bilayers. *Biochim. Biophys. Acta* 600:655–665.

Wisniewska, A., J. Draus, and W. K. Subczynski. 2006. Is fluid mosaic model of biological membranes fully relevant? Studies on lipid organization in model and biological membranes. *Cell. Mol. Biol. Lett.* 8:147–154.

Wisniewska, A., Y. Nishimoto, J. S. Hyde, A. Kusumi, and W. K. Subczynski. 1996. Depth dependence of the perturbing effect of placing a bulky group (oxazoline ring spin labels) in the membrane on the membrane phase transition. *Biochim. Biophys. Acta* 1278:68–72.

Wisniewska, A. and W. K. Subczynski. 1998. Effects of polar carotenoids on the shape of the hydrophobic barrier of phospholipid bilayers. *Biochim. Biophys. Acta* 1368:235–246.

Wisniewska, A. and W. K. Subczynski. 2006a. Accumulation of macular xanthophylls in unsaturated membrane domines. *Free Radic. Biol. Med.* 40:1820–1826.

Wisniewska, A. and W. K. Subczynski. 2006b. Distribution of macular xanthophylls between domains in model of photoreceptor outer segment membranes. *Free Radic. Biol. Med.* 4:1257–1265.

Wisniewska, A., J. Widomska, and W. K. Subczynski. 2006. Carotenoid-membrane interactions in liposomes: Effect of dipolar, monopolar, and nonpolar carotenoids. *Acta Biochim. Polonica* 53:475–484.

Yin, J.-J. and W. K. Subczynski. 1996. Effect of lutein and cholesterol on alkyl chain bending in lipid bilayers: A pulse electron paramagnetic resonance spin labeling study. *Biophys. J.* 71:832–839.

Yokoyama, A., G. Sandmann, T. Hoshino, K. Adachi, M. Sakai, and Y. Shizuri. 1995. Thermozeaxanthins, new carotenoid-glycoside-esters from thermophilic eubacterium *Thermus thermophilus*. *Tetrahedron Lett.* 36:4901–4904.

Part IV

Chemical Breakdown of Carotenoids In Vitro *and* In Vivo

11 Formation of Carotenoid Oxygenated Cleavage Products

Catherine Caris-Veyrat

CONTENTS

11.1 INTRODUCTION

Molecules formed from carotenoids are given different names in the literature, for instance, carotenoid-derived products, degraded carotenoids (Walberg and Eklund 1998), carotenoid decomposition products (Wang 2004), carotenoid oxidation products, carotenoid oxidative/degradative products (Wang 2004), carotenoid oxidative breakdown products (Bonnie and Choo 1999), oxidative cleavage products, apocarotenoids, and lycopenoids (Lindshield et al. 2007). The use of each term is justified in the relevant context of each cited article, making it very difficult or even impossible to choose one of these terms for general use.

In this chapter, we focus on a category of molecules obtained from carotenoids in which at least one of the carbon–carbon bonds has been cleaved and at least one oxygen atom has been introduced. These products are referred to as carotenoid oxygenated cleavage products. According to the accepted rules of carotenoid nomenclature (Weedon and Moss 1995), "derivatives in which the carbon skeleton has been shortened by the formal removal of fragments from one end or both ends of a carotenoid" are called, respectively, apo- or diapocarotenoids. When fission occurs on a cyclic bond, the C40 carbon skeleton is retained, and the products are called seco-carotenoids. In most cases, the organic functional group replacing the lost end of the carotenoid contains at least one oxygen atom and is often an alcohol, aldehyde, ketone, carboxylic acid, or ester function (Figure 11.1). Like apocarotenoids, norcarotenoids have fewer than 40 carbon atoms. However, those that have been eliminated come from within the carotenoid skeleton, and, as such, they do not fit our definition of cleavage compounds. Two types of oxygenated cleavage products of carotenoids can be distinguished: volatiles and nonvolatiles. We concentrate our review on nonvolatile products, mentioning studies on volatile compounds either when nonvolatiles have been studied at the same time or when the effects of food thermal processing on carotenoids are described.

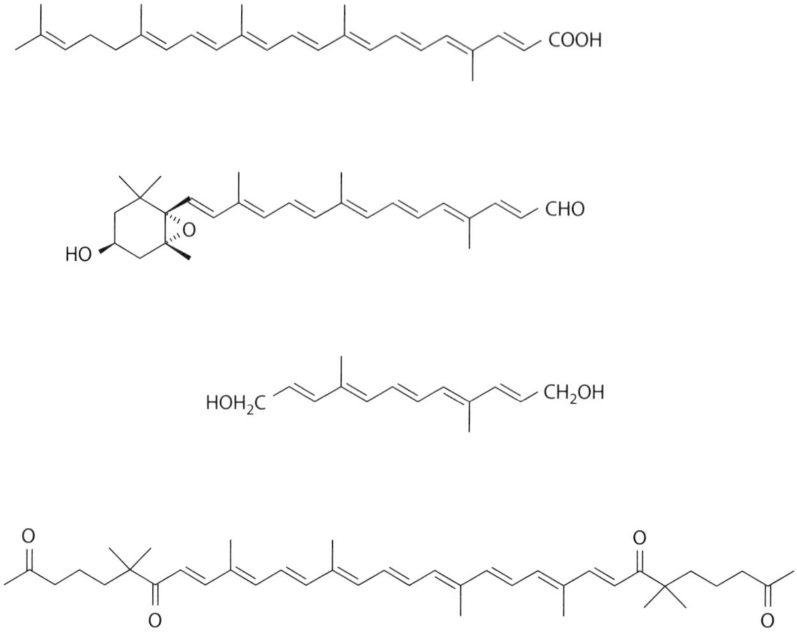

504.4

502

547.2

562

FIGURE 11.1 Chemical structures of carotenoid oxidation products occurring in nature: apocarotenoids:10′-apolycopen-10′-oic acid (504.4), apo-10′-violaxanthal (502), diapocarotenoid:rosafluin (547.2), and seco-carotenoid:β-carotenone (562). The compound number corresponds to those in Britton et al. (2004).

After a short presentation of naturally occurring oxygenated cleavage compounds, we describe different ways by which they can be formed starting from the parent carotenoid, and we give some information on their mechanisms of formation when available in the literature.

11.2 OCCURRENCE IN NATURE AND FORMATION IN BIOCHEMICAL SYSTEMS

In nature, some 117 apocarotenoids have been reported, 88 of which have been fully identified. Another six naturally occurring seco-carotenoids have been referenced as carotenoids (Britton et al. 2004). Apo- and seco-carotenoids represent around 15% of the carotenoids so far reported. This subfamily of carotenoids would be even larger if one considers retinoids and norisoprenoids, but these compounds are excluded by nomenclature rules (IUPAC 1971, 1975) that dictate that they are not deemed to be carotenoids because of the absence of the two central methyl groups (at C20 and C20′). Retinoic acid, retinal, and retinol (vitamin A) can be considered as carotenoid oxygenated cleavage products of the provitamin A carotenoids, such as β-carotene or β-cryptoxanthin, and are formed in humans by enzymatic cleavage. The theories of their mechanism of formation were for many years controversial, with two hypotheses based on a central and/or excentric cleavage. Krinsky and coworkers have shown that the excentric cleavage of β-carotene occurs, giving rise to a series of apocarotenals and even retinoic acid, when β-carotene is incubated in various biochemical systems (Tang et al. 1991, Wang et al. 1991, 1992, Yeum et al. 1995). It is only recently that cleavage enzymes have been identified. The first central cleavage enzyme was partially purified via cloning of its encoding cDNAs from different organisms (von Lintig and Vogt 2000, Wyss et al. 2000, Paik et al. 2001, Redmond et al. 2001) and was shown to be a monooxygenase-type enzyme (Leuenberger et al. 2001). Another enzyme that catalyzes the excentric cleavage of β-carotene in the 9′,10′ position was shown to occur in humans and mice producing apo-10′-carotenal (Kiefer et al. 2001). A similar enzyme from ferret, a model used to study carotenoid metabolism in humans, was shown to oxidatively cleave not only β-carotene but also 5-(Z) and 13-(Z)-lycopene *in vitro* at the 9′,10′ carbon–carbon double bond (Hu et al. 2006), thus producing the corresponding apocarotenals and apolycopenals. And, for the first time, lycopene oxygenated cleavage compounds, apo-8′-lycopenal and apo-12′-lycopenal, were found to occur *in vivo* in rat liver (Gajic et al. 2006). These findings on the biosynthetic route to the formation of apocarotenals in animals and the discovery of an enzyme catalyzing the asymmetric cleavage of carotenoids has generated heightened interest in carotenoid oxygenated cleavage products and their possible biological role *in vivo*.

Apocarotenoids are also found in plants, where they are bioactive mediators. They can act as visual or volatile signals to attract pollinating and seed dispersal agents, and are also key players in allelopathic interactions, plant defense, and even plant architecture (Bouvier et al. 2005). Abscisic acid is an essential plant metabolite that can be considered a carotenoid oxygenated cleavage product. It is formed via the specific oxidative cleavage of the 11′,12′ carbon–carbon double bond of 9′-(Z)-neoxanthin. As already mentioned, in this chapter we focus on nonvolatile compounds, but it is worth noting that a large structural diversity is found among apocarotenoids with 9 or 13 carbons present in fruits, wines, and tobacco, many of which possess aromatic properties, making them popular for use in commercial flavoring and as fragrances.

11.3 FORMATION BY AUTOXIDATION IN MODEL SYSTEMS

Autoxidation has been defined as "a spontaneous oxidation in air of a substance, not requiring a catalyst." (Miller et al. 1990) However, because ground state molecular oxygen is in the triplet form and most biomolecules exist in the singlet form, reactions between them are spin forbidden, although they do occur very slowly (bimolecular rate constant $k \leq 10^{-5}\ M^{-1}s^{-1}$), i.e., over the time frame of days (Miller et al. 1990). As a consequence, direct reactions between biomolecules, such as carotenoids and dioxygen, are either very slow or, when quicker, probably catalyzed or accelerated by trace metal ions, light, or heat. Because of their long, conjugated polyenic chain, carotenoids are highly susceptible to autoxidation.

Researchers have studied the products formed and their mechanism of formation through the nonradical and nonmetal initiated autoxidation of carotenoids in experimental models by using only organic solvent and a flow of oxygen. In 1970, El-Tinay and Chichester (1970) first studied the reaction between β-carotene and oxygen in toluene at 60°C in the dark. The products of the reaction were tentatively identified as 5,6- and 5,8-epoxides, 5,6; 5′,6′- and 5,8; 5′,8′-diepoxides of β-carotene and "polyene carbonyl," which was not further identified. The authors deduced that the site of "initial attack" of oxygen was on the terminal carbon–carbon double bond with the highest electron density in the polyene chain. They also found overall zero-order reaction kinetics and activation energy of 10.20 kcal/mol. The authors concluded that there is an "associated intermediate complex between β-carotene and oxygen" with "free radical character." More than 20 years later, a similar experimental model (β-carotene, toluene, 60°C, oxygen, 120 min) was used by Handelman et al. (1991). Using HPLC and mass analysis, the authors could tentatively identify the 5,6-epoxide of β-carotene and apocarotenals, but some compounds remained unidentified. Using comparable experimental conditions but lower temperatures and longer times (β-carotene, benzene or tetrachloromethane, 30°C, oxygen, dark, 48 and 77 h), Mordi et al. (1993) published the identification of the mono- and diepoxides of β-carotene, as previously detected, together with *Z*-isomers, apocarotenals of different lengths, volatile short compounds, and minor or oligomeric compounds not previously identified. To explain the formation of these compounds, the authors proposed a radical-mediated reaction in which the initiation process involves the formation of a diradical of β-carotene. The results of the three studies are not completely comparable since some experimental conditions are different and some are not indicated (time, absence of light). However, very similar types of products were found: mono- and diepoxides of β-carotene and oxygenated cleavage compounds such as apocarotenals. Other researchers studied the mechanism of autoxidation of β-carotene in organic solutions: Takahashi et al. (1999) proposed a kinetic model on the basis of an autocatalytic free radical chain reaction mechanism; Martin et al. (1999) proposed that triplet oxygen adds to an "undisturbed" carotene and calculated an energy needed for the reaction of 18 kcal/mol, which is in agreement with the experimental value of E_a = 16 kcal/mol.

Using an aqueous model system, the speed of autoxidation was compared for different carotenoids (Henry et al. 2000). Carotenoids were adsorbed onto a C18 solid phase and exposed to a continuous flow of water saturated with oxygen or ozone at 30°C. The major reaction products of β-carotene were identified as 13-(*Z*)-, 9-(*Z*)-isomers, a di-(*Z*)-isomer, the oxygenated cleavage products β-apo-13-carotenone and β-apo-14-carotenal, and also the β-carotene 5,8-epoxide and β-carotene 5,8-endoperoxide. The degradation of all the carotenoids followed zero-order reaction kinetics with the following relative rates: lycopene > β-cryptoxanthin > (*E*)-β-carotene > 9-(*Z*)-β-carotene. Recently, the autoxidation of β-carotene in an aqueous model system was studied in the presence of light during a long period (30 days) (Rodriguez and Rodriguez-Amaya 2007). The main products were *Z*-isomers, hydroxylated compounds in position 4 and epoxide-containing compounds in positions 5,6; 5′,6′; 5,8; and 5′,8′. Oxygenated cleavage compounds (apo-8′, apo-10′, apo-12′, apo-14′, and apo-15-carotenals) were also detected but in very small amounts, probably due to the limited concentration of oxygen available. The compounds identified were very similar when a low-moisture model system (β-carotene impregnated into starch) was used in the presence of light or in the dark over 21 days. Some of these compounds were also detected in very low levels in processed food products (mango and acerola juices, dried apricots). In an aqueous system, using Tween 40 to solubilize lycopene, autoxidation at 37°C for 72 h produced oxygenated cleavage compounds, some of which were identified as apolycopenals, among which acycloretinal and one as apolycopenone (Kim et al. 2001).

Studies on the autoxidation of carotenoids in liposomal suspensions have also been performed since liposomes can mimic the environment of carotenoids *in vivo*. Kim et al. have studied the autoxidation of lycopene (Kim et al. 2001), ζ-carotene (Kim 2004), and phytofluene (Kim et al. 2005) in liposomal suspensions and identified oxygenated cleavage compounds. The stability to oxidation at room temperature of various carotenoids has also been studied when incorporated in pig liver microsomes (Socaciu et al. 2000), and taking into account membrane dynamics. After 3 h of reaction, β-carotene and lycopene had completely degraded, whereas the xanthophylls tested were shown to be more stable.

Interestingly, early examples of carotenoid autoxidation in the literature described the influence of lipids or other antioxidants on the autoxidation of carotenoids (Lisle 1951, Budowski and Bondi 1960). In the study by Budowski and Bondi (1960), the influence of fat was found to be a "prooxidant." In this case the oxidation of carotenoids was probably caused not only by molecular oxygen but also by lipid oxidation products, a now well-known phenomena called "co-oxidation," which has been studied in lipid solution, in aqueous solution catalyzed by enzymes (Grosch and Laskawy 1979), and even in food systems in relation to carotenoid oxidation (Perez-Galvez and Minguez-Mosquera 2001). The influence of α-tocopherol on the autoxidation of carotenoids was also studied, for instance, by Takahashi et al. (2003), who showed that carotene oxidation was suppressed as long as the tocopherol remained in the system, and thus that α-tocopherol protected β-carotene from autoxidation. The oxidative cleavage compounds of β-carotene were also found to be formed after reaction with alkylperoxides generated by 2,2′-azobis (2,4-dimethylvaleronitrile) (AMVN) (Yamauchi et al. 1993) and during the peroxyl radical-initiated peroxidation of methyl linoleate and its autoxidation in the bulk phase (Yamauchi et al. 1998). The products contained formyl or cyclic ether groups in the chain of carbon–carbon double bonds. The authors obtained similar compounds with canthaxanthin when it reacted either with peroxyl radicals generated by thermolysis of AMVN in benzene or when it reacted as an antioxidant during the peroxidation of methyl linoleate initiated by AMVN in bulk phase (Yamauchi and Kato 1998). The authors of the paper conclude that these products are formed from the decomposition of oxygenated products which themselves were formed by the trapping of lipid peroxyl radicals by β-carotene or canthaxanthin.

The autoxidation of carotenoids in cell medium is highly probable when experiments are conducted over periods of a few hours. The autoxidation of canthaxanthin in a cell culture medium was shown to give all-(*E*)- and 13-(*Z*)-4-oxoretinoic acid, both of which were shown to induce gap junction communication (Hanusch et al. 1995).

The interaction of carotenoids with cigarette smoke has become a subject of interest since the results of the Alpha-Tocopherol Beta-Carotene Cancer Prevention Study Group 1994 (ATBC) and CARET (Omenn et al. 1996) studies were released. β-Carotene has been hypothesized to promote lung carcinogenesis by acting as a prooxidant in the smoke-exposed lung. Thus, the autoxidation of β-carotene in the presence of cigarette smoke was studied in model systems (toluene) (Baker et al. 1999). The major product was identified as 4-nitro-β-carotene, but apocarotenals and β-carotene epoxides were also encountered.

In conclusion, the oxidation of carotenoids by molecular oxygen, so-called "autoxidation," is a complex phenomenon that is very probably initiated by an external factor (radical, metal, etc.) and for which different mechanisms have been proposed. The autoxidation of a carotenoid is important to take into account when working with this molecule even for short periods of time, for example, in cell cultures or studying antioxidant activity, since it can lower the apparent antioxidant activity of a carotenoid (Vulcain et al. 2005).

11.4 FORMATION BY CHEMICAL OXIDATION

Carotenoid oxygenated cleavage products were first produced in order to help in the structural identification of carotenoids (Karrer and Jucker 1950). Carotenoids were oxidized expressly to form small fragments that could be analyzed with the techniques available at the time. The chemical structure of the parent carotenoids was deduced from those of its oxidation products. For example, stepwise degradation by oxidation with alkaline potassium permanganate or chromic acid or ozonolysis was used to obtain large fragments of carotenoids that could be used to deduce the carotenoid structure (Karrer and Jucker 1950). Another example is the oxidation by manganese dioxide used as a chemical derivatization in microscale tests to elucidate the presence of allylic primary- and secondary-hydroxy groups in carotenoids, with the allylic aldehyde or ketone formed exhibiting a bathochromic shifted UV/Vis spectrum (Uebelhart and Eugster 1988).

Carotenoids act as antioxidants in photosynthetic tissues by inactivating singlet oxygen through a physical reaction. However, concomitant chemical reactions can occur, consuming the carotenoids. The photosensitized oxidation of β-carotene has been studied in a model system using rose bengal as the photosensitizer in toluene/methanol. The solution was bubbled with dioxygen and illuminated with a quartz-halogen lamp (Stratton et al. 1993). Apocarotenoids, β-ionone, and β-carotene 5,8-endoperoxide were found as the products of the reaction.

In order to obtain insights into the mechanism of excentric cleavage of carotenoids in human gastric mucosal homogenate, Yeum et al. (1995) incubated β-carotene with a lipid hydroperoxide, mainly (13*S*)-hydroperoxy-*cis*, *trans*-9,11-octadecadienoic acid (13-LOOH), the primary product of lipoxygenase, and linoleic acid. Apo-8′, apo-10′, apo-12′, apo-14′, apo-carotenals, apo-13-carotenone, retinoic acid, and retinol were identified as products of the reaction of β-carotene and the hydroperoxide. The same products were obtained when β-carotene was incubated with lipoxygenase and linoleic acid and also with human gastric mucosal homogenate, suggesting that lipoxygenase is involved in carotenoid metabolism in the human gastric compartment.

Few other examples of the use of chemical oxidation of carotenoids in order to obtain carotenoid oxygenated cleavage products have been described in the literature. Different reagents have been used in order to obtain carotenoid oxygenated derivatives such as epoxides (Rodriguez and Rodriguez-Amaya 2007), dihydrooxepin (Zurcher et al. 1997), ozonides (Zurcher and Pfander 1999), or oxo-carotenoids (Molnar et al. 2006). These reactions sometimes also produced carotenoid oxygenated cleavage compounds as by-products (Molnar et al. 2006). Our focus being oxygenated cleavage products, we concentrate on the presentation of the reactions that aimed to produce these target compounds.

The ozonolysis of carotenoids was employed in order to obtain oxygenated cleavage products for biological tests, for example, for lycopene. In this case, among a series of products, one product formed by a double oxidative cleavage was purified and characterized as (*E*,*E*,*E*)-4-methyl-8-oxo-2,4,6-nonatrienal, and it was shown to be active in the induction of apoptosis in HL-60 cells (Zhang et al. 2003).

Osmium tetroxide/hydrogen peroxide was used to oxidatively cleave β-carotene chemically (Wendler and Rosenblum 1950). This reagent was later used to produce oxygenated cleavage compounds from lycopene. Indeed, the acyclic analogue of retinal, i.e., acycloretinal, also named apo-15-lycopenal, is of interest for its potential biological activity. Acycloretinal was obtained by the oxidation of lycopene using osmium tetroxide/hydrogen peroxide; subsequent oxidation or reduction gave, respectively, acycloretinoic acid and acycloretinol (Wingerath et al. 1999). Using similar experimental conditions to those of Wendler et al., Aust et al. obtained several oxidation products, among which one, diapocarotenoid (2,7,11-trimethyltetradecahexaene-1,14-dial), was identified and shown to stimulate gap junction communication (Aust et al. 2003).

The oxidation of β-carotene with potassium permanganate was described in a dichloromethane/water reaction mixture (Rodriguez and Rodriguez-Amaya 2007). After 12 h, 20% of the carotenoid was still present. The products of the reaction were identified as apocarotenals (apo-8′- to apo-15-carotenal = retinal), semi-β-carotenone, monoepoxides, and hydroxy-β-carotene-5,8-epoxide.

We have developed a biphasic oxidation protocol using the hydrophilic oxidant potassium permanganate (Caris-Veyrat et al. 2003), which we applied to lycopene. Cetyltrimethylammonium bromide was used as a phase transfer agent to achieve the contact of the hydrophilic oxidant with the lipophilic carotenoid lycopene dissolved in methylene chloride/toluene (50/50, v/v). Analysis of the reaction mixture by HPLC-DAD-MS revealed the presence of (1) apolycopenals and apolycopenones derived from a single oxidative cleavage and (2) diapocarotenedials derived from a double oxidative cleavage of lycopene which had lost the two Ψ-end groups of lycopene (Figure 11.2). No apolycopenoic acids were found in the reaction mixture, indicating that, in our experimental conditions, there was no further oxidation of apolycopenals by potassium permanganate. This oxidation

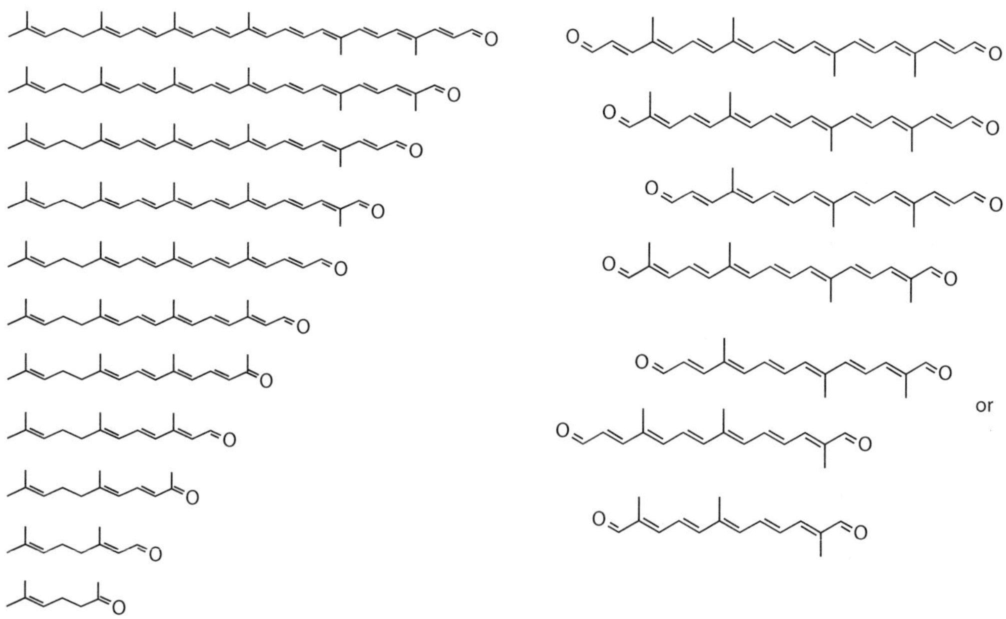

FIGURE 11.2 Lycopene oxygenated cleavage compounds produced by the reaction of potassium permanganate on lycopene in a biphasic medium: apolycopenals and apolycopenones (left column), and diapocarotendials (right column).

method allowed the production of the complete range of the possible apolycopenals formed by oxidative cleavage of conjugated carbon–carbon double bonds of lycopene and also six diapocarotenedials, which opened up the possibility of preparing these compounds by preparative HPLC for further use.

Potassium permanganate is a versatile reagent that can react with carbon–carbon double bonds by different mechanisms in order to produce different types of compounds (Fatiadi 1987). In the conditions used by Rodriguez and Rodriguez-Amaya (2007) and ourselves, the use of potassium permanganate generated the oxidative cleavage of the double bonds of the studied carotenoids and gave apocarotenoids without further oxidation to carboxylic acid functions. In these reactions, the initial step of the reaction may involve a [3 + 2] electrocyclic addition of permanganate ion to the pi bond, thus forming the cyclic hypomanganate Mn(V) ester (Figure 11.3). Intramolecular electron transfer may then occur to give the oxidized cyclic manganate Mn(VI) ester, which, in turn, by rearrangement and fragmentation, will give the final products: cleavage products bearing aldehyde groups.

The chemically catalyzed oxidation of carotenoids by metalloporphyrins has also been described in the literature. In 2000, French et al. described a central cleavage mimic system (ruthenium porphyrin linked to cyclodextrins) that exhibited a 15,15′-regioselectivity of about 40% in the oxidative cleavage of β-carotene by *tert*-butyl hydroperoxide in a biphasic system (French et al. 2000).

Simultaneously, we used ruthenium porphyrin in order to catalyze the oxidation of a carotenoid by molecular oxygen. Our focus was on the experimental modeling of the eccentric cleavage of β-carotene (Caris-Veyrat et al. 2001) and lycopene (Caris-Veyrat et al. 2003). Different types of products were found in the reaction mixture and tentatively characterized by HPLC-DAD-MS: (*Z*)-isomers, epoxides, apolycopenals, and apolycopenones. When lycopene was allowed to react in the presence of ruthenium tetraphenyl porphyrin and oxygen, it was slowly oxidized and disappeared completely within 24 h. The reaction mixture continued to evolve for up to 96 h. Different types of products could be detected and tentatively attributed using HPLC-DAD-MS analysis. *Z*-isomers, among which the 9-, 13-, and 15-(*Z*) were tentatively attributed, together with “in chain” epoxides, as well as apolycopenals, both long or short, were found in the reaction mixture. Following these reactions, over 96 h allowed us to trace the appearance/disappearance of each class of compounds and even each individual compound in the case of apolycopenals. (*Z*)-isomers of lycopene were detected after 1 h of reaction and had almost disappeared after 24 h, so as (*E*)-lycopene, whereas other reaction products (epoxides and apolycopenals) either were still present after 24 h or even continued to be formed after 24 h and until 96 h. Moreover, it is known that a (*Z*)-olefin is at least 10 times more reactive than the (*E*)-isomer in a competitive oxidation by a metalloporphyrin catalytic

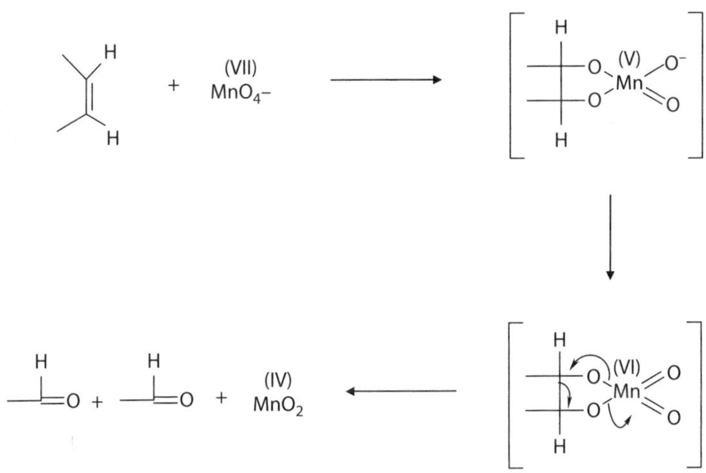

FIGURE 11.3 Mechanism for oxidative cleavage of carbon–carbon double bonds by potassium permanganate by which apocarotenals are produced (as described in Fatiadi [1987]).

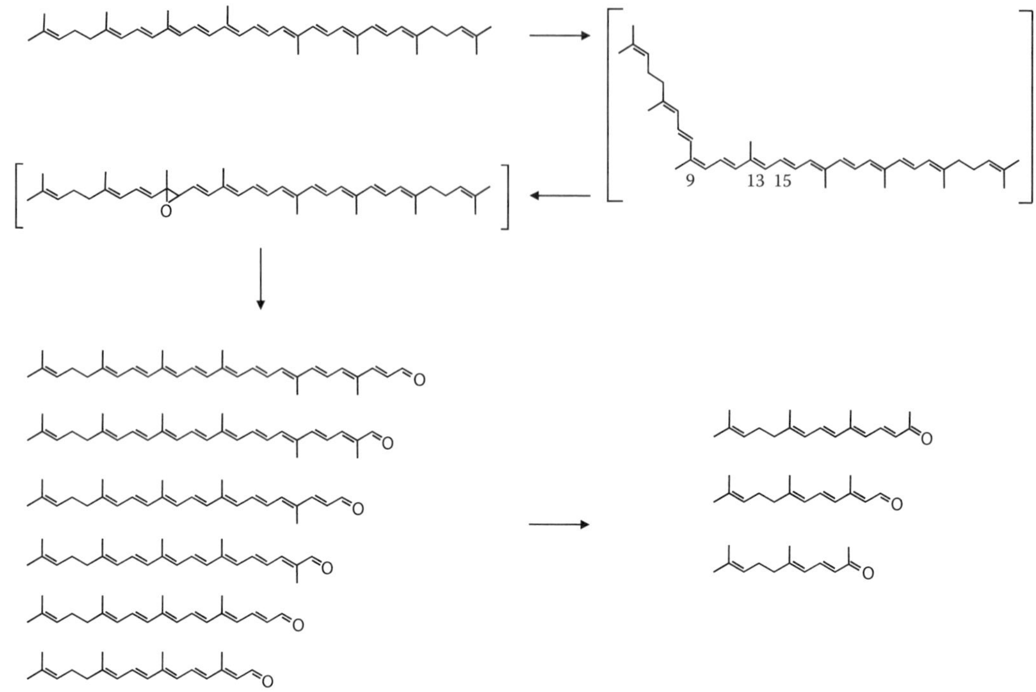

FIGURE 11.4 Hypothesis of the sequence of events when lycopene is oxidized by molecular oxygen in the presence of ruthenium tetraphenylporphyrin.

system similar to the one we used (Groves and Quinn 1985). These results allowed us to hypothesize a mechanism in which the *Z*-isomers would be the first products to appear, which then would be transformed into "in-chain" epoxides, which, in turn, would undergo an oxidative cleavage to give apolycopenals. Moreover, long apolycopenals could possibly be converted into shorter ones by a similar sequence of events (Figure 11.4).

A similar catalytic system, but with a more hindered porphyrin (tetramesitylporphyrin = tetraphenylporphyrin bearing three methyl substituents in *ortho* and *para* positions on each phenyl group), was tested for β-carotene oxidation by molecular oxygen (Caris-Veyrat et al. 2001). This system was chosen to slow down the oxidation process and thus make it possible to identify possible intermediates by HPLC-DAD-MS analysis. After just 1 h of reaction, the first products of the reaction could be seen, mainly *Z*-isomers. After 6 h, the chromatogram became more complex (Figure 11.5), and we could tentatively identify three families of compounds: *Z*-isomers, epoxides, and apocarotenals. After 24 h of reaction, β-carotene almost completely disappeared, but many reaction products were still visible. A detailed analysis of the chromatograms revealed the presence of a series of monooxygenated cleavage compounds, i.e., apocarotenals and also some epoxides of these apocarotenals. Moreover, diapocarotendials were also detected and tentatively identified. It is important to note that these last compounds were not detected in the similar model with lycopene. The oxidation mechanism thus appears more complex in this setup. In Figure 11.6, we propose a sequence of events that could occur in the reaction mixture. As we have observed with lycopene (Caris-Veyrat, Schmid et al. 2003), we hypothesize that β-carotene may be first isomerized and then oxidized and cleaved to form apocarotenals, which themselves may either undergo a second cleavage to produce diapocarotendials or which may be oxidized into 5,6-epoxide. This latter product could either isomerize to give an apocarotenal with a 5,8-furanoxide function, which could, in turn, be cleaved into diapocarotendials, or it may be directly cleaved to produce a diapocarotendial. Apocarotenals bearing epoxide or furanoxide functions may also be formed by the cleavage of the corresponding epoxide/furanoxide β-carotene.

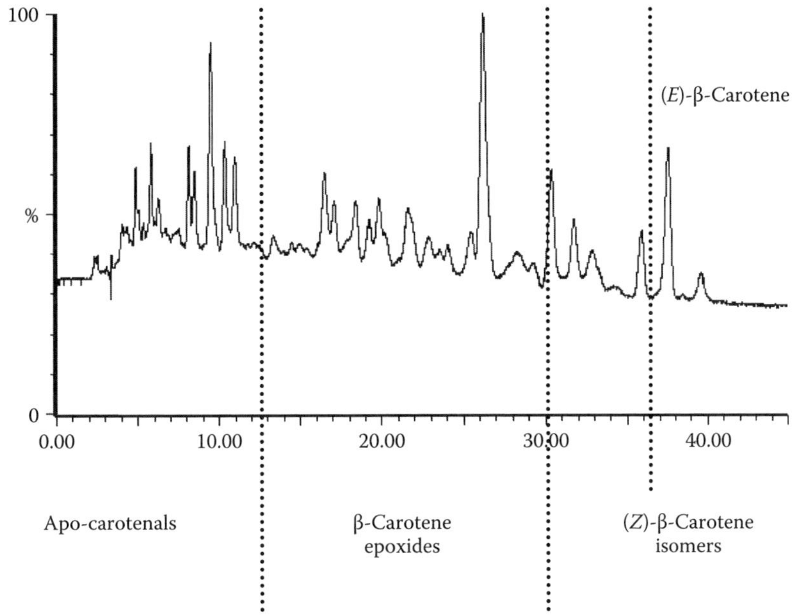

FIGURE 11.5 Chromatograms at 450 nm of the reaction mixture at 6 h of catalytic oxidation of β-carotene by dioxygen catalyzed by ruthenium mesitylporphyrin.

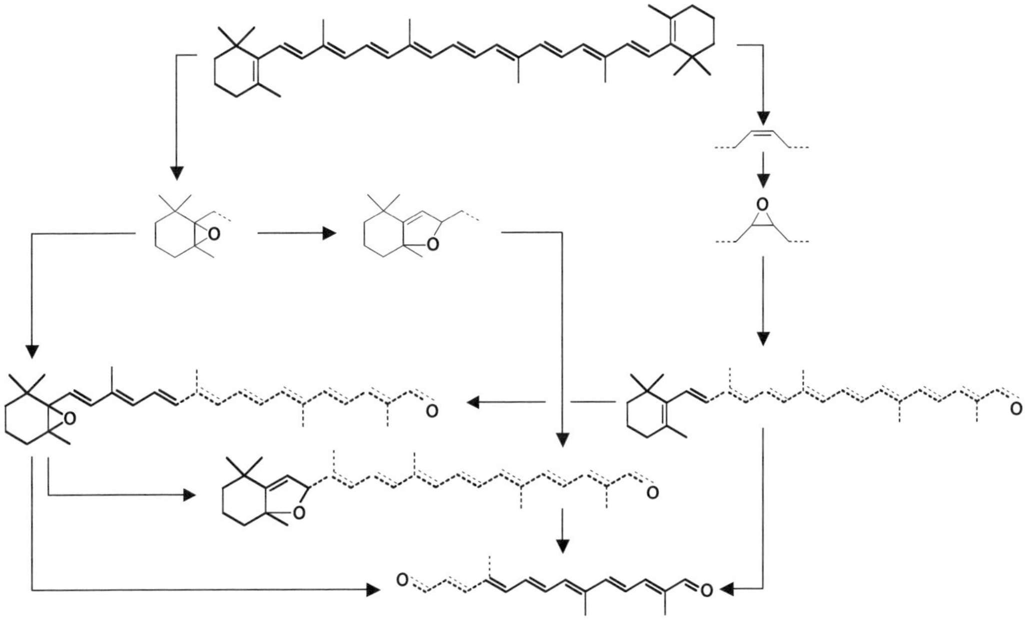

FIGURE 11.6 Hypothesis of the sequence of events when β-carotene is oxidized by molecular oxygen in the presence of ruthenium tetramesitylporphyrin. Parts of the chemical formula in dotted line indicate that length of the carbon chain may vary.

The literature contains other examples of the chemical oxidation of carotenoids that aim to mimic oxidation processes that potentially occur *in vivo*. For example, hypochlorous acid, an oxidant produced by polymorphonuclear leukocytes during inflammatory processes, was shown to oxidatively cleave β-carotene into apocarotenals and shorter chain compounds (Sommerburg et al. 2003).

It should be noted that partial or total organic synthesis was used to produce carotenoid oxygenated cleavage products such as, for example, apo-8′-lycopenal (Surmatis et al. 1966).

The ready availability of carotenoid oxidation products through chemical methods will facilitate their use as standard identification tools in complex media such as biological fluids, and it will enable *in vitro* investigation of their biological activity. Moreover, these studies can help in understanding the mechanisms by which carotenoids can be either chemically or biochemically cleaved *in vivo*.

11.5 FORMATION DURING FOOD PROCESSING OR MODEL FOOD SYSTEMS

Elevated temperature is the main factor affecting the integrity of carotenoids during food processing. Numerous studies have been made in order to quantify carotenoid degradation, some of which analyzed the products formed in detail, commonly oxygenated cleavage compounds. A review on the thermal degradation of carotenoids, which produces volatile and nonvolatile compounds, was published by Bonnie and Choo (1999). Some articles mentioned in this chapter dealing with carotenoid oxygenated cleavage compounds are discussed here along with other articles published at that time.

Thermal treatments generate not only oxygenated cleavage compounds of carotenoids but also oxidation compounds that do not necessarily undergo a cleavage reaction of the hydrocarbon chain, such as epoxides or furanoxides of the parent carotenoids, most often in positions 5,6 and 5′,6′, because the electron density of the double bonds is the highest at the extremities of the conjugated carbonated chain. Their rearrangement products possessing 5,8- and 5′,8′-furanoxide groups can also be found. These compounds can be generated from a 5,6-epoxy-carotenoid, itself produced from a nonepoxy carotenoid during the thermal treatment (Kanasawud and Crouzet 1990a,b), or from a 5,6 (5′,6′)-epoxy-carotenoid already present in the product before heating (Dhuique-Mayer et al. 2007). Thermal treatments can also transform carotenoids into compounds formed by cleavage of the polyenic chain followed by a rearrangement, by means of a radical mechanism (Edmunds and Johnstone 1965), without the introduction of an oxygen atom.

The degradation of β-carotene during different heat treatments and extrusion cooking, a widely used processing technique in the food industry, has been studied by Marty and Berset (1988, 1990). Several apocarotenals were identified by HPLC together with β-carotene epoxides in *E* and *Z* forms. The authors of the articles propose that chain breaks progress from the end of the molecule to the center with increasing strength of the treatments since the longest chain compounds (apo-8′- and apo-10′-carotenals) were obtained for all treatments, whereas shorter ones (apo-12′- and apo-14′-carotenals) were obtained for the more severe treatments, except for the shortest compound, i.e., apo-15-carotenal, which was detected for each heating treatment. Thus, a direct attack on all double bonds, and particularly on the central C15=C15′, cannot be excluded.

These results are confirmed by two studies on the effect of high temperatures (170°C –250°C) used for the deodorization of palm oil, which led to the oxygenated cleavage of β-carotene, forming apo-13-carotenone, apo-14′- and apo-15-carotenals, i.e., relatively short chain length apocarotenoids (Ouyang et al. 1980). A “dioxetane mechanism” was suggested to explain the formation of these products. The effects of similar treatments applied to palm oil deodorization for deep frying were tested on β-carotene by Onyewu et al. (1986) After 4 h of heating at 210°C, more than 70 nonvolatile compounds were detected: 7 of them were identified, including 2 apocarotenoids (apo-13-carotenone and apo-14-carotenal). Other products included hydrocarbons of shorter chains formed from cleavage and rearrangement reactions, but without the addition of oxygen atoms. Three volatile compounds were also identified.

Most of the studies on the thermal degradation of carotenoids analyzed the volatile fraction, as the identification of nonvolatile fractions was probably more complex to analyze. A study was published recently on the volatile compounds generated by the thermal degradation of carotenoids in different oleoresins of paprika, tomato, and marigold (Rios et al. 2008). Two groups of compounds were distinguished: cyclic olefins with or without oxygen atoms and compounds qualified as "linear ketones." In the first group, some of the compounds identified were hydrocarbons, such as m-xylene, toluene, or 2,6-dimethylnaphtalene; others were oxygenated, such as methylbenzaldehyde, isophorone, loliolide or ethanone, and 1-methylphenyl was identified for the first time as a carotenoid-thermodegraded compound. Two "linear ketone" type compounds were identified as 6-methyl-3,5-heptadien-2-one and 6-methyl-5-hepten-2-one. Intramolecular cyclization followed by an elimination reaction in the chain or a heterocyclic fragmentation reaction and oxidation reactions are mechanisms proposed to explain the occurrence of detected compounds.

Kanasawud and Crouzet have studied the mechanism for formation of volatile compounds by thermal degradation of β-carotene and lycopene in aqueous medium (Kanasawud and Crouzet 1990a,b). Such a model system is considered by the authors to be representative of the conditions found during the treatment of vegetable products. In the case of lycopene, two of the compounds identified, 2-methyl-2-hepten-6-one and citral, have already been found in the volatile fraction of tomato and tomato products. New compounds have been identified: 5-hexen-2-one, hexane-2,5-dione, and 6-methyl-3,5-heptadien-2-one, possibly formed from transient pseudoionone and geranyl acetate. According to the kinetics of their formation, the authors concluded that most of these products are formed mainly from all-(*E*)-lycopene and not (*Z*)-isomers of lycopene, which are also found as minor products in the reaction mixture.

11.6 CONCLUSIONS

Carotenoid oxygenated cleavage compounds include many different chemical structures and can be formed in various ways. Their influence on the organoleptic quality of food is well known, at least for volatile compounds, and some of them have been identified as aroma compounds (e.g., pseudoionone). Except for retinoids, their occurrence in humans has not been proven to date, but their biological effects, which could be either beneficial or detrimental for health, are well documented *in vitro* and strongly suspected *in vivo* (Wang 2004). Further research is needed to localize them *in vivo* and to determine if they contain significant biological activity.

ACKNOWLEDGMENTS

I thank my collaborators Michel Carail and Eric Reynaud for their participation in the work described and scientific discussions.

REFERENCES

Aust, O. et al. (2003). Lycopene oxidation product enhances gap junctional communication. *Food Chem. Toxicol.* **41**(10): 1399–1407.

Baker, D. L. et al. (1999). Reactions of beta-carotene with cigarette smoke oxidants. Identification of carotenoid oxidation products and evaluation of the prooxidant antioxidant effect. *Chem. Res. Toxicol.* **12**(6): 535–543.

Bonnie, T. Y. P. and Y. M. Choo (1999). Oxidation and thermal degradation of carotenoids. *J. Oil Palm Res.* **11**(1): 62–78.

Bouvier, F. et al. (2005). Oxidative tailoring of carotenoids: A prospect towards novel functions in plants. *Trends Plant Sci.* **10**(4): 187–194.

Britton, G. et al. (2004). *Carotenoids Handbook*. Basel, Switzerland: Birkhäuser Verlag.

Budowski, P. and A. Bondi (1960). Autoxidation of carotene and vitamin A. Influence of fat and antioxidants. *Arch. Biochem. Biophys.* **89**: 66–73.

Caris-Veyrat, C. et al. (2001). Mild oxidative cleavage of beta,beta-carotene by dioxygen induced by a ruthenium porphyrin catalyst: Characterization of products and of some possible intermediates. *New J. Chem.* **25**(2): 203–206.

Caris-Veyrat, C. et al. (2003). Cleavage products of lycopene produced by in vitro oxidations: Characterization and mechanisms of formation. *J. Agric. Food Chem.* **51**(25): 7318–7325.

Dhuique-Mayer, C. et al. (2007). Thermal degradation of antioxidant micronutrients in citrus juice: Kinetics and newly formed compounds. *J. Agric. Food Chem.* **55**(10): 4209–4216.

Edmunds, F. S. and R. A. W. Johnstone (1965). Constituents of cigarette smoke. Part IX. The pyrolysis of polyenes and the formation of aromatic hydrocarbons. *J. Chem. Soc*: 2892–2897.

El-Tinay, A. H. and C. O. Chichester (1970). Oxidation of beta-carotene. Site of initial attack. *J. Org. Chem.* **56**(7): 2290–2293.

Fatiadi, A. J. (1987). The classical permanganate ion: A novel antioxidant in organic chemistry. *Synthesis* **85**: 85–127.

French, R. R. et al. (2000). A supramolecular enzyme mimic that catalyzes the 15,15′ double bond scission of beta,beta-carotene. *Angew. Chem. Int. Ed.* **39**(7): 1267–1269.

Gajic, M. et al. (2006). Apo-8′-lycopenal and apo-12′-lycopenal are metabolic products of lycopene in rat liver. *J. Nutr. Biochem.* **136**(6): 1552–1557.

Grosch, W. and G. Laskawy (1979). Co-oxidation of carotenes requires one soybean lipoxygenase isoenzyme. *Biochim. Biophys. Acta* **575**(3): 439–445.

Handelman, G. J. et al. (1991). Characterization of products formed during the autoxidation of β-carotene. *Free Rad. Biol. Med.* **10**: 427–437.

Hanusch, M. et al. (1995). Induction of gap junctional communications by 4-oxoretinoic acid generated from its precursor canthaxanthin. *Arch. Biochem. Biophys.* **317**(2): 423–428.

IUPAC Commission on the Nomenclature of Organic Chemistry and IUPAC-IUB Commission on Biochemical Nomenclature (1971). *Carotenoids*. O. Isler. Basel, Switzerland: Birkhäuser Verlag, pp. 851–864.

IUPAC Commission on the Nomenclature of Organic Chemistry (CNOC) and IUPAC-IUB Commission on Biochemical Nomenclature (CBN) (1975). Nomenclature of carotenoids (Rules approved 1974). *Pure Appl. Chem.* **41**: 107–131.

Henry, L. K. et al. (2000). Effects of ozone and oxygen on the degradation of carotenoids in an aqueous model system. *J. Agric. Food Chem.* **48**(10): 5008–5013.

Hu, K. Q. et al. (2006). The biochemical characterization of ferret carotene-9′,10′-monooxygenase catalyzing cleavage of carotenoids *in vitro* and *in vivo*. *J. Biol. Chem.* **281**(28): 19327–19338.

Kanasawud, P. and J. C. Crouzet (1990a). Mechanism of formation of volatile compounds by thermal degradation of carotenoids in aqueous medium. 1. Beta-carotene degradation. *J. Agric. Food Chem.* **38**(1): 237–243.

Kanasawud, P. and J. C. Crouzet (1990b). Mechanism of formation of volatile compounds by thermal degradation of carotenoids in aqueous medium. 2. Lycopene degradation. *J. Agric. Food Chem.* **38**: 1238–1242.

Karrer, P. and E. Jucker (1950). *Carotenoids*. New York: Elsevier.

Kiefer, C. et al. (2001). Identification and characterization of a mammalian enzyme catalyzing the asymmetric oxidative cleavage of provitamin A. *J. Biol. Chem.* **276**(17): 14110–14116.

Kim, S. J. (2004). Cleavage products formed through autoxidation of zeta-carotene in liposomal suspension. *Food Sci. Biotech.* **13**(2): 202–207.

Kim, S. J. et al. (2001). Formation of cleavage products by autoxidation of lycopene. *Lipids* **36**(2): 191–199.

Kim, S. J. et al. (2005). Oxidative cleavage products derived from phytofluene by pig liver homogenate. *Food Sci. Biotech.* **14**(3): 424–427.

Leuenberger, M. G. et al. (2001). The reaction mechanism of the enzyme-catalyzed central cleavage of beta-carotene to retinal. *Ang. Chem. Int. Ed.* **40**(14): 2614–2617.

Lindshield, B. L. et al. (2007). Lycopenoids: Are lycopene metabolites bioactive? *Arch. Biochem. Biophys.* **458**(2): 136–140.

Lisle, E. B. (1951). The effect of carcinogenic and other related compounds on the autoxidation of carotene and other autoxidizable systems. *Cancer Res.* **11**(3): 153–156.

Martin, H. D. et al. (1999). Chemistry of carotenoid oxidation and free radical reactions. *Pure Appl. Chem.* **71**(12): 2253–2262.

Marty, C. and C. Berset (1988). Degradation products of *trans*-beta-carotene produced during extrusion cooking. *J. Food Sci.* **53**: 1880–1886.

Marty, C. and C. Berset (1990). Factors Affecting the thermal degradation of all-*trans*-beta-carotene. *J. Agric. Food Chem.* **38**(4): 1063–1067.

Miller, D. M. et al. (1990). Transition metals as catalysts of "autoxidation" reactions. *Free Rad. Biol. Med.* **8**: 95–108.

Molnar, P. et al. (2006). Preparation and spectroscopic characterization of 3′-oxolutein. *Lett. Org. Chem.* **3**(10): 723–734.

Mordi, R. C. et al. (1993). Oxidative degradation of beta-carotene and β-apo-8′-carotenal. *Tetrahedron* **49**(4): 911–928.

Omenn, G. S. et al. (1996). Effects of a combination of beta-carotene and vitamin A on lung cancer and cardiovascular disease. *N. Engl. J. Med.* **334**: 1150–1155.

Onyewu, P. N. et al. (1986). Characterization of β-carotene thermal degradation products in a model food system. *J. Am. Oil Chem. Sci.* **63**(11): 1437–1441.

Ouyang, J. et al. (1980). Formation of carbonyl compounds from beta-carotene during palm oil deodorization. *J. Food Sci.* **43**: 1214–1222.

Paik, J. et al. (2001). Expression and characterization of a murine enzyme able to cleave beta-carotene—The formation of retinoids. *J. Biol. Chem.* **276**(34): 32160–32168.

Perez-Galvez, A. and M. I. Minguez-Mosquera (2001). Structure-reactivity relationship in the oxidation of carotenoid pigments of the pepper (*Capsicum annuum* L.). *J. Agric. Food Chem.* **49**(10): 4864–4869.

Redmond, T. M. et al. (2001). Identification, expression, and substrate specificity of a mammalian beta-carotene 15,15′-dioxygenase. *J. Biol. Chem.* **276**(9): 6560–6565.

Rios, J. J. et al. (2008). Description of volatile compounds generated by the degradation of carotenoids in paprika, tomato and marigold oleoresins. *Food Chem.* **106**(3): 1145–1153.

Rodriguez, E. B. and D. B. Rodriguez-Amaya (2007). Formation of apocarotenals and epoxycarotenoids from beta-carotene by chemical reactions and by autoxidation in model systems and processed foods. *Food Chem.* **101**(2): 563–572.

Socaciu, C. et al. (2000). Carotenoid incorporation into microsomes: Yields, stability and membrane dynamics. *Spectrochim. Acta A Mol. Biomol. Spect.* **56**(14): 2799–2809.

Sommerburg, O. et al. (2003). Beta-carotene cleavage products after oxidation mediated by hypochlorous acid–A model for neutrophil-derived degradation. *Free Rad. Biol. Med.* **35**(11): 1480–1490.

Stratton, S. P. et al. (1993). Isolation and identification of singlet oxygen oxidation products of beta-carotene. *Chem. Res. Toxicol.* **6**(4): 542–547.

Surmatis, J. D. et al. (1966). Total synthesis of rhodovibrin (OH-P481), anhydrorhodovibrin (P481), and rhodopin. *J. Org. Chem.* **31**(1): 186–188.

Takahashi, A. et al. (1999). Kinetic model for autoxidation of beta-carotene in organic solutions. *J. Am. Oil Chem. Sci.* **76**(8): 897–903.

Takahashi, A. et al. (2003). A rigorous kinetic model for beta-carotene oxidation in the presence of an antioxidant, alpha-tocopherol. *J. Am. Oil Chem. Soc.* **80**(12): 1241–1247.

Tang, G. et al. (1991). Characterization of β-apo-13 carotenone and β-apo-14′-carotenal as enzymatic products of the excentric cleavage of β-carotene. *Biochemistry* **30**: 9829–9834.

The Alpha-Tocopherol Beta-Carotene Cancer Prevention Study Group (1994). The effect of vitamin E and beta carotene on the incidence of lung cancer and other cancers in male smokers. *N. Engl. J. Med.* **330**: 1029–1035.

Uebelhart, P. and C. H. Eugster (1988). Synthesen von enantiomerenreinen apoviolaxanthin-säuren, -olen- und -alen (persicaxanthin, sinensiaxanthin und β-Citraurin-epoxid) und ihrer furanoiden umlagerungsprodukte. *Helv. Chim. Acta* **71**: 1983.

von Lintig, J. and K. Vogt (2000). Filling the gap in vitamin A research—Molecular identification of an enzyme cleaving beta-carotene to retinal. *J. Biol. Chem.* **275**(16): 11915–11920.

Vulcain, E. et al. (2005). Inhibition of the metmyoglobin-induced peroxidation of linoleic acid by dietary antioxidants: Action in the aqueous vs. lipid phase. *Free Rad. Res.* **39**(5): 547–563.

Walberg, I. and A.-M. Eklund (1998). Degraded carotenoids. *Carotenoids, Volume 3: Biosynthesis and Metabolism*. G. Britton, S. Liaaen-Jensen and H. Pfander, eds. Basel, Switzerland: Birkhäuser Verlag, pp. 195–216.

Wang, X. D. (2004). Carotenoid oxidative/degradative products and their biological activities. *Carotenoids in Health and Disease*. N. I. Krinsky, S. T. Mayne and H. Sies, eds. New York: Marcel Dekker, pp. 313–335.

Wang, X. D. et al. (1991). Enzymatic conversion of beta-carotene into beta-apo-carotenals and retinoids by human, monkey, ferret and rat tissue. *Arch. Biochem. Biophys.* **285**(1): 8–16.

Wang, X. D. et al. (1992). Retinoic acid can be produced from excentric cleavage of beta-carotene in human intestinal mucosa. *Arch. Biochem. Biophys.* **293**(2): 298–304.

Weedon, B. C. L. and G. P. Moss (1995). Structure and nomenclature. *Carotenoids, Isolation and Analysis*. G. Britton, S. Liaaen-Jensen and H. Pfander, eds. Basel, Switzerland: Birkäuser Verlag, pp. 1A.

Wendler, N. L. and C. Rosenblum (1950). The oxidation of beta-carotene. *J. Am. Chem. Soc.* **72**: 234–239.

Wingerath, T. et al. (1999). Analysis of cyclic and acyclic analogs of retinol, retinoic acid, and retinal by laser desorption ionization-, matrix-assisted laser desorption ionization-mass spectrometry, and UV/Vis. spectroscopy. *Anal. Biochem.* **272**(2): 232–242.

Wyss, A. et al. (2000). Cloning and expression of β,β-carotene 15,15′-dioxygenase. *Biochem. Biophys. Res. Commun.* **271**(2): 334–336.

Yamauchi, R. and K. Kato (1998). Products formed by peroxyl radical-mediated oxidation of canthaxanthin in benzene and in methyl linoleate. *J. Agric. Food Chem.* **46**(12): 5066–5071.

Yamauchi, R. et al. (1993). Products formed by peroxyl radical-mediated oxidation of beta-carotene. *J. Agric. Food Chem.* **41**: 708–713.

Yamauchi, R. et al. (1998). Oxidation products of beta-carotene during the peroxidation of methyl linoleate in the bulk phase. *Biosci. Biotechnol. Biochem.* **62**(7): 1301–1306.

Yeum, K.-J. et al. (1995). Similar metabolites formed from beta-carotene by human gastric mucosal homogenates, lipoxygenase, or linoleic acid hydroperoxyde. *Arch. Biochem. Biophys.* **321**(1): 167–174.

Zhang, H. et al. (2003). A novel cleavage product formed by autoxidation of lycopene induces apoptosis in HL-60 cells. *Free Radic. Biol. Med.* **35**(12): 1653–1663.

Zurcher, M. and H. Pfander (1999). Oxidation of carotenoids-II: Ozonides as products of the oxidation of canthaxanthin. *Tetrahedron* **55**(8): 2307–2310.

Zurcher, M. et al. (1997). Oxidation of carotenoids-I. Dihydrooxepin derivatives as products of oxidation of canthaxanthin and beta,beta-carotene. *Tet. Lett.* **38**(45): 7853–7856.

12 Thermal and Photochemical Degradation of Carotenoids

Claudio D. Borsarelli and Adriana Z. Mercadante

CONTENTS

12.1 INTRODUCTION

The most characteristic feature of the carotenoid structure is the presence of several conjugated double bonds in the chain. The polyene chain is responsible for the light absorption properties and also for the susceptibility of carotenoids to degradation under high temperature, low pH, light, and reactive oxygen species, among other factors. Nevertheless, heat processing has become an important part of the food chain, and both processed and fresh foods are often exposed under fluorescent light in supermarkets.

Although carotenoids are naturally stabilized by the plant matrix, cutting or disrupting of fruit and vegetable tissues favors their exposure to oxygen and endogenous oxidative enzymes, thus provoking their isomerization and oxidation. Differences between fruit and vegetable species, such as the localization of carotenoids in the tissue and its physical state, may be crucial factors for the susceptibility of these pigments to *trans* to *cis* isomerization and oxidation reactions. In food systems, the mechanisms involved in both thermal and photochemical degradations are much more complex than in model systems. Along with the environmental factors and the carotenoid structure that play important known roles, the physical state and location in cellular organelles, and interactions between different naturally occurring food compounds are much more difficult to predict. Therefore, comparison of published data regarding the extension of carotenoid degradation is a difficult task since different foods are processed and stored under different combinations of temperature, light and time, etc. and such conditions are sometimes only partially described.

12.2 THERMALLY INDUCED DEGRADATION

Although *trans* to *cis* isomerization *per se* is not expected to cause major changes in color, it is the first step for intramolecular cyclization to form cyclic volatile compounds under conditions of high temperature. The oxidation of carotenoid is also required for subsequent reorganization and fragmentation to aldehydes and ketones with low molecular weight, such as those observed in oleoresins (Rios et al. 2008). Kanasawud and Crouzet (1990) had proposed a reaction mechanism with mono- and di-epoxides of β-carotene as intermediates for the formation of carotenoid-derived volatiles from β-carotene in heated aqueous medium.

Carotenoid degradation kinetics and visual color changes in model and food systems submitted to heating processes are complex phenomena although simple first-order kinetics models have been widely applied in the reports available in the literature (see Sections 12.2.1 and 12.2.2). Considering that thermal degradation includes reversible *trans* to *cis* isomerization, the formation of oxidation products (e.g., epoxide and apo-carotenal), epoxy to furanoid rearrangement, and degradation to volatile compounds, the latter three are all irreversible reactions, the simplest mechanism for overall carotenoid changes expected to occur in model and food systems submitted to heating is shown in Figure 12.1. Taking into consideration these overall carotenoid changes, we should expect at least a bi-exponencial or two-stage first-order decay to explain the different simultaneous reactions, e.g., reversible isomerization and irreversible degradation (Capellos and Bielski 1972, Rios et al. 2005).

Furthermore, the mechanism shown in Figure 12.1 considers only the all-*trans*-carotenoid form as the initial compound; however, although the all-*trans*-isomer predominates, *cis*-isomers are also commonly found in model solutions and even more frequently in food systems, since these isomers are in equilibrium in the solution. Therefore, the initial carotenoid system often contains a mixture of isomers, whose composition changes according to the carotenoid structure, solvent, and heat treatment. For example, the isomerization rate of β-carotene is higher in nonpolar solvents, e.g., petroleum ether and toluene, than in polar solvents (Zechmeister 1944).

Since spontaneous isomerization occurs in solution, the difficulty lies in the mathematical calculations for the determination of the kinetic constant rates for all the compounds found in this complex mechanism. In fact, the different initial amounts of *cis*-isomers in the system can lead to misinterpretation when analyzing real world data. For example, a simulated cashew–apple juice system (water:ethanol, 8:2), containing initially 22.0% of all-*trans*-β-carotene, 2.0% of β-carotene *cis* isomers, 57.4% of all-*trans*-β-cryptoxanthin, and 4.2% β-cryptoxanthin *cis* isomers in relation to the total carotenoid content, changed, respectively, to 17.5%, 8.0%, 37.1%, and 17.5% after heating at 90°C for 240 min (Zepka and Mercadante 2009). These different final percentages obtained for the *cis*-isomers of β-carotene and β-cryptoxanthin were expected because changes should occur toward the isomeric equilibrium for each carotenoid. In fact, the initial ratios for the all-*trans*:*cis* isomers of β-carotene and β-cryptoxanthin were similar, *ca.* 92:8 at room temperature, and both changed to *ca.* 69:31 after heating. Similar results for β-carotene and lutein in toluene solutions were observed, with *cis*-isomers increasing at different rates to yield a *trans*:*cis* ratio of approximately 65:35, when the equilibrium had been reached and further thermal processing did not affect the final isomeric proportion (Aman et al. 2005a).

Taking into account that heat treatment inactivates some oxidative enzymes and causes the rupture of some cellular structures, greater extractability of carotenoids is expected to occur in processed foods. Therefore, when mild temperatures are applied, it is very common to obtain higher carotenoid content in a processed food as compared to its fresh counter part. For example, total

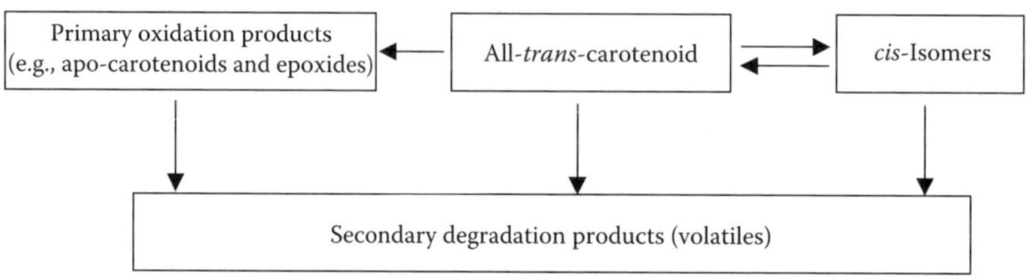

FIGURE 12.1 Overall structural changes occurring in carotenoids during heating.

carotenoid content increased from 10.9 μg/g in unblanched pumpkin puree to 12.5 μg/g after 2 min blanching and to 14.1 μg/g after further treatment at 60°C for 2 h (Dutta et al. 2006).

12.2.1 Thermal Degradation in Model Systems

The spontaneous isomerization of all-*trans*- carotenoids at room temperature is a slow process, and its rate depends on the solvent and the pigment structure. For example, the initial solutions of β-carotene in a mixture of tetrahydrofuran (THF), methanol, and acetonitrile containing *ca.* 95% of all-*trans*- and 5% of 9-*cis*- plus 13-*cis*-isomers was transformed to 90% all-*trans*-β-carotene and 9% of 9-*cis*- plus 13-*cis*-carotene after 24 h of spontaneous isomerization at 25°C (Pesek et al. 1990). However, in chloroform, the amounts of 13-*cis*- and 9-*cis*-β-carotene, respectively, increased to 15.6% and 13.6% of the total β-carotene content (Pesek et al. 1990). Moreover, thermal isomerization of β-carotene in hexane was reported to be <1.3% for 1 h at room temperature (Kuki et al. 1991). In addition, as can be seen in Table 12.1, increased temperature accelerated this process, and equilibrium was reached within 15–60 min under reflux (Zechmeister 1944).

Based on high performance liquid chromatography (HPLC) studies regarding the equilibration of isomeric fractions of β-carotene isomers at 45°C, a model consisting of two reversible concurrent isomerization reactions was developed by Pesek et al. 1990. Under dark storage conditions at 45°C, a β-carotene solution reached an equilibrium after 4–6 days yielding approximately 66% all-*trans*-, 8% of 9-*cis*-, and 25% of 13-*cis*-β-carotene. The observed rate constant (k) for the formation of the 13-*cis*- isomer was faster than that of the 9-*cis*-β-carotene isomer, and the back rate constants toward the all-*trans*- isomer were intrinsically faster as compared to the formation of *cis*-isomers of β-carotene (Chart 12.1).

A more complex isomerization pattern induced by heating was proposed for β-carotene isomers (all-*trans*-, 7-*cis*-, 9-*cis*-, 13-*cis*-, and 15-*cis*-) in *n*-hexane solution heated at 80°C (Kuki et al. 1991). Starting from each β-carotene isomer, the following isomerization products were observed

TABLE 12.1
Percentage of β-Carotene Stereoisomers Formed at Different Temperatures in Petroleum Ether–Benzene Solution

T (°C)	Experimental Time (h)	*cis*-Isomers (%)
20	24	1.0
	168	5.5
	1176	11.1
40	1	4.0
	3	5.4
	24	11.2
60	1	7.5
	3	9.7
80	1	8.5
	3	31.9
	24	34.1

Sources: Data from Carter, G.P. and Gillam, A.E., *Biochem. J.*, 33, 1325, 1939; Zechmeister, L., *Chem. Rev.*, 34, 267, 1944.

CHART 12.1
Observed Rate Constants (k_{obs}) for the Equilibration of All-*trans*-β-Carotene in the Dark at 45°C

$$\text{9-}cis \underset{k_{-2}}{\overset{k_2}{\rightleftharpoons}} \text{All-}trans \underset{k_{-1}}{\overset{k_1}{\rightleftharpoons}} \text{13-}cis$$

Reactions	k_{obs} (min^{-1})
all-*trans* → 13-*cis*	1.6×10^{-4}
13-*cis* → all-*trans*	4.2×10^{-4}
All-*trans* → 9-*cis*	6.9×10^{-6}
9-*cis* → all-*trans*	7.6×10^{-5}

Source: Data from Pesek, C.A. and Warthesen, J.J., *J. Agric. Food Chem.*, 38, 1313, 1990.

under heating: (a) all-*trans*- isomerized into 13-*cis*- > 15-*cis*- ≈ 9-*cis*-; (b) 7-*cis*- isomerized into 7,13′-di-*cis*- > 7,15-di-*cis*- > 7,13-di-*cis*-; (c) 9-*cis*- isomerized into 9,13′-di-*cis*- > 9,15-di-*cis*- > all-*trans*- > 9,13-di-*cis*-; (d) 13-*cis*- isomerized into all-*trans*- >> 15-*cis*-; and (e) 15-*cis*- isomerized into all-*trans*- > 13-*cis*-. Major differences in the isomerization patterns found for thermal isomerization when compared to direct photoisomerization and sensitized photoisomerization (see Section 12.3) were (a) starting from 7-*cis*-, only di-*cis*- isomers were formed and (b) starting from 13-*cis*- and 15-*cis*- mutual isomerization between the central-*cis* isomers took place (Kuki et al. 1991). Moreover, calculated π bond orders for docosaundecaene, a model for β-carotene, in the S_0 state showed that the C–C bond order decreases from both ends (0.922) toward the center (0.833) indicating that thermal isomerization (S_0-state) can take place more easily in the central part (Kuki et al. 1991).

The isomerization and degradation of dried β-carotene were evaluated in an oven heated at temperatures between 50°C and 150°C up to 30 min, as well as by reflux heating at 70°C during 140 min, using first-order kinetic decay (Chen and Huang 1998). Although the degradation of all-*trans*-β-carotene became significant after heating at 50°C and 100°C for 25 and 10 min, respectively, no significant changes were found in the amounts of 9-*cis*-, 13-*cis*-, or 15-*cis*-β-carotenes (Table 12.2). When all-*trans*-β-carotene was heated under reflux in hexane, its concentration decreased with increasing heating time; however, after 70 min levels remained constant indicating that the whole system approached an equilibrium (*trans*:*cis* ≈ 49:51). The major isomers formed during heating were 13-*cis*-β-carotene, both under oven and reflux, while the 13,15-di-*cis*-β-carotene was only found at temperatures higher than 120°C (Chen and Huang 1998). These results also indicated that reflux heating is more likely to induce β-carotene isomerization, while oven-heating is more likely to cause β-carotene degradation. This phenomenon can be attributed either to differences in degradation mechanisms as affected by the temperature, the oxygen access, and the physical state of the reaction system or by the highest activation energy required for the formation of di-*cis*- as compared to mono-*cis*- carotenoid (Zechmeister 1944).

The degradation of α- and β-carotene crystals upon heating at 150°C fitted a reversible first-order model, *trans*- to *cis*- conversion occurred two- to threefold slower than that observed for the backward reaction; in other words, the equilibrium toward the all-*trans*- isomer was favored (Chen et al. 1994). Four *cis*- isomers of β-carotene (13,15-di-*cis*-, 15-*cis*-, 13-*cis*-, and 9-*cis*-) and three isomers of α-carotene (15-*cis*-, 13-*cis*-, and 9-*cis*-) were formed during the heating of their respective all-*trans*- carotene crystals. The 13-*cis* isomer of both carotenes was found in greater amounts (Chen et al. 1994). In this system, α-carotene degraded faster than β-carotene (Table 12.2).

A dry, thin lycopene layer heated at 50°C, 100°C, and 150°C showed first-order kinetic decay (Lee and Chen 2002). At 50°C, isomerization dominated in the first 9 h; however, degradation was favored afterward. On the other hand, at 100°C and 150°C degradation proceeded faster than isomerization. Although *cis* isomer identification was not confirmed by standards, the mono-*cis* lycopene isomers, 5-*cis*-, 9-*cis*-, 13-*cis*-, and 15-*cis*-, degraded at the same rate as did all-*trans*-lycopene,

TABLE 12.2
Observed Rate Constant (k_{obs}) and Activation Energy (E_a) Values Found for Carotenoid Thermal Degradation in Model Systems

Model Systems	Carotenoid	T (°C)	k_{obs}(min^{-1})	E_a (kcal/mol)	Reference
Crystal	All-*trans*-α-carotene	150	4.3×10^{-2}	n.r.	Chen et al. (1994)
Crystal	All-*trans*-β-carotene	100	2.0×10^{-3}	9.3	Chen and Huang (1998)
Crystal	All-*trans*-β-carotene	150	1.7×10^{-2}	n.r.	Chen et al. (1994)
Dry thin layer	Lycopene	100	12.4×10^{-3}	14.6	Lee and Chen (2002)
		150	16.5×10^{-2}		
Safflower seed oil	All-*trans*-β-carotene	95	5.4×10^{-3}	26.2	Henry et al. (1998)
	9-*cis*-β-carotene	95	5.9×10^{-3}	25.1	
	Lycopene lutein	95	8.6×10^{-3}	19.8	
		95	4.5×10^{-3}	24.9	
Chlorophyll *a* + methyl stearate (hexane)	All-*trans*-β-carotene	60	2.2×10^{-2}	n.r.	Liu and Chen (1998)
		120	8.2×10^{-2}		
Chlorophyll *a* + methyl oleate	All-*trans*-β-carotene	60	1.3×10^{-2}	n.r.	Liu and Chen (1998)
		120	4.2×10^{-2}		
Chlorophyll *a* + methyl linoleate	All-*trans*-β-carotene	60	6.0×10^{-3}	n.r.	Liu and Chen (1998)
		120	1.9×10^{-2}		
Ethanol/water (2:8)	Bixin	98	2.0×10^{-2}	36.9	Rios et al. (2005)
Amorphous powder	Norbixin	n.r.	n.r.	36.8	Silva et al. (2007)

Note: n.r., not reported.

whereas the rates of formation of two di-*cis* lycopene isomers showed increasing trends during heating. At 150°C lycopene degraded almost ten-times faster than β-carotene crystals (Chen et al. 1994), as compared in Table 12.2.

The thermal degradation of all-*trans*-β-carotene, 9-*cis*-β-carotene, lycopene, and lutein was studied in an oil model system, safflower seed oil, at 75°C, 85°C, and 95°C (Henry et al. 1998). The kinetic data was fitted as first-order reaction for all carotenoids; and the k_{obs} value calculated for lycopene was about twice as high as those found for the other carotenoids, whereas no significant difference was found between the stability of β-carotene isomers (Table 12.2). The calculated E_a values were similar for all-*trans*-β-carotene, 9-*cis*-β-carotene, and lutein, while lycopene with lower E_a was found to be less affected by temperature. Heating β-carotene at several temperatures formed 13-*cis*-carotene in higher amounts, followed by 9-*cis*-β-carotene and an unidentified *cis* isomer. Although several degradation products were formed during lycopene heating and lutein heating, they were not identified (Henry et al. 1998).

In toluene solution, 84.7% and 83.4% of the initial contents of β-carotene and lutein were, respectively, retained after heating at 98°C for 60 min. In chloroplast preparations, a similar degradation rate of β-carotene (83.2%) was observed, whereas 72.0% of lutein remained (Aman et al. 2005a). The addition of fat to chloroplast did not affect the retention of total β-carotene (82.0%), whereas an enhancement of lutein stability was found (93.0%). In these systems, apart from degradation, all-*trans*-β-carotene and all-*trans*-lutein were partially converted into its *cis*- isomers. After heat treatment at 98°C for 1 h, the predominant *cis*- isomers were 13-*cis*-β-carotene, 13-*cis*-lutein, and 13′-*cis*-lutein in toluene, whereas 9-*cis*-β-carotene, 9-*cis*-lutein, and 9′-*cis*-lutein were found as the major *cis*- isomers in chloroplasts. The different isomeric profile after heating may result from the interactions of chlorophylls in the chloroplast enhancing the formation of the 9-*cis* isomers. It is remarkable that these effects, occurring in well-organized chlorophyll–protein complexes, were still observed after

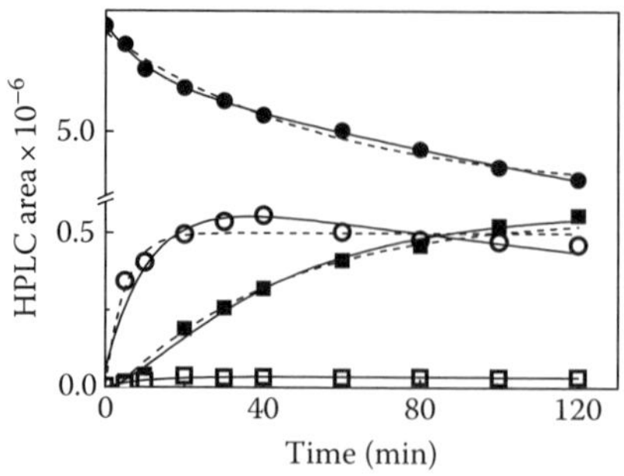

FIGURE 12.2 Mono exponential (dashed lines) and bi-exponential fitting (solid lines) of the kinetic HPLC data for the thermal degradation of bixin in a 20% ethanolic solution at 98°C: (●) Bixin; (○) sum of di-*cis* bixin peaks; (■) all-*trans*-bixin; (□) oxidation compound (C17). (From Rios, A.O. et al., *J. Agric. Food Chem.*, 53, 2307, 2005. With permission.)

the denaturation of the chlorophyll–carotenoid complexes by heat treatment at 98°C for 60 min. The addition of fat to the chloroplasts had a negligible effect on the isomerization rates of both carotenoids indicating the absence of crystalline carotenoids in such an organelle (Aman et al. 2005a).

On the other hand, the type of methyl fatty acid added to a system containing all-*trans*-β-carotene and chlorophyll *a* heated at 60°C and 120°C was significant (Liu and Chen 1998), Table 12.2. Since the systems were maintained in the dark, although in the presence of air, the addition of chlorophyll was not expected to photocatalyze the isomerization reaction. The first-order degradation rate of β-carotene significantly decreased with the increased number of double bonds in the methyl fatty acid; e.g., methyl linoleate < methyl oleate < methyl stearate. The authors claimed that methyl linoleate can compete with β-carotene for molecular oxygen, and thus less oxygen was available to react with β-carotene; and that methyl linoleate is more susceptible to react with free radicals than β-carotene. At 60°C, 13-*cis*-β-carotene was the predominant isomer formed, whereas besides 13-*cis*- isomer, 15-*cis*- and 13,15-di-*cis*-β-carotene were also found in larger amounts at 120°C (Liu and Chen 1998).

The thermal degradation kinetics of bixin, along with the products formed, in a water/ethanol (8:2) solution was studied as a function of temperature (70°C–125°C) (Rios et al. 2005). During heating, the consumption of the visible band of bixin (400–500 nm) was accompanied by an increase in the absorbance below 400 nm, without the presence of clear isosbestic points, indicating that degradation rate was strongly dependent on the monitoring wavelength due to the formation of bixin isomers and degradation products at different rate constants and blue-shifted absorption spectra. The decay of bixin and the formation of several products were confirmed by HPLC. At all temperatures, although the decay curves could be adjusted to a first-order rate law (exponential fitting) as indicated by the dashed lines in Figure 12.2, much better fits (solid lines in Figure 12.2) of the kinetic data were obtained using the bi-exponential Equation 12.1:

$$A_t - A_\infty = a_1 \exp(-k_{obs,1}t) + a_2 \exp(-k_{obs,2}t) \tag{12.1}$$

where

A_t and A_∞ are the transitory and final HPLC areas
a_1 and a_2 are the pre-exponential factors
$k_{obs,1}$ and $k_{obs,2}$ are, respectively, the observed fast and slow first-order rate constants

This bi-exponential behavior confirms the presence of reversible isomerization steps coupled with irreversible degradation steps and accounts for the role of the di-*cis* isomers as reaction intermediates, according to the general reaction scheme presented in Figure 12.1. The dependence of the rate constant of each elementary step on temperature allowed the calculation of the respective activation

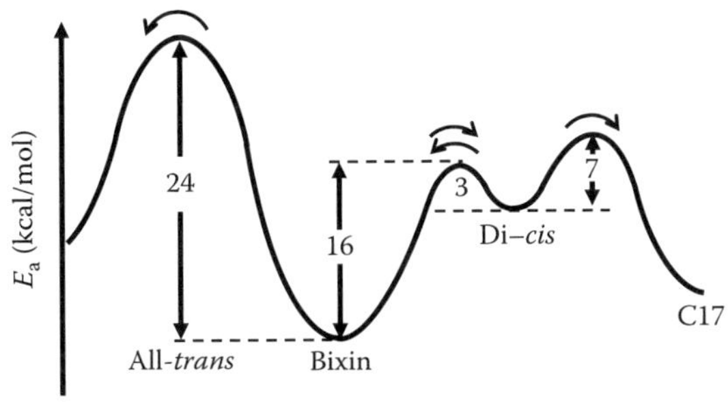

FIGURE 12.3 Coupled reaction scheme proposed for the degradation of bixin and the formation of its primary products and the respective activation energy E_a.

energy (E_a). Thus, the isomerization of bixin to all-*trans*-bixin was very slow with E_a = 24 kcal/mol, in good agreement with the value reported by Zechmeister and Escue (1944). The di-*cis* isomers were formed faster (E_a = 16 kcal/mol), both di-*cis*-isomers easily revert to bixin by a low activated pathway (E_a = 3 kcal/mol) or irreversibly react to yield 4,8-dimethyltetradecahexaenedioic acid monomethyl ester (C17) as oxidation product, with an energy barrier of 7 kcal/mol, Figure 12.3. These transformations were accompanied by the formation of *m*-xylene as the major volatile compound (Scotter et al. 2001), involving the di-*cis* isomer as an intermediate in the mechanism of the thermal degradation of bixin (Scotter 1995).

The thermal decomposition of norbixin powder was analyzed by thermogravimetric analysis at heating rates of 5°C, 10°C, and 20°C in the range of 25°C–900°C (Silva et al. 2007). Differential scanning calorimetry (DSC) curves showed that thermal decomposition reactions occurred in the solid phase (<280°C). Considering a first-order reaction mechanism, the kinetic parameters for the first stage of the thermal decomposition was calculated using integral and approximate methods. The Coats–Redfern model gave decreased E_a values as the heating rates increased: 36.8 kcal/mol at 5°C/min, 31.4 kcal/mol at 10°C/min, and 23.7 at 20°C/min (Silva et al. 2007). Interestingly, the E_a = 36.8 kcal/mol calculated for norbixin by DSC was the same as that obtained for the global activation energy (i.e., 24 + 16 − 3 = 37 kcal/mol) of bixin (Rios et al. 2005), calculated according to coupled reversible isomerization with irreversible degradation reactions.

Recently, Mercadante (2008) summarized some characteristics observed in carotenoid model systems submitted to heating, such as isomerization is the main reaction that occurs during heating at atmospheric pressure and at temperatures lower than 100°C; the 13-*cis*- is formed at higher rates than the 9-*cis*-carotenoid isomer; formation of oxidation products from β-carotene, such as epoxides and apo-carotenals, as well as di-*cis*- isomers occurs under stronger conditions, e.g., high temperature, long time or high pressure. In addition, it is remarkable that under heating, the products formed, detected by HPLC, are usually found in much smaller amounts than the amount of carotenoid destroyed. This fact indicates that both noncolored and volatiles compounds are also formed under heat processes (Zepka et al. 2009).

12.2.2 Thermal Degradation in Food Systems

The changes that occur after the heat processing of food systems are often monitored by different parameters, such as total carotenoid content (and therefore isomerization and oxidation are underestimated), individual carotenoids (overall changes may be missed), and CIELAB color parameters (no information on carotenoid degradation mechanism). The data given in Table 12.3 reflects the influence of matrix composition, food state (liquid or solid), and measured parameter on the carotenoid degradation kinetics.

The degradation kinetics of total carotenoid contents and visual color of papaya puree were investigated at temperatures between 70°C and 105°C (Ahmed et al. 2002). The thermal degradation of total carotenoids and color change parameters (Hunter, $a \times b$ values) followed first-order reaction

TABLE 12.3
Observed Rate Constant (k_{obs}) and Activation Energy (E_a) Values Found for Carotenoid Thermal Degradation in Food Systems

Food Systems	Parameters Measured	T (°C)	k_{obs} (min^{-1})	E_a (kcal/mol)	Reference
Papaya puree	Total carotenoids	90	6.0×10^{-3}	4.91	Ahmed et al. (2002)
	$a\times b$		4.2×10^{-3}	7.78	
Pumpkin puree	Total carotenoids	90	7.0×10^{-3}	6.51	Dutta et al. (2006)
	L		7.0×10^{-3}	8.04	
	ΔE		n.r.	7.26	
Orange juice	β-Carotene	90	5.3×10^{-3}	26.27	Dhuique-Mayer et al. (2007)
	β-cryptoxanthin		4.3×10^{-3}	37.26	
	Zeinoxanthin		3.9×10^{-3}	29.33	
Tomato pulp	Lycopene	100 (increased TS)	2.3×10^{-3}	n.r.	Sharma and Le Maguer (1996)
		100 (constant TS)	1.7×10^{-3}		
Insoluble tomato solids in water	Lycopene	100	1.3×10^{-3}	n.r.	Sharma and Le Maguer (1996)
Oleoresin from tomato cv. Roma	Total lycopene	100	5.9×10^{-1}	11.5	Hackett et al. (2004)
Oleoresin from tomato cv. High lycopene	Total lycopene	100	1.97	11.7	Hackett et al. (2004)
Oleoresin from tomato cv. Tangerine	Total lycopene	100	6.27	15.0	Hackett et al. (2004)

Notes: n.r., not reported and TS, total solids.

kinetics, and the dependence of the rate constant (k) followed the Arrhenius relationship. Similar kinetic behavior of both total carotenoids and color parameters (L and ΔE) was verified for the thermal degradation of pumpkin puree blanched for 2 min in 1% NaCl solution (Dutta et al. 2006). As a result of the heating, the CIELAB color parameters a, b, and L decreased indicating losses on yellowness color and reduced luminosity due to the degradation of carotenoids and the formation of darker compounds by parallel reactions, such as Maillard reaction (Dutta et al. 2006).

As can be seen in Table 12.3, higher activation energy E_a was calculated from the kinetic analysis of the color parameters as compared to those from the decay of the total carotenoid contents. This mismatch is a consequence of the complex thermal degradation mechanism and the type of information that can be extracted from the analyzed parameter. For instance, color parameters are global properties of the reacting mixture and can represent an average change between the consumption of the starting material with the formation of new colored carotenoid species, e.g., *cis*-isomers, epoxides, apo-carotenoids, and also brown-pigments formed by parallel Maillard reaction between reductive sugars and free amino groups of proteins (Machiels and Istasse 2002). On the other hand, total carotenoid concentration also considers the increasing concentration of products, which could have an important contribution after 10% of reactant conversion. Thus, the kinetic decay analysis at longer time and larger conversion values underestimate the degradation rate of the starting carotenoid (usually the all-*trans* isomer) because of the formation of the other colored carotenoid products.

Mango puree was produced on a laboratory scale, mimicking typical operations in continuous and small-size batches, applying pasteurization between 85°C and 93°C up to 16 min (Vásquez-Caicedo et al. 2007). Although significant *trans*- to *cis*- isomerization of β-carotene occurred, especially by the formation of 13-*cis*-β-carotene, provitamin A (*trans*- + *cis*-β-carotene) losses were similar in all applied pasteurization treatments. In mango nectar, the relative proportions of the β-carotene stereoisomers altered during processing from 85.5%, 9.8%, and 4.7% for all-*trans*-, 13-*cis*-, and 9-*cis*-β-carotene, respectively, in the raw mesocarp to 67.5%, 22.4%, and 15.2% in the pasteurized nectar. These small changes did not allow the calculation of the degradation kinetics. Color changes induced by pasteurization were verified by the marked losses of color intensity (C^*), which correlated with heating time, but only a slight increase in hue (H) was observed. These changes were also observed in the UV–visible absorption spectra of the total carotenoid mango extract, with the formation of a band between 360 and 380 nm, corresponding to the formation of *cis*-isomers, and a hypsochromic shift of 20 nm in the maximum absorbance band, probably due to epoxy–furanoide rearrangement. In commercially processed mango juice, auroxanthin, not present in fresh mangoes, appeared at an appreciable level due to the conversion of the 5,6-epoxide groups of violaxanthin to the 5,8-furanoxide groups of auroxanthin resulting in a hypsochromic shift of 40 nm (Mercadante and Rodriguez-Amaya 1998).

The thermal degradation kinetics of carotenoids in citrus juice during isothermal treatment at temperatures between 75°C and 100°C fitted with a first-order rate law (Dhuique-Mayer et al. 2007). Differences in stability among the main provitamin A carotenoids, and between these and xanthophylls were found, e.g., the main provitamin A carotenoids were not significantly affected during conventional thermal processing (Table 12.3). The HPLC-diode array detection-mass spectrometry analysis of degradation products showed that the rearrangement of the epoxide function in positions 5,6 into a furanoxide function in positions 5,8 was a common reaction for several xanthophylls (Dhuique-Mayer et al. 2007). Similar results were previously reported, no changes in β-carotene and β-cryptoxanthin levels were observed after the pasteurization of orange juice at 90°C for 30 s, whereas violaxanthin decreased 46%, *cis*-violaxanthin 20%, and antheraxanthin 25% (Lee and Coates 2003).

Contrary to the carotenoid behavior during orange juice pasteurization, losses of 46%–54% in the all-*trans*-α- and all-*trans*-β-carotene contents and the formation of *cis*-isomers were also verified for the pasteurization of carrot juice at 110°C and at 120°C, both for 30 s (Chen et al. 1995). In addition, all *cis*- isomer levels increased, with 13-*cis*-β-carotene and 15-*cis*-α-carotene formed in the largest amount. Heating at 121°C for 30 min caused further losses of 61% in all-*trans*-α-carotene and 55% in all-*trans*-β-carotene (Chen et al. 1995). However, minor effects on the amounts of *trans*- and *cis*- isomers of α- and β-carotenes were observed after the acidification and the heating of carrot juice at 105°C for 25 s (Chen et al. 1995).

Unheated carrot juices produced from carrots blanched at 80°C for 10 min were devoid of *cis*-isomers, and further pasteurization or sterilization process formed only 13-*cis*-β-carotene, respectively, at 2% and 5% (Marx et al. 2003). However, extensive carrot blanching (100°C for 60 min) caused the losses of 26%–29% in total β-carotene content, along with an increased 13-*cis*-β-carotene content up to 10% after pasteurization (T_{max} 95°C, $F = 3$) and to 14% after sterilization (T_{max} 121°C, $F = 5$) (Marx et al. 2003). The addition of grape oil to carrot juice before heat treatment enhanced the 13-*cis*-β-carotene formation (18.8%) as compared to the control (6.0%) (Marx et al. 2003). This fact is probably due to the partial dissolution of crystalline carotene, present in the intact carrot in lipid droplets, since the solubilization of carotenes during blanching is a prerequisite for the formation of *cis*-isomers.

Lycopene degradation was higher (24%) when tomato pulp was concentrated at 100°C for 120 min, as compared to heating at the same conditions, but without concentration (18.6%), and was also higher than water-insoluble tomato pulp solids heated in distilled water without concentration (15%) indicating that acids and sugars contributed to the loss of lycopene (Sharma and Le Maguer 1996). The authors reported that some kinetic models were studied, and the best fit model was found for zero-order kinetics; however, it was not suitable because the rate of lycopene loss was dependent on the initial lycopene concentration. Therefore, the authors considered a pseudo first-order model to explain the lycopene degradation during the tomato pulp heating. The features of the UV–visible spectra did not change after heating and no extra peaks were detected by the HPLC, although all peak areas were smaller than the control sample (Sharma and Le Maguer 1996).

The thermal stability and the isomerization of lycopene in oleoresins prepared from three different tomato varieties, Roma, High Lycopene, and Tangerine, were studied at temperatures ranging from 25°C to 100°C (Hackett et al. 2004). The first-order model was applied to determine the total lycopene degradation rates, although the correlation values (R^2) varied from 0.60–0.86. The lycopene degradation rates in oleoresins from Roma and High Lycopene were slower than that found in Tangerine oleroresin due to the high content of prolycopene (tetra-*cis*-lycopene) in this tomato variety (Table 12.3). The authors reported that at 25°C and at 50°C, lycopene degraded mainly through oxidation without isomerization, while lycopene isomerization increased at 75°C and 100°C reaching the formation of eight unidentified lycopene geometrical isomers. The addition of antioxidants, α-tocopherol or butylated hydroxytoluene (BHT), slowed by half the degradation rate of lycopene at 50°C (Hackett et al. 2004).

In five different tomato varieties submitted to thermal treatment in water or water/oil (8:2) at 100°C for 30 min, all-*trans*-lycopene, prolycopene, all-*trans*-δ-carotene, and all-*trans*-γ-carotene did not undergo isomerization, though heat treatment imparted changes to the physical ultrastructure, such as cell wall and organelle deformations (Nguyen et al. 2001). In contrast, on an average 21% of β-carotene and 27% of lutein were found to be present as *cis* isomers. These differences can be explained by the structural specificities of the carotenoids, such as molecular shape, ease of crystal formation, and further organization in multilayers or aggregates, and their storage at different locations in the cell. Once in the aggregated form, lycopene molecules might be able to resist further structural changes. On the contrary, β-carotene with two bulky β-ionone rings may not be able to easily assemble into an ordered and a stable structure, as do lycopene molecules. However, the presence of vegetable oil did not alter the thermal stability of all carotenoids evaluated (Nguyen et al. 2001).

Similar results were obtained during hot-break processing of tomato juice and even of tomato paste, the amounts of *trans*- and *cis*- lycopene isomers remained almost unchanged, whereas increased levels of *cis*-β-carotene were found during these processes (Abushita et al. 2000).

The sterilization ($T_{max} = 121°C$, $F = 5$) of sweet corn resulted in decreased amounts of total lutein by 26% and total zeaxanthin by 29% accompanied by increased amounts of *cis*- lutein from 12% to 30% and of *cis*-zeaxanthin from 7% to 25% (Aman et al. 2005b). The relative amounts of 13-*cis*-lutein increased by 11%, 13′-*cis*-lutein by 10%, and 13-*cis*-zeaxanthin by 21%, whereas the levels of the corresponding 9-*cis*- isomers remained practically unchanged after corn canning. The total lutein content decreased by 17% after spinach blanching with vapor ($T \approx 100°C$, 2 min) and the amount of lutein *cis*-stereoisomers of processed spinach decreased from 21% to 14%. While the contents of 9-*cis*-lutein and 9′-*cis*-lutein decreased, 13-*cis*-lutein and 13′-*cis*-lutein practically remained unaffected (Aman et al. 2005b). The differences in xanthophyll isomerization in corn and spinach upon thermal treatment are so far not understood. It may be assumed that the localization of carotenoids in different plastids of vegetables plays an important role, since lutein and zeaxanthin are localized in the chromoplasts in sweet corn, and lutein is exclusively localized in the chloroplasts of spinach (Aman et al. 2005b).

Another interesting difference regarding the formation of *cis* isomers, either at C-13 or C-9, was reported for mango slices dried in an overflow tray dryer (75°C, for 3–3.5 h) or in a solar-tunnel-dryer (60°C–62°C, for 7–8 h) (Pott et al. 2003). Both drying processes resulted in the complete degradation of xanthophylls and the partial degradation of all-*trans*-β-carotene, as previously mentioned for mango pureé (Vásquez-Caicedo et al. 2007) and mango juice (Mercadante and Rodriguez-Amaya 1998). The isomerization was shown to depend on the drying process; conventionally dried mangoes were characterized by the elevated amounts of 13-*cis*-β-carotene (from *ca*. 25% to 37%) and the negligible formation of 9-*cis*-isomers, whereas solar-dried mango slices contained additional amounts of the 9-*cis*-isomer (Pott et al. 2003).

In general, the major consequences of food thermal processing, either at laboratory or commercial scales, on carotenoids are the transformation of the 5,6-epoxy to the 5,8-furanoid rings, *trans*- to *cis*- isomerization and oxidation. In addition, independently of the food matrix or thermal process, the predominant *cis*- isomer of β-carotene, lutein, zeaxanthin, and β-cryptoxanthin, formed in processed red, yellow, and orange fruits and vegetables, is the 13-*cis*- form (and 13′-*cis*- for asymmetric carotenoids) followed by the smaller quantities of 9-*cis*- and 15-*cis*- isomers. However, in processed green vegetables, the 9-*cis*- isomers of β-carotene and lutein (in this case also the 9′-*cis*-) are predominantly formed. The degradation rates of different carotenoids depend not only on the carotenoid structure but also on their localization within the cell organelles, e.g., the slower degradation rates of lycopene were observed in food systems as compared to those of β-carotene, in contrast to what happens in model systems.

12.3 DIRECT AND SENSITIZED LIGHT-INDUCED DEGRADATIONS

In addition to food systems, carotenoid pigments are ubiquitous in the photosynthetic membranes of plants and bacteria where they are involved in the collection of sunlight for photosynthetic work, the dissipation of excited triplet-state energy, and the regulation of singlet energy flow to the photosynthetic reaction centers (Cogdell and Frank 1987). Most of the photobiological properties of carotenoids, such as antenna and photoprotector pigments, are related with their peculiar photophysical behavior and the characteristics of their excited states, which in turn are strongly determined by the extension of the one-dimensional π-electron conjugated system (Gust et al. 1993). The characteristic intense absorption band in the visible region of carotenoids (400–550 nm, ε_{max}~10^5 M^{-1} cm^{-1}), produced by the strong electric dipole, allowed transition from the ground state (S_0) to the second excited singlet state S_2. The electronic transition to the lowest excited singlet state, S_1, is forbidden by symmetry and thus its absorption band is undetected under normal conditions. As expected for upper excited states, the lifetime of the S_2 state is very short (<200 fs) and therefore the fluorescence and intersystem crossing quantum yields are extremely low ($<10^{-4}$) (Gust et al. 1993). However, despite their very fast unimolecular deactivation pathways, carotenoids show photochemical activity either by intra- (or direct photolysis) or intermolecular (photosensitized) populations of their singlet and triplet excited state manifolds.

12.3.1 PHOTOLYSIS IN MODEL AND FOOD SYSTEMS

Direct photolysis or steady state photolysis of carotenoids in model systems, e.g., homogenous solvents and microheterogeneous solutions (micelles, liposomes, etc.), has been studied under different conditions by several groups. In general, irradiation with UV or visible light produces the bleaching of the characteristic intense absorption band of the carotenoid at 400–550 nm together with a progressive blueshifting and absorbance increment in the region of 300–400 nm due to the formation of products with shorter chromophores, as depicted for the photodegradation of lycopene in Triton X-100 aqueous micelles, Figure 12.4.

The extent of the photodegradation reaction is measured by the photodegradation quantum yield, Φ_{pd}, which is defined as the fraction of molecules degraded in relation to how many photons were absorbed, and quantifies the light sensitivity of the molecule (Turro 1978). Usually, $\Phi_{pd} \leq 1$ but values higher than unity can indicate more complex processes, such as radical chain reactions.

Skibsted and coworkers have determined the dependence of solvent, oxygen partial pressure, temperature, and irradiation wavelength on the Φ_{pd} of several C_{40} carotenoids, such as astaxanthin, canthaxanthin, zeaxanthin, lutein, and β-carotene (Jørgensen and Skibsted 1990, Christophersen et al. 1991, Nielsen et al. 1996, Mortensen and Skibsted 1999, Hansen and Skibsted 2000). In general, it was observed that shorter irradiation wavelength, higher solvent polarity, and higher oxygen concentration all give rise to higher Φ_{pd} values. For instance, the Φ_{pd} value for β-carotene in air-saturated toluene solution was strongly reduced from 2.1×10^{-3} at 313 nm to 1.7×10^{-6} at 436 nm, whereas the Φ_{pd} values of canthaxanthin were almost one order of magnitude higher in chloroform and acetone than in toluene and vegetable oil (Christophersen et al. 1991, Nielsen et al. 1996). In addition, the Φ_{pd} increased between two and three times by changing the oxygen partial pressure from zero to unity

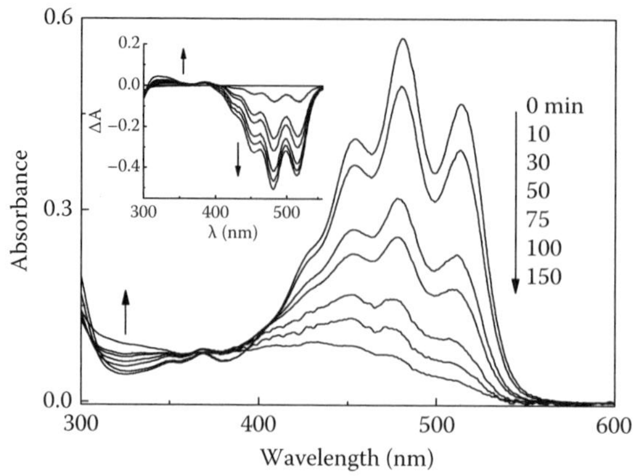

FIGURE 12.4 Lycopene photodegradation in 0.02 M Triton X-100 aqueous solutions illuminated with a 150 W (> 380 nm) filament lamp. Inset: evolution of the difference absorption spectrum (ΔA).

indicating that the photodegradation proceeded both through both oxygen-independent and oxygen-dependent pathways (Christophersen et al. 1991, Nielsen et al. 1996).

In view of these facts, the photodegradation mechanism was explained by occurring either from closely spaced vibrationally excited electronic singlet states or from the triplet manifold (Nielsen et al. 1996), Chart 12.2. In the absence of oxygen and low carotenoid concentrations, unimolecular photodegradation is expected. It has been shown that carotenoids with pure π,π^* excited states degrade almost from the singlet excited state due to their very low Φ_{isc} to populate the triplet state. Thus the unimolecular degradation was proposed to occur by breaking one of the central carbon–carbon bonds of the excited singlet state leading to radical products (Nielsen et al. 1996). On the other hand, carotenoids with carbonyl substituents (n,π^* states), such as canthaxanthin and astaxanthin, the Φ_{isc} increased about 20 times as compared with β-carotene. Therefore, the photobleaching can be significant from both singlet (${}^1\Phi_{pd}$) and triplet (${}^3\Phi_{pd}$) states (Nielsen et al. 1996), Chart 12.2.

CHART 12.2
Photophysical and Photochemical Pathways, and Quantum Yields (Φ) of β-Carotene and Canthaxanthin in Deaerated Toluene Solution at 25°C

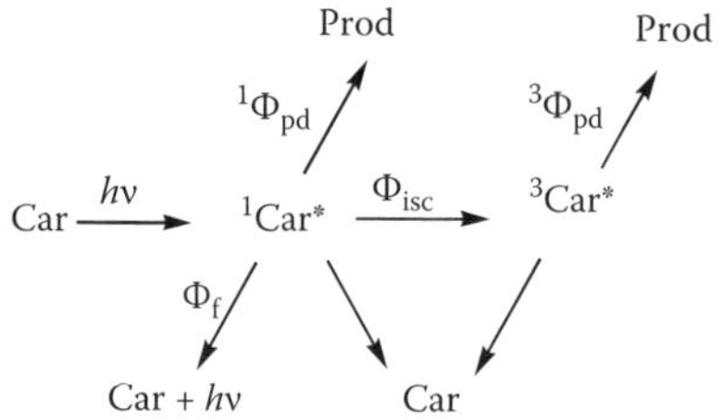

Quantum Yield	β-Carotene	Canthaxanthin
Φ_f[a]	9.6×10^{-4}	1.1×10^{-4}
Φ_{isc}[a]	5.4×10^{-4}	9.7×10^{-3}
${}^1\Phi_{pd}$[b]	3.8×10^{-5}	4.9×10^{-6}
${}^3\Phi_{pd}$[b]	2.7×10^{-7}	7.0×10^{-6}

Source: Nielsen, B.R. et al., *J. Agric. Food Chem.*, 44, 2106, 1996.

[a] Excitation at 355 nm.

[b] Excitation at 366 nm.

However, in spite of the higher Φ_{isc} for canthaxanthin, in both degassed and aerated solutions, β-carotene showed lower photostability than canthaxanthin (Nielsen et al. 1996). This effect was ascribed to the longer lifetime of the singlet state of β-carotene (10 ps) than for canthaxanthin (5 ps) (Wasielewski and Kispert 1986) allowing the increase in the bimolecular interaction with oxygen (Christophersen et al. 1991, Nielsen et al. 1996). However, the same authors suggested that the reactivity difference may be due to the nature of excited states, since carbonyl-containing carotenoids may be expected to show less diradical character in the central part of the excited molecule (Nielsen et al. 1996). Probably, the oxygen effect is due to its interaction with these diradical species rather than with the very short-lived excited species, since the quenching efficiency of singlet excited carotenoids by molecular oxygen in aerated organic solvents can be expected to be $<10^{-4}$.

Temperature did not affect the photodegradation upon photolysis with wavelengths longer than 350 nm. However, an activation energy of *ca.* 6.7 kcal/mol was observed in photolysis experiments at 313 nm, and it was also found that thermal degradation (or ground state degradation) was insignificant in UV (<400 nm) photolysis experiments, but was competing if visible light was used, because of the lower Φ_{pd} value under illumination in this spectral region (Mortensen and Skibsted 1999). However, the distribution of degradation products is highly dependent on the photolysis conditions and in many cases confusing results can be observed. For example, β-carotene showed that under mild photolysis conditions the basic carotenoid carbon skeleton was retained, and several *cis* and oxygenated isomers were formed. However, under exhaustive photolysis conditions carotenoid fragmentation produces volatile compounds, such as β-ionone and 6-hydroxy-2,2,6-trimethylcyclohexanone (Isoe et al. 1969, 1972).

Petersen et al. (1999) have studied the light stability of two commercial carotenoids extracts, e.g., annatto extract and β-carotene preparation, which are used as colorants for Cheddar cheese preparation. The apparent quantum yields for photodegradation by monochromatic light were determined at 25°C for buffer solutions containing sodium caseinate at pH 5.2 and 5.4. The solutions of commercial β-carotene were more photostable than the solutions of annatto, and although the quantum yields for photodegradation for both solutions depended significantly on both pH and irradiation wavelength, exposure to UV light (313 and 366 nm) caused more photobleaching than exposure to visible light (436 nm), consistent with the photodegradation model proposed by Skibsted and coworkers (Nielsen et al. 1996, Mortensen and Skibsted 1999, Hansen and Skibsted 2000).

The photodegradation rate of β-carotene and all-*trans*-8′-apo-β-caroten-8′-al in DMPC (dimyristoyl-L-α-phosphatidylcholine) liposomes was higher than in ethanol solutions, indicating that the vibrational relaxation of the carotenoid excited states in the membrane model system is less efficient than in homogeneous solution (He et al. 2000). The lifetime of the lowest excited singlet state of the carotenal in DMPC liposomes is 27 ps, longer than in ethanol (17 ps), while for β-carotene its lifetime was almost independent of the microenvironment, e.g., ≈10 ps (He et al. 2000). This result indicated that the carotenal was located in a more rigid region of the liposome membrane than is the case of β-carotene.

In good electron acceptor solvents, such as carbon tetrachloride and chloroform, the photodegradation of carotenoids is significantly increased as compared to other solvents (Christophersen et al. 1991, Mortensen and Skibsted 1999), because of a direct photoinduced electron-transfer reaction from the excited singlet state of the carotenoids to the solvent, as determined by transient absorption spectroscopy (Jeevarajan et al. 1996, Mortensen and Skibsted 1996, 1997a,b, El-Agamey et al. 2005), Equation 12.2:

$$\text{Car} + \text{CHCl}_3 \xrightarrow{h\nu} \left[\text{Car}^{\bullet+} \cdots \text{CHCl}_3^{\bullet-}\right] \rightarrow \text{Car}^{\bullet+} + \text{CHCl}_2^{\bullet} + \text{Cl}^- \tag{12.2}$$

The bleaching of carotenoids was simultaneous with the formation of near-infrared absorbing intermediates in the microsecond timescale. The formation of an adduct ion-pair is instantaneous during the laser pulse (<10 ns) with maximum absorption in the region 830–950 nm, depending on the carotenoid. This transient decays to form the solvated carotenoid radical cation with transient absorption at 850–1050 nm. The second order rate constant for photobleaching was in the range of 10^8–10^9 M^{-1} s^{-1} depending on the carotenoid structure, following the reactivity order: carotenes > hydroxyxanthophylls > ketoxanthophylls (Jeevarajan et al. 1996, Mortensen and Skibsted 1996, 1997a,b, El-Agamey et al. 2005).

The photodegradation of β-carotene by UVA irradiation in the presence of a series of mono-, di-, tri-, and tetrasulfides volatile compounds was investigated in ethanol solutions (Arita et al. 2005). The reaction was accelerated as the number of sulfur atoms was increased, and also by increasing either the light intensity or the initial concentration of the reactants, indicating the second-order nature of the photochemical reaction. The photodegradation rate of zeaxanthin was similar to that observed for β-carotene showing that the reaction was independent of the carotenoid structure (Arita et al. 2005).

The photostability of β-carotene and canthaxanthin was also evaluated in the presence of semiconductor particles, e.g., CdS or ZnO, irradiated with light at >350 nm (Gao et al. 1998). All-*trans*-β-carotene in dichloromethane solutions showed a rapid degradation in the presence of the semiconductors, but canthaxanthin underwent significant photocatalyzed degradation only on ZnO, not on CdS. HPLC studies indicated that CdS catalyzed the *trans–cis* photoisomerization of both carotenoids. As in the photoisomerization in the absence of semiconductor, the major *cis* isomers have the 9-*cis* and 13-*cis* configurations, but, under otherwise the same condition, the ratio of *cis/trans* isomers has doubled. In contrast to CdS, ZnO did not catalyze the photoisomerization of the carotenoids, although it enhanced their rate of degradation. A photoisomerization mechanism involving carotenoid radicals formed by reaction with interstitial sulfur on the CdS surface was proposed, with the formation of a S^+–Car bond favoring the geometrical isomerization of the carotenoid radicals to various *cis* configurations. In the case of ZnO, this attacking interaction is not available and photoisomerization does not compete with the photodegradation pathway (Gao et al. 1998).

The photobleaching of β-carotene by fluorescent light in fatty acid ester solutions showed an autoxidation kinetic profile with the rate of degradation of β-carotene in the order laurate > oleate > linoleate (Carnevale et al. 1979). The presence of a radical scavenger retarded the autoxidation, thus leading to the view that protection against autoxidation is built into the system by the unsaturation in the fatty acid.

Pesek and Wathersen (1990) used HPLC to study the photodegradation kinetics of *all-trans*-β-carotene in organic solvent mixture (acetonitrile/methanol/THF, 42:58:1 v/v/v) at 28°C by illumination with standard fluorescence lamp. They observed a first-order kinetic decay for the degradation of the carotenoid with $k_{pd} = 0.0018\,h^{-1}$, together with the formation of the 9-*cis* and 13-*cis* isomers as main photoisomerization products. The proportion of both *cis*-isomers was increased at the beginning of the illumination and both were later consumed during the process. In fact, the amount of *cis*-isomers formed (<20%) did not equal the amount of all-*trans*-carotenoid degraded (*ca.* 70%), indicating that photodegradation of all isomers to form volatile and nonvolatile compounds with shorter conjugated double bond chains occurred in parallel degradation channels. The thermal (dark) isomerization of the carotenoid was also parallel but less extensive than that under illumination and no degradation products were detected, indicating that under dark conditions mainly *trans–cis* isomerization occurred.

In addition, Chen and Huang (1998) have studied the photodegradation of *all-trans*-β-carotene in hexane solution at −5.4°C to minimize the thermal contribution, observing first-order decay for the carotenoid degradation with $k_{pd} = 1 \times 10^{-3}\,h^{-1}$. After 20 h of illumination the carotenoid degraded *ca.* 60%, together with the simultaneous formation of relatively small amounts of several *cis* isomers in the proportions of 13-*cis* > 15-*cis* > 9-*cis* ≥ 13,15-di-*cis*. 13,15-di-*cis* was the last isomer to be detected during the first 4 h of illumination, indicating that the 13-*cis* and/or 15-*cis* isomers are the precursors of this di-*cis*-isomer. All *cis*-isomers were degraded after prolonged illumination indicating the formation of degradation products from all isomers (Chen and Huang 1998). In spite of the different solvents used in both studies (Pesek and Warthesen 1990, Chen and Huang 1998), an activation energy $E_a \approx 3$ kcal/mol for the photodegradation of β-carotene can be estimated, which is a much lower value when compared to those observed for thermal degradation processes (Table 12.2).

The photoisomerization of the all-*trans*-zeaxanthin in a solvent mixture of methyl tertiary butyl ether (MTBE):methanol (5:95, v/v) at 25°C was evaluated upon illumination at four different wavelengths, e.g., 450, 540, 580 and 670 nm, corresponding to the electronic transitions of zeaxanthin from the ground state to the singlet excited states: $1^1B_u^+$, $3^1A_g^-$, $1^1B_u^-$, and $2^1A_g^-$, respectively (Milanowska and Gruszecki 2005). The photoisomerization quantum efficiency, Φ_{iso}, of the all-*trans*-zeaxanthin was found to differ considerably, in the ratio of 1:15:160:29 at 450, 540, 580, and 670 nm, respectively. The sequence of the quantum efficiency values suggests that the carotenoid triplet state 1^3B_u, populated via the internal conversion from the 1^3A_g triplet state that is generated by the intersystem crossing from the $1^1B_u^-$ state, may be involved in the light-induced isomerization (Milanowska and Gruszecki 2005).

The photodegradation of solid crystals of lycopene produced upon illumination with 20 W fluorescent lamps at 25°C was studied by Lee and Chen (2002). The degradation of all-*trans*-lycopene showed first-order kinetics with an observed degradation rate of 0.018 h^{-1}, a value similar to that previously reported in a vegetable juice system (Pesek and Warthesen 1987). The loss of lycopene after 144 h of illumination was *ca.* 13.1%, and it was almost transformed in several mono-*cis*-isomers (e.g., 5-*cis*-, 9- *cis*-, 13- *cis*-, and 15-*cis*-lycopene) and an unidentified di-*cis*-isomer of lycopene, which represented *ca.* 22% of the formed isomer products. The amount of the total mono-*cis* isomers increased initially and then decreased during prolonged illumination together with the formation of the di-*cis* isomer (Lee and Chen 2002).

The photostability of lycopene in commercial tomato powders was evaluated during storage under fluorescent light (38,500 lux) at room temperature for up to 6 weeks (Anguelova and Warthesen 2000). HPLC and spectral analysis were used to determine lycopene losses and the formation of *cis* isomers and degradation products. The lycopene isomer content of the starting material was *ca.* 93.5% of all-*trans*-lycopene, 5.3% of 5,5′-di-*cis*-lycopene, and 1.2% of 15-*cis*-lycopene. During illumination a color fading together with the development of hay or grassy odors, characteristic of the odors due to oxidation products, were observed. Decreases of all-*trans*-lycopene (*ca.* 30%) were accompanied by increases in the contents of the 5,5′-di-*cis* isomer and 5,6-dihydroxy-5,6-dihydrolycopene as a proportion of the total lycopene present in the sample after storage. These facts suggested that the degradation of all-*trans* lycopene proceeded through isomerization and autoxidation (Anguelova and Warthesen 2000).

The photostability of the natural occurring 9′-*cis* carotenoids bixin and norbixin, Figure 12.5, has received the attention of several research groups (Najar et al. 1988, Pimentel and Stringheta

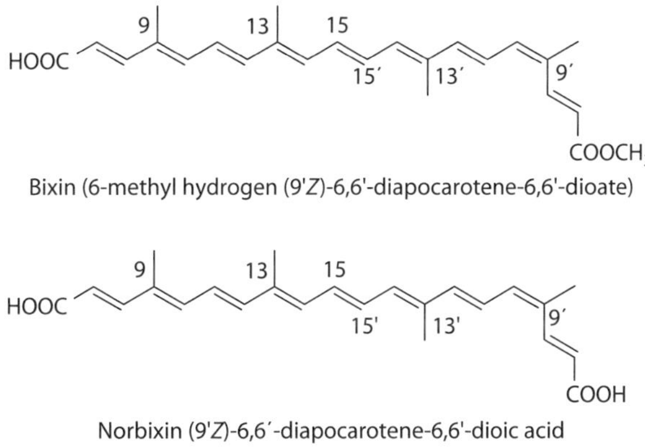

FIGURE 12.5 Structures of bixin and norbixin, *cis* carotenoids from annatto.

TABLE 12.4
Observed Rate Constant for the Photodegradation of Carotenoids (k_{pd}) in Some Model and Food Systems

Carotenoid	Model System	k_{pd} (h^{-1})	Reference
β-Carotene	ACN:MeOH:THF 42:58:1 v/v/v (28°C)	1.8×10^{-3}	Pesek and Warthesen (1990)
	n-Hexane (−5.4°C)	1.0×10^{-3}	Chen and Huang (1998)
Lycopene	Solid crystals (25°C)	1.0×10^{-2}	Lee and Chen (2002)
Bixin	$ClCH_3$ (24°C, air saturated, 1380 lux)	0.73	Najar et al. (1988)
	$ClCH_3$ (24°C, N_2 saturated, 1380 lux)	0.50	Najar et al. (1988)
	$ClCH_3$ (24°C, air saturated, 430 lux)	0.10	Najar et al. (1988)
	$ClCH_3$ (24°C, air saturated, 430 lux, 20% w/v ascorbyl palmitate)	0.05	Najar et al. (1988)
	Food System		
β-Carotene	Freeze-dried carrot pulp powder (25°C)	1.8×10^{-3}	Tang and Chen (2000)
α-Carotene	Freeze-dried carrot pulp powder (25°C)	1.6×10^{-3}	Tang and Chen (2000)
Lutein	Freeze-dried carrot pulp powder (25°C)	5.4×10^{-4}	Tang and Chen (2000)

1999, Prentice-Hernández and Rusig 1999, Barbosa et al. 2005), because of their larger solubility in polar solvents or in alkaline media. Chloroform solutions of annatto extracts containing *ca.* 0.26 g/L of bixin also showed a first-order bleaching kinetics upon excitation with a tungsten filament lamp (Najar et al. 1988), Table 12.4. In air-saturated solutions both the k_{pd} and the final percentage of degraded bixin increased with the light intensity. However, in N_2-saturated solutions the reaction was similar to that under aerated conditions, and the presence of ascorbyl palmitate as an antioxidant reduced both the rate and the extent of the photobleaching reaction (Najar et al. 1988). Considering the electron-acceptor ability of chloroform, these results could be explained by a photoinduced electron-transfer mechanism as indicated in Equation 12.2 (see above), as demonstrated by several groups using transient absorption spectroscopy for other carotenoids (Jeevarajan et al. 1996, Mortensen and Skibsted 1996, 1997a,b, El-Agamey et al. 2005).

The photostability of these *cis*-carotenoids (bixin and norbixin) was evaluated in the presence of edible biopolymers, such as gum arabic and maltodextrin (MD) (Pimentel and Stringheta 1999, Prentice-Hernández and Rusig 1999, Barbosa et al. 2005). The absorbance changes at 453 nm (ΔA_{453}) of alkaline aqueous extracts (pH 8.5) of annatto (containing norbixin) showed complex decay behavior when illuminated with standard fluorescence lamps (40 W) and no effect of oxygen was observed (Pimentel and Stringheta 1999). The complex kinetic behavior of ΔA_{453} can be ascribed to the progressive formation of blue edge-absorbing intermediate products as it is shown for the photobleaching of lycopene in Figure 12.4. The lack of an oxygen effect may indicate that the photobleaching reaction is through the very short-lived singlet state of the carotenoid. The addition of MD to the aqueous solutions did not show any change on the photobleaching of norbixin in the presence or the absence of light.

However, microencapsulation with edible biopolymers by spray-drying increased the photostability of both extract or pure carotenoids (Prentice-Hernández and Rusig 1999, Barbosa et al. 2005). In the microencapsulation process the interest core molecule is coated by a wall material, such as edible biopolymers, increasing the shelf-life of the core material, and/or improving its solubility in suitable solvents, and/or controlling its delivery into the solution (Gharsallaoui et al. 2007). The photoprotective effect depends on the microencapsulation material and conditions, but similar kinetic behavior was observed among different systems (Prentice-Hernández and Rusig 1999, Barbosa et al. 2005). Figure 12.6 compares the kinetic profile for the photodegradation of non- and microencapsulated bixin solutions with MD obtained from two laboratories (Prentice-Hernández and Rusig

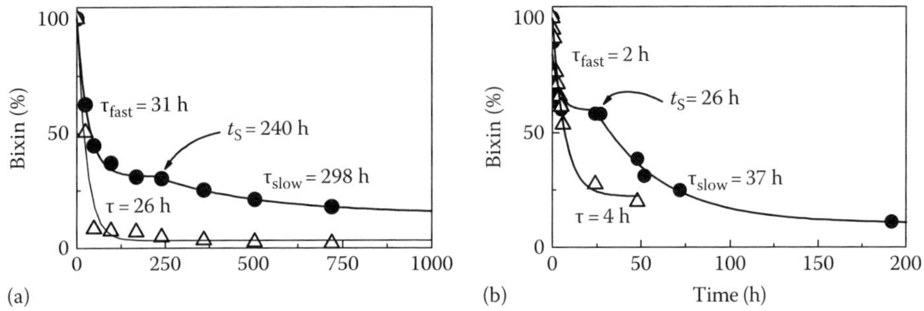

FIGURE 12.6 Comparison of photodegradation kinetics of bixin in aqueous solutions containing maltodextrin (MD) under different conditions: (●) microencapsulated and (Δ) not encapsulated. (From Barbosa, M.I.M.J. et al., *Food Res. Int.*, 38, 989, 2005. With permission.)

1999, Barbosa et al. 2005). Despite the sample heterogeneity, a characteristic that is intrinsic of the spray-drying technique (Gharsallaoui et al. 2007), both sets of experiments showed similar kinetic profiles for the degradation of bixin. In principle, bixin dissolved in MD solutions (not microencapsulated) was very labile under illumination conditions, and its decay showed a simple first-order behavior. By contrast, the photodegradation of bixin microencapsulated solutions showed biphasic first-order decay. In both cases, it was observed that the lifetime of the fast decay (τ_{fast}) was similar to that observed for the photodegradation of non-microencapsulated bixin in solution (Barbosa et al. 2005). Therefore, the fast decay component was considered to result from the photodegradation of bixin located outside the microcapsules, where the carotenoid molecules are highly exposed to the surrounding environment. In turn, the slower decay component (τ_{slow}), which started after the lag period, t_S, should correspond to the degradation of bixin molecules incorporated into the microcapsules, which are slowly released as the microcapsule is swollen by the aqueous solvent. Thus, these core bixin molecules are more protected from both oxidative and photochemical degradations. The efficiency of encapsulation of core bixin molecules and its photostability were larger in GA than in MD, and it was also shown that depending on the choice of the wall material and coemulsifier the effective lifetime of the carotenoid can be tuned (Barbosa et al. 2005). Table 12.4 summarizes the observed rate constant values for the photodegradation of some selected carotenoids under several conditions.

The influence of light exposure on the degradation and the isomerization of pure carotenoids and chloroplast-bound carotenoids were compared by Aman et al. (2005a). The illumination of freshly prepared chloroplast isolates caused an initial increase in the level of lutein (9.6%) and β-carotene (29.8%), while pure carotenoids exhibited time-dependent degradation as described above. These authors claimed that carotenoid stability has to be evaluated for every individual pigment in its genuine environment, since stability data based on model systems (e.g., pure carotenoids in homogeneous solvents) may not be transferred to complex food matrices without an intensive investigation (Aman et al. 2005a).

Changes in carotenoid contents and antioxidant activities of three tomato genotypes, labeled DRW 5981, HP 1, and Esperanza, grown inside of a greenhouse either covered with polyethylene transparent to UV-B or depleted of UV-B by a special covering film was evaluated by Giuntini et al. (2005). The results indicated that the genotype Esperanza showed low capacity for accumulating carotenoids and a great susceptibility to the detrimental effects of UV-B. Conversely, the DRW genotype shows high carotenoid levels under sunlight conditions and a further promotion by UV-B (Giuntini et. al. 2005).

The photostability of carotenoids during the storage of acidified and pasteurized carrot juice was evaluated at several storage temperatures, e.g., at 4°C, 25°C, and 35°C during 3 months illuminated with fluorescent light (20 W, 1500 lux), (Chen et al. 1996). The isomerization and the degradation of carotenoids were monitored by HPLC with diode-array detection and the results showed that the amounts of lutein, α-carotene, and β-carotene in carrot juice decreased with increasing storage temperature and that 9-*cis*- isomers were the major types of isomers formed under light storage (Chen et al. 1996).

The photostability of carotenoids in freeze-dried powder from carrot pulp waste under light at 25°C was analyzed by HPLC with photodiode-array detection upon illumination with fluorescence light (1500 lux) (Tang and Chen 2000). Results showed that the amounts of all-*trans*- forms of main components, α-carotene, β-carotene, and lutein, decreased with increasing illumination time with the formation of 9-*cis* derivatives as main isomers. The degradation rates of the total amount of all-*trans* plus *cis* forms of each pigment at 25°C were 5.4×10^{-4}, 1.6×10^{-3}, and 1.8×10^{-3} h^{-1} for lutein, α-carotene, and β-carotene, respectively. The degradation rate of β-carotene was identical to that observed in homogeneous solvents (Pesek and Warthesen 1990), Table 12.4. Additionally the Hunter L and b values of the powder decreased with increasing storage time and temperature, while the a (red) value showed an insignificant change ($p > 0.05$).

The stability of carotenoids in tomato juice during storage under fluorescent light (2500 lux) at 4°C, 25°C, and 35°C for 12 weeks was studied by Lin and Chen (2005). Light enhanced the degradation and the isomerization of all-*trans*-lutein; with more formation of 13-*cis*-lutein than 9-*cis*-lutein. Similar trends were observed for β-carotene but also the formation of di-*cis* isomers was observed. For lycopene, 15-*cis*-lycopene was the major isomer formed during dark storage at 4°C, while 9-*cis*- and 13-*cis*-lycopene were favored at 25°C and 5-*cis*- as well as 13-*cis*-lycopene dominated at 35°C. Under light storage, both 9-*cis* and di-*cis*-lycopene were the main isomers generated at 35°C, whereas 13-*cis*- and 15-*cis*-lycopene were the most abundant at 4°C and 25°C. Therefore, by increasing the storage temperature larger losses of the all-*trans*- and *cis*- forms of lutein, β-carotene, and lycopene occurred during illumination. All-*trans*-lycopene showed the highest degradation efficiency, followed by all-*trans*-β-carotene and all-*trans*-lutein. More *cis* isomers of lycopene than lutein or β-carotene were generated during storage. However, the major type of isomers formed may vary, depending on storage conditions (Lin and Chen 2005).

12.3.2 Photosensitized Degradation in Model and Food Systems

The photosensitized transformation of carotenoids has been studied using several sensitizer molecules, such as chlorophylls, iodine, rose bengal (RB), and methylene blue (MB) and in general terms isomerization is the major pathway of reaction.

The products of dye-sensitized photoisomerization (excitation at 337 nm of anthracene as sensitizer) and direct photoisomerization (excitation at 488 and 337 nm) of all-*trans*-, 7-*cis*-, 9-*cis*-, 13-*cis*-, and 15-*cis*- isomers of β-carotene in deaerated (by N_2 bubbling) *n*-hexane were analyzed by HPLC (Kuki et al. 1991). The following isomerization patterns were found for each starting isomer: (a) all-*trans* ⇄ 13- *cis* > 9- *cis* > 15- *cis*, (b) 7- *cis* ⇄ all-*trans* > 9-*cis* > 7,15-di-*cis* ≈7,13′-di- *cis*, (c) 9- *cis* ⇄ all-*trans* >> 9,15-di- *cis* > 9,13′-di- *cis* > 9,13-di-*cis* > 13- *cis*, (d) 13- *cis* ⇄ all-*trans* >> 9-*cis*, and (e) 15- *cis* ⇄ all-*trans*. The isomerization quantum yields (Φ_{iso}) in the photosensitized reaction is between two and four orders of magnitude higher when compared with the direct photolysis at 488 and 337 nm photolysis, respectively (Chart 12.3).

In addition, the results indicated that the efficiency of *cis* → *trans* increased as the initial *cis* double bond configuration is shifted from the center of the polyenic chain, consistent with the T_1, triplet excited state potential curve that has a very shallow minimum at the 15-*cis* position compared to the deep minima at the all-*trans* position. The results strongly suggest that isomerization takes place via the T_1 state of the carotenoid even in the case of direct photoexcitation, with their Φ_{iso} much lower than in photosensitized process because of the very low intersystem crossing quantum yield, Φ_{isc} ($<10^{-3}$) (Nielsen et al. 1996).

CHART 12.3
Comparison of Photosensitized and Direct Photolysis Isomerization Quantum Yields (Φ_{iso}) of All-*trans* and several *cis*-Isomers of β-Carotene in *n*-Hexane

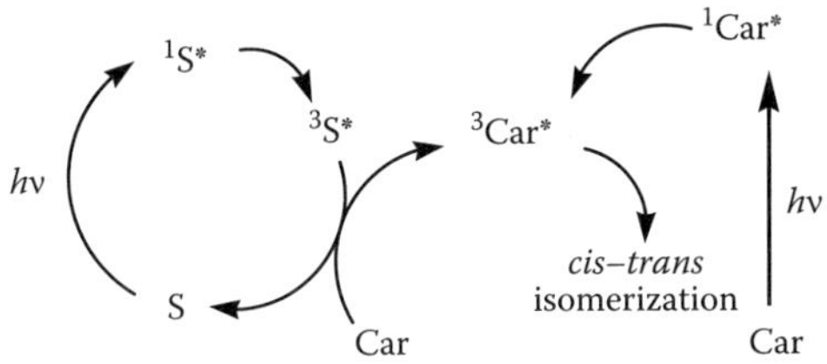

Carotenoid	Photosensitized	Direct Photolysis	
		488 nm (×10[6])	337 nm (×10[4])
all-*trans*	0.044	4.0	1.0
7-*cis*	0.117	14.8	8.9
9-*cis*	0.145	18.5	10.6
13-*cis*	0.865	85.3	21.8
15-*cis*	0.976	98.7	28.1

Source: Data from Kuki, M. et al., *J. Phys. Chem.*, 95, 7171, 1991.

The iodine-catalyzed photoisomerization of all-*trans*- α- and β-carotenes in hexane solutions produced by illumination with 20 W fluorescence light (2000 lux) and monitored by HPLC with diode-array detection yielded a different isomer distribution (Chen et al. 1994). Four *cis* isomers of β-carotene (9-*cis*, 13-*cis*, 15-*cis*, and 13,15-di-*cis*) and three *cis* isomers of α-carotene (9-*cis*, 13-*cis*, and 15-*cis*) were separated and detected. The kinetic data fit into a reversible first-order model. The major isomers formed during the photosensitized reaction of each carotenoid were 13,15-di-*cis*-β-carotene and 13-*cis*-α-carotene (Chen et al. 1994).

The isomerization of all-*trans*-β-carotene under N_2 atmosphere by photosensitization action of eight chlorophyll compounds naturally present in the extracts of green vegetables was investigated by illumination with fluorescent white light (3000 lux) at 12°C to minimize the thermal degradation (O'Neil and Schwartz 1995). All chlorophylls showed similar isomeric distribution and efficiency, 9-*cis*-β-carotene is the main isomer formed. On the other hand, the illumination of all-*trans*-β-carotene without chlorophylls indicated that the main isomer formed was 13-*cis*-β-carotene, probably due to the population of a different triplet state manifold by direct photolysis.

In presence of a dye sensitizer (S) and ground state oxygen, $^3O_2$, two pathways lead to the population of $^3Car^*$ that can be produced according to the following energy-transfer mechanism, Figure 12.7 (Foote et al. 1970, Montenegro et al. 2004). Therefore, either in absence or presence of oxygen, the formation of $^3Car^*$ is principally mediated by an electronic energy-transfer mechanism. The carotenoid isomerization occurring from the $^3Car^*$ state favors the *cis* → *trans* process (Foote et al. 1970, Montenegro et al. 2004). Under aerobic conditions, $^3O_2$ competes with the carotenoid molecules for the deactivation of the triplet state of the sensitizer, $^3S^*$. The relative efficiency of each

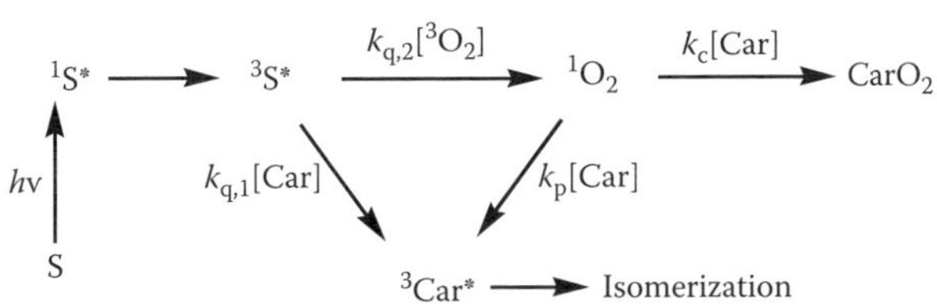

FIGURE 12.7 Photosensitized generation of singlet molecular oxygen and carotenoid triplet state by energy-transfer process. S, $^1S^*$, and $^3S^*$ represents, respectively, the ground state, singlet and triplet excited states of the sensitizer molecule. $^3O_2$ and $^1O_2$ are ground and singlet molecular oxygen, and Car represents carotenoid molecule. In turn, $k_{q,1}$ and $k_{q,2}$ are the bimolecular quenching rate constant of $^3S^*$ by Car and $^3O_2$, respectively. Finally, k_c and k_p represent the chemical reaction rate constant and the physical quenching rate constant of $^1O_2$ by Car, respectively. The unimolecular decays of $^3S^*$ and $^1O_2$ were not indicated, since at moderated carotenoid concentration [Car] are negligible.

pathway is given by the ratio $k_{q,1}[Car]/k_{q,2}[O_2]$. Considering that in air-saturated organic solvents the $^3O_2$ concentration is *ca.* 2 mM (Murov et al. 1993), and both $k_{q,1}$ and $k_{q,2}$ are near diffusion controlled ($\geq 1 \times 10^9$ M^{-1}s^{-1}, Montenegro et al. 2004), the energy-transfer from $^3S^*$ to $^3O_2$ is more efficient and the formation of singlet oxygen ($^1O_2$) is favored.

However, carotenoids with more than nine conjugated double bonds are known as diffusional quenchers of $^1O_2$, with bimolecular quenching rate constant $k_Q \approx 1$–2×10^{10} M^{-1}s^{-1} (Farmillo and Wilkinson 1973, Baltschun et al. 1997, Edge et al. 1997, Montenegro et al. 2002, Schweitzer and Schmidt 2003), which involves both physical (k_p) and chemical (k_c) deactivation processes, Figure 12.7. For most carotenoids, the physical interaction with $^1O_2$ is the principal quenching pathway (i.e., $k_c/k_p \approx 10^{-3}$–10^{-4}) yielding as products the carotenoid triplet excited state, $^3Car^*$, and the triplet ground state molecular oxygen, $^3O_2$.The large efficiency of energy-transfer from $^1O_2$ to forming $^3Car^*$ is based on the down-hill energy cascade driving force. The triplet energy, E_T, for carotenoids that quench $^1O_2$ with $k_Q > 1 \times 10^{10}$ M^{-1}s^{-1} lies below the energy level of $^1O_2$ (22.5 kcal/mol), as confirmed by experimental measurements using laser-induced optoacoustic spectroscopy for β-carotene (E_T = 19.5 kcal/mol, Lambert and Redmond 1994), and bixin (E_T = 18.0 kcal/mol, Rios et al. 2007). The biological relevance of a predominant physical quenching pathway is that almost none of the quencher molecules are consumed during the process, thus allowing its repeated participation in consecutive interactions with $^1O_2$.

Recently, Montenegro et al. (2004) described the photosensitized isomerization mechanism of bixin, a naturally occurring 9-*cis* carotenoid, in acetonitrile:methanol (1:1) solution using RB or MB as a sensitizer. HPLC-diode array detector analysis showed that bixin was almost quantitatively transformed into its all-*trans* isomer, with identical activation energy (E_a = 6 kcal/mol) for both N_2- and air-saturated solutions. This activation value is four times lower than that observed for the dark (thermal) *cis* → *trans* isomerization in water:ethanol (8:2) mixtures (Rios et al. 2005), suggesting the participation of the excited triplet of bixin ($^3Bix^*$) as the precursor state of the photosensitized process. The participation of the $^3Bix^*$ was confirmed using laser-flash photolysis experiments by the detection of the typical carotenoid triplet absorption band at 520 nm. In addition, the $^3Bix^*$ was quenched by the bixin ground state (self-quenching) and by ground state oxygen, $^3O_2$. In this case, the oxidative degradation was observed by the reaction of $^1O_2$ with all-*trans*-bixin. The respective rate constant values describing the individual steps in the MB-mediated photosensitization of bixin are summarized in Table 12.5 (Montenegro et al. 2004).

Despite the high physical quenching efficiency of long conjugated carotenoids, $^1O_2$-mediated carotenoid oxidation is produced in long-term photosensitized processes, due to the chemical quenching pathway (Stratton et al. 1993, Montenegro et al. 2002). Interestingly, it has been observed that the oxidative quenching rate is independent of the carotenoid nature and/or extension of the polyenic chain, with $k_c \approx 1 \times 10^6$ M^{-1}s^{-1} (Montenegro et al. 2002, Borsarelli et al. 2007). This result differs from k_p, which is strongly dependent on the number of conjugated double bonds because of the decrease in the E_T value of the $^3Car^*$ (Baltschun et al. 1997, Edge et al. 1997, Montenegro et al. 2002).

The product distribution resulting from β-carotene oxidization by $^1O_2$ was studied by Stratton et al. (1993) using reverse-phase HPLC, UV-vis spectrophotometry, and mass spectrometry .The oxidation products were identified as β-ionone, β-apo-14′-carotenal, β-apo-10′-carotenal, β-apo-8′-carotenal, and β-carotene-5,8-endoperoxide. The formation of 5,8-endoperoxide derivative by a [4+2] Diels–Alder addition mechanism was also reported in the $^1O_2$-mediated oxidation of β-carotene in reverse micelles (Montenegro et al. 2002), β-ionone (Borsarelli et al. 2007), and of the A1E retinoid derivative (Jockusch et al. 2004).

The bacteriopheophytin *a*-photosensitized oxygenation of β-carotene was also studied in air-saturated acetone (Fiedor et al. 2001). The carotenoid was rapidly oxygenated under strong illumination of the sensitizer with red light ($\lambda_{exc} \geq 630$ nm). At the same time the photosensitizer undergoes only a slight (<10%) photodegradation. Seven major oxygen-containing products of carotenoids

TABLE 12.5
Unimolecular and Bimolecular Rate Constants for the Different Elementary Steps Involved in the MB Photosensitized Degradation of Bixin in Acetonitrile: Methanol (1:1) Solutions at 25°C

Process	Rate Constant
$^3MB^* \rightarrow MB$	$2.9 \times 10^4\ s^{-1}$
$^3MB^* + Bix \rightarrow MB + {}^3Bix^*$	$7.0 \times 10^9\ M^{-1}s^{-1}$
$^3MB^* + {}^3O_2 \rightarrow MB + {}^1O_2$	$2.0 \times 10^9\ M^{-1}s^{-1}$
$^1O_2 \rightarrow {}^3O_2$	$6.7 \times 10^4\ s^{-1}$
$^1O_2 + Bix \rightarrow {}^3O_2 + {}^3Bix^*$	$1.3 \times 10^{10}\ M^{-1}s^{-1}$
$^3Bix^* \rightarrow$ all-*trans*	$4.2 \times 10^4\ s^{-1}$
$^3Bix^* + Bix \rightarrow Bix + Bix$	$1.5 \times 10^9\ M^{-1}s^{-1}$
$^3Bix^* + {}^3O_2 \rightarrow Bix + {}^3O_2$	$7.0 \times 10^8\ M^{-1}s^{-1}$
$^1O_2$ + all-*trans* → Oxidation products	$1.0 \times 10^6\ M^{-1}s^{-1}$

Source: Data from Montenegro, M.A. et al., *J. Agric. Food Chem.*, 52, 367, 2004.

were isolated by preparative HPLC chromatography. Their molecular weight analysis indicated that the sequential accumulation of up to six oxygen atoms is produced on the C40-skeleton. Two possible mechanisms were envisaged: One was the formation of epoxides and perhaps their subsequent decomposition (hydrolysis) to vicinal diols. The other proposed mechanism was sequential [4+2] photocycloadditions at the β-rings to form 5,8-endoperoxide intermediates (Fiedor et al. 2001).

12.4 CONCLUSIONS

Carotenoid degradation by thermal and photochemical processes may produce both isomers (*trans*, mono-*cis* and di-*cis*) and degradation products (oxidation, volatile, and nonvolatile chain breaking compounds). While *cis*- isomers retain most of the color properties of the parent carotenoid, degradation products are less colored because of the resulting shorter chromophore. Depending on the degradation treatment, the formation of volatile compounds (such as β-ionone) and autoxidation products can occur. Both the carotenoid stability and the product distribution are dependent on the media properties, such as solvent polarity, electron-acceptor and electron-donor abilities, pH, and microviscosity. Since in most photosensitized processes the triplet excited state of the carotenoid $^3Car^*$ acts as an intermediate in the *cis–trans* isomerization, the photosensitized activation energy is almost four to five times lower than for thermally activated isomerization.

Finally, due to the simultaneous reversible isomerization and coupled irreversible degradation reactions, as depicted in Figure 12.1, the kinetic profile for the degradation of the starting parent carotenoid is not fitted well by a first-order equation, and a bi-exponential equation set is required. The use of simple first-order model in both thermal and light-induced degradations of carotenoids can produce miscalculation and/or misinterpretation of both rate constants and activation energies. Special care must be taken on the use of global physicochemical properties, e.g., color parameters changes and total carotenoid concentration, for the calculation of activation parameters, since these parameters account for global change and their kinetic profile can be strongly dependent on the reaction system, products formed, and parallel reactions.

ACKNOWLEDGMENTS

We thank Consejo Nacional de Investigaciones Científicas y Técnicas (CONICET) from Argentina and Fundação de Amparo a Pesquisa do Estado de São Paulo (FAPESP), CNPq, and FAEPEX-UNICAMP from Brazil for financial support. C.D.B is a member of the Research Career of CONICET.

REFERENCES

Abushita, A. A., H. G. Daood, and P. A. Biacs. 2000. Change in carotenoids and antioxidant vitamins in tomato as a function of varietal and technological factors. *J. Agric. Food Chem.* 48: 2075–2081.

Ahmed, J., U. S. Shivhare, and K. S. Sandhu. 2002. Thermal degradation kinetics of carotenoids and visual color of papaya puree. *J. Food Sci.* 67: 2692–2695.

Aman, R., A. Schieber, and R. Carle. 2005a. Effects of heating and illumination on *trans-cis* isomerization and degradation of β-carotene and lutein in isolated spinach chloroplasts. *J. Agric. Food Chem.* 53: 9512–9518.

Aman, R., J. Biehl, R. Carle et al. 2005b. Application of HPLC coupled with DAD, APcI-MS and NMR to the analysis of lutein and zeaxanthin stereoisomers in thermally processed vegetables. *Food Chem.* 92: 753–763.

Anguelova, T. and J. J. Warthesen. 2000. Lycopene stability in tomato powders. *J. Food Sci.* 65: 67–70.

Arita, S., S. Ando, H. Hosoda et al. 2005. Acceleration effect of sulfides on photodegradation of carotenoids by UVA irradiation. *Biosci. Biotechnol. Biochem.* 69: 1786–1789.

Baltschun, D., S. Beutner, K. Briviba et al. 1997. Singlet oxygen quenching abilities of carotenoids. *Liebigs Ann. Recueil.* 1887–1893.

Barbosa, M. I. M. J., C. D. Borsarelli, and A. Z. Mercadante. 2005. Light stability of spray-dried bixin encapsulated with different edible polysaccharide preparations. *Food Res. Int.* 38: 989–994.

Borsarelli, C. D., M. Michne, A. La Venia et al. 2007. UVA self-photosensitized oxygenation of β-ionone. *Photochem. Photobiol.* 83: 1313–1318.

Capellos, C. and B. H. J. Bielski. 1972. *Kinetic Systems. Mathematical Description of Chemical Kinetics in Solution*. New York: Wiley-Interscience.

Carnevale, J., E. R. Cole, and G. Crank. 1979. Fluorescent light catalyzed autoxidation of β-carotene. *J. Agric. Food Chem.* 27: 462–463.

Carter, G. P. and A. E. Gillam. 1939. CLXI: The isomerisation of carotenes. III. Reconsideration of the change β-carotene to α-carotene. *Biochem. J.* 33: 1325–1331.

Chen, B. H. and J. H. Huang. 1998. Degradation and isomerization of chlorophyll *a* and β-carotene as affected by various heating and illumination treatments. *Food Chem.* 62: 299–307.

Chen, B. H., T. M. Chen, and J. T. Chien. 1994. Kinetic model for studying the isomerization of α- and β-carotene during heating and illumination. *J. Agric. Food Chem.* 42: 2391–2397.

Chen, B. H., H. Y. Peng, and H. E. Chen. 1995. Changes of carotenoids, color, and vitamin A contents during processing of carrot juice. *J. Agric. Food Chem.* 43: 1912–1918.

Chen, H. E., H. Y. Peng, and B. H. Chen. 1996. Stability of carotenoids and vitamin A during storage of carrot juice. *Food Chem.* 57: 497–503.

Christophersen, A.G., H. Jun, K. Jørgensen et al. 1991. Photobleaching of astaxanthin and canthaxanthin. *Z. Lebensm. Unters. Forsch.* 192: 433–439.

Cogdell, R. J. and H. A. Frank. 1987. The function of carotenoids in photosynthesis. *Biochim. Biophys. Acta* 815: 63–79.

Dhuique-Mayer, C., M. Tbatou, M. Carail et al. 2007. Thermal degradation of antioxidant micronutrients in citrus juice: Kinetics and newly formed compounds. *J. Agric. Food Chem.* 55:4209–4216.

Dutta, D., A. Dutta, U. Raychaudhuri et al. 2006. Rheological characteristics and thermal degradation kinetics of beta-carotene in pumpkin puree. *J. Food Eng.* 76: 538–546.

Edge R., D. J. McGarvey, and T. G. Truscott. 1997. The carotenoids as anti-oxidants—A review. *J. Photochem. Photobiol. B: Biol.* 41: 189–200.

El-Agamey, A., M. Burke, R. Edge et al. 2005. Photolysis of carotenoids in chloroform: Enhanced yields of carotenoid radical cations in the presence of a tryptophan ester. *Rad. Phys. Chem.* 72: 341–345.

Farmillo, A. and F. Wilkinson. 1973. On the mechanism of quenching of singlet oxygen in solution. *Photochem. Photobiol.* 18: 447–450.

Fiedor, J., L. Fiedor, J. Winkler et al. 2001. Photodynamics of the bacteriochlorophyll–carotenoid system. 1. Bacteriochlorophyll-photosensitized oxygenation of β-carotene in acetone. *Photochem. Photobiol.* 74: 64–71.

Foote, C. S., Y. C. Chang, and R. W. Denny. 1970. Chemistry of singlet oxygen. XI. *Cis–trans* isomerization of carotenoids by singlet oxygen and a probable quenching mechanism. *J. Am. Chem. Soc.* 92: 5218–5219.

Gao, G., Y. Deng, and L. D. Kispert. 1998. Semiconductor photocatalysis: Photodegradation and *trans-cis* photoisomerization of carotenoids. *J. Phys. Chem. B.* 102: 3897–3901.

Gharsallaoui, A., G. Roudaut, O. Chambin et al. 2007. Applications of spray-drying in microencapsulation of food ingredients: An overview. *Food Res. Int.* 40: 1107–1121.

Giuntini, D., G. Graziani, B. Lercari et al. 2005. Changes in carotenoid and ascorbic acid contents in fruits of different tomato genotypes related to the depletion of UV-B radiation. *J. Agric. Food Chem.* 53: 3174–3181.

Gust, D., T. A. Moore, A. L. Moore et al. 1993. The photochemistry of carotenoids. *Ann. N. Y. Acad. Sci.* 691: 32–47.

Hackett, M. M., J. H. Lee, D. Francis et al. 2004. Thermal stability and isomerization of lycopene in tomato oleoresins from different varieties. *J. Food Sci.* 69: C536–C541.

Hansen, E. and L. H. Skibsted. 2000. Light-induced oxidative changes in a model dairy spread. Wavelength dependence of quantum yields. *J. Agric. Food Chem.* 48: 3090–3094.

He, Z., L. D. Kispert, R. M. Metzger et al. 2000. Carotenoids in liposomes: Photodegradation, excited state lifetimes, and energy transfer. *J. Phys. Chem. B.* 104: 6302–6307.

Henry, L. K., G. L. Catignani, and S. J. Schwartz. 1998. Oxidative degradation kinetics of lycopene, lutein and 9-*cis* and all-*trans* β-carotene. *J. Am. Oil Chem. Soc.* 75: 823–829.

Isoe, S., S. B. Hyeon, S. Katsumura et al. 1972. Photo-oxygenation of carotenoids. II. The absolute configuration of ioliolide and dihydroactinidioline. *Tetrahed. Lett.* 13: 2517–2519.

Isoe, S., S. B. Hyeon, and T. Sakan. 1969. Photo-oxygenation of carotenoids. I. The formation of dihydroactinidiolide and β-ionone from β-carotene. *Tetrahed. Lett.* 10, 279–281.

Jeevarajan, A. S., L. D. Kispert, N. I. Avdievich et al. 1996. Role of excited singlet state in the photooxidation of carotenoids: A time-resolved Q-band EPR study. *J. Phys. Chem.* 103: 669–671.

Jockusch, S., R. X. Ren, Y. P. Jang et al. 2004. Photochemistry of A1E, a retinoid with a conjugated pyridinium moiety: Competition between pericyclic photooxygenation and pericyclization. *J. Am. Chem. Soc.* 126: 4646–4652.

Jørgensen, K. and L. H. Skibsted. 1990. Light sensitivity of carotenoids used as food colours. *Z. Lebensm. Unters. Forsch.*, 190: 306–313.

Kanasawud, P. and J. C. Crouzet. 1990. Mechanism of formation of volatile compounds by thermal degradation of carotenoids in aqueous medium. 1. β-Carotene degradation. *J. Agric. Food Chem.* 38: 237–243.

Kuki, M., Y. Koyama, and H. Nagae. 1991. Triplet-sensitized and thermal isomerization of all-*trans*, 7-*cis*, 9-*cis*, 13-*cis*, and 15-*cis* isomers of β-carotene: Configurational dependence of the quantum yield of isomerization via the T1 state. *J. Phys. Chem.*, 95: 7171–7180.

Lambert, C. and R. W. Redmond. 1994. Triplet energy level of β-carotene. *Chem. Phys. Lett.* 228: 495–498.

Lee, H. S. and G. A. Coates. 2003. Effect of thermal pasteurization on Valencia orange juice color and pigments. *Lebensm.-Wiss. U.-Technol.* 36, 153–156.

Lee, M. T. and B. H. Chen. 2002. Stability of lycopene during heating and illumination in a model system. *Food Chem.* 78: 425–432.

Lin, C. H. and B. H. Chen. 2005. Stability of carotenoids in tomato juice during storage. *Food Chem.* 90: 837–846.

Liu, M. H. and B. H. Chen. 1998. Relationship between chlorophyll *a* and β-carotene in a lipid-containing model system during heating. *Food Chem.* 61: 41–47.

Machiels, D. and L. Istasse. 2002. La rèaction de Maillard: Importance et applications en chimie des aliments. *Ann. Med. Vet.* 146: 347–352.

Marx, M., M. Stuparic, A. Schieber et al. 2003. Effects of thermal processing on *trans-cis*-isomerization of β-carotene in carrot juices and carotene-containing preparations. *Food Chem.* 83: 609–617.

Mercadante, A. Z. 2008. Carotenoids in foods: Sources and stability during processing and storage. In *Food Colorants: Chemical and Functional Properties*, ed. C. Socaciu, pp. 213–240. Boca Raton, FL: CRC Press.

Mercadante, A. Z. and D. B. Rodriguez-Amaya. 1998. Effects of ripening, cultivar differences, and processing on the carotenoid composition of mango. *J. Agric. Food Chem.* 46: 128–130.

Milanowska, J. and W. I. Gruszecki. 2005. Heat-induced and light-induced isomerization of the xanthophyll pigment zeaxanthin. *J. Photochem. Photobiol. B: Biol.* 80: 178–186.

Montenegro, M. A., A. de O. Rios, M. A. Nazareno et al. 2004. Model studies on the photosensitized isomerization of bixin. *J. Agric. Food Chem.* 52: 367–373.
Montenegro, M. A., M. A. Nazareno, E. N. Durantini et al. 2002. Singlet oxygen quenching ability of carotenoids in a reverse micelle membrane mimetic system. *Photochem. Photobiol.* 75: 353–362.
Mortensen, A. and L. H. Skibsted. 1996. Kinetics of photobleaching of β-carotene in chloroformand formation of transient carotenoid species absorbing in the near infrared. *Free Rad. Res.* 25: 355–368.
Mortensen, A. and L. H. Skibsted. 1997a. Free radical transients in photobleaching of xanthophylls and carotenes. *Free Rad. Res.* 26: 549–563.
Mortensen, A. and L. H. Skibsted. 1997b. Real time detection of reactions between radicals of lycopene and tocopherol homologues. *Free Rad. Res.* 27: 229–234.
Mortensen, A. and L. H. Skibsted. 1999. Carotenoid photobleaching. *Meth. Enzymol.* 299: 408–421.
Murov, S. L., I. Carmichael, and G. L. Hug. 1993. In *Handbook of Photochemistry*, 2nd ed., pp. 289–291. New York: Marcel Dekker.
Najar, S. V., F. O. Bobbio, and P. A. Bobbio. 1988. Effects of light, air, anti-oxidants and pro-oxidants on annatto extracts (*Bixa orellana*). *Food Chem.* 29: 283–289.
Nguyen, M., D. Francis, and S. Schwartz. 2001. Thermal isomerisation susceptibility of carotenoids in different tomato varieties. *J. Sci. Food Agric.* 81: 910–917.
Nielsen, B. R., A. Mortensen, K. Jørgensen et al. 1996. Singlet versus triplet reactivity in photodegradation of C_{40} carotenoids. *J. Agric. Food Chem.* 44: 2106–2113.
O'Neil, C. A. and S. J. Schwartz. 1995. Photoisomerization of β-carotene by photosensitization with chlorophyll derivatives as sensitizers. *J. Agric. Food Chem.* 43: 631–635.
Pesek, C. A. and J. J. Warthesen. 1987. Photodegradation of carotenoids in vegetable juice system. *J. Food Sci.* 53: 744–746.
Pesek, C. A. and J. J. Warthesen. 1990. Kinetic model for photoisomerization and concomitant photodegradation of β-carotenes. *J. Agric. Food Chem.* 38: 1313–1315.
Pesek, C. A., J. J. Warthesen, and P. S. Taoukis. 1990. A kinetic model for equilibration of isomeric β-carotenes. *J. Agric. Food Chem.* 38: 41–45.
Petersen, M., L. Wiking, and H. Stapelfeldt. 1999. Light sensitivity of two colorants for Cheddar cheese. Quantum yields for photodegradation in an aqueous model system in relation to light stability of cheese in illuminated display. *J. Dairy Res.* 66: 599–607.
Pimentel, F. A. and P. C. Stringheta. 1999. Influência da luz e do oxigênio sobre a estabilidade do norbixinato de potássio em presença do maltodextrina. *Rev. Bras. Cor. Nat.* 3: 21–26.
Pott, I., M. Marx, S. Neidhart et al. 2003. Quantitative determination of β-carotene stereoisomers in fresh, dried, and solar-dried mangoes (*Mangifera indica* L.). *J. Agric. Food Chem.* 51: 4527–4531.
Prentice-Hernández, C. and O. Rusig. 1999. Effect of light on the stability of a microencapsulated extract obtained from anatto (*Bixa orellana* L.). *Braz. J. Food Technol.* 2: 185–189.
Rios, A. de O., C. D. Borsarelli, and A. Z. Mercadante. 2005. Thermal degradation kinetics of bixin in an aqueous model system. *J. Agric. Food Chem.* 53: 2307–2311.
Rios, J. J., E. Fernández-García, M. I. Mínguez-Mosquera et al. 2008. Description of volatile compounds generated by the degradation of carotenoids in paprika, tomato and marigold oleoresins. *Food Chem.* 106: 1145–1153.
Rios, A. de O., A. Z. Mercadante, and C. D. Borsarelli. 2007. Triplet state energy of the carotenoid bixin determined by photoacoustic calorimetry. *Dyes Pigments* 74: 561–565.
Schweitzer, C. and R. Schmidt. 2003. Physical mechanisms of generation and deactivation of singlet oxygen. *Chem. Rev.* 103: 1685–1757.
Scotter, M. J. 1995. Characterization of the coloured thermal degradation products of bixin from annatto and a revised mechanism for their formation. *Food Chem.* 53: 177–185.
Scotter, M., J. L. Castle, and G. P. Appleton. 2001. Kinetics and yields for the formation of coloured and aromatic thermal degradation products of annatto in foods. *Food Chem.* 74: 365–375.
Sharma, S. K. and M. Le Maguer. 1996. Kinetics of lycopene degradation in tomato pulp solids under different processing and storage conditions. *Food Res. Int.* 29: 309–315.
Silva, G. S., A. G. Souza, J. R. Botelho et al. 2007. Kinetics study of norbixin's first stage thermal decomposition, using dynamic method. *J. Thermal Anal. Calorim.* 87: 871–874.
Stratton, S. P., W. H. Schaefer, and D. C. Liebler. 1993. Isolation and identification of singlet oxygen oxidation products of β-carotene. *Chem. Res. Toxicol.* 6: 542–547.
Tang, Y. C. and B. H. Chen. 2000. Pigment change of freeze-dried carotenoid powder during storage. *Food Chem.* 69: 11–17.
Turro, N. J. 1978. *Modern Molecular Photochemistry*. Menlo Park, CA: Benjamin/Cummings.

Vásquez-Caicedo, A. L., S. Schilling, R. Carle et al. 2007. Effects of thermal processing and fruit matrix on β-carotene stability and enzyme inactivation during transformation of mangoes into purée and nectar. *Food Chem*. 102: 1172–1186.

Wasielewski, M. R. and L. D. Kispert. 1986. Direct measurement of the lowest excited state lifetime of all-*trans*-β-carotene and related carotenoids. *Chem. Phys. Lett*. 128: 238–243.

Zechmeister, L. 1944. *Cis–trans* isomerization and stereochemistry of carotenoids and diphenylpolyenes. *Chem. Rev*. 34: 267–344.

Zechmeister, L. and R. B. Escue. 1944. A stereochemical study of methylbixin. *J. Am. Chem. Soc*. 66: 322–330.

Zepka, L. Q. and A. Z. Mercadante. 2009. Degradation compounds of carotenoids formed during heating of a simulated cashew apple juice. *Food Chem*. 117: 28–34.

Zepka, L. Q., C. D. Borsarelli, M. A. A. P. da Silva et al. 2009. Thermal degradation kinetics of carotenoids in a cashew apple juice model and its impact on the system color. *J. Agric. Food Chem*. in press. doi: 10.1021/jf900558a.

Part V

Antioxidant and Photoprotection Functions and Reactions Involving Singlet Oxygen and Reactive Oxygen Species

13 The Functional Role of Xanthophylls in the Primate Retina

Wolfgang Schalch, Richard A. Bone, and John T. Landrum

CONTENTS

13.1 INTRODUCTION

In autumn, as chlorophyll gradually disappears from leaves, the colorful world of carotenoids unfolds with colors ranging from deep red to light yellow. The prime reason for the presence of carotenoids in plants, however, is not to generate the beauty of the autumnal colors; it is their capability to drive nonphotochemical quenching reactions and to dissipate the energy of excess absorbed light as heat to protect the photosynthetic reaction centers from damage that makes carotenoids so important (Nayak et al. 2001). For millennia, photosynthetic organisms have been using these important properties of carotenoids (Frank and Cogdell 1996). As evolution proceeded, at least one other system that is simultaneously exposed to light and oxygen has adopted these principles as well: the eye, where xanthophylls have been used not only as active antioxidants but also as passive intraocular (blue)-light filters. The occurrence of such intraocular color filters in the vertebrate kingdom was comprehensively reviewed in 1933 by Walls and Judd (1933). The yellow corneae of some fish species or the oil droplets in the retina of reptiles and birds are examples of these intraocular color filters based on the presence of carotenoids. An alternative carotenoid system, not based on oil droplets, is present in humans and nonhuman primates. This is the "macula lutea," also called the yellow spot at the location of highest visual acuity within the retina. Two xanthophylls, lutein and zeaxanthin, are responsible for its distinctive yellow color. These molecules in the macula lutea are concentrated up to a combined concentration of about 1 mM, the highest concentration of carotenoids found anywhere in the primate body (Landrum et al. 1999b).

13.2 CAROTENOIDS: CAROTENES AND XANTHOPHYLLS

Carotenoids are a group of more than 750 naturally occurring molecules (Britton et al. 2004) of which about 50 occur in the normal human food chain. Of these, only 24 have, so far, been detected in human plasma and tissues (Khachik et al. 1995), with only six molecules being abundant in normal human plasma (for chemical formulas see Figure 13.1). Carotenoids are subdivided into two main classes: the carotenes, cyclized (e.g., β-carotene) or uncyclized (e.g., lycopene) hydrocarbons, and the xanthophylls, which have hydroxyl groups (e.g., lutein and zeaxanthin), keto-groups (e.g., canthaxanthin), or both (e.g., astaxanthin) as functional groups.

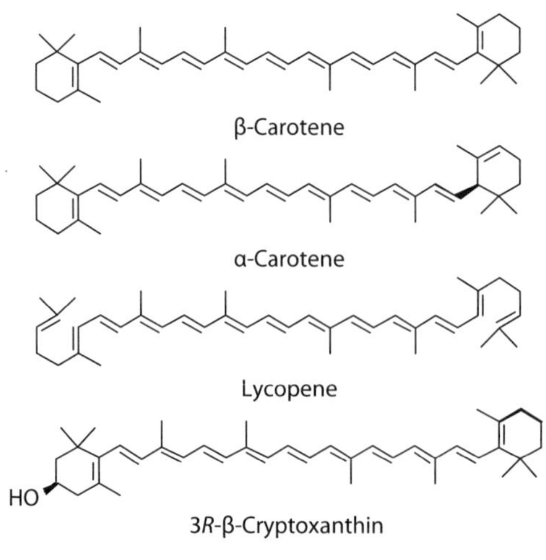

FIGURE 13.1 Chemical formulas of "major plasma carotenoids."

13.3 HISTORICAL BACKGROUND

Anatomically, the presence of the macula lutea, the distinctively yellow spot in the center of the retina, was first described in 1782 by Buzzi (1782) and again in 1799, based on an independent discovery, by Soemmering (1799). An interesting account of this discovery and related observations in monkeys is available from Home (1798), who mentions that the yellow spot is absent in fetuses and infants under one year of age and that it appears brighter in young people and paler in the elderly. Much later, Belloni (1983) provided interesting additional details of the discovery and many references to historic articles that are relevant to the subject. In 1981, almost 200 years after the original report, the history of the discovery of the macular yellow pigment was comprehensively reviewed by Nussbaum et al. (1981), who eloquently describe the 200 years of research and controversies. The authors mention that the existence of the macular yellow pigment was not generally accepted to be an anatomical feature *in vivo* for many years. Even as recently as 1958, some authors maintained that the macula lutea was a postmortem artifact (Nordenson 1958) although it had been photographed in the living eye using red-free illumination more than two decades earlier (Fincham 1936).

The chemical composition of the macular yellow pigment was originally characterized by Wald as being a "leaf xanthophyll" carotenoid (Wald 1945). It took another 40 years until the molecules lutein and zeaxanthin were identified as its main constituents by Bone et al. (1985), and in 1993 the same authors reported that macular zeaxanthin is itself comprised of two stereoisomers, (3*R*,3′*R*)-zeaxanthin and (3*R*,3′*S*)-zeaxanthin (Bone et al. 1993), this compound will be called "(*meso*)-zeaxanthin" throughout this article. These three molecules are often collectively called the "macular xanthophylls" (for their chemical formulas see Figure 13.2).

The first hypotheses to explain the physiological importance and function of the yellow macular pigment (MP) dealt exclusively with its putative effects on visual acuity, which were believed to be

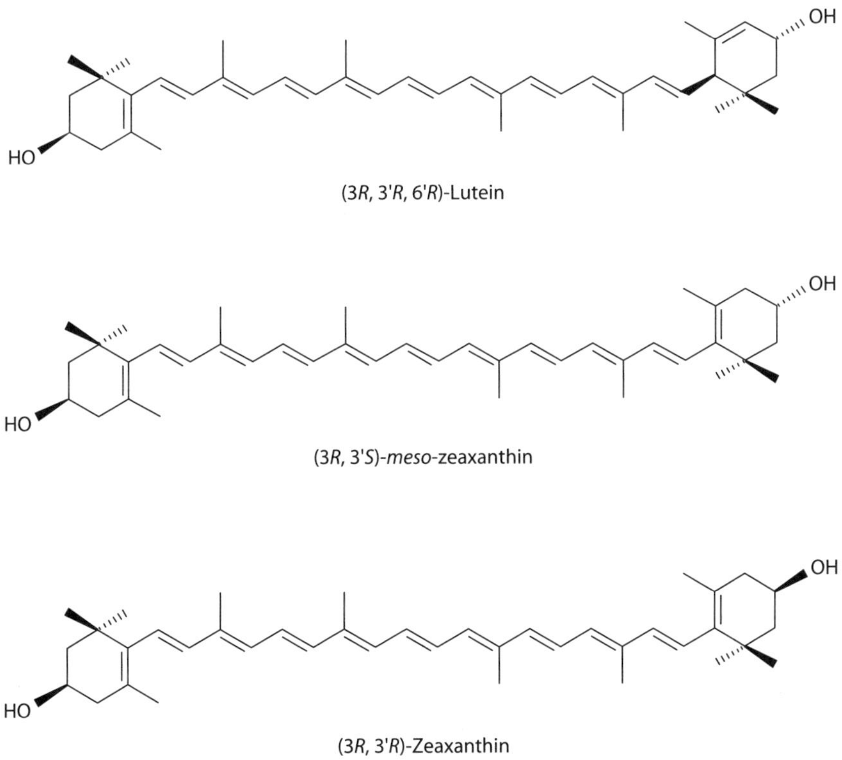

FIGURE 13.2 Chemical formulas of "macular xanthophylls." It can be noted that the chemical structures of (meso)-zeaxanthin and lutein differ only by the position of a single double bond.

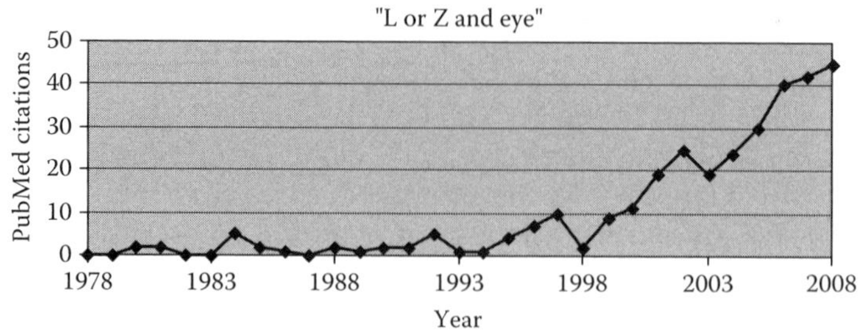

FIGURE 13.3 Time-course of publications "L or Z and eye." The graph was produced by entering the search term "lutein or zeaxanthin and eye," searching in all fields of the international biomedical research article database Pubmed for the years 1950–2008. The abstracts of the retrieved articles were checked to ensure that they were indeed relevant to the search term. The number of articles found was noted and plotted against the year of publication.

mediated by the MP's ability to absorb the strongly scattered blue light and in turn this was expected to ameliorate chromatic aberration. Even Nussbaum et al. (1981) only transiently mentioned the possibility that xanthophylls, because of these properties, could also contribute to risk reduction of ophthalmic diseases.

In contrast, a review by Kirschfeld (1982) published one year later specifically discussed the potential protective properties of macular yellow pigment, and the author pointed out that the effect of light on animal tissues is ambivalent. On the one side, light is necessary for multiple functions: for vision in animals and in photosynthetic organisms to gain energy. On the other side, light is potentially dangerous: it is capable of inducing damage by photooxidation, especially in the presence of sensitizing pigments that are ubiquitous in aerobic cells such as haems, cytochromes, or lipofuscin. Kirschfeld discusses several examples that illustrate how a compromise was achieved to cope with this dichotomy. This theme was taken up by a later review in 1992 (Schalch 1992) pinpointing the importance of lutein and zeaxanthin in the retinal environment, which is characterized by the simultaneous presence of light and oxygen, and how the macular xanthophylls could contribute to risk reduction of age-related macular degeneration (AMD). This latter suggestion first received scientific support by epidemiological findings, which demonstrated that plasma concentrations (EDCC Study Group 1993) or dietary intake levels (Seddon et al. 1994) of lutein and zeaxanthin are lower in patients with neovascular AMD, which may have increased their risk of developing AMD.

Based on this hypothesis, from the year 1990 onward, lutein and zeaxanthin have received increasing attention in the scientific literature as possible contributors to risk reduction of AMD, an attention that continues as indicated by the still increasing number of hits in PubMed for the keywords "eye" and "lutein" or "zeaxanthin" (Figure 13.3) and by a steady flux of review articles on the subject (some of the more recent reviews are O'Connell et al. [2006], Trumbo and Ellwood [2006], Whitehead et al. [2006], Bhosale and Bernstein [2007], Coleman and Chew [2007], Loughman et al. [2007], Renzi and Johnson [2007], Afzal and Afzal [2008], and Loane et al. [2008]).

13.4 OCCURRENCE OF CAROTENOIDS IN THE EYE

Apart from lutein and zeaxanthin, a number of other xanthophylls are present in the retina at low concentrations, including 3′-epi-lutein, lactucaxanthin, 3′-dehydrolutein, and β,β-carotene-3,3′-dione (Khachik et al. 1997a, Bernstein et al. 2001). Interestingly, β-carotene, the provitamin A precursor of the chromophore of the visual pigments, has only been identified in traces in the retina, if at all.

In the eye, carotenoids occur mainly in the retina, but the xanthophylls, lutein and zeaxanthin, can also be detected in the lens and the ciliary body. The human lens contains lutein and zeaxanthin in roughly equal amounts but no other carotenoids. Their amount in the lens is several orders of magnitude smaller than that in the macula lutea and they mainly appear to be concentrated in the epithelium/cortex tissue (Yeum et al. 1999).

Lutein and zeaxanthin are the dominant carotenoids in nonretinal eye tissue, and lycopene and β-carotene have been found in the ciliary body, which after the retina and the retinal pigment epithelium (RPE) contains the highest quantity of carotenoids (Bernstein et al. 2001). The orbital adipose tissue also contains measurable quantities of lutein and β-carotene, and possibly other carotenoids as minor constituents (Sires et al. 2001). It is also interesting to note that lutein was recently identified in the vitreous body of human fetuses, 15–28 weeks old (Yakovleva et al. 2007). However, these results may have to be considered with caution, because the vitreous bodies were described as substantially being penetrated with hyaloid blood vessels, which could have contaminated the vitreous with blood.

A xanthophyll that is structurally similar to zeaxanthin, canthaxanthin (β,β-carotene-4,4′-dione) is not normally identified in retinal tissues. However, in cases when canthaxanthin has been ingested over a long time and in elevated doses as an oral tanning agent or as a treatment for light sensitivity, the formation of crystalloid deposits in the human retina has been observed. These deposits preferentially occurred around and within the macular region, a condition that was termed "canthaxanthin retinopathy" (Boudreault et al. 1893), although it was found to be reversible (Harnois et al. 1989, Leyon et al. 1990) and without clinical consequences (Arden and Barker 1991, Koepcke et al. 1995, Goralczyk et al. 2000). The retina of a human subject who had been taking high doses of a combination of canthaxanthin and β-carotene could comprehensively be analyzed postmortem (Daicker et al. 1987). While canthaxanthin was readily identifiable in this retina by HPLC, no β-carotene could be detected although it had also been ingested in substantial quantities.

13.5 TOPOGRAPHY OF THE MACULAR PIGMENT

13.5.1 Cross-Sectional Distribution in the Retina

The cross-sectional distribution (Figure 13.4, bottom) of the yellow MP in the retina has been evaluated in retinal cross-sections. There, the yellow pigmentation is seen to be concentrated primarily within a layer called Henle's fiber layer (Snodderly et al. 1984), Light must consequently first pass through the yellow MP before reaching the photoreceptor outer segments. Lutein and zeaxanthin are also present in the rod outer segments (Rapp et al. 2000), although at lower concentrations, and may be present in cone outer segments as well. Furthermore, Müller cells have been suggested to be a reservoir of macular xanthophylls (Gass 1999).

In summary, the xanthophylls in the Henle's fiber layer can act as a blue-light filter, passively shielding the fragile macula against potentially damaging blue light, whereas the xanthophylls in the outer segments are directly and actively involved in quenching of reactive oxygen species generated by the simultaneous presence of light and oxygen.

13.5.2 Horizontal Distribution in the Retina

The horizontal distribution (Figure 13.4, top) of the macular xanthophylls across the retina has been studied in detail by measuring concentrations in postmortem eyes via HPLC (Bone et al. 1997). The macular xanthophylls are detectable across the entire retina but have their highest concentration in the center of the macula. The local zeaxanthin to lutein ratio depends on the distance from the fovea and decreases from about 2:1 at its center to a low of near 1:2 in the peripheral retina. The variation in the zeaxanthin/lutein ratio across the retina suggests that the chemical and biochemical influences operating on the xanthophylls in the peripheral retina are different from those in the central macula. This is an area about which not much is known and would constitute an interesting field of research.

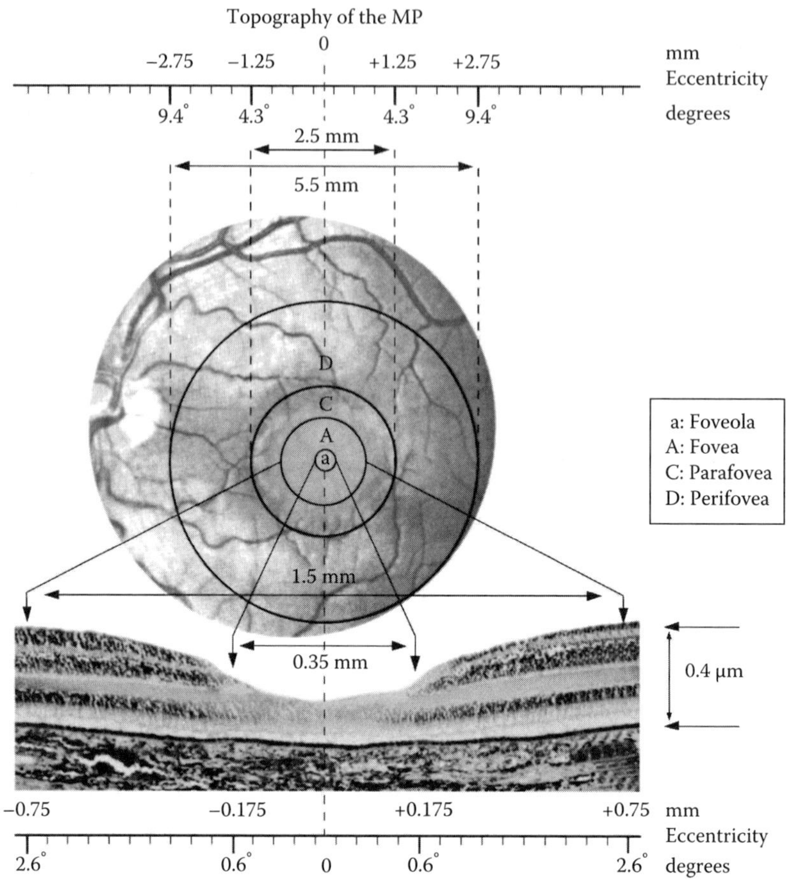

FIGURE 13.4 (See color insert following page 336.) Topography of the MP. This figure schematically shows the distribution of the yellow MP across the retina: horizontally (top) and vertically (bottom). (From Gass, J.D., *Stereoscopic Atlas of Macular Diseases Diagnosis and Treatment*, Vol. 1, Mosby – Year Book Inc., 3, 1997. With permission.)

As already mentioned, macular zeaxanthin comprises two stereoisomers, the normal dietary (3*R*,3′*R*)-zeaxanthin and (3*R*,3′*S*)-zeaxanthin(=(*meso*)-zeaxanthin), of which the latter is not normally a dietary component (Bone et al. 1993) and is not found in any other compartment of the body except in the retina. The concentration of (*meso*)-zeaxanthin in the retina decreases from a maximum within the central fovea to a minimum in the peripheral retina, similar to the situation with (3*R*,3′*R*)-zeaxanthin. This distribution inversely reflects the relative concentration of lutein in the retina and gave rise to a hypothesis (Bone et al. 1997) that (*meso*)-zeaxanthin is formed in the retina from lutein. This was confirmed by an experiment in which xanthophyll-depleted monkeys had been supplemented with chemically pure lutein or (3*R*,3′*R*)-zeaxanthin (Johnson et al. 2005). (*Meso*)-Zeaxanthin was exclusively detected in the retina of lutein-fed monkeys but not in retinas of zeaxanthin-fed animals, demonstrating that it is a retina-specific metabolite of lutein only. The mechanism of its formation has not been established but may involve oxidation–reduction reactions that are mediated photochemically, enzymatically, or both. Thus, (*meso*)-zeaxanthin is a metabolite unique to the primate macula.

The observation that lutein in the retina is converted to (*meso*)-zeaxanthin appears to be physiologically plausible: due to the presence of, in comparison to lutein, one additional conjugated double bond, it has a stronger singlet oxygen quenching capability (Cantrell et al. 2003). Furthermore, it has been reported that (*meso*)-zeaxanthin provides a somewhat better protection against the oxidation of lipid constituents of membranes than zeaxanthin (Bhosale and Bernstein 2005). Like zeaxanthin, (*meso*)-zeaxanthin appears to span the cell membrane in a perpendicular orientation, whereas lutein tends to lie close to the membrane surface being positioned parallel to the liposome phospholipids (Gabrielska and Gruszecki 1996, Krinsky 2002). The perpendicular orientation results in a closer spatial association of zeaxanthin with the polyunsaturated fatty acids of the core of the membrane, which may be relevant to its oxidation protection potential. Recent results with xanthophylls using model membranes enriched with polyunsaturated fatty acids appear to support these ideas (McNulty et al. 2008).

13.6 ABSORPTION AND TRANSPORT INTO THE RETINA

13.6.1 General

Humans and nonhuman primates are not able to synthesize carotenoids *de novo* and are therefore exclusively dependent on the diet or dietary supplementation for a continuous supply of preformed carotenoids. It is obvious that for the accumulation of the macular xanthophylls in the retina, it is essential that they are first absorbed from the food or from the nutritional supplements in the gastrointestinal tract and then transported into the retina. This entire topic of bioavailability, including the involvement of lipoproteins, has been reviewed numerous times (one of the more recent reviews being Loane et al. [2008]) and, therefore, will not be discussed in detail here. However, it has to be acknowledged that for fulfilling any function in the macula one of the prerequisites is that ingested xanthophylls do indeed reach the retina, a process called the specific bioavailability at the target organ. Therefore, it is important to briefly review the knowledge on the accumulation of MP, as measured by MP optical density (MPOD) in response to dietary or supplemental ingestion of xanthophylls.

13.6.2 Carotenoid-Binding Proteins

The existence of the "macula lutea" as a localized feature within the retina suggests a specific function for the macular xanthophylls as well as an active mechanism for their accumulation. The occurrence of carotenoid-binding proteins in vertebrates and invertebrates was recently reviewed (Bhosale and Bernstein 2007). Bernstein et al. (1997) first reported that tubulin, a protein found preferentially in neural tissue, has carotenoid-binding properties, although it has been found to be relatively unspecific, suggesting that it may only be a "high capacity site for the passive deposition of accumulated carotenoids." Later, a protein was isolated from postmortem human retinae that exhibited specificity for xanthophylls as opposed to carotenes. This protein was identified as the Pi isoform of gluthathione S transferase (GSTP1) (Bhosale et al. 2004) and appears to be specific for zeaxanthin and (*meso*)-zeaxanthin while lutein is only weakly bound. By immunocytochemical labeling with an antibody against GSTP1, the authors have demonstrated its primary localization to the Henle's fiber region of the macula. In an experiment published later, they presented evidence that zeaxanthin in the presence of its binding protein underwent oxidation at a slower rate than in its absence (Bhosale and Bernstein 2005). Involvement of the RPE in the retinal accumulation of xanthophylls was long suspected but only recently was evidence supporting it presented in a publication, which reported that RPE cells take up xanthophylls preferentially compared to β-carotene and that uptake could be inhibited by antibodies against the scavenger receptor class B1 (SR-B1) (During et al. 2008).

13.7 RESPONSES TO SUPPLEMENTATION WITH XANTHOPHYLLS

13.7.1 General

In 1941, Wald et al. reported that excreted xanthophylls were entirely of dietary origin and that only negligible amounts could originate from intestinal organisms (Wald et al. 1941, Wald 1945). Forty years later, this dietary origin was confirmed by results obtained in monkeys raised on a carotenoid-free diet (Malinow et al. 1980). In this experiment, the carotenoid-deprived macaques did not have measurable plasma concentrations of carotenoids including lutein and zeaxanthin and no yellow MP. Over the years, the xanthophyll deprivation of the macula led to distinct ophthalmic consequences, so-called window defects, visible during fluorescein angiography, which signal a malfunction in the cells of the RPE. The observed defects were similar to those seen in human AMD. This experiment was probably the first controlled attempt to modulate MPOD in nonhuman primates by dietary means (Malinow et al. 1980) and the first indication that a long-lasting xanthophyll deficiency can have ophthalmic consequences. Under special conditions, deprivation of carotenoids can also occur in humans. The human disease cystic fibrosis, a consequence of which is that the absorption of fat-soluble nutrients including carotenoids is severely impaired, provides a model for this. Schupp et al. (2004) have evaluated MPOD and plasma levels of lutein and zeaxanthin in 10 cystic fibrosis patients. Their results indicated that in comparison to age- and gender-matched healthy subjects these patients had about 50% lower MPOD levels along with total plasma xanthophyll concentrations that on average were as much as 57% lower than those in the control group. This indicates that MP density decreases if the supply of lutein and zeaxanthin is not maintained. The reverse situation, namely, the response of the human organism to supplementation with lutein and zeaxanthin has also been studied, mostly in terms of lutein and zeaxanthin plasma concentrations and MPOD.

13.7.2 Responses of Plasma Concentration

If xanthophylls are ingested, they appear in the plasma within a few hours and definitive responses of the plasma concentrations of lutein and zeaxanthin can normally be observed. These responses have systematically been evaluated by many authors with Thürmann et al. (2005) being one of the more recent evaluations for lutein and Hartmann et al. (2004) for zeaxanthin. To study the pharmacokinetic characteristics of the xanthophylls, the authors supplemented 16 subjects with daily lutein doses of 4.1 or 20.5 mg/day of lutein for 42 days and concomitantly measured their plasma concentrations. This supplementation led to about 3.5- and 10-fold increases of lutein plasma concentrations, respectively. The study also provided evidence that the concentration of the other plasma carotenoids measured remained unchanged, despite the drastic increase in lutein plasma concentrations. Later, the same formulation of lutein and zeaxanthin was studied in the LUXEA Study by Schalch et al. (2007), using doses of 10 and 20 mg/day. Combining the data of these two studies gave the opportunity to evaluate the general pharmacokinetic characteristics of the xanthophylls and it could be deduced that the response of plasma concentrations to supplementation with xanthophylls was dose-dependent, following saturable kinetics, so that the highest plasma concentration that can be attained by xanthophyll supplementation is limited at about 3 μmol/L (v_{max} in Figure 13.5). Plasma responses to supplementation with xanthophylls including (meso)-zeaxanthin were recently measured by Thurnham et al. (2008), who supplemented 19 subjects with a mixture of lutein, (meso)-zeaxanthin, and zeaxanthin for 3 weeks. The authors concluded that (meso)-zeaxanthin was less well absorbed than (3*R*,3′*R*)-zeaxanthin from the administered mixture but they did not measure MPOD levels as Bone et al. (2007) did (see Section 13.7.3).

While the presence of lutein and zeaxanthin in plasma is obviously an important condition for the accumulation of these substances in the retina, the question is whether higher plasma levels lead also to higher macular levels. This relationship between plasma concentrations of xanthophylls and MPOD has been studied by many investigators. Two recent publications (Mares et al. 2006, Nolan et al. 2007a) have analyzed this relationship in two different cohorts (one from the United States and one from Ireland, each with about 700 subjects), and have reported statistically significant direct correlations between lutein and zeaxanthin plasma concentrations and MPOD, confirming independent earlier reports from other authors (i.e., Bone et al. [2003]), that higher plasma levels of lutein and zeaxanthin are correlated with a denser MP. The next strongest predictor for MPOD was dietary intake of xanthophylls.

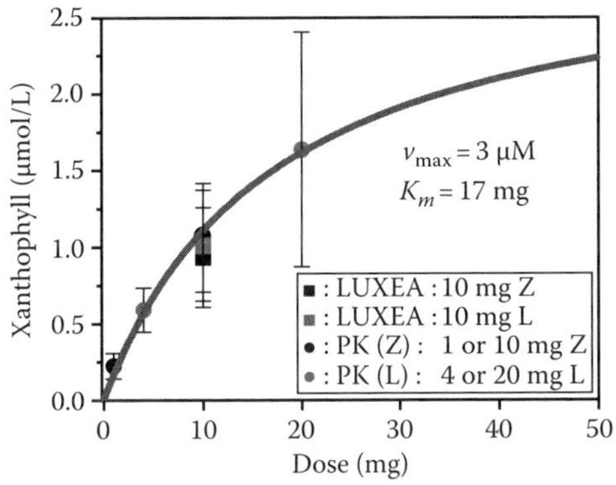

FIGURE 13.5 Plasma responses to supplementation with xanthophylls. This figure is a compilation of steady-state plasma concentrations reached after supplementation with lutein or zeaxanthin. The data are from three human studies that used the xanthophylls in identically formulated preparations. (Schalch et al. 2007: LUXEA study, black and grey squares, 10 mg zeaxanthin or lutein, respectively; Thürmann et al. 2005: grey circles, 4 and 20 mg lutein; Hartmann et al. 2004: black circles, 1 and 10 mg zeaxanthin) The curve perfectly follows Michaelis–Menton kinetics, with v_{max} (3 μM) being the highest steady-state concentration that theoretically could be reached and K_m (17 mg) being the xanthophyll dose at which half of this steady-state level would be reached. The data indicate that from a pharmacokinetic perspective lutein and zeaxanthin appear to be identical. (Courtesy of Dr. W. Cohn.)

13.7.3 Responses of MPOD

Since the original work in monkeys (Malinow et al. 1980), numerous articles have presented evidence that the plasma concentrations of lutein and zeaxanthin as well as MPOD can be modulated by diet, especially by the intake of fruits and vegetables (Hammond et al. 1997), eggs (Handelman et al. 1999), or lutein and zeaxanthin supplements (Schalch et al. 2007).

In terms of MPOD responses, human supplementation studies with xanthophylls have yielded a wide range of results, which are characterized by a substantially larger variability than that of plasma responses: MPOD values can vary by one order of magnitude or more. Differences between measuring techniques, study length, and subject training are all factors that could contribute to this variability, as well as inherent differences between individuals including genetic factors and differences in the specific xanthophyll product used for supplementation.

Xanthophyll supplementation was primarily studied using lutein. The first supplementation study with lutein during which the time course of MPOD was also followed was reported by Landrum et al. (1997). Although it was a rather small study, the supplement of 30 mg for 140 days showed increases of around 40%. A substantial number of other studies have been done since then. In these studies, maximal MPOD increases of over (Trieschmann et al. 2007) or almost 50% (Stringham and Hammond 2008), around 40% (Landrum et al. 1997), or no responses (Yolton et al. 2002, Cardinault et al. 2003) were reported. Trieschmann et al. (2007) supplemented about 100 subjects with 12 mg/day of lutein in ester form (together with 1 mg zeaxanthin) and reported MPOD increases of over 50%, while Stringham et al. (2008) found an MPOD increase of almost 50% when supplementing 40 subjects with 10 mg nonesterified lutein and 2 mg zeaxanthin for 6 months. More modest increases of 15%–23% were reported by Aleman et al. (2001) with 20 mg lutein for 6 months, Berendschot et al. (2000) with 10 mg lutein for 3 months, Duncan et al. (2002) with 20 mg for 6 months, Schweitzer et al. (2002) with 6 mg for 40 days, and Schalch et al. (2007) with 10–20 mg for up to 1 year.

In summary, lutein supplementation has been shown to raise MPOD in the majority of subjects supplemented. In the few cases where no increases of MPOD could be observed, longer periods of supplementation (>6 months) and higher dosages (>20 mg/day) may be necessary to cause significant responses. The question of whether nonresponders exist or whether these individuals merely respond more slowly to supplementation than normal remains an unanswered question.

In contrast, supplementation with zeaxanthin or (meso)-zeaxanthin has received much less attention and there is a paucity of published data. Modest increases of plasma concentrations and MPOD increases after supplementation with (meso)-zeaxanthin were reported by Bone et al. (2007). In another study, the authors supplemented two subjects with 30 mg/day of pure (*R*,*R*)-zeaxanthin extracted from *Flavobacteria* for 4 months and reported statistically significant MPOD increases of about 10%. These MPOD increases were smaller than those observed with lutein in an earlier study of the same authors (Landrum et al. 1997). However, this most probably was due to differences in formulation of the zeaxanthin and lutein. In another slightly larger study, eight subjects were supplemented with pure zeaxanthin. MPOD increases could be identified by heterochromatic flicker photometry (HFP) in five of the subjects, whereas at the end of supplementation, MPOD values below baseline and thus a decrease of MPOD, were reported in the other three (Garnett et al. 2002). Schalch et al. (2007) have supplemented pure, chemically synthesized zeaxanthin and reported a corrected MPOD increase of 15% as measured by HFP; the correction of MPOD was for the increase in pigment concentration in the parafoveal region. Pigment increases such as these in the parafoveal location that HFP uses as reference can cause MPOD to appear to decline, which was observed. This may indicate a similar situation occurred in the study reported above (Garnett et al. 2002). This observation of parafoveal pigment increases upon supplementation with xanthophylls is consistent with results from several other supplementation studies. In one of them (Wenzel et al. 2007), three subjects consumed 30 mg lutein and 2.7 mg zeaxanthin/day for 120 days. The authors recorded MPOD by HFP at four discrete eccentricities from 20′ to 120′. In all three subjects MPOD increased significantly at the two most central measurement loci. However, a trend of increasing pigment at the reference location at 7° eccentricity was observed as well, suggesting that parafoveal pigment increases may not be specific to zeaxanthin but can also be observed with lutein in special situations in particular when supplementing at higher doses (Johnson et al. 2008b). This phenomenon was also reported in an epidemiological study that investigated the age dependency of MPOD (Berendschot and van Norren 2005).

13.7.4 Modulators of MPOD

In addition to supplementation or dietary intake of the xanthophylls, several other modulators that influence the MPOD response of subjects to supplementation with xanthophylls were reported (Mares et al. 2006). Larger waist circumference and the presence of diabetes predicted a decrease of MPOD. In contrast to earlier findings, iris color was not related to MPOD. No dependence of MPOD on age was revealed in this study but this may be because of its lower age limit of 53 years.

Among the possible determinants of MPOD, age as the most evident risk factor for AMD is probably the most important. The earliest report on an observation relevant to the presence or absence of the yellow MP at birth is from Schwalbe (1874), who stated that the pigment is rarely present at birth. This is consistent with the observation of Bone et al. (1988), based on HPLC determination of xanthophylls in the central retina of postmortem eye, that the xanthophylls are present in prenatal eyes but that they do not form a visible yellow spot until about 6 months after birth. Bone et al. (1988) also report that the youngest eyes have lutein as the predominating xanthophyll and that it is only in older eyes that zeaxanthin becomes more dominant. Their data suggest that the retinal content of xanthophylls is independent of age. In contrast to this conclusion, Nolan et al. (2007b) have reported a decline of MPOD with age in an Irish population ($n = 800$). In the same article they also have reviewed 23 studies published earlier and pointed out that in 14 of these studies a decline of MPOD was found. Together with their own study and a more recent study from Japan (Obana et al. 2008), it appears that the majority of studies do indeed suggest a decline of MPOD with age, but a final conclusion on this topic has not yet been reached, mainly because the results may be dependent on the method chosen for MPOD measurement.

MPOD may, at least partly, be dependent on genetic factors, as suggested by a recent study (Liew et al. 2005). A genetic linkage may not be the primary determining factor, however: MPOD appeared to be different in monozygotic twins (Hammond et al. 1995), who apparently can have different levels of MPOD depending on differences in their specific environment, particularly with regard to their diet.

From 2005 onward, several research groups have independently identified genes that appear to be strongly linked with the risk for AMD (Marx 2006). Whether and how the presence or absence of these genes is linked to an individual's MPOD remains to be established. A recent publication suggests that plasma concentration of only lycopene and β-cryptoxanthin, but not lutein and zeaxanthin, differ in subjects bearing different single nucleotide polymorphisms of genes involved in lipid metabolism (Borel et al. 2007). Furthermore, ethnicity seems to influence plasma levels of carotenoids (Kant and Graubard 2007) as well as MPOD levels and distribution (Wolf-Schnurrbusch et al. 2007). The question remains whether dietary or supplemental intake of the macular xanthophylls can influence the course of the disease in subjects who possess one or more genes that have been identified as risk factors for AMD.

13.7.5 Supplementation Experiments in Monkeys

A series of publications has reported results of supplementation experiments with monkeys. Motivated by an earlier investigation (Malinow et al. 1980), the authors supplemented groups of carotenoid-depleted rhesus monkeys with either pure lutein or pure zeaxanthin at doses of 2.2 mg/kg/day (equivalent to 12–24 mg of carotenoid/day and animal) for 6–12 months (Neuringer et al. 2004). Plasma concentrations of lutein rose faster, to higher initial levels, than those of zeaxanthin but by approximately 16 weeks both had stabilized at comparable levels of about 0.8 μmol/L. This was equivalent to a 10-fold increase compared with plasma levels of normal chow-fed animals. MPOD increased gradually and variably in both groups. However, by 16 months MPOD had approached levels of only around 50% of that seen in monkeys that were fed normal monkey chow throughout their lives. The lifelong carotenoid deprivation may have impaired the retina's natural ability to accumulate xanthophylls to its full extent during the supplementation period.

13.8 THE FUNCTIONAL ROLE OF XANTHOPHYLLS

The observation that lutein and zeaxanthin occur in the highest concentration in the macula soon raised expectations that the macular xanthophylls may be essential in maintaining structure and function of the retina by contributing not only to risk reduction of macular diseases but also to improving visual performance of the healthy eye, which was the original hypothesis to explain the presence of the macular yellow pigment as mentioned previously.

13.8.1 First Human Supplementation Studies with Xanthophylls

Starting in the late 1940s, just after it was realized that xanthophylls occur in the retina and that they are provided by dietary intake, a number of supplementation studies were conducted with "Helenien," a lutein–dipalmitate ester that had been discovered in the flower *Helenium autumnale* by Nobel Laureate R. Kuhn (Kuhn and Winterstein 1930). The helenien used for supplementation was extracted from the marigold flower *Tagetes patula flore pleno*, and not from *Tagetes erecta*, the main commercial source of lutein today. Under the name "Adaptinol," helenien was commercialized by Bayer from the late 1940s on (Cüppers and Wagner 1950, Tarpo and Cucu 1961). As the name implies, the effect on dark adaptation was the main target of its application. The mechanistic basis of this effect had been evaluated in frogs using electroretinograms (ERG) (Mueller-Limmroth et al. 1958), measuring retinal oxygen consumption (Schmitt et al. 1959), and the determination of retinal sodium and potassium contents (Berges et al. 1959). However, the scientific basis of this application remained weak and was derived from the erroneous idea that the action of lutein was similar to vitamin A in the visual process (von Studnitz and Loevenich 1947). Lutein at that time was believed to be a precursor of vitamin A like β-carotene. Today, however, we know that lutein and zeaxanthin do not have any provitamin A activity (Weiser and Kormann 1993). Nevertheless, in a substantial number of studies improving effects on dark adaptation could indeed be demonstrated (Monje 1948, Cüppers and Wagner 1950, Klaes and Riegel 1951, von Studnitz 1952, Andreani and Volpi 1956, Cuccagna 1956, Mosci 1956, Mueller-Limmroth and Schmidt 1961b, Cilotti 1963), while other authors (Wuestenberg 1951, De Ferreira and Da Maia 1956, Pfeifer 1957) were not able to confirm these effects. The most frequently used doses ranged from 5 to 20 mg of the ester and were taken over periods of 2–6 weeks. Hayano et al. (1959) appear to have been the first and only scientists who followed adaptinol treatment with measuring plasma concentrations of lutein. They did this first in frogs and presented evidence that parenteral administration of helenien increased its levels in liver and blood. In humans, they found that adaptinol supplementation increased the lutein plasma level in normal subjects and that dark adaptation improved proportionally. Interestingly, the plasma lutein levels of patients with retinitis pigmentosa (RP) were initially very low and clinical improvement in dark adaptation could only be demonstrated in patients who showed an increase of plasma lutein levels. Adaptinol was also tested in subjects with various other ophthalmic diseases, in particular night blindness (Oka 1955, Andreani and Volpi 1956, Cuccagna 1956, De Ferreira and Da Maia 1956, Mosci 1956, Hayano, Koide et al. 1959, Sole et al. 1984), but also myopia (Asciano and Bellizzi 1974, Sole et al. 1984) and tapeto-retinal degenerations (Mueller-Limmroth and Kueper 1961a), with mixed results being reported. After 1984, interest in helenien and adaptinol appears to have vanished as no respective publications can be retrieved after then.

The above-mentioned early supplementation studies with xanthophylls did not measure the changes in MPOD associated with supplementation probably because an easy-to-use technique for its noninvasive measurement in the human retina eye was not available at that time. The first apparatus for this purpose had been described in 1953 by deVries et al. (1953) and it is only since the 1970s that publications can be found that report on systematic measurements of MPOD in the human retina.

13.8.2 Lutein's and Zeaxanthin's Role in Risk Reduction of AMD

For hundreds of years, the dried fruit of the Chinese wolfberry (also called *Fructus lycii*), "Gou Qi Zi" (*Lycium barbarum*) has been a constituent of traditional Chinese herbal medicine for the treatment of visual disorders (Huang 1993, Benzie et al. 2006). This probably was the first recorded "medicinal use" of one of the macular xanthophylls. The dried fruit contains high levels of zeaxanthin–dipalmitate, up to 1.1 g/kg (Inbaraj et al. 2008), making zeaxanthin a logical lead compound for this plant, which is not only prescribed as a medicine but also commonly used in home cooking in China. Plasma levels of zeaxanthin increase when ingesting this berry (Breithaupt et al. 2004). Furthermore, MPOD levels increased significantly in 7 volunteers who received daily doses of about 20 mg zeaxanthin via ingesting this berry for 3 months (Leung et al. 2001).

AMD is, as the name implies, an age-related degenerative condition of the macula. If the macula becomes dysfunctional, visual tasks requiring high resolution such as recognizing faces or reading become progressively more difficult until, in the late stages of advanced AMD, they become impossible. Advanced AMD is the leading cause of legal blindness in the United States and other developed countries, and it is expected that the prevalence of this disease will drastically increase, and may reach close to 3 million individuals within less than 20 years in the United States alone (Friedman et al. 2004). Evidence of AMD is first observable for most individuals between the ages of 55 and 65 with the build up of characteristic yellow deposits within and around the macular area. These deposits, called drusen, contain lipofuscin and its derivatives. Most people with these early changes still have satisfactory vision but they are at risk of developing advanced AMD. Advanced AMD, which is responsible for profound vision loss, has two forms: dry and wet. Central geographic atrophy, the "dry" form of advanced AMD, causes these problems through loss of photoreceptors and cells supporting the photoreceptors in the central part of the retina. Currently, no treatment is available for this condition, but recently a probable gene may have been identified (Yang et al. 2008). Neovascular or exudative AMD, the "wet" form of AMD, causes vision loss due to abnormal blood vessel growth (angiogenesis) beneath and into the macula. These newly formed blood vessels are imperfect and blood leaks from them causing blood accumulation under the retina, which leads to irreversible damage to the functional layers of the macula. Finally, vision is completely lost if the condition is left untreated. An effective but very expensive treatment regimen for this neovascular ("wet") form of AMD has recently become available. However, an intervention that would prevent or at least slow the progression of this disease would certainly be a welcome alternative (de Jong 2006).

The etiology of AMD is not completely understood, but some ideas regarding its pathogenesis have been developed. Photoreceptors are constantly exposed to (photo)-oxidative damage in the environment of the retina, which is characterized by the simultaneous presence of light and oxygen. As a consequence, they are damaged and become dysfunctional. Before new photoreceptors can be formed, dysfunctional photoreceptors must be disposed of. This task is accomplished by the highly metabolically active RPE cells. It is estimated that during a period of about 10 days, each RPE cell has to phagocytose, digest, and eliminate into the blood flow about 50 photoreceptors. Thus, during 60 years, more than 100,000 photoreceptors are to be processed by a single RPE cell. It is not surprising that during this very dynamic metabolic activity, digestion, and elimination of spent photoreceptors is not always complete and cell debris accumulates, mostly in the form of lipofuscin and its derivates, causing a progressive malfunctioning and eventual death of not only the RPE but also of the photoreceptor cells (Sun and Nathans 2001).

Logical targets for risk reduction and prevention of AMD appear to include support to the RPE cells so that they are better able to cope with their exceptional metabolic burden, the reduction of the generation of new but imperfect blood vessels by inhibiting angiogenesis, reduction of blue light which has the highest damage potential of the visible light reaching the macula, and reduction of oxidative damage by antioxidants.

The evidence available to date indicates that lutein and zeaxanthin could contribute to achieving the last two objectives, namely, the reduction of actinic insults caused by blue light and quenching reactive oxygen species. This follows from the dual presence of xanthophylls in the macula: their prereceptoral location and their presence within the outer segments themselves, as discussed in Section 13.5.

Recent experimental evidence indicates that lutein and zeaxanthin may be instrumental in maintaining a healthy RPE. Rhesus monkeys raised on a xanthophyll-free diet since birth exhibited a distorted profile of the RPE cells in the macula, with a reduced cell density in the center of the fovea, whereas normally the maximum density of RPE cells is to be found there (Leung et al. 2004). Supplementation of the animals with lutein or zeaxanthin altered the RPE cell profile in a way that is consistent with a migration of RPE cells toward the fovea, and appears to have induced a "normalization" of the RPE cell profile. In a recent publication (Izumi-Nagai et al. 2007), a state of choroidal neovascularization was induced in mice by laser photocoagulation and it was shown that mice pretreated with lutein were protected from this neovascularization and that a number of inflammatory biomarkers were suppressed. Furthermore, in diabetic mice treated with zeaxanthin, the diabetes-induced retinal oxidative damage could be reduced along with a decrease of VEGF (Kowluru et al. 2008).

The main parameter used to assess the amount of xanthophylls in the retina is the MPOD. Recently, a comprehensive review (Nolan et al. 2007b), which demonstrated that age, smoking, and a family history of AMD were all correlated with a reduced MPOD in a statistically significant manner, was published. Although these correlations do not necessarily signal a causal relationship they provide suggestive evidence for the contribution of xanthophylls to risk reduction of AMD. However, the possible contribution of lutein and zeaxanthin to risk reduction of AMD is supported by experimental, epidemiological, and clinical evidence as described in the following sections.

13.8.2.1 Experimental and Epidemiological Evidence

The contribution of lutein and zeaxanthin to the risk reduction of AMD is mainly based on two properties of the xanthophylls: one is their blue-light absorption and the other is their antioxidant property. It was estimated that the MP can attenuate up to 40% of the blue light that hits the macula (Krinsky et al. 2003). The antioxidant properties of the macular xanthophylls have been demonstrated many times. They can quench singlet oxygen as well as other reactive oxygen intermediates (Krinsky and Deneke 1982) and an oxidized metabolite of lutein, 3′-dehydro-lutein, has been identified in plasma and in the retina (Khachik et al. 1997a).

Xanthophylls can further inhibit the peroxidation of membrane phospholipids (Lim et al. 1992) and reduce photooxidation of lipofuscin fluorophores (Kim et al. 2006), which are implicated in the pathogenesis of AMD (Sparrow and Boulton 2005). Furthermore, it was shown that light-induced damage to photoreceptors was reduced in quails fed zeaxanthin, with the number of apoptotic photoreceptor cells being inversely related to the concentration of zeaxanthin in the retina (Thomson et al. 2002).

Results of human epidemiological studies investigating the relationship of MPOD and dietary or supplemental intake of lutein and zeaxanthin with the risk of AMD are somewhat variable, presenting a mixed picture and not all studies were able to generate supportive evidence (van den Langenberg et al. 1998, Flood et al. 2002, Cho et al. 2004). This is not surprising in view of the fact that AMD is a degenerative disease that develops over a lifetime with many confounding factors prevailing making an epidemiological assessment difficult. Early data indicated that subjects with low dietary intake (Seddon et al. 1994) or plasma levels (EDCC Study Group 1993) of macular xanthophylls had a higher risk for neovascular AMD. These results are consistent with a more recent evaluation by Snellen et al. (2002) of the prevalence of AMD in relation to antioxidant and xanthophyll intake. They reported a dose–response relationship with higher xanthophyll intakes exhibiting lower prevalence rates. A more recent epidemiological study investigating the relationship of plasma levels of xanthophylls and the risk for AMD (Delcourt et al. 2006) indicated that subjects in the south of France had a lower risk for AMD if they had higher plasma concentrations particularly of zeaxanthin, confirming results of Gale et al. (2003) in a U.K. population. An epidemiological evaluation of data from the Age-Related Eye Disease Study (AREDS) indicated that subjects in the highest quintile of dietary lutein and zeaxanthin intake had a statistically significant lower risk of developing different manifestations of AMD (AREDS Research Group et al. 2007).

Some studies have reported lower MPOD in AMD eyes or in eyes at risk of developing AMD (Schweitzer et al. 2000, Beatty et al. 2001, Bernstein et al. 2002, Obana et al. 2008) by using different noninvasive measuring techniques in the living eye. In contrast, Bone et al. (2001) have determined lutein and zeaxanthin directly in postmortem retinal tissue samples by HPLC from normal subjects and subjects with AMD. The results demonstrated that the average lutein and zeaxanthin levels were lower in the AMD retinas than in the normal retinas. Those individuals with the highest quartile of xanthophyll concentration in the outer annulus had an 82% lower risk for AMD when compared to those in the lowest quartile (Landrum et al. 1999b). Because this relationship was found in the outer annulus which is relatively unaffected by AMD, this observation lends support to the conclusion that the observed reduction of MPOD may be preceding the disease rather than resulting from the disease. Thus, low carotenoid concentrations in the retina can be a risk factor for AMD. How and to what extent the quantitative amount of carotenoids in the macula modulates an individual's AMD risk is still open to debate.

The question of how exposure to sunlight contributes to the etiology of AMD was recently investigated together with plasma concentration of antioxidants including lutein and zeaxanthin. This was done in course of the EUREYE study conducted in 4750 subjects older than 65 years from across Europe. The participants were interviewed for their lifetime sunlight exposure and gave a plasma sample for biochemical analyses. The results of the study indicated a strong inverse association of sunlight exposure and neovascular AMD, particularly in subjects with low antioxidant plasma levels with odds ratios being as high as 3.72 for subjects low in vitamins E and C and zeaxanthin (Fletcher et al. 2008). Furthermore, odds ratios for AMD in this study were generally increased for almost every combination of lower lutein and zeaxanthin plasma concentrations.

Overall, a substantial number of epidemiological and experimental studies suggests that lutein and zeaxanthin could contribute to risk reduction of AMD. Two recent articles in this respect appear to be particularly supporting because they outline that many AMD risk factors are associated either with a relative dietary lack of key nutrients including lutein and zeaxanthin (Nolan et al. 2006), or with a reduced MPOD (Nolan et al. 2007b).

13.8.2.2 Clinical Evidence

The question whether lutein and zeaxanthin can contribute to lowering the risk for AMD cannot be answered unequivocally by epidemiological studies. Only randomized controlled trials (RCTs) during the course of which xanthophylls are supplemented in a double-blind, placebo-controlled, and randomized manner, and in which results are evaluated according to clear predefined efficacy criteria (Seddon and Hennekens 1994) have the potential to provide definitive answers. The specific long-term time-course and intricate nature of AMD make the design of such studies difficult, however.

To date, no results from large RCTs evaluating whether supplementation with lutein and/or zeaxanthin influences disease-specific endpoints has been published. One reason for this is that lutein and zeaxanthin supplements have only recently become available for human consumption. In 1992, the National Eye Institute initiated the AREDS in 3600 people (AREDS Research Group 1999). The results indicated that regular ingestion of a dietary antioxidant supplement containing vitamins E and C, β-carotene, zinc, and copper could reduce the progression of advanced AMD relative to controls (AREDS Research Group 2001). Recently, another RCT (AREDS II) was initiated by the NEI. Early in 2007, this trial began recruiting the planned 4000 subjects. The supplements for this study provide daily doses of 10 mg lutein and 2 mg zeaxanthin in combination with long-chain polyunsaturated fatty acids (LCPUFAs). The effects of combining LCPUFAs with xanthophylls has been evaluated by at least two research groups in the meantime (Huang et al. 2008, Johnson et al. 2008a,b) with results indicating that addition of LCPUFA did not change the plasma levels of the supplemented xanthophylls.

What have been published are small-scale lutein supplementation studies. One (Dagnelie et al. 2000) reported significantly improved visual function in 16 patients with congenital retinal degenerations who were supplemented with 20–40 mg lutein/day for 26 weeks. A case–control study found improvements in a number of visual function tests, including contrast sensitivity in patients who consumed lutein-rich spinach at an intake level of 30 mg of lutein/day for 26 weeks (Richer 1999). A larger and longer double-blind, placebo-controlled supplementation study with lutein and an antioxidant mixture in 90 subjects, showed statistically significant improvements in selected visual functions of AMD patients who took either 10 mg/day of lutein alone, or 10 mg/day of lutein incorporated in an antioxidant formula, compared with those taking placebo (Richer et al. 2004). In another study, 21 patients diagnosed with RP and 8 normal subjects were supplemented with a daily dose of 20 mg of lutein for a period of 6 months (Aleman et al. 2001). Plasma lutein concentrations increased in all participants but MPOD, as measured by HFP, increased significantly only in half of them. These "retinal responders" had a less severe course of the disease than the nonresponders. Inner retinal thickness, measured by optical coherence tomography, correlated positively with the level of MP density at 0.5° eccentricity, a relationship that was significant for patients, but not for healthy controls. In contrast, results of a recent study indicate that central retinal thickness is indeed directly correlated with MPOD in healthy subjects (Liew et al. 2006). Parisi et al. (2008) supplemented 15 early AMD patients with an antioxidant mixture providing, among other substances, daily amounts of 10 mg lutein and 1 mg zeaxanthin for 12 months. When comparing the patients' multifocal ERG recordings with those of an untreated control group they noted that supplementation had induced an improvement of retinal function, which was specific for the central retina but was not noted in peripheral retinal areas. While this is a preliminary finding in a small group of patients, it indicates that lutein and zeaxanthin supplementation may have had a "therapeutic" effect that could be measured by a functionally important parameter. The authors conclude that the improved ERG signal is indicative of a functional improvement of preganglionic elements. If this were true, a retinal element other than the photoreceptors and the RPE would have been positively altered by lutein supplementation for the first time. These results are consistent with an earlier study (Falsini et al. 2003) using a higher daily dose of lutein (15 mg/day) in a similar antioxidant combination. The main measurement parameter in this study was the macular, cone-mediated focal electroretinogram (FERG), another assay of retinal function. The results indicated a significant improvement of the FERG variables in supplemented subjects when compared to nonsupplemented control individuals. For both studies mentioned above, it would have been interesting to relate the measured effects to the patients' MPOD responses, however, neither plasma nor retinal levels of xanthophylls were measured in these studies.

In summary, while only the NEI initiated AREDS II study is a large enough RCT to have the potential to provide definitive evidence as to whether the macular xanthophylls can indeed reduce the risk of AMD, the evidence available to date that lutein and zeaxanthin could contribute to this is not only biologically plausible but also supported by various experimental, epidemiological, and small-scale clinical studies. Although the benefits of lutein and zeaxanthin in this respect may be moderate to small, their safety is well documented.

A research direction based on a hypothesis over 200 years old, but only recently starting to emerge, proposes to evaluate the role of the MP for optimal visual performance, thus investigating lutein's and zeaxanthin's effects beyond risk reduction of retinal diseases.

13.8.3 The Xanthophyll's Emerging Roles in Optimizing Visual Performance

13.8.3.1 General

Long before the chemical identity of the "macular yellow" was determined, there were hypotheses about its role in vision. As early as 1866, it was conjectured that the color of the "macular yellow" might be physiologically important for human vision (Schultze 1866), purely on grounds of considerations that it was a prereceptoral blue-light filter which could reduce chromatic aberration and would thereby improve visual acuity, reduce blue haze, glare, and dazzle, and enhance contrast (Holm 1922). Thus, ideas about the functions of MP in the healthy eye originated much earlier than the ideas that it may contribute to risk reduction of AMD. From an evolutionary perspective, it is hard to understand why nature should have prepared to control a disease that only becomes overt far after the reproductive phase. In contrast, if MP contributed to improving visual performance, this could indeed have been essential for the survival of an early hunter/gatherer population, particularly at twilight when hunting activities predominantly occurred. Based on the similarities of the light-absorption characteristics of MP and the action spectrum of rhodopsin, a recent hypothesis suggests that increased MP could contribute to better visual acuity by reducing the activation of rod photoreceptors. This effect would be most important in the mesopic (twilight) range when the photoreceptors are adapting from photopic (high light intensity) to scotopic (low light intensity) conditions. During this transition, both rods and cones are active but the quality of the image is degraded by the contribution of rods, with their poor contrast sensitivity and resolving power (Kvansakul et al. 2006) as compared to the contribution of cones. Reducing the rod contribution would increase visual performance.

The current ideas about the action of the MP on visual performance, which were recently reviewed (Loughman et al. 2007), have previously been grouped into three separate hypotheses: the "acuity hypothesis" and "visibility hypothesis" (Wooten and Hammond 2002), and the "glare hypothesis" (Stringham and Hammond 2007).

13.8.3.2 The Acuity Hypothesis

The idea that MP could improve visual acuity was first mentioned by Max Schultze (1866) and systematically investigated by Reading and Weale, who demonstrated that an ideal intraocular filter, which would eliminate chromatic aberration has light-absorption characteristics identical to those of the MP (Reading and Weale 1974). The focal length of the eye's optic media decreases with wavelength. This effect, chromatic aberration, results in an imperfect retinal image having prismatic, colored fringes. In other words, if the eye is in focus for green light, the blue parts of an image are focused in front of the retina, whereas the red parts are focused behind the retina (Reading and Weale 1974). Chromatic aberration is much greater for blue light than for the longer wavelengths of the spectrum. At 460 nm, the dominant wavelength of the blue sky and the peak absorption of MP, the aberration of blue light amounts to −1.2 dioptres (Hammond et al. 2001). Visual acuity and contrast sensitivity are related parameters that both contribute to the resolving power of the eye. Visual acuity is a measure of the smallest angle between two points subtended at the retina, or the distance at which two lines can be distinguished as separate. In a contrast sensitivity test, the subject views sinusoidal gratings covering a range of spatial frequencies and the contrast ratio is adjusted for each until the bars can only just be discriminated. The ability of subjects to demonstrate high visual acuity or contrast sensitivity, assuming their refractive errors have been corrected, will depend on a variety of factors such as pupil size, cone density, and clarity of the optic media. Not surprisingly, visual acuity and contrast sensitivity in healthy eyes tend to decrease with age. Since carotenoids, in particular lutein and zeaxanthin, may also be associated with a reduction in the incidence of cataracts (Moeller et al. 2000) and therefore a preservation of the clarity of the lens, supplementation with lutein or zeaxanthin may additionally assist in the maintenance of visual acuity.

Supplementation with lutein at 20 mg/day for up to 1 year was shown to significantly improve contrast acuity, a combined parameter from visual acuity and contrast sensitivity (Kvansakul et al. 2006). This was the first controlled supplementation study with lutein and zeaxanthin that systematically studied the effects of supplementation on visual performance in healthy subjects. Although the study was small, the results support the classical hypotheses that MP may influence vision. Furthermore, the study provided evidence for a reduction of intraocular scatter by supplementation with lutein. In addition to these data in healthy subjects, there is also limited evidence that lutein supplementation can improve visual acuity as measured by visual-acuity charts in subjects with degenerative ocular diseases, such as reported by Richer et al. for AMD patients with 10 mg lutein/day (Richer et al. 2004), by Dagnelie et al. for RP patients with 40 and 20 mg/day (Dagnelie et al. 2000), and Olmedilla et al. for cataract patients with 6 mg/day (Olmedilla et al. 2001). In contrast, an epidemiological study investigating the relationship of MPOD to gap resolution acuity concluded that it is unlikely that MP can improve visual acuity by reducing the effects of chromatic aberration (Engles et al. 2007) seemingly supporting an opinion of Weale (2007) based on theoretical considerations. However, the investigation was done in photopic conditions and did not use supplementation. Therefore, its results cannot be generalized to mesopic lighting situations (Kvansakul et al. 2006) where data, which were generated by specifically supplementing lutein and zeaxanthin, indeed support the acuity hypothesis. Furthermore, as pointed out by Lougham et al. (2007), there are numerous other limitations in the Engles et al. study, weakening its conclusion that MP cannot influence visual acuity.

13.8.3.3 The Visibility Hypothesis

In outdoor situations, the scattering of light by large ("Mie" scatter) and small ("Rayleigh" scatter) particles is responsible for the generation of a phenomenon called "blue haze." The blue color of this haze arises because blue light is more heavily scattered than the other wavelengths of visible light. The consequence is that targets viewed outdoors often appear reddish being surrounded by a blue background, thus being reduced in contrast with a resulting reduced visibility. Reduction of the surrounding blue haze by MP could increase the target's contrast and in turn its visibility (Wooten and Hammond 2002). Wooten and Hammond have modeled this situation quantitatively and concluded that subjects with high MP levels could see up to 30% further than subjects with low levels, which could be important for, among others, civil and military aviation applications.

13.8.3.4 The Glare Hypothesis

When subjects are exposed to light, they can report pain and discomfort, particularly when the intensity of light changes quickly from dim to bright. This response is called photophobia.

Photophobia is not limited to changing levels of brightness but can even be chronic as in migraine headaches, for example. A variety of clinical conditions such as RP and AMD can also cause photophobia. Stringham et al. (2003) investigated its dependency on wavelength and found that photophobia was predominantly induced by light of shorter wavelengths (blue light). Wenzel et al. (2006) have measured MPOD and its relationship to photophobia light threshold and reported that the measured thresholds are inversely correlated with MPOD. In addition to photophobia, bright light can induce the sensation of glare. Sensitivity to glare is often exacerbated by increasing age and by diseases of the lens that result in increased light scattering within the eye. Glare sensitivity may be assessed by measuring contrast sensitivity in the presence of a nearby glare source, for example, a pair of halogen lamps that simulate the headlights of an oncoming car. In 36 healthy non-supplemented subjects, MPOD was measured by HFP and sensitivity to glare was measured by assessing their photostress recovery time, the time span until vision returns after the subjects had been "blinded" by a bright glare light. It was found that photostress recovery time was significantly shorter for subjects with higher MPOD levels (Stringham and Hammond 2007).

These correlational data were later extended by supplementing 40 healthy subjects with a mixture of 10 mg lutein and 2 mg zeaxanthin for 6 months and again measuring photostress recovery time. Supplementation increased MPOD levels on average by 35% and along with this MPOD increase photostress recovery time was significantly ($p = 0.01$) reduced (Stringham and Hammond 2008). Although the study was not placebo-controlled or randomized, together with the results of the correlational study mentioned above, its data strongly support an inverse relationship of MPOD and photostress recovery time. It is possible that increasing the level of MP would diminish the amount of scattered blue light reaching the photoreceptors, and this might also result in lowered sensitivity to glare (Hammond et al. 2001). However, light scatter within the eye has been demonstrated to be independent of wavelength (Whittaker et al. 1993). Thus, the scattered longer wavelengths would not be removed. This may be the reason why supplementation with lutein, zeaxanthin, or a combination of both carotenoids was consistently shown to reduce intraocular light scatter in healthy eyes, but not at a level of statistical significance (Kvansakul et al. 2006).

13.8.4 Possible Actions of Lutein and Zeaxanthin Beyond the Retina

The retina had been named an "approachable part of the brain" (Dowling 1987) and indeed emerging data suggest that lutein and zeaxanthin supplementation can have effects on the brain and on cognitive performance. Generally, this appears plausible because of the natural occurrence of lutein and zeaxanthin throughout the nervous system, particularly in locations relevant for cognitive and visual processing (Craft et al. 2004). In this context, Johnson et al. (2008b) have recently supplemented 11 elderly subjects with 12 mg lutein/day for 4 months and reported statistically significant improvements in verbal fluency and memory scores along with marked increases of MPOD. In an epidemiological study, Renzi et al. have investigated the relationship of MPOD and cognitive function in 118 older adults. MPOD turned out to be the strongest and most consistent correlate of cognitive function across all tested indices, although in absolute terms the xanthophylls accounted for only small but significant proportions of variance (Renzi et al. 2008b).

More specific to the events after electrical signals have been generated in the retina are observations that critical flicker fusion thresholds, a classical measure of central processing speed relevant to the dynamic functioning of the visual system, are directly proportional ($p < 0.001$) to MPOD, as first reported by Hammond and Wooten (2005) and confirmed by Renzi et al. (2008a) in a larger population.

13.8.5 Xanthophylls and the Developing Eye

Two recent articles (Hammond and Frick 2007, Zimmer and Hammond 2007) review and discuss the potential importance that lutein and zeaxanthin have for the developing retina. Indeed, the protective properties of the macular xanthophylls may be of particular importance early in life and it was mentioned above that lutein in the eye is already present before birth (Bone et al. 1988, Yakovleva et al. 2007). The lens of infants is virtually clear and therefore transmits unfiltered blue light to the retina (Dillon et al. 2004). It is possible that initial actinic insults to the retina occurring during early childhood and adolescence may lead to retinal diseases later in life and could be reduced if the MPOD of infants were increased.

Breastfed infants are exclusively dependent on the lutein and zeaxanthin content of mother's milk because lutein and zeaxanthin cannot be biosynthesized by the human body as mentioned earlier. In comparison to other carotenoids present in mother's milk, lutein and zeaxanthin were reported to constitute the highest relative amount (Khachik et al. 1997b, Azeredo and Trugo 2008). Their concentrations in mother's milk approximately reflect maternal intake levels of these carotenoids (Canfield et al. 2003, Jackson and Zimmer 2007). Currently, most commercially available infant formulas either do not contain lutein and zeaxanthin at all or only in trace amounts. In this context, an earlier publication (Johnson and Norkus 1995) documented decreasing lutein and zeaxanthin plasma levels in infants who were formula-fed for 1 month after birth.

13.9 THE SAFETY OF SUPPLEMENTED LUTEIN AND ZEAXANTHIN

The safety of supplementation with xanthophylls has been well established. Lutein and zeaxanthin are a natural part of the diet and intake from natural sources is in the range 1–6 mg (Koushik et al. 2006). Furthermore, two trials with focus on safety have been conducted in nonhuman primates with FloraGLO® lutein formulated by DSM (Goralczyk et al. 2002, Khachik et al. 2006), which documented an excellent safety profile of lutein and the smaller (6%–8%) amount of zeaxanthin normally present in natural lutein preparations. Crystalline lutein, the main ingredient of lutein beadlets was given generally recognized as safe status in the year 2001 based on the results of toxicology data. Later, Shao and Hathcock (2006) conducted a formal risk assessment of lutein by analyzing all published human studies during which lutein was supplemented and determined an observed safe level of 20 mg/day while noting that much higher levels of lutein have been used without adverse effects. The safety of lutein and its esterified form was again confirmed recently by a systematic toxicological comparison of the two substances (Harikumar et al. 2008).

Ocular safety of lutein and zeaxanthin supplementation in humans was documented in a human supplementation study involving almost 100 subjects who were exposed to daily doses of 10–20 mg over 6–12 months (Schalch and Barker 2005, Schalch et al. 2007). In the year 2004, the Joint FAO/WHO Expert Committee on Food Additives (JECFA) has set a group ADI of 2 mg/kg body weight/day for lutein and zeaxanthin taken together, which is equivalent to 120 mg of xanthophylls/day for a 60 kg person (summary and conclusions of the 63th meeting of the Joint FAO/WHO Expert Committee on food additives (JECFA), June 8–17, 2004, Geneva, Switzerland).

ACKNOWLEDGMENTS

Julia Bird's valuable input for compiling Figure 13.3 and for reviewing the manuscript is appreciated. We also thank Willy Cohn for providing Figure 13.5.

REFERENCES

Afzal, A. and M. Afzal (2008). Photoprotective carotenoids lutein and zeaxanthin: Their role in AMD. *Curr. Nutr. Food Sci.* **4**(2): 127–134.

Aleman, T. S., J. L. Duncan et al. (2001). Macular pigment and lutein supplementation in retinitis pigmentosa and Usher syndrome. *Invest. Ophthalmol. Vis. Sci.* **42**(8): 1873–1881.

Andreani, D. and U. Volpi (1956). Effect of helenien (adaptinol) on light sense in normal subjects and patients with retinitis pigmentosa (Italian). *G. Ital. Oftal.* **9**: 565–573.

Arden, G. B. and F. M. Barker (1991). Canthaxanthin and the eye: A critical ocular and toxicological assessment. *J. Toxicol. Cutaneous Ocular Toxicol.* **10**: 115–155.

AREDS Research Group (1999). The age-related eye disease study (AREDS): Design implications, AREDS Report No. 1. *Contr. Clin. Trials* **20**: 573–600.

AREDS Research Group (2001). A randomized, placebo-controlled, clinical trial of high-dose supplementation with vitamins C and E, β-carotene, and zinc for age-related macular degeneration and vision loss, AREDS Report No. 8. *Arch. Ophthalmol.* **119**(10): 1417–1436.

AREDS Research Group, J. P. SanGiovanni et al. (2007). The relationship of dietary carotenoid and vitamin A, E, and C intake with age-related macular degeneration in a case-control study, AREDS Report No. 22. *Arch. Ophthalmol.* **125**(9): 1225–1232.

Asciano, F. and M. Bellizzi (1974). Light and colour sensitivity in malignant myopia before and after treatment with helenien and vitamin A (Italian). *Annali Otal.* **100**: 191–202.

Azeredo, V. B. and N. M. Trugo (2008). Retinol, carotenoids, and tocopherols in the milk of lactating adolescents and relationships with plasma concentrations. *Nutrition* **24**(2): 133–139.

Beatty, S., I. J. Murray et al. (2001). Macular pigment and risk for age-related macular degeneration in subjects from a Northern European population. *Invest. Ophthalmol. Vis. Sci.* **42**(2): 439–446.

Belloni, L. (1983). Contribution of Francesco Buzzi to the discovery of the macula lutea and the fovea centralis of the human eye (German). *Gesnerus* **40**(1–2): 23–30.

Benzie, I. F., W. Y. Chung et al. (2006). Enhanced bioavailability of zeaxanthin in a milk-based formulation of wolfberry (Gou Qi Zi, *Fructus barbarum* L.). *Br. J. Nutr.* **96**(1): 154–160.

Berendschot, T. T., R. A. Goldbohm et al. (2000). Influence of lutein supplementation on macular pigment, assessed with two objective techniques. *Invest. Ophthalmol. Vis. Sci.* **41**(11): 3322–3326.

Berendschot, T. T. and D. van Norren (2005). On the age dependency of the macular pigment optical density. *Exp. Eye Res.* **81**(5): 602–609.

Berges, D., G. Schmitt et al. (1959). The effects of helenien and vitamin A on the primary sight process. III. A flame photometric determination of the potassium and sodium content of the retina (German). *Z. Biol.* **111**: 220–227.

Bernstein, P. S., N. A. Balashov et al. (1997). Retinal tubulin binds macular carotenoids. *Invest. Ophthalmol. Vis. Sci.* **38**: 167–175.

Bernstein, P. S., F. Khachik et al. (2001). Identification and quantitation of carotenoids and their metabolites in the tissues of the human eye. *Exp. Eye Res.* **72**(3): 215–223.

Bernstein, P. S., D. Y. Zhao et al. (2002). Resonance Raman measurement of macular carotenoids in normal subjects and in age-related macular degeneration patients. *Ophthalmology* **109**(10): 1780–1787.

Bhosale, P. and P. S. Bernstein (2005). Synergistic effects of zeaxanthin and its binding protein in the prevention of lipid membrane oxidation. *Biochim. Biophys. Acta* **1740**(2): 116–121.

Bhosale, P. and P. S. Bernstein (2007). Vertebrate and invertebrate carotenoid-binding proteins. *Arch. Biochem. Biophys.* **458**(2): 121–127.

Bhosale, P., A. J. Larson et al. (2004). Identification and characterization of a Pi isoform of glutathione S-transferase (GSTP1) as a zeaxanthin-binding protein in the macula of the human eye. *J. Biol. Chem.* **279**(47): 49447–49454.

Bone, R. A., J. T. Landrum et al. (1985). Preliminary identification of the human macular pigment. *Vis. Res.* **25**(11): 1531–1535.

Bone, R. A., J. T. Landrum et al. (1988). Analysis of the macular pigment by HPLC: Retinal distribution and age study. *Invest. Ophthalmol. Vis. Sci.* **29**(6): 843–849.

Bone, R. A., J. T. Landrum et al. (1993). Stereochemistry of the human macular carotenoids. *Invest. Ophthalmol. Vis. Sci.* **34**: 2033–2040.

Bone, R. A., J. T. Landrum et al. (1997). Distribution of lutein and zeaxanthin stereoisomers in the human retina. *Exp. Eye Res.* **64**: 211–218.

Bone, R. A., J. T. Landrum et al. (2001). Macular pigment in donor eyes with and without AMD: A case-control study. *Invest. Ophthalmol. Vis. Sci.* **42**(1): 235–242.

Bone, R. A., J. T. Landrum et al. (2003). Lutein and zeaxanthin dietary supplements raise macular pigment density and serum concentrations of these carotenoids in humans. *J. Nutr.* **133**(4): 992–998.

Bone, R. A., J. T. Landrum et al. (2007). Macular pigment response to a supplement containing meso-zeaxanthin, lutein and zeaxanthin. *Nutr. Metab. (London)* **4**: 12.

Borel, P., M. Moussa et al. (2007). Human plasma levels of vitamin E and carotenoids are associated with genetic polymorphisms in genes involved in lipid metabolism. *J. Nutr.* **137**(12): 2653–2659.

Boudreault, G., P. Cortin et al. (1893). Canthaxanthine retinopathy: 1. Clinical study in 51 consumers (French). *Can. J. Ophthalmol.* **18**(7): 325–328.

Breithaupt, D. E., P. Weller et al. (2004). Comparison of plasma responses in human subjects after the ingestion of 3R,3R′-zeaxanthin dipalmitate from wolfberry (*Lycium barbarum*) and non-esterified 3R,3R′-zeaxanthin using chiral high-performance liquid chromatography. *Br. J. Nutr.* **91**(5): 707–713.

Britton, G., S. Liaaen-Jensen et al. (2004). *Handbook of Carotenoids*. Birkhäuser AG, Basel, Switzerland.

Buzzi, F. (1782). Nuove sperienze fatte sull' occhio umano dal sig. Fancesco Buzzi chirurgo oculista ed ajutante chriurgo dello spedale maggiore di milano. *Opuscoli Scelti Sulle Scienze e Sulle Arti* **5**: 87–95.

Canfield, L. M., M. T. Clandinin et al. (2003). Multinational study of major breast milk carotenoids of healthy mothers. *Eur. J. Nutr.* **42**(3): 133–141.

Cantrell, A., D. J. McGarvey et al. (2003). Singlet oxygen quenching by dietary carotenoids in a model membrane environment. *Arch. Biochem. Biophys.* **412**(1): 47–54.

Cardinault, N., J. Gorrand et al. (2003). Short-term supplementation with lutein affects biomarkers of lutein status similarly in young and elderly subjects. *Exp. Gerontol.* **38**(5): 573–582.

Cho, E., J. M. Seddon et al. (2004). Prospective study of intake of fruits, vegetables, vitamins, and carotenoids and risk of age-related maculopathy. *Arch. Ophthalmol.* **122**(6): 883–892.

Cilotti, P. (1963). The action of thioctic acid and helenien on visual acuity in reduced lighting in normal subjects (Italian). *Ann. Ottalmol. Clin. Ocul.* **89**: 1000–1004.

Coleman, H. and E. Chew (2007). Nutritional supplementation in age-related macular degeneration. *Curr. Opin. Ophthalmol.* **18**(3): 220–223.

Craft, N. E., T. B. Haitema et al. (2004). Carotenoid, tocopherol, and retinol concentrations in elderly human brain. *J. Nutr. Health Aging* **8**(3): 156–162.

Cuccagna, F. (1956). Effect of helenien on retinal adaptation in myopia and retinitis pigmentosa (Italian). *Arch. Ottal.* **60**: 311–321.

Cüppers, C. and E. Wagner (1950). Pharmacological effect on retinal function; normal dark adaptation (German). *Klin Monatsblätter Augenheilkd* **117**(1): 59–69.

Dagnelie, G., I. S. Zorge et al. (2000). Lutein improves visual function in some patients with retinal degeneration: A pilot study via the Internet. *Optometry* **71**(3): 147–164.

Daicker, B., K. Schiedt et al. (1987). Canthaxanthin retinopathy–An investigation by light and electron microscopy and physicochemical analysis. *Graefe's Arch. Clin. Exp. Ophthalmol.* **225**(3): 189–197.

De Ferreira, C. and C. Da Maia (1956). Effect of adaptinol in primary pigmentary retinosis (Portugese). *Gaz. Med. Port.* **9**(5): 579–581.

de Jong, P. T. (2006). Age-related macular degeneration. *N. Engl. J. Med.* **355**(4): 1474–1485.

Delcourt, C., I. Carriere et al. (2006). Plasma lutein and zeaxanthin and other carotenoids as modifiable risk factors for age-related maculopathy and cataract: The POLA study. *Invest. Ophthalmol. Vis. Sci.* **47**(6): 2329–2335.

deVries, H., R. Jielof et al. (1953). Properties of the eye with respect to polarized light. *Physica* **19**: 419–432.

Dillon, J., L. Zheng et al. (2004). Transmission of light to the aging human retina: possible implications for age related macular degeneration. *Exp. Eye Res.* **79**(6): 753–759.

Dowling, J. E. (1987). *The Retina, an Approachable Part of the Brain*. The Belknap Press of Harvard University Press, Cambridge, MA and London.

Duncan, J. L., T. S. Aleman et al. (2002). Macular pigment and lutein supplementation in choroideremia. *Exp. Eye Res.* **74**(3): 371–381.

During, A., S. Doraiswamy et al. (2008). Xanthophylls are preferentially taken up compared to beta-carotene by retinal cells via a scavenger receptor BI-dependent mechanism. *J. Lipid Res.* **49**: 1715–1724.

EDCC Study Group (1993). Antioxidant status and neovascular age-related macular degeneration. *Arch. Ophthalmol.* **111**: 104–109.

Engles, M., B. Wooten et al. (2007). Macular pigment: A test of the acuity hypothesis. *Invest. Ophthalmol. Vis. Sci.* **48**(6): 2922–2931.

Falsini, B., M. Piccardi et al. (2003). Influence of short-term antioxidant supplementation on macular function in age-related maculopathy: A pilot study including electrophysiologic assessment. *Ophthalmology* **110**(1): 51–60.

Fincham, E. F. (1936). The photography of the human eye. *J. Photogr.* **76**: 268–274.

Fletcher, A. E., G. C. Bentham et al. (2008). Sunlight exposure, antioxidants, and age-related macular degeneration. *Arch. Ophthalmol.* **126**(10): 1396–1403.

Flood, V., W. Smith et al. (2002). Dietary antioxidant intake and incidence of early age-related maculopathy: The blue mountains eye study. *Ophthalmology* **109**(12): 2272–2278.

Frank, H. A. and R. J. Cogdell (1996). Carotenoids in photosynthesis. *Photochem. Photobiol.* **63**(3): 257–264.

Friedman, D. S., B. J. O'Colmain et al. (2004). Prevalence of age-related macular degeneration in the United States. *Arch. Ophthalmol.* **122**(4): 564–572.

Gabrielska, J. and W. I. Gruszecki (1996). Zeaxanthin (dihydroxy-beta-carotene) but not beta-carotene rigidifies lipid membranes: A 1H-NMR study of carotenoid-egg phosphatidylcholine liposomes. *Biochim. Biophys. Acta* **1285**(2): 167–174.

Gale, C. R., N. F. Hall et al. (2003). Lutein and zeaxanthin status and risk of age-related macular degeneration. *Invest. Ophthalmol. Vis. Sci.* **44**(6): 2461–2465.

Garnett, K. M., L. H. Guerra et al. (2002). Serum and macular pigment responses to supplementation with lutein or zeaxanthin. *ARVO 2002*: Abstracts Volume #1, Program #2820.

Gass, J. D. (1997). *Stereoscopic Atlas of Macular Diseases Diagnosis and Treatment*. Vol. 1, Mosby – Year Book Inc., p. 3.

Gass, J. D. M. (1999). Müller cell cone, an overlooked part of the anatomy of the fovea centralis. *Arch. Ophthalmol.* **117**: 821–823.

Goralczyk, R., F. M. Barker et al. (2000). Dose dependency of canthaxanthin crystals in monkey retina and spatial distribution of its metabolites. *Invest. Ophthalmol. Vis. Sci.* **41**(6): 1513–1522.

Goralczyk, R., F. M. Barker et al. (2002). Ocular safety of lutein and zeaxanthin in a long-term study in cynomolgous monkeys. *ARVO 2002* Abstracts Volume #1, Program #2546.

Hammond, B. R. and J. E. Frick (2007). Nutritional protection of the developing retina. *Hong Kong Pract.* **29**: 1–8.

Hammond, B. R., K. Fuld et al. (1995). Macular pigment density in monozygotic twins. *Invest. Ophthalmol. Vis. Sci.* **36**(12): 2531–2541.

Hammond, B. R., E. J. Johnson et al. (1997). Dietary modification of human macular pigment density. *Invest. Ophthalmol. Vis. Sci.* **38**(9): 1795–801.

Hammond, B. R., B. R. Wooten et al. (2001). Carotenoids in the retina and lens: Possible acute and chronic effects on human visual performance. *Arch. Biochem. Biophys.* **385**(1): 41–46.

Hammond, B. R. and B. R. Wooten (2005). CFF thresholds: Relation to macular pigment optical density. *Ophthal. Physiol. Opt.* **25**(4): 315–319.

Handelman, G. J., Z. D. Nightingale et al. (1999). Lutein and zeaxanthin concentrations in plasma after dietary supplementation with egg yolk. *Am. J. Clin. Nutr.* **70**: 247–251.

Harikumar, K. B., C. V. Nimita et al. (2008). Toxicity profile of lutein and lutein ester isolated from marigold flowers (*Tagetes erecta*). *Int. J. Toxicol.* **27**(1): 1–9.

Harnois, C., J. Samson et al. (1989). Canthaxanthin retinopathy. Anatomic and functional reversibility. *Arch. Ophthalmol.* **107**(4): 538–540.

Hartmann, D., P. A. Thürmann et al. (2004). Plasma kinetics of zeaxanthin and 3′-dehydro-lutein after multiple oral doses of synthetic zeaxanthin. *Am. J. Clin. Nutr.* **79**(3): 410–417.

Hayano, S., Y. Koide et al. (1959). Actions of adaptinol observed from the blood lutein pattern (Japanese). *J. Clin. Ophthalmol.* **13**(6): 931–937.

Holm, E. (1922). Das gelbe maculapigment und seine optische bedeutung. *Graefe's Arch. Clin. Exp. Ophthalmol.* **108**(1–2): 1–85.

Home, E. (1798). An account of the orifice in the retina of the human eye, discovered by Professor Soemmering. *Philos. Trans. R. Soc. London.* **Pt 2**: 332–345.

Huang, K. C. (1993). *The Pharmacology of Chinese Herbs*. CRC Press, New York, pp. 221–215.

Huang, L. L., H. R. Coleman et al. (2008). Oral supplementation of lutein/zeaxanthin and omega-3 long chain polyunsaturated fatty acids in persons aged 60 years and older, with or without age-related macular degeneration. *Invest. Ophthalmol. Vis. Sci.* **49**(9): 3864–3869.

Inbaraj, B. S., H. Lu et al. (2008). Determination of carotenoids and their esters in fruits of *Lycium barbarum* Linnaeus by HPLC-DAD-APCI-MS. *J. Pharm. Biomed. Anal.* **47**(4–5): 812–818.

Izumi-Nagai, K., N. Nagai et al. (2007). Macular pigment lutein is antiinflammatory in preventing choroidal neovascularization. *Arterioscler. Thromb. Vasc. Biol.*, Oct. 11 [Epub ahead of print].

Jackson, J. G. and J. P. Zimmer (2007). Lutein and zeaxanthin in human milk independently and significantly differ among women from Japan, Mexico, and the United Kingdom. *Nutr. Res.* **27**(8): 449–453.

Johnson, E. J., H. Y. Chung et al. (2008a). The influence of supplemental lutein and docosahexaenoic acid on serum, lipoproteins, and macular pigmentation. *Am. J. Clin. Nutr.* **87**(5): 1521–1529.

Johnson, E. J., K. A. McDonald et al. (2008b). Cognitive findings of an exploratory trial of docosahexaenoic acid and lutein supplementation in older women. *Nutr. Neurosci.* **11**(2): 75–83.

Johnson, E. J., M. Neuringer et al. (2005). Nutritional manipulation of primate retinas. III. Effects of lutein or zeaxanthin supplementation on adipose and retina of xanthophyll-free monkeys. *Invest. Ophthalmol. Vis. Sci.* **46**(2): 692–702.

Johnson, E. J. and E. Norkus (1995). Contribution of beta-carotene from beta-carotene enriched formulas to individual and total serum carotenoids in term infants. *FASEB J.* **9**(4 Pt 3): 1869.

Kant, A. K. and B. I. Graubard (2007). Ethnicity is an independent correlate of biomarkers of micronutrient intake and status in American adults. *J. Nutr.* **137**(11): 2456–2463.

Khachik, F., G. R. Beecher et al. (1995). Lutein, lycopene, and their oxidative metabolites in chemoprevention of cancer. *J. Cell. Biochem. Suppl.* **22**: 236–246.

Khachik, F., P. S. Bernstein et al. (1997a). Identification of lutein and zeaxanthin oxidation products in human and monkey retinas. *Invest. Ophthalmol. Vis. Sci.* **38**(9): 1802–1811.

Khachik, F., E. London et al. (2006). Chronic ingestion of (3R,3′R,6′R)-lutein and (3R,3′R)-zeaxanthin in the female rhesus macaque. *Invest. Ophthalmol. Vis. Sci.* **47**(12): 5476–5486.

Khachik, F., C. J. Spangler et al. (1997b). Identification, quantification, and relative concentrations of carotenoids and their metabolites in human milk and serum. *Anal. Chem.* **69**: 1873–1881.

Kim, S. R., K. Nakanishi et al. (2006). Photooxidation of A2-PE, a photoreceptor outer segment fluorophore, and protection by lutein and zeaxanthin. *Exp. Eye Res.* **82**(5): 828–839.

Kirschfeld, K. (1982). Carotenoid pigments: Their possible role in protecting against photooxidation in eyes and photoreceptor cells. *Proc. R. Soc. Lond. B Biol. Sci.* **216**(1202): 71–85.

Klaes, H. and H. Riegel (1951). Effect of adaptinol on dark adaptation of the human eye (German). *Medizinische Monatsschrift* **5**(5): 334–337.

Koepcke, W., F. M. Barker et al. (1995). Canthaxanthin deposition in the retina: A biostatistical evaluation of 411 patients. *J. Toxicol. Cutaneous Ocular Toxicol.* **14**(2): 89–104.

Koushik, A., D. J. Hunter et al. (2006). Intake of the major carotenoids and the risk of epithelial ovarian cancer in a pooled analysis of 10 cohort studies. *Int. J. Cancer* **119**(9): 2148–2154.

Kowluru, R. A., B. Menon et al. (2008). Beneficial effect of zeaxanthin on retinal metabolic abnormalities in diabetic rats. *Invest. Ophthalmol. Vis. Sci.* **49**(4): 1645–1651.

Krinsky, N. I. (2002). Possible biologic mechanisms for a protective role of xanthophylls. *J. Nutr.* **132**(3): 540S–542S.

Krinsky, N. I. and S. M. Deneke (1982). Interaction of oxygen and oxy-radicals with carotenoids. *J. Natl. Cancer Inst.* **69**(1): 205–210.

Krinsky, N. I., J. T. Landrum et al. (2003). Biologic mechanisms of the protective role of lutein and zeaxanthin in the eye. *Annu. Rev. Nutr.* **23**: 171–201.

Kuhn, R. and A. Winterstein (1930). About the occurrence of lutein in the plant kingdom (German). *Naturwissenschaften* **18**: 754.

Kvansakul, J., M. Rodriguez-Carmona et al. (2006). Supplementation with the carotenoids lutein and zeaxanthin improves human visual performance. *Ophthal. Physiol. Opt.* **26**(4): 362–371.

Landrum, J. T., R. A. Bone et al. (1997). A one year study of the macular pigment: The effect of 140 days of a lutein supplement. *Exp. Eye Res.* **65**(1): 57–62.

Landrum, J. T., R. A. Bone et al. (1999a). Carotenoids in the human retina. *Pure Appl. Chem.* **71**(12): 2237–2244.

Landrum, J. T., R. A. Bone et al. (1999b). Analysis of zeaxanthin distribution within individual human retinas. *Methods Enzymol.* **299**: 457–467.

Leung, I. Y., M. M. Sandstrom et al. (2004). Nutritional manipulation of primate retinas, II: Effects of age, n-3 fatty acids, lutein, and zeaxanthin on retinal pigment epithelium. *Invest. Ophthalmol. Vis. Sci.* **45**(9): 3244–3256.

Leung, I. Y. F., M. O. M. Tso et al. (2001). Absorption of zeaxanthin in human subjects after ingestion of fructus lycii (gou qi zi). *Invest. Ophthalmol. Vis. Sci.* **42**(4): S359, Program #1942.

Leyon, H., A. M. Ros et al. (1990). Reversibility of canthaxanthin deposits within the retina. *Acta Ophthalmol (Copenh)* **68**(5): 607–611.

Liew, S. H., C. E. Gilbert et al. (2005). Heritability of macular pigment: A twin study. *Invest. Ophthalmol. Vis. Sci.* **46**(12): 4430–4436.

Liew, S. H., C. E. Gilbert et al. (2006). Central retinal thickness is positively correlated with macular pigment optical density. *Exp. Eye Res.* **82**(5): 915–920.

Lim, B. P., A. Nagao et al. (1992). Anti-oxidant activity of xanthophylls on peroxyl radical-mediated phospholipid peroxidation. *Biochim. Biophys. Acta* **1126**: 178–184.

Loane, E., J. M. Nolan et al. (2008). Transport and retinal capture of lutein and zeaxanthin with reference to age-related macular degeneration. *Surv. Ophthalmol.* **53**(1): 68–81.

Loughman, J., P. Davison et al. (2007). Impact of macular pigment on visual performance. *Optician* **234**(6116): 23–28.

Malinow, M. R., L. Feeney-Burns et al. (1980). Diet-related macular anomalies in monkeys. *Invest. Ophthalmol. Vis. Sci.* **19**(8): 857–863.

Mares, J. A., T. L. LaRowe et al. (2006). Predictors of optical density of lutein and zeaxanthin in retinas of older women in the carotenoids in age-related eye disease study, an ancillary study of the women's health initiative. *Am. J. Clin. Nutr.* **84**(5): 1107–1122.

Marx, J. (2006). Genetics–A clearer view of macular degeneration. *Science* **311**(5768): 1704–1705.

McNulty, H., R. F. Jacob et al. (2008). Biologic activity of carotenoids related to distinct membrane physicochemical interactions. *Am. J. Cardiol.* **101**(10A): 20D–29D.

Moeller, S. M., P. F. Jacques et al. (2000). The potential role of dietary xanthophylls in cataract and age-related macular degeneration. *J. Am. Coll. Nutr.* **19**(5 Suppl): 522S–527S.

Monje, M. (1948). On the effect of helenien on the darkness adjustment of humans with normal night vision (German). *von Graefe's Arch. Ophthalmol.* **148**: 679–705.

Mosci, L. (1956). Effect of heleniene (adaptinol) on sensitivity to light in man, experimental study (Italian). *Ann. Ottalmol. Clin. Ocul.* **82**(5): 237–254.

Mueller-Limmroth, W., D. Berges et al. (1958). The effects of helenien and vitamin A on the primary visual processes. I. An electroretinographic study on the problem of the sites of helenien and vitamin A actions (German). *Z. Biol.* **110**(5–6): 457–475.

Mueller-Limmroth, W. and J. Kueper (1961a). On the effect of adaptinol on the electroretinogram in tapetoretinal degenerations (German). *Klin Monatsblätter Augenheilkd* **138**: 37–41.

Mueller-Limmroth, W. and B. Schmidt (1961b). The human ERG after glare in the course of early dark adaptation and its modification by adaptinol (German). *Med. Welt* **27**: 1413–1421.

Nayak, L., M. K. Raval et al. (2001). Photoprotection of green leaves by zeaxanthin, a two-channel process. *Curr. Sci.* **81**(9): 1165–1166.

Neuringer, M., M. M. Sandstrom et al. (2004). Nutritional manipulation of primate retinas, I: Effects of lutein or zeaxanthin supplements on serum and macular pigment in xanthophyll-free rhesus monkeys. *Invest. Ophthalmol. Vis. Sci.* **45**(9): 3234–3243.

Nolan, J. M., J. Stack et al. (2006). Monthly consistency of macular pigment optical density and serum concentrations of lutein and zeaxanthin. *Curr. Eye Res.* **31**(2): 199–213.

Nolan, J. M., J. Stack et al. (2007a). The relationships between macular pigment optical density and its constituent carotenoids in diet and serum. *Invest. Ophthalmol. Vis. Sci.* **48**(2): 571–582.

Nolan, J. M., J. Stack et al. (2007b). Risk factors for age-related maculopathy are associated with a relative lack of macular pigment. *Exp. Eye Res.* **84**(1): 61–74.

Nordenson, J. W. (1958). Neue argumente zur frage der makulafarbe. *Graefe's Arch. Clin. Exp. Ophthalmol.* **160**: 43–46.

Nussbaum, J. J., R. C. Pruett et al. (1981). Historic perspectives. Macular yellow pigment. The first 200 years. *Retina* **1**(4): 296–310.

Obana, A., T. Hiramitsu et al. (2008). Macular carotenoid levels of normal subjects and age-related maculopathy patients in a Japanese population. *Ophthalmology* **115**(1): 147–157.

O'Connell, E., K. Neelam et al. (2006). Macular carotenoids and age-related maculopathy. *Ann. Acad. Med. Singapore* **35**(11): 821–830, Nov.

Oka, S. (1955). Experience with adaptinol (Bayer) in night blindness (Japanese). *Ophthalmol. Japonica Folia* **6**(4): 156–160.

Olmedilla, B., F. Granado et al. (2001). Lutein in patients with cataracts and age-related macular degeneration: a long term supplementation study. *J. Sci. Food Agric.* **81**(9): 904–909.

Parisi, V., M. Tedeschi et al. (2008). Carotenoids and antioxidants in age-related maculopathy Italian study: Multifocal electroretinogram modifications after 1 year. *Ophthalmology* **115**(2): 324–333.

Pfeifer, H. (1957). Experiments on the effect of helenien on normal and reduced dark-adaptation (German). *von Graefe's Arch. Ophthalmol.* **159**(3): 311–322.

Rapp, L. M., S. S. Maple et al. (2000). Lutein and zeaxanthin concentrations in rod outer segment membranes from perifoveal and peripheral human retina. *Invest. Ophthalmol. Vis. Sci.* **41**(5): 1200–1209.

Reading, V. M. and R. A. Weale (1974). Macular pigment and chromatic aberration. *J. Opt. Soc. Am.* **64**(2): 231–234.

Renzi, L. M., B. R. Hammond et al. (2008a). The effects of macular pigment on temporal vision. *ARVO 2008:* Abstract Book on CDROM: Program #4958.

Renzi, L. M., A. Iannaccone et al. (2008b). Relationships among macular pigment optical density (MPOD), serum lutein (L), zeaxanthin (Z), and n3 fatty acids (FAs). *ARVO 2008*: Abstract Book on CDROM: Program #2123.

Renzi, L. M. and E. J. Johnson (2007). Lutein and age-related ocular disorders in the older adult: A review. *J. Nutr. Elder* **26**(3–4): 139–157.

Richer, S. (1999). Part II: ARMD–pilot (case series) environmental intervention data. *J. Am. Optom. Assoc.* **70**(1): 24–36.

Richer, S., W. Stiles et al. (2004). Double-masked, placebo-controlled, randomized trial of lutein and antioxidant supplementation in the intervention of atrophic age-related macular degeneration: The veterans LAST study (lutein antioxidant supplementation trial). *Optometry* **75**(4): 216–229.

Schalch, W. (1992). Carotenoids in the retina–A review of their possible role in preventing or limiting damage caused by light and oxygen. *EXS (Experientia Suppl. Ser.)* **62**: 280–298.

Schalch, W. and F. M. Barker (2005). Ocular and general safety of supplementation with zeaxanthin and lutein; plasma exposure levels of carotenoids and 3′-dehydro-lutein—Results of the LUXEA study. ARVO 2005: Abstract Book on CDROM: Program #1765.

Schalch, W., W. Cohn et al. (2007). Xanthophyll accumulation in the human retina during supplementation with lutein or zeaxanthin–the LUXEA (lutein xanthophyll eye accumulation) study. *Arch. Biochem. Biophys.* **458**(2): 128–135.

Schmitt, G., D. Berges et al. (1959). The effects of helenien and vitamin A on the primary sight process. II. Tissue respiration of the isolated retina under the influence of helenien and vitamin A (German). *Z. Biol.* **111**: 213–219.

Schultze, M. (1866). Ueber den gelben fleck der retina, seinen einfluss auf normales sehen und auf farbenblindheit. *Bonn, Verlag von Max Cohen & Sohn Vortrag gehalten in der Sitzung der Medicinischen Section der Niederrheinischen Gesellschaft für Natur-und Heilkunde zu Bonn* am 9. Mai **1866**: 1–16.

Schupp, C., E. Olano-Martin et al. (2004). Lutein, zeaxanthin, macular pigment, and visual function in adult cystic fibrosis patients. *Am. J. Clin. Nutr.* **79**(6): 1045–1052.

Schwalbe, G. (1874). *Microscopische Anatomie des Sehnerven, der Netzhaut und des Glaskörpers*. Leipzig, Engelmann, Germany.

Schweitzer, D., G. E. Lang et al. (2000). Age-related maculopathy. Comparative studies of patients, their children and healthy controls (German). *Ophthalmology* **97**(2): 84–90.

Schweitzer, D., G. E. Lang et al. (2002). Objective determination of the optical density of xanthophyll after supplementation with lutein (original German title: Objektive bestimmung der optischen dichte von xanthophyll nach supplementation von lutein). *Ophthalmology* **99**: 270–275.

Seddon, J. M., U. A. Ajani et al. (1994). Dietary carotenoids, vitamins A, C, and E, and advanced age-related macular degeneration. Eye disease case-control study group. *JAMA* **272**(18): 1413–1420.

Seddon, J. M. and C. H. Hennekens (1994). Vitamins, minerals, and macular degeneration, promising but unproven hypotheses. *Arch. Ophthalmol.* **112**: 176–179.

Shao, A. and J. N. Hathcock (2006). Risk assessment for the carotenoids lutein and lycopene. *Regul. Toxicol. Pharmacol.* **45**(3): 289–298.

Sires, B. S., J. C. Saari et al. (2001). The color difference in orbital fat. *Arch. Ophthalmol.* **119**(6): 868–871.

Snellen, E. L., A. L. Verbeek et al. (2002). Neovascular age-related macular degeneration and its relationship to antioxidant intake. *Acta Ophthalmol. Scand.* **80**(4): 368–371.

Snodderly, D. M., P. K. Brown et al. (1984). The macular pigment. I. Absorbance spectra, localization, and discrimination from other yellow pigments in primate retinas. *Invest. Ophthalmol. Vis. Sci.* **25**(6): 660–673.

Soemmering, S. T. (1799). De foramine centrali limbo luteo cincto retinae humanae (Latin). *Commentationes Societatis Regiae Scientiarum Göttingensis* **13**: 3–13.

Sole, P., D. Rigal et al. (1984). Effects of cyaninoside chloride and heleniene on mesopic and scotopic vision in myopia and night blindness (French). *J. Fr. Ophthalmol.* **7**(1): 35–39.

Sparrow, J. R. and M. Boulton (2005). RPE lipofuscin and its role in retinal pathobiology. Review. *Exp. Eye Res.* **80**(5): 595.

Stringham, J. M., K. Fuld et al. (2003). Action spectrum for photophobia. *J. Opt. Soc. Am. A Opt. Image Sci. Vis.* **20**(10): 1852–1858.

Stringham, J. M. and B. R. Hammond (2007). The glare hypothesis of macular pigment function. *Optom. Vis. Sci.* **84**(9): 859–864.

Stringham, J. M. and B. R. Hammond (2008). Macular pigment and visual performance under glare conditions. *Optom. Vis. Sci.* **85**(2): 82–88.

Sun, H. and J. Nathans (2001). The challenge of macular degeneration. *Sci. Am.* (Oct. Issue): 69–75.

Tarpo, E. and V. Cucu (1961). The blossoms of the genus Tagetes as the raw material for the production of helenien (German). *Pharmazie* **16**: 486–488.

Thomson, L. R., Y. Toyoda et al. (2002). Elevated retinal zeaxanthin and prevention of light-induced photoreceptor cell death in quail. *Invest. Ophthalmol. Vis. Sci.* **43**(11): 3538–3549.

Thürmann, P. A., W. Schalch et al. (2005). Plasma kinetics of lutein, zeaxanthin, and 3′-dehydro-lutein after multiple oral doses of a lutein supplement. *Am. J. Clin. Nutr.* **82**(1): 88–97.

Thurnham, D. I., A. Tremel et al. (2008). A supplementation study in human subjects with a combination of meso-zeaxanthin, (3R,3′R)-zeaxanthin and (3R,3′R,6′R)-lutein. *Br. J. Nutr.* **100**(6): 1307–1314.

Trieschmann, M., S. Beatty et al. (2007). Changes in macular pigment optical density and serum concentrations of its constituent carotenoids following supplemental lutein and zeaxanthin: The LUNA study. *Exp. Eye Res.* **84**(4): 718–728.

Trumbo, P. R. and K. C. Ellwood (2006). Lutein and zeaxanthin intakes and risk of age-related macular degeneration and cataracts: An evaluation using the Food and Drug Administration's evidence-based review system for health claims. *Am. J. Clin. Nutr.* **84**(5): 971–974.

van den Langenberg, G. M., J. A. Mares-Perlman et al. (1998). Associations between antioxidant and zinc intake and the 5-year incidence of early age-related maculopathy in the beaver dam eye study. *Am. J. Epidemiol.* **148**(2): 204–214.

von Studnitz, G. (1952). Die Steigerung der Dunkeladaptation durch Adaptinol [Increase of dark adaptation by adaptinol] (German). *Klin Monatsblätter Augenheilkd* **120**: 632–636.

von Studnitz, G. and H. K. Loevenich (1947). About the increase in human darkness adaptation by carotenoids (German). *Klin Monatsblätter Augenheilkd* **111**: 193–210.

Wald, G., W. R. Carroll et al. (1941). The human excretion of carotenoids and vitamin A. *Science* **94**(2430): 95–96.

Wald, G. L. (1945). Human vision and the spectrum. *Nature (London)* **101**: 653–658.

Walls, G. L. and H. D. Judd (1933). The intra-ocular colour-filters of vertebrates. *Br. J. Ophthalmol.* **17**: 641–675, 705–725.

Weale, R. A. (2007). Guest editorial: Notes on the macular pigment. *Ophthal. Physiol. Opt.* **27**(1): 1–10.

Weiser, H. and A. W. Kormann (1993). Provitamin A activities and physiological functions of carotenoids in animals. *Ann. N.Y. Acad. Sci.* **691**: 213–215.

Wenzel, A. J., K. Fuld et al. (2006). Macular pigment optical density and photophobia light threshold. *Vis. Res.* **46**(28): 4615–4622.

Wenzel, A. J., J. P. Sheehan et al. (2007). Macular pigment optical density at four retinal loci during 120 days of lutein supplementation. *Ophthal. Physiol. Opt.* **27**(4): 329–335.

Whitehead, A. J., J. A. Mares et al. (2006). Macular pigment: A review of current knowledge. *Arch. Ophthalmol.* **124**(7): 1038–1045.

Whittaker, D., R. Steen et al. (1993). Light scatter in the normal young, elderly, and cataractous eye demonstrates little wavelength dependency. *Optom. Vis. Sci.* **70**: 963–968.

Wolf-Schnurrbusch, U. E. K., N. Röösli et al. (2007). Ethnic differences in macular pigment density and distribution. *Invest. Ophthalmol. Vis. Sci.* **48**(8): 3783–3787.

Wooten, B. R. and B. R. Hammond (2002). Macular pigment: Influences on visual acuity and visibility. *Prog. Retin. Eye Res.* **21**(2): 225–240.

Wuestenberg, W. (1951). The effect of adaptinol on normal dark adaptation (German). *Klin Monatsblätter Augenheilkd* **119**(5): 524–528.

Yakovleva, M. A., I. G. Panova et al. (2007). Identification of carotenoids in the vitreous body of the prenatal human eye. *Ontogenez* **38**(5): 380–385.

Yang, Z., C. Stratton et al. (2008). Toll-like receptor 3 and geographic atrophy in age-related macular degeneration. *N. Engl. J. Med.*, Aug. 27 [Epub ahead of print] Links.

Yeum, K. J., F. M. Shang et al. (1999). Fat-soluble nutrient concentrations in different layers of human cataractous lens. *Curr. Eye Res.* **19**(6): 502–505.

Yolton, D., B. DeRuyter et al. (2002). Failure of oral supplement containing lutein to change macular pigment density. *Optom. Vis. Sci.* **79**(12): 104.

Zimmer, J. P. and B. R. Hammond (2007). Possible influences of lutein and zeaxanthin on the developing retina. *Clin. Ophthalmol.* **1**(1): 25–35.

14 Properties of Carotenoid Radicals and Excited States and Their Potential Role in Biological Systems

Ruth Edge and George Truscott

CONTENTS

14.1 INTRODUCTION

The C_{40} carotenoids (CARs) and their oxygenated derivatives xanthophylls (XANs) are one of nature's major antioxidant pigments and they efficiently quench singlet oxygen [$^1O_2$] and interact with damaging free radicals. Indeed, carotenoids protect bacterial and green plant photosynthetic systems and the skin from $^1O_2$ damage. XANs protect the macula of the eye and the interaction/quenching of free radicals can be observed in photosynthetic systems and are also believed to be linked to the protective role of CARs against the initiation of chronic disease.

The overall process of $^1O_2$ quenching simply converts the excess energy of singlet oxygen to heat via the carotenoid [CAR] lowest excited triplet state [^{3}CAR].

$$^1O_2 + CAR \rightarrow O_2 + {}^3CAR \quad (14.1)$$

$$^3CAR \rightarrow CAR + heat \quad (14.2)$$

The reaction of CARs with free radicals is much more complex and depends mostly on the nature of the free radical [$RO^{\bullet}$] rather than on the CAR. Certainly, at least four processes have been reported. Of course, in all four processes, the unpaired electron of the free radical is transferred to the CAR so that a new, carotenoid radical (or CAR adduct radical) is produced.

$$RO^{\bullet} + CAR \rightarrow RO^{-} + CAR^{\bullet +} \quad (14.3)$$

$$RO^{\bullet} + CAR \rightarrow RO^{+} + CAR^{\bullet -} \quad (14.4)$$

$$RO^{\bullet} + CAR \rightarrow ROH^{\bullet} + CAR(-H)^{\bullet} \quad (14.5)$$

$$RO^{\bullet} + CAR \rightarrow (RO - CAR)^{\bullet} \quad (14.6)$$

where

- $CAR^{\bullet +}$ and $CAR^{\bullet -}$ are the radical cations and anions of CARs generated by electron transfer to or from the radical $RO^{\bullet}$
- CAR $(-H)^{\bullet}$ is the radical formed via H-atom transfer to $RO^{\bullet}$
- $(RO-CAR)^{\bullet}$ is an adduct radical

The reactivity of the resulting CAR radical [$CAR^{\bullet +}$, $CAR^{\bullet -}$, $CAR(-H)^{\bullet}$, $RO-CAR^{\bullet}$] depends, of course, on the nature of this species.

Strong oxidizing radicals (such as peroxyl radicals $RO_2^{\bullet}$) generate $CAR^{\bullet +}$ via electron transfer and, because the radical $CAR^{\bullet +}$ are themselves strong oxidizing agents (see Section 14.4.3.2 and Table 14.12), this species may well be the most important of the CAR radicals formed.

14.2 REACTIONS BETWEEN CAROTENOIDS AND SINGLET OXYGEN

In biological systems, sensitizers such as porphyrins, chlorophylls, and riboflavin can sensitize $^1O_2$ production and this can lead to deleterious effects including DNA damage and lipid peroxidation.

The quenching of $^1O_2$ by carotenoids and how this reaction protects against $^1O_2$ mediated photooxidation reactions has been much discussed. In this chapter, the older literature on singlet oxygen quenching is collated with newer (including some previously unpublished) results. All the dietary carotenoids studied are extremely efficient $^1O_2$ quenchers and there is little difference in their individual efficiencies, in homogeneous environments (e.g., organic solvents) for this important function. Results in microheterogeneous environments such as liposomes (as cell membrane models) are more complex and this is, at least in part, due to the aggregation of the carotenoids. A useful model of aggregation effects comes from the studies of the $^1O_2$ quenching in alcohol/water mixtures (Gruszecki 1999, Burke 2001).

The first demonstration that β-carotene could inhibit photosensitized oxidation and was, therefore, an efficient quencher of $^1O_2$ was reported by Foote and Denny (1968). Subsequently, Farmilo and Wilkinson (1973) showed that electron exchange energy transfer quenching producing the carotenoid triplet state (^3CAR) is the principal mechanism of carotenoid photoprotection against $^1O_2$: although, chemical quenching also occurs leading to the destruction of the carotenoid.

Once produced, ^3CAR can easily return to the ground state dissipating the energy as heat or it can be quenched physically via enhanced intersystem crossing by ground state oxygen, Scheme

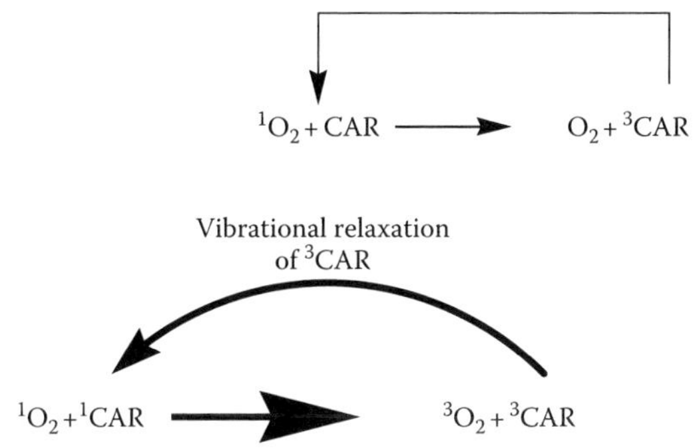

SCHEME 14.1

14.1. Thus, the carotenoid acts as a catalyst deactivating $^1O_2$. Many different carotenoids have been studied to investigate the influence of different carotenoid structural characteristics on the ability to quench $^1O_2$. Much of this work has been carried out in organic solvents with some typical results, taken from Conn et al. (1991), Rodgers and Bates (1980), and Edge et al. (1997) as shown in Table 14.1.

The three unsymmetrical carotenoids such as asteroidenone, adonixanthin, and adonirubin are not well known and their structures are shown in Figure 14.1. However, they have been studied in detail as $^1O_2$ quenchers both in benzene and methanol as shown in Table 14.2.

TABLE 14.1
Singlet Oxygen Quenching Rate Constants for Carotenoids in Benzene

Carotenoid	*N*	k_q (×10^9 M^{-1} s^{-1})
Dodecapreno-β-carotene	19	23.0
Decapreno-β-carotene (DECA)	15	20.0
Tetradehydrolycopene	15	10.7
Rhodoxanthin	12 (+2, C=O)	12.0
Astaxanthin (ASTA)	11 (+2, C=O)	14.0
Canthaxanthin (CAN)	11 (+2, C=O)	12.0
Lycopene (LYC)	11	17.0
Dihydroxylycopene	11	5.1
All-*trans*-β-carotene (β-CAR)	11	13.0
15-*cis*-β-carotene	11	11.0
9-*cis*-β-carotene	11	11.0
Zeaxanthin (ZEA)	11	12.0
α-carotene	10	12.0
β-apo-8′-carotenal (APO)	10	5.27
Lutein (LUT)	10	6.64
Violaxanthin	9	16.0
Septapreno-β-carotene (SEPTA)	9	1.38
7,7′dihydro-β-carotene (77DH)	8	0.3

Note: *N*, Number of conjugated double bonds.

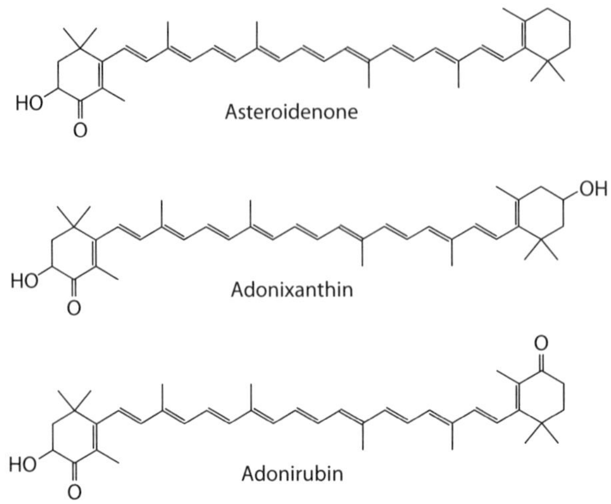

FIGURE 14.1 Structures of three asymmetrical xanthophylls.

TABLE 14.2
Singlet Oxygen Quenching Rate Constants for Asymmetric Carotenoids in Benzene and Methanol

Carotenoid	*n*	k_q ($\times 10^9$ M^{-1} s^{-1}) (Benzene)	k_q ($\times 10^9$ M^{-1} s^{-1}) (MeOD)
Asteroidenone	11 (+1 C=O)	14.8	18.2
Adonixanthin	11 (+1 C=O)	12.3	18.2
Adonirubin	11 (+2 C=O)	10.4	13.2

Source: Burke, M., Pulsed radiation studies of carotenoid radicals and excited states, PhD thesis, University of Keele, Keele, U.K., 2001.

As can be seen in homogeneous environments such as benzene quenching of singlet oxygen by carotenoids is near to diffusion controlled ($k_q \sim 1 \times 10^{10}$ M^{-1} s^{-1}) and the rate constants given in Table 14.1 indicate that the ability of the carotenoids to quench singlet oxygen increases with the increasing number of conjugated double bonds (*n*). This data are in agreement with Devasagayam et al. (1992) who noted that the quenching efficiency increases with increasing wavelength of $\pi\pi^*$ absorption maximum. This principle suggests that the energy transfer from excited $^1O_2$ becomes more exothermic as the conjugation of the carotenoid increases. Of course, simple Hückel theory predicts a lowering of singlet state energy (of the carotenoid) on increasing conjugation, accompanied by a decrease in the triplet energy level. In fact, several research groups have demonstrated a linear relationship between λ_{max} of the ground state and the triplet state, which is the state involved in the quenching process. Farmilo and Wilkinson (1973), and Wilkinson and Ho (1978) demonstrated that electron exchange energy transfer is the principal mechanism by which carotenoids accept excitation energy from $^1O_2$, producing the carotenoid triplet state (Equation 14.1).

The three unsymmetrical carotenoids have also been studied in methanol (Burke 2001) and all are very efficient singlet oxygen quenchers. This may be attributable to the polarity of the molecules. These asymmetrical XANs will possess a permanent dipole and their solvent interaction will be increased compared to symmetrical carotenoids. This enhanced solvent interaction will lower the energy of the triplet state, making energy transfer from $^1O_2$ to the carotenoid faster.

As noted earlier, environments such as water/methanol mixtures are useful models of membrane environments. These mixed solvents lead to a reduced efficiency of $^1O_2$ quenching and the quenching becomes negligible at high water concentrations. Figure 14.2 shows an example of this behavior for zeaxanthin (ZEA), as the aggregation of ZEA is increased.

At 70% methanol (30% D_2O), very little quenching is observed and this correlates with the formation of a new band in the ground state spectrum in methanol/water mixtures as shown in Figure 14.3.

In general, when water is added to homogeneous organic solutions containing carotenoids, spectral changes indicate that carotenoid aggregation occurs. The absorption band attributed to the monomer decreases with the addition of water (>15%) with the concomitant increase in a new absorption band at lower wavelength attributed to a carotenoid dimer/aggregate. The spectral shift of the carotenoid dimer/aggregate to shorter wavelength is attributed to exciton coupling interactions. This splitting leads to a forbidden lower energy transition and an allowed higher energy transition leading to a blue shift. Overall, the stacking of carotenoids occurs in order to reduce the exposure of the hydrophobic system to the polar aqueous environment.

Cantrell et al. (2003) studied the quenching of $^1O_2$ by several dietary carotenoids in dipalmitoyl phosphatidylcholine (DPPC) unilamellar liposomes. These workers used water soluble and lipid soluble $^1O_2$ sensitizers so that a comparison of the efficiencies of quenching $^1O_2$ generated within and outside the membrane model could be made. Perhaps surprisingly there was little difference in the efficiency of quenching in either situation. Typical results are presented in Table 14.3 (taken from Cantrell et al. (2003 and 2006)).

This implies that the rate-determining step is the migration of the $^1O_2$ through the membrane rather than through the water to the membrane surface. However, as can be seen, there was a marked difference in the behavior of the different dietary carotenoids with all-*trans*-β-carotene (β-CAR) and lycopene (LYC) being the most efficient and the XANs, especially lutein (LUT), being rather inefficient. For ZEA, a pivotal XAN in the protection of the macular, a particularly unexpected result was reported. It is instructive to compare β-cryptoxanthin (β-CRYP), with only one terminal hydroxyl group, and ZEA, with two such groups.

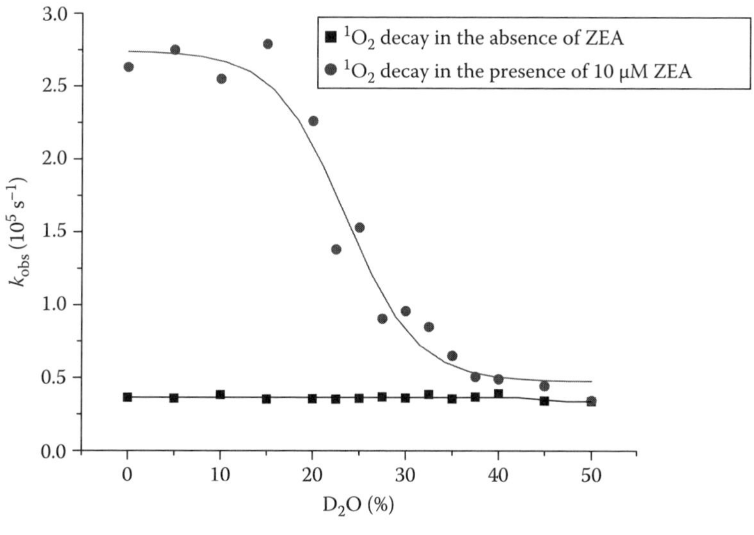

FIGURE 14.2 The effect of increasing D_2O (inducing zeaxanthin aggregation) on the singlet oxygen deactivation efficiency of zeaxanthin.

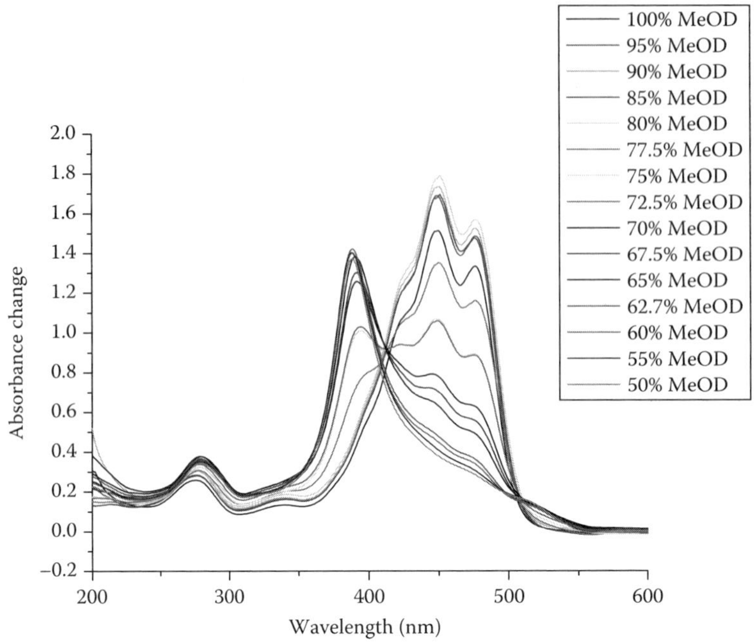

FIGURE 14.3 (See color insert following page 336.) Ground state absorption spectra of 1×10^{-5} M zeaxanthin in various MeOD/D_2O mixtures.

TABLE 14.3
Second-Order Quenching Rate Constants for the Quenching of $^1O_2$ by Carotenoids in Unilamellar DPPC Liposomes, Benzene, and Triton X-100/405 Micelles

		k_q ($\times10^8$ M^{-1} s^{-1})			
		DPPC Liposomes			
Carotenoid	*n*	RB Sensitization	PBA Sensitization	Benzene	Micelles
LYC	11	24.0	23	170	20
β-CAR	11	23	25	130	24
CAN	11	23	16	120	30
ASTA	11	5.9	—	110	29
ZEA*	11	2.3	1.7	160	25
β-CRYP	11	1.8	1.4	130	—
LUT	10	1.1	0.82	66	33

Note: RB, rose bengal and PBA, 4-(1-pyrene)butyric acid.
* Values obtained at low concentrations from linear portion of curve.

Figure 14.4 shows that β-CRYP (like all carotenoids in homogeneous solution and all except ZEA in liposomes) exhibits a linear plot with the quenching of $^1O_2$ increasing as the concentration of the carotenoid increases. While ZEA shows a bell-shaped plot and zero singlet oxygen quenching at concentrations >70 μM (see Figure 14.5). Such behavior of ZEA is symptomatic of its unique

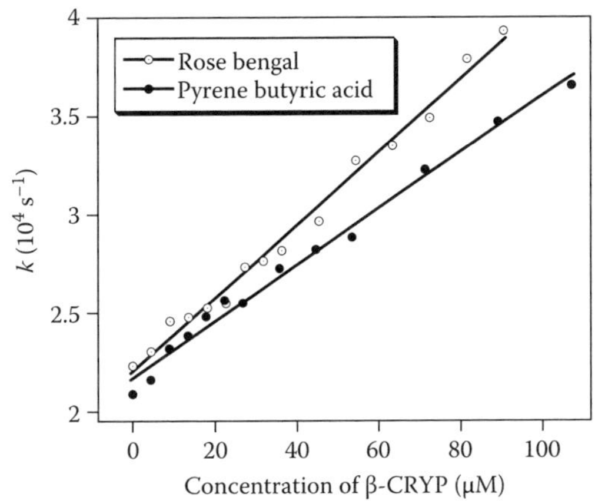

FIGURE 14.4 Rate of decay of $^1O_2$ against β-CRYP concentration in air-saturated solutions of DPPC unilamellar liposomes using either RB or PBA as $^1O_2$ sensitizer.

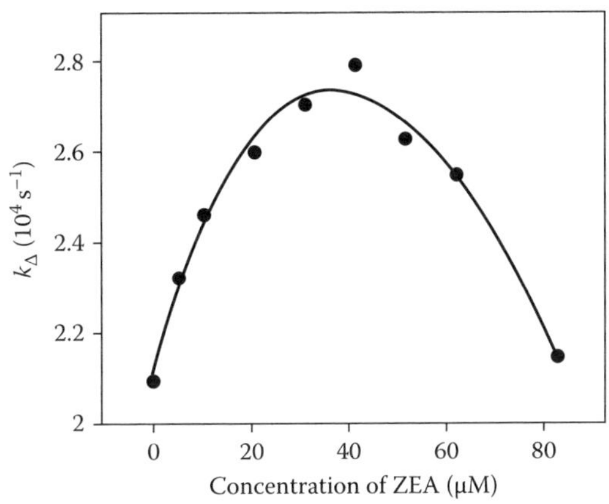

FIGURE 14.5 Rate of decay of $^1O_2$ against ZEA concentration in air-saturated solutions of DPPC unilamellar liposomes using RB as singlet oxygen sensitizer (a similar, but less marked, effect is observed with PBA as sensitizer).

properties and its location and orientation within the membrane. ZEA is a dihydroxy-carotenoid with a rodlike structure and has a tendency to form aggregates within a liposomal environment. The polar hydroxyl groups of ZEA are likely to form hydrogen bonds with the polar head groups of the lipid, and ZEA is therefore anchored to the lipid bilayer. The biophysical interactions of ZEA with the lipid membrane result in ZEA exerting a major influence on the properties of the bilayer (Okulski et al. 2000) in such a way as to rigidify the membrane and inhibit the penetration of small molecules. Such effects are likely to influence the interactions of ZEA with other molecules/species present in the aqueous phase or within the membrane and may restrict radical and excited state scavenging, particularly at higher concentrations. However, β-CRYP that contains only one hydroxy group is likely to have greater freedom and may be less prone to form aggregates. Furthermore, the ground state spectra of ZEA and β-CRYP differ; ZEA shows a sharp, blue-shifted spectrum in methanol:water mixtures (see earlier), thought to be caused by a "card-pack" H-type aggregate (Okulski et al. 2000). That is, ZEA behaves quite like the carotenoids that aggregated in water/methanol solutions while the other carotenoids in the DPPC liposomes do not exhibit this behavior.

For comparison, included in Table 14.3 are the k_q values obtained in detergent micelles along with k_q values obtained in homogeneous solvent benzene. As can be seen, the second-order rate constant for $^1O_2$ quenching in a liposomal environment is a factor of ~4 lower for β-CAR compared to the second-order rate constant obtained in the aromatic solvent. While, there is a marked ~80–130 fold difference between the k_q values determined in liposomal environments compared to the k_q values determined in the aromatic solvent for the XANs.

The present results for β-CAR incorporated into DPPC vesicles compare favorably with those of β-CAR in detergent micelles, this is to be expected because carotene molecules reside in the hydrophobic core of the micelle and likewise they reside in the hydrophobic region of the phospholipid bilayer of liposomes (between the two lipid layers) away from the water interface as depicted in Figure 14.6 (taken from Burke (2001)). Although the two types of vesicles have somewhat different structures, $^1O_2$ penetration into each type of vesicles is required before β-CAR is able to quench $^1O_2$. It is known that nonpolar carotenoids, in particular the carotenes, decrease the penetration barrier for small molecules to the membrane headgroup region of phospholipid vesicles. Most probably, due to the additional space in the headgroup region, resulting from the pigment–lipid interaction in the hydrophobic region of the phospholipid bilayer, there is a greater permeability in the head group region, which aids $^1O_2$ diffusion throughout the entire lipid bilayer, by acting as a portal of entry for $^1O_2$.

The second-order quenching rate constants for the two XANs in DPPC liposomes are quite different from those reported in micelles. In micelles, where XANs are accommodated in a similar manner to carotenes, very little variation in the second-order quenching rate constant is observed (see Table 14.3), but in contrast, a ~26-fold difference in reactivity is observed between the XANs, LUT, ZEA, and β-CAR in a liposomal environment. There are two possible explanations for this; polar carotenoids such as ZEA and LUT incorporated into liposome bilayers have been shown to limit molecular oxygen penetration within the lipid bilayer as demonstrated by the pigment-related decrease of oxygen diffusion-concentration product (Subczynski et al. 1992). Due to their transmembrane orientation with both polar end groups anchored at the inner and the outer lipid–water interface respectively (Gruszecki and Sielewiesiuk 1990), they act as "molecular rivets" rigidifying the lipid membrane by restricting many molecular motions of individual lipid molecules. This type of interaction reinforces the lipid bilayer and thus restricts the diffusion of small molecules

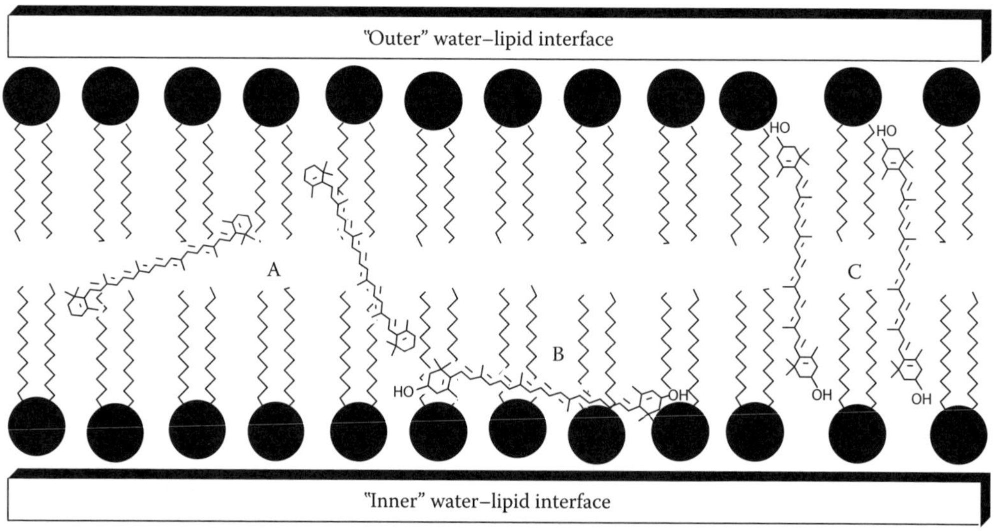

FIGURE 14.6 Typical orientations of carotenoids within a lipid bilayer (A denotes β-CAR, B LUT, and C ZEA).

such as excited state oxygen through the lipid bilayer. Secondly, polar carotenoids aggregate more effectively than their nonpolar counterparts, within phospholipid bilayers (Gruszecki 1990) and it has been shown that the efficiency of $^1O_2$ deactivation decreases the XAN aggregation to a greater extent. These two points may in part explain the huge difference between the determined second-order rate constants for β-CAR and the XANs ZEA and LUT in liposomes. The second-order rate constant for the quenching of $^1O_2$ by LUT embedded within the lipid bilayer of unilamellar liposomes is slightly lower than the value observed for ZEA. This may reflect the orientation of XANs within the bilayer; LUT unlike ZEA has the potential to rotate an entire terminal ring round about the 6′–7′ single bond. This provides the possibility of interaction of both hydroxyl groups located at the 3 and 3′ positions with the same water–lipid interface (Gruszecki 1999). However, conformational (high level) calculations on the barriers to ring rotation indicate rather low values (4–8 kcal mol^{-1}) and it is likely that a combination of flex/extension and rotational barriers taken together with the extent of H-bonding that controls the site selection in the membrane (J. Landrum, personal communication). This possible conformation of LUT allows for the existence of two essentially different pools of pigment molecules, one orientated perpendicular to the plane of the membrane and the second having the same orientation as ZEA (almost parallel to the plane of the membrane). If the "twisted" LUT molecules anchor themselves to the inner water–lipid interface of the liposome, via the two-hydroxyl groups, then if quenching of $^1O_2$ is to occur, $^1O_2$ must traverse the entire lipid bilayer in order to relinquish its excitation energy to a LUT molecule anchored to the inner water–lipid interface.

In summary, it should be noted that the two XANs pivotal in the macular protection, LUT and ZEA (together with β-CRYP), while efficient $^1O_2$ quenchers in solvents such as benzene are the most inefficient in the cell membrane models. Singlet oxygen quenching efficiency is dependent upon the environment. In organic solvents, such as benzene, quenching is near to diffusion controlled but in mixed solvent systems, such as water/methanol, quenching may approach zero as the carotenoids tend to form aggregates.

The aggregation and the orientation of a carotenoid in the lipid bilayer may be major factors in determining the efficiency of $^1O_2$ quenching, for example, ZEA may span the membrane and aggregate while β-CAR, β-CRYP, and LYC are more randomly ordered.

14.3 INTERACTIONS OF CAROTENOIDS WITH FREE RADICALS

14.3.1 Sulfur-Containing Radicals

The reaction of β-CAR with thiyl (RS$^•$) and thiyl sulfonyl (RSO$_2^•$) radicals have both been reported using pulse radiolysis (Everett et al. 1995, 1996). It was found that radical addition to β-CAR occurred and that β-CAR scavenges the thiyl radical, including that derived from glutathione, only via this mechanism, whereas it reacts with thiyl sulfonyl radicals by electron transfer as well.

Mortensen et al. (1997) and Mortensen (2000) used pulse radiolysis to generate RS$^•$ from RSH via H atom transfer to a carbon-centered radical ($^•CH_2(CH_3)_2COH$). The two thiyl radicals studied were the glutathione radical and the $HOCH_2CH_2S^•$ (2-mercaptoethanol thiyl) radical. In each case, there was a loss of ground state absorption due to the parent carotenoid but no corresponding absorption was detected at wavelengths longer than 600 nm. The adduct was found to absorb in a similar spectral region as the ground state of the parent carotenoid and the bleaching in this spectral region was biphasic with a fast step due to an addition process:

$$\mathrm{CAR} + \mathrm{RS}^{\bullet} \rightarrow [\mathrm{RS{-}CAR}]^{\bullet} \tag{14.7}$$

and a slower bimolecular step proposed as the decay of this adduct:

$$2[\text{RS–CAR}]^{\bullet} \rightarrow \text{products per second} \tag{14.8}$$

The rate of reaction was found to be virtually independent of the carotenoid structure, which is in contrast to electron transfer reactions (see Section 14.3.2).

14.3.2 NO_x

It has been suggested (Gabr et al. 1995) that nitric oxide ($NO^{\bullet}$), which is, of course, a radical, bleaches β-CAR presumably by forming addition complexes. However, when we completely exclude oxygen from the system we found no evidence of an interaction between NO and β-CAR (unpublished). Therefore, the observed reaction by Gabr et al. may have been due to nitrogen dioxide ($NO_2^{\bullet}$).

In fact, Everett et al. (1995, 1996) have reported the scavenging of $NO_2^{\bullet}$ by β-CAR, and their results indicate that the reaction proceeds via electron transfer only and no radical addition occurs. The electron transfer was shown to proceed with a rate constant of $1.1 \times 10^8\ M^{-1}\ s^{-1}$ in *tert*-butanol/water mixtures (50% v/v). This study was extended by the same workers (Mortensen et al. 1997) to include five other carotenoids, with canthaxanthin (CAN) having the lowest rate constant of reaction with $NO_2^{\bullet}$ ($1.2 \times 10^7\ M^{-1}\ s^{-1}$), and LYC having the second highest ($1.9 \times 10^7\ M^{-1}\ s^{-1}$) after ZEA ($2.1 \times 10^7\ M^{-1}\ s^{-1}$). All the rate constants obtained were an order of magnitude below that for β-CAR. However, the experiments were carried out in 60:40%, v/v *tert*-butanol/water mixture (80:20%, v/v for LYC due to aggregation) rather than the 50% (v/v) mixture used for β-CAR and the $NO_2^{\bullet}$ was generated in a different way.

Böhm et al. (1995) have studied the protective effect of β-CAR and LYC against cell membrane damage by $NO_2^{\bullet}$, showing that LYC is more than twice as effective as β-CAR. These authors observe two species from the reaction, both in the infrared, assigning them to the radical cation and a radical addition product.

A possible explanation is that at the high concentrations of $NO_2^{\bullet}$ addition across a carotenoid double bond could occur. This reaction has been observed by Pryor and Lightsey (1981) for cyclohexene when concentrations of 1% $NO_2^{\bullet}$ (10,000 ppm) were used, and Kikugawa et al. (1997) have shown that β-CAR in hexane is completely destroyed by two equimolar amounts of $NO_2^{\bullet}$, with the absorption spectra gradually decreasing and blue-shifting, possibly indicating a gradual decrease in conjugation.

We have studied the effect of the combinations of antioxidants loaded onto cells *in vivo* via supplementation as well as via *in vitro* incubation with human lymphocytes. These studies were also extended to include peroxynitrite-induced cell membrane damage as well as $NO_2^{\bullet}$-induced damage. Both peroxynitrite (ONOO–) and $NO_2^{\bullet}$ can be formed from $NO^{\bullet}$ (Beckman and Crow 1993), which is a radical with a wide range of important *in vivo* roles, such as the control of systemic blood pressure and acting as a messenger molecule and it is present in cigarette smoke, at up to 500 ppm (Cueto and Pryor 1994). Of course, $NO_2^{\bullet}$ is also a major environmental air pollutant and it can initiate lipid peroxidation. Peroxynitrite also initiates lipid peroxidation (Radi et al. 1991) and it has been shown to oxidize proteins (Lacsamana and Gebicki 1996).

The results of the lymphocyte experiments with $NO_2^{\bullet}$ and ONOO– are given in Tables 14.4 and 14.5. The major finding is that cells that are treated with the β-CAR in addition to vitamins E and C *in vivo* and exposed to $NO_2^{\bullet}$ show the cell staining of 6.0% whereas, without the antioxidants, the cell staining was 61.4%. That is, the presence of all three of the antioxidants leads to a protection factor (PF) of 10.2. the protection by β-CAR alone gave a PF of only 2.0, for α-tocopherol alone it was 1.8 and for ascorbic acid 1.2.

For *in vitro* treatment, the antioxidant combination leads to a PF of 10.0. With β-carotene alone as the antioxidant the PF was only 3.5, while for α-tocopherol alone it was 3.6, and for ascorbic acid alone there was no significant protection.

TABLE 14.4
Lymphocyte Membrane Protection by Antioxidants against NO_2 Cell Membrane Destruction Is Shown by Cell Staining with Eosin

Cells Incubated with	Percentage of Stained Cells	Protection Factor
β-CAR + vitamins E + C *in vivo*	6.0 (without 61.4)	10.2
β-CAR in vivo	26.9 (without 53.2)	2.0
Vitamin E *in vivo*	28.4 (without 50.1)	1.8
Vitamin C *in vivo*	41.0 (without 51.0)	1.2
β-CAR + vitamins E + C *in vitro*	5.3 (without 52.9)	10.0
β-CAR *in vitro*	14.6 (without 51.6)	3.5
Vitamin E *in vitro*	14.8 (without 53.0)	3.6
Vitamin C *in vitro*	48.0 (without 48.9)	1.0

Note: For the in vitro experiments the corresponding cell staining was 5.3% and 52.9%.

TABLE 14.5
Lymphocyte Membrane Protection by Antioxidants against ONOO– Cell Membrane Destruction Is Shown by Cell Staining with Eosin

Cells Incubated with	Percentage of Stained Cells	Protection Factor
β-CAR + vitamins E + C *in vivo*	5.2 (without 43.3)	8.3
β-CAR *in vivo*	32.4 (without 55.1)	1.7
Vitamin E *in vivo*	27.1 (without 53.8)	2.0
Vitamin C *in vivo*	36.1 (without 50.9)	1.4
β-CAR + vitamins E + C *in vitro*	7.3 (without 59.5)	8.2
β-CAR *in vitro*	38.1 (without 49.1)	1.3
Vitamin E *in vitro*	14.0 (without 48.0)	3.4
Vitamin C *in vitro*	34.9 (without 47.9)	1.4

The second major finding is that cell protection was also observed against the peroxynitrite anion. Thus, *in vivo*, the staining increased from 5.2% with the three antioxidants to 43.3% without the antioxidants (giving a PF of 8.3). For the *in vitro* experiments, the corresponding cell staining was 7.3% and 59.5%, that is, a PF against ONOO– of 8.2 as shown in Table 14.5.

Hence, for both of the oxidants, $NO_2^{\bullet}$ and ONOO–, a marked synergism in cell protection by the antioxidant combination of β-CAR with vitamins E and C was observed for both *in vivo* and *in vitro* experiments, although the synergistic effect was more pronounced in protection from $NO_2^{\bullet}$.

The results on the cellular protection against $NO_2^{\bullet}$ can be interpreted as the $NO_2^{\bullet}$ reacting with the three antioxidants to produce their radicals, with ascorbic acid reacting least efficiently, probably due to the lower reduction potential of its radical. Moreover, Arroyo et al. (1992) reported that $NO^{\bullet}$- and $NO_2^{\bullet}$-induced mutations in *Salmonella typhimurium* TA1535 were inhibited efficiently by β-CAR and tocopherols, but not at all by ascorbic acid.

The synergistic effect observed in the presence of all three antioxidants implies that there is an interaction between the individual antioxidant components. The direct interaction of the α-tocopherol radical and ascorbic acid is already well established (Bisby and Parker 1995) and a study by Mayne and Parker (1989) on chicks deficient in vitamin E and selenium showed that the addition of CAN to their diet increased their resistance to lipid peroxidation mainly by increasing membrane α-tocopherol levels, and only weakly by a direct antioxidant effect. Moreover, Li et al. (1995) reported synergism between α-tocopherol and β-CAR in inhibiting the peroxidation of linoleic acid, observing a diminished consumption of vitamin E in the presence of β-CAR. In addition to these results, later there will be a discussion on the direct interaction between α-tocopherol radical cation and carotenoids, as well as between carotenoid radical cations and vitamin C.

The protection of cells against ONOO– is difficult to interpret since no β-$CAR^{\bullet+}$ formation is observed when ONOO– is generated in water, with the β-CAR in micelles. However, a slow reaction may occur, and indeed Kikugawa et al. (1997) have shown that ONOO–/ONOOH (prepared from H_2O_2 and $NO_2^{\bullet}$) reacts with β-CAR, observing ground state–bleaching in a dose-dependent manner. They also found that the loss of the β-CAR absorption was only partially inhibited by both α-tocopherol and ascorbic acid (50% and 70%, respectively) indicating that β-CAR is a better scavenger.

14.3.3 Peroxyl Radicals

14.3.3.1 Arylperoxyl Radicals

Arylperoxyl radicals ($ArO_2^{\bullet}$) (9-phenanthryl peroxyl, 1-naphthyl peroxyl, and 2-naphthyl peroxyl) were generated via the pulse radiolysis of arylbromides in methanol (see reactions below) (Edge 1998):

$$ArBr + e^-_{sol} \rightarrow Ar^{\bullet} \tag{14.9}$$

$$Ar^{\bullet} + O_2 \rightarrow ArO_2^{\bullet} \tag{14.10}$$

The arylperoxyl radicals produced have absorbtion maxima at 750, 800, and 550 nm for 9-phenanthryl peroxyl, 1-naphthyl peroxyl, and 2-naphthyl peroxyl radicals, respectively, and are not observed in argon-saturated solutions, supporting their assignments as peroxyl radicals.

The rate constants for the reactions of the arylperoxyl radicals with carotenoids were determined from the first-order kinetics of the formation of the carotenoid radicals produced (using a range of carotenoid concentrations). The three arylperoxyl radicals were all observed to react with carotenoids to yield the carotenoid radical cations via electron transfer.

From Table 14.6 it can be seen that, with the exception of astaxanthin (ASTA), the rate constants for the electron transfer reactions decrease for each carotenoid in the order 9-phenanthryl peroxyl > 1-naphthyl peroxyl > 2-naphthyl peroxyl. This order of reactivity should be related to the reduction potentials of the radicals, with 9-phenanthryl peroxyl having the highest reduction potential. The same order of reactivity for these three arylperoxyl radicals reacting with Trolox was shown by Neta and coworkers (Alfassi et al. 1995). The reactivities of all the carotenoids studied are similar

TABLE 14.6
Second-Order Rate Constants for the Reaction of $ArO_2^{\bullet}$ with Carotenoids

	k (×10^8 M^{-1} s^{-1}) ±20% for Reaction with Carotenoids			
Peroxyl Radical	**Lutein**	**Zeaxanthin**	**Astaxanthin**	**β-Carotene**
9-phenanthryl peroxyl	4.0	3.0	—	8.8
1-naphthyl peroxyl	0.9	1.3	0.8	0.3
2-naphthyl peroxyl	0.5	0.2	1.4	—

for each arylperoxyl radical, indicating that the nature of the carotenoid does not have a significant effect upon these electron transfer reactions. This was also the conclusion of Mortensen et al. (1997), who found that as well as the rate of scavenging, the mechanism of the scavenging (i.e., radical addition, electron transfer, or both) is strongly dependent on the nature of the oxidizing species and much less dependent on the carotenoid structure. Their work was also undertaken in a polar solvent, hence it could be that significant differences in carotenoid scavenging abilities are more easily observed in hydrocarbon solvents, as used in other studies by us reported later and in another study by Mortensen and Skibsted where large differences in the carotenoid antioxidant activity has been reported (Mortensen and Skibsted 1997a).

14.3.3.2 Chlorinated Peroxyl Radicals

Packer et al. generated the trichloromethyl peroxyl radical $CCl_3O_2^{\bullet}$ via pulse radiolysis (Packer et al. 1981). In the presence of β-CAR there was a fast bleaching of the carotene ground state with a rate constant of $1.5 \times 10^9\ M^{-1}\ s^{-1}$. The loss of ground state absorption was accompanied by an increase in absorption in the near infrared, indicating that interaction between the peroxyl radical and the β-CAR produces $\beta\text{-CAR}^{\bullet+}$. Hill et al. further extended this work, studying the interaction of the $CCl_3O_2^{\bullet}$ radical with six carotenoids in aqueous TX-100 micelles at pH 7 (Hill et al. 1995). They observed two peaks with different λ_{max}, in the near-infrared spectral region for all carotenoids studied with different kinetics at the two wavelengths. The species absorbing at the shorter wavelength decayed into the other species, which was assigned to the radical cation. The species absorbing at the shorter wavelength was suggested to be an addition radical similar to that proposed by Burton and Ingold (1984) that subsequently "falls apart" to yield the radical cation. Hill et al. also suggest that oxygen-centered radicals are required for the production of adducts, since without the presence of oxygen, the $CCl_3^{\bullet}$ radical reacts with the carotenoids yielding the carotenoid radical cation only. We have also observed (unpublished results) similar reactions upon the pulse radiolysis of β-CAR in chloroform or carbon tetrachloride, with a species absorbing at lower wavelengths than the radical cation only observable when oxygen is present. However, in dichloromethane only the radical cation spectrum was observed. In addition, Adhikari et al. (2000) have also observed similar reactions with $CCl_3O_2^{\bullet}$ and $CBr_3O_2^{\bullet}$ in a quaternary microemulsion and found that retinol is formed as a stable product.

The work by Hill et al. also noted differences for ASTA compared with the other carotenoids studied. Its radical cation was not formed initially from $CCl_3O_2^{\bullet}$, but was formed solely through the proposed addition radical. Unfortunately, LYC could not be studied due to its insolubility in TX 100 micelles. However, since LYC appears, from its quenching of $^1O_2$ and its protection against $NO_2^{\bullet}$, to be the most efficient natural carotenoid antioxidant, we repeated this work using 4% TX 405:TX 100 (4:1) mixed micelles for both β-CAR and LYC (unpublished) and have observed LYC behaving in a different manner to the other carotenoids as there appears to be no conversion of the "adduct" to the radical cation.

Skibsted and coworkers (Mortensen and Skibsted 1996) have shown that upon the laser flash photolysis of carotenoids in chloroform bleaching of the ground state absorption is observed and there is formation of two near infrared–absorbing species ($\lambda_{max} \approx 920$ and 1000 nm for β-CAR). The species absorbing at about 1000 nm is $\beta\text{-CAR}^{\bullet+}$ and, as with the carotenoid/$CCl_3O_2^{\bullet}$ system noted earlier, the $\beta\text{-CAR}^{\bullet+}$ is formed from the other species. The nature of the other species is not defined although an adduct or a neutral carotenoid radical is proposed.

This work was extended to carotenoids containing keto, hydroxy, and aldehyde groups in halogenated solvents (Mortensen and Skibsted 1997b). All the XANs produce a transient species in $CHCl_3$ absorbing in the 850–960 nm region following laser excitation and this transient decays by first-order kinetics to the radical cation absorbing at longer wavelengths (870–1040 nm). In contrast, the authors note that, while carotenoids are also bleached in CCl_4, no near infrared–absorbing species arise on laser excitation in this solvent. Possibly the neutral radical, $CAR^{\bullet}$, is produced via hydrogen atom transfer, and this may not absorb in the near infrared.

14.3.3.3 Acylperoxyl Radicals

The reactions of carotenoids with several acylperoxyl radicals have been undertaken (Mortensen 2001, El-Agamey and McGarvey 2003) (e.g., the phenylacetylperoxyl radical) and different reactions have been observed depending on the polarity of the solvent. The initial product observed in all solvents absorbs in the visible and is proposed to be an addition radical. In hexane or benzene, there is no formation of near infrared–absorbing species and the addition radical decays forming epoxides or cyclic ethers. In polar solvents, two near infrared–absorbing species are observed, with the longer wavelength absorber being assigned as the radical cation. The relative amounts of the two species have been shown to change with dielectric constant, with the radical cation absorption decreasing relative to that of the shorter wavelength absorber as the solvent polarity decreases. In fact, in 1-decanol for 7,7′-dihydro-β-carotene the radical cation cannot be observed and the absorption due to the radical addition product at 450 nm is larger than that due to the shorter wavelength near–infrared absorber. This observation suggested that the shorter wavelength near-infrared absorber is either an ion-pair of the carotenoid radical cation and the peroxide anion or an isomer of the carotenoid radical cation that thermally isomerizes.

14.3.4 Reducing Radicals

The radical anions of a variety of carotenoids have been shown to absorb in the infrared (like the radical cations). The anions typically absorb at wavelengths around 120 nm shorter than their respective radical cations in nonpolar solvents, such as benzene and hexane. However, for carotenoids containing carbonyl groups on the rings, the order is switched and it is the anions that absorb farthest to the red (Dawe and Land 1975, Lafferty et al. 1977, Hill 1994).

The radical anions have been shown by electrochemistry to be strongly reducing, with E(β-CAR/β-CAR$^{\bullet-}$) = −1.63 V in tetrahydrofuran (Park 1978) or −1.68 V in a mixed aprotic solvent (Mairanovsky et al. 1975) against the standard calomel electrode. Edge et al. (2007) recently studied the one electron reduction of carotenoids in aqueous micellar solutions and by comparing the reactivity of various carotenoids with $CO_2^{\bullet-}$, the acetone ketyl radical (AC$^{\bullet-}$), and ACH$^{\bullet}$ a range between −1950 and −2100 mV (against the normal hydrogen electrode [NHE]) was obtained for the reduction potential of both β-CAR and ZEA and a value more positive than −1450 mV (vs. NHE) for ASTA, CAN, and β-*apo*-8′-carotenol (APO). Surprisingly, in polar solvents the anions of the carbonyl-containing carotenoids absorb to the blue of their respective radical cations, unlike in nonpolar solvents, suggesting that in polar solvents the ground state is stabilized relative to the excited state.

We have also observed (El-Agamey et al. 2006, Edge et al. 2007) that the radical anions of carotenoids containing carbonyl groups abstract a proton from water (or methanol) forming the corresponding neutral radical absorbing at a much shorter wavelength, only just to the red of the neutral carotenoid absorption. (e.g., 580 nm for CANH$^{\bullet}$ in TX-100), see Equation 14.11 and Figure 14.7 as an example of this proton abstraction reaction.

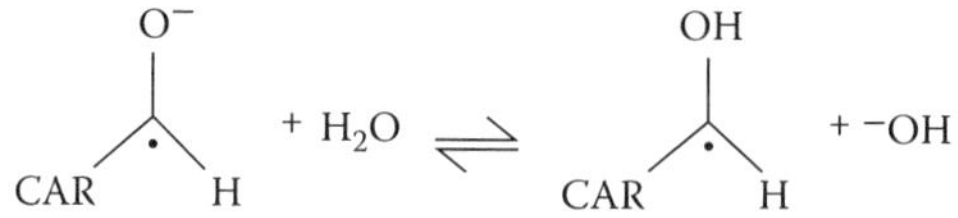

(14.11)

At high pH (~13), the equilibrium is shifted to the left and only CAR$^{\bullet-}$ is observed and upon lowering the pH the amount of CAR$^{\bullet-}$ decreases and CARH$^{\bullet}$ increases. By plotting pH versus the yields of CAR$^{\bullet-}$ or CARH$^{\bullet}$, the pKa of each neutral radical could be determined. These were found to be 10.6 ± 0.2 for ASTAH$^{\bullet}$, 11.7 ± 0.2 for CANH$^{\bullet}$, and 10.2 ± 0.1 for APOH$^{\bullet}$.

The second-order decays of the uncharged neutral radicals are very similar to those of the radical anions so that, perhaps surprisingly, the negative charge does not hinder the radical–radical interaction.

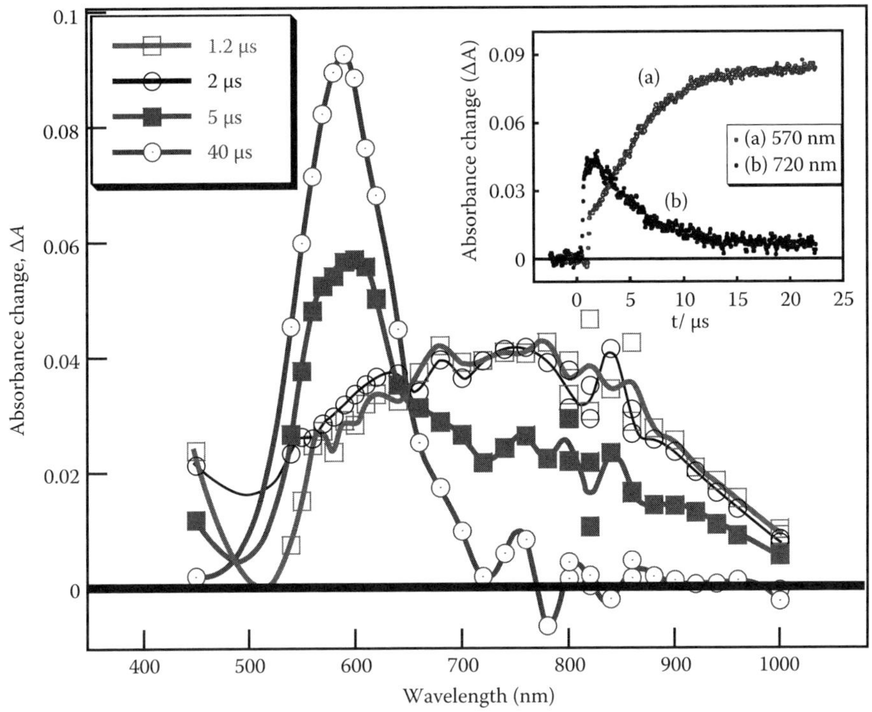

FIGURE 14.7 Transient absorption spectra observed following pulse radiolysis of CAN and formate in argon-saturated aqueous 2% TX-100 (pH=7.1). Inset: Kinetic traces of $CANH^{\bullet}$ at 570 nm and $CAN^{\bullet-}$ at 720 nm, showing the decay of the radical anion and concomitant formation of the neutral radical.

14.4 REACTIVITY OF CAROTENOID RADICALS

14.4.1 Interaction with Oxygen

Carotenoid radical anions contrast with radical cations in that they have been shown to react with oxygen at diffusion-controlled rates (Conn et al. 1992) whereas the radical cations do not react with oxygen (Dawe and Land 1975) at all.

For the neutral addition radicals of carotenoids, with acylperoxyl radicals, it was shown (El-Agamey and McGarvey 2003) that no reaction could be observed with up to 0.01M oxygen, giving an upper limit of $\approx 10^5$ M^{-1} s^{-1} for the rate constant. However, more recently, the same authors (El-Agamey and McGarvey 2005) have reported a reversible oxygen addition to a neutral carbon-centered carotenoid addition radical from the reaction of carotenoids with phenylthiyl radicals. In the absence of oxygen, these radicals decay over hundreds of milliseconds, and the decay was shown to increase with the addition of oxygen. For PhS-77DH$^{\bullet}$, the rate of oxygen addition was shown to be 4.3×10^4 M^{-1} s^{-1}, that is, below their previously suggested limit. This work has been recently extended (El-Agamey and McGarvey 2007) to a wide range of carotenoid-phenylthiyl addition radicals leading to the rate constants of $0.32–4.3 \times 10^4$ M^{-1} s^{-1}.

14.4.2 Interaction with Other Carotenoids

14.4.2.1 Radical Anions

In hexane, β-CAR, LYC, septareno-β-carotene (SEPTA), and decapreno-β-carotene (DECA) were studied and Table 14.7 gives the electron transfer second-order rate constants for various pairs, with

TABLE 14.7
Bimolecular Rate constants for Electron Transfer between Carotenoid Pairs in Argon Saturated Hexane ($CAR_1^{\bullet-} + CAR_2 \rightarrow CAR_1 + CAR_2^{\bullet-}$)

	Rate Constant (±10%)/(×10^9 M^{-1} s^{-1}) for Reaction with CAR_2		
$CAR_1^{\bullet-}$	DECA	LYC	β-CAR
SEPTA$^{\bullet-}$	63	12	20
β-CAR$^{\bullet-}$	11	14	—

Note: The limit of the rate constants for all the back reactions is ≤1 × 10^9 M^{-1} s^{-1}.

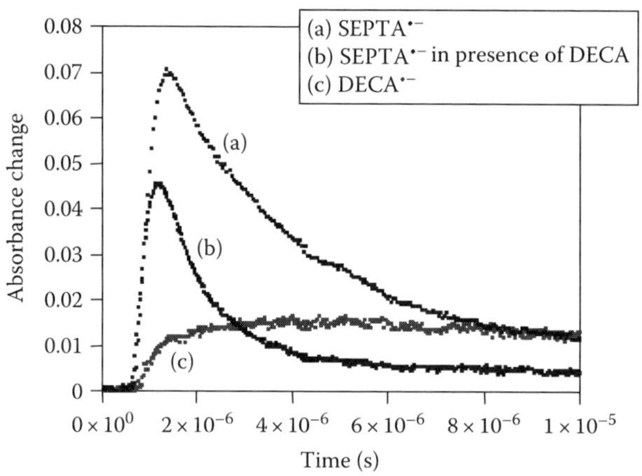

FIGURE 14.8 Decay trace of SEPTA$^{\bullet-}$ with and without 1×10^{-5} M DECA and formation of DECA$^{\bullet-}$ from SEPTA$^{\bullet-}$.

Figure 14.8 showing, as an example, the decay of SEPTA$^{\bullet-}$ in the absence and the presence of DECA and also the growth of the DECA$^{\bullet-}$ as it is formed from SEPTA$^{\bullet-}$.

These results produce an ordering of the one-electron reduction potentials as shown in Figure 14.9. This order is consistent with results on the reactions of oxygen and porphyrins with carotenoids (McVie at al. 1979, Conn et al. 1992), for example, β-CAR$^{\bullet-}$ reacts much more efficiently with oxygen than LYC$^{\bullet-}$ and DECA$^{\bullet-}$. Comparative studies have been made in benzene due to the decreased solubility of XANs in hexane and Table 14.8 gives the corresponding bimolecular rate constants for electron transfer. Overall, the one-electron reduction potentials increase in the order ZEA < β-CAR ≈ LUT < LYC < APO ≈ CAN < ASTA.

These results suggest that hydroxyl groups on the rings of the XANs (as in ZEA and LUT) decrease the reduction potential and that carbonyl groups significantly increase the reduction potential. This is again consistent with results on the reactions of oxygen and porphyrins with carotenoids (McVie at al. 1979, Conn et al. 1992), for example, CAN$^{\bullet-}$ reacts with oxygen at only 1.0 × 10^8 M^{-1} s^{-1} compared with 24 × 10^8 M^{-1} s^{-1} for β-CAR$^{\bullet-}$.

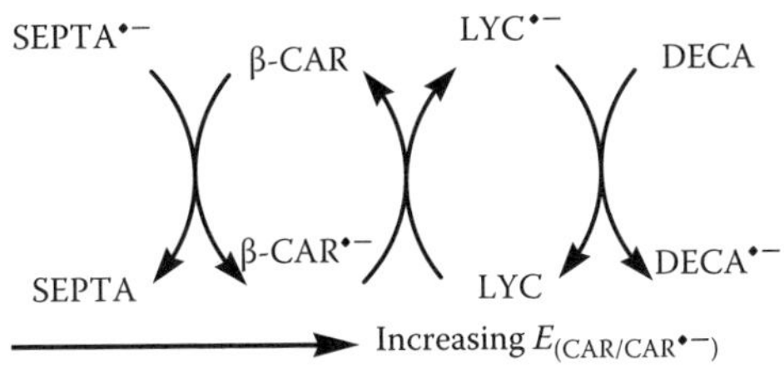

FIGURE 14.9 Relative ordering of the one-electron reduction potentials (E(CAR/CAR•-)) of several carotenoids in hexane.

TABLE 14.8
Bimolecular Rate Constants for Electron Transfer between Carotenoid Pairs in Argon Saturated Benzene ($CAR_1^{•-} + CAR_2 \rightarrow CAR_1 + CAR_2^{•-}$)

	Rate Constant (±10%) ($\times 10^9$ $M^{-1}s^{-1}$) for Reaction with CAR_2					
$CAR_1^{•-}$	ASTA	CAN	APO	LYC	LUT	β-CAR
ZEA•−	15	15	10	3.0	3.8	3.7
β-CAR•−	14	7.7	13	6.2	≤0.5	
LUT•−	13	7.5	10	2.5		
LYC•−	12	10	10			
APO•−	1.1	≤0.2				
CAN•−	1.9					

Note: The limit of the rate constants for all the back reactions is $\leq 5 \times 10^8$ M^{-1} s^{-1}.

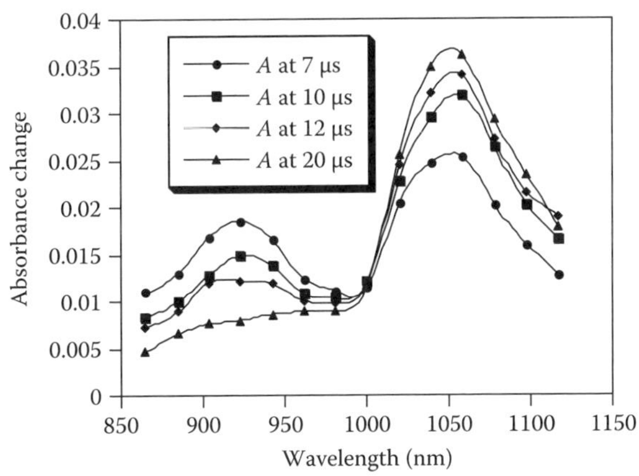

FIGURE 14.10 Transient absorption spectra observed following pulse radiolysis of 1×10^{-4} M ASTA with 1×10^{-5} M LYC in argon flushed benzene.

14.4.2.2 Radical Cations

Figure 14.10 shows the spectral changes over time on the pulse radiolysis of ASTA in the presence of LYC. Similar data were observed for 11 pairs of carotenoids and have allowed the electron transfer second-order rate constants to be determined (see Table 14.9). These pulse radiolysis kinetic studies show LYC efficiently quenches the radical cations of all the XANs studied, whereas β-CAR reduces only $ASTA^{\bullet+}$, $CAN^{\bullet+}$, and $APO^{\bullet+}$ (XANs containing carbonyl groups) and give an order for the ease of electron transfer as shown in Figure 14.11.

It is interesting that $CAR^{\bullet+}$ arising from the three carotenoids present in the human macular (LUT, ZEA, and MZEA [mesozeaxanthin]) are all repaired efficiently by LYC but not by β-CAR. The retina is the only organ in the human body, which is continually exposed to the high levels of focused radiation and is in a highly oxygenated environment and this combination means there is a high likelihood of oxy-radical and $^1O_2$ generation. LUT, ZEA, and MZEA all contain terminal hydroxyl groups, and, as discussed in Section 14.2, this allows them to span membranes. If this is the case, then those XAN containing hydroxyl groups will probably be more accessible to species in the extracellular environment, such as vitamin C, which may be able to regenerate these XAN from their radical cations.

The retina does not contain high concentrations of hydrocarbon carotenoids but Mares-Perlman et al. (1995) have shown a correlation between age-related macular degeneration and low levels of serum LYC and this apparent contradiction is discussed in Section 14.5.

TABLE 14.9
Bimolecular Rate Constants for Electron Transfer between Carotenoid Pairs ($CAR_1^{\bullet-} + CAR_2 \rightarrow CAR_1 + CAR_2^{\bullet-}$)

	Rate Constant (±10%) ($\times 10^9$ M^{-1} s^{-1}) for Reaction with CAR_2		
$CAR_1^{\bullet+}$	LYC	β-CAR	ZEA
$ASTA^{\bullet+}$	9.2	8.0	4.6
$APO^{\bullet+}$	11.2	6.3	7.7
$CAN^{\bullet+}$	7.9	4.8	<1.0
$LUT^{\bullet+}$	5.2	<1.0	<1.0
$MZEA^{\bullet+}$	7.8	<1.0	—
$ZEA^{\bullet+}$	6.9	<1.0	—

Note: All other pairs and back reactions have rate constants $<1 \times 10^9$ M^{-1} s^{-1}.

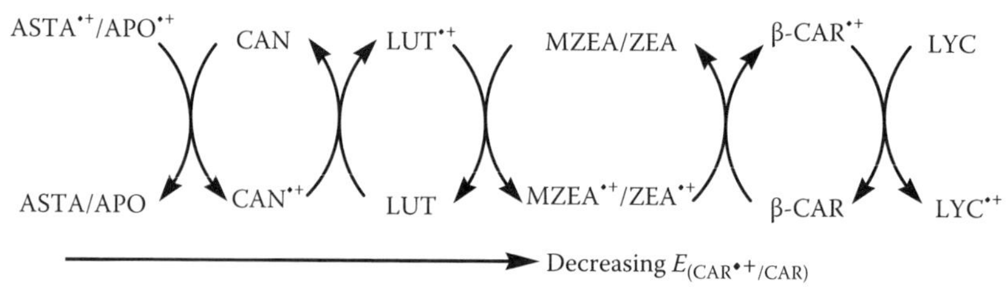

FIGURE 14.11 Relative ordering of the one-electron reduction potentials ($E(CAR^{\bullet+}/CAR)$) of several carotenoid radical cations in benzene.

The ordering of the oxidation potentials correlates well with the order of the λ_{max} of $CAR^{\bullet+}$, the lower the λ_{max} the higher the reduction potential. Hence, a decrease in reduction potential is dependent on extending the chromophore, better overlap of the C=C π-orbitals, and increasing the electron density in the conjugated chain. The results of two electochemical studies are consistent with these results (Mairanovsky et al. 1975, Grant et al. 1988) as well as with studies on scavenging several radical species (Miller et al. 1996, Mortensen and Skibsted 1997a, Woodall et al. 1997a). For example, Mortensen and Skibsted studied the reaction of carotenoids with phenoxyl radicals in di-tert-butyl peroxide/benzene solutions and the fastest rate of reaction was seen for LYC, followed by β-CAR, then the hydroxy-substituted XANs ZEA and LUT. However, this cannot be taken to indicate that the carotenes and the LYC in particular, will necessarily be most effective as antioxidants since they may well be destroyed more quickly. Indeed, in a study by Woodall et al. (1997b) on the inhibition of egg yolk phosphatidylcholine lipid peroxidation by carotenoids, LYC was destroyed fastest and afforded the least protection.

14.4.3 Interaction with Biological Substrates

14.4.3.1 Water-Soluble Antioxidants

The bimolecular rate constants were determined (Burke 2001) for the repair of carotenoid radical cations by trolox, ascorbic, ferrulic, and uric acids from the pulse radiolysis studies of carotenoids in aqueous micellar solutions (see Table 14.10).

As can be seen, all the water-soluble compounds are capable of quenching $CAR^{\bullet+}$ efficiently, with rate constants of the order of 10^6, 10^7, and 10^8 M^{-1} s^{-1} for ferrulic acid, both ascorbic and uric acids, and Trolox, respectively.

Burke et al. (2001a) have also demonstrated that the radical cations of carotenoids are quenched by vitamin C in liposomal environments and Figure 14.12 shows quenching plots for the reaction of the β-$CAR^{\bullet+}$ with a range of vitamin C concentrations and corresponds to a second-order rate constant for the quenching of 1.1×10^7 M^{-1} s^{-1}.

Perhaps surprisingly, due to ZEA and β-CAR having different orientations in membranes (as discussed in Section 14.2), approximately the same value for the quenching rate constant was obtained for $ZEA^{\bullet+}$ (9.7×10^6 M^{-1} s^{-1}) as for β-$CAR^{\bullet+}$.

This suggests that either the β-$CAR^{\bullet+}$, which is more polar than the parent β-CAR, can efficiently reorientate so as to interact with the vitamin C in the aqueous phase, or that the ascorbic

TABLE 14.10
Second-Order Rate Constants for the Repair of Carotenoid Radical Cations by Four Biologically Relevant Molecules in Triton Detergent Micelles

		k ($\times 10^7$ M^{-1} s^{-1})			
$CAR^{\bullet+}$	λ_{max}(nm)	Ascorbic Acid	Trolox	Ferrulic Acid	Uric Acid
β-CAR	940	1.00	19.1	0.082	1.12
LYC	970	1.76	18.6	0.108	1.01
ASTA	880	5.50	51.8	0.583	12.1
CAN	890	4.29	47.2	0.517	1.03
LUT	900	1.76	30.7	0.144	1.48
ZEA	940	1.50	25.6	0.147	0.96
SEPTA	850	3.07	31.3	0.229	3.81
77DH	770	1.17	30.1	0.244	6.17

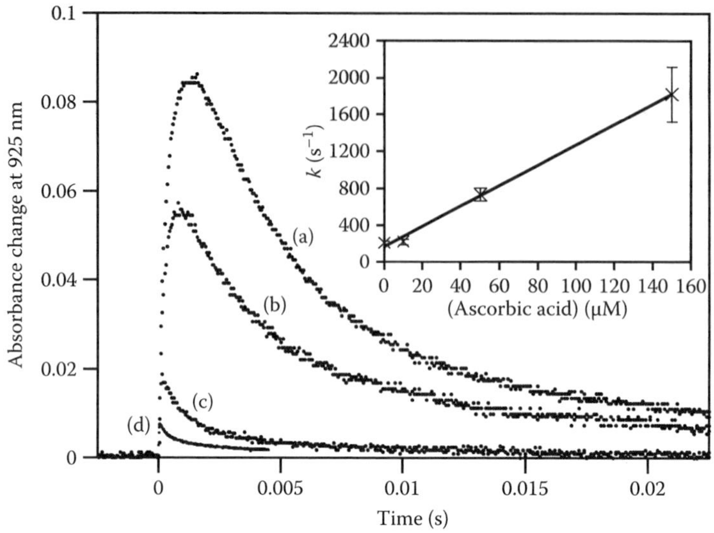

FIGURE 14.12 Decay of β-$CAR^{•+}$ in unilamellar DPPC liposomes (a) without ascorbic acid, (b) with 10 μM ascorbic acid, (c) with 50 μM ascorbic acid, (d) with 150 μM ascorbic acid. Inset: Quenching plot.

acid can penetrate far into the hydrophobic regions of the model membranes. It is known that nonpolar carotenoids, in particular the carotenes, can decrease the penetration barrier for small molecules to the membrane headgroup region of phospholipid vesicles (Chaturvedi and Kurup 1986, Strzalka and Gruszecki 1994). This is most probably due to additional space in the headgroup region resulting from the pigment–lipid interaction in the hydrophobic region of the phospholipid bilayer. This greater permeability in the head group region may in fact aid ascorbic acid diffusion throughout the entire lipid leaflet, by acting as a portal of entry for ascorbic acid. In addition, the fact that ZEA has a rigidifying effect as it spans the membranes may slow the diffusion of small molecules, such as vitamin C, into the membrane.

The second-order rate constant for the repair of $LUT^{•+}$ is approximately half the value observed for $ZEA^{•+}$ (5.2×10^6 M^{-1} s^{-1}). This lower second-order rate constant for LUT may reflect LUT's orientation within the bilayer, as discussed in Section 14.2 for the quenching of singlet oxygen by LUT.

The aforementioned results refer to unilamellar membrane models but essentially similar results are obtained in multilamellar vesicles, though the kinetics are more complex in such systems. The numerical values observed in these model membranes simply show that one or more of the aforementioned factors arise; however, in the *in vivo* situation, the preeminent effect is unknown but may well be the proximity of the hydroxyl group to the water interface.

14.4.3.2 Amino Acids

The studies of tyrosine and cysteine show that at pH 7, both of these amino acids react with $CAR^{•+}$ (see Table 14.11), thus oxidizing these amino acids to their corresponding radicals:

$$CAR^{\bullet+} + TyrOH \rightarrow CAR + TyrO^{\bullet} + H^{+} \tag{14.12}$$

$$CAR^{\bullet+} + CysH \rightarrow CAR + Cys^{\bullet} + H^{+} \tag{14.13}$$

This suggests the possible deleterious effects of carotenoids, for example, on membrane proteins, if, following a radical scavenging reaction, the radical cations so formed are not efficiently repaired.

TABLE 14.11
Second-Order Rate Constants for the Quenching of Carotenoid Radical Cations by Tyrosine and Cysteine in Detergent Micelles

	k (×10^5 M^{-1} s^{-1})	
CAR$^{•+}$	Tyrosine	Cysteine
β -CAR in TX-100	0.22	16.9
CAN in TX-100	0.41	10.9
ZEA in TX-100	0.81	7.8
ASTA in TX-100	0.94	10.1

There is evidence that carotenoid radical cations can persist for up to one second in some micellar environments (Burke et al. 2001a) so that, in the absence of a "repair" process, the radical may survive until it reacts with another biomolecule.

For tryptophan, there seems to be an equilibrium, which could also lead to some tryptophan radical formation and subsequent protein damage:

$$\text{TrpH}^{\bullet+} + \text{CAR} \rightleftharpoons \text{TrpH} + \text{CAR}^{\bullet+} \tag{14.14}$$

The studies of this equilibrium as a function of pH enabled the estimation of the absolute one-electron reduction potentials of CAR$^{•+}$ in an aqueous micellar environment (Edge et al. 2000, Burke et al. 2001b) (see Table 14.12 for typical results). As can be seen, the potentials of all the dietary carotenoid radical cations are very similar but LYC$^{•+}$ has the lowest potential implying that it is the best carotenoid antioxidant against free radicals (of course, this is an oversimplification, see above).

14.5 BIOMEDICAL CONSEQUENCES

A number of speculations can arise from the photo-physical data presented earlier. The major carotenoids that protect the macular are the XANs, ZEA, and LUT, yet, at least in the model membranes studies, these are rather poor $^1O_2$ quenchers (compared, e.g., with β-CAR and LYC). However, free

TABLE 14.12
One-Electron Reduction Potentials for CAR$^{•+}$

CAR$^{•+}$	E^0 (mV) ± 25 mV
β -CAR in TX-100	1060
CAN in TX-100	1041
ZEA in TX-100	1031
ASTA in TX-100	1030
β -CAR in TX-405/TX-100	1028
LYC in TX-405/TX-100	980

Note: Solubility problems require mixed micelles for LYC.

radical quenching is also a pivotal requirement for an efficient antioxidant. The antioxidant capacity does not only depend on the efficiency of quenching/removal of an oxidizing free radical but also on the reactivity and the lifetime of the products of the quenching reaction. For a strong oxidizing radical such as $NO_2^{\bullet}$ the product is the radical cation of the carotenoid:

$$\text{CAR} + \text{NO}_2^{\bullet} \rightarrow \text{CAR}^{+} + \text{NO}_2^{-} \quad (14.15)$$

and carotenoid radical cations are themselves strong oxidizing species and can have a relatively long lifetime. However, water-soluble antioxidants such as vitamin C can efficiently reconvert a carotenoid radical cation to the parent carotenoid:

$$\text{CAR}^{\bullet+} + \text{Vitamin C} \rightarrow \text{CAR} + \text{Vitamin C}^{\bullet+} \quad (14.16)$$

and this process would preclude the damaging (pro-oxidative) effects of a $CAR^{\bullet+}$. For such processes to be extremely efficient, as would certainly be necessary to protect the macular, the $CAR^{\bullet+}$ and the vitamin C need to be in close proximity. Because ZEA and LUT have terminal –OH groups they are fixed to the lipid–water interface leading to a "super-efficient" antioxidant system. Other hydrocarbon carotenoids such as LYC and β-CAR would need to reorientate to interact with vitamin C possibly reducing the efficiency of the antioxidant system (although this is not evident from the comparison of β-CAR and ZEA discussed in Section 14.4.3.1).

A somewhat related situation can be used to explain the well-publicized lung-cancer inducing effects of β-carotene in heavy smokers. This subpopulation will have low vitamin C levels and hence damage due to smoke components, such as $NO_2^{\bullet}$, can produce β-$CAR^{\bullet+}$ which will reach the lung and initiate damage. In nonsmokers, the vitamin C (or other water-soluble antioxidant) is likely to be present in sufficient concentration to preclude this damaging process. Indeed, this speculation has been promoted by the American Chemical Society as the subject of a "press release" in 1997 (Böhm et al. 1997).

One final, perhaps, extreme speculation concerns the claim that LYC can protect the macular even though it does not accumulate significantly in the eye (Mares-Pearlman et al. 1995). Possibly, dietary LUT and ZEA are protected from being retained as the corresponding radical cations by LYC, for example:

$$\text{ZEA}^{\bullet+} + \text{LYC} \rightarrow \text{ZEA} + \text{LYC}^{\bullet+} \quad (14.17)$$

Interestingly, β-CAR does not have the ability to so react with LUT and ZEA radical cations (see above) and does not appear to have any beneficial effects on macular protection.

REFERENCES

Adhikari, S., Kapoor, S., Chattopadhyay, S., and Mukherjee, T. 2000. Pulse radiolytic oxidation of β-carotene with halogenated alkylperoxyl radicals in a quaternary microemulsion: Formation of retinal. *Biophys. Chem.* 88:111–117.

Alfassi, Z.B., Khaikin, G.I., and Neta, P. 1995. Arylperoxyl radicals. Formation, absorption spectra, and reactivity in aqueous alcohol solutions. *J. Phys. Chem.* 99:265–268.

Arroyo, P.L., Hatch-Pigott, V., Mower, H.F., and Cooney, R.V. 1992. Mutagenicity of nitric oxide and its inhibition by antioxidants. *Mutation Res.* 281:193–202.

Beckman, J.S. and Crow, J.P. 1993. Pathological implications of nitric oxide, superoxide and peroxynitrite formation. *Biochem. Soc. Trans.* 21:330–334.

Bisby, R.H. and Parker, A.W. 1995. Reaction of ascorbate with the α-tocopheroxyl radical in micellar and bilayer membrane systems. *Arch. Biochem. Biophys.* 317:170–178.

Böhm, F., Tinkler, J.H., and Truscott, T.G. 1995. Carotenoids protect against cell membrane damage by the nitrogen dioxide radical. *Nat. Med.* 1:98–99.

Böhm, F., Edge, R., Land, E.J., McGarvey, D.J., and Truscott, T.G. 1997. Carotenoids enhance vitamin E antioxidant efficiency. *J. Am. Chem. Soc.* 119:621–622.

Burke, M. 2001. Pulsed radiation studies of carotenoid radicals and excited states, PhD thesis, University of Keele, Keele, U.K.

Burke, M., Edge, R., Land, E.J., and Truscott, T.G. 2001a. Characterisation of carotenoid radical cations in liposomal environments: interaction with vitamin C. *J. Photochem. Photobiol. B: Biol.* 60:1–6.

Burke, M., Edge, R., Land, E.J., McGarvey, D.J., and Truscott, T.G. 2001b. One-electron reduction potentials of dietary carotenoid radical cations in aqueous micellar environments. *FEBS Lett.* 500:132–136.

Burton, G.W. and Ingold, K.U. 1984. β-carotene: An unusual type of lipid antioxidant. *Science* 224:569–573.

Cantrell, A., McGarvey, D.J., Truscott, T.G., Rancan, F., and Boehm, F. 2003. Singlet oxygen quenching by dietary carotenoids in a model membrane environment. *Arch. Biochem. Biophys.* 412:47–54.

Cantrell, A., Land, E.J., and Truscott, T.G. 2006. Pulsed radiation studies of xanthophylls. In *The Life and Scientific Legacy of George Porter*, eds. D. Phillips and J. Barber, pp. 438–447. London: Imperial College Press.

Chaturvedi, V.K. and Kurup, C.K.R. 1986. Interaction of lutein with phosphotidylcholine bilayers. *Biochim. Biophys. Acta* 860:286–292.

Conn, P.F., Schalch, W., and Truscott, T.G. 1991. The singlet oxygen and carotenoid interaction. *J. Photochem. Photobiol. B: Biol.* 11:41–47.

Conn, P.F., Lambert, C., Land, E.J., Schalch, W., and Truscott, T.G. 1992. Carotene-oxygen radical interactions. *Free Rad. Res. Commun.* 16:401–408.

Cueto, R. and Pryor, W.A. 1994. Cigarette smoke chemistry: conversion of nitric oxide to nitrogen dioxide and reactions of nitrogen oxides with other smoke components as studied by fourier transform infrared spectroscopy. *Vib. Spectrosc.* 7:97–111.

Dawe E.A. and Land E.J. 1975. Radical ions derived from photosynthetic polyenes. *J. Chem. Soc. Faraday Trans. I* 71:2162–2169.

Devasagayam, T.P.A., Werner, T., Ippendorf, H., Martin H.-D., and Sies, H. 1992. Synthetic carotenoids, novel polyene polyketones and new capsorubin isomers as efficient quenchers of singlet molecular oxygen. *Photochem. Photobiol.* 55:511–514.

Edge, R. 1998. Spectroscopic and kinetic investigations of carotenoid radical ions and excited states, PhD thesis, Keele University, Keele, U.K.

Edge, R., McGarvey, D.J., and Truscott, T.G. 1997. The carotenoids as antioxidants – A review. *J. Photochem. Photobiol. B: Biol.* 41:189–200.

Edge, R., Land, E.J., McGarvey, D.J., Burke, M., and Truscott, T.G. 2000. The reduction potential of the β-carotene$^{\bullet+}$/β-carotene couple in an aqueous micro heterogeneous environment. *FEBS Lett.* 471:125–7.

Edge, R., El-Agamey, A., Land E.J., Navaratnam, S.,and Truscott, T.G. 2007. Studies of carotenoid one-electron reduction radicals. *Arch. Biochem. Biopys.* 458:104–110.

El-Agamey, A. and McGarvey, D.J. 2003. Evidence for a lack of reactivity of carotenoid radicals towards oxygen: A laser flash photolysis study of the reactions of carotenoids with acylperoxyl radicals in polar and non-polar solvents. *J. Am. Chem. Soc.* 125:3330–3340.

El-Agamey, A. and McGarvey, D.J. 2005. First direct observation of reversible oxygen addition to a carotenoid-derived carbon-centered neutral radical. *Org. Lett.* 18:3957–3960.

El-Agamey, A., Edge, R., Navaratnam, S., Land, E.J., and Truscott, T.G. 2006. Carotenoid radical anions and their protonated derivatives. *Org. Lett.* 8:4255–4258.

El-Agamey, A. and McGarvey, D.J. 2007. The reactivity of carotenoid radicals with oxygen. *Free Rad. Res.* 41:295–302.

Everett, S.A., Kundu, S.C., Maddix, S., and Willson, R.L. 1995. Mechanisms of free-radical scavenging by the nutritional antioxidant β-carotene. *Biochem. Soc. Trans.* 23:230S.

Everett, S.A., Dennis, M.F., Patel, K.B., Maddix, S., Kundu, S.C., and Willson, R.L. 1996. Scavenging of nitrogen dioxide, thiol, and sulphonyl free radicals by the nutritional antioxidant β-carotene. *J. Biol. Chem.* 271:3988–3994.

Farmilo, A. and Wilkinson, F. 1973. On the mechanism of quenching of singlet oxygen in solution. *Photochem. Photobiol.* 18:447–450.

Foote, C.S. and Denny R.W. 1968. Chemistry of singlet oxygen. VIII. Quenching by β-carotene. *J. Am. Chem. Soc.* 90:6233–6235.

Gabr, I., Patel, R.P., Symons, M.C.R., and Wilson, M.T. 1995. Novel reactions of nitric oxide in biological systems. *J. Chem. Soc., Chem. Commun.* 915–916.

Grant, J.L., Kramer, V.J., Ding, R., and Kispert, L.D. 1988. Carotenoid cation radicals: Electromechanical, optical and EPR study. *J. Am. Chem. Soc.* 110:2151–2157.

Gruszecki, W.I. 1990. Violaxanthin and zeaxanthin aggregation in the lipid-water system. *Studia Biophysica* 139:95–101.

Gruszecki, W.I. 1999. Carotenoids in Membranes. In *The Photochemistry of Carotenoids*, eds. H. A. Frank, A.J. Young, G. Britton, and R. J. Cogdell, pp. 363–379. Dordrecht, the Netherlands: Kluwer Academic Publishers.

Gruszecki, W.I. and Sielewiesiuk, J. 1990. Orientation of xanthophylls in phosphatidylcholine multibilayers. *Biochim. Biophys. Acta* 1023:405–412.

Hill, T.J. 1994. Molecular mechanisms of photoprotection, PhD thesis, University of Keele, Kelle, U.K.

Hill, T.J., Land, E.J., McGarvey, D.J., Schalch, W., Tinkler, J.H., and Truscott, T.G. 1995. Interactions between carotenoids and the $CCl_3O_2^{\bullet}$ radical. *J. Am. Chem. Soc.* 117:8322–8326.

Kikugawa, K., Hiramoto, K., Tomiyama, S., and Asano, Y. 1997. β-Carotene effectively scavenges toxic nitrogen oxides: Nitrogen dioxide and peroxynitrous acid. *FEBS Lett.* 404:175–178.

Lacsamana, M. and Gebicki, J.M. 1996. Peroxidation of proteins by peroxynitrite. In *Natural Antioxidants and Food Quality in Atherosclerosis and Cancer Prevention,* ed. J. Kampulainen, pp. 55–59. Cambridge, U.K.: Royal Society of Chemistry.

Lafferty, J., Roach, A.C., Sinclair, R.S., Truscott, T.G., and Land, E.J., 1977. Absorption spectra of radical ions of polyenes of biological interest. *J. Chem. Soc. Faraday Trans. I* 73:416–429.

Li, Z.-L., Wu, L.-M., Ma, L.-P., Liu, Y.-C., and Liu, Z.-L. 1995. Antioxidant synergism and mutual protection of α-tocopherol and β-carotene in the inhibition of radical-initiated peroxidation of linoleic acid in solution. *J. Phys. Org. Chem.* 8:774–780.

Mares-Perlman, J.A., Bracy, W.E., Klein, R. et al. 1995. Serum antioxidants and age-related macular degeneration in a population based control study. *Arch. Opthalmol.* 113:1518–1523.

Mairanovsky, V.G., Engovatov, A.A., Ioffe, N.T., and Samokhvalov, G.I. 1975. Electron-donor and electron-acceptor properties of carotenoids: Electrochemical study of carotenes. *J. Electroanal. Chem.* 66:123–137.

Mayne, S.T. and Parker, R.S. 1989. Antioxidant activity of dietary canthaxanthin. *Nutr. Cancer* 12:225–236.

McVie, J., Sinclair, R.S., Tait, D., Truscott, T.G., and Land E.J. 1979. Electron transfer reactions involving porphyrins and carotenoids. *J. Chem. Soc. Faraday Trans. I* 75:2869–2872.

Miller, N.J., Sampson, J., Candeias, L.P., Bramley, P.M., and Rice-Evans, C.A. 1996. Antioxidant activities of carotenes and xanthophylls. *FEBS Lett.* 384:240–242.

Mortensen, A. 2000. Mechanism and kinetics of scavenging of the phenylthiyl radical by carotenoids. A laser flash photolysis study. *Asian Chem. Lett.* 4:135–143.

Mortensen, A. 2001. Scavenging of acetylperoxyl radicals and quenching of triplet diacetyl by β-carotene: mechanisms and kinetics. *J. Photochem. Photobiol. B: Biol.* 61:62–67.

Mortensen, A. and Skibsted, L.H. 1996. Kinetics of photobleaching of β-carotene in chloroform and formation of transient carotenoid species absorbing in the near infrared. *Free Rad. Res.* 25:355–368.

Mortensen, A. and Skibsted, L.H. 1997a. Importance of carotenoid structure in radical scavenging reactions. *J. Agric. Food. Chem.* 45:2970–2977.

Mortensen, A. and Skibsted, L.H. 1997b. Free radical transients in photobleaching of xanthophylls and carotenes. *Free Rad. Res.* 26:549–563.

Mortensen, A., Skibsted, L.H., Sampson, J., Rice-Evans, C., and Everett, S.A. 1997. Comparative mechanisms and rates of free radical scavenging by carotenoid antioxidants. *FEBS Lett.* 418:91–97.

Okulski, W., Sujak, A., and Gruszecki, W.I. 2000. Dipalmitoylphosphatidylcholine membranes modified with zeaxanthin: Numeric study of membrane organisation. *Biochim. Biophys. Acta* 1509:216–228.

Packer, J.E., Mahood, J.S., Mora-Arellano, V.O., Slater, T.F., Willson, R.L., and Wolfenden, B.S. 1981. Free radicals and singlet oxygen scavengers: Reaction of a peroxy-radical with β-carotene, diphenyl furan and 1,4-diazobicyclo (2,2,2)-octane. *Biochem. Biophys. Res. Commun.* 98:901–906.

Park, S.-M. 1978. Electrochemical studies of β-carotene, all-trans-retinal and all-trans-retinol in tetrahydrofuran. *J. Electrochem. Soc.* 125:216–222.

Pryor, W.A. and Lightsey, J.W. 1981. Mechanisms of nitrogen dioxide reactions: Initiation of lipid peroxidation and the production of nitrous acid. *Science* 214:435–437.

Radi, R., Beckman, J.S., Bush, K.M., and Freeman, B.A. 1991. Peroxynitrite-induced membrane lipid peroxidation: The cytotoxic potential of superoxide and nitric oxide. *Arch. Biochem. Biophys.,* 288:481–7.

Rodgers, M.A.J. and Bates A.L. 1980. Kinetic and spectroscopic features of some carotenoid triplet states: Sensitization by singlet oxygen. *Photochem. Photobiol.* 31:533–537.

Strzalka, K. and Gruszecki, W.I. 1994. Effect of β-carotene on structural and dynamic properties of model phosphtidylcholine membranes. 1. An EPR spin-label study. *Biochim. Biophys. Acta* 1194:138–142.

Subczynski, W.K., Markowska, E., Gruszecki, W.I., and Sielewiesiuk, J. 1992. Effects of polar carotenoids on dimyristoylphosphatidylcholine membranes – Spin-label study. *Biochim. Biophys. Acta* 1105:97–108.
Wilkinson, F. and Ho, W.-T. 1978. Electronic energy transfer from singlet molecular oxygen to carotenoids. *Spectrosc. Lett.* 11:455–463.
Woodall, A.A., Lee, S.W.-M., Weesie, R.J., Jackson, M.J., and Britton, G. 1997a. Oxidation of carotenoids by free radicals: Relationship between structure and reactivity. *Biochim. Biophys. Acta* 1336:33–42.
Woodall, A.A., Britton, G., and Jackson, M.J. 1997b. Carotenoids and protection of phospholipids in solution or in liposomes against oxidation by peroxyl radicals: Relationship between carotenoid structure and protective ability. *Biochim. Biophys. Acta* 1336:575–586.

15 Carotenoid Uptake and Protection in Cultured RPE

Małgorzata Różanowska and Bartosz Różanowski

CONTENTS

15.1 INTRODUCTION

Carotenoids accumulating in the human body are obtained exclusively from our diet. Out of almost 50 carotenoids present in a typical human diet, about 14 are absorbed into the blood (Khachik et al., 1997), and only two of them—lutein and zeaxanthin (Figure 15.1)—accumulate in the retina (Bernstein et al., 2001; Bone and Landrum, 1992; Bone et al., 1988, 1997; Davies and Morland, 2004; Khachik et al., 1997, 2002). Lutein and zeaxanthin are particularly concentrated in photoreceptor axons and inner plexiform layer in the area including and surrounding

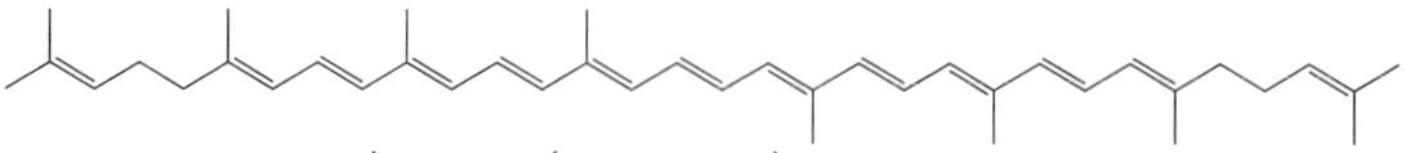

Lycopene (γ,γ-carotene)

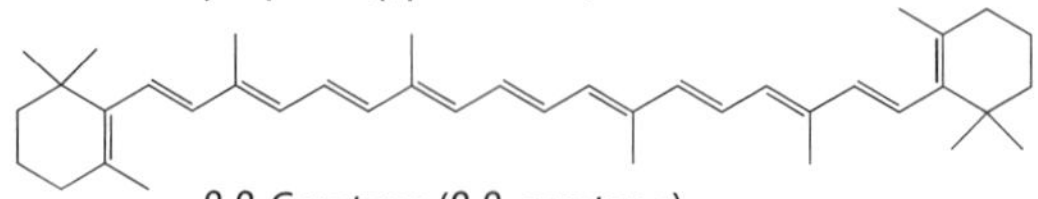

β,β-Carotene (β,β-carotene)

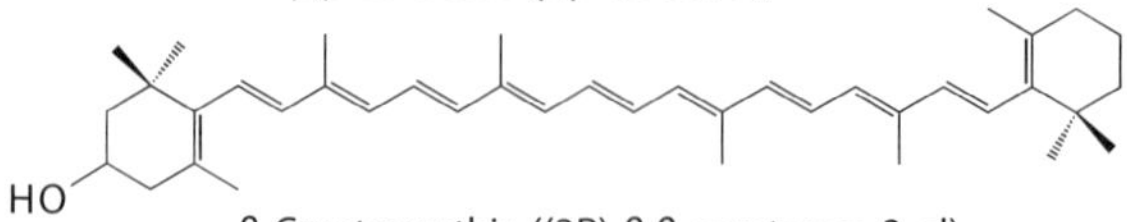

β-Cryptoxanthin ((3R)-β,β-carotenen-3-ol)

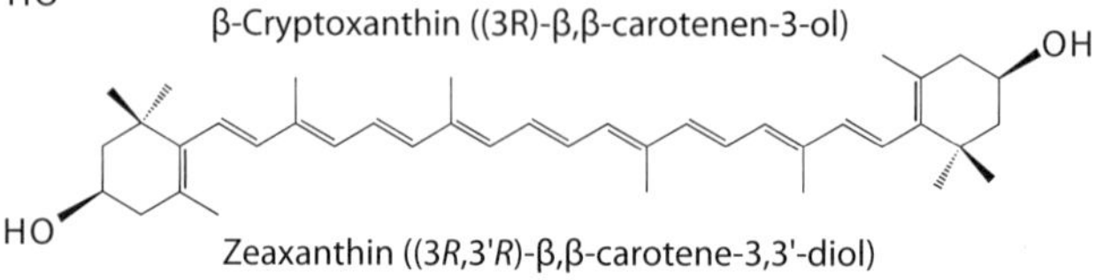

Zeaxanthin ((3*R*,3'*R*)-β,β-carotene-3,3'-diol)

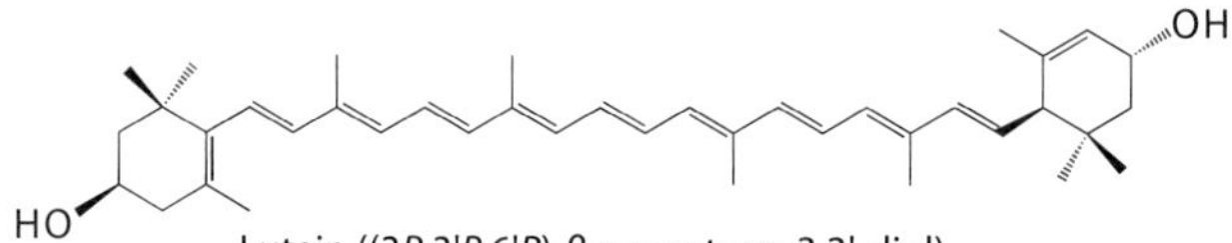

Lutein ((3*R*,3'*R*,6'*R*)-β,ε-carotene-3,3'-diol)

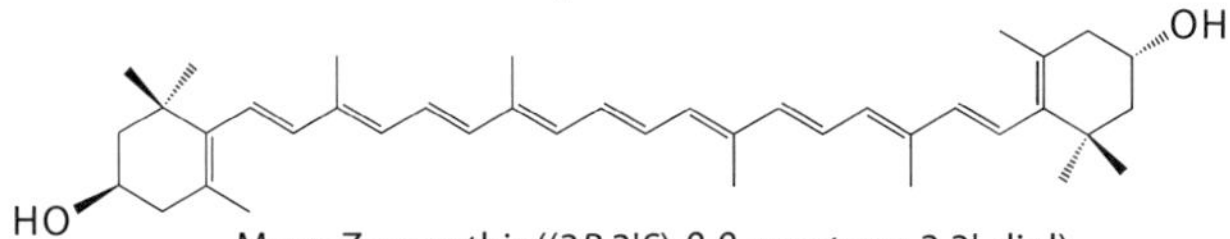

Meso-Zeaxanthin((3*R*,3'*S*)-β,β-carotene-3,3'-diol)

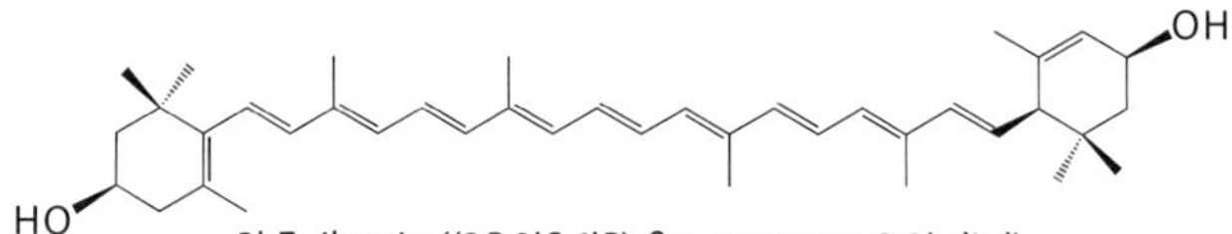

3'-Epilutein ((3*R*,3'*S*,6'*R*)-β,ε-carotene-3,3'-diol)

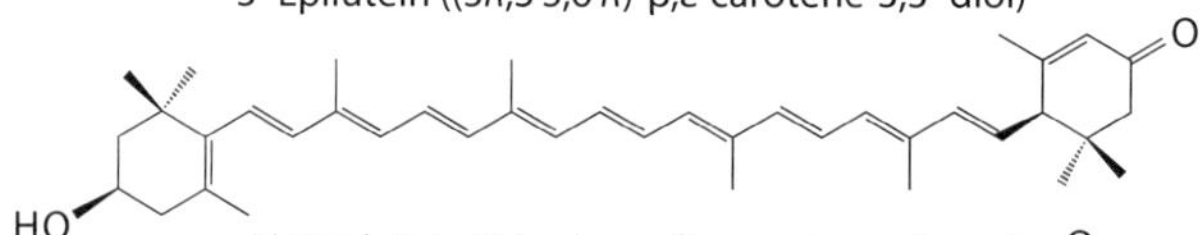

3'-Oxolutein (3-hydroxy-β,ε-carotene-3'-one)

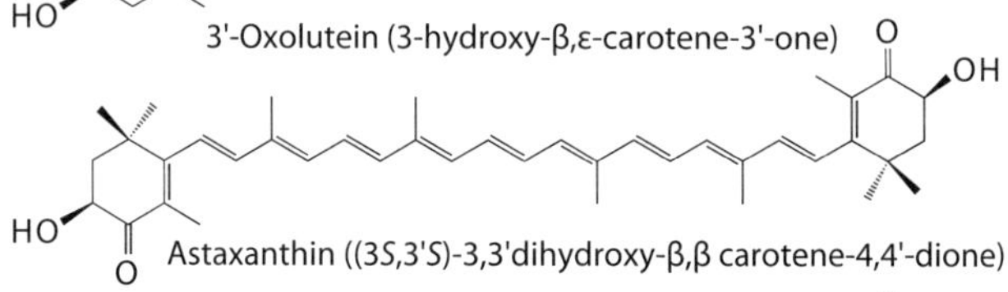

Astaxanthin ((3*S*,3'*S*)-3,3'dihydroxy-β,β carotene-4,4'-dione)

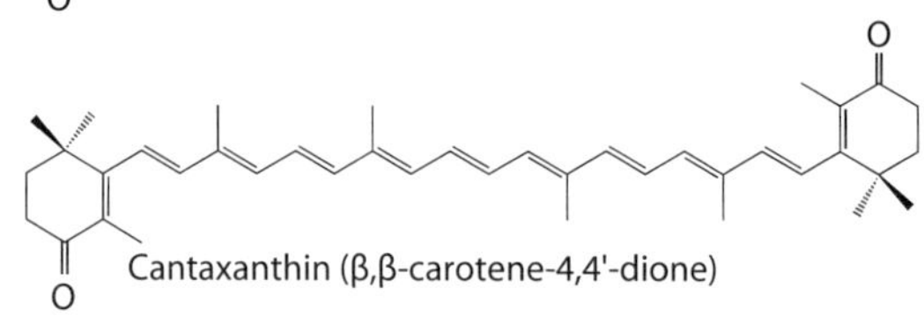

Cantaxanthin (β,β-carotene-4,4'-dione)

FIGURE 15.1 Structures of carotenoids important for vision. Oxygen-containing carotenoids belong to a subclass of carotenoids known as xanthophylls.

the area with the highest density of photoreceptors, *fovea centralis*, responsible for acute vision. Due to the high optical density of accumulated carotenoids, this area can be visible as a yellow spot, *macula lutea*, and therefore lutein and zeaxanthin are often referred to as the macular pigment (Berendschot and van Norren, 2006; Bernstein et al., 2001; Bone et al., 1988, 1993, 1997; Snodderly et al., 1984a,b). About 25% of total retinal lutein and zeaxanthin is present in the outer retina—photoreceptor outer segments (POS) and retinal pigment epithelium (RPE) (Rapp et al., 2000; Sommerburg et al., 1999).

Some observational epidemiological studies have shown a reduced risk of age-related macular degeneration (AMD) or rate of its progression in people with a higher intake and/or higher plasma concentrations of lutein and zeaxanthin (Delcourt et al., 2006; Goldberg et al., 1988; Moeller et al., 2006; Nolan et al., 2007; SanGiovanni et al., 2007; Seddon et al., 1994; Snellen et al., 2002; Sperduto, 1993; Tan et al., 2008; van Leeuwen et al., 2003). AMD is the leading cause of blindness in people above 60 years old in the developed countries and becomes an increasingly important socio-economic problem due to ageing populations of extended lifespans.

AMD is a progressive disease that affects RPE and photoreceptors in the central part of the retina. Initial changes include formation of deposits between RPE and Bruch's membrane, so-called drusen, and/or pigmentary abnormalities in the RPE. Even though at that stage the vision is not affected, the changes are a strong predictor of developing of the advanced form of AMD and therefore, the stage when they are present is usually referred to as early AMD. The disease progresses when drusen become more numerous, and eventually confluent, and RPE cells die leaving depigmented areas in a process called geographic atrophy, which are often surrounded by areas with an increased pigmentation. This form of the disease, called the "dry" form, is present in about 90% of AMD victims. In 10% of AMD patients, new blood vessels from the choroid invade the retina leading to the rapid progression of vision loss in the so-called "wet" form of the disease.

Despite recent findings of several genes associated with AMD (Edwards, 2008; Lotery and Trump, 2007; Moshfeghi and Blumenkranz, 2007; Mullins, 2007; Scholl et al., 2007; Swaroop et al., 2007), the etiology of the disease still remains largely unknown and involves a complex interaction of genetic and environmental factors (Gorin, 2007). Current anti-angiogenic treatments target only the so-called "wet" form of the disease (Owens et al., 2006). There is no effective prevention or treatment for the atrophic form of the disease affecting the majority of AMD victims (reviewed by Chong et al. (2007), Coleman and Chew (2007), Donaldson and Pulido (2006), Eter et al. (2006), Guymer and Chong (2006), and Yeoh et al. (2006)). The only widely used approach is supplementing AMD patients with a combination of zinc, vitamin C, vitamin E and β-carotene. This mixture has been tested in a large clinical trial, the Age-Related Eye Disease Study (AREDS), where the effects of supplementation on the progression of AMD and vision loss were followed for up to seven years (AREDS, 2001). The participants were randomly allocated placebo, zinc and/or antioxidant mixture of β-carotene with vitamin C and E. Supplementation with zinc alone reduced the risk of progression to advanced AMD by 20%, while antioxidant mixture composed of about 15 mg of β-carotene, 400 IU vitamin E and 500 mg vitamin C reduced the risk by 17%. Combined zinc and vitamin C, vitamin E and β-carotene reduced the risk by 25%, and only these results were statistically significant. While in the published report from the study it is mentioned that during the trial some participants assigned to antioxidant supplements opted for mixtures without β-carotene, no further data on that group of patients were provided. Therefore, the data available are not conclusive whether β-carotene is an essential component of the AREDS mixture to be protective against progression of AMD. Further evaluation of whether β-carotene is needed as a component of the AREDS mixture will be tested in a multicenter controlled, randomized trial—the Age-Related Eye Disease Study 2 (AREDS2). Other trials, testing the effects of supplementation with β-carotene alone or in a mixture with other antioxidants, did not show any statistically significant differences on AMD development between supplemented and non-supplemented groups of people (Evans, 2006; West et al., 2006).

As the therapy of AMD is very limited, there is an urgent need to develop an intervention to prevent vision loss. The epidemiological data together with the well-documented antioxidant properties of carotenoids in studies *in vitro* and with proven increases in macular pigment density in most people via dietary supplementation (Beatty et al., 2004; Berendschot et al., 2000; Bone et al., 2003; Hammond et al., 1997; Iannaccone et al., 2007; Landrum et al., 1997), including patients with early AMD (Koh et al., 2004; Obana et al., 2008; Richer et al., 2007; Trieschmann et al., 2007), have prompted researchers to consider lutein and zeaxanthin as modifiable risk factors for AMD. Evaluation of safety and efficacy of supplementation with lutein and zeaxanthin against AMD will be tested in the recently launched trial AREDS2 (Seddon, 2007). Yet, numerous dietary supplements containing carotenoids are marketed and advertised as being beneficial for the eye, and not-surprisingly the use of those supplements is growing (Jones, 2007; Kiser and Dagnelie, 2008).

However, several other epidemiological studies found no statistically significant relationship of susceptibility to AMD and dietary intake of lutein and zeaxanthin, or optical density of macular pigment (Cho et al., 2004; Gale et al., 2003; Kanis et al., 2007; LaRowe et al., 2008; Mares-Perlman et al., 1995, 2001; Morris et al., 2007; Sanders et al., 1993; Trumbo and Ellwood, 2006). Moreover, a recent epidemiological study of dietary intake of fat and carotenoids in Australia found an increased prevalence of late AMD in patients consuming a high fat diet rich in xanthophylls (Robman et al., 2007; Vu et al., 2006), while in another study a higher β-carotene intake was associated with an increased risk of AMD (Tan et al., 2008).

Furthermore, several studies have shown that in some individuals an increased intake of xanthophylls does not lead to increased levels of xanthophylls in their plasmas and/or retinas, and macular pigment densities do not exhibit a positive correlation with plasma levels of lutein and zeaxanthin (Aleman et al., 2001; Bernstein et al., 2002b; Bone et al., 2000, 2001, 2003; Hammond et al., 1995, 1997). These apparently conflicting epidemiological results need to be interpreted with caution as a diet rich in fruit and vegetables includes a great variety of phytochemicals that may independently, or in cooperation with lutein or zeaxanthin, and other dietary components affect carotenoid uptake and function in the retina.

Clearly, there is an urgent need to understand the roles of carotenoids in the retina and the mechanism of carotenoid uptake. The RPE cells are present at the blood–retina barrier and are the primary site of damage in AMD as well as in some other retinal degenerations. In this chapter, we discuss current understanding of carotenoid uptake into cells and the hypothetical roles of the RPE in selective uptake and retention of carotenoids in the retina. We discuss the advantages of cultured RPE cells as an appropriate model to test many of the hypothetical pathways of carotenoid transport. We also review the multiple bioactivities of carotenoids in cellular systems and their potential in protection of the RPE. As carotenoids are prone to oxidative damage themselves, and their degradation products exhibit several deleterious effects, we also discuss the mechanisms protecting the carotenoids from oxidative degradation applicable to the RPE and testable *in vitro*.

15.2 POTENTIAL PROTECTIVE ROLE OF CAROTENOIDS IN THE RETINA AS ANTIOXIDANTS

Carotenoids exhibit several potent antioxidant properties—filtering out blue light, quenching excited states of photosensitizers and oxygen, as well as scavenging of free radicals, and these properties are well documented in solution, micelles, liposomes, cell culture, and experimental animals, and therefore carotenoids are believed to play a protective role in the retina (summarized recently in several elegant reviews: Bahrami et al., 2006; Davies and Morland, 2004; El-Agamey et al., 2004a,b; Krinsky et al., 2003; Stahl and Sies, 2005).

Lutein and zeaxanthin can act as efficient blue light filters in the *macula lutea* due to their high millimolar concentration in the axons of photoreceptor cells, and their high absorption coefficients (Bone et al., 1992; Britton, 1995). The optical density of macular pigment varies between individuals from about 0.1 to 0.8, meaning that they are responsible for preventing 20%–85% of the incident blue light from reaching photoreceptor inner and outer segments and the RPE (Werner et al., 1987). Blue light, covering the shortest wavelengths reaching the retina in the adult eye, is the most susceptible to scatter and refraction. Thus, eliminating blue light before reaching photoreceptive part of the retina may reduce chromatic aberration and improve visual acuity (Reading and Weale, 1974; Wooten and Hammond, 2002). Consistently, some trials indicate that supplementation with xanthophylls improves certain parameters of visual function (Bahrami et al., 2006; Cangemi, 2007; Kvansakul et al., 2006; Parisi et al., 2008; Richer et al., 2004; Rodriguez-Carmona et al., 2006).

Blue light is the most efficient part of the visible spectrum reaching the adult human retina to trigger light-induced damage (Boulton et al., 2001; Rozanowska and Sarna, 2005). It has been demonstrated that blue light initiates production of reactive oxygen species by mitochondria, all-*trans*-retinal, lipofuscin and melanosomes (Boulton et al., 2001; Godley et al., 2005; Rozanowska and Sarna, 2005; Rozanowska et al., 2002)—molecules and organelles particularly abundant in the inner and outer segments of photoreceptors and the RPE. Thus, reducing blue light irradiance levels of these parts of the retina can minimize photoactivation of photosensitizers and subsequent photodamage.

Moreover, carotenoids may quench electronically excited states and scavenge free radicals formed in the retina, and therefore protect biomolecules from oxidative damage. Due to the low energy level of the first excited triplet state (^{3}Car), carotenoids (Car) can act as efficient acceptors of triplet state energy from photosensitizers (S) (Equation 15.1), such as all-*trans*-retinal, the photosensitizers of lipofuscin (Rozanowska et al., 1998), or singlet oxygen ($^1O_2$) (Equation 15.2) (Cantrell et al., 2003):

$$\mathrm{Car} + {}^3\mathrm{S} \rightarrow {}^3\mathrm{Car} + \mathrm{S} \tag{15.1}$$

$$\mathrm{Car} + {}^1\mathrm{O}_2 \rightarrow {}^3\mathrm{Car} + \mathrm{O}_2 \tag{15.2}$$

Carotenoids are particularly valuable as singlet oxygen quenchers. They accept the energy from the excited state of molecular oxygen, singlet oxygen ($^1O_2$) and, as a result, the oxygen molecule returns to its ground state, while the carotenoid is left in a triplet state that thermally deactivates to the ground state (Equation 15.2). Thus, it is a safe physical mode of quenching of $^1O_2$ and results in no chemical modification to any of the interacting molecules. The bimolecular rate constants of interactions of carotenoids with singlet oxygen are close to the diffusion controlled limits, with zeaxanthin being about twice more effective than lutein (Cantrell et al., 2003).

Carotenoids interact with a number of free radicals either via electron (Equation 15.3) or hydrogen (Equation 15.4) transfer, or forming an addition complex (Equation 15.5) (El-Agamey et al., 2004b):

$$\mathrm{Car} + \mathrm{R}^\bullet \rightarrow \mathrm{Car}^\bullet + \mathrm{R}^- \tag{15.3}$$

$$\mathrm{Car} + \mathrm{R}^\bullet \rightarrow \mathrm{Car}^\bullet + \mathrm{RH} \tag{15.4}$$

$$\mathrm{Car} + \mathrm{R}^\bullet \rightarrow \left[\mathrm{R} - \mathrm{Car}\right]^\bullet \tag{15.5}$$

In case of scavenging of lipid-derived peroxyl radicals (LOO$^\bullet$), the radical adduct formed [LOO-Car]$^\bullet$ is less reactive than the LOO$^\bullet$, so carotenoids act as chain-breaking antioxidants in lipid peroxidation (Equation 15.6):

$$\mathrm{Car} + \mathrm{LOO}^\bullet \rightarrow [\mathrm{LOO} - \mathrm{Car}]^\bullet \tag{15.6}$$

15.3 RPE AS A MEDIATOR OF SPECIFIC UPTAKE OF CAROTENOIDS INTO THE RETINA

RPE plays numerous functions essential for proper structure and function of retinal photoreceptors. They include the maintenance of the blood–retina barrier, selective uptake and transport of nutrients from the blood to the retina and removal of waste products to the blood, enzymatic cleavage of β-carotene into vitamin A, storage of vitamin A and its metabolic transformations, phagocytosis and molecular renewal of POS, expression and secretion of growth factors and immunomodulatory cytokines (Aizman et al., 2007; Aleman et al., 2001; Crane et al., 2000a,b; Elner et al., 2006; Holtkamp et al., 2001; Leuenberger et al., 2001; Lindqvist and Andersson, 2002; Maminishkis et al., 2006; Momma et al., 2003; Strauss, 2005).

Interestingly, carotenoids more abundant in the blood plasma than zeaxanthin, such as lycopene, β-carotene, and β-cryptoxanthin, do not accumulate in the retina. RPE cells express β,β-carotene 15,15′-monooxygenase (BCO), formerly known as β-carotene 15,15′-dioxygenase, an enzyme that catalyzes the oxidative cleavage of β-carotene into two molecules of all-*trans*-retinal (Aleman et al., 2001; Bhatti et al., 2003; Chichili et al., 2005; Leuenberger et al., 2001; Lindqvist and Andersson, 2002). Therefore it may be suggested that β-carotene transported into RPE-cells is efficiently cleaved into retinal molecules. BCO cleaves also β-cryptoxanthin (Lindqvist and Andersson, 2002), and its absence in the retina may also be explained by its efficient cleavage to retinoids. However, lycopene, often the most abundant carotenoid in human plasma, cannot serve as a substrate for BCO, and yet it is not detectable in the neural retina (Khachik et al., 2002).

15.3.1 RPE as the Blood–Retina Barrier

The presence of only two dietary carotenoids in the retina, lutein and zeaxanthin, out of about 14 normally present in the plasma indicates their highly specific uptake and retention (Bernstein et al., 2001; Bone and Landrum, 1992; Bone et al., 1988, 1997, 1993; Davies and Morland, 2004; Khachik et al., 1997, 2002). The retina–blood barrier is formed by the tight *zonulae occludentes* of the endothelial cells in the inner retina and of the RPE, a monolayer of cells which separates the outer retina from its choroidal blood supply (Strauss, 2005).

The greatest concentration of the macular pigment is present in the avascular part of the retina. This suggests that the RPE may play the predominant role in uptake and transport of xanthophylls to the photoreceptors. Moreover, about 25% of the total retinal xanthophylls are present in the POS (Rapp et al., 2000; Sommerburg et al., 1999), which, under normal conditions, are intimately associated with the RPE. This proximity lends further support to the hypothesis of a role for the RPE in the selective uptake of carotenoids into the retina.

15.3.2 Carotenoid Delivery to the RPE from Blood

The RPE separates the neural retina from the fenestrated vascular bed of choriocapillaris. Carotenoids are transported in the blood serum mainly bound to lipoproteins (Parker, 1996; Wang et al., 2007). Lipoproteins are multimolecular assemblies of noncovalently bound lipids and proteins. The outer layer of a lipoprotein is comprised of a single layer of phospholipids, cholesterol and apo-lipoproteins, which surround a central core composed of neutral lipids, mainly triacylglycerol and esterified cholesterol. Nonpolar carotenoids such as β-carotene and lycopene tend to exhibit similar pattern of distribution among lipoproteins to that of cholesterol, namely, 58%–73% are carried by low density lipoproteins (LDL), 17%–26% by high density lipoproteins (HDL), and 10%–16% by very low density lipoproteins (VLDL). About 53% of lutein and zeaxanthin are bound to HDL, 31% to LDL, and 16% to VLDL. Cryptoxanthin is distributed almost equally between LDL and HDL and the remaining 20% is bound to VLDL. Within each type of lipoprotein there is a further variation in carotenoid distribution depending on lipoprotein composition (Brown and Fragoso, 1994). For example, the HDL fraction rich in apo-lipoprotein E (ApoE) contains about a threefold greater amount of β-carotene than the ApoE-poor HDL fraction (Brown and Fragoso, 1994).

It has been well documented *in vivo* that the RPE takes up lipoproteins from serum (Elner, 2002; Gordiyenko et al., 2004; Wang and Anderson, 1993). For example, it was demonstrated that an intravenous injection of LDLs in rodents leads to formation of some LDL deposits at Bruch's membrane followed by internalization of this LDL within the RPE during a 24 h period (Gordiyenko et al., 2004). Thus, it may be suggested that carotenoids are taken up from serum into the RPE together with lipoproteins.

15.3.2.1 Lipoprotein Receptors Expressed by the RPE

RPE expresses several receptors responsible for lipoprotein uptake, including lipoprotein receptors: LDL receptor (LDLR) (Tserentsoodol et al., 2006b), VLDL receptor (VLDLR) (Hu et al., 2008), as well as scavenger receptors: cluster determinant 36 (CD36), scavenger receptor class B type I (SR-BI) and type II (SR-BII) (Calvo et al., 1997, 1998; Duncan et al., 2002; Provost et al., 2003; Ryeom et al., 1996b). CD36, an 88 kDa integral membrane glycoprotein, preferentially binds oxidized LDL, but also native HDL, LDL, and VLDL, which are subsequently endocytosed and degraded in cells to release lipids (Calvo et al., 1998). In RPE cells, CD36 also participates in phagocytosis of shed parts of POS (Ryeom et al., 1996a,b).

SR-BI, a 82 kDa glycoprotein with two cytoplasmic C- and N-terminal domains separated by a large extracellular domain, acts mainly as a multiligand HDL receptor (Acton et al., 1996). In particular, SR-BI acts as receptor for anionic lipids and mediates selective uptake of HDL cholesteryl esters and other lipids, vitamin E, and carotenoids, as well as transport from cells of free cholesterol to lipoprotein and non-lipoprotein acceptors (Ji et al., 1997; Krieger, 1999; 2001; Reboul et al., 2005, 2006; Uittenbogaard et al., 2002). The major role of SR-BI in carotenoid uptake has been demonstrated by comparison of β-carotene uptake in wild-type and SR-BI knockout mice and *in vitro* in a polarized monolayer of the Caco-2 intestinal cell line, used as a model of human intestinal epithelium (During and Harrison, 2007; Reboul et al., 2005, 2006). The polarized Caco-2 cells express SR-BI mainly on their apical site. The incubation of these cells with SR-BI blocking antibodies or other specific inhibitors of SR-BI reduces the uptake of lutein solubilized in micelles by about 80% and partly inhibits the lutein efflux from cells into the basal side (Reboul et al., 2005). Interestingly, lutein absorption from a micellar suspension of 900 nM lutein is reduced by 28% in the presence of 200 nM β-carotene but is not affected in the presence of 130 nM lycopene (Reboul et al., 2005). It needs to be stressed, however, that in the experiments described, due to the solubility limitations, the lycopene was tested at lower concentrations than β-carotene. Nevertheless, these data indicate that there is a competition at least between some carotenoids for their uptake to cultured intestinal cells.

Recent data indicate that SR-BI is a nonspecific receptor for many lipophilic molecules (Lorenzi et al., 2008; Reboul et al., 2007b). Apart from HDLs, rodent SR-BI also binds to LDL, VLDL, acetylated LDL, oxidized LDL, and maleylated bovine serum albumin. SR-BII has a similar ligand specificity and function to that of SR-BI (Webb et al., 1998). However, it has been shown that vitamin E (which like carotenoids is carried in the bloodstream mainly by LDL and HDL) is transported more efficiently into the endothelial cells from HDLs than from LDLs (Balazs et al., 2004; Kaempf-Rotzoll et al., 2003; Mardones and Rigotti, 2004). This is in striking contrast to cholesterol, which is taken up much more efficiently from LDLs than HDLs by the RPE to the retina (Tserentsoodol et al., 2006b). It remains to be shown which lipoproteins are the main carriers for carotenoids transported from blood into the RPE.

There are several factors which may be responsible for the variability in the uptake of different carotenoids. These include differences in the distribution of carotenoids among serum lipoproteins and differences in the expression on cell membranes of the various receptors/transporters responsible for uptake of different types of lipoproteins. It may be further expected that the efficiencies for uptake of different carotenoids may vary for different types of receptors/transporters. Also, the further processing of internalized different carotenoids may vary.

The roles of lipoprotein and scavenger receptors, particularly SR-BI/II and CD36, in carotenoid uptake in the RPE cells still awaits exhaustive investigation.

15.3.2.2 Metabolic Pathways in the RPE of Pro-Vitamin A Carotenoids

β-carotene does not accumulate in the neural retina in detectable amounts—its concentration in the neural retina is at most 1% of total carotenoid content (Handelman et al., 1991a). However, it is present in substantial concentrations in the combined RPE-choroid tissues isolated from human cadaver eyes (Bernstein et al., 2001). RPE highly expresses the BCO enzyme and actively converts β-carotene into all-*trans*-retinal (Bhatti et al., 2003; Chichili et al., 2005). BCO has been also detected in bovine neural retina (Chichili et al., 2005). Thus it may be suspected that conversion to vitamin A is the typical fate of β-carotene in the retina. Vitamin A is important for vision as a precursor of the visual pigment chromophore, 11-*cis*-retinal (Figure 15.2a).

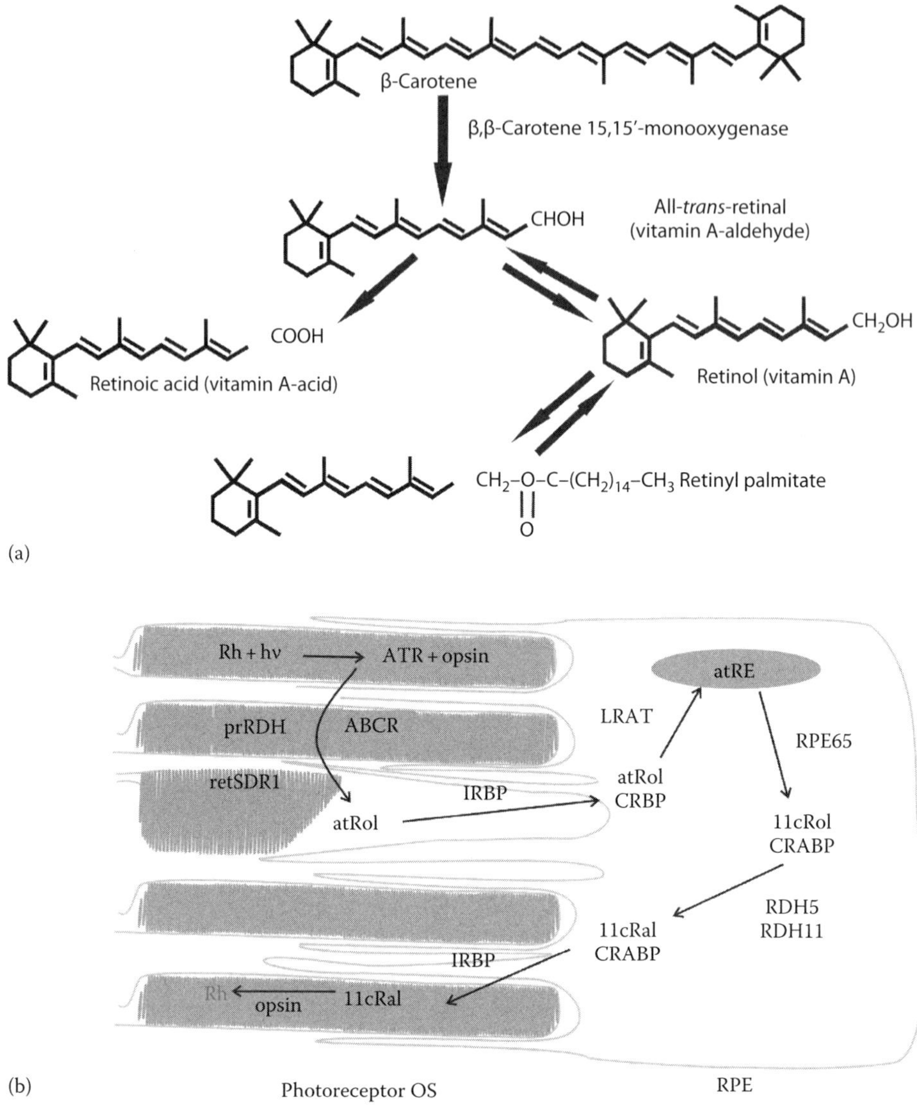

FIGURE 15.2 Schematic diagrams depicting the fate of β-carotene and its metabolite, all-*trans*-retinal in the retina: (a) β-Carotene is converted by β,β-carotene 15,15′-monooxygenase to all-*trans*-retinal (Bhatti et al., 2003; Chichili et al., 2005). (b) All-*trans*-retinal (ATR) can be reversibly reduced to form all-*trans*-retinol (atRol) which itself may be esterified with fatty acids to form retinyl esters (atRE). atRE can undergo transformations in the retina which are collectively referred to as the retinoid cycle. atRE is accumulated in so called retinosomes in the RPE. It has been determined in *post mortem* human eyes that retinoids present in RPE/choroid, mainly retinyl esters, account for 2.5 mol equiv of the rhodopsin present in the outer segments. Activated RPE65 is responsible for isomerization and hydrolysis of atRE and formation of 11-*cis*-retinol (11cRol). 11cRol bound to a chaperone cellular retinoid binding protein (CRABP) is a substrate for retinol dehydrogenases RDH5 or RDH11 and becomes oxidized to 11-*cis*-retinal (11cRal). 11cRal is transported from the RPE to photoreceptor outer segment (OS) and this transport is facilitated by a chaperone, interphotoreceptor retinoid binding protein (IRBP). Binding of 11cRal to opsin regenerates rhodopsin (Rh). Rh is the photosensitive chromophore which upon photoactivation by absorption of a photon initiates a visual cascade leading to visual perception. The primary physical event of visual perception, upon absorption of light, is isomerization of the 11cRal of the Rh chromophore to all-*trans*-retinal (ATR). Following isomerization, ATR is hydrolysed from opsin protein and is subsequently reduced to atRol by photoreceptor retinol dehydrogenases (prRDH in rods, and both prRDH and retSDR1 in cones). This process may be facilitated by the ABCR protein.

It has been demonstrated that supplementation with vitamin A improves dark adaptation in elderly patients including those with early AMD (Owsley et al., 2006). Presumably this effect is related to increased stores of vitamin A in the RPE which in turn supports transport of larger fluxes of 11-*cis*-retinal to the POS for rhodopsin regeneration (Figure 15.2b) (Lamb and Pugh, 2004). Characteristic features of the aging retina include inadequately supported and kinked POS (Marshall et al., 1979), which may affect transport of retinoids between photoreceptors and RPE (Jackson et al., 2002). In particular, compromised delivery of 11-*cis*-retinal needed to regenerate the inactive state of rhodopsin may affect the termination of phototransduction and dark adaptation. Indeed, aging is associated with delayed termination of phototransduction and dark adaptation and further delays in both processes are associated with late AMD (Jackson et al., 1999, 2006; Owsley et al., 2001).

However, the impairment in transport of retinoids within the compromised POS may increase the risk of accumulation of free retinal within photoreceptor membranes (Różanowska and Różanowski, 2008; Rozanowska and Sarna, 2005). Free all-*trans*-retinal can induce oxidative damage to photoreceptor lipids and proteins, which may then contribute to the formation of lipofuscin in the RPE (Różanowska and Różanowski, 2008; Rozanowska and Sarna, 2005). Lipofuscin accumulates with ageing and its elevated levels are observed in several retinal degenerations. Within lipofuscin potent photosensitizers are present and therefore its accumulation in the RPE increases the risk of photodamage and phototoxicity. It has been suggested that high concentrations of lipofuscin contribute to RPE cell death and development of atrophic areas in retinas of AMD patients (Holz et al., 2007; Katz, 2002; Schmitz-Valckenberg et al., 2006).

Moreover, efficient rhodopsin regeneration may precede enzymatic reduction of all-*trans*-retinal to all-*trans*-retinol in the aged retina (Figure 15.2c) (Schadel et al., 2003). Upon rhodopsin regeneration, all-*trans*-retinal is released from the "exit" site of the protein into the lipid membrane (Figure 15.2c) (Schadel et al., 2003). From here the removal of all-*trans*-retinal to the outer leaflet of the disc membrane is dependent on activity of ATP-binding cassette trasporter A4 (ABCA4) present in the rim of photoreceptor disc, known also as ABCR protein.

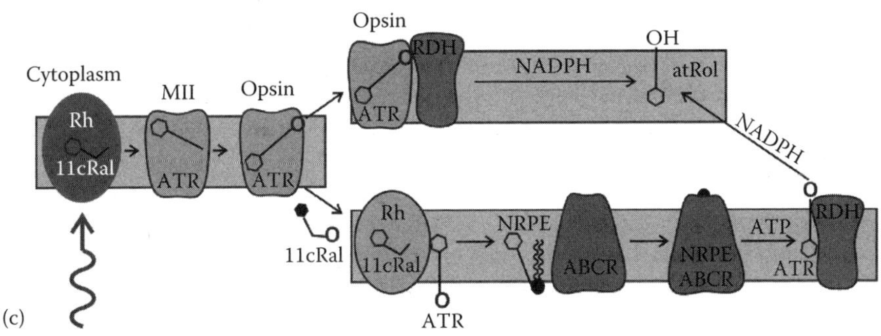

FIGURE 15.2 (continued) Following reduction atRol is transported to the RPE chaperoned by IRBP or cellular retinol binding protein (CRBP). Once inside the RPE atRol is esterified by lecithin:retinol acyltransferase (LRAT) to form atRE. (c) ATR which is of vital importance to the photoreceptor OS is a potentially damaging molecule capable of photosensitizing molecular oxygen and therefore is a carefully regulated species. Photoactivation of rhodopsin (Rh) leads to formation of biochemically active metarhodopsin II (MII) from which ATR is hydrolyzed. There are two pathways leading to enzymatic ATR reduction to atRol upon hydrolysis from opsin. ATR may either be reduced by NADPH-dependent RDH while bound to the opsin "exit site," or after it is released to the inner leaflet of the disk membrane upon binding to opsin of 11cRal and rhodopsin regeneration. The reduction of ATR is accelerated by ABCR, which transports ATR complexes with phosphatidylethanolamine, *N*-retinylidene phosphatidylethanolamine (NRPE) to the outer leaflet of disc membrane. Finally, ATR is enzymatically reduced to atRol. (Modified from Rozanowska, M. and Sarna, T., *Photochem. Photobiol.*, 81, 1305, 2005.)

The importance of ABCR protein in retinal degeneration is underscored by epidemiologic data showing that certain variants of *ABCR* gene are a causative factor for Stargardt's disease and are associated with an increased susceptibility to AMD in a subpopulation of AMD patients (Allikmets, 2000; Allikmets et al., 1997; Bernstein et al., 2002a). ABCR is responsible for transport of all-*trans*-retinal-phosphatidylethanolamine conjugate from the inner leaflet of photoreceptor disc membrane to the outer leaflet enabling its enzymatic reduction to all-*trans*-retinol (Beharry et al., 2004) (Figure 15.2c). In the absence of *ABCR* in knockout animals or in cases of ABCR dysfunction as observed in Stargardt's disease patients with mutated *ABCR*, elevated amounts of all-*trans*-retinal accumulate in POS upon exposure to light and photobleaching of visual pigments (Mata et al., 2000, 2001; Weng et al., 1999). As a result, early in life affected people and animals accumulate large amounts of lipofuscin.

Therefore, it may be speculated that elevating levels of vitamin A through supplementation in the retina with fully functional synthesis of 11-*cis*-retinal from all-*trans* retinol but dysfunctional transport of retinoids, such as in the aged retina or retina with dysfunctional ABCR, may increase the risk of retinal-mediated damage. At present, several drugs are being tested or developed with the aim to decrease the efficiency of the retinoid cycle (Golczak et al., 2005a, b; Maiti et al., 2006; Radu et al., 2003, 2005; Travis et al., 2007).

15.3.2.3 Expression and Secretion of Lipoproteins by the RPE

Once internalized within the RPE, there must be a mechanism for carotenoid transport to photoreceptors. The RPE metabolizes lipids from phagocytosed POS and provides a constant supply of lipids to photoreceptors for the synthesis of new discs and molecular renewal of lipids within existing discs (Strauss, 2005). Thus there is a constant transfer of lipids from the RPE to photoreceptors. It has been shown in the rabbit and monkey that intraveneous administration of lipophilic benzoporphyrin bound to LDLs results in an efficient delivery of the fluorescent photosensitizer not only to the RPE but also to photoreceptors; this occurs within 20 min following injection (Haimovici et al., 1997; Miller et al., 1995).

These data suggest that one of possible mechanisms of carotenoid delivery to the neural retina may involve lipoprotein uptake from the basal side of the RPE followed by its retro-endocytosis on the apical site (Lorenzi et al., 2008). Alternatively, the endocytosed lipoprotein may be degraded in the RPE followed by secretion of certain lipophilic components from the lipoprotein at the apical site. Due to low solubility of carotenoids in aqueous solutions, it may be suggested that they are secreted already bound to a protein or that an acceptor protein is available in the interphotoreceptor matrix and/or POS.

In the neural retina there are at least two types of proteins with high affinity and specificity for binding of lutein and zeaxanthin (Bhosale and Bernstein, 2007; Bhosale et al., 2004; Loane et al., 2008; Yemelyanov et al., 2001). Bernstein and colleagues have isolated two xanthophyll binding proteins (XBPs) from human retina having molecular weights of 25 and 55 kDa (Yemelyanov et al., 2001). They have shown that these XBPs have high affinity for lutein, 3′-epilutein, meso-zeaxanthin, β-cryptoxanthin, and zeaxanthin, substantially smaller for a diketodihydroxycarotenoid, astaxanthin, while binding of β-carotene or the diketocarotenoid, canthaxanthin, was negligible (Yemelyanov et al., 2001). In their subsequent study of XBPs, Bernstein et al. have purified from human retina a protein fraction corresponding to 23 kDa which then has been resolved into four components using two-dimensional gel electrophoresis (Bhosale et al., 2004). The most prominent component has been identified as glutathione-S-transferase class pi (GSTP1) and has been shown to bind zeaxanthin and meso-zeaxanthin with high affinity, but not lutein (Bhosale et al., 2004).

GSTP1 belongs to a superfamily of phase II detoxification enzymes, glutathione transferases (GSTs), which become upregulated in response to oxidative stress. Mammalian GSTs are divided into three major families: cytosolic, mitochondrial, and microsomal GSTs (Hayes et al., 2005). Cytosolic and mitochondrial GSTs are soluble enzymes, whereas microsomal GSTs are membrane associated. The best-known enzymatic function of many GSTs is conjugation of glutathione to various nonpolar electrophilic metabolites, such as epoxides or unsaturated aldehydic products of lipid peroxidation (Ellis, 2007). However, GSTs may exhibit other functions, including peroxidase activity and isomerase activity toward various unsaturated compounds.

GSTP1 is one of cytosolic enzymes. Interestingly, Bernstein et al. have demonstrated that the human recombinant isoforms, GSTM1 and GSTA1, which are also cytosolic enzymes, exhibit very low affinity binding for zeaxanthin in comparison to GSTP1 (Bhosale et al., 2004). Other potential XBPs, namely, tubulin, albumin, HDL, LDL, and β-lactoglobulin, have not exhibited high-affinity saturable binding of zeaxanthin comparable with that of GSTP1.

Interestingly, none of those three isoforms of GSTs tested has exhibited high-affinity binding of lutein. Out of all other candidate proteins tested, only albumin has exhibited saturable high-affinity binding for lutein (Bhosale et al., 2004). It may be suggested that XBPs may be involved in specific accumulation of xanthophylls as well as in xanthopyll transport inside the cells or even in between cells.

It needs to be noted that apart from expression of lipoprotein receptors, RPE itself expresses several apolipoproteins (Bartl et al., 2001; Ishida et al., 2004; Li et al., 2006; Malek et al., 2003; Tserentsoodol et al., 2006a). So far, six apolipoproteins have been identified as being expressed by the RPE, namely, apolipoprotein A-I (ApoA-I), ApoB, ApoC-I, ApoC-II, ApoE, and ApoJ (clusterin) (Bailey et al., 2004; Bartl et al., 2001; Ishida et al., 2004; Li et al., 2006; Malek et al., 2003; Tserentsoodol et al., 2006a). In addition to their functions as lipid transporters and receptor ligands, apo-lipoproteins can act as modulators of several enzymes. The basic characteristics of apo-lipoproteins expressed by the RPE are described below.

ApoA-I is the major apo-lipoprotein of HDL (Tserentsoodol et al., 2006a; Zannis et al., 2006), which is involved in lipid efflux from peripheral cells and its transport to the liver. Also, free ApoA-I has been shown to act as an acceptor for cholesterol secreted from cells and to stimulate its efflux.

ApoB is a component of chylomicrons and VLDL and is essential in liver, intestine, and heart for assembly of large lipoproteins containing triglyceride and esterified cholesterol (Bjorkegren et al., 2001). To assemble lipoproteins containing ApoB, a microsomal triglyceride transfer protein (MTP) is required (Li et al., 2005). Significantly, Curcio and colleagues have demonstrated that human RPE and neural retina synthesizes MTP, and the RPE assembles and secretes ApoB-containing lipoproteins *in vitro* (Li et al., 2005; Malek et al., 2003). They have suggested that these lipoproteins secreted by the RPE may then accumulate on the basal side of the RPE as age-related deposits in Bruch's membrane (Li et al., 2005; Malek et al., 2003).

ApoC-I is expressed mainly in liver but also in lung, skin, testis, spleen, neural retina, and RPE. Its multiple functions include the activation of lecithin cholesterol acyltransferase (LCAT) and the inhibition, among others, of lipoprotein and hepatic lipases that hydrolyze triglycerides in particle cores. Notably, both LCAT and lipoprotein lipases are expressed in RPE and choroid (Li et al., 2006). Moreover ApoC-I has been shown to displace ApoE on the VLDL and LDL and thus hinder their binding and uptake via their corresponding receptors (Li et al., 2006).

ApoC-II is expressed in liver and intestine, and both the neural retina and RPE (Li et al., 2006). In contrast to ApoC-I, it can function as an activator of lipoprotein lipase. Similar to ApoA-I, ApoA-II, and ApoE, in the absence of lipid to stabilize its structure, ApoC-II forms amyloid assemblies.

ApoE is a component of VLDL, HDL, and chylomicrons and is expressed in many tissues including the RPE and neural retina. It is found in atherosclerotic intima and sub-RPE deposits (Anderson et al., 2001; Malek et al., 2003). Interestingly, free ApoE is a ligand for the SR-BI receptor and stimulates selective uptake of cholesterol esters from HDLs (Bultel-Brienne et al., 2002). ApoE has anti-angiogenic, anti-inflammatory, and antioxidant effects (Browning et al., 1994; Kelly et al., 1994; Tangirala et al., 2001). Its importance in the retina is highlighted by the fact that certain isoforms of ApoE, ApoE-2, and ApoE-4 are associated with an increased and decreased risk of AMD, respectively (Thakkinstian et al., 2006). In human eyes, *post mortem* ApoE has been found to be localized predominantly at the basal surface of the RPE (Anderson et al., 2001). In polarized, cultured RPE cells, ApoE has been secreted both at the apical and basal sides, and its secretion has been stimulated by the presence of HDL (Ishida et al., 2004).

ApoJ is another protein component of HDL which is highly expressed by the RPE and neural retina, especially under oxidative stress conditions (Wong et al., 2000, 2001). It can act as a complement regulatory protein, which by binding to and inactivating the membrane-attack complex can prevent cytolysis (Bartl et al., 2001). ApoJ accumulation was identified in drusen in AMD patients (Sakaguchi et al., 2002; Wong et al., 2000).

The expression of all these apo-lipoproteins by the RPE, and its ability to form lipoprotein particles suggest that these newly formed lipoproteins may be involved in the transport of lipophilic molecules, including carotenoids, from the RPE to the neural retina and/or to the choroidal blood supply. Testing the roles of apolipoproteins and lipoprotein particles in carotenoid secretion from the RPE is another subject awaiting experimental investigation.

15.3.2.4 Transporters Potentially Involved in Carotenoid Movement in the Retina

While it may be speculated that in the RPE both lipoprotein and/or scavenger receptors are likely to be involved in carotenoid uptake from the blood, it is not clear what mechanism(s) are responsible for carotenoid transport through the RPE into the neural retina. Also, it is not clear what mechanism(s) are responsible for selective accumulation in the retina of only two carotenoids.

Intracellular transport and efflux from cells of lipophilic molecules can be mediated by several members of the ATP-binding cassette (ABC) transporters family, some of which have been identified in the brain, including the retina (Kim et al., 2008; Sarkadi et al., 2006).

One of those transporters is an ABC transporter A1 (ABCA1) which is widely expressed, with particularly high expression in the adrenal gland and uterus and moderate expression in the liver and brain (Kim et al., 2008). In human neuronal tissue ABCA1 is expressed by multiple cell types: isolated human fetal neurons, microglia, astrocytes, and oligodendrocytes (Kim et al., 2008). ABCA1 has been identified in both the RPE and neural retina (Bailey et al., 2004; Lakkaraju et al., 2007; Tserentsoodol et al., 2006a). The most extensively investigated function of ABCA1 is its role in reverse transport of lipids from cells via HDLs to the liver (Faulkner et al., 2008; Lakkaraju et al., 2007; Lorenzi et al., 2008). Importantly, ABCA1 also mediates intracellular efflux of cholesterol from late endosomes/lysosomes (Chen et al., 2001).

The importance of ABCA1 in carotenoid uptake into the retina is evident from studies on the Wisconsin hypo-alpha mutant (WHAM) chicken having a recessive sex-linked mutation in the ABCA1 transporter gene (Connor et al., 2007). Proper function of ABCA1 is essential for the formation of HDLs (Faulkner et al., 2008; Tserentsoodol et al., 2006a). The mutant chicken exhibits a severe deficiency of HDLs, having levels as much as 18.8 times smaller than in the control chicken (Connor et al., 2007). Normally, HDLs are the predominant lipoproteins in chicken plasma, accounting for 88% of total lipoproteins in the control chicken. In the WHAM chicken, HDLs account for only 15% of total lipoproteins. The level of total lipoproteins in the mutant chicken is 3.2 times smaller than in the control chicken. The partial compensation for the lipoprotein concentration in the mutant chicken is due to a 1.5-fold increased level of LDLs in comparison to the control chicken.

Importantly, the mutant chicken exhibits lower levels of lutein and zeaxanthin in plasma and several other tissues in comparison with the control chicken, and that difference is already apparent in 1-day-old chickens and remains in 28-day-old chickens fed the same diet (Connor et al., 2007). In the WHAM chickens, the levels of lutein in the plasma, retina, skin, adipose tissue, liver and heart, respectively, have been found to be only 8%, 10%, 18%, 33%, 52%, and 60% of the corresponding levels in control chickens. Even though the diet in these chickens included three times more lutein than zeaxanthin and these ratios have been present in the plasma of both the control and WHAM chickens, there was a preferential accumulation of zeaxanthin over lutein in their retinas.

The ratio of zeaxanthin to lutein was found to be 1.6 and 1.3 in the retinas of the control and WHAM chickens, respectively.

Interestingly, a 5.2-fold increase of lutein in the diet of the chickens for 4 weeks led to a substantial 4.4-fold increase in lutein in the plasma of the WHAM chickens, but only a 2.5-fold increase in control chickens (Connor et al., 2007). Overall, the plasma level of lutein was still 4.8 times greater in the control chickens than in WHAM chickens fed a lutein-rich diet. Furthermore, both types of chickens on lutein-rich diet reached similar levels of lutein in their heart and liver. Yet, the difference in the levels of lutein in the retina of the control and WHAM chickens on lutein-rich diet became even greater than in chickens on the control diets. The retinas of WHAM chickens accumulated only 6% as much lutein as was accumulated in retinas the control chickens.

Comparisons on WHAM chickens fed a lutein-rich diet with control chickens on control diet indicate dysfunctional uptake of lutein into the retinas of WHAM chickens (Connor et al., 2007). Even though the plasma levels of lutein in WHAM chickens on lutein-rich diet was 2.7 times smaller than its levels in the control chickens on the control diet, the lutein levels in the hearts and livers of WHAM chickens were 2.1- and 1.4-fold greater than the controls. The concentration of lutein in the retinas of WHAM chickens on lutein rich diet was 6.1-fold smaller than in the retina of the control chickens on control diet.

It has been suggested that the smaller accumulation of lutein in the retinas of WHAM chickens relative to the control chickens was due to a preferential uptake of HDLs into the retina (Connor et al., 2007): the WHAM chickens exhibiting a relatively lower concentration of HDLs in the plasma accumulate less lutein in their retinas than the control chickens.

However, lutein is transported in the plasma bound not only to HDL but also to other lipoproteins which are taken up by the RPE, such as LDL and VLDL (Calvo et al., 1997, 1998; Duncan et al., 2002; Elner, 2002; Gordiyenko et al., 2004; Hu et al., 2008; Parker, 1996; Provost et al., 2003; Ryeom et al., 1996b; Tserentsoodol et al., 2006b; Wang and Anderson, 1993; Wang et al., 2007). Therefore, it can be suggested that xanthophyll deficiency in the retina of WHAM chickens is due not only to deficiency in the HDL but also due to dysfunctional ABCA1 in the retina. In this hypothetical scenario ABCA1 is responsible for transport of xanthophylls endocytosed by the RPE into the neural retina. Thus, the dysfunction of a mutated ABCA1 in WHAM chickens may be directly responsible for the low levels of xanthophylls in their retinas.

In addition to its presence in the RPE, ABCA1 has been found to be localized in the neural retina, particularly in the ganglion cell layer and rod photoreceptor inner segments (Tserentsoodol et al., 2006a), suggesting it may be involved in carotenoid transport throughout the retina.

ABCA1 interacts with at least two apo-lipoproteins expressed by the RPE, ApoA1, and ApoE (Faulkner et al., 2008; Von Eckardstein et al., 2001). The neural retina expresses all apo-lipoproteins, which are expressed by the RPE, namely, ApoA-I, ApoC-I, ApoC-II, ApoE, and ApoJ (Li et al., 2006; Tserentsoodol et al., 2006a). ApoA1 was identified in the ganglion cell layer, the rod photoreceptor inner segment layer, and the rod photoreceptor outer segment layer, presumably localized to the interphotoreceptor matrix (Tserentsoodol et al., 2006a). In addition, the neural retina expresses ApoA-II that has not been identified in the RPE (Li et al., 2006). It may be speculated that at least some of these apo-lipoproteins within the neural retina have the potential to act as transporters of xanthophylls moving from the RPE into the retina.

The neural retina expresses several lipoprotein receptors including SR-BI, SR-BII (Tserentsoodol et al., 2006a,b), and VLDL receptor (VLDLR) (Hu et al., 2008). Thus, carotenoid flow through the RPE and further transport in the neural retina may also be mediated by lipoprotein receptors (Tserentsoodol et al., 2006a,b). SR-BI and SR-BII have been found to be localized mainly to the ganglion cell layer and POS in the monkey retina (Tserentsoodol et al., 2006a).

Apart from SR-BI, SR-BII, CD36, and ABCA1, a microarray analysis of genc expression in human RPE reveals some additional lipid transporters that might potentially be involved in intracellular transport of carotenoids and/or their efflux from the RPE cells into the neural retina or out of the retina into the choroidal blood (van Soest et al., 2007). These include other ABC transporters: ABCA6, ABCA8, and ABCA10 (van Soest et al., 2007). The functions of those transporters are not well understood. ABCA6 is also expressed in liver, lung, heart, and brain and is believed to be involved in lipid homeostasis (Kaminski et al., 2001). ABCA8 is also expressed in human brain and in isolated fetal brain neurons and astrocytes, and it is involved in transport of lipophilic molecules, including the bioactive lipid, leukotriene C4 (Kim et al., 2008). In the mouse brain, ABCA8 was identified in the choroid plexus (Matsumoto et al., 2003), which is responsible for the formation of the blood–cerebrospinal fluid barrier. ABCA10 is ubiquitously expressed, with the highest gene expression levels detectable in the heart, brain, and the gastrointestinal tract, and it is believed to be involved in lipid homeostasis (Wenzel et al., 2003).

It may be suggested that as a result of lipoprotein uptake a variety of carotenoids enter the RPE and subsequently lutein and zeaxanthin are transported into the neural retina, provitamin A carotenoids are metabolized to form vitamin A within the RPE, and all others are secreted at the basal side and back into the blood (Figure 15.3). It can be further argued that other ABC transporters known also as multidrug resistance (MDR) proteins may be involved in efflux of carotenoids and/or their metabolites out of the retina. In humans, the three major types of MDR proteins include members of the ABCB, ABCC, and ABCG subfamily, and they take part in the maintenance of the blood–brain barrier (Sarkadi et al., 2006). MDR proteins act as efflux transporters for a wide variety of hydrophobic compounds. MDR1 (ABCB1/P-glycoprotein) is one of MDR proteins and is expressed in the RPE (Aukunuru et al., 2001; Constable et al., 2006; Esser et al., 1998; Kennedy and Mangini, 2002). MDR1 transports amphipatic compounds with a molecular mass of 300–2000 Da including anticancer drugs and antibiotics (Sarkadi et al., 2006) from cells. Studies on patients suffering from proliferative vitreopathy show that MDR1 protein expression in the RPE is strongly upregulated upon exposure to the drug used for its treatment, daunomycin (Esser et al., 1998).

Altogether, the role of transporters of lipophilic molecules regulating the movement of carotenoids through the RPE and into the neural retina is another area awaiting experimental investigation.

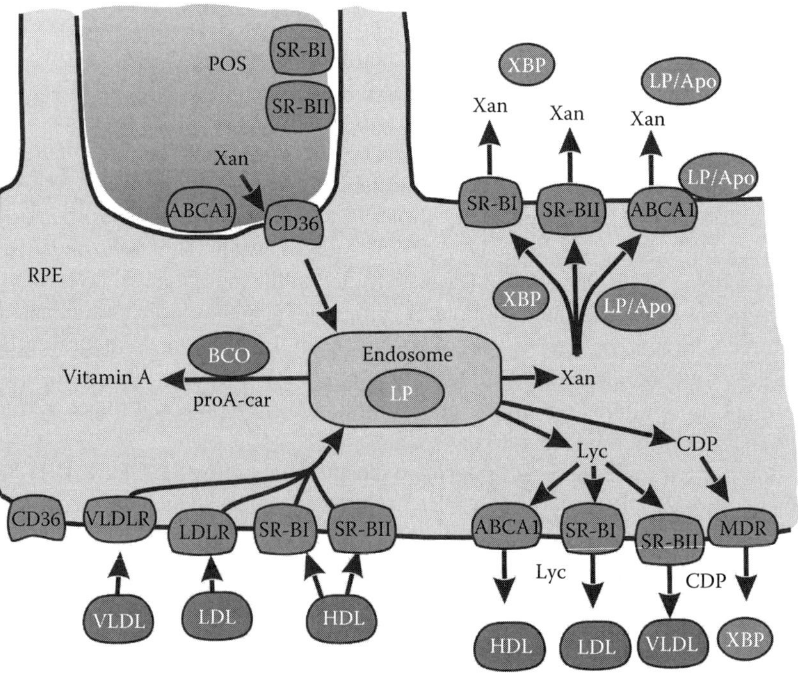

FIGURE 15.3 Hypothetical pathways responsible for carotenoid uptake, metabolic transformations, transcytosis to the neural retin, or secretion to the blood.

15.4 CULTURED RPE AS A MODEL OF PHYSIOLOGICAL RPE FUNCTIONS

Despite recent advances in understanding carotenoid uptake in the intestines, the specific uptake and retention of carotenoids in the retina is still poorly understood (Loane et al., 2008). In contrast to photoreceptors, which in culture rapidly lose their morphology and physiological function, RPE cells, cultured under appropriate conditions can sustain several features characteristic of them under physiological conditions (Figure 15.4). These include phagocytosis, ability to form polarized monolayers, expression of growth factors, cytokines, and certain components of the complement (Chen et al., 2007; Crane et al., 2000a; Cui et al., 2006; Ebihara et al., 2007; Ershov and Bazan, 2000; Ishida et al., 2004; Joffre et al., 2007; Karl et al., 2007; Maminishkis et al., 2006; Momma et al., 2003; Mukherjee et al., 2007). When isolated and cultured under appropriate conditions, RPE cells retain the ability to synthesize melanin and sustain the retinoid cycle by synthesizing 11-*cis*-retinal from all-*trans*-retinol (Maminishkis et al., 2006; von Recum et al., 1999).

Cultures of primary RPE and some RPE cell lines express scavenger receptors and transporters that are deemed likely participants in carotenoid uptake and further transport. These include CD36 (Ryeom et al., 1996a,b), SR-BI and SR-BII (Duncan et al., 2002), ABCA1 (Lakkaraju et al., 2007), and ABCB1 (MDR1/P-glycoprotein) (Aukunuru et al., 2001; Constable et al., 2006; Kennedy and Mangini, 2002). Human RPE cell line, D407, has been shown to express fully functional BCO (Chichili et al., 2005). Moreover cultured RPE cells express similar apo-lipoproteins to those of the RPE *in situ*, including ApoA-I, ApoB, ApoC-I, ApoC-II, ApoE, ApoJ, and microsomal triglyceride transfer protein (MTP), which is required for assembly of lipoproteins containing ApoB (Li et al., 2005, 2006; Suuronen et al., 2007). In particular, cultures of ARPE-19, a spontaneously immortalized human RPE cell line derived from 19-year-old human male (Dunn et al., 1996), exhibits differentiated properties and expresses all of these scavenger receptors, transporters, apo-lipoproteins, and MTP. Thus RPE cultures appear as an excellent model to study carotenoid uptake, dynamics of transport, and effects of carotenoids on the RPE.

15.4.1 Carotenoid Uptake, Accumulation, and Secretion in Cultured RPE Cells

The methodology to study *in vitro* drug uptake, metabolism, and removal is well established in extensive studies of blood–brain barrier, intestinal drug metabolism, and drug resistance in cancer cells (Cecchelli et al., 2007; Reboul et al., 2006; Sarkadi et al., 2006; van de Kerkhof et al., 2007). In particular, the mechanisms responsible for intestinal absorption of carotenoids have recently been a subject of intensive investigation where the Caco-2 cell line has been used as a model of intestinal enterocytes (O'Sullivan et al., 2007; Reboul et al., 2005; Yonekura and Nagao, 2007). In contrast to previous beliefs of the diffusion of carotenoids and tocopherols into the enterocytes being a passive process, it has now been shown that their uptake is mediated by SR-BI and possibly other receptors (During and Harrison, 2007; van Bennekum et al., 2005). Study of competitive uptake of α-tocopherol by Caco-2 cells in the presence of a mixture of carotenoids (lycopene, β-carotene, and lutein) demonstrated that carotenoids significantly reduce α-tocopherol uptake, thus providing further support for an important role of SR-BI in the uptake of both carotenoids and vitamin E

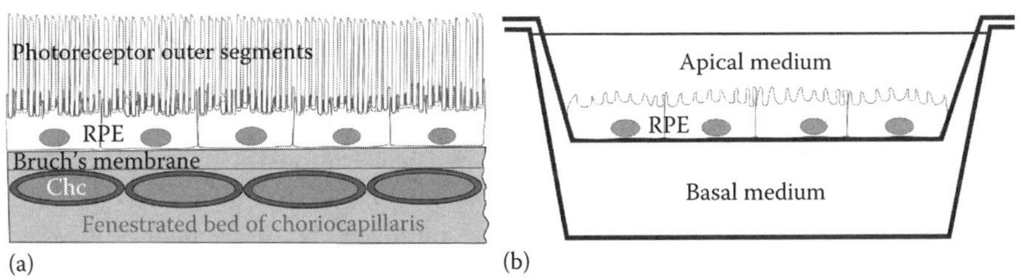

FIGURE 15.4 Polarized cultures of RPE as a model of blood–retina barrier: (a) schematic diagram of the RPE at the blood–retina barrier and (b) culture of polarized RPE cells as a model of the blood–retina barrier.

(Reboul et al., 2007a,b). As mentioned earlier the competitive uptake occurs also in the presence of a mixture of carotenoids where absorption of lutein is inhibited by β-carotene but not by lycopene (Reboul et al., 2005). This indicates that the presence of a mixture of different lipophilic substrates can strongly influence the uptake of certain carotenoids. It has also been demonstrated that cultured Caco-2 cells secrete β-carotene, preferentially within micelles rich in long fatty acids (Yonekura et al., 2006), suggesting that carotenoids can be stored in the cell or secreted depending on the absence or presence of appropriate carotenoid acceptors.

Despite the feasibility of using cultured RPE cells for studies similar to those performed using Caco-2 cells, the role of the RPE in carotenoid uptake and dynamic regulation has only just begun to be investigated. As carotenoids are carried in blood by lipoproteins, lipoprotein-rich serum seems to be the most appropriate vehicle for carotenoid delivery to cultured RPE cells. Indeed, recent studies comparing carotenoid delivery from fetal calf serum and from organic solvents showed that delivery in the presence of serum was superior to tetrahydrofuran (Shafaa et al., 2007).

Our experiments employing the ARPE-19 cell line demonstrate that these cells in culture exhibit selectivity in accumulation of β-carotene and hydroxy-carotenoids–lutein, zeaxanthin, over a lutein metabolite containing a keto group, (3R, 6′R)-3-hydroxy-β,ε-carotene-3′-one (3′-oxolutein) (Figure 15.1) (Rozanowska et al., 2004b). 3′-oxo-lutein has been identified in human and monkey retina as well as in human blood plasma particularly upon long-term supplementation with lutein (Bernstein et al., 2001; Bhosale et al., 2007a,b; Khachik et al., 1997, 2006). In our experimental setup, the ARPE-19 cells have been fed for a period of up to 3 weeks with culture medium containing 10% fetal calf serum, 2 μM carotenoids and 0.2% dimethylsulphoxide used for solubilization of carotenoids in their stock solutions. The cells gradually accumulated increasing amounts of zeaxanthin, lutein, and β-carotene. After three weeks, concentrations of up to 165 pmol/million cells were observed. Yet, the accumulation of 3′-oxolutein remained unchanged at low levels (below 20 pmol/million cells) throughout the experiment (Rozanowska et al., 2004b). This selective discrimination against accumulation of 3′-oxolutein is intriguing. It may be argued that 3′-oxolutein must have entered the cells in a similar way to the other carotenoids—either bound to serum lipoproteins or directly through solubilization in the lipid plasma membrane. Therefore, an efficient efflux mechanism must operate to remove it from the cells.

It is of interest to determine if it is the keto group of 3′-oxolutein that accounts for this observation. Comparison of the uptake by ARPE-19 cells, and ultimately by RPE *in situ*, of other keto containing carotenoids, such as astaxanthin and canthaxanthin, to find out whether ARPE-19 cells exhibit similar behavior toward all keto carotenoids may provide insights into transport mechanisms. Canthaxanthin and astaxanthin are naturally occurring carotenoids and are used in the food industry to add color to foods such as sausage and fish and as such are part of the human food chain.

Astaxanthin has been suggested as a potential dietary supplement to improve retinal function (Hussein et al., 2006; Parisi et al., 2008). While normally astaxanthin is not present in human plasma, a high dose of dietary astaxanthin does result in its appearance in appreciable concentrations within the plasma where it is bound mainly to the lipoproteins, VLDL, LDL, and HDL (Coral-Hinostroza et al., 2004). To our knowledge, to date there is no report identifying astaxanthin in human retina.

Canthaxanthin has been used in the treatment of various light-sensitive dermatoses and in over-the-counter "tanning pills" in the United States and Europe (Haught et al., 2007; Leyon et al., 1990). It is no longer approved for the tanning purpose because of its adverse effects (Haught et al., 2007). These include hepatitis, aplastic anemia, urticaria, and a retinopathy (Chan et al., 2006; Espaillat et al., 1999; Haught et al., 2007; Leyon et al., 1990). The canthaxanthin retinopathy has been observed in humans after high canthaxanthin intake (more than 30 mg/day) and is characterized by yellow crystal-like deposits in the inner retina, usually with no effects on vision except for one case where visual field defect was reported. In experimental studies of canthaxanthin effects in cynomolgus monkeys fed daily for 2.5 years with up to 48.6 mg canthaxanthin/kg of body weight, no retinal changes were detected *in vivo* by ophthalmoscopy, biomicroscopy, or electroretinography (Goralczyk et al., 1997). However, postmortem examination using polarization microscopy showed that animals on canthaxanthin doses of at least 0.2 mg/kg of body weight per day exhibited a circular zone in the peripheral retina containing birefringent, polymorphous red, orange, or white inclusions. These inclusions were located mainly in the nerve fiber layer, ganglion cell layer, inner plexiform layer, and inner nuclear layer. There were no other apparent histological changes observed in the retinas (Goralczyk et al., 1997, 2000). Subsequent study revealed that, in monkeys, a long-term daily administration of extremely high doses of 0.2–0.5 g of canthaxanthin/kg of body weight are required to induce retinal crystal detectable by ophthalmoscopy but still the visual function has not been affected (Goralczyk et al., 2000). Similarly, in cats, canthaxanthin supplementation with doses of between 2 and 16 mg/kg/day for up to 6 months resulted in canthaxanthin accumulation in the retina overlying the *tapetum lucidum* but no ophthalmological evidence of crystal formation was found (Scallon et al., 1988). Electroretinograms performed after one and two months of canthaxanthin supplementation showed no significant change compared to baseline. Markedly however, in the cat RPE canthaxanthin induced pronounced changes including disruption of some phagolysosomes (Scallon et al., 1988).

Interestingly, our recent results show that both astaxanthin and canthaxanthin accumulate in ARPE-19 cells *in vitro* in concentrations similar to those of zeaxanthin (Rozanowska and Rozanowski, unpublished). Therefore, at least in the ARPE-19 cell line the keto groups of astaxanthin and canthaxanthin do not prevent from their efficient accumulation.

In another study of carotenoid accumulation, cultured ARPE-19 cells were treated with a lipophilic extract from tomatoes solubilized in ethanol and injected into the culture medium for 24 h. The extract, containing β-carotene, lycopene, and lutein at relative ratios of 23, 13, and 1, respectively, led to internalization of carotenoids at ratios of 9, 1.3, and 1, respectively (Chichili et al., 2006). These results indicate preferential accumulation of β-carotene and lutein over lycopene in ARPE-19 cells.

Considering the expression of proteins involved (or potentially involved) in carotenoid transport, binding, and metabolism, and drawing on analogy to other cells involved in uptake and transfer of carotenoids, a hypothetical scenario of carotenoid uptake by the RPE can be suggested (Figure 15.3). It may be hypothesized that scavenger receptors SR-BI, SR-BII, and CD36 and lipoprotein receptors are involved in carotenoid uptake from the serum lipoproteins. Next, endocytosed lipoproteins release lipophilic molecules, which are transported from the endolysosomal compartment via ABCA1 to apo-lipoproteins, XBPs (GSTP1, albumin) or to BCO. BCO in the RPE appears to convert β-carotene and cryptoxanthin into vitamin A, which may explain their absence in the retina. Other carotenoids are transported either into the neural retina or outward flow back through the basal side. As albumin has been identified in the interphotoreceptor matrix (Adler and Edwards, 2000), it may be argued that it may serve as a potential acceptor for lutein transported from the RPE on the apical side into the neural retina. The transport of carotenoids within the retina may be facilitated by ABCA1, SR-BI/SR-BII, or MDR1. The presence in the retina of apo-lipoproteins and XBPs with differing affinities for different types of carotenoids may function in the selective retention of lutein and zeaxanthin.

SR-BI is highly abundant in POS (Tserentsoodol et al., 2006a). Therefore it may be speculated that it plays a role in further uptake of carotenoid-enriched (lipo)proteins and their transport from the outer segment and then into deeper layers of the retina.

So far, we have focused mainly on the potential pathways of carotenoid uptake by the RPE from the choroidal blood supply. As mentioned earlier, POS contain lutein and zeaxanthin, and their distal tips are phagocytosed by the RPE. Therefore POS is another source of xanthophylls in the RPE and this pathway of carotenoid delivery and their further fate can be easily tested in cultured RPE. It has been shown that exposure of RPE cells *in vitro* to HDL stimulates efflux of phospholipids from phagocytosed POS out of the cell (Ishida et al., 2006). Thus it is of interest to determine whether that transport may potentially include xanthophylls and whether other types of lipoproteins may stimulate the carotenoid efflux from RPE cells. If such a transport pathway exists, then, how that transport depends on the type of the carotenoid and whether it is greater on the basal side or the apical side in polarized RPE monolayers would be informative.

Altogether, there are many unknowns about carotenoid transport in the retina. However, present knowledge on carotenoid uptake in other cell types and the finding of multiple proteins potentially involved in carotenoid transport in the RPE and adjacent neural retina leads to the suggestion that several hypothetical pathways exist (Figure 15.3). Many such pathways can be easily tested in cultured RPE.

15.5 CAROTENOID PROTECTION IN THE RPE

In the retina, RPE cells are under constant oxidative stress: (1) they are exposed to high oxygen tensions; (2) include abundant intracellular polyunsaturated lipids, including docosahexaenoate with six unsaturated double bonds extremely susceptible to peroxidation and are exposed to polyunsaturated lipids present in POS and Bruch's membrane from both apical and basal sides, respectively; (3) are involved in transport of iron between the retina and choroidal blood supply; (4) are exposed to a high intensity of visible light; and (5) potent photosensitizers present both inside and outside the cell that can activate oxygen in the presence of light (Boulton et al., 2001; He et al., 2007; Różanowska and Różanowski, 2008; Rozanowska and Sarna, 2005; Wong et al., 2007).

As discussed earlier, due to their antioxidant properties carotenoids have a potential to provide protection against oxidative damage. However, it has been demonstrated that at least a part of the lutein and zeaxanthin in the retina is bound to glutathione transferase GSTP1 and other as yet to be identified XBPs with high affinity for xanthophylls, and tubulin (Bernstein et al., 1997; Bhosale et al., 2004; Yemelyanov et al., 2001). It has been shown that the binding of zeaxanthin or meso-zeaxanthin to GSTP1 enhances their antioxidant action against lipid peroxidation induced by thermolabile azo-compounds as a source of peroxyl radicals (Bhosale and Bernstein, 2005). It remains to be determined how the binding to these proteins affects xanthophyll ability to quench excited states of photosensitizers and singlet oxygen.

Moreover, carotenoids themselves are very susceptible to oxidative damage and their oxidation products include deleterious aldehydes (Failloux et al., 2003; Hurst et al., 2005; Rozanowski and Rozanowska, 2005; Siems et al., 2000, 2002; Sommerburg et al., 2003). Therefore it is of interest to find out how carotenoids can offer antioxidant protection in cellular systems, how stable the carotenoids are within cells, and what the fate of the carotenoid degradation products is.

15.5.1 Effects of Carotenoids on Oxidative Stress in Cultured RPE Cells

As mentioned previously, the ability of carotenoids to inhibit oxidative stress was tested *in vitro* in many different cell types. In the retina only lutein and zeaxanthin accumulate in sufficient concentrations to exert direct antioxidant effects, therefore our further discussion of these antioxidant effects will be focused mainly on those two xanthophylls.

Importantly, several studies have shown that cultured cells such as dermal fibroblasts, melanoma cells, or ARPE-19 cells can accumulate substantial amounts of lutein and zeaxanthin without deleterious effects on cell viability (Lornejad-Schafer et al., 2007; Philips et al., 2007; Roberts et al., 2002; Rozanowska et al., 2004b). However, the exposure of ARPE-19 cells to higher concentrations of lutein, such as 10 μM over a period of 19 h which has led to internalization of 470 pmol of lutein/million cells, has been shown to result in a small 9% decrease in their mitochondrial activity (Kanofsky and Sima, 2006). This effect increased upon exposure of lutein-laden cells to blue (430 nm; 8.4 J/cm^2) or green light (502 nm; 32 8.4 J/cm^2), where the mitochondrial activity has decreased by about 15% (Kanofsky and Sima, 2006). It should be noted that lutein used in these studies was of low purity of 70% and therefore it is unclear whether the cytotoxic effects can be ascribed to lutein or other contaminants.

In another study of protective effects of lutein, cultures of primary human RPE cells were incubated for 2h with 40μM lutein, after which any remaining lutein in the culture medium was washed away and the medium was replaced with phosphate buffered saline (Roberts et al., 2002). Lutein did not cause any changes in cell morphology or any increase in DNA damage assessed by the comet assay (Roberts et al., 2002). Cultures with and without lutein were irradiated either with visible light (>400nm; 2.5 J/cm^2) to mimic light reaching the adult human retina or with UVB light above 300nm (0.08 J/cm^2) to mimic the UV light reaching the retina in young children. Visible light increased DNA damage only slightly and it has been prevented completely by lutein. UVB irradiation caused extensive DNA damage which was also completely prevented in the presence of lutein (Roberts et al., 2002).

Carotenoids are excellent singlet oxygen quenchers, so it may be expected that they will be particularly efficient at protecting cells against photosensitized damage involving singlet oxygen. However, to be able to act as a singlet oxygen quencher, a carotenoid must be in the immediate proximity of the photosensitizer (within 220nm) (Kuimova et al., 2009; Redmond and Kochevar, 2006). Thus subcellular localization of the source of singlet oxygen and carotenoids are likely to play a major role in the effectiveness of carotenoids in protection against photosensitized $^1O_2$. Interestingly, we and others have demonstrated that despite accumulation of large concentrations of lutein or zeaxanthin inside ARPE-19 cells, no significant protection against photosensitized damage could be observed (Kanofsky and Sima, 2006; Rozanowska et al., 2004b; Wrona et al., 2004). The photosensitizers tested included rose bengal (Rozanowska et al., 2004b), merocyanine 540 (Wrona et al., 2004), acridine orange, and *cis*-di(4-sulfonatophenyl)diphenylporphine (Kanofsky and Sima, 2006). Rose bengal exists in an anionic form at neutral pH and, despite association with membrane lipids, it is not permeable through the plasma membrane. Merocyanine 540 binds to cellular membranes and the pattern of its distribution varies in different cell lines between the plasma membrane, mitochondria, and lysosomes (Chen et al., 2000). Acridine orange and *cis*-di(4-sulfonatophenyl) diphenylporphine are localized mainly in lysosomes (Kanofsky and Sima, 2006). It may be suggested that lutein or zeaxanthin have not exerted any protective effects on photosensitized damage induced by these photosensitizers because they were not colocalized in the same subcellular compartment and/or were bound to XBPs which limited their ability to act as singlet oxygen quenchers. In contrast, synthetic carotenoid derivatives that colocalize with photosensitizers in lysosomes or mitochondria of cultured ARPE-19 cells have been shown to offer a substantial protective effect against photosensitized damage (Kanofsky and Sima, 2006). Clearly, the subcellular localization of lutein and zeaxanthin and determination of whether they are present in free forms or are bound to proteins requires elucidation.

Lutein and zeaxanthin have also been tested as potential protection against formation of lipofuscin in cultured RPE (Sundelin and Nilsson, 2001). Lipofuscin is a complex aggregate of lipids and proteins including several fluorophores and photosensitizers which accumulate in the RPE mainly as a result of incomplete lysosomal digestion of POS (Różanowska and Różanowski, 2008; Rozanowska and Sarna, 2005). Zeaxanthin and lutein have been tested in primary cultures of rabbit and bovine RPE cells fed POS under 40% oxygen, conditions leading to rapid accumulation of lipofuscin-like inclusions (Sundelin and Nilsson, 2001). The cultured cells were supplemented with antioxidants 4 days before the first feeding with POS, and every 48h thereafter. Xanthophylls were injected into the culture medium directly from their 2mM stock solutions in tetrahydrofuran in the presence of butylated hydroxytoluene to give a final concentration of 10μM. Administration of zeaxanthin or lutein led to a substantial, up to ~60% inhibition of accumulation of fluorescent inclusions in rabbit RPE. However, a morphometric analysis of inclusion bodies in bovine RPE showed that lipofuscin accumulation was diminished by 16% or less in xanthophyll supplemented cells (Sundelin and Nilsson, 2001). This discrepancy may be explained if xanthophylls interrupt the pathway leading to the formation of some lipofuscin fluorophores, while not preventing from overall POS oxidation and formation of products nonsusceptile to lysosomal digestion.

Altogether, studies in cultured RPE indicate that lutein and zeaxanthin may provide antioxidant protection in the RPE but more research is required to determine the exact mechanisms responsible for the observed protective effects or the lack thereof.

15.6 PRO-OXIDANT EFFECTS OF CAROTENOIDS

It has been shown that carotenoids can act as a pro-oxidant at high oxygen tensions (Burton and Ingold, 1984). These observations can be explained considering the mechanism of the antioxidant action of carotenoids. When lipid-derived peroxyl radicals ($LOO^•$) react with carotenoids, the radical adduct formed, $[LOO\text{-}Car]^•$, is less reactive than the $LOO^•$ and so carotenoids act as chain breaking antioxidants in lipid peroxidation (Equation 15.6). However, in the presence of high concentrations of oxygen, the oxygen molecule can add to $[LOO\text{-}Car]^•$ generating another peroxyl radical, $LOO\text{-}Car\text{-}OO^•$, which readily propagates lipid peroxidation (Equation 15.7):

$$\mathrm{Car} + \mathrm{LOO}^{\bullet} \rightarrow [\mathrm{LOO} - \mathrm{Car}]^{\bullet} \quad (15.6)$$

$$\left[\mathrm{LOO} - \mathrm{Car}\right]^{\bullet} + \mathrm{O}_2 \rightarrow \mathrm{LOO} - \mathrm{Car} - \mathrm{OO}^{\bullet} \quad (15.7)$$

This mechanism explains the pro-oxidant behavior of carotenoids observed when oxygen partial pressures are higher than 150 mmHg (Burton and Ingold, 1984; Palozza et al., 1995, 1997). It is believed that oxygen tensions encountered under physiological conditions are not high enough to induce this pro-oxidant action of carotenoids.

It has been shown in many studies that protective effects of carotenoids can be observed only at small carotenoid concentrations, whereas at high concentrations carotenoids exert pro-oxidant effects via propagation of free radical damage (Chucair et al., 2007; Lowe et al., 1999; Palozza, 1998, 2001; Young and Lowe, 2001). For example, supplementation of rat retinal photoreceptors with small concentrations of lutein and zeaxanthin reduces apoptosis in photoreceptors, preserves mitochondrial potential, and prevents cytochrome *c* release from mitochondria subjected to oxidative stress induced by paraquat or hydrogen peroxide (Chucair et al., 2007). However, this protective effect has been observed only at low concentrations of xanthophylls, of 0.14 and 0.17 μM for lutein and zeaxanthin, respectively. Higher concentrations of carotenoids have led to deleterious effects (Chucair et al., 2007).

Carotenoid-radical adducts, such as $LOO\text{-}Car\text{-}OO^•$, are not the only products of carotenoid-free radical interactions which may exhibit pro-oxidant properties. Carotenoid cation radicals can be damaging to biomolecules. It has been shown by pulse radiolysis that $Car^{•+}$ can oxidize amino acids such as tyrosine and cysteine (Burke et al., 2001; Edge et al., 2000a). Carotenoid cation radicals may be generated as a result of interaction with the nitrogen dioxide radical, $NO_2^•$, a product of interaction of nitric oxide with oxygen, both molecules being highly abundant in the retina (Bohm et al., 1995; Everett et al., 1996) (Equation 15.8):

$$\mathrm{Car} + \mathrm{NO}_2^{\bullet} \rightarrow \mathrm{Car}^{\bullet+} + \mathrm{NO}_2^{-} \quad (15.8)$$

Moreover, carotenoid cation radicals can be formed as a result of oxidation of carotenoids by iron ions, Fe(III) (Equation 15.9) (Polyakov et al., 2001):

$$\mathrm{Car} + \mathrm{Fe(III)} \rightarrow \mathrm{R} - \mathrm{Car}^{\bullet+} + \mathrm{Fe(II)} \quad (15.9)$$

It should be stressed that in the RPE transport of iron ions between the photoreceptors and choroidal blood supply is constantly occurring (He et al., 2007; Wong et al., 2007). Iron is essential for the proper function and survival of every cell as it serves as a co-factor for vital mitochondrial enzymes.

Moreover in the retina, iron is a cofactor of a number of other enzymes, including nitric oxide synthase, β-carotene monooxygenase, and RPE65-isomerohydrolase converting all-*trans*-retinol to 11-*cis*-retinol in the visual cycle.

The reduction of Fe(III) by carotenoids may have deleterious consequences. The reduced iron Fe(II) can react with hydrogen peroxide leading to the formation of hydroxyl radical, the most reactive free radical encountered in biological systems (Equation 15.10):

$$Fe(II) + H_2O_2 \rightarrow Fe(III) + OH^- + {}^{\bullet}OH \tag{15.10}$$

Moreover, redox cycling of free or weekly chelated iron may decompose a lipid hydroperoxide and thus initiate a chain of lipid peroxidation (Halliwell and Gutteridge, 2000).

In summation, as a result of interaction with free radicals, carotenoids can themselves become a source of free radicals and may induce further damaging reactions.

15.7 PRO-OXIDANT AND CYTOTOXIC PROPERTIES OF THE DEGRADATION PRODUCTS OF CAROTENOIDS

In most assays designed to study antioxidant action of carotenoids, the effects of carotenoids were followed for a relatively short periods of time, while carotenoids were still present at substantial concentrations. Carotenoids, such as β-carotene, lutein, and zeaxanthin, undergo rapid degradation upon exposure to oxidants or irradiation with ultraviolet and visible light (Ojima et al., 1993; Siems et al., 1999, 2005).

The retinal environment puts the carotenoids under constant threat of oxidative damage due to very active metabolism, high oxygen tension, abundant polyunsaturated lipids, and continual exposure to high fluxes of visible light in the presence of potent photosensitizers (Beatty et al., 2000; Różanowska and Różanowski, 2008; Rozanowska and Sarna, 2005). With ageing, the risk of oxidative damage is further elevated due to accumulation of iron, lipofuscin, and age-related changes in melanosomes (He et al., 2007; Rozanowska et al., 1995, 2002, 2008c; Wong et al., 2007). Both lipofuscin and aged melanosomes can act as photoinducible generators of reactive oxygen species and exhibit cytotoxicity (Boulton et al., 2004; Rozanowska and Sarna 2005; Rozanowski et al., 2004a, 2008a,b,c).

Increased age and smoking are associated with an increased oxidative stress and depletion of low-molecular weight antioxidants in many tissues including the retina (Beatty et al., 2000). Epidemiological studies indicate that on average, smokers develop late stage AMD 10 years earlier than nonsmokers (Kelly et al., 2004; Klein et al., 1998; Mitchell et al., 2002; Thornton et al., 2005; Tomany et al., 2004).

In the AMD retina, the risk of oxidative damage is elevated even further in comparison to the healthy age-matched retina (Beatty et al., 2000; Donoso et al., 2006; He et al., 2007; Różanowska and Różanowski, 2008; Wong et al., 2007). The AMD retina exhibits five-fold higher content of iron (Hahn et al., 2003) and even though the proteins involved in iron transport and storage, such as transferrin, ferritin, and ferroportin, are upregulated (Chowers et al., 2006; Dentchev et al., 2005), it may be speculated that this is not sufficient to prevent iron-mediated damage (Chowers et al., 2006). Moreover, there is growing evidence suggesting that chronic inflammation is involved in AMD (Anderson et al., 2002; Dasch et al., 2005; Donoso et al., 2006; Hageman et al., 2001; Holtkamp et al., 2001; Johnson et al., 2000, 2001; Kijlstra et al., 2005; Kuehn, 2005; Nozaki et al., 2006; Wiggs, 2006; Zarbin, 2004), suggesting that oxidative stress of the AMD retina can be further exacerbated by production of reactive oxygen and nitrogen species by phagocytic cells.

Several markers of oxidative stress have been identified in AMD retinas, including proteins modified by products of lipid peroxidation (Crabb et al., 2002; Gu et al., 2003; Hollyfield et al., 2003). Overall, there is growing body of evidence implicating oxidative stress in the development and progression of AMD (Anderson et al., 2002; Beatty et al., 2000; Seddon et al., 2004).

Thus it may be suggested that, particularly in AMD, retinal carotenoids are at risk of oxidative damage. The free radical pathways leading to carotenoid oxidation have been already discussed. Markedly, while overall singlet oxygen quenching by carotenoids occurs mainly via a physical route of energy transfer followed by a thermal deactivation of the excited state of carotenoid molecule, a small fraction of interactions with singlet oxygen do lead to oxidation of carotenoids (Fiedor et al., 2005; Stratton et al., 1993). Moreover, hypochlorite a potent oxidant, produced under physiological conditions by activated neutrophils and macrophages, also leads to rapid carotenoid degradation (Sommerburg et al., 2003).

The degradation of β-carotene has become the subject of extensive studies especially after two large clinical trials indicated that β-carotene supplementation substantially increased the risk of lung cancer in smokers and asbestos workers (Albanes et al., 1995; Omenn, 1996; Omenn et al., 1996). Oxidative degradation of β-carotene induced by free radicals or singlet oxygen leads to the formation of several different endoperoxides, epoxides, and apo-carotenals, including all-*trans*-retinal (Fiedor et al., 2005; Handelman et al., 1991b; Kennedy and Liebler, 1991; McClure and Liebler, 1995; Mordi et al., 1991, 1993; Sommerburg et al., 2003; Stratton et al., 1993). All-*trans*-retinal is a potent photosensitizer that upon photoexcitation with blue light photogenerates singlet oxygen and free radicals (Rozanowska and Sarna, 2005). These properties indicate that all-*trans*-retinal may exert damaging effects upon the retina as a consequence of irradiation with blue light, an action opposite to protective action of its precursor, β-carotene.

It has been also shown in numerous studies that degradation products of β-carotene can exert an action opposite to their parent compound and induce damage to biomolecules independent on light (Klamt et al., 2003; Marques et al., 2004; Murata and Kawanishi, 2000; Siems et al., 2002). For instance, incubation of retinal, β-apo-8′-carotenal or β-carotene with 2′-deoxyguanosine for 72 h under aerobic conditions leads to the formation of a mutagenic adduct, 1,N^2-entheno-2′-deoxyguanosine (Marques et al., 2004), and the yields of those adducts are up to 12-fold increased when hydrogen peroxide is included in the incubation mixture. While β-carotene also leads to the adduct formation in this assay, it may be argued that under the conditions employed, β-carotene becomes at least partly degraded to apo-carotenoids by the end of the incubation period.

Nucleic acids are not the only biomolecules susceptible to damage by carotenoid degradation products. Degradation products of β-carotene have been shown to induce damage to mitochondrial proteins and lipids (Siems et al., 2002), to inhibit mitochondrial respiration in isolated rat liver mitochondria, and to induce uncoupling of oxidative phosphorylation (Siems et al., 2005). Moreover, it has been demonstrated that the degradation products of β-carotene, which include various aldehydes, are more potent inhibitors of Na-K ATPase than 4-hydroxynonenal, an aldehydic product of lipid peroxidaton (Siems et al., 2000).

Numerous studies have demonstrated that degradation products of β-carotene exhibit deleterious effects in cellular systems (Alija et al., 2004, 2006; Hurst et al., 2005; Salerno et al., 2005; Siems et al., 2003). A mixture of β-carotene degradation products exerts pro-apoptotic effects and cytotoxicity to human neutrophils (Salerno et al., 2005; Siems et al., 2003), and enhances the genotoxic effects of oxidative stress in primary rat hepatocytes (Alija et al., 2004, 2006), as well as dramatically reduces mitochondrial activity in a human leukaemic cell line, K562, and RPE 28 SV4 cell line derived from stably transformed fetal human retinal pigmented epithelial cells (Hurst et al., 2005). As a result of degradation or enzymatic cleavage of β-carotene, retinoids are formed, which are powerful modulators of cell proliferation, differentiation, and apoptosis (Blomhoff and Blomhoff, 2006).

In some studies it was shown that β-carotene decomposes more rapidly than lutein and zeaxanthin when exposed to oxidants or light in the presence and absence of rose bengal as a photosensitizer (Hurst et al., 2004; Ojima et al., 1993; Siems et al., 1999). However, it is not a rule, as lutein and zeaxanthin are depleted faster than β-carotene during methylene blue photosensitized oxidation of human plasma (Ojima et al., 1993).

While degradation products of carotenoids often exhibit completely different properties from their parent compounds, they are often similar in their abilities to reduce Fe(III) and undergo subsequent oxidative degradation (Panzella et al., 2004).

Relatively little is known about metabolic pathways of carotenoids other than β-carotene and the cellular effects they (and their degradation products) may be responsible for. Several studies have been performed on lycopene, an acyclic carotenoid, which is believed to protect against prostate cancer. It has been demonstrated that rats fed with a lycopene-enriched diet accumulated a number of different metabolites in their livers, including the aldehydic products, apo-8′-lycopenal and apo-12′-lycopenal (Gajic et al., 2006). In two other studies it was shown that lycopene easily undergoes oxidative cleavage and its oxidation products induce apoptosis in several cancer cell lines (Kotake-Nara et al., 2002; Nagao, 2004).

However, the metabolic pathways of lutein and zeaxanthin are only beginning to be discovered. Several derivatives of dietary xanthophylls have been identified in the retina, such as 3′-epilutein, meso-zeaxanthin, 3′-oxolutein, and 3-methoxyzeaxanthin, and it has been suggested that they may be formed as a result of nonenzymatic oxidative modifications (Bernstein et al., 2001, 2002b; Bhosale et al., 2007b; Khachik et al., 1997). The *macula lutea* contains predominantly meso-zeaxanthin (Figure 15.1), which is believed to originate from either oxidative modification or double bond isomerization of dietary lutein (Khachik et al., 1997, 2002).

Because of structural similarities between β-carotene and lutein and zeaxanthin, it may be expected that as a consequence of the oxidative degradation of these xanthophylls, products analogous to oxidation products of β-carotene will be formed. Indeed, in post mortem of human retinas two aldehydic products were identified, 3-hydroxy-β-ionone and 3-hydroxy-14′-apocarotenal, both are likely derived from oxidative cleavage of lutein or zeaxanthin (Prasain et al., 2005). We have recently shown that the degradation of lutein *in vitro* during iron ion-mediated lipid peroxidation in liposomes yields numerous products that include potent photosensitizers, which, upon absorption of blue light, photosensitize the generation of singlet oxygen, superoxide, and hydroxyl radicals (Rozanowski and Rozanowska, 2005). Interestingly, the quantum yield of singlet oxygen generation by the degradation products of lutein is similar to that of all-*trans*-retinal.

The susceptibility of carotenoids to degradation is also important with regard to production and storage of carotenoids in dietary supplements. It needs to be noted that there is often a great discrepancy between the declared and real content of lutein in commercially available supplements (Breithaupt and Schlatterer, 2005). In a study of 14 products from local supermarkets and pharmacies, seven contained smaller amounts of lutein than specified, and varied from 11% to 93% of the stated values (Breithaupt and Schlatterer, 2005). It may be suspected that part of the xanthophylls degraded during storage. Carotenoids do not require any special oxidants to undergo degradation. It is enough to leave carotenoids exposed to air to induce their autooxidative degradation to apocarotenals and short-chain carbonyl compounds (Kim, 2004; Kim et al., 2001; Mordi et al., 1991, 1993). It is not clear whether the degradation products of carotenoids are absorbed from the gastrointestinal track and, if so, whether they can accumulate in toxic levels in human tissues, such as the retina. Clearly, this is another area demanding experimental investigation. In particular, it is of interest to determine the role of RPE in carotenoid protection, the metabolism of carotenoid oxidation products and the possible ways of their removal from the retina.

15.8 PRO-OXIDANT AND CYTOTOXIC EFFECTS OF CAROTENOIDS AND THEIR DEGRADATION PRODUCTS IN CULTURED RPE CELLS

As mentioned previously, β-carotene oxidation products substantially reduce mitochondrial activity in an RPE cell line, 28 SV4, derived from fetal human RPE cells (Hurst et al., 2005). In these experiments β-carotene was degraded in dichloromethane/methanol/water solution by hypochlorite, NaOCl. Exposure of the 28 SV4 cells to a mixture of the degradation products corresponding to 0.1 mM β-carotene led to ~75% decrease in mitochondrial activity. The absorption spectra of the extracted degradation products of β-carotene solubilized in phosphate buffered saline exhibited a major absorption peak at 220 nm which can be ascribed to a mixture of low-molecular-weight short-chain aldehydes and ketones, including β-ionone. In addition, the absorption spectra show a broad shoulder extending from 270 to 345 nm and attributed to longer-chain carotenoid degradation products, such as β-apo-carotenals. The absorption spectra of the degradation products suggest that they have shorter chains than all-*trans*-retinal. Thus, it may be argued that while β-carotene does not accumulate in the RPE in sufficient concentrations to impose a risk of generation of harmful concentrations of its degradation products, the abundant retinoids accumulated in the RPE may become a plentiful source of these degradation products.

In their subsequent studies, van Kuijk and colleagues (Kalariya et al., 2008) tested the effects of oxidation products of β-carotene, lutein, and zeaxanthin on cultured ARPE-19 cells. Consistent with previous results, β-carotene degradation products, as well as lutein and zeaxanthin degradation products, induced a dose dependent loss of mitochondrial activity. The highest concentrations of the degradation products tested, corresponding to 0.1 mM of the parent carotenoid, induced up to 90% loss of mitochondrial activity. Degradation products of β-carotene induced a dose dependent increase in the intracellular level of reactive oxygen species measured by a fluorescent dye, 2′,7′-dichlorofluorescin diacetate (DCF-DA). Using annexin V staining of phosphatidylserine exposed to the outer leaflet of the plasma membrane as an early indicator of apoptosis, it was also shown that the degradation products of all three carotenoids induced apoptotic cell death. In the case of β-carotene degradation products the investigations included measurements of mitochondrial membrane potential and morphological assessment of cellular nuclei. Upon treatment with the β-carotene degradation products, the mitochondrial membrane potential substantially decreased while the nuclei exhibited characteristic features of apoptosis—condensation and fragmentation. Degradation products of all three carotenoids induced activation of redox-sensitive transcription factors, nuclear factor kappaB (NF-κB), and activating protein 1 (AP-1). The damaging effects of degradation products of all three carotenoids tested were ameliorated by pretreatment of cells with 1 mM *N*-acetylcysteine (NAC)—an effective free radical scavenger and a precursor of glutathione.

It may be speculated that in the previously mentioned study on ARPE-19 cells exposed to lutein of low purity (Kanofsky and Sima, 2006), the apparent cytotoxic effects observed in dark that are exacerbated upon irradiation of lutein-laden cells with blue or green light may have been caused by (some) degradation products of lutein.

Certainly, the identification of the degradation products responsible for the cytotoxic effects and their metabolic pathways require a thorough elucidation, and a cultured RPE offers a good model for these investigations.

15.9 EFFECT OF BINDING TO PROTEINS ON CAROTENOID SUSCEPTIBILITY TO DEGRADATION

Altogether, it is clear that the degradation products of carotenoids can exert deleterious effects on cells. Minimizing the risk of oxidative damage to carotenoids seems to be the most effective way to avoid their harmful actions.

It is believed that in the retina, xanthophylls remain bound to proteins (Bhosale and Bernstein, 2007; Loane et al., 2008). Thus, it is of interest to determine how the binding affects the action of carotenoids as antioxidants and their susceptibility to oxidative degradation. As mentioned earlier, it has been demonstrated that binding of zeaxanthin or meso-zeaxanthin to GSTP1 can slightly inhibit carotenoid degradation and enhances their antioxidant action against lipid peroxidation induced by thermolabile azo-compounds (Bhosale and Bernstein, 2005). The effects of zeaxanthin and meso-zeaxanthin on the formation of thiobarbituric acid reactive species (TBARS) were compared in the presence and absence of GSTP1. It was shown that in the presence of GSTP1 the degradation of xanthophylls was slightly inhibited, particularly at the initial phase of the lipid peroxidation assay

(Bhosale and Bernstein, 2005). The synergistic antioxidant effect of xanthophylls and GSTP1 can be ascribed at least in part to the protective effects of the binding protein against xanthophyll degradation (Bhosale and Bernstein, 2005). It remains to be established how GSTP1 affects the ability of xanthophylls to quench singlet oxygen and how other XBPs affect xanthophyll antioxidant actions and susceptibility to degradation.

15.10 COOPERATION OF CAROTENOIDS WITH OTHER ANTIOXIDANTS

Another way to protect carotenoids from oxidative degradation is to ensure an adequate antioxidant system to minimize the oxidative stress. If endogenous antioxidants are not sufficient, it may be needed to add exogenously an ancillary antioxidant. Indeed, it has been shown that the degradation of xanthophylls can be slightly slowed down in liposomes or in cultured ARPE-19 cells in the presence of low-molecular weight antioxidants such as α-tocopherol or vitamin C (ascorbate) (Figure 15.5) (Wrona et al., 2003, 2004). While the effects of vitamins E and C are relatively small (Wrona et al., 2003, 2004), they encourage further attempts to search for a more effective combination of antioxidants to protect against carotenoid degradation.

Several studies demonstrate that a combination of a carotenoid with another antioxidant can increase the protection in a synergistic way or induce protective effects where a single antioxidant does not cause a significant difference (Bohm et al., 1998a,b, 2001; Sanz et al., 2007; Wrona et al., 2003, 2004). For example, experiments *in vivo* and *in vitro* on an animal model of retinitis pigmentosa, the rd1 mouse, have shown that a combination of lutein, zeaxanthin, α-lipoic acid, and reduced glutathione, but none of the single antioxidants, diminished oxidative damage to DNA and prevented apoptosis of photoreceptors, whereas single antioxidants were ineffective (Sanz et al., 2007).

One possible mechanism responsible for cooperative action of antioxidants is reduction of a semi-oxidized carotenoid by another antioxidant. Carotenoid cation radicals can be reduced, and therefore recycled to the parent molecule, by α-tocopherol, ascorbate, and melanins (Edge et al., 2000b; El-Agamey et al., 2004b) (Figure 15.5). Interestingly, lycopene can reduce radical cations of other carotenoids, such as astaxanthin, β-carotene, lutein, and zeaxanthin (Edge et al., 1998).

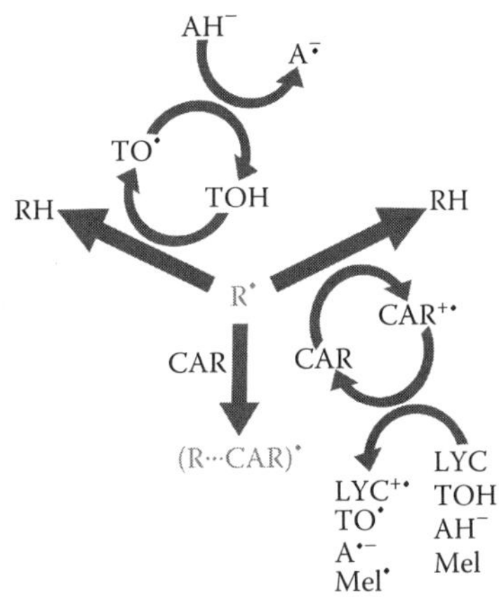

FIGURE 15.5 A scheme of the interactions of carotenoids (Car) with free radicals (R•) and carotenoid cation radical (Car+•) with other antioxidants—tocopherol (TOH), ascorbate (AH−), melanin (Mel). Car can add R• and form a resonance-stabilized radical adduct ([R-Car]•) or reduce R• by electron transfer or hydrogen donation to produce RH. The carotenoid cation radical (Car+•) is directly formed when Car acts as an electron donor. Car+• can be recycled to regenerate Car through a reduction reaction involving TOH, AH−, and/or Mel. The benign tocopheryl radical (TO•), ascorbyl radical (A•−), and/or melanin radical (Mel•) are formed. Thus, in the presence of other antioxidants, such as TOH, AH−, and Mel, carotenoids can be spared from degradation and therefore may provide longer lived protection as singlet oxygen quenchers.

It is noteworthy that some epidemiological studies have found that lycopene, but not lutein nor zeaxanthin, is substantially decreased in serum of AMD patients compared with age-matched control subjects (Cardinault et al., 2005).

In the case of photosensitized oxidation where both singlet oxygen and free radicals are involved, an even simpler explanation of synergistic action is credible—inhibition of the free radical chain reaction by α-tocopherol or ascorbate can protect the carotenoid from free radical-mediated degradation so it can function longer as a singlet oxygen quencher (Wrona et al., 2003, 2004). Interactions such as these may explain synergistic protection offered by a combination of antioxidants observed in many systems (Bohm et al., 1998a,b, 2001; Wrona et al., 2003, 2004).

These possible cooperative effects of antioxidant mixtures have been tested in ARPE-19 cells supplemented with zeaxanthin and α-tocopherol or ascorbate and exposed to photosensitized action of merocyanine 540 for up to 60 min (Wrona et al., 2004). To assess cell viability mitochondrial activity was determined and endogenous cholesterol was employed as a reporter of the damage pathway (Girotti and Korytowski, 2000). Interaction of cholesterol with singlet oxygen leads to formation of a specific product, 5α-cholesterol hydroperoxide, which can be detected by high performance liquid chromatography using electrochemical detection. Decomposition of 5α-cholesterol hydroperoxides or interaction of cholesterol with free radicals leads to the formation of other cholesterol hydroperoxides, such as 7α,β-cholesterol hydroperoxides. Supplementation of cells solely with zeaxanthin or α-tocopherol has provided no significant protection of ARPE-19 cells from cell death even though α-tocopherol alone exerted a significant inhibitory effect on photoformation of 7α,β-cholesterol hydroperoxides and zeaxanthin alone inhibited photoformation of 5α-cholesterol hydroperoxide (Wrona et al., 2004). The supplementation of cells with 0.5 mM ascorbate alone offered a small protection against cell death but was detectable only after 10 min of irradiation with visible light. Interestingly, ascorbate significantly inhibited photoformation of 5α-cholesterol hydroperoxide, and this effect was significant even after 60 min of irradiation. Combinations of zeaxanthin with either α-tocopherol or ascorbate provided a significant synergistic protection of cell viability for up to 30 min of irradiation and an inhibitory effect against photoformation of both, 5α- and 7α,β -cholesterol hydroperoxides for up to 60 min of irradiation.

As mentioned previously, in the AMD retina iron metabolism is compromised (He et al., 2007; Wong et al., 2007). Thus, it is of interest to determine the effects of potential antioxidants in the presence of iron. In an *in vitro* study of ARPE-19 cells, addition of a lipophilic iron complex led to about a ninefold increase in the photosensitized yield of 7α,β-cholesterol hydroperoxides (Wrona et al., 2004). In the presence of the iron, ascorbate exerted pro-oxidant effects, while the effects of α-tocopherol, zeaxanthin, or their combination were still protective (Wrona et al., 2004). Thus, it appears that the effects of potential antioxidants are strongly dependent on the sources of oxidative damage. The same antioxidant may be protective under certain conditions and exert deleterious effects when the conditions are changed. Therefore a detailed understanding of the sources of the oxidative damage is required in order to design an adequate antioxidant mixture.

Another study looking at the effect of a combination of antioxidants on protection of ARPE-19 cells against oxidative damage used a lipophilic extract of tomatoes containing carotenoids at concentrations of 1.2 mg of β-carotene, 0.75 mg of lycopene, 0.05 mg of lutein, and 0.17 mg of α-tocopherol/g of dry weight of tomato powder (Chichili et al., 2006). The mixture offered a substantial protection against oxidative damage induced by hydrogen peroxide in the absence and presence of sodium nitrate. Hydrogen peroxide induced extensive carbonylation of cellular proteins and formation of thiobarbituric acid reactive substances. Exposure of cells to both, H_2O_2 and $NaNO_2$, led to tyrosine nitration. All these effects were substantially diminished upon supplementation of cells with the tomato extract. Further studies are needed to determine whether the same outcome can be achieved upon supplementation of cells with a mixture of those carotenoids without possible additional components from tomatoes, which even at trace concentrations might upregulate cellular antioxidant defense mechanisms (Baur and Sinclair, 2006; Dinkova-Kostova and Talalay, 2008).

15.11 BIOACTIVITIES OF CAROTENOIDS OTHER THAN DIRECT ANTIOXIDANTS

Apart from their direct action as antioxidants, quenching electronically excited states and scavenging free radicals, carotenoids play many diverse roles in modulating inflammatory, angiogenic, apoptotic, and transcription pathways (Ben-Dor et al., 2005; Chew and Park, 2004; Chew et al., 2003; Maccarrone et al., 2005; Santos et al., 1996, 1998; Selvaraj and Klasing, 2006; Selvaraj et al., 2006; Sharoni et al., 2004; Sumantran et al., 2000). These various bioactivities have been investigated in numerous studies in animals and *in vitro* on different cell types. Even though many of these bioactivities are of great relevance for the retina, studies in that area are scarce. Next we shortly discuss indirect carotenoid bioactivities focusing on those aspects which are particularly relevant to the RPE and can be tested *in vitro*.

15.11.1 Modulation of Inflammatory Pathways

RPE cells can express a number of anti-inflammatory and pro-inflammatory cytokines, and complement factors constitutively and/or upon stimulation (Chen et al., 2007; Crane et al., 2000a; Ebihara et al., 2007; Holtkamp et al., 2001; Joffre et al., 2007). Therefore the effects of carotenoids on inflammatory responses are of great relevance to the retina.

There is a growing body of evidence that carotenoids such as lutein and zeaxanthin exert potent immunomodulatory effects in humans, animals, and in isolated macrophages (Chew and Park, 2004; Ekam et al., 2006; Fuller et al., 1992; Hozawa et al., 2007; Izumi-Nagai et al., 2007; Jin et al., 2006; Koutsos et al., 2007; Lidebjer et al., 2007; Santos et al., 1996, 1998; Seddon et al., 2006; Selvaraj and Klasing, 2006; Selvaraj et al., 2006; Walston et al., 2006). Lutein has been shown to exert anti-inflammatory effects on endotoxin-induced uveitis in the rat by inhibiting the nuclear factor (NF)-κB-dependent signaling pathway and the subsequent production of pro-inflammatory mediators (Jin et al., 2006). An intravenous administration of lutein at doses of 10 mg/kg and 100 mg/kg effectively prevents infiltration of macrophages into the aqueous humor in the rat eye induced by lipopolysaccharide, substantially inhibits activation of NF-κB responsible for production of some inflammatory cytokines, and reduces the expression of several markers of inflammation, such as nitric oxide, interleukin 6 (IL-6), monocyte chemotactic protein 1 (MCP-1), and tumor necrosis factor α (TNF-α).

In the rat retina, ischemia upregulates expression of the neuronal nitric oxide synthase and cyclo-oxygenase-2; these effects can be effectively inhibited by lutein (Choi et al., 2006).

Lutein can also effectively inhibit infiltration of macrophages and upregulation of pro-inflammatory proteins, such as vascular endothelial growth factor, MCP-1, and intercellular adhesion molecule-1 upon laser photocoagulation of a murine retina (Izumi-Nagai et al., 2007). In this model the potent antiangiogenic action of lutein against choroidal neovascularization is probably related to suppression of NF-κB pathway, including IκB-α degradation and p65 nuclear translocation (Izumi-Nagai et al., 2007).

Interestingly, opposing immunomodulatory effects can be observed using different carotenoid concentrations, and the effects on up- or downregulation of the inflammatory response by lutein can be further modified by fatty acids (Selvaraj and Klasing, 2006; Selvaraj et al., 2006). For example, it has been shown that lutein affects expression of inducible nitric oxide synthase (iNOS) in liposaccharide-stimulated macrophages isolated from chickens or HD11 cells *in vitro* (Selvaraj et al., 2006). At low concentration, 0.01 mM, lutein induces an upregulation of iNOS mRNA. Increasing the concentration of lutein 10 times results in a downregulation of iNOS mRNA. The action of lutein is modulated by eicosapentaenoate and, through the peroxisome proliferator activated receptor γ and retinoid X receptor (RXR) pathways, they can modulate iNOS expression in macrophages (Rafi and Shafaie, 2007; Selvaraj et al., 2006).

Also degradation products of carotenoids exhibit immunomodulatory effects. For example, β-carotene degradation products can stimulate production of superoxide by activated neutrophils at low micromolar concentrations but exhibit inhibitory effects at concentrations above 20 μM (Siems et al., 2003).

Considering the propensity of RPE cells to express a range of anti-inflammatory and proinflammatory proteins, the proven role of inflammation in uveitis and AMD, as well as potent immunomodulatory effects of carotenoids, it is of importance to determine the immunomodulatory effects of carotenoids and their degradation products in the RPE.

15.11.2 Remodeling of Extracellular Matrix

Expression of matrix metalloproteinases (MMPs) and their inhibitors is an important function of the RPE, particularly with respect to the maintenance of appropriate permeability of the Bruch's membrane (Ahir et al., 2002). This function can be tested *in vitro* (Marin-Castano et al., 2006). For example, it has been shown that the expression of MMP-2, TIPM-2s (tissue inhibitor of MMP-2), and type IV collagen by cultured ARPE-19 cells is affected by repetitive exposures to nonlethal oxidant injury with hydroquinone (Marin-Castano et al., 2006). Oxidative stress decreases MMP-2 activity and increases collagen type IV accumulation.

It has been demonstrated in other cell types that lutein can inhibit expression of MMPs and/or activity (Philips et al., 2007). For example, in dermal fibroblasts lutein inhibits expression of MMP-1 and decreases levels of MMP-2 protein (Philips et al., 2007). In melanoma cells, lutein inhibits MMP-1 expression while stimulating TIMP-2 (Philips et al., 2007). Moreover it has been shown that lutein inhibits elastin expression in fibroblasts subjected to oxidative stress by exposure to ultraviolet light (Philips et al., 2007). These results clearly indicate that lutein can play an important role in remodeling of the extracellular matrix.

Therefore it is of interest to determine whether carotenoids can modulate the turnover of the extracellular matrix by the RPE by affecting the expression of MMPs, elastin, and/or collagen. Cultured RPE cells are a suitable model for such investigations.

15.11.3 Modulation of Lipid Metabolism and Transport

Interestingly, it has been shown that supplementation of greenfinches with lutein and zeaxanthin at a ratio of 20:1 increases plasma levels of triglycerides and bird body mass (Horak et al., 2006). These data suggest that xanthophylls may affect lipid metabolism.

Another indication of involvement of certain xanthophylls in lipid metabolism comes from studies on macrophages *in vitro*. Supplementation of macrophages with lutein or β-cryptoxanthin has been found to upregulate the lipid transporter ABCA1 expression (Matsumoto et al., 2007). ABCA1 expression is regulated at the transcriptional level by the liver X receptors, RXR, and retinoic acid receptor (RAR) (Costet et al., 2003; Koldamova et al., 2003; Venkateswaran et al., 2000). These receptors have been known to be activated by ligands such as 22(*R*)-hydroxycholesterol, 9-*cis*-retinoic acid, and all-*trans*-retinoic acid (Costet et al., 2003; Koldamova et al., 2003; Venkateswaran et al., 2000). It has been recently determined that β-cryptoxanthin and lutein, but not β-carotene, zeaxanthin, astaxanthin, or lycopene can serve as ligands for RAR (Matsumoto et al., 2007). Moreover, it has been shown that β-cryptoxanthin can upregulate ABCA1 and ABCG1 expression in macrophages and this effect is inhibited by the RAR pan-antagonist, LE540 (Matsumoto et al., 2007).

As the RPE plays an important role in lipid metabolism and regulation of dynamic transport between the choriocapillaris and photoreceptors, it is important to determine whether carotenoids affect these pathways in the RPE.

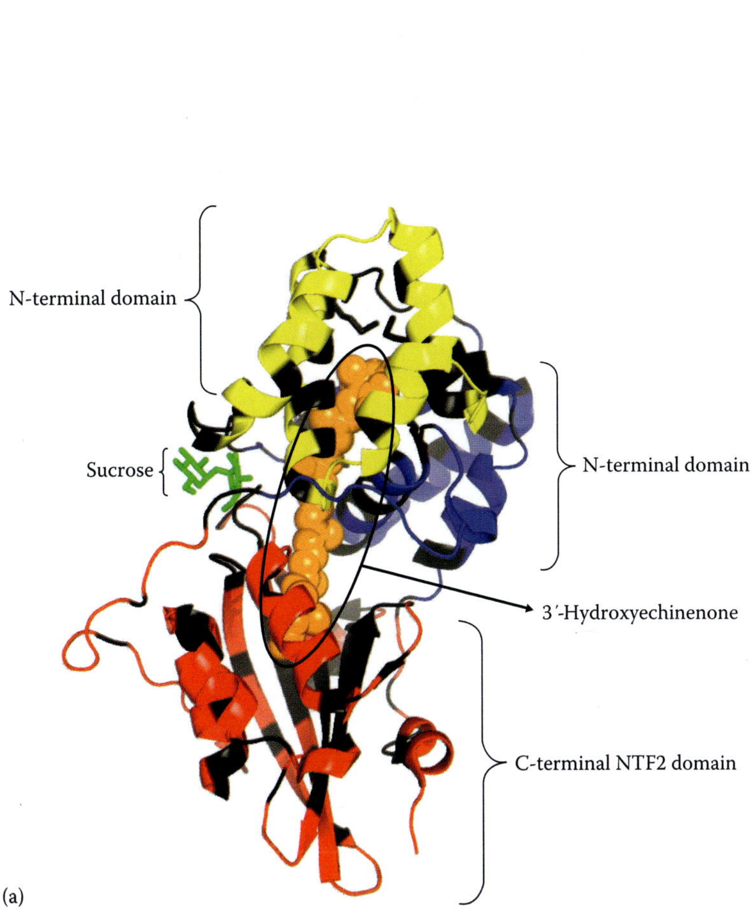

FIGURE 1.3 The structure of the OCP. (a) Ribbon diagram of the *A. maxima* OCP structure. The two helical bundles making up the N-terminal domain are uppermost; the C-terminal NTF2 domain is shown in red. The 3′-hydroxyechinenone molecule is shown in space-filling representation and the sucrose molecule and the side chains of conserved Met residues are shown in sticks. Absolutely conserved amino acids are shown in black (Shading as in (a); tubes correspond to alpha-helices; arrows, beta-strands; the amino acids comprising each element of secondary structure are indicated). (Created using Pymol, http://www.pymol.org.)

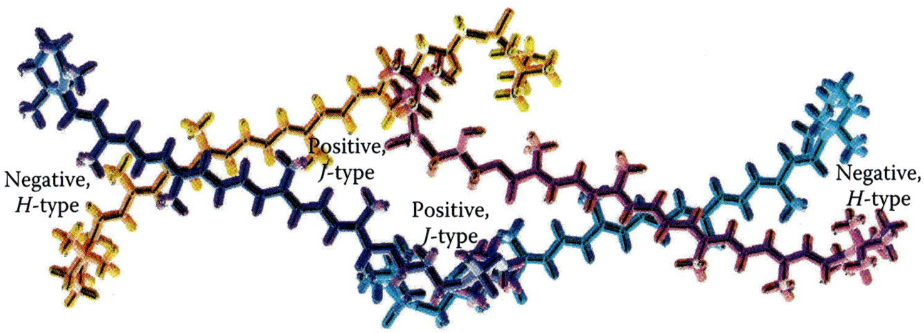

FIGURE 3.7 Molecular arrangements for *H*- and *J*-aggregates. Tetrameric lysophospholipid (*R*)-**3.15** forms predominantly *H*-aggregates in addition to a small percentage of *J*-aggregates. The calculated VIS absorption of the tetramer (*R*)-**3.15** is in accordance with the experimental VIS spectra. (Reprinted from Foss, B.J. et al., *Chem. Eur. J.*, 11, 4103, 2005b. With permission.)

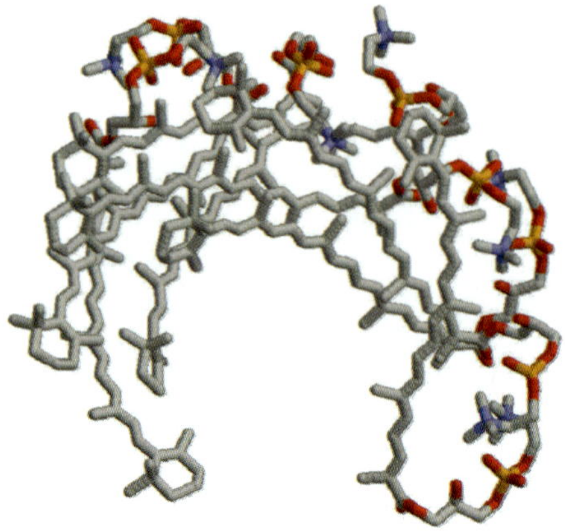

FIGURE 3.11 Optically active *P*-oligomer unit, built from eight optically inactive (*R*)-**3.15** monomers. The calculated spectra of this octamer is in accordance with both the experimental VIS and CD spectra. (Reprinted from Foss, B.J. et al., *Chem. Eur. J.*, 11, 4103, 2005b. With permission.)

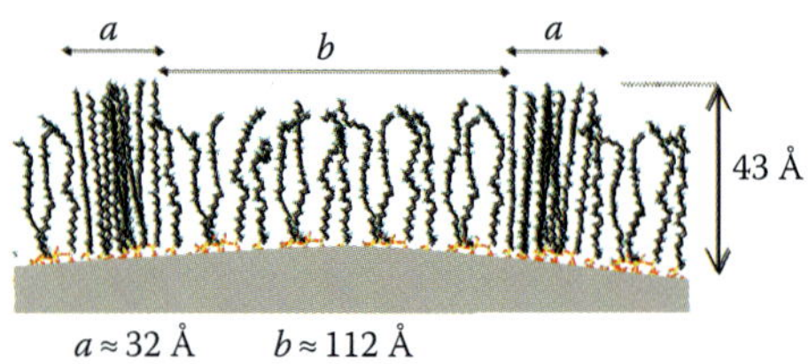

FIGURE 4.2 Alkyl chain organization of a C_{30} phase. (From Raitza, M. et al., *Investigating the Surface Morphology of Triacontyl Phases with Spin-Diffusion Solid-State NMR Spectroscopy*, John Wiley & Sons Ltd., 3489, 2000. With permission.)

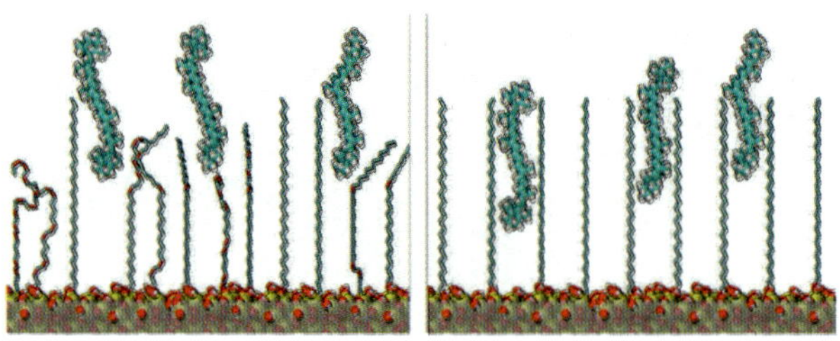

FIGURE 4.3 Slot model. (From Meyer, C. et al., *Nuclear Magnetic Resonance and High Performance Liquid Chromatography Evaluation of Polymer Based Stationary Phases Immobilized on Silica*, Springer-Verlag GmbH, 686, 2005. With permission.)

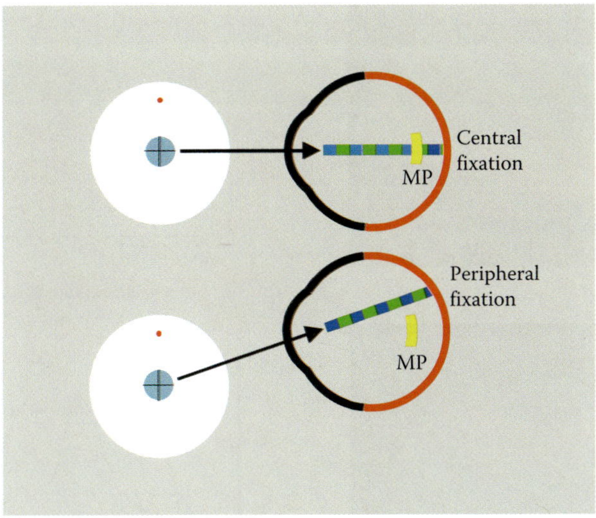

FIGURE 5.2 Illustration of the method of HFP. On viewing the stimulus directly (upper), MP attenuates the blue component of the stimulus whereas with peripheral viewing (lower), no such attenuation occurs. In each case, the subject adjusts the luminance of the blue component until it matches the luminance of the green component, which is unaffected by MP.

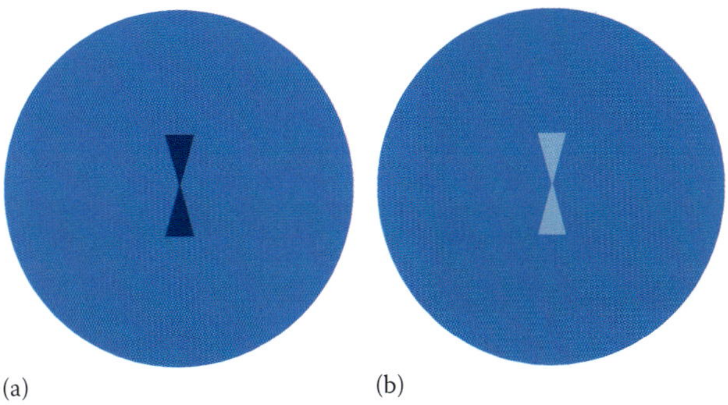

FIGURE 5.3 Appearance of the visual field in dichroism-based photometry. The background field is unpolarized and is of wavelength 460 nm. The triangles are also of wavelength 460 nm, but are polarized. In (a), the plane of polarization is horizontal causing the triangles to appear darker; in (b) the plane of polarization is vertical causing them to appear lighter. In each case, the subject adjusts the luminance of the triangles until they match the luminance of the unpolarized background.

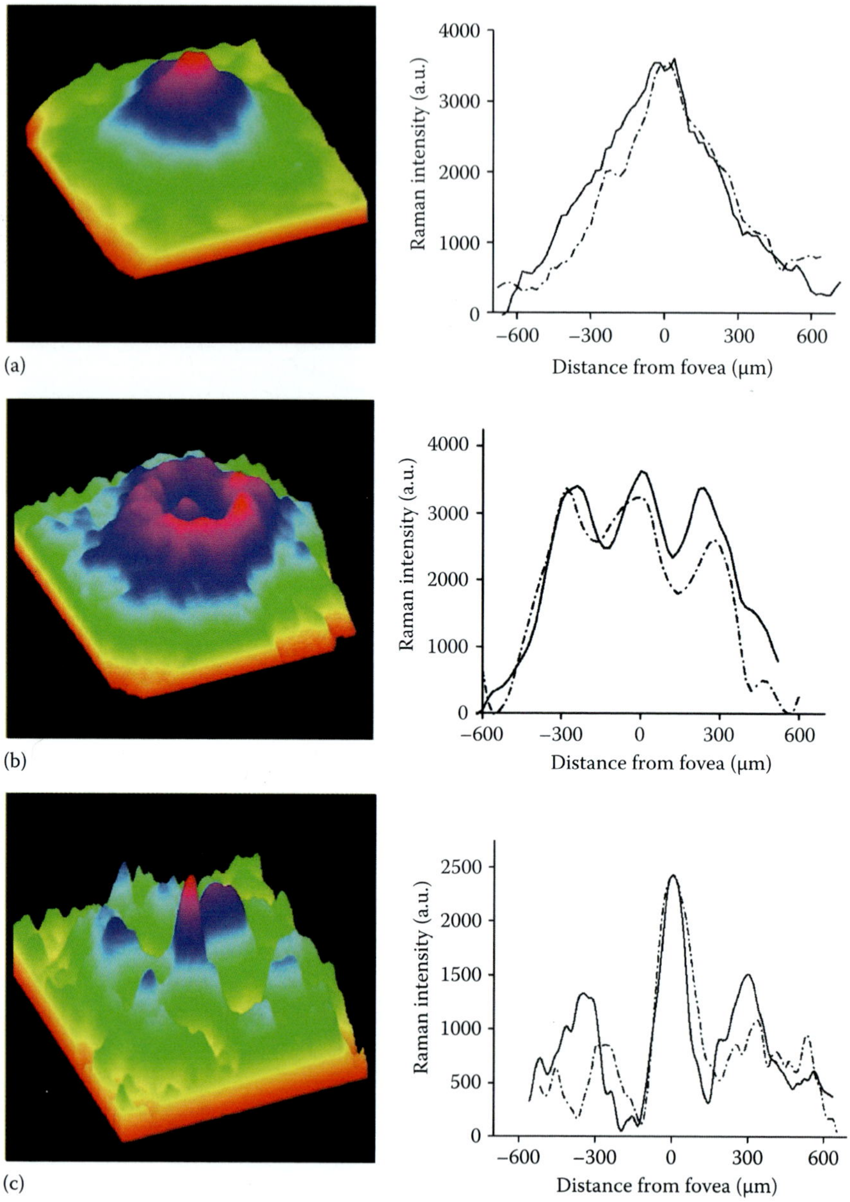

FIGURE 6.8 Pseudocolor scaled, three-dimensional MP RRI images of three volunteer subjects, along with related line plot profiles, derived for each distribution along nasal–temporal (solid line) and inferior–superior meridians (dashed line), both running through the center of the macula. Distribution (a), which is representative for most healthy subjects, features a nearly rotationally symmetric MP distribution with monotonous decrease of concentration levels from the center to the periphery. Distribution (b) features a small central peak with a strong, surrounding, ring-like component. Varying in relative strength of central and ring components, this "ring-like" pattern is encountered in about 30% of the population. Distribution (c) is an example for a fragmented distribution with narrow central peak and broken-up ring structure, measured in a subject with mild form of dry macular degeneration. All images are color coded with the same intensity scale (not shown).

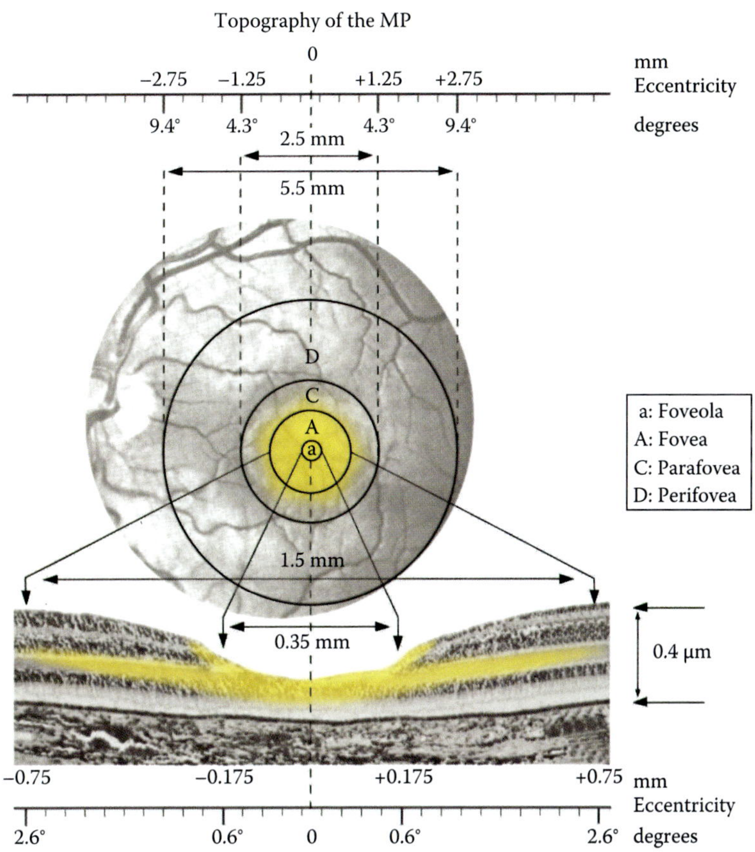

FIGURE 13.4 Topography of the MP. This figure schematically shows the distribution of the yellow MP across the retina: horizontally (top) and vertically (bottom). (From Gass, J.D., *Stereoscopic Atlas of Macular Diseases Diagnosis and Treatment*, Vol. 1, Mosby – Year Book Inc., 3, 1997. With permission.)

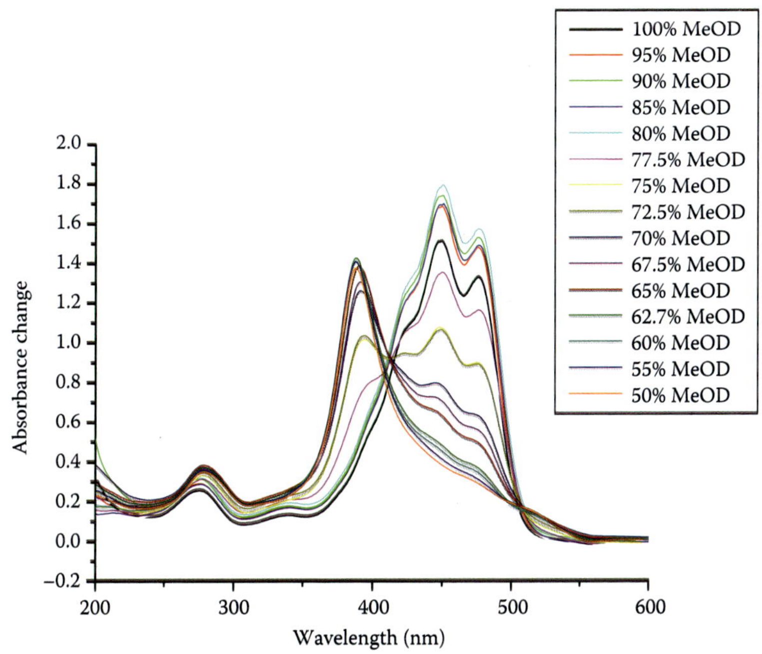

FIGURE 14.3 Ground state absorption spectra of 1×10^{-5} M zeaxanthin in various MeOD/D_2O mixtures.

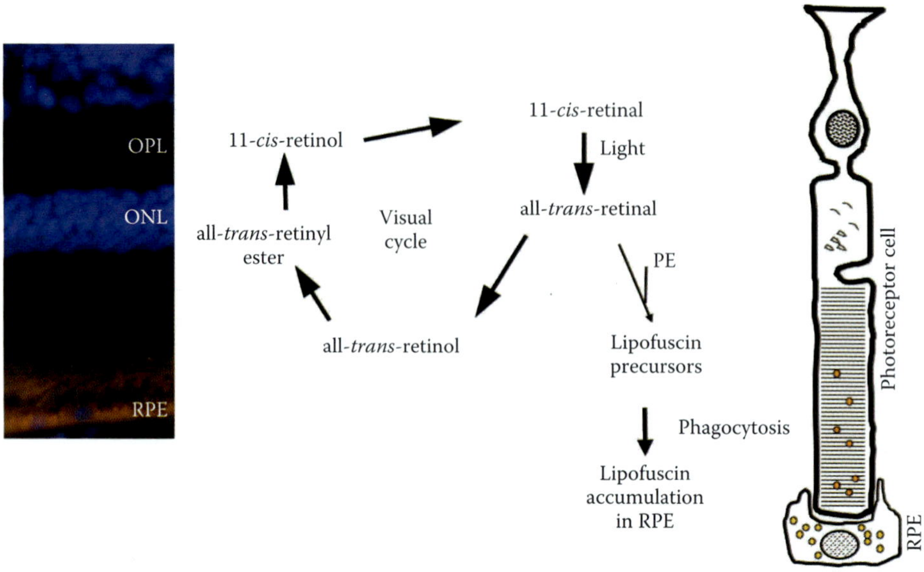

FIGURE 16.2 Intersection of the visual (retinoid) cycle and pathway for RPE lipofuscin formation. The photoisomerization of 11-*cis*-retinal leads to the release of all-*trans*-retinal from rhodopsin. All-*trans*-retinal for the most part is reduced to all-*trans*-retinol and remains in the visual cycle for reconversion to 11-*cis*-retinal. All-*trans*-retinal can also react inaptly, leave the visual cycle, and forming lipofuscin precursors. At least some of the reactions leading to the lipofuscin pathway are between all-*trans*-retinal and PE in a 2:1 ratio. After the phagocytosis of shed outer segment membrane by RPE, lipofuscin accumulates in the latter cells. RPE lipofuscin detected as autofluorescence in monkey retina imaged by fluorescence microscopy (left). Nuclei are stained with DAPI. The autofluorescence adjacent to the RPE is at the level of outer segments and is likely attributable to lipofuscin precursors that form in outer segments. Outer nuclear layer (ONL); outer plexiform layer (OPL).

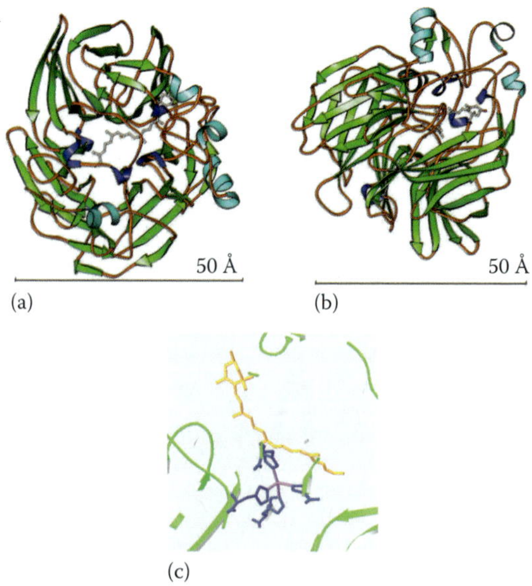

FIGURE 19.5 Structure of SynACO (PDB ID 2BIW). (a) SynACO has a rare seven-bladed propeller structure with the substrate tunnel perpendicular to the propellers. The Fe^{2+} metal center (shown in orange) is coordinated by four histidines. The substrate β-apo-8′-carotenal is shown in gray. (b) Side view showing the α-helical region of the entrance loops (shown in blue). A hydrophobic patch lies over the substrate tunnel. (c) Slab view of the active site histidine residues (HIS183, HIS238, HIS304, HIS404) and the apocarotenoid substrate. The substrate is isomerized from an all-*trans* configuration to a *cis* configuration at the 13,14- and 13′,14′-double bonds. Cleavage occurs at the 15,15′-double bond.

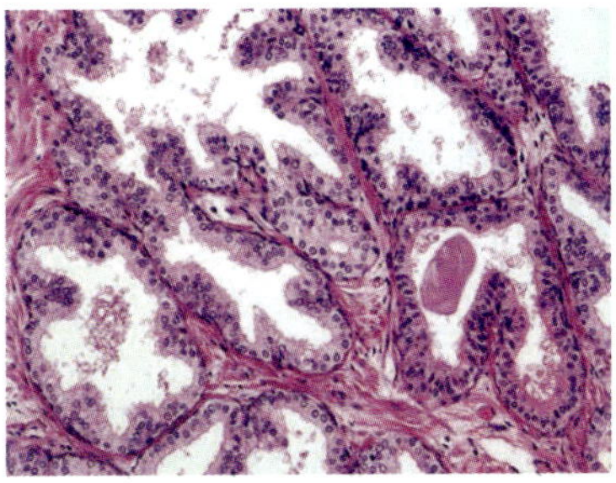

FIGURE 21.1 Normal tissue from human prostate showing secretory section—hematoxilin and eosin staining showing epithelial cells lining secretory ducts backed by basal and stromal cells. (Courtesy of A. Brollo, Wikimedia Commons, 2005.)

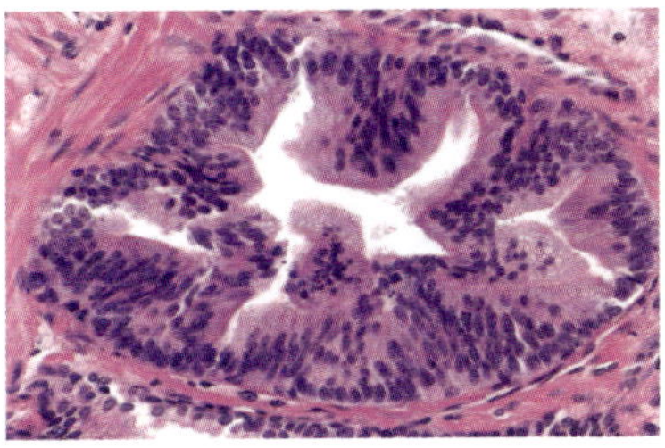

HGPIN—possible precursor of neoplasia, proliferative with enlarged nuclei but normal glands

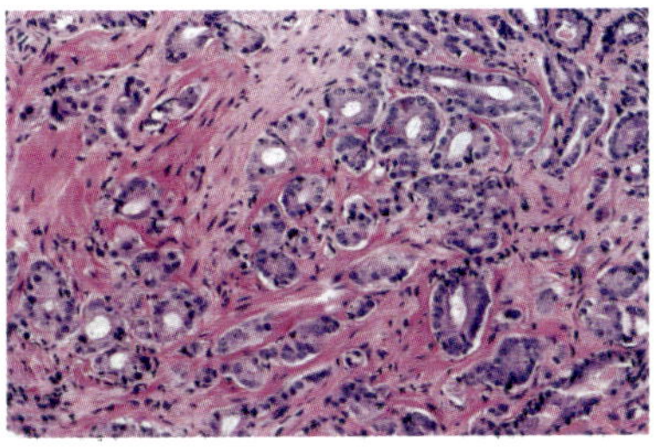

Gleason pattern 3—most common, glands vary in shape with haphazardly infiltrating stroma

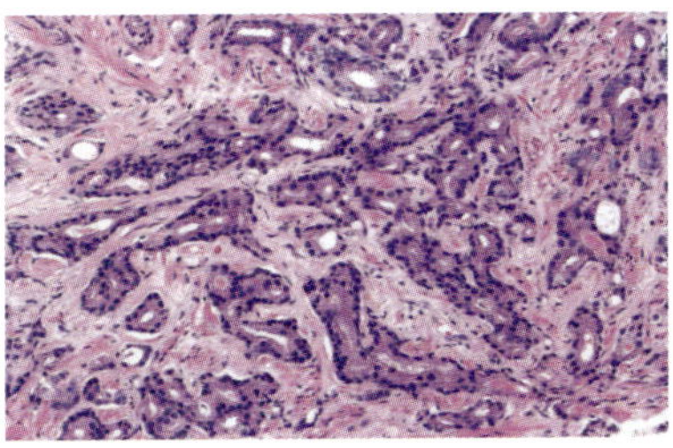

Gleason pattern 4—glands have fused

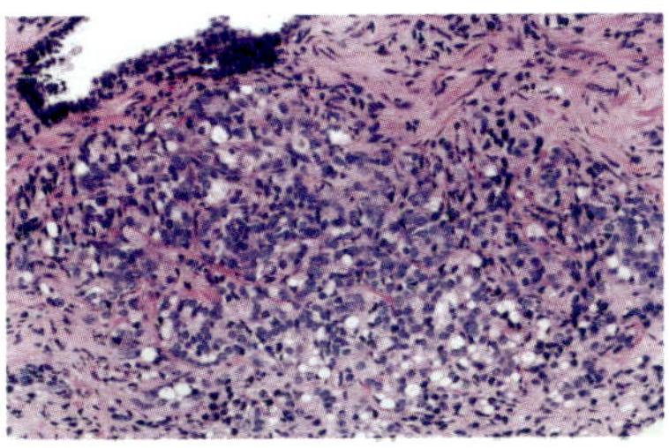

Gleason pattern 5—loss of glandular structure

FIGURE 21.2 Human prostate showing progression of cancer grading from benign to severe—hemotoxilin and eosin staining showing changes in the organization of epithelial cells forming ducts and surrounding stromal cells.

FIGURE 25.1 Larval butterflies have distinctive coloration patterns: (a) Monarch (*Danaus plexippus*; yellow, white, and black), (b) Swallowtail (*Papilio polyxenes asterius*; yellow, black, and green), (c) Queen (*Danaus gillipus*; black, white, and yellow, some red coloration is also observable in this specimen), and (d) Atala (*Eurnaceus atala florida*; red with yellow spots).

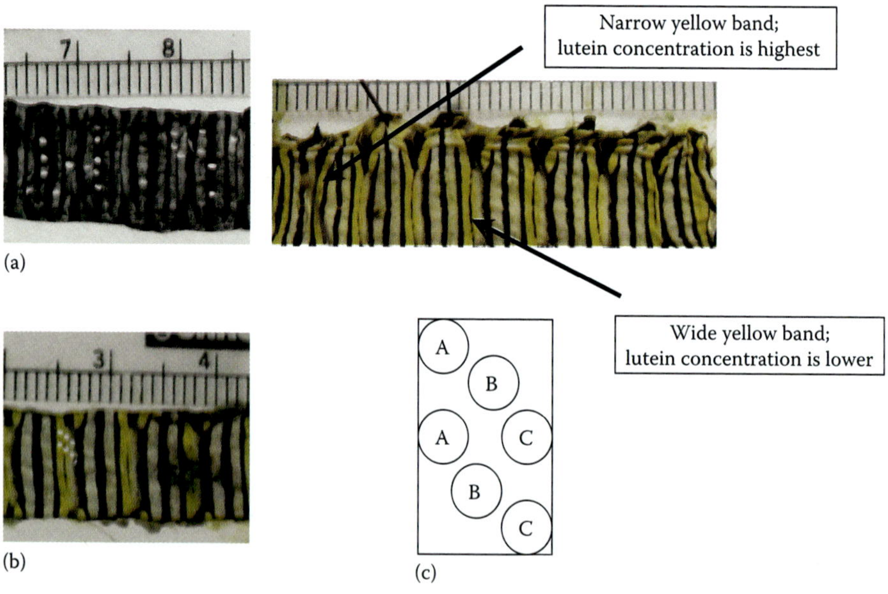

FIGURE 25.2 The sampling of colored bands in the epidermis of the Monarch was accomplished using the trephines of different sizes suited for providing samples of separate regions. (a) The dissected epidermis and cuticle showing the complete banding pattern of the Monarch and the positions where punches were obtained from one specimen. (b) Sampling of wider yellow bands using a smaller (0.39 mm) diameter trephine is shown. (c) Drawing showing how the sampling positions using the small trephine from across the wide yellow band were combined.

15.11.4 Other Effects of Carotenoids

Carotenoids have been found to exert numerous other effects of potential importance for the RPE. Carotenoids can activate transcription pathways (Ben-Dor et al., 2005; Kalariya et al., 2008; Palozza et al., 2006; Sharoni et al., 2004); for example, by activation of the antioxidant response element (ARE) (Ben-Dor et al., 2005; Sharoni et al., 2004). The ARE is an enhancer sequence responsible for the expression of many phase-II detoxification and antioxidant genes. Thus carotenoids may upregulate cellular antioxidant defenses.

As already mentioned, van Kuijk and colleagues (Kalariya et al., 2008) tested the effects of oxidation products of β-carotene, lutein, and zeaxanthin on the activation of redox-sensitive transcription factors, NF-κB, and AP-1 in cultured ARPE-19 cells. Degradation products of all three carotenoids induced activation of NF-κB and AP-1, and these effects were ameliorated by pretreatment of cells with 1 mM NAC. NF-κB is a major transcription factor that binds to promoter sites of many pro-inflammatory cytokines such as IL-1, IL-6, TNF-α, and iNOS. These results indicate that the degradation products of carotenoids can stimulate a pro-inflammatory pathway.

Intercellular communication can be affected by different carotenoids and their oxidation products, and opposing effects can be observed depending on their concentrations (Stahl et al., 1998). Carotenoids play a role in the induction and stimulation of intercellular communication via gap junctions, which in turn play an important role in the regulation of cell growth, differentiation, and apoptosis (Tapiero et al., 2004).

Carotenoids can strongly modulate apoptotic pathways (Palozza et al., 2006). For example, it has been demonstrated that lutein and zeaxanthin modulate the expression of anti-and pro-apoptotic factors and can selectively induce apoptosis in cancer cells but not in normal cells (Chew et al., 2003; Maccarrone et al., 2005).

It has been shown that carotenoids, such as lycopene, β-carotene, zeaxanthin, lutein, or astaxanthin, can inhibit oxidative damage to DNA at low concentrations or short incubation times, but higher concentrations or longer incubation times may enhance DNA damage and/or inhibit the DNA repair mechanisms so the net effect of carotenoids on oxidative stress-induced DNA damage may become deleterious (Astley et al., 2002, 2004a,b; Santocono et al., 2006). Numerous studies point to antiproliferative effects of some carotenoids in cancer cells and monocytes (Cheng et al., 2007; Gunasekera et al., 2007; McDevitt et al., 2005; Sun and Yao, 2007).

Lutein and zeaxanthin seem to exert distinct effects on distribution of the RPE cells in the retina (Leung et al., 2004). In animal studies, Leung and colleagues demonstrated that animals supplemented with either lutein and zeaxanthin on a low *n*-3 fatty acid diet had a lower RPE cell density than unsupplemented animals on the same diet (Leung et al., 2004). The authors suggested that macular xanthophylls could stimulate the movement of RPE cells away from the central retina (Leung et al., 2004).

Altogether, there is strong evidence that carotenoids can exert multiple effects via modulation of signaling pathways. Because oxidative stress can easily degrade carotenoids, it is not enough to investigate the effects of intact carotenoids, but it is necessary to elucidate the effects of their degradation products as well.

15.12 SUMMARY

Carotenoids seem to have a great therapeutic potential; they are superior singlet oxygen quenchers and can modulate a variety of cellular processes. However, the effects of carotenoids may vary depending on their concentrations and the presence of other nutrients, such as lipids. Carotenoids are susceptible to oxidative degradation and their degradation products exhibit different properties than their parent compounds, and include numerous potentially toxic components. The mechanism of specific accumulation of lutein and zeaxanthin in the retina, their roles, and their metabolic transformations are just being discovered. Elucidating these processes is particularly important to developing a clear understanding of retinal degenerations, such as AMD. The retina affected by AMD is under increased oxidative stress in comparison to the healthy retina (Anderson et al., 2002; Beatty et al., 2000; Seddon et al., 2004), thus there is an increased risk that lutein and zeaxanthin could be degraded. It is not known what the fate of the xanthophyll degradation products is in the retina. It is an open question whether and how xanthophyll degradation products affect the AMD retina. At present, there is no direct evidence that supplementation with carotenoids is likely to prevent or slow down the progression of AMD. Yet, in every food store and pharmacy there is a host of dietary supplements containing β-carotene, lutein, and/or zeaxanthin with labels implying a positive effect on eye health (Arora et al., 2004). Clearly, there is an urgent need to understand the many roles carotenoids play in the retina as well as the effects of their degradation products. Cultured RPE cells provide a good model to investigate at least some of these processes, including carotenoid uptake and secretion, effects on antioxidant, inflammatory, angiogenic and apoptotic pathways, lipid metabolism, and remodeling of the extracellular matrix.

REFERENCES

Acton, S, Rigotti, A, Landschulz, KT, Xu, S, Hobbs, HH, and Krieger, M, 1996. Identification of scavenger receptor SR-BI as a high density lipoprotein receptor. *Science* 271, 518–520.

Adler, AJ and Edwards, RB, 2000. Human interphotoreceptor matrix contains serum albumin and retinol-binding protein. *Exp Eye Res* 70, 227–234.

Ahir, A, Guo, L, Hussain, AA, and Marshall, J, 2002. Expression of metalloproteinases from human retinal pigment epithelial cells and their effects on the hydraulic conductivity of Bruch's membrane. *Invest Ophthalmol Vis Sci* 43, 458–465.

Aizman, A, Johnson, MW, and Elner, SG, 2007. Treatment of acute retinal necrosis syndrome with oral antiviral medications. *Ophthalmology* 114, 307–312.

Albanes, D, Heinonen, OP, Huttunen, JK, Taylor, PR, Virtamo, J, Edwards, BK, Haapakoski, J, Rautalahti, M, Hartman, AM, Palmgren, J, and Greenwald, P, 1995. Effects of alpha-tocopherol and beta-carotene supplements on cancer incidence in the alpha-tocopherol beta-carotene cancer prevention study. *Am J Clin Nutr* 62, S1427–S1430.

Aleman, TS, Duncan, JL, Bieber, ML, de Castro, E, Marks, DA, Gardner, LM, Steinberg, JD, Cideciyan, AV, Maguire, MG, and Jacobson, SG, 2001. Macular pigment and lutein supplementation in retinitis pigmentosa and Usher syndrome. *Invest Ophthalmol Vis Sci* 42, 1873–1881.

Alija, AJ, Bresgen, N, Sommerburg, O, Siems, W, and Eckl, PM, 2004. Cytotoxic and genotoxic effects of beta-carotene breakdown products on primary rat hepatocytes. *Carcinogenesis* 25, 827–831.

Alija, AJ, Bresgen, N, Sommerburg, O, Langhans, CD, Siems, W, and Eckl, PM, 2006. β-Carotene breakdown products enhance genotoxic effects of oxidative stress in primary rat hepatocytes. *Carcinogenesis* 27, 1128–1133.

Allikmets, R, Singh, N, Sun, H, Shroyer, NE, Hutchinson, A, Chidambaram, A, Gerrard, B et al. 1997. A photoreceptor cell-specific ATP-binding transporter gene (ABCR) is mutated in recessive Stargardt macular dystrophy. *Nat Genet* 15, 236–246.

Allikmets, R, 2000. Simple and complex ABCR: Genetic predisposition to retinal disease. *Am J Hum Genet* 67, 793–799.

Anderson, DH, Ozaki, S, Nealon, M, Neitz, J, Mullins, RF, Hageman, GS, and Johnson, LV, 2001. Local cellular sources of apolipoprotein E in the human retina and retinal pigmented epithelium: implications for the process of drusen formation. *Am J Ophthalmol* 131, 767–781.

Anderson, DH, Mullins, RF, Hageman, GS, and Johnson, LV, 2002. Perspective—A role for local inflammation in the formation of drusen in the aging eye. *Am J Ophthalmol* 134, 411–431.

AREDS, 2001. A randomized, placebo-controlled, clinical trial of high-dose supplementation with vitamins C and E, beta carotene, and zinc for age-related macular degeneration and vision loss, AREDS Report No. 8. *Arch Ophthalmol* 119, 1417–1436.

Arora, S, Musadiq, M, Mukherji, S, and Yang, YC, 2004. Eye nutrient products for age-related macular degeneration: what do they contain? *Eye* 18, 470–473.

Astley, SB, Elliott, RM, Archer, DB, and Southon, S, 2002. Increased cellular carotenoid levels reduce the persistence of DNA single-strand breaks after oxidative challenge. *Nutr Cancer* 43, 202–213.

Astley, SB, Elliott, RM, Archer, DB, and Southon, S, 2004a. Evidence that dietary supplementation with carotenoids and carotenoid-rich foods modulates the DNA damage: repair balance in human lymphocytes. *Br J Nutr* 91, 63–72.

Astley, SB, Hughes, DA, Wright, AJA, Elliott, RM, and Southon, S, 2004b. DNA damage and susceptibility to oxidative damage in lymphocytes: Effects of carotenoids *in vitro* and *in vivo*. *Br J Nutr* 91, 53–61.

Aukunuru, JV, Sunkara, G, Bandi, N, Thoreson, WB, and Kompella, UB, 2001. Expression of multidrug resistance-associated protein (MRP) in human retinal pigment epithelial cells and its interaction with BAPSG, a novel aldose reductase inhibitor. *Pharm Res* 18, 565–572.

Bahrami, H, Melia, M, and Dagnelie, G, 2006. Lutein supplementation in retinitis pigmentosa: PC-based vision assessment in a randomized double-masked placebo-controlled clinical trial [NCT00029289]. *BMC Ophthalmol* 6, 23.

Bailey, KR, Ishida, BY, Duncan, KG, Kane, JP, and Schwartz, DM, 2004. Basal reverse cholesterol transport of retinal pigment epithelium cell digested photoreceptor outer segment lipids. *Invest Ophthalmol Vis Sci* 45, U721.

Balazs, Z, Panzenboeck, U, Hammer, A, Sovic, A, Quehenberger, O, Malle, E, and Sattler, W, 2004. Uptake and transport of high-density lipoprotein (HDL) and HDL-associated alpha-tocopherol by an *in vitro* blood-brain barrier model. *J Neurochem* 89, 939–950.

Bartl, MM, Luckenbach, T, Bergner, O, Ullrich, O, and Koch-Brandt, C, 2001. Multiple receptors mediate apoJ-dependent clearance of cellular debris into nonprofessional phagocytes. *Exp Cell Res* 271, 130–141.

Baur, JA and Sinclair, DA, 2006. Therapeutic potential of resveratrol: The *in vivo* evidence. *Nat Rev Drug Discov* 5, 493–506.

Beatty, S, Koh, HH, Henson, D, and Boulton, M, 2000. The role of oxidative stress in the pathogenesis of age-related macular degeneration. *Surv Ophthalmol* 45, 115–134.

Beatty, S, Nolan, J, Kavanagh, H, and O'Donovan, O, 2004. Macular pigment optical density and its relationship with serum and dietary levels of lutein and zeaxanthin. *Arch Biochem Biophys* 430, 70–76.

Beharry, S, Zhong, M, and Molday, RS, 2004. N-retinylidene-phosphatidylethanolamine is the preferred retinoid substrate for the photoreceptor-specific ABC transporter ABCA4 (ABCR). *J Biol Chem* 279, 53972–53979.

Ben-Dor, A, Steiner, M, Gheber, L, Danilenko, M, Dubi, N, Linnewiel, K, Zick, A, Sharoni, Y, and Levy, J, 2005. Carotenoids activate the antioxidant response element transcription system. *Mol Cancer Ther* 4, 177–186.

Berendschot, TT and van Norren, D, 2006. Macular pigment shows ringlike structures. *Invest Ophthalmol Vis Sci* 47, 709–714.

Berendschot, TTJM, Goldbohm, RA, Klopping, WAA, van de Kraats, J, van Norel, J, and van Norren, D, 2000. Influence of lutein supplementation on macular pigment, assessed with two objective techniques. *Invest Ophthalmol Vis Sci* 41, 3322–3326.

Bernstein, PS, Balashov, NA, Tsong, ED, and Rando, RR, 1997. Retinal tubulin binds macular carotenoids. *Invest Ophthalmol Vis Sci* 38, 167–175.

Bernstein, PS, Khachik, F, Carvalho, LS, Muir, GJ, Zhao, DY, and Katz, NB, 2001. Identification and quantitation of carotenoids and their metabolites in the tissues of the human eye. *Exp Eye Res* 72, 215–223.

Bernstein, PS, Leppert, M, Singh, N, Dean, M, Lewis, RA, Lupski, JR, Allikmets, R, and Seddon, JM, 2002a. Genotype-phenotype analysis of ABCR variants in macular degeneration probands and siblings. *Invest Ophthalmol Vis Sci* 43, 466–473.

Bernstein, PS, Zhao, DY, Wintch, SW, Ermakov, IV, McClane, RW, and Gellermann, W, 2002b. Resonance Raman measurement of macular carotenoids in normal subjects and in age-related macular degeneration patients. *Ophthalmology* 109, 1780–1787.

Bhatti, RA, Yu, S, Boulanger, A, Fariss, RN, Guo, Y, Bernstein, SL, Gentleman, S, and Redmond, TM, 2003. Expression of beta-carotene 15,15′ monooxygenase in retina and RPE-choroid. *Invest Ophthalmol Vis Sci* 44, 44–49.

Bhosale, P, Larson, AJ, Frederick, JM, Southwick, K, Thulin, CD, and Bernstein, PS., 2004. Identification and characterization of a Pi isoform of glutathione S-transferase (GSTP1) as a zeaxanthin-binding protein in the macula of the human eye. *J Biol Chem* 279, 49447–49454.

Bhosale, P and Bernstein, PS, 2005. Synergistic effects of zeaxanthin and its binding protein in the prevention of lipid membrane oxidation. *Biochim Biophys Acta* 1740, 116–121.

Bhosale, P and Bernstein, PS, 2007. Vertebrate and invertebrate carotenoid-binding proteins. *Arch Biochem Biophys* 458, 121–127.

Bhosale, P, Zhao da, Y, and Bernstein, PS, 2007a. HPLC measurement of ocular carotenoid levels in human donor eyes in the lutein supplementation era. *Invest Ophthalmol Vis Sci* 48, 543–549.

Bhosale, P, Zhao da, Y, Serban, B, and Bernstein, PS, 2007b. Identification of 3-methoxyzeaxanthin as a novel age-related carotenoid metabolite in the human macula. *Invest Ophthalmol Vis Sci* 48, 1435–1440.

Bjorkegren, J, Veniant, M, Kim, SK, Withycombe, SK, Wood, PA, Hellerstein, MK, Neese, RA, and Young, SG, 2001. Lipoprotein secretion and triglyceride stores in the heart. *J Biol Chem* 276, 38511–38517.

Blomhoff, R and Blomhoff, HK, 2006. Overview of retinoid metabolism and function. *J Neurobiol* 66, 606–630.

Bohm, F, Tinkler, JH, and Truscott, TG, 1995. Carotenoids protect against cell membrane damage by the nitrogen dioxide radical. *Nat Med* 1, 98–99.

Bohm, F, Edge, R, Lange, L, and Truscott, TG, 1998a. Enhanced protection of human cells against ultraviolet light by antioxidant combinations involving dietary carotenoids. *J Photochem Photobiol B* 44, 211–215.

Bohm, F, Edge, R, McGarvey, DJ, and Truscott, TG, 1998b. Beta-carotene with vitamins E and C offers synergistic cell protection against NOx. *FEBS Lett* 436, 387–389.

Bohm, F, Edge, R, Foley, S, Lange, L, and Truscott, TG, 2001. Antioxidant inhibition of porphyrin-induced cellular phototoxicity. *J Photochem Photobiol B* 65, 177–183.

Bone, RA, Landrum, JT, Fernandez, L, and Tarsis, SL, 1988. Analysis of the macular pigment by HPLC: Retinal distribution and age study. *Invest Ophthalmol Vis Sci* 29, 843–849.

Bone, RA and Landrum, JT, 1992. Distribution of macular pigment components, zeaxanthin and lutein, in human retina. *Methods Enzymol* 213, 360–366.

Bone, RA, Landrum, JT, and Cains, A, 1992. Optical density spectra of the macular pigment *in vivo* and *in vitro*. *Vision Res* 32, 105–110.

Bone, RA, Landrum, JT, Hime, GW, Cains, A, and Zamor, J, 1993. Stereochemistry of the human macular carotenoids. *Invest Ophthalmol Vis Sci* 34, 2033–2040.

Bone, RA, Landrum, JT, Friedes, LM, Gomez, CM, Kilburn, MD, Menendez, E, Vidal, I., and Wang, W., 1997. Distribution of lutein and zeaxanthin stereoisomers in the human retina. *Exp Eye Res* 64, 211–218.

Bone, RA, Landrum, JT, Dixon, Z, Chen, Y, and Llerena, CM, 2000. Lutein and zeaxanthin in the eyes, serum and diet of human subjects. *Exp Eye Res* 71, 239–245.

Bone, RA, Landrum, JT, Mayne, ST, Gomez, CM, Tibor, SE, and Twaroska, EE, 2001. Macular pigment in donor eyes with and without AMD: A case-control study. *Invest Ophthalmol Vis Sci* 42, 235–240.

Bone, RA, Landrum, JT, Guerra, LH, and Ruiz, CA, 2003. Lutein and zeaxanthin dietary supplements raise macular pigment density and serum concentrations of these carotenoids in humans. *J Nutr* 133, 992–998.

Boulton, M, Rozanowska, M, and Rozanowski, B, 2001. Retinal photodamage. *J Photochem Photobiol B* 64, 144–161.

Boulton, M, Rozanowska, M, Rozanowski, B, and Wess, T, 2004. The photoreactivity of ocular lipofuscin. *Photochem Photobiol Sci* 3, 759–764.

Breithaupt, DE and Schlatterer, J, 2005. Lutein and zeaxanthin in new dietary supplements—Analysis and quantification. *Eur Food Res Technol* 220, 648–652.

Britton, G, 1995. UV/visible spectroscopy. In: Britton, G, Liaaen-Jensen, S, and Pfander, H. (Eds.), *Carotenoids*. Birkhauser, Basel, Switzerland, pp. 13–62.

Brown, AJ and Fragoso, YD, 1994. Tocopherols and carotenes are differently distributed in subfractions of high-density lipoprotein. *Biochim Biophys Acta* 1210, 373–376.

Browning, PJ, Roberts, DD, Zabrenetzky, V, Bryant, J, Kaplan, M, Washington, RH, Panet, A, Gallo, RC, and Vogel, T, 1994. Apolipoprotein E (ApoE), a novel heparin-binding protein inhibits the development of Kaposi's sarcoma-like lesions in BALB/c nu/nu mice. *J Exp Med* 180, 1949–1954.

Bultel-Brienne, S, Lestavel, S, Pilon, A, Laffont, I, Tailleux, A, Fruchart, JC, Siest, G, and Clavey, V, 2002. Lipid free apolipoprotein E binds to the class B type I scavenger receptor I (SR-BI) and enhances cholesteryl ester uptake from lipoproteins. *J Biol Chem* 277, 36092–36099.

Burke, M, Edge, R, Land, EJ, McGarvey, DJ, and Truscott, TG, 2001. One-electron reduction potentials of dietary carotenoid radical cations in aqueous micellar environments. *FEBS Lett* 500, 132–136.

Burton, GW and Ingold, KU, 1984. Beta-carotene: An unusual type of lipid antioxidant. *Science* 224, 569–573.

Calvo, D, Gomez-Coronado, D, Lasuncion, MA, and Vega, MA, 1997. CLA-1 is an 85-kD plasma membrane glycoprotein that acts as a high-affinity receptor for both native (HDL, LDL, and VLDL) and modified (OxLDL and AcLDL) lipoproteins. *Arterioscler Thromb Vasc Biol* 17, 2341–2349.

Calvo, D, Gomez-Coronado, D, Suarez, Y, Lasuncion, MA, and Vega, MA, 1998. Human CD36 is a high affinity receptor for the native lipoproteins HDL, LDL, and VLDL. *J Lipid Res* 39, 777–788.

Cangemi, FE, 2007. TOZAL Study: An open case control study of an oral antioxidant and omega-3 supplement for dry AMD. *BMC Ophthalmol* 7, 3.

Cantrell, A, McGarvey, DJ, Truscott, TG, Rancan, F, and Bohm, F, 2003. Singlet oxygen quenching by dietary carotenoids in a model membrane environment. *Arch Biochem Biophys* 412, 47–54.

Cardinault, N, Abalain, JH, Sairafi, B, Coudray, C, Grolier, P, Rambeau, M, Carre, JL, Mazur, A, and Rock, E, 2005. Lycopene but not lutein nor zeaxanthin decreases in serum and lipoproteins in age-related macular degeneration patients. *Clinica Chimica Acta* 357, 34–42.

Cecchelli, R, Berezowski, V, Lundquist, S, Culot, M, Renftel, M, Dehouck, MP, and Fenart, L, 2007. Modelling of the blood-brain barrier in drug discovery and development. *Nat Rev Drug Discov* 6, 650–661.

Chan, A, Ko, TH, and Duker, JS, 2006. Ultrahigh-resolution optical coherence tomography of canthaxanthine retinal crystals. *Ophthalmic Surg Lasers Imaging* 37, 138–139.

Chen, JY, Cheung, NH, Fung, MC, Wen, JM, Leung, WN, and Mak, NK, 2000. Subcellular localization of merocyanine 540 (MC540) and induction of apoptosis in murine myeloid leukemia cells. *Photochem Photobiol* 72, 114–120.

Chen, M, Forrester, JV, and Xu, H, 2007. Synthesis of complement factor H by retinal pigment epithelial cells is down-regulated by oxidized photoreceptor outer segments. *Exp Eye Res* 84, 635–645.

Chen, W, Sun, Y, Welch, C, Gorelik, A, Leventhal, AR, Tabas, I, and Tall, AR, 2001. Preferential ATP-binding cassette transporter A1-mediated cholesterol efflux from late endosomes/lysosomes. *J Biol Chem* 276, 43564–43569.

Cheng, HC, Chien, H, Liao, CH, Yang, YY, and Huang, SY, 2007. Carotenoids suppress proliferating cell nuclear antigen and cyclin D1 expression in oral carcinogenic models. *J Nutr Biochem* 18, 667–675.

Chew, BP, Brown, CM, Park, JS, and Mixter, PF, 2003. Dietary lutein inhibits mouse mammary tumor growth by regulating angiogenesis and apoptosis. *Anticancer Res* 23, 3333–3339.

Chew, BP and Park, JS, 2004. Carotenoid action on the immune response. *J Nutr* 134, 257S–261S.

Chichili, GR, Nohr, D, Schaffer, M, von Lintig, J, and Biesalski, HK, 2005. Beta-carotene conversion into vitamin A in human retinal pigment epithelial cells. *Invest Ophthalmol Vis Sci* 46, 3562–3569.

Chichili, GR, Nohr, D, Frank, J, Flaccus, A, Fraser, PD, Enfissi, EM, and Biesalski, HK, 2006. Protective effects of tomato extract with elevated beta-carotene levels on oxidative stress in ARPE-19 cells. *Br J Nutr* 96, 643–649.

Cho, E, Seddon, JM, Rosner, B, Willett, WC, and Hankinson, SE, 2004. Prospective study of intake of fruits, vegetables, vitamins, and carotenoids and risk of age-related maculopathy. *Arch Ophthalmol* 122, 883–892.

Choi, JS, Kim, D, Hong, YM, Mizuno, S, and Joo, CK, 2006. Inhibition of nNOS and COX-2 expression by lutein in acute retinal ischemia. *Nutrition* 22, 668–671.

Chong, EW, Wong, TY, Kreis, AJ, Simpson, JA, and Guymer, RH, 2007. Dietary antioxidants and primary prevention of age related macular degeneration: Systematic review and meta-analysis. *BMJ* 335, 755.

Chowers, I, Wong, R, Dentchev, T, Farkas, RH, Iacovelli, J, Gunatilaka, TL, Medeiros, NE, Presley, JB, Campochiaro, PA, Curcio, CA, Dunaief, JL, and Zack, DJ, 2006. The iron carrier transferrin is upregulated in retinas from patients with age-related macular degeneration. *Invest Ophthalmol Vis Sci* 47, 2135–2140.

Chucair, AJ, Rotstein, NP, Sangiovanni, JP, During, A, Chew, EY, and Politi, LE, 2007. Lutein and zeaxanthin protect photoreceptors from apoptosis induced by oxidative stress: Relation with docosahexaenoic acid. *Invest Ophthalmol Vis Sci* 48, 5168–5177.

Coleman, H and Chew, E, 2007. Nutritional supplementation in age-related macular degeneration. *Curr Opin Ophthalmol* 18, 220–223.

Connor, WE, Duell, PB, Kean, R, and Wang, Y, 2007. The prime role of HDL to transport lutein into the retina: Evidence from HDL-deficient WHAM chicks having a mutant ABCA1 transporter. *Invest Ophthalmol Vis Sci* 48, 4226–4231.

Constable, PA, Lawrenson, JG, Dolman, DE, Arden, GB, and Abbott, NJ, 2006. P-Glycoprotein expression in human retinal pigment epithelium cell lines. *Exp Eye Res* 83, 24–30.

Coral-Hinostroza, GN, Ytrestoyl, T, Ruyter, B, and Bjerkeng, B, 2004. Plasma appearance of unesterified astaxanthin geometrical E/Z and optical R/S isomers in men given single doses of a mixture of optical 3 and 3′ R/S isomers of astaxanthin fatty acyl diesters. *Comp Biochem Physiol C-Toxicol Pharmacol* 139, 99–110.

Costet, P, Lalanne, F, Gerbod-Giannone, MC, Molina, JR, Fu, X, Lund, EG, Gudas, LJ, and Tall, AR, 2003. Retinoic acid receptor-mediated induction of ABCA1 in macrophages. *Mol Cell Biol* 23, 7756–7766.

Crabb, JW, Miyagi, M, Gu, X, Shadrach, K, West, KA, Sakaguchi, H, Kamei, M, Hasan, A, Yan, L, Rayborn, ME, Salomon, RG, and Hollyfield, JG, 2002. Drusen proteome analysis: an approach to the etiology of age-related macular degeneration. *Proc Natl Acad Sci U S A* 99, 14682–14687.

Crane, IJ, Wallace, CA, McKillop-Smith, S, and Forrester, JV, 2000a. Control of chemokine production at the blood–retina barrier. *Immunology* 101, 426–433.

Crane, IJ, Wallace, CA, McKillop-Smith, S, and Forrester, JV, 2000b. CXCR4 receptor expression on human retinal pigment epithelial cells from the blood–retina barrier leads to chemokine secretion and migration in response to stromal cell-derived factor 1 alpha. *J Immunol* 165, 4372–4378.

Cui, HS, Hayasaka, S, Zhang, XY, Hayasaka, Y, Chi, ZL, and Zheng, LS, 2006. Effect of berberine on interleukin 8 and monocyte chemotactic protein 1 expression in a human retinal pigment epithelial cell line. *Ophthalmic Res* 38, 149–157.

Dasch, B, Fuhs, A, Behrens, T, Meister, A, Wellmann, J, Fobker, M, Pauleikhoff, D, and Hense, HW, 2005. Inflammatory markers in age-related maculopathy—Cross-sectional analysis from the Muenster Aging and Retina Study. *Arch Ophthalmol* 123, 1501–1506.

Davies, NP and Morland, AB, 2004. Macular pigments: Their characteristics and putative role. *Prog Retin Eye Res* 23, 533–559.

Delcourt, C, Carriere, I, Delage, M, Barberger-Gateau, P, and Schalch, W, 2006. Plasma lutein and zeaxanthin and other carotenoids as modifiable risk factors for age-related maculopathy and cataract: The POLA Study. *Invest Ophthalmol Vis Sci* 47, 2329–2335.

Dentchev, T, Hahn, P, and Dunaief, JL, 2005. Strong labeling for iron and the iron-handling proteins ferritin and ferroportin in the photoreceptor layer in age-related macular degeneration. *Arch Ophthalmol* 123, 1745–1746.

Dinkova-Kostova, AT and Talalay, P, 2008. Direct and indirect antioxidant properties of inducers of cytoprotective proteins, *Mol Nutr Food Res* 52 Suppl, S128–S138.

Donaldson, MJ and Pulido, JS, 2006. Treatment of nonexudative (dry) age-related macular degeneration. *Curr Opin Ophthalmol* 17, 267–274.

Donoso, LA, Kim, D, Frost, A, Callahan, A, and Hageman, G, 2006. The role of inflammation in the pathogenesis of age-related macular degeneration. *Surv Ophthalmol* 51, 137–152.

Duncan, KG, Bailey, KR, Kane, JP, and Schwartz, DM, 2002. Human retinal pigment epithelial cells express scavenger receptors BI and BII. *Biochem Biophys Res Commun* 292, 1017–1022.

Dunn, KC, Aotaki-Keen, AE, Putkey, FR, and Hjelmeland, LM, 1996. ARPE-19, a human retinal pigment epithelial cell line with differentiated properties. *Exp Eye Res* 62, 155–169.

During, A, and Harrison, EH, 2007. Mechanisms of provitamin A (carotenoid) and vitamin A (retinol) transport into and out of intestinal Caco-2 cells. *J Lipid Res* 48, 2283–2294.

Ebihara, N, Chen, L, Tokura, T, Ushio, H, Iwatsu, M, and Murakami, A, 2007. Distinct functions between toll-like receptors 3 and 9 in retinal pigment epithelial cells. *Ophthalmic Res* 39, 155–163.

Edge, R, Land, EJ, McGarvey, D, Mulroy, L, and Truscott, TG, 1998. Relative one-electron reduction potentials of carotenoid radical cations and the interactions of carotenoids with the vitamin E radical cation. *J Am Chem Soc* 120, 4087–4090.

Edge, R, Land, EJ, McGarvey, DJ, Burke, M, and Truscott, TG, 2000a. The reduction potential of the beta-carotene+ /beta-carotene couple in an aqueous micro-heterogeneous environment. *FEBS Lett* 471, 125–127.

Edge, R, Land, EJ, Rozanowska, M, Sarna, T, and Truscott, TG, 2000b. Carotenoid radical-melanin interactions. *J Phys Chem B* 104, 7193–7196.

Edwards, AO, 2008. Genetics of age-related macular degeneration. *Recent Advances in Retinal Degeneration* 613, 211–219.

Ekam, VS, Udosen, EO, and Chigbu, AE, 2006. Comparative effect of carotenoid complex from Golden Neo-Life Dynamite (GNLD) and carrot extracted carotenoids on immune parameters in albino Wistar rats. *Niger J Physiol Sci* 21, 1–4.

El-Agamey, A, Cantrell, A, Land, EJ, McGarvey, DJ, and Truscott, TG, 2004a. Are dietary carotenoids beneficial? Reactions of carotenoids with oxy-radicals and singlet oxygen. *Photochem Photobiol Sci* 3, 802–811.

El-Agamey, A, Lowe, GM, McGarvey, DJ, Mortensen, A, Phillip, DM, Truscott, TG, and Young, AJ, 2004b. Carotenoid radical chemistry and antioxidant/pro-oxidant properties. *Arch Biochem Biophys* 430, 37–48.

Ellis, EM, 2007. Reactive carbonyls and oxidative stress: Potential for therapeutic intervention. *Pharmacol Ther* 115, 13–24.

Elner, SG, Delmonte, D, Bian, ZM, Lukacs, NW, and Elner, VM, 2006. Differential expression of retinal pigment epithelium (RPE) IP-10 and interleukin-8. *Exp Eye Res* 83, 374–379.

Elner, VM, 2002. Retinal pigment epithelial acid lipase activity and lipoprotein receptors: Effects of dietary omega-3 fatty acids. *Trans Am Ophthalmol Soc* 100, 301–338.

Ershov, AV and Bazan, NG, 2000. Photoreceptor phagocytosis selectively activates PPARgamma expression in retinal pigment epithelial cells. *J Neurosci Res* 60, 328–337.

Espaillat, A, Aiello, LP, Arrigg, PG, Villalobos, R, Silver, PM, and Cavicchi, RW, 1999. Canthaxanthine retinopathy. *Arch Ophthalmol* 117, 412–413.

Esser, P, Tervooren, D, Heimann, K, Kociok, N, Bartz-Schmidt, KU, Walter, P, and Weller, M, 1998. Intravitreal daunomycin induces multidrug resistance in proliferative vitreoretinopathy. *Invest Ophthalmol Vis Sci* 39, 164–170.

Eter, N, Krohne, TU, and Holz, FG, 2006. New pharmacologic approaches to therapy for age-related macular degeneration. *BioDrugs* 20, 167–179.

Evans, JR, 2006. Antioxidant vitamin and mineral supplements for slowing the progression of age-related macular degeneration. *Cochrane Database Syst Rev* 19(2): CD000254.

Everett, SA, Dennis, MF, Patel, KB, Maddix, S, Kundu, SC, and Willson, RL, 1996. Scavenging of nitrogen dioxide, thiyl, and sulfonyl free radicals by the nutritional antioxidant beta-carotene. *J Biol Chem* 271, 3988–3994.

Failloux, N, Bonnet, I, Baron, MH, and Perrier, E, 2003. Quantitative analysis of vitamin A degradation by Raman spectroscopy. *Appl Spectrosc* 57, 1117–1122.

Faulkner, LE, Panagotopulos, SE, Johnson, JD, Woollett, LA, Hui, DY, Witting, SR, Maiorano, JN, and Davidson, WS, 2008. An analysis of the role of a retroendocytosis pathway in ATP-binding cassette transporter (ABCA1)—Mediated cholesterol efflux from macrophages. *J Lipid Res*, M800048–JLR800200.

Fiedor, J, Fiedor, L, Haessner, R, and Scheer, H, 2005. Cyclic endoperoxides of beta-carotene, potential pro-oxidants, as products of chemical quenching of singlet oxygen. *Biochim Biophys Acta-Bioenerg* 1709, 1–4.

Fuller, CJ, Faulkner, H, Bendich, A, Parker, RS, and Roe, DA, 1992. Effect of beta-carotene supplementation on photosuppression of delayed-type hypersensitivity in normal young men. *Am J Clin Nutr* 56, 684–690.

Gajic, M, Zaripheh, S, Sun, FR, and Erdman, JW, 2006. Apo-8′-lycopenal and apo-12′-lycopenal are metabolic products of lycopene in rat liver. *J Nutr* 136, 1552–1557.

Gale, CR, Hall, NF, Phillips, DI, and Martyn, CN, 2003. Lutein and zeaxanthin status and risk of age-related macular degeneration. *Invest Ophthalmol Vis Sci* 44, 2461–2465.

Girotti, AW and Korytowski, W, 2000. Cholesterol as a singlet oxygen detector in biological systems. *Methods Enzymol* 319, 85–100.

Godley, BF, Shamsi, FA, Liang, FQ, Jarrett, SG, Davies, S, and Boulton, M, 2005. Blue light induces mitochondrial DNA damage and free radical production in epithelial cells. *J Biol Chem* 280, 21061–21066.

Golczak, M, Imanishi, Y, Kuksa, V, Maeda, T, Kubota, R, and Palczewski, K, 2005a. Lecithin: Retinol acyltransferase is responsible for amidation of retinylamine, a potent inhibitor of the retinoid cycle. *J Biol Chem* 280, 42263–42273.

Golczak, M, Kuksa, V, Maeda, T, Moise, AR, and Palczewski, K, 2005b. Positively charged retinoids are potent and selective inhibitors of the *trans–cis* isornerization in the retinoid (visual) cycle. *Proc Natl Acad Sci U S A* 102, 8162–8167.

Goldberg, J, Flowerdew, G, Smith, E, Brody, JA, and Tso, MO, 1988. Factors associated with age-related macular degeneration. An analysis of data from the first National Health and Nutrition Examination Survey. *Am J Epidemiol* 128, 700–710.

Goralczyk, R, Buser, S, Bausch, J, Bee, W, Zuhlke, U, and Barker, FM, 1997. Occurrence of birefringent retinal inclusions in cynomolgus monkeys after high doses of canthaxanthin. *Invest Ophthalmol Vis Sci* 38, 741–752.

Goralczyk, R, Barker, FM, Buser, S, Liechti, H, and Bausch, J, 2000. Dose dependency of canthaxanthin crystals in monkey retina and spatial distribution of its metabolites. *Invest Ophthalmol Vis Sci* 41, 1513–1522.

Gordiyenko, N, Campos, M, Lee, JW, Fariss, RN, Sztein, J, and Rodriguez, IR, 2004. RPE cells internalize low-density lipoprotein (LDL) and oxidized LDL (oxLDL) in large quantities in vitro and in vivo. *Invest Ophthalmol Vis Sci* 45, 2822–2829.

Gorin, MB, 2007. A clinician's view of the molecular genetics of age-related maculopathy. *Arch Ophthalmol* 125, 21–29.

Gu, XR, Meer, SG, Miyagi, M, Rayborn, ME, Hollyfield, JG, Crabb, JW, and Salomon, RG, 2003. Carboxyethylpyrrole protein adducts and autoantibodies, biomarkers for age-related macular degeneration. *J Biol Chem* 278, 42027–42035.

Gunasekera, RS, Sewgobind, K, Desai, S, Dunn, L, Black, HS, McKeehan, WL, and Patil, B, 2007. Lycopene and lutein inhibit proliferation in rat prostate carcinoma cells. *Nutr Cancer* 58, 171–177.

Guymer, RH and Chong, EW, 2006. Modifiable risk factors for age-related macular degeneration. *Med J Aust* 184, 455–458.

Hageman, GS, Luthert, PJ, Chong, NHV, Johnson, LV, Anderson, DH, and Mullins, RF, 2001. An integrated hypothesis that considers drusen as biomarkers of immune-mediated processes at the RPE-Bruch's membrane interface in aging and age-related macular degeneration. *Prog Retin Eye Res* 20, 705–732.

Hahn, P, Milam, AH, and Dunaief, JL, 2003. Maculas affected by age-related macular degeneration contain increased chelatable iron in the retinal pigment epithelium and Bruch's membrane. *Arch Ophthalmol* 121, 1099–1105.

Haimovici, R, Kramer, M, Miller, JW, Hasan, T, Flotte, TJ, Schomacker, KT, and Gragoudas, ES, 1997. Localization of lipoprotein-delivered benzoporphyrin derivative in the rabbit eye. *Curr Eye Res* 16, 83–90.

Halliwell, B and Gutteridge, JMC, 2000. *Free Radicals in Biology and Medicine*. Oxford University Press, Oxford.

Hammond, BR, Fuld, K, and Currancelentano, J, 1995. Macular pigment density in monozygotic twins. *Invest Ophthalmol Vis Sci* 36, 2531–2541.

Hammond, BR, Johnson, EJ, Russell, RM, Krinsky, NI, Yeum, KJ, Edwards, RB, and Snodderly, DM, 1997. Dietary modification of human macular pigment density. *Invest Ophthalmol Vis Sci* 38, 1795–1801.

Handelman, GJ, Snodderly, DM, Krinsky, NI, Russett, MD, and Adler, AJ, 1991a. Biological control of primate macular pigment—Biochemical and densitometric studies. *Invest Ophthalmol Vis Sci* 32, 257–267.

Handelman, GJ, Vankuijk, F, Chatterjee, A, and Krinsky, NI, 1991b. Characterization of products formed during the autoxidation of beta-carotene. *Free Radic Biol Med* 10, 427–437.

Haught, JM, Patel, S, and English, JC, 3rd, 2007. Xanthoderma: A clinical review. *J Am Acad Dermatol* 57, 1051–1058.

Hayes, JD, Flanagan, JU, and Jowsey, IR, 2005. Glutathione transferases. *Annu Rev Pharmacol Toxicol* 45, 51–88.

He, X, Hahn, P, Iacovelli, J, Wong, R, King, C, Bhisitkul, R, Massaro-Giordano, M, and Dunaief, JL, 2007. Iron homeostasis and toxicity in retinal degeneration. *Prog Retin Eye Res* 26, 649–673.

Hollyfield, JG, Salomon, RG, and Crabb, JW, 2003. Proteomic approaches to understanding age-related macular degeneration, In: LaVail, MM, Anderson, RE, and Hollyfield, JG (Eds.), *Retinal Degenerations: Mechanisms and Experimental Therapy*. Kluwer Academic/Plenum Publishers, New York, pp. 83–89.

Holtkamp, GM, Kijlstra, A, Peek, R, and de Vos, AF, 2001. Retinal pigment epithelium-immune system interactions: Cytokine production and cytokine-induced changes. *Prog Retin Eye Res* 20, 29–48.

Holz, FG, Bindewald-Wittich, A, Fleckenstein, M, Dreyhaupt, J, Scholl, HP, and Schmitz-Valckenberg, S, 2007. Progression of geographic atrophy and impact of fundus autofluorescence patterns in age-related macular degeneration. *Am J Ophthalmol* 143, 463–472.

Horak, P, Zilmer, M, Saks, L, Ots, I, Karu, U, and Zilmer, K, 2006. Antioxidant protection, carotenoids and the costs of immune challenge in greenfinches. *J Exp Biol* 209, 4329–4338.

Hozawa, A, Jacobs, DR, Jr., Steffes, MW, Gross, MD, Steffen, LM, and Lee, DH, 2007. Relationships of circulating carotenoid concentrations with several markers of inflammation, oxidative stress, and endothelial dysfunction: The Coronary Artery Risk Development in Young Adults (CARDIA)/Young Adult Longitudinal Trends in Antioxidants (YALTA) study. *Clin Chem* 53, 447–455.

Hu, W, Jiang, A, Liang, J, Meng, H, Chang, B, Gao, H, and Qiao, X, 2008. Expression of VLDLR in the retina and evolution of subretinal neovascularization in the knockout mouse model's retinal angiomatous proliferation. *Invest Ophthalmol Vis Sci* 49, 407–415.

Hurst, JS, Contreras, JE, Siems, WG, and van Kuijk, F, 2004. Oxidation of carotenoids by heat and tobacco smoke. *Biofactors* 20, 23–35.

Hurst, JS, Saini, MK, Jin, GF, Awasthi, YC, and van Kuijk, FJ, 2005. Toxicity of oxidized beta-carotene to cultured human cells. *Exp Eye Res* 81, 239–243.

Hussein, G, Sankawa, U, Goto, H, Matsumoto, K, and Watanabe, H, 2006. Astaxanthin, a carotenoid with potential in human health and nutrition. *J Nat Prod* 69, 443–449.

Iannaccone, A, Mura, M, Gallaher, KT, Johnson, EJ, Todd, WA, Kenyon, E, Harris, TL, Harris, T, Satterfield, S, Johnson, KC, and Kritchevsky, SB, 2007. Macular pigment optical density in the elderly: Findings in a large biracial Midsouth population sample. *Invest Ophthalmol Vis Sci* 48, 1458–1465.

Ishida, BY, Bailey, KR, Duncan, KG, Chalkley, RJ, Burlingame, AL, Kane, JP, and Schwartz, DM, 2004. Regulated expression of apolipoprotein E by human retinal pigment epithelial cells. *J Lipid Res* 45, 263–271.

Ishida, BY, Duncan, KG, Bailey, KR, Kane, JP, and Schwartz, DM, 2006. High density lipoprotein mediated lipid efflux from retinal pigment epithelial cells in culture. *Br J Ophthalmol* 90, 616–620.

Izumi-Nagai, K, Nagai, N, Ohgami, K, Satofuka, S, Ozawa, Y, Tsubota, K, Umezawa, K, Ohno, S, Oike, Y, and Ishida, S, 2007. Macular pigment lutein is antiinflammatory in preventing choroidal neovascularization. *Arterioscler Thromb Vasc Biol* 27, 2555–2562.

Jackson, GR, Owsley, C, and McGwin, G, 1999. Aging and dark adaptation. *Vision Res* 39, 3975–3982.

Jackson, GR, Owsley, C, and Curcio, CA, 2002. Photoreceptor degeneration and dysfunction in aging and age-related maculopathy. *Ageing Res Rev* 1, 381–396.

Jackson, GR, McGwin, G, Phillips, JM, Klein, R, and Owsley, C, 2006. Impact of aging and age-related maculopathy on inactivation of the a-wave of the rod-mediated electroretinogram. *Vision Res* 46, 1422–1431.

Ji, Y, Jian, B, Wang, N, Sun, Y, Moya, ML, Phillips, MC, Rothblat, GH, Swaney, JB, and Tall, AR, 1997. Scavenger receptor BI promotes high density lipoprotein-mediated cellular cholesterol efflux. *J Biol Chem* 272, 20982–20985.

Jin, XH, Ohgami, K, Shiratori, K, Suzuki, Y, Hirano, T, Koyama, Y, Yoshida, K, Ilieva, I, Iseki, K, and Ohno, S, 2006. Inhibitory effects of lutein on endotoxin-induced uveitis in Lewis rats. *Invest Ophthalmol Vis Sci* 47, 2562–2568.

Joffre, C, Leclere, L, Buteau, B, Martine, L, Cabaret, S, Malvitte, L, Acar, N, Lizard, G, Bron, A, and Creuzot-Garcher, C, 2007. Oxysterols induced inflammation and oxidation in primary porcine retinal pigment epithelial cells. *Curr Eye Res* 32, 271–280.

Johnson, LV, Ozaki, S, Staples, MK, Erickson, PA, and Anderson, DH, 2000. A potential role for immune complex pathogenesis in drusen formation. *Exp Eye Res* 70, 441–449.

Johnson, LV, Leitner, WP, Staples, MK, and Anderson, DH, 2001. Complement activation and inflammatory processes in drusen formation and age related macular degeneration. *Exp Eye Res* 73, 887–896.

Jones, AA, 2007. Age related macular degeneration—Should your patients be taking additional supplements? *Aust Fam Physician* 36, 1026–1028.

Kaempf-Rotzoll, DE, Traber, MG, and Arai, H, 2003. Vitamin E and transfer proteins. *Curr Opin Lipidol* 14, 249–254.

Kalariya, NM, Ramana, KV, Srivastava, SK, and van Kuijk, FJ, 2008. Carotenoid derived aldehydes-induced oxidative stress causes apoptotic cell death in human retinal pigment epithelial cells. *Exp Eye Res* 86, 70–80.

Kaminski, WE, Wenzel, JJ, Piehler, A, Langmann, T, and Schmitz, G, 2001. ABCA6, a novel a subclass ABC transporter. *Biochem Biophys Res Commun* 285, 1295–1301.

Kanis, MJ, Berendschot, TT, and van Norren, D, 2007. Influence of macular pigment and melanin on incident early AMD in a white population. *Graefes Arch Clin Exp Ophthalmol* 245, 767–773.

Kanofsky, JR and Sima, PD, 2006. Synthetic carotenoid derivatives prevent photosensitised killing of retinal pigment epithelial cells more effectively than lutein. *Exp Eye Res* 82, 907–914.

Karl, MO, Valtink, M, Bednarz, J, and Engelmann, K, 2007. Cell culture conditions affect RPE phagocytic function. *Graefes Arch Clin Exp Ophthalmol* 245, 981–991.

Katz, ML, 2002. Potential role of retinal pigment epithelial lipofuscin accumulation in age-related macular degeneration. *Arch Gerontol Geriatr* 34, 359–370.

Kelly, ME, Clay, MA, Mistry, MJ, Hsieh-Li, HM, and Harmony, JA, 1994. Apolipoprotein E inhibition of proliferation of mitogen-activated T lymphocytes: Production of interleukin 2 with reduced biological activity. *Cell Immunol* 159, 124–139.

Kelly, SP, Thornton, J, Lyratzopoulos, G, Edwards, R, and Mitchell, P, 2004. Smoking and blindness—Strong evidence for the link, but public awareness lags. *Br Med J* 328, 537–538.

Kennedy, BG and Mangini, NJ, 2002. P-glycoprotein expression in human retinal pigment epithelium. *Mol Vis* 8, 422–430.

Kennedy, TA and Liebler, DC, 1991. Peroxyl radical oxidation of beta-carotene—Formation of beta-carotene epoxides. *Chem Res Toxicol* 4, 290–295.

Khachik, F, Bernstein, PS, and Garland, DL, 1997. Identification of lutein and zeaxanthin oxidation products in human and monkey retinas. *Invest Ophthalmol Vis Sci* 38, 1802–1811.

Khachik, F, de Moura, FF, Zhao, DY, Aebischer, CP, and Bernstein, PS, 2002. Transformations of selected carotenoids in plasma, liver, and ocular tissues of humans and in nonprimate animal models. *Invest Ophthalmol Vis Sci* 43, 3383–3392.

Khachik, F, de Moura, FF, Chew, EY, Douglass, LW, Ferris, FL, 3rd, Kim, J, and Thompson, DJ, 2006. The effect of lutein and zeaxanthin supplementation on metabolites of these carotenoids in the serum of persons aged 60 or older. *Invest Ophthalmol Vis Sci* 47, 5234–5242.

Kijlstra, A, La Heij, EC, and Hendrikse, F, 2005. Immunological factors in the pathogenesis and treatment of age-related macular degeneration. *Ocul Immunol Inflamm* 13, 3–11.

Kim, SJ, Nara, E, Kobayashi, H, Terao, J, and Nagao, A, 2001. Formation of cleavage products by autoxidation of lycopene. *Lipids* 36, 191–199.

Kim, SJ, 2004. Cleavage products formed through autoxidation of zeta-carotene in liposomal suspension. *Food Sci Biotechnol* 13, 202–207.

Kim, WS, Weickert, CS, and Garner, B, 2008. Role of ATP-binding cassette transporters in brain lipid transport and neurological disease. *J Neurochem* 104, 1145–1166.

Kiser, AK and Dagnelie, G, 2008. Reported effects of non-traditional treatments and complementary and alternative medicine by retinitis pigmentosa patients. *Clin Exp Optom* 91, 166–176.

Klamt, F, Dal-Pizzol, F, Roehrs, R, de Oliveira, RB, Dalmolin, R, Henriques, JAP, de Andrades, HHR, Ramos, A, Saffi, J, and Moreira, JCF, 2003. Genotoxicity, recombinogenicity and cellular preneoplasic transformation induced by vitamin A supplementation. *Mutat Res Genet Toxicol Environ Mutagen* 539, 117–125.

Klein, R, Klein, BEK, and Moss, SE, 1998. Relation of smoking to the incidence of age-related maculopathy—The Beaver Dam Eye Study. *Am J Epidemiol* 147, 103–110.

Koh, HH, Murray, IJ, Nolan, D, Carden, D, Feather, J, and Beatty, S, 2004. Plasma and macular responses to lutein supplement in subjects with and without age-related maculopathy: A pilot study. *Exp Eye Res* 79, 21–27.

Koldamova, RP, Lefterov, IM, Ikonomovic, MD, Skoko, J, Lefterov, PI, Isanski, BA, DeKosky, ST, and Lazo, JS, 2003. 22R-hydroxycholesterol and 9-*cis*-retinoic acid induce ATP-binding cassette transporter A1 expression and cholesterol efflux in brain cells and decrease amyloid beta secretion. *J Biol Chem* 278, 13244–13256.

Kotake-Nara, E, Kim, SJ, Kobori, M, Miyashita, K, and Nagao, A, 2002. Acyclo-retinoic acid induces apoptosis in human prostate cancer cells. *Anticancer Res* 22, 689–695.

Koutsos, EA, Garcia Lopez, JC, and Klasing, KC, 2007. Maternal and dietary carotenoids interactively affect cutaneous basophil responses in growing chickens (Gallus gallus domesticus). *Comp Biochem Physiol B Biochem Mol Biol* 147, 87–92.

Krieger, M, 1999. Charting the fate of the "good cholesterol": Identification and characterization of the high-density lipoprotein receptor SR-BI. *Annu Rev Biochem* 68, 523–558.

Krieger, M, 2001. Scavenger receptor class B type I is a multiligand HDL receptor that influences diverse physiologic systems. *J Clin Invest* 108, 793–797.

Krinsky, NI, Landrum, JT, and Bone, RA, 2003. Biologic mechanisms of the protective role of lutein and zeaxanthin in the eye. *Annu Rev Nutr* 23, 171–201.

Kuehn, BM, 2005. Inflammation suspected in eye disorders. *J Am Med Assoc* 294, 31–32.

Kuimova, MK, Yahioglu, G, and Ogilby, PR, 2009. Singlet oxygen in a cell: Spatially dependent lifetimes and quenching rate constants. *J Am Chem Soc* 131, 332–340.

Kvansakul, J, Rodriguez-Carmona, M, Edgar, DF, Barker, FM, Kopcke, W, Schalch, W, and Barbur, JL, 2006. Supplementation with the carotenoids lutein or zeaxanthin improves human visual performance. *Ophthalmic Physiol Opt* 26, 362–371.

Lakkaraju, A, Finnemann, SC, and Rodriguez-Boulan, E, 2007. The lipofuscin fluorophore A2E perturbs cholesterol metabolism in retinal pigment epithelial cells. *Proc Natl Acad Sci U S A* 104, 11026–11031.

Lamb, TD and Pugh, EN, 2004. Dark adaptation and the retinoid cycle of vision. *Prog Retin Eye Res* 23, 307–380.

Landrum, JT, Bone, RA, Joa, H, Kilburn, MD, Moore, LL, and Sprague, KE, 1997. A one year study of the macular pigment: The effect of 140 days of a lutein supplement. *Exp Eye Res* 65, 57–62.

LaRowe, TL, Mares, JA, Snodderly, DM, Klein, ML, Wooten, BR, and Chappell, R, 2008. Macular pigment density and age-related maculopathy in the Carotenoids in Age-Related Eye Disease Study. An ancillary study of the women's health initiative. *Ophthalmology* 115, 876–883 e871.

Leuenberger, MG, Engeloch-Jarret, C, and Woggon, WD, 2001. The reaction mechanism of the enzyme-catalyzed central cleavage of beta-carotene to retinal. *Angew Chem Int Ed Engl* 40, 2613–2617.

Leung, IY, Sandstrom, MM, Zucker, CL, Neuringer, M, and Snodderly, DM, 2004. Nutritional manipulation of primate retinas, II: Effects of age, n-3 fatty acids, lutein, and zeaxanthin on retinal pigment epithelium. *Invest Ophthalmol Vis Sci* 45, 3244–3256.

Leyon, H, Ros, AM, Nyberg, S, and Algvere, P, 1990. Reversibility of canthaxanthin deposits within the retina. *Acta Ophthalmol (Copenh)* 68, 607–611.

Li, CM, Presley, B, Zhang, XM, Dashti, N, Chung, BH, Medeiros, NE, Guidry, C, and Curcio, CA, 2005. Retina expresses microsomal triglyceride transfer protein: Implications for age-related maculopathy. *J Lipid Res* 46, 628–640.

Li, CM, Clark, ME, Chimento, MF, and Curcio, CA, 2006. Apolipoprotein localization in isolated drusen and retinal apolipoprotein gene expression. *Invest Ophthalmol Vis Sci* 47, 3119–3128.

Lidebjer, C, Leanderson, P, Ernerudh, J, and Jonasson, L, 2007. Low plasma levels of oxygenated carotenoids in patients with coronary artery disease. *Nutr Metab Cardiovasc Dis* 17, 448–456.

Lindqvist, A and Andersson, S, 2002. Biochemical properties of purified recombinant human beta-carotene 15,15′-monooxygenase. *J Biol Chem* 277, 23942–23948.

Loane, E, Nolan, JM, O'Donovan, O, Bhosale, P, Bernstein, PS, and Beatty, S, 2008. Transport and retinal capture of lutein and zeaxanthin with reference to age-related macular degeneration. *Surv Ophthalmol* 53, 68–81.

Lorenzi, I, von Eckardstein, A, Cavelier, C, Radosavljevic, S, and Rohrer, L, 2008. Apolipoprotein A-I but not high-density lipoproteins are internalised by RAW macrophages: Roles of ATP-binding cassette transporter A1 and scavenger receptor BI. *J Mol Med* 86, 171–183.

Lornejad-Schafer, MR, Lambert, C, Breithaupt, DE, Biesalski, HK, and Frank, J, 2007. Solubility, uptake and biocompatibility of lutein and zeaxanthin delivered to cultured human retinal pigment epithelial cells in tween40 micelles. *Eur J Nutr* 46, 79–86.

Lotery, A and Trump, D, 2007. Progress in defining the molecular biology of age related macular degeneration. *Hum Genet* 122, 219–236.

Lowe, GM, Booth, LA, Young, AJ, and Bilton, RF, 1999. Lycopene and beta-carotene protect against oxidative damage in HT29 cells at low concentrations but rapidly lose this capacity at higher doses. *Free Radic Res* 30, 141–151.

Maccarrone, M, Bari, M, Gasperi, V, and Demmig-Adams, B, 2005. The photoreceptor protector zeaxanthin induces cell death in neuroblastoma cells. *Anticancer Res* 25, 3871–3876.

Maiti, P, Kong, J, Kim, SR, Sparrow, JR, Allikmets, R, and Rando, RR, 2006. Small molecule RPE65 antagonists limit the visual cycle and prevent lipofuscin formation. *Biochemistry* 45, 852–860.

Malek, G, Li, CM, Guidry, C, Medeiros, NE, and Curcio, CA, 2003. Apolipoprotein B in cholesterol-containing drusen and basal deposits of human eyes with age-related maculopathy. *Am J Pathol* 162, 413–425.

Maminishkis, A, Chen, S, Jalickee, S, Banzon, T, Shi, G, Wang, FE, Ehalt, T, Hammer, JA, and Miller, SS, 2006. Confluent monolayers of cultured human fetal retinal pigment epithelium exhibit morphology and physiology of native tissue. *Invest Ophthalmol Vis Sci* 47, 3612–3624.

Mardones, P and Rigotti, A, 2004. Cellular mechanisms of vitamin E uptake: Relevance in alpha-tocopherol metabolism and potential implications for disease. *J Nutr Biochem* 15, 252–260.

Mares-Perlman, JA, Brady, WE, Klein, R, Klein, BE, Bowen, P, Stacewicz-Sapuntzakis, M, and Palta, M, 1995. Serum antioxidants and age-related macular degeneration in a population-based case-control study. *Arch Ophthalmol* 113, 1518–1523.

Mares-Perlman, JA, Fisher, AI, Klein, R, Block, G, Millen, AE, and Wright, JD, 2001. Lutein and zeaxanthin in the diet and serum and their relation to age-related maculopathy in the Third National Health and Nutrition Examination Survey. *Am J Epidemiol* 153, 424–432.

Marin-Castano, ME, Striker, GE, Alcazar, O, Catanuto, P, Espinosa-Heidmann, DG, and Cousins, SW, 2006. Repetitive nonlethal oxidant injury to retinal pigment epithelium decreased extracellular matrix turnover *in vitro* and induced sub-RPE deposits in vivo. *Invest Ophthalmol Vis Sci* 47, 4098–4112.

Marques, SA, Loureiro, APM, Gornes, OF, Garcia, CCM, Di Mascio, P, and Medeiros, MHG, 2004. Induction of 1,N-2-etheno-2′-deoxyguanosine in DNA exposed to beta-carotene oxidation products. *FEBS Lett* 560, 125–130.

Marshall, J, Grindle, J, Ansell, PL, and Borwein, B, 1979. Convolution in human rods—Aging process. *Br J Ophthalmol* 63, 181–187.

Mata, NL, Weng, J, and Travis, GH, 2000. Biosynthesis of a major lipofuscin fluorophore in mice and humans with ABCR-mediated retinal and macular degeneration. *Proc Natl Acad Sci U S A* 97, 7154–7159.

Mata, NL, Tzekov, RT, Liu, XR, Weng, J, Birch, DG, and Travis, GH, 2001. Delayed. dark-adaptation and lipofuscin accumulation in abcr+/– mice: Implications for involvement of ABCR in age-related macular degeneration. *Invest Ophthalmol Vis Sci* 42, 1685–1690.

Matsumoto, A, Mizukami, H, Mizuno, S, Umegaki, K, Nishikawa, J, Shudo, K, Kagechika, H, and Inoue, M, 2007. Beta-cryptoxanthin, a novel natural RAR ligand, induces ATP-binding cassette transporters in macrophages. *Biochem Pharmacol* 74, 256–264.

Matsumoto, N, Kitayama, H, Kitada, M, Kimura, K, Noda, M, and Ide, C, 2003. Isolation of a set of genes expressed in the choroid plexus of the mouse using suppression subtractive hybridization. *Neuroscience* 117, 405–415.

McClure, TD and Liebler, DC, 1995. Electron capture negative chemical ionization mass spectrometry and tandem mass-spectrometry analysis of beta-carotene, alpha-tocopherol and their oxidation products. *J Mass Spectrom* 30, 1480–1488.

McDevitt, TM, Tchao, R, Harrison, EH, and Morel, DW, 2005. Carotenoids normally present in serum inhibit proliferation and induce differentiation of a human monocyte/macrophage cell line (U937). *J Nutr* 135, 160–164.

Miller, JW, Walsh, AW, Kramer, M, Hasan, T, Michaud, N, Flotte, TJ, Haimovici, R, and Gragoudas, ES, 1995. Photodynamic therapy of experimental choroidal neovascularization using lipoprotein-delivered benzoporphyrin. *Arch Ophthalmol* 113, 810–818.

Mitchell, P, Wang, JJ, Smith, W, and Leeder, SR, 2002. Smoking and the 5-year incidence of age-related maculopathy—The Blue Mountains Eye Study. *Arch Ophthalmol* 120, 1357–1363.

Moeller, SM, Parekh, N, Tinker, L, Ritenbaugh, C, Blodi, B, Wallace, RB, and Mares, JA, 2006. Associations between intermediate age-related macular degeneration and lutein and zeaxanthin in the Carotenoids in Age-related Eye Disease Study (CAREDS): Ancillary study of the Women's Health Initiative. *Arch Ophthalmol* 124, 1151–1162.

Momma, Y, Nagineni, CN, Chin, MS, Srinivasan, K, Detrick, B, and Hooks, JJ, 2003. Differential expression of chemokines by human retinal pigment epithelial cells infected with cytomegalovirus. *Invest Ophthalmol Vis Sci* 44, 2026–2033.

Mordi, RC, Walton, JC, Burton, GW, Hughes, L, Ingold, KU, and Lindsay, DA, 1991. Exploratory study of beta-carotene autoxidation. *Tetrahedron Lett* 32, 4203–4206.

Mordi, RC, Walton, JC, Burton, GW, Hughes, L, Ingold, KU, Lindsay, DA, and Moffatt, DJ, 1993. Oxidative degradation of beta-carotene and beta-apo-8′-carotenal. *Tetrahedron* 49, 911–928.

Morris, MS, Jacques, PF, Chylack, LT, Hankinson, SE, Willett, WC, Hubbard, LD, and Taylor, A, 2007. Intake of zinc and antioxidant micronutrients and early age-related maculopathy lesions. *Ophthalmic Epidemiol* 14, 288–298.

Moshfeghi, DM and Blumenkranz, MS, 2007. Role of genetic factors and inflammation in age-related macular degeneration. *Retina* 27, 269–275.

Mukherjee, PK, Marcheselli, VL, de Rivero Vaccari, JC, Gordon, WC, Jackson, FE, and Bazan, NG, 2007. Photoreceptor outer segment phagocytosis attenuates oxidative stress-induced apoptosis with concomitant neuroprotectin D1 synthesis. *Proc Natl Acad Sci U S A* 104, 13158–13163.

Mullins, RF, 2007. Genetic insights into the pathobiology of age-related macular degeneration. *Int Ophthalmol Clin* 47, 1–14.

Murata, M and Kawanishi, S, 2000. Oxidative DNA damage by vitamin A and its derivative via superoxide generation. *J Biol Chem* 275, 2003–2008.

Nagao, A, 2004. Oxidative conversion of carotenoids to retinoids and other products. *J Nutr* 134, 237S–240.

Nolan, JM, Stack, J, O, OD, Loane, E, and Beatty, S, 2007. Risk factors for age-related maculopathy are associated with a relative lack of macular pigment. *Exp Eye Res* 84, 61–74.

Nozaki, M, Raisler, BJ, Sakurai, E, Sarma, JV, Barnum, SR, Lambris, JD, Chen, Y, Zhang, K, Ambati, BK, Baffi, JZ, and Ambati, J, 2006. Drusen complement components C3a and C5a promote choroidal neovascularization. *Proc Natl Acad Sci U S A* 103, 2328–2333.

O'Sullivan, L, Ryan, L, and O'Brien, N, 2007. Comparison of the uptake and secretion of carotene and xanthophyll carotenoids by Caco-2 intestinal cells. *Br J Nutr* 98, 38–44.

Obana, A, Hiramitsu, T, Gohto, Y, Ohira, A, Mizuno, S, Hirano, T, Bernstein, PS, Fujii, H, Iseki, K, Tanito, M, and Hotta, Y, 2008. Macular carotenoid levels of normal subjects and age-related maculopathy patients in a Japanese population. *Ophthalmology* 115, 147–157.

Ojima, F, Sakamoto, H, Ishiguro, Y, and Terao, J, 1993. Consumption of carotenoids in photosensitized oxidation of human plasma and plasma low-density-lipoprotein. *Free Radic Biol Med* 15, 377–384.

Omenn, GS, 1996. Antioxidant vitamins, cancer, and cardiovascular disease. *N Engl J Med* 335, 1067–1068.

Omenn, GS, Goodman, GE, Thornquist, MD, Balmes, J, Cullen, MR, Glass, A, Keogh, JP, Meyskens, FL, Valanis, B, Williams, JH, Barnhart, S, Cherniack, MG, Brodkin, CA, and Hammar, S, 1996. Risk factors for lung cancer and for intervention effects in CARET, the beta-carotene and retinol efficacy trial. *J Natl Cancer Inst* 88, 1550–1559.

Owens, SL, Bunce, C, Brannon, AJ, Xing, W, Chisholm, IH, Gross, M, Guymer, RH, Holz, FG, and Bird, AC, 2006. Prophylactic laser treatment hastens choroidal neovascularization in unilateral age-related maculopathy: Final results of the drusen laser study. *Am J Ophthalmol* 141, 276–281.

Owsley, C, Jackson, GR, White, M, Feist, R, and Edwards, D, 2001. Delays in rod-mediated dark adaptation in early age-related maculopathy. *Ophthalmology* 108, 1196–1202.

Owsley, C, McGwin, G, Jackson, GR, Heimburger, DC, Piyathilake, CJ, Klein, R, White, MF, and Kallies, K, 2006. Effect of short-term, high-dose retinol on dark adaptation in aging and early age-related maculopathy. *Invest Ophthalmol Vis Sci* 47, 1310–1318.

Palozza, P, Calviello, G, and Bartoli, GM, 1995. Prooxidant activity of beta-carotene under 100-percent oxygen pressure in rat liver microsomes. *Free Radic Biol Med* 19, 887–892.

Palozza, P, Luberto, C, Calviello, G, Ricci, P, and Bartoli, GM, 1997. Antioxidant and prooxidant role of beta-carotene in murine normal and tumor thymocytes: Effects of oxygen partial pressure. *Free Radic Biol Med* 22, 1065–1073.

Palozza, P, 1998. Prooxidant actions of carotenoids in biologic systems. *Nutr Rev* 56, 257–265.

Palozza, P, Calviello, G, Serini, S, Maggiano, N, Lanza, P, Ranelletti, FO, and Bartoli, GM, 2001. Beta-carotene at high concentrations induces apoptosis by enhancing oxy-radical production in human adenocarcinoma cells. *Free Radic Biol Med* 30, 1000–1007.

Palozza, P, Serini, S, and Calviello, G, 2006. Carotenoids as modulators of intracellular signaling pathways. *Curr Signal Transduct Ther* 1, 325–335.

Panzella, L, Manini, P, Napolitano, A, and d'Ischia, M, 2004. Free radical oxidation of (E)-retinoic acid by the Fenton reagent: Competing epoxidation and oxidative breakdown pathways and novel products of 5,6-epoxyretinoic acid transformation. *Chem Res Toxicol* 17, 1716–1724.

Parisi, V, Tedeschi, M, Gallinaro, G, Varano, M, Saviano, S, and Piermarocchi, S, 2008. Carotenoids and antioxidants in age-related maculopathy italian study: multifocal electroretinogram modifications after 1 year. *Ophthalmology* 115, 324–333 e322.

Parker, RS, 1996. Absorption, metabolism, and transport of carotenoids. *FASEB J* 10, 542–551.

Philips, N, Keller, T, Hendrix, C, Hamilton, S, Arena, R, Tuason, M, and Gonzalez, S, 2007. Regulation of the extracellular matrix remodeling by lutein in dermal fibroblasts, melanoma cells, and ultraviolet radiation exposed fibroblasts. *Arch Dermatol Res* 299, 373–379.

Polyakov, NE, Leshina, TV, Konovalova, TA, and Kispert, LD, 2001. Carotenoids as scavengers of free radicals in a Fenton reaction: Antioxidants or pro-oxidants? *Free Radic Biol Med* 31, 398–404.

Prasain, JK, Moore, R, Hurst, JS, Barnes, S, and van Kuijk, F, 2005. Electrospray tandem mass spectrometric analysis of zeaxanthin and its oxidation products. *J Mass Spectrom* 40, 916–923.

Provost, AC, Pequignot, MO, Sainton, KM, Gadin, S, Salle, S, Marchant, D, Hales, DB, and Abitbol, M, 2003. Expression of SR-BI receptor and StAR protein in rat ocular tissues. *C R Biol* 326, 841–851.

Radu, RA, Mata, NL, Nusinowitz, S, Liu, X, Sieving, PA, and Travis, GH, 2003. Treatment with isotretinoin inhibits lipofuscin accumulation in a mouse model of recessive Stargardt's macular degeneration. *Proc Natl Acad Sci U S A* 100, 4742–4747.

Radu, RA, Han, Y, Bui, TV, Nusinowitz, S, Bok, D, Lichter, J, Widder, K, Travis, GH, and Mata, NL, 2005. Reductions in serum vitamin A arrest accumulation of toxic retinal fluorophores: A potential therapy for treatment of lipofuscin-based retinal diseases. *Invest Ophthalmol Vis Sci* 46, 4393–4401.

Rafi, MM and Shafaie, Y, 2007. Dietary lutein modulates inducible nitric oxide synthase (iNOS) gene and protein expression in mouse macrophage cells (RAW 264.7). *Mol Nutr Food Res* 51, 333–340.

Rapp, LM, Maple, SS, and Choi, JH, 2000. Lutein and zeaxanthin concentrations in rod outer segment membranes from perifoveal and peripheral human retina. *Invest Ophthalmol Vis Sci* 41, 1200–1209.

Reading, VM and Weale, RA, 1974. Macular pigment and chromatic aberration. *J Opt Soc Am* 64, 231–234.

Reboul, E, Abou, L, Mikail, C, Ghiringhelli, O, Andre, M, Portugal, H, Jourdheuil-Rahmani, D, Amiot, MJ, Lairon, D, and Borel, P, 2005. Lutein transport by Caco-2 TC-7 cells occurs partly by a facilitated process involving the scavenger receptor class B type I (SR-BI). *Biochem J* 387, 455–461.

Reboul, E, Klein, A, Bietrix, F, Gleize, B, Malezet-Desmoulins, C, Schneider, M, Margotat, A, Lagrost, L, Collet, X, and Borel, P, 2006. Scavenger receptor class B type I (SR-BI) is involved in vitamin E transport across the enterocyte. *J Biol Chem* 281, 4739–4745.

Reboul, E, Thap, S, Perrot, E, Amiot, MJ, Lairon, D, and Borel, P, 2007a. Effect of the main dietary antioxidants (carotenoids, gamma-tocopherol, polyphenols, and vitamin C) on alpha-tocopherol absorption. *Eur J Clin Nutr* 61, 1167–1173.

Reboul, E, Thap, S, Tourniaire, F, Andre, M, Juhel, C, Morange, S, Amiot, MJ, Lairon, D, and Borel, P, 2007b. Differential effect of dietary antioxidant classes (carotenoids, polyphenols, vitamins C and E) on lutein absorption. *Br J Nutr* 97, 440–446.

Redmond, RW and Kochevar, IE, 2006. Spatially resolved cellular responses to singlet oxygen. *Photochem Photobiol* 82, 1178–1186.

Richer, S, Stiles, W, Statkute, L, Pulido, J, Frankowski, J, Rudy, D, Pei, K, Tsipursky, M, and Nyland, J, 2004. Double-masked, placebo-controlled, randomized trial of lutein and antioxidant supplementation in the intervention of atrophic age-related macular degeneration: the Veterans LAST study (Lutein Antioxidant Supplementation Trial). *Optometry* 75, 216–230.

Richer, S, Devenport, J, and Lang, JC, 2007. LAST II: Differential temporal responses of macular pigment optical density in patients with atrophic age-related macular degeneration to dietary supplementation with xanthophylls. *Optometry* 78, 213–219.

Roberts, JE, Kukielczak, BM, Hu, DN, Miller, DS, Bilski, P, Sik, RH, Motten, AG, and Chignell, CF, 2002. The role of A2E in prevention or enhancement of light damage in human retinal pigment epithelial cells. *Photochem Photobiol* 75, 184–190.

Robman, L, Vu, H, Hodge, A, Tikellis, G, Dimitrov, P, McCarty, C, and Guymer, R, 2007. Dietary lutein, zeaxanthin, and fats and the progression of age-related macular degeneration. *Can J Ophthalmol* 42, 720–726.

Rodriguez-Carmona, M, Kvansakul, J, Harlow, JA, Kopcke, W, Schalch, W, and Barbur, JL, 2006. The effects of supplementation with lutein and/or zeaxanthin on human macular pigment density and colour vision. *Ophthalmic Physiol Opt* 26, 137–147.

Rozanowska, M, Jarvis-Evans, J, Korytowski, W, Boulton, ME, Burke, JM, and Sarna, T, 1995. Blue light-induced reactivity of retinal age pigment. In vitro generation of oxygen-reactive species. *J Biol Chem* 270, 18825–18830.

Rozanowska, M, Wessels, J, Boulton, M, Burke, JM, Rodgers, MA, Truscott, TG, and Sarna, T, 1998. Blue light-induced singlet oxygen generation by retinal lipofuscin in non-polar media. *Free Radic Biol Med* 24, 1107–1112.

Rozanowska, M, Korytowski, W, Rozanowski, B, Skumatz, C, Boulton, ME, Burke, JM, and Sarna, T, 2002. Photoreactivity of aged human RPE melanosomes: A comparison with lipofuscin. *Invest Ophthalmol Vis Sci* 43, 2088–2096.

Rozanowska, M, Pawlak, A, Rozanowski, B, Skumatz, C, Zareba, M, Boulton, ME, Burke, JM, Sarna, T, and Simon, JD, 2004a. Age-related changes in the photoreactivity of retinal lipofuscin granules: Role of chloroform-insoluble components. *Invest Ophthalmol Vis Sci* 45, 1052–1060.

Rozanowska, M and Sarna, T, 2005. Light-induced damage to the retina: Role of rhodopsin chromophore revisited. *Photochem Photobiol* 81, 1305–1330.

Różanowska, M and Różanowski, B, 2008. Visual transduction and age-related changes in lipofuscin. In: Tombran-Tink, J and Barnstable, CJ (Eds.), *Ophthalmology Research: The Visual Transduction Cascade*. The Humana Press Inc., Totowa, NJ, pp. 405–446.

Rozanowska, MB, Bigaj, J, Boulton, ME, Czuba-Pelech, B, Landrum, J, Rozanowski, B, and Zareba, M, 2004b. Uptake of carotenoids and their antioxidant action in ARPE-19 cells in culture. *Invest Ophthalmol Vis Sci* 45, U531.

Rozanowski, B and Rozanowska, M, 2005. Degradation products of xanthophyll include blue-light absorbing photosensitizers. *Invest Ophthalmol Vis Sci* 46, E-Abstract 3032.

Rozanowski, B, Burke, J, Sarna, T, and Rozanowska, M, 2008a. The pro-oxidant effects of interactions of ascorbate with photoexcited melanin fade away with aging of the retina. *Photochem Photobiol* 84, 658–670.

Rozanowski, B, Burke, JM, Boulton, ME, Sarna, T, and Rozanowska, M, 2008b. Human RPE melanosomes protect from photosensitized and iron-mediated oxidation but become pro-oxidant in the presence of iron upon photodegradation. *Invest Ophthalmol Vis Sci* 49, 2838–2847.

Rozanowski, B, Cuenco, J, Davies, S, Shamsi, FA, Zadlo, A, Dayhaw-Barker, P, Rozanowska, M, Sarna, T, and Boulton, ME, 2008c. The phototoxicity of aged human retinal melanosomes. *Photochem Photobiol* 84, 650–657.

Ryeom, SW, Silverstein, RL, Scotto, A, and Sparrow, JR, 1996a. Binding of anionic phospholipids to retinal pigment epithelium may be mediated by the scavenger receptor CD36. *J Biol Chem* 271, 20536–20539.

Ryeom, SW, Sparrow, JR, and Silverstein, RL, 1996b. CD36 participates in the phagocytosis of rod outer segments by retinal pigment epithelium. *J Cell Sci* 109, 387–395.

Sakaguchi, H, Miyagi, M, Shadrach, KG, Rayborn, ME, Crabb, JW, and Hollyfield, JG, 2002. Clusterin is present in drusen in age-related macular degeneration. *Exp Eye Res* 74, 547–549.

Salerno, C, Crifo, C, Capuozzo, E, Sommerburg, O, Langhans, CD, and Siems, W, 2005. Effect of carotenoid oxidation products on neutrophil viability and function. *Biofactors* 24, 185–192.

Sanders, TA, Haines, AP, Wormald, R, Wright, LA, and Obeid, O, 1993. Essential fatty acids, plasma cholesterol, and fat-soluble vitamins in subjects with age-related maculopathy and matched control subjects. *Am J Clin Nutr* 57, 428–433.

SanGiovanni, JP, Chew, EY, Clemons, TE, Ferris, FL, 3rd, Gensler, G, Lindblad, AS, Milton, RC, Seddon, JM, and Sperduto, RD, 2007. The relationship of dietary carotenoid and vitamin A, E, and C intake with age-related macular degeneration in a case-control study, AREDS Report No. 22. *Arch Ophthalmol* 125, 1225–1232.

Santocono, M, Zurria, M, Berrettini, M, Fedeli, D, and Falcioni, G, 2006. Influence of astaxanthin, zeaxanthin and lutein on DNA damage and repair in UVA-irradiated cells. *J Photochem Photobiol B* 85, 205–215.

Santos, MS, Meydani, SN, Leka, L, Wu, DY, Fotouhi, N, Meydani, M, Hennekens, CH, and Gaziano, JM, 1996. Natural killer cell activity in elderly men is enhanced by beta-carotene supplementation. *Am J Clin Nutr* 64, 772–777.

Santos, MS, Gaziano, JM, Leka, LS, Beharka, AA, Hennekens, CH, and Meydani, SN, 1998. Beta-carotene-induced enhancement of natural killer cell activity in elderly men: An investigation of the role of cytokines. *Am J Clin Nutr* 68, 164–170.

Sanz, MM, Johnson, LE, Ahuja, S, Ekstrom, PA, Romero, J, and van Veen, T, 2007. Significant photoreceptor rescue by treatment with a combination of antioxidants in an animal model for retinal degeneration. *Neuroscience* 145, 1120–1129.

Sarkadi, B, Homolya, L, Szakacs, G, and Varadi, A, 2006. Human multidrug resistance ABCB and ABCG transporters: Participation in a chemoimmunity defense system. *Physiol Rev* 86, 1179–1236.

Scallon, LJ, Burke, JM, Mieler, WF, Kies, JC, and Aaberg, TM, 1988. Canthaxanthine-induced retinal pigment epithelial changes in the cat. *Curr Eye Res* 7, 687–693.

Schadel, SA, Heck, M, Maretzki, D, Filipek, S, Teller, DC, Palczewski, K, and Hofmann, KP, 2003. Ligand channeling within a G-protein-coupled receptor—The entry and exit of retinals in native opsin. *J Biol Chem* 278, 24896–24903.

Schmitz-Valckenberg, S, Bindewald-Wittich, A, Dolar-Szczasny, J, Dreyhaupt, J, Wolf, S, Scholl, HP, and Holz, FG, 2006. Correlation between the area of increased autofluorescence surrounding geographic atrophy and disease progression in patients with AMD. *Invest Ophthalmol Vis Sci* 47, 2648–2654.

Scholl, HPN, Fleckenstein, M, Issa, PC, Keilhauer, C, Holz, FG, and Weber, BHF, 2007. An update on the genetics of age-related macular degeneration. *Mol Vis* 13, 196–205.

Seddon, JM, Ajani, UA, Sperduto, RD, Hiller, R, Blair, N, Burton, TC, Farber, MD, Gragoudas, ES, Haller, J, Miller, DT, Yannuzzi, LA, and Willett, W, 1994. Dietary carotenoids, vitamin A, vitamin C, and vitamin E, and advanced age-related macular degeneration. *J Am Med Assoc* 272, 1413–1420.

Seddon, JM, Gensler, G, Milton, RC, Klein, ML, and Rifai, N, 2004. Association between C-reactive protein and age-related macular degeneration. *J Am Med Assoc* 291, 704–710.

Seddon, JM, Gensler, G, Klein, ML, and Milton, RC, 2006. C-reactive protein and homocysteine are associated with dietary and behavioral risk factors for age-related macular degeneration. *Nutrition* 22, 441–443.

Seddon, JM, 2007. Multivitamin-multimineral supplements and eye disease: age-related macular degeneration and cataract. *Am J Clin Nutr* 85, 304S–307S.

Selvaraj, RK and Klasing, KC, 2006. Lutein and eicosapentaenoic acid interact to modify iNOS mRNA levels through the PPAR gamma/RXR pathway in chickens and HD11 cell lines. *J Nutr* 136, 1610–1616.

Selvaraj, RK, Koutsos, EA, Calvert, CC, and Klasing, KC, 2006. Dietary lutein and fat interact to modify macrophage properties in chicks hatched from carotenoid deplete or replete eggs. *J Anim Physiol Anim Nutr (Berl)* 90, 70–80.

Shafaa, MW, Diehl, HA, and Socaciu, C, 2007. The solubilisation pattern of lutein, zeaxanthin, canthaxanthin and beta-carotene differ characteristically in liposomes, liver microsomes and retinal epithelial cells. *Biophys Chem* 129, 111–119.

Sharoni, Y, Danilenko, M, Dubi, N, Ben-Dor, A, and Levy, J, 2004. Carotenoids and transcription. *Arch Biochem Biophys* 430, 89–96.

Siems, W, Sommerburg, O, Schild, L, Augustin, W, Langhans, CD, and Wiswedel, I, 2002. Beta-carotene cleavage products induce oxidative stress in vitro by impairing mitochondrial respiration. *Faseb J* 16, U335–U353.

Siems, W, Capuozzo, E, Crifo, C, Sommerburg, O, Langhans, CD, Schlipalius, L, Wiswedel, I, Kraemer, K, and Salerno, C, 2003. Carotenoid cleavage products modify respiratory burst and induce apoptosis of human neutrophils. *Biochim Biophys Acta* 1639, 27–33.

Siems, W, Wiswedel, I, Salerno, C, Crifo, C, Augustin, W, Schild, L, Langhans, CD, and Sommerburg, O, 2005. Beta-carotene breakdown products may impair mitochondrial functions—Potential side effects of high-dose beta-carotene supplementation. *J Nutr Biochem* 16, 385–397.

Siems, WG, Sommerburg, O, and van Kuijk, F, 1999. Lycopene and beta-carotene decompose more rapidly than lutein and zeaxanthin upon exposure to various pro-oxidants in vitro. *Biofactors* 10, 105–113.

Siems, WG, Sommerburg, O, Hurst, JS, and van Kuijk, F, 2000. Carotenoid oxidative degradation products inhibit Na^+-K^+-ATPase. *Free Radic Res* 33, 427–435.

Snellen, EL, Verbeek, AL, Van Den Hoogen, GW, Cruysberg, JR, and Hoyng, CB, 2002. Neovascular age-related macular degeneration and its relationship to antioxidant intake. *Acta Ophthalmol Scand* 80, 368–371.

Snodderly, DM, Auran, JD, and Delori, FC, 1984a. The macular pigment.2. Spatial distribution in primate retinas. *Invest Ophthalmol Vis Sci* 25, 674–685.

Snodderly, DM, Brown, PK, Delori, FC, and Auran, JD, 1984b. The macular pigment.1. Absorbance spectra, localization, and discrimination from other yellow pigments in primate retinas. *Invest Ophthalmol Vis Sci* 25, 660–673.

Sommerburg, O, Siems, WG, Hurst, JS, Lewis, JW, Kliger, DS, and van Kuijk, F, 1999. Lutein and zeaxanthin are associated with photoreceptors in the human retina. *Curr Eye Res* 19, 491–495.

Sommerburg, O, Langhans, CD, Arnhold, J, Leichsenring, M, Salerno, C, Crifo, C, Hoffmann, GF, Debatin, KM, and Siems, WG, 2003. Beta-carotene cleavage products after oxidation mediated by hypochlorous acid—A model for neutrophil-derived degradation. *Free Radic Biol Med* 35, 1480–1490.

Sperduto, RD, 1993. Antioxidant status and neovascular age-related macular degeneration. *Arch Ophthalmol* 111, 104–109.

Stahl, W, Junghans, A, de Boer, B, Driomina, ES, Briviba, K, and Sies, H, 1998. Carotenoid mixtures protect multilamellar liposomes against oxidative damage: Synergistic effects of lycopene and lutein. *FEBS Lett* 427, 305–308.

Stahl, W and Sies, H, 2005. Bioactivity and protective effects of natural carotenoids. *Biochim Biophys Acta-Mol Basis Dis* 1740, 101–107.

Stratton, SP, Schaefer, WH, and Liebler, DC, 1993. Isolation and identification of singlet oxygen oxidation-products of beta-carotene. *Chem Res Toxicol* 6, 542–547.

Strauss, O, 2005. The retinal pigment epithelium in visual function. *Physiol Rev* 85, 845–881.

Sumantran, VN, Zhang, R, Lee, DS, and Wicha, MS, 2000. Differential regulation of apoptosis in normal versus transformed mammary epithelium by lutein and retinoic acid. *Cancer Epidemiol Biomarkers Prev* 9, 257–263.

Sun, Z and Yao, H, 2007. The influence of di-acetylation of the hydroxyl groups on the anti-tumor-proliferation activity of lutein and zeaxanthin. *Asia Pac J Clin Nutr* 16(Suppl 1), 447–452.

Sundelin, SP and Nilsson, SE, 2001. Lipofuscin-formation in retinal pigment epithelial cells is reduced by antioxidants. *Free Radic Biol Med* 31, 217–225.

Suuronen, T, Nuutinen, T, Ryhanen, T, Kaarniranta, K, and Salminen, A, 2007. Epigenetic regulation of clusterin/apolipoprotein J expression in retinal pigment epithelial cells. *Biochem Biophys Res Commun* 357, 397–401.

Swaroop, A, Branham, KEH, Chen, W, and Abecasis, G, 2007. Genetic susceptibility to age-related macular degeneration: A paradigm for dissecting complex disease traits. *Hum Mol Genet* 16, R174–R182.

Tan, JS, Wang, JJ, Flood, V, Rochtchina, E, Smith, W, and Mitchell, P, 2008. Dietary antioxidants and the long-term incidence of age-related macular degeneration: The Blue Mountains Eye Study. *Ophthalmology* 115, 334–341.

Tangirala, RK, Pratico, D, FitzGerald, GA, Chun, S, Tsukamoto, K, Maugeais, C, Usher, DC, Pure, E, and Rader, DJ, 2001. Reduction of isoprostanes and regression of advanced atherosclerosis by apolipoprotein E. *J Biol Chem* 276, 261–266.

Tapiero, H, Townsend, DM, and Tew, KD, 2004. The role of carotenoids in the prevention of human pathologies. *Biomed Pharmacother* 58, 100–110.

Thakkinstian, A, Bowe, S, McEvoy, M, Smith, W, and Attia, J, 2006. Association between apolipoprotein E polymorphisms and age-related macular degeneration: A HuGE review and meta-analysis. *Am J Epidemiol* 164, 813–822.

Thornton, J, Edwards, R, Mitchell, P, Harrison, RA, Buchan, I, and Kelly, SP, 2005. Smoking and age-related macular degeneration: A review of association. *Eye* 19, 935–944.

Tomany, SC, Wang, HJ, van Leeuwen, R, Klein, R, Mitchell, P, Vingerling, JR, Klein, BEK, Smith, W, and de Jong, P, 2004. Risk factors for incident age-related macular degeneration—Pooled findings from 3 continents. *Ophthalmology* 111, 1280–1287.

Travis, GH, Golczak, M, Moise, AR, and Palczewski, K, 2007. Diseases caused by defects in the visual cycle: Retinoids as potential therapeutic agents. *Annu Rev Pharmacol Toxicol* 47, 469–512.

Trieschmann, M, Beatty, S, Nolan, JM, Hense, HW, Heimes, B, Austermann, U, Fobker, M, and Pauleikhoff, D, 2007. Changes in macular pigment optical density and serum concentrations of its constituent carotenoids following supplemental lutein and zeaxanthin: The LUNA study. *Exp Eye Res* 84, 718–728.

Trumbo, PR and Ellwood, KC, 2006. Lutein and zeaxanthin intakes and risk of age-related macular degeneration and cataracts: An evaluation using the Food and Drug Administration's evidence-based review system for health claims. *Am J Clin Nutr* 84, 971–974.

Tserentsoodol, N, Gordiyenko, NV, Pascual, I, Lee, JW, Fliesler, SJ, and Rodriguez, IR, 2006a. Intraretinal lipid transport is dependent on high density lipoprotein-like particles and class B scavenger receptors. *Mol Vis* 12, 1319–1333.

Tserentsoodol, N, Sztein, J, Campos, M, Gordiyenko, NV, Fariss, RN, Lee, JW, Fliesler, SJ, and Rodriguez, IR, 2006b. Uptake of cholesterol by the retina occurs primarily via a low density lipoprotein receptor-mediated process. *Mol Vis* 12, 1306–1318.

Uittenbogaard, A, Everson, WV, Matveev, SV, and Smart, EJ, 2002. Cholesteryl ester is transported from caveolae to internal membranes as part of a caveolin-annexin II lipid-protein complex. *J Biol Chem* 277, 4925–4931.

van Bennekum, A, Werder, M, Thuahnai, ST, Han, CH, Duong, P, Williams, DL, Wettstein, P, Schulthess, G, Phillips, MC, and Hauser, H, 2005. Class B scavenger receptor-mediated intestinal absorption of dietary ss-carotene and cholesterol. *Biochemistry* 44, 4517–4525.

van de Kerkhof, EG, de Graaf, IAM, and Groothuis, GMM, 2007. In vitro methods to study intestinal drug metabolism. *Curr Drug Metab* 8, 658–675.

van Leeuwen, R, Klaver, CC, Vingerling, JR, Hofman, A, and de Jong, PT, 2003. Epidemiology of age-related maculopathy: A review. *Eur J Epidemiol* 18, 845–854.

van Soest, SS, de Wit, GM, Essing, AH, ten Brink, JB, Kamphuis, W, de Jong, PT, and Bergen, AA, 2007. Comparison of human retinal pigment epithelium gene expression in macula and periphery highlights potential topographic differences in Bruch's membrane. *Mol Vis* 13, 1608–1617.

Venkateswaran, A, Laffitte, BA, Joseph, SB, Mak, PA, Wilpitz, DC, Edwards, PA, and Tontonoz, P, 2000. Control of cellular cholesterol efflux by the nuclear oxysterol receptor LXR alpha. *Proc Natl Acad Sci U S A* 97, 12097–12102.

Von Eckardstein, A, Langer, C, Engel, T, Schaukal, I, Cignarella, A, Reinhardt, J, Lorkowski, S, Li, Z, Zhou, X, Cullen, P, and Assmann, G, 2001. ATP binding cassette transporter ABCA1 modulates the secretion of apolipoprotein E from human monocyte-derived macrophages. *FASEB J* 15, 1555–1561.

von Recum, HA, Okano, T, Kim, SW, and Bernstein, PS, 1999. Maintenance of retinoid metabolism in human retinal pigment epithelium cell culture. *Exp Eye Res* 69, 97–107.

Vu, HT, Robman, L, McCarty, CA, Taylor, HR, and Hodge, A, 2006. Does dietary lutein and zeaxanthin increase the risk of age related macular degeneration? The Melbourne Visual Impairment Project. *Br J Ophthalmol* 90, 389–390.

Walston, J, Xue, Q, Semba, RD, Ferrucci, L, Cappola, AR, Ricks, M, Guralnik, J, and Fried, LP, 2006. Serum antioxidants, inflammation, and total mortality in older women. *Am J Epidemiol* 163, 18–26.

Wang, N and Anderson, RE, 1993. Transport of 22:6n-3 in the plasma and uptake into retinal pigment epithelium and retina. *Exp Eye Res* 57, 225–233.

Wang, W, Connor, SL, Johnson, EJ, Klein, ML, Hughes, S, and Connor, WE, 2007. Effect of dietary lutein and zeaxanthin on plasma carotenoids and their transport in lipoproteins in age-related macular degeneration. *Am J Clin Nutr* 85, 762–769.

Webb, NR, Connell, PM, Graf, GA, Smart, EJ, de Villiers, WJ, de Beer, FC, and van der Westhuyzen, DR, 1998. SR-BII, an isoform of the scavenger receptor BI containing an alternate cytoplasmic tail, mediates lipid transfer between high density lipoprotein and cells. *J Biol Chem* 273, 15241–15248.

Weng, J, Mata, NL, Azarian, SM, Tzekov, RT, Birch, DG, and Travis, GH, 1999. Insights into the function of Rim protein in photoreceptors and etiology of Stargardt's disease from the phenotype in abcr knockout mice. *Cell* 98, 13–23.

Wenzel, JJ, Kaminski, WE, Piehler, A, Heimerl, S, Langmann, T, and Schmitz, G, 2003. ABCA10, a novel cholesterol-regulated ABCA6-like ABC transporter. *Biochem Biophys Res Commun* 306, 1089–1098.

Werner, JS, Donnelly, SK, and Kliegl, R, 1987. Aging and human macular pigment density. *Vision Res* 27, 257–268.

West, AL, Oren, GA, and Moroi, SE, 2006. Evidence for the use of nutritional supplements and herbal medicines in common eye diseases. *Am J Ophthalmol* 141, 157–166.

Wiggs, JL, 2006. Complement factor H and macular degeneration—The genome yields an important clue. *Arch Ophthalmol* 124, 577–578.

Wong, P, Pfeffer, BA, Bernstein, SL, Chambers, ML, Chader, GJ, Zakeri, ZF, Wu, YQ, Wilson, MR, and Becerra, SP, 2000. Clusterin protein diversity in the primate eye. *Mol Vis* 6, 184–191.

Wong, P, Ulyanova, T, Organisciak, DT, Bennett, S, Lakins, J, Arnold, JM, Kutty, RK, Tenniswood, M, vanVeen, T, Darrow, RM, and Chader, G, 2001. Expression of multiple forms of clusterin during light-induced retinal degeneration. *Curr Eye Res* 23, 157–165.

Wong, RW, Richa, DC, Hahn, P, Green, WR, and Dunaief, JL, 2007. Iron toxicity as a potential factor in AMD. *Retina* 27, 997–1003.

Wooten, BR and Hammond, BR, 2002. Macular pigment: influences on visual acuity and visibility. *Prog Retin Eye Res* 21, 225–240.

Wrona, M, Korytowski, W, Rozanowska, M, Sarna, T, and Truscott, TG, 2003. Cooperation of antioxidants in protection against photosensitized oxidation. *Free Radic Biol Med* 35, 1319–1329.

Wrona, M, Rozanowska, M, and Sarna, T, 2004. Zeaxanthin in combination with ascorbic acid or alpha-tocopherol protects ARPE-19 cells against photosensitized peroxidation of lipids. *Free Radic Biol Med* 36, 1094–1101.

Yemelyanov, AY, Katz, NB, and Bernstein, PS, 2001. Ligand-binding characterization of xanthophyll carotenoids to solubilized membrane proteins derived from human retina. *Exp Eye Res* 72, 381–392.

Yeoh, J, Sims, J, and Guymer, RH, 2006. A review of drug options in age-related macular degeneration therapy and potential new agents. *Expert Opin Pharmacother* 7, 2355–2368.

Yonekura, L, Tsuzuki, W, and Nagao, A, 2006. Acyl moieties modulate the effects of phospholipids on beta-carotene uptake by Caco-2 cells. *Lipids* 41, 629–636.

Yonekura, L and Nagao, A, 2007. Intestinal absorption of dietary carotenoids. *Mol Nutr Food Res* 51, 107–115.

Young, AJ and Lowe, GM, 2001. Antioxidant and prooxidant properties of carotenoids. *Arch Biochem Biophys* 385, 20–27.

Zannis, VI, Chroni, A, and Krieger, M, 2006. Role of apoA-I, ABCA1, LCAT, and SR-BI in the biogenesis of HDL. *J Mol Med* 84, 276–294.

Zarbin, MA, 2004. Current concepts in the pathogenesis of age-related macular degeneration. *Arch Ophthalmol* 122, 598–614.

16 The Carotenoids of Macular Pigment and Bisretinoid Lipofuscin Precursors in Photoreceptor Outer Segments

Janet R. Sparrow and So Ra Kim

CONTENTS

16.1 INTRODUCTION

The oxygen atom–containing carotenoids (xanthophylls), zeaxanthin and lutein, Figure 16.1, are obtained by humans through the dietary intake of fruits and vegetables and become incorporated in the retina as macroscopically visible macular pigment. These yellow-colored pigments are particularly abundant in the fovea, their concentration declining steeply toward the peripheral retina. Of the two carotenoids, zeaxanthin is more concentrated in the central 10° of the retina while lutein dominates at eccentricities greater than 35° (Bone et al., 1988; Snodderly et al., 1991). The specificity of this distribution indicates the selective uptake of macular pigments by specific binding proteins (Bhosale et al., 2004). Nevertheless, the concentration of lutein and zeaxanthin in the macula varies among individuals (Bone et al., 1997) and it is likely that the extent of oral intake is responsible for these differences (Hammond et al., 1997; Landrum et al., 1997). Indeed, the long-term intake of the dietary supplements of lutein increases the levels of macular pigment (Bhosale et al., 2007). The highest levels of lutein and zeaxanthin are present in photoreceptor cell axonal processes (Henle's fibers) (Snodderly et al., 1984) but 25% of total retinal carotenoids are present within photoreceptor outer segments (Rapp et al., 2000; Sommerburg et al., 1999). Given their hydrophobicity, lutein and zeaxanthin readily integrate into the lipophilic compartment of cell membranes (Landrum and Bone, 2001).

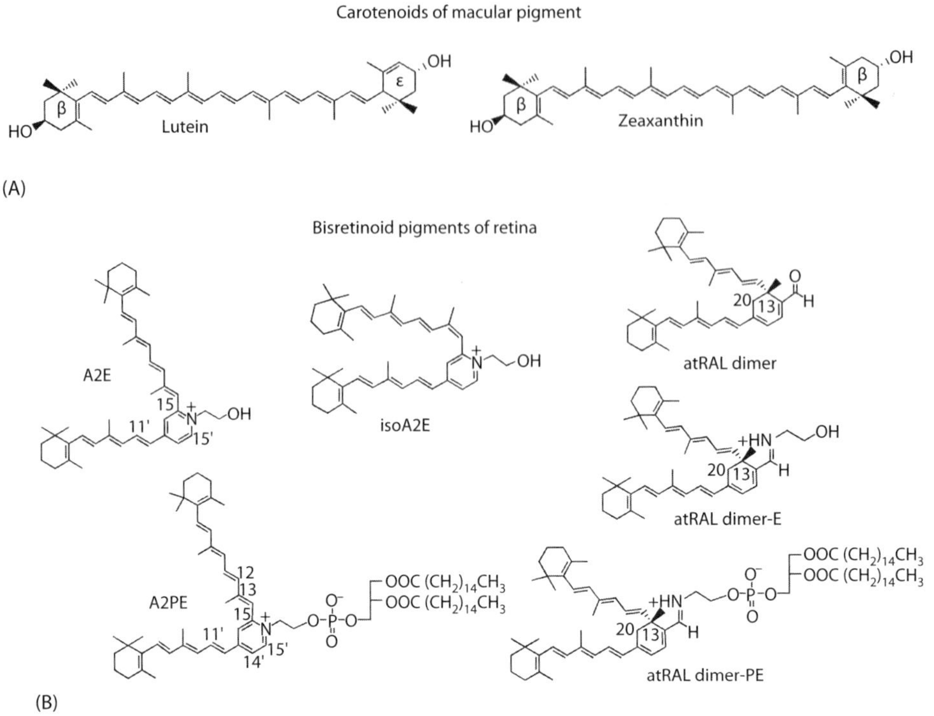

FIGURE 16.1 Carotenoids of macular pigment and bisretinoid pigments of retina. (A) Structures of the carotenoids lutein and zeaxanthin. Due to positioning of the double bonds in the ionone rings, zeaxanthin has 11 conjugated double bonds and lutein has 10. As compared to purely hydrocarbon carotenoids such as β-carotene, hydroxyl (OH) substitution within the ionone rings of lutein and zeaxanthin (xanthophylls class) confer greater polarity. (B) The known bisretinoid pigments include A2E, isoA2E, other *cis*-isomers (not shown) and their precursor A2PE and the atRAL dimer series of pigments, unconjugated atRAL dimer, atRAL dimer-PE, and atRAL dimer-E. All of these molecules have a polyene structure consisting of a conjugated system of alternating double and single carbon–carbon bonds with methyl groups (CH_3) attached to the carbon backbone as side groups and with additional conjugation into ionone rings situated at each end of the carbon chains. A2E is generated from A2PE and atRAL dimer-E is generated form atRAL dimer-PE, by enzyme-mediated phosphate hydrolysis. Both A2E and A2PE have a unique pyridinium ring from which two retinoid-derived side-arms extend, with six double-bond conjugations on the long arm and five on the short arm. The pigments atRAL dimer-E and atRAL dimer-PE are protonated Schiff base conjugates with seven double-bond conjugations on the long arm and four on the short arm.

16.2 LIGHT FILTERING AND ANTIOXIDANT PROPERTIES OF MACULAR PIGMENT

The macular pigment (λ_{max} ~ 450) that is present in Henle's fibers is in a position to filter short wavelength visible light, a function that serves to decrease chromatic aberration and scatter in the foveal image (Reading and Weale, 1974). Indeed, due to their broad band absorbance centered around 450 nm (Junghans et al., 2001), both lutein and zeaxanthin can attenuate short wavelength light. It is suggested that this function is aided by hydrophilic substituents (OH) on the ionone rings that allow lutein and zeaxanthin to assume appropriate positions by forming hydrogen bonds with polar head groups at the membrane surface (Krinsky, 2002; Sujak et al., 2002). Studies utilizing carotenoid-containing unilamellar liposomes indicate that the filtering efficiency of lutein may be better than that of zeaxanthin, β-carotene, and lycopene (Junghans et al., 2001; Sujak et al., 2002).

The antioxidant properties of carotenoids are facilitated by the presence of multiple, closely spaced energy levels between the excited states and the ground state of the molecules. Thus, triplet–triplet energy transfer from photosensitizer to carotenoid occurs, thereby preventing energy transfer from photosensitizer to oxygen and the formation of singlet oxygen (Martin et al., 1999). Carotenoids are also the excellent physical quenchers of singlet oxygen, the energy of singlet molecular oxygen being transferred to the carotenoid molecule to yield ground state oxygen, and a triplet-excited carotenoid. Energy transfer in this way is possible because the triplet energy level of the carotenoid is lower than the energy level of singlet oxygen.

16.3 PHOTOREACTIVE BISRETINOID COMPOUNDS IN PHOTORECEPTOR OUTER SEGMENTS

Given the presence of lutein and zeaxanthin in photoreceptor outer segments and because of the known antioxidant properties of carotenoids (Khachik et al., 1997; Landrum and Bone, 2001), questions arise as to whether there are specific molecules in photoreceptor outer segments toward which carotenoid protection might be directed. Photoreactive molecules that could be important in this regard are the bisretinoid molecules that form in photoreceptor outer segments as the precursors of retinal pigment epithelial (RPE) lipofuscin (Figure 16.1). One of these molecules is A2PE (Figure 16.1), a phosphatidyl-pyridinium bisretinoid pigment that forms through a biogenic cascade (Liu et al., 2000; Parish et al., 1998) that involves reactions between the membrane phospholipid phosphatidylethanolamine (PE) and all-*trans*-retinal, the latter being generated upon the photoisomerization of 11-*cis*-retinal (Figure 16.2). Intermediates in biogenic pathway include

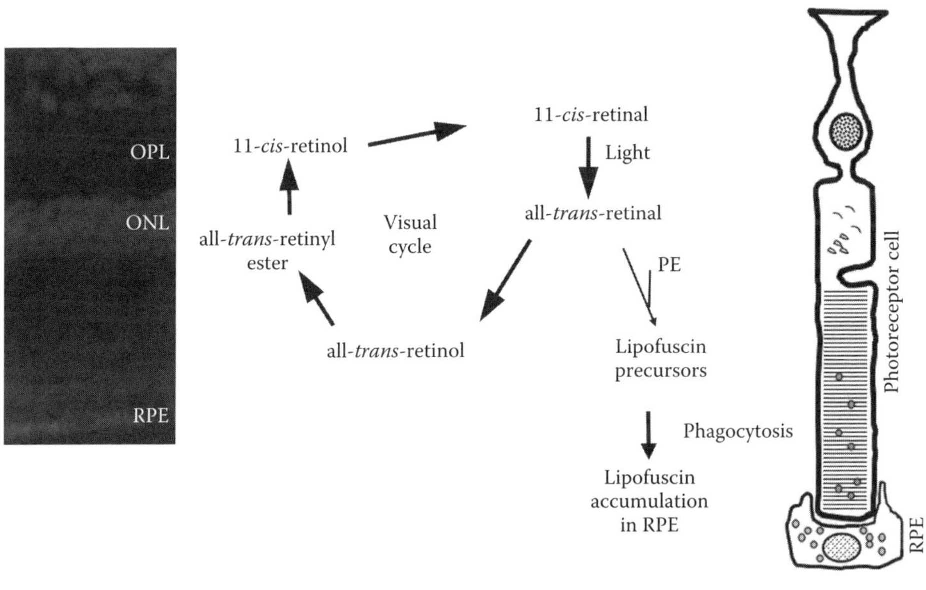

FIGURE 16.2 (See color insert following page 336.) Intersection of the visual (retinoid) cycle and pathway for RPE lipofuscin formation. The photoisomerization of 11-*cis*-retinal leads to the release of all-*trans*-retinal from rhodopsin. All-*trans*-retinal for the most part is reduced to all-*trans*-retinol and remains in the visual cycle for reconversion to 11-*cis*-retinal. All-*trans*-retinal can also react inaptly, leave the visual cycle, and forming lipofuscin precursors. At least some of the reactions leading to the lipofuscin pathway are between all-*trans*-retinal and PE in a 2:1 ratio. After the phagocytosis of shed outer segment membrane by RPE, lipofuscin accumulates in the latter cells. RPE lipofuscin detected as autofluorescence in monkey retina imaged by fluorescence microscopy (left). Nuclei are stained with DAPI. The autofluorescence adjacent to the RPE is at the level of outer segments and is likely attributable to lipofuscin precursors that form in outer segments. Outer nuclear layer (ONL); outer plexiform layer (OPL).

the Schiff base conjugate, *N*-retinylidene-phosphatidyl-ethanolamine (NRPE), and a phosphatidyl dihydropyridinium molecule (dihydro-A2PE) that undergoes automatic oxidative aromatization to yield A2PE (Figure 16.3) (Kim et al., 2007a; Liu et al., 2000; Parish et al., 1998). By fast atom bombardment tandem mass spectrometry with collision-induced dissociation mass spectrometric analysis, the structure of A2PE was confirmed, the detection of the permanent positive charge of the quaternary nitrogen of A2PE by positive ionization was definitive for identification (Liu et al., 2000). Furthermore, it was shown that A2PE is the immediate precursor of A2E since phosphate cleavage by the lysosomal enzyme phospholipase D yields peaks in the HPLC profile that can be identified as A2E and isoA2E on the basis of UV-visible absorbance and retention time (Ben-Shabat et al., 2002; Liu et al., 2000). Several lines of investigation also demonstrated that A2PE forms within photoreceptor outer segments. For instance, A2PE was generated in isolated outer segments irradiated to release endogenous all-*trans*-retinal or outer segments incubated with exogenous all-*trans*-retinal (Ben-Shabat et al., 2002; Liu et al., 2000). By mass spectrometric analysis, A2PE was also detected in the orange-colored photoreceptor outer segment debris that accumulates in the subretinal space in Royal College of Surgeon rats due to the failure of RPE cells to phagocytose shed outer segment membrane (Ben-Shabat et al., 2002) (Figure 16.4). An autofluorescent material that could be attributed to lipofuscin precursors such as A2PE, has also been described within the

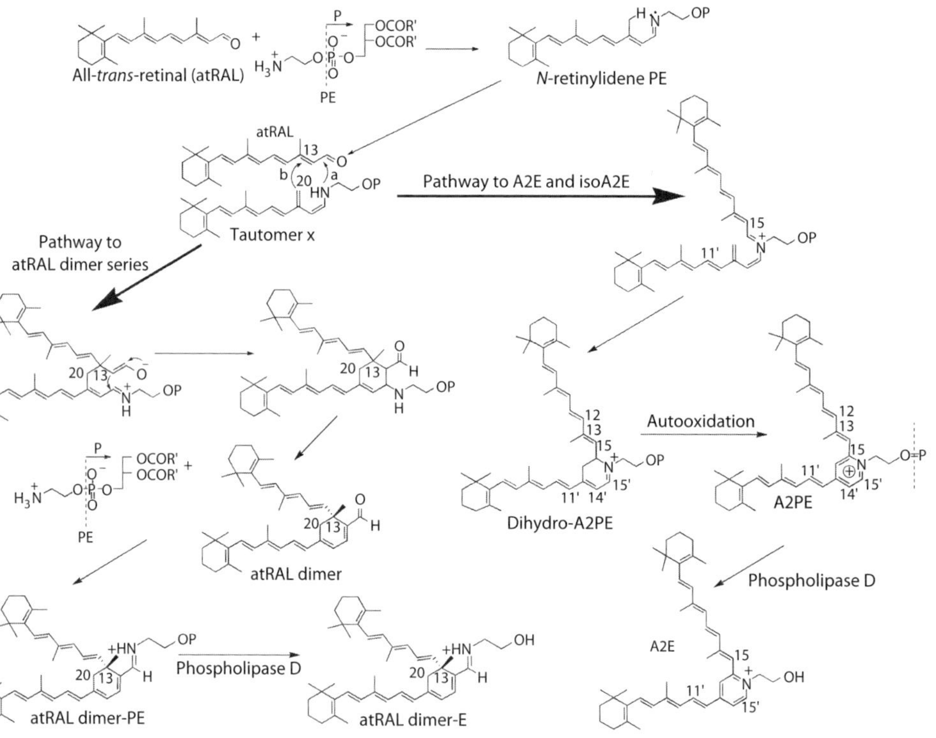

FIGURE 16.3 Proposed pathways for biosynthesis of A2E/isoA2E and pigments of the atRAL dimer series. All-*trans*-retinal that is released from opsin when 11-*cis*-retinal photoisomerizes reacts with PE to generate the Schiff base NRPE. The pathway to lipofuscin formation continues with the reaction of a second molecule of all-*trans*-retinal. Both atRAL dimer and A2PE may form from the same tautomer X. A2PE is produced via the intermediate dihydro-A2PE that undergoes automatic oxidation. Within RPE cell lysosomes, A2PE undergoes phosphate cleavage to release A2E. Phospholipase D can mediate this hydrolysis. atRAL dimer reacts with PE to form the protonated Schiff base conjugate atRAL dimer-PE; atRAL dimer-E forms from the enzyme-mediated hydrolysis of atRAL dimer-PE.

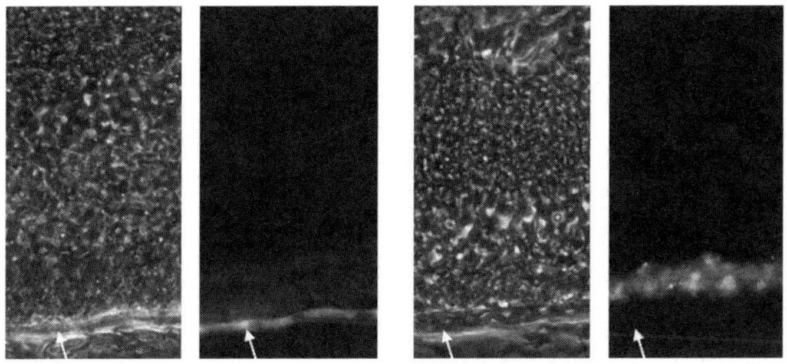

FIGURE 16.4 Precursors of RPE cell lipofuscin form in the outer segments of photoreceptor cells. The retina of normal rat (A, B) and the Royal College of Surgeons (RCS) rat (C, D) viewed under the phase contrast (A, C) and the epifluorescence microscopy (B, D). In the normal rat, autofluorescent material accumulates as lipofuscin in RPE cells (arrows). In the RCS, due to a defect in RPE cell phagocytosis, shed outer segment membrane builds up at the photoreceptor-RPE interface; the autofluorescence in this debris is attributable to lipofuscin precursors that form in photoreceptor outer segments.

photoreceptor cell membrane in patients with Stargardt disease and retinitis pigmentosa (Birnbach et al., 1994; Bunt-Milam et al., 1983; Szamier and Berson, 1977). A2PE is not the only lipofuscin precursor in photoreceptor outer segments; however, since molecules of all-*trans*-retinal also condense to form unconjugated (all-*trans*-retinal dimer, atRAL dimer) and conjugated (all-*trans*-retinal dimer-phosphatidylethanolamine [atRAL dimer-PE] and all-*trans*-retinal dimer-ethanolamine [atRAL dimer-E]) forms of the atRAL dimer series of lipofuscin pigments that are deposited in RPE cells with outer segment phagocytosis (Fishkin et al., 2005; Kim et al., 2007b).

16.4 LUTEIN AND ZEAXANTHIN ATTENUATE A2PE PHOTOOXIDATION

When A2PE is irradiated at 430 nm, examined by fast atom bombardment ionization mass spectrometry (FAB-MS) and compared to the unirradiated sample, molecular ion peaks greater than the mass-to-charge ratio (*m*/*z* 1223 for dipalmitoyl-A2PE) of A2PE are generated (Figure 16.5A). Each of these higher *m*/*z* peaks (*m*/*z* 1239, 1255, 1271, 1287) differs from its neighbors by mass 16, the series of *m*/*z* peaks representing the sequential addition of oxygen at carbon–carbon double bonds of the retinoid-derived side-arms of A2PE (photooxidation). With analysis by reverse phase HPLC, the irradiation of A2PE also causes a pronounced decrease in the absorbance of the A2PE peak due to the loss of A2PE as photooxidation proceeds. However, when A2PE is irradiated in the presence of lutein or zeaxanthin, the FAB-MS data demonstrate that A2PE photooxidation is inhibited (Figure 16.5A). Specifically, we found that molecular ion peaks at *m*/*z* 1271 and 1287 were absent and the *m*/*z* peaks at 1239 and 1255 exhibited reduced intensity. Moreover, analysis by reverse phase HPLC, showed that the consumption of A2PE that accompanies photooxidation, was reduced (Figure 16.5B). The inhibition afforded by zeaxanthin was also more pronounced than that associated with lutein and for both lutein and zeaxanthin, the effects were greater than with α-tocopherol.

16.5 ZEAXANTHIN AND LUTEIN QUENCH SINGLET OXYGEN

The evidence that lutein and zeaxanthin attenuate the photooxidation of A2PE by quenching singlet oxygen, came from studies utilizing an aromatic compound (endoperoxide of 1,4-dimethyl naphthalene) that decomposes to release singlet oxygen (Turro et al., 1981). For these experiments A2E were used, since its polyene structure and singlet oxygen quenching ability is comparable to A2PE

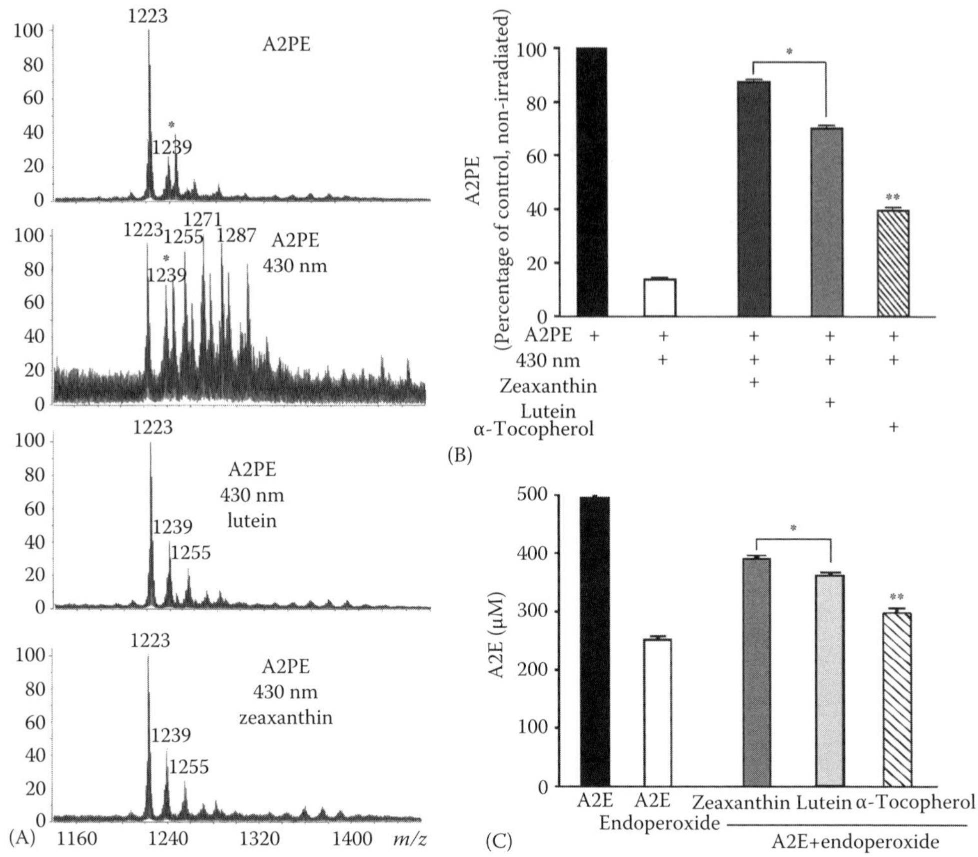

FIGURE 16.5 Photooxidation of A2PE and is decreased by lutein and zeaxanthin. The singlet oxygen quenching activity of lutein and zeaxanthin. (A) FAB-MS of nonirradiated A2PE, A2PE irradiated at 430 nm, and A2PE illuminated at 430 nm in the presence of lutein or zeaxanthin. The molecular ion peak at mass-to-charge (*m/z*) ratio 1223 corresponds to the molecular mass of A2PE. A2PE photooxidation is reflected by the presence of additional higher molecular weight peaks for example, *m/z* 1239, 1255, 1271, 1287 in the irradiated samples. Illumination in the presence of lutein and zeaxanthin reduces the formation of these photooxidation products. *, matrix peak at *m/z* 1245 {M + Na}. (B) Lutein and zeaxanthin protect against A2PE photooxidation. A2PE (200 μM) with and without lutein and zeaxanthin (200 μM) or α-tocopherol (200 μM) was irradiated at 430 nm. A2PE was quantified by reverse-phase HPLC and integrated peak areas were normalized to an external standard of A2E. The loss of A2PE after 430 nm irradiation is indicative of A2PE photooxidation; the attenuation of this loss in the presence of lutein and zeaxanthin indicates protection against A2PE photooxidation. +, the presence of compound/irradiation. Values are mean ± SD of 4–7 experiments, 3 replicates per experiment. $^{*}p < 0.01$, for the indicated comparison; $^{**}p < 0.001$ for α-tocopherol versus zeaxanthin and lutein values; ANOVA followed by the Newman Keul Multiple Comparison test. (C) Zeaxanthin, lutein, and α-tocopherol quench singlet oxygen thus reducing A2E oxidation. The consumption of A2E that accompanies oxidation was quantified by HPLC after A2E (500 μM) was exposed to singlet oxygen generated from the endoperoxide of 1,4-dimethyl-naphthalene (1 mM) in the absence and the presence of zeaxanthin, lutein, and α-tocopherol (1 mM). Bar height in the presence of antioxidant is positively correlated with quenching ability. Values are mean ± SD of 3 experiments. $^{*}p < 0.05$, zeaxanthin versus lutein; $^{**}p < 0.01$ α-tocopherol versus lutein and zeaxanthin. ANOVA followed by the Newman Keul Multiple Comparison test.

(Figure 16.5). By analyzing peak areas in reverse phase HPLC chromatograms to quantify the loss of A2E as it reacts with singlet oxygen, it was shown that both zeaxanthin and lutein were able to compete with A2E for the quenching of singlet oxygen. Again, zeaxanthin was a more efficient quencher than lutein and both more effectively protected A2E than did α-tocopherol. The ability of zeaxanthin to serve as a better quencher of singlet oxygen under specified conditions, is significant since similar differences in the quenching activity of lutein and zeaxanthin have been previously reported (Mascio et al., 1991). The quenching activity of carotenoids is principally dependent on the number of conjugate double bonds in the molecule (Stahl et al., 1997). Thus, it is of interest that zeaxanthin has 11 conjugated double bonds (9 conjugated double bonds in the polyene chain and 2 double bonds of the β-ionone rings) (Figure 16.1), while lutein has 10 conjugated double bonds (9 conjugated double bonds in the polyene chain and 1 double bond in the β-ionone ring). Perhaps, it is also significant that of lutein and zeaxanthin, the latter is present at higher concentration in the fovea, an area that is resistant to degenerative changes in early AMD (foveal sparing) and in retinal degenerations associated with photosensitizing drugs (Bull's eye maculopathy) (Weiter et al., 1988).

The quenching of singlet oxygen by carotenoids is known to occur, for the most part, by direct energy transfer between the molecules (physical quenching) with chemical reaction between the singlet oxygen and the carotenoid molecule (chemical quenching) accounting for only a minor portion of the overall quenching rate (Stahl and Sies, 2003). Consistent with this, it was found that under conditions of 430 nm irradiation, the addition of lutein or zeaxanthin protected against A2E/A2PE photooxidation without any evidence of oxidation of the carotenoids. Similarly, analysis by both HPLC and FAB-MS to compare A2E, lutein, and zeaxanthin in terms of susceptibility to oxidation in the presence of the singlet oxygen generator 1,4-dimethyl naphthalene or MCPBA (*meta*-chloroperoxybenzoic acid), a strong oxidizing agent, showed that little carotenoid was consumed relative to A2E. For antioxidant functioning, this feature could be important, as it would allow lutein and zeaxanthin to participate in multiple quenching cycles with only slow turnover.

16.6 STRUCTURAL FEATURES OF ZEAXANTHIN AND LUTEIN VERSUS A2PE

Despite the structural similarities between lutein and zeaxanthin on the one hand and A2E/A2PE on the other (Figure 16.1), lutein and zeaxanthin quench singlet oxygen by physical mechanisms while A2E/A2PE quench singlet oxygen by chemical reaction. An important structural feature of carotenoids is the extended conjugation system (Figure 16.1) along which π-electrons become delocalized. This arrangement may reduce the electron density such that the polyene chain is less likely to be subject to electrophilic chemical reactions involving singlet oxygen. By contrast, A2PE/A2E have separate conjugated systems of double bonds extending along each of two side-arms: a long arm extends *ortho* to the pyridinium nitrogen and a short arm extends *para* to the pyridinium nitrogen (Figure 16.1). It is evident that the side-arms constitute separate conjugation systems since each side-arm generates a separate absorbance peak: the absorbance generated by the short arm has λ_{max} ~337 nm while the absorbance of the long arm exhibits a λ_{max} of ~439 (Jang et al., 2005). Thus, as compared to lutein and zeaxanthin, the polyene side-arms of A2PE/A2E with only three and four conjugated double bonds, constitute more electron-rich systems that are highly susceptible to reaction with electrophilic singlet oxygen. The importance of the extent of conjugation is further illustrated by the observation that the shorter arm of A2E is even more susceptible to the electrophilic attack than the longer arm (Jang et al., 2005).

16.7 SUMMARY

The molecules responsible for light damage to photoreceptors have not been identified but light damage is known to be dependent on the presence of 11-*cis*-retinal, the chromophore of rods, and cones (Grimm et al., 2001; Wenzel et al., 2005). Exposure to blue light (430 nm) is also more damaging than exposure to green light (550 nm), even when light at these wavelengths is delivered at the same luminosity. The experiments presented here demonstrate that A2PE, which forms subsequent to 11-*cis*-retinal photoisomerization and release of all-*trans*-retinal, absorbs maximally in the short wavelength region of the spectrum (λ_{max} ~ 450 nm) and can serve as a photosensitizer. Through absorption of high-energy, short-wavelength light, lutein and zeaxanthin may reduce the amount of light in the blue region of the spectrum that reaches the photosensitizers that are responsible for light damage to the retina (Bone et al., 1997). More directly, these carotenoids may also serve as antioxidants (Khachik et al., 1997).

ACKNOWLEDGMENTS

This work was supported by NEI grant EY12951, the Kaplen Fund, and unrestricted funds to the Department of Ophthalmology, Columbia University from Research to Prevent Blindness.

REFERENCES

Ben-Shabat, S., Parish, C.A., Vollmer, H.R., Itagaki, Y., Fishkin, N., Nakanishi, K., Sparrow, J.R., 2002. Biosynthetic studies of A2E, a major fluorophore of RPE lipofuscin. *J Biol Chem*. 277, 7183–7190.

Bhosale, P., Larson, A.J., Frederick, J.M., Southwick, K., Thulin, C.D., Bernstein, P.S., 2004. Identification and characterization of a Pi isoform of glutathioine S-transferase (GSTP1) as a zeaxanthin-binding protein in the macula of the human eye. *J Biol Chem*. 279, 49447–49454.

Bhosale, P., Zhao da, Y., Bernstein, P.S., 2007. HPLC measurement of ocular carotenoid levels in human donor eyes in the lutein supplementation era. *Invest Ophthalmol Vis Sci*. 48, 543–549.

Birnbach, C.D., Jarvelainen, M., Possin, D.E., Milam, A.H., 1994. Histopathology and immunocytochemistry of the neurosensory retina in fundus flavimaculatus. *Ophthalmology*. 101, 1211–1219.

Bone, R.A., Landrum, J.T., Friedes, L.M., Gomez, C.M., Kilburn, M.D., Menendez, E., Vidal, I., Wang, W., 1997. Distribution of lutein and zeaxanthin stereoisomers in the human retina. *Exp Eye Res*. 64, 211–218.

Bone, R.A., Landrum, J.T., Fernandez, L., Tarsis, S.L., 1988. Analysis of the macular pigment by HPLC: Retinal distribution and age study. *Invest Ophthalmol Vis Sci*. 29, 843–849.

Bunt-Milam, A.H., Kalina, R.E., Pagon, R.A., 1983. Clinical–ultrastructural study of a retinal dystrophy. *Invest Ophthalmol Vis Sci*. 24, 458–469.

Fishkin, N., Sparrow, J.R., Allikmets, R., Nakanishi, K., 2005. Isolation and characterization of a retinal pigment epithelial cell fluorophore: An all-*trans*-retinal dimer conjugate. *Proc Natl Acad Sci U S A*. 102, 7091–7096.

Grimm, C., Wenzel, A., Williams, T.P., Rol, P.O., Hafezi, F., Reme, C.E., 2001. Rhodopsin-mediated blue-light damage to the rat retina: Effect of photoreversal of bleaching. *Invest Ophthalmol Vis Sci*. 42, 497–505.

Hammond, B.R., Johnson, E.J., Russell, R.M., Krinsky, N.I., Yeum, K.J., Edwards, R.B., Snodderly, D.M., 1997. Dietary modification of human macular pigment density. *Invest Ophthalmol Vis Sci*. 38, 1795–1801.

Jang, Y.P., Matsuda, H., Itagaki, Y., Nakanishi, K., Sparrow, J.R., 2005. Characterization of peroxy-A2E and furan-A2E photooxidation products and detection in human and mouse retinal pigment epithelial cells lipofuscin. *J Biol Chem*. 280, 39732–39739.

Junghans, A., Sies, H., Stahl, W., 2001. Macular pigments lutein and zeaxanthin as blue light filters studied in liposomes. *Arch Biochem Biophys*. 391, 160–164.

Khachik, F., Bernstein, P.S., Garland, D.L., 1997. Identification of lutein and zeaxanthin oxidation products in human and monkey retinas. *Invest Ophthalmol Vis Sci*. 38,1802–1811.

Kim, S.R., He, J., Yanase, E., Jang, Y.P., Berova, N., Sparrow, J.R., Nakanishi, K., 2007a. Characterization of dihydro-A2PE: An intermediate in the A2E biosynthetic pathway. *Biochemistry* 46, 10122–10129.

Kim, S.R., Jang, Y.P., Jockusch, S., Fishkin, N.E., Turro, N.J., Sparrow, J.R., 2007b. The all-trans-retinal dimer series of lipofuscin pigments in retinal pigment epithelial cells in a recessive Stargardt disease model. *Proc Natl Acad Sci U S A*. 104, 19273–19278.

Krinsky, N.I., 2002. Possible biologic mechanisms for a protective role of zanthophylls. *J Nutr*. 132, 540S–542S.

Landrum, J.T., Bone, R.A., 2001. Lutein, zeaxanthin and the macular pigment. *Arch Biochem Biophys*. 385, 28–40.

Landrum, J.T., Bone, R.A., Joa, H., Kilburn, M.D., Moore, L.L., Sprague, K.E., 1997. A one year study of the macular pigment: The effect of 140 days of a lutein supplement. *Exp Eye Res*. 65, 57–62.

Liu, J., Itagaki, Y., Ben-Shabat, S., Nakanishi, K., Sparrow, J.R., 2000. The biosynthesis of A2E, a fluorophore of aging retina, involves the formation of the precursor, A2-PE, in the photoreceptor outer segment membrane. *J Biol Chem*. 275, 29354–29360.

Martin, H.D., Ruck, C., Schmidt, M., Sell, S., Beutner, S., Mayer, B., Walsh, R., 1999. Chemistry of carotenoid oxidation and free radical reactions. *Pure Appl Chem*. 71, 2253–2262.

Mascio, P.D., Murphy, M.E., Sies, H., 1991. Antioxidant defense systems: The role of carotenoids, tocopherols, and thiols. *Am J Clin Nutr*. 53, 194S–200S.

Parish, C.A., Hashimoto, M., Nakanishi, K., Dillon, J., Sparrow, J.R., 1998. Isolation and one-step preparation of A2E and iso-A2E, fluorophores from human retinal pigment epithelium. *Proc Natl Acad Sci U S A*. 95, 14609–14613.

Rapp, L.M., Maple, S.S., Choi, J.H., 2000. Lutein and zeaxanthin concentrations in rod outer segment membranes from perifoveal and peripheral human retina. *Invest Ophthalmol Vis Sci*. 41, 1200–1209.

Reading, V.M., Weale, R.A., 1974. Macular pigment and chromatic aberration. *J Opt Soc Am*. 64, 231–234.

Snodderly, D.M., Auran, J.D., Delori, F.C., 1984. The macular pigment. II: Spatial distribution in primate retinas. *Invest Ophthalmol Vis Sci*. 25, 674–685.

Snodderly, D.M., Handelman, G.J., Adler, A.J., 1991. Distribution of individual macular pigment carotenoids in central retina of macaque and squirrel monkeys. *Invest Ophthalmol Vis Sci*. 32, 268–279.

Sommerburg, O.G., Siems, W.G., Hurst, J.S., Lewis, J.W., Kliger, D.S., van Kuijk, F.J., 1999. Lutein and zeaxanthin are associated with photoreceptors in the human retina. *Curr Eye Res*. 19, 491–495.

Stahl, W., Sies, H., 2003. Antioxidant activity of carotenoids. *Mol Aspect Med*. 24, 345–351.

Stahl, W., Nicolai, S., Briviba, K., Hanusch, M., Broszeit, G., Peters, M., Martin, H.D., Sies, H., 1997. Biological activities of natural and synthetic carotenoids: Induction of gap junctional communication and singlet oxygen quenching. *Carcinogenesis*. 18, 89–92.

Sujak, A., Mazurek, P., Gruszecki, W.I., 2002. Xanthophyll pigments lutein and zeaxanthin in lipid multilayers formed with dimyristoylphosphatidylcholine. *J Photochem Photobiol B*. 68, 39–44.

Szamier, R.B., Berson, E.L., 1977. Retinal ultrastructure in advanced retinitis pigmentosa. *Invest Ophthalmol Vis Sci*. 16, 947–962.

Turro, N.J., Chow, M.-F., Rigaudyo, J., 1981. Mechanism of thermolysis of endoperoxides of aromatic compounds. Activation parameters, magnetic field, and magnetic isotope effects. *J Am Chem Soc*. 103, 7218–7222.

Weiter, J.J., Delori, F.C., Dorey, C.K., 1988. Central sparing in annular macular degeneration. *Am J Ophthalmol*. 106, 286–290.

Wenzel, A., Grimm, C., Samardzija, M., Reme, C.E., 2005. Molecular mechanisms of light-induced photoreceptor apoptosis and neuroprotection for retinal degeneration. *Prog Ret Eye Res*. 24, 275–306.

Part VI

Cell Culture Methods Applied to Understanding Carotenoid Recognition and Action

17 Mechanisms of Intestinal Absorption of Carotenoids: Insights from *In Vitro* Systems

Earl H. Harrison

CONTENTS

17.1 INTRODUCTION

Carotenoids are synthesized in plants and in certain microorganisms such as some bacteria, algae, and fungi. They are a group of pigments that are widespread in nature and responsible for the yellow/orange/red/purple colors of many fruits, flowers, birds, insects, and marine animals. Over 600 carotenoids have been isolated from natural sources; nearly 60 of them have been detected in the human diet (Mangels et al., 1993) and ~20 of them in human blood and tissues (Parker, 1989). β-Carotene (β-C), α-carotene (α-C), lycopene (LYC), lutein (LUT), and β-cryptoxanthin are the five most prominent carotenoids present in the human body. In the human diet, plant food sources are the major contributors of carotenoids: carrots, squash, and dark-green leafy vegetables for β-C, carrots for α-C, tomatoes, and watermelon for LYC, kale, peas, spinach, and broccoli for LUT, and sweet red peppers, oranges, and papaya for β-cryptoxanthin.

All carotenoids are derived from the basic linear polyisoprenoid structure of LYC that contains 40 carbon atoms and an extended system of 13 conjugated double bonds. Carotenoids are derived from this parent structure by cyclization (i.e., formation of β- or ε-ionone rings) at one (i.e., γ-carotene) or two ends (i.e., β-C and α-C) of the polyene chain and by dehydrogenation and/or oxidation. The structures of several major carotenoids are shown in Figure 17.1. The carotenoid group is divided into the carotenes, hydrocarbon carotenoids with unsubstituted rings, and the xanthophylls, carotenoids with at least one oxygen atom. They exist mostly in the all-*trans* configuration, but they can be subject to a *cis* isomerization at any double bond of their polyene chain, resulting in a large number of mono- and poly-*cis* isomers (in theory) (Britton, 1995).

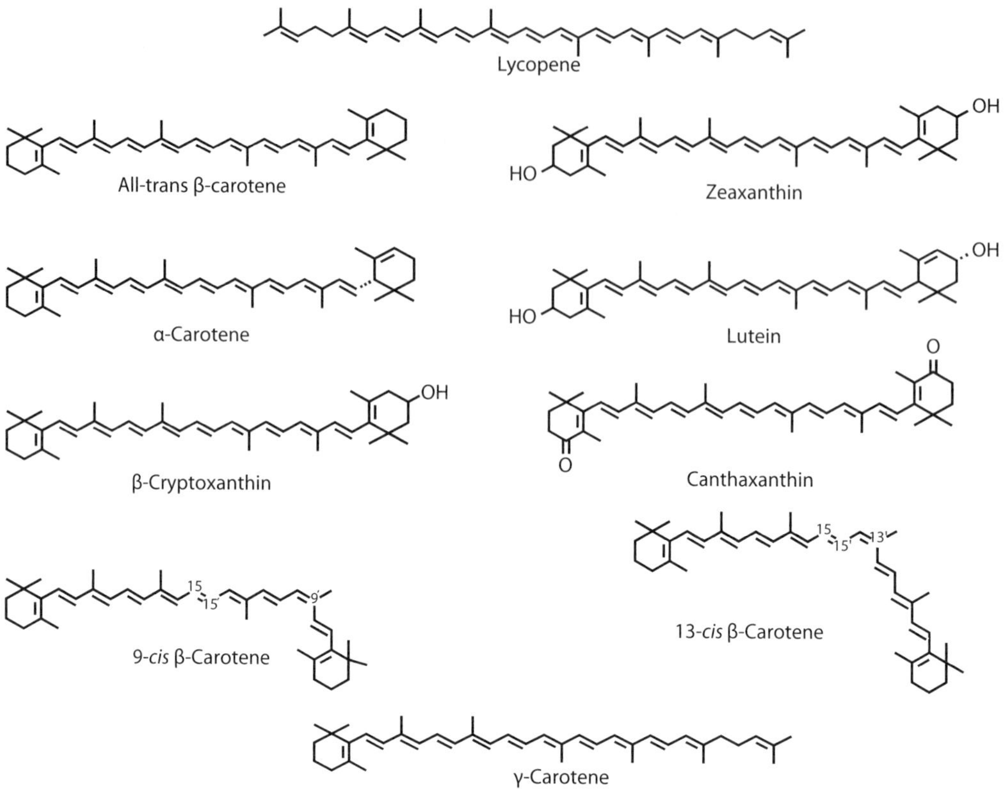

FIGURE 17.1 Structures of some of the major carotenoids.

Carotenoids are hydrophobic molecules and thus are located in lipophilic sites of cells, such as bilayer membranes. Their hydrophobic character is decreased with an increased number of polar substitutents (mainly hydroxyl groups free or esterified with glycosides), thus affecting the positioning of the carotenoid molecule in biological membranes. For example, the dihydroxycarotenoids such as LUT and zeaxanthin (ZEA) may orient themselves perpendicular to the membrane surface as "molecular rivet" in order to expose their hydroxyl groups to a more polar environment. In contrast, the carotenes such as β-C and LYC could position themselves parallel to the membrane surface to remain in a more lipophilic environment in the inner core of the bilayer membranes (Parker, 1989; Britton, 1995). Thus, carotenoid molecules can have substantial effects on the thickness, strength, and fluidity of membranes and thus affect many of their functions.

To move through an aqueous environment *in vivo*, carotenoids must form complexes with proteins. For example, the ketocarotenoids (i.e., canthaxanthin and astaxanthin) interact with proteins by the formation of Schiff's bases between their keto groups and specific lysine residues of the proteins, while the other carotenoids (e.g., the carotenes) form mostly hydrophobic interactions in amphipathic areas of the proteins or with the lipid components of lipoproteins. Specific carotenoid–protein complexes have been reported mainly in plants and in invertebrates (e.g., cyanobacteria, crustaceans, and silkworm) (Bullerjahn et al., 1986; Zagalsky et al., 1991; Jouni and Wells, 1993). In vertebrates, data on the existence of carotenoproteins are limited. Although no intracellular β-carotene-binding protein was found in bovine liver and intestine (Gugger and Erdman, 1996), a cellular carotenoid-binding protein with a high specificity for the carotenes was reported in ferret liver (Lakshman and Rao, 1999) and a specific xanthophyll-binding protein was reported in the human retina and macula (Yemelyanov et al., 2001). As an alternative mechanism for their water solubilization, carotenoids could use small cytosolic carrier vesicles (Gugger and Erdman, 1996). In nature, carotenoids can be also present in very fine physical dispersions (or crystalline aggregates) in aqueous media; oranges, tomatoes, and carrots are well-known examples of sources that contain such aggregates (Klaui and Bauernfiend, 1981). These differential physicochemical characteristics, that is, chemical structure, positioning in biological membranes, and interaction with proteins, may account for the differences observed among carotenoids in their absorption and metabolism as well as their biological activities.

Several epidemiological studies have shown that the consumption of carotenoid-rich foods is associated with a reduced risk of certain cancers, cardiovascular disease, and age-related macular degeneration (Peto et al., 1981; Ziegler, 1991; Seddon et al., 1994; van Poppel, 1996). These preventive effects of carotenoids could be related to their major function as vitamin A precursors and/or their actions as antioxidants, modulators of the immune response, and inducers of gap-junction communications (Olson, 1998). Not all carotenoids do have a protective effect against a specific disease. They are, however, generally recognized as safe for human health in contrast to vitamin A, which has the potential for toxicity at high doses. Thus, the potential use of carotenoids, as supplements or from natural food sources, to prevent certain chronic diseases in addition to their use for preventing vitamin A deficiency has stimulated a renewed interest in the carotenoid field.

Carotenoid absorption and metabolism have been comprehensively reviewed (Erdman et al., 1993; Parker, 1996; van Vliet, 1996; Furr and Clark, 1997; Yeum and Russell, 2002) and this chapter will focus only on recent advances in these areas. A particular emphasis will be placed on studies that used *in vitro* and cell culture models as tools to understand better the mechanisms of absorption on the molecular level.

17.2 INTESTINAL CAROTENOID ABSORPTION

Knowledge about human carotenoid absorption is mostly derived from studies conducted with β-C. Rodents, because of their high efficiency of cleaving provitamin A carotenoids in intestine, are not a good animal model for studying human carotenoid absorption. As alternatives, ferrets, preruminant calves, and gerbils have been used (Poor et al., 1992; Wang et al., 1992; Pollack et al., 1994). However, none of these animal models completely mimic carotenoid metabolism in humans (Lee et al., 1999). There are different methods to quantify the intestinal absorption of carotenoids in humans, such as the intake-excretion "balance" approach and the total plasma "carotenoid response" approach. Both of these methods give only a rough estimate of intestinal absorption per se. Recent approaches using stable isotopes, coupled with mass spectral analysis of the carotenoid and its newly synthesized metabolites isolated from the postprandial triglycerides (TG)-rich lipoprotein plasma fraction, are the most promising methods in terms of accurate measurement of carotenoid absorption. However, such studies are costly and complex, and the data generated are currently limited and difficult to compare due to the use of different experimental designs (Novotny et al., 1995; Lin et al., 2000; Tang et al., 2000; Van Lieshout et al., 2001). Although such methods have a great promise in assessing carotenoid bioavailability and bioefficacy from different food sources in humans (Edwards et al., 2001; You et al., 2002; Van Lieshout et al., 2003), they do not provide mechanistic information about the carotenoid absorption process itself.

The *in vivo* intestinal absorption of carotenoids involves several crucial steps: (1) release of carotenoids from the food matrix, (2) solubilization of carotenoids into mixed lipid micelles in the lumen, (3) cellular uptake of carotenoids by intestinal mucosal cells, (4) incorporation of carotenoids into chylomicrons (CM), and (5) secretion of carotenoids and their metabolites associated with CM into the lymph (Figure 17.2). In this overall process, several basic aspects still remain to be clarified such as the absolute absorption efficiencies of the different carotenoids, the nature of luminal and intracellular factors regulating the process of absorption, the mechanisms of intracellular transport of carotenoids and of their incorporation into CM, and the nature of interactions between carotenoids occurring during their intestinal absorption. Given the limitations of using human subjects for these kinds of investigations, a simple alternative model for studying intestinal carotenoid absorption on

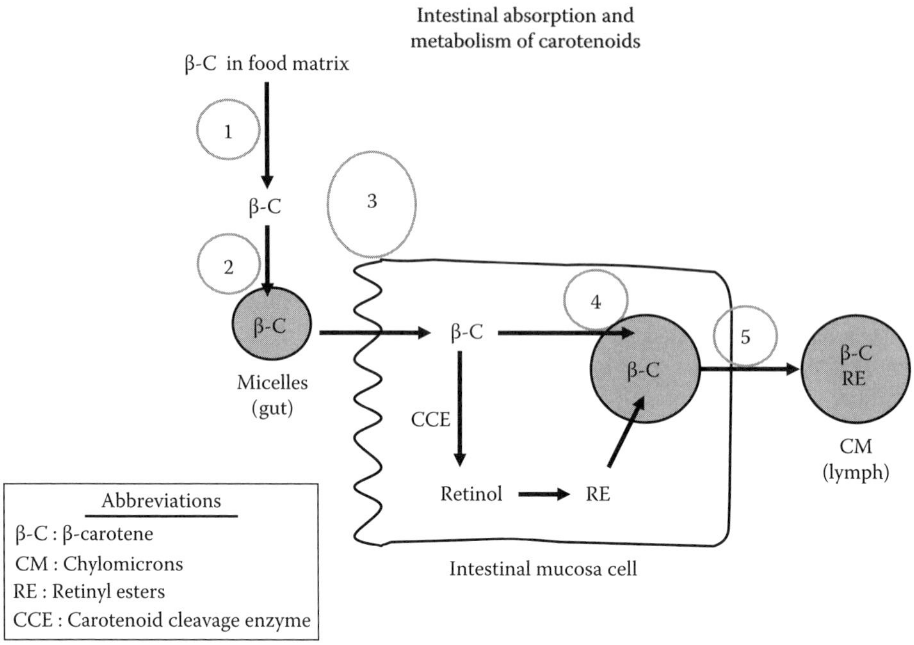

FIGURE 17.2 Intestinal absorption and metabolism of β-C. Numbered steps are explained in the text.

the molecular level would be useful. An *in vitro* intestinal cell culture system mimicking the *in vivo* intestinal absorption of carotenoids was recently proposed (Steps 3–5 of the above-mentioned steps) (During et al., 2002). The description of this model and several applications are presented below.

17.2.1 An *In Vitro* Model to Study Intestinal Absorption of Carotenoids

One obligate step for fat-soluble nutrients, such as carotenoids, to cross the intestinal barrier is their incorporation into CM assembled in the enterocytes. Under normal cell culture conditions, human intestinal Caco-2 cells are unable to form CM. However, when supplemented with oleic acid (OA) and taurocholate (TC) as described earlier (Luchoomun and Hussain, 1999), highly differentiated parent Caco-2 cells (without BCO activity) and the derived TC7 cells (with BCO activity) cultured on membranes were able to form and secrete CM. The high OA concentration is necessary to induce intracellular TG synthesis and thus CM assembly. Because Caco-2 cells are more efficient than TC7 cells in terms of both CM formation and β-C transport, and because β-C cleavage might complicate studies on provitamin A carotenoid absorption per se, the parent Caco-2 cell line was chosen in most studies. CM secreted by Caco-2 cells are characterized as particles rich in (newly synthesized) TG (~90% of total secreted) containing apolipoprotein B (~30% of total secreted) and phospholipids (~20% of total secreted) and with an average diameter of ~60 nm (determined by laser light scattering) (During et al., 2002). These characteristics are similar to those of CM secreted *in vivo* by the enterocytes.

Thus, in contrast to previous *in vivo* models, this *in vitro* model provides the possibility of dissociating experimentally two important processes of the intestinal carotenoid absorption: cellular uptake and secretion. Under conditions mimicking the postprandial state (TC:OA supplementation), differentiated Caco-2 cells were able (1) to take up carotenoids at the apical side and to incorporate them into CM and (2) to secrete them at the basolateral side, associated with CM fractions. In this model, no attempt has yet been made to reproduce the *in vivo* physiochemical conditions occurring in the intestinal lumen, such as carotenoid release from the food matrix and solubilization into mixed lipid micelles. Carotenoids were delivered to Caco-2 cells in aqueous suspension with Tween 40 (During et al., 2002). Using this cell culture system in conjunction with an *in vitro* digestion procedure (steps 1 and 2 of the above-mentioned steps) (Garrett et al., 1999), carotenoids are transferred from the food to bile salt micelles, could be useful to assess the bioavailability of carotenoids from different types of food matrices *in vitro*. These first two steps of carotenoid absorption have been mimicked using Caco-2 cells cultured on plastic (Garrett et al., 2000).

17.2.2 Kinetics of β-C Transport through Intestinal Cells

Only a few studies have been done on the kinetics of carotenoid absorption. Based on earlier rat studies (El-Gorab et al., 1975; Hollander and Ruble, 1978), the intestinal absorption of carotenoids was thought to be a passive diffusion process determined by the concentration gradient of the carotenoid across the intestinal cell membrane. The kinetics of β-C transport through Caco-2 cell monolayers, characterized for both steps (cellular uptake and secretion in CM), showed curvilinear, time-dependent (Figure 17.3a), and saturable, concentration-dependent (apparent Km of 7–10 μM; Figure 17.3b) processes (During et al., 2002). Thus, these data suggest that the intestinal transport of carotenoids might be facilitated by the participation of a specific epithelial transporter; a hypothesis that contrasts with previous investigations. The contrast between studies could be due to the use of different models, human cells and rats, respectively, and possibly due to the use of different β-C concentration ranges, 0.5–23 μM (During et al., 2002) and 0.5–11 μM (Hollander and Ruble, 1978) or 6–60 μM (El-Gorab et al., 1975), respectively.

The saturation of β-C transport through Caco-2 cell monolayers occurred at β-C concentrations of 15–20 μM (equivalent to a daily β-C intake of 100 mg or more); a concentration range far higher than the "physiological" concentration range. It was estimated that the β-C concentration of 1 μM at the apical side of cells (or 400 pmol β-C/cm^2 of Caco-2 cell monolayer) was close to the physiological level of β-C found in the gut (200 pmol β-C/cm^2 of surface of absorption) after ingestion of a daily β-C dose of 5 mg (During et al., 2002).

Under linear concentration conditions (for a β-C concentration range of 0.12–6 μM) at 16 h incubation and under cell culture conditions mimicking the *in vivo* postprandial state, the extent of absorption of all-*trans* β-C through Caco-2 cell monolayers was 11%; a value similar to that reported from different human studies. In humans, the bioavailability of a single dose of β-C

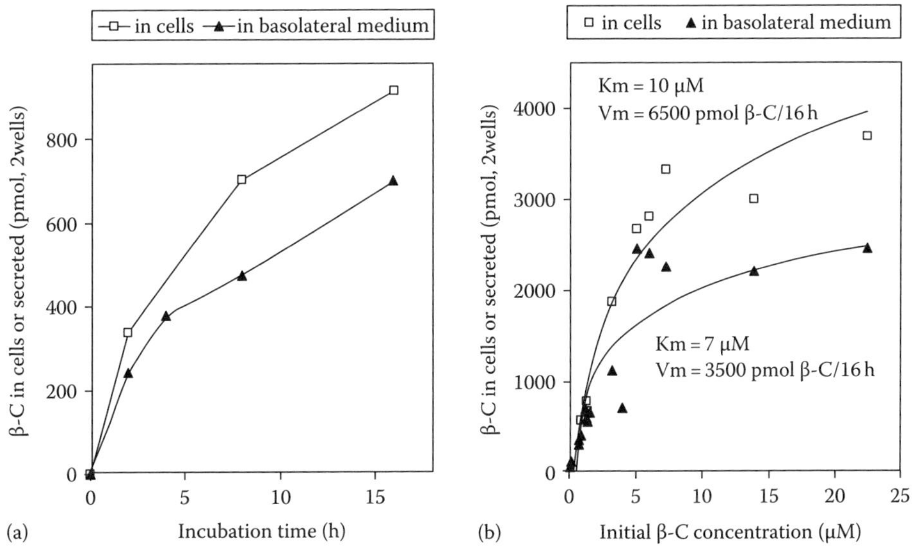

FIGURE 17.3 Kinetics of β-C transport through Caco-2 cell monolayers as a function of (a) the incubation time at a fixed β-C concentration (1 μM) and (b) the initial β-C concentration for 16 h incubation. (Modified from During, A. et al., *J. Lipid Res.*, 43, 1086, 2002.)

(in oil or in capsule) was 9%–17% using the lymph-cannulation approach (Goodman et al., 1966), 11% using carotenoid and retinyl ester response in the TG-rich lipoprotein plasma fraction approach (van Vliet et al., 1995), and 3%–22% using the most recent isotopic tracer approaches (Novotny et al., 1995; Lin et al., 2000). The fact that the extent of β-C absorption obtained with Caco-2 cells falls within the range observed *in vivo* adds confidence to the *in vitro* model for studying human intestinal absorption of carotenoids. Finally, of the total β-C secreted by Caco-2 cells, 80% was associated with CM, 10% with VLDL, and 10% with the nonlipoprotein fraction (During et al., 2002), pointing to the importance of CM assembly for β-C secretion into the lymph *in vivo*.

17.2.3 Selective Uptake of All-*trans* β-C versus Its *cis* Isomers by Intestinal Cells

Human studies (Jensen et al., 1987; Gaziano et al., 1995; Stahl et al., 1995; You et al., 1996; Johnson et al., 1997) have consistently reported a preferential accumulation of all-*trans* β-C in total plasma, and in the postprandial TG-rich lipoprotein plasma fraction, compared to its 9-*cis* isomer. These differences in plasma response between the two geometrical isomers suggested either a selective intestinal transport of all-*trans* β-C versus its 9-*cis* isomer or an intestinal *cis–trans* isomerization of 9-*cis* β-C into all-*trans* β-C. This later possibility was brought up by a study (You et al., 1996) showing a significant accumulation of [^{13}C]-all-*trans* β-C in plasma of subjects who ingested only [^{13}C]-9-*cis* β-C. Starting with an initial concentration (1 μM) for the three geometrical isomers of β-C applied separately to the *in vitro* system described above, it was demonstrated that both 9-*cis* and 13-*cis* β-C were taken up by Caco-2 cells to only one-fifth of the extent of all-*trans* β-C (During et al., 2002). The extent of absorption of the two *cis* isomers through Caco-2 cell monolayers was less than 3.5% (compared to 11% for all-*trans* β-C) (Table 17.1), indicating that the discrimination between β-C isomers occurred at the cellular uptake level of the intestinal absorption process.

The β-C isomer selectivity seems to be tissue-specific; a preferential uptake of the all-*trans* isomer was shown in hepatic stellate HSC-T6 cells and in cell-free system from rat liver microsomes, but not in endothelial EAHY cells or U937 monocyte-macrophages (During et al., 2002). When Caco-2 cells were incubated with only 9-*cis* β-C, all-*trans* β-C did not increase in cells or in the basolateral medium, indicating that there is no *cis–trans* isomerization occurring in intestinal cells. Thus, the isomerization of 9-*cis* β-C observed *in vivo* (You et al., 1996) could take place in the

TABLE 17.1
Differential Absorption of Individual Carotenoids through Caco-2 Cell Monolayers

Carotenoid Name	Extent of Absorption (%)[a]
All-*trans* β-carotene	11.2 ± 2.4
13-*cis* β-Carotene	3.1 ± 1.9[b]
9-*cis* β-Carotene	1.9 ± 1.5[b]
α-Carotene	9.6 ± 1.5
Lutein	6.7 ± 1.9[b]
Lycopene	2.3 ± 2.0[b]

Source: Modified from During, A. et al., *J. Lipid Res.*, 43, 1086, 2002.

[a] Values are means ± SD, $n = 3$ or more independent experiments, at 16 h incubation for 1 μM carotenoid.

[b] $P < 0.0005$ compared to all-trans β-C value.

gastrointestinal lumen before the cellular uptake, probably under the action of enzymes related to gut microflora since the spontaneous isomerization of 9-*cis* β-C to all-*trans* β-C is not thermodynamically favored (von Doering et al., 1995). Taken together, these data on the selective uptake of β-C isomers by Caco-2 cells support the idea of a specific transporter involved in the intestinal absorption process of carotenoids.

17.2.4 Differential Intestinal Transport of Individual Carotenoids

Data on the intestinal absorption of carotenoids other than β-C are limited. It seems that the more polar carotenoids (xanthophylls) are absorbed better than the carotenes. Supporting this idea, it was reported that the plasma response for LUT was twice as high as it was for β-C when single doses of those carotenoids were given in oil (Kostic et al., 1995) and that both LUT and ZEA versus β-C were preferentially increased in CM after ingestion of a carotenoid mixture, "Betatene" (Gartner et al., 1996). In addition, the relative bioavailability of LUT from vegetables was reported to be five times higher than that of β-C (van het Hof et al., 1999), but in the same study the plasma response of LUT was substantially smaller than that of β-C after simultaneous ingestion of pure LUT and β-C dissolved in oil. In fact, in these different human studies, the difference in plasma response between carotenoids may not reflect a difference in true absorption. There are several factors for which each carotenoid seems to follow a different pattern such as (a) the differential transfer of carotenoids from food matrices to the lipid micelles: LYC (from tomato puree) was reported to be less efficiently transferred to the micellar phase of the duodenum than β-C (from carrot puree) and LUT (from chopped spinach) *in vivo* (Tyssandier et al., 2003), (b) the differential stability of carotenoids: LYC and β-C decomposed more rapidly than LUT and ZEA upon exposure to various pro-oxidants *in vitro* (Siems et al., 1999), (c) the differential metabolism of carotenoids: at least 35% (up to 75%) of the absorbed β-C is converted to retinyl esters in intestinal cells (Goodman et al., 1966; van Vliet et al., 1995; O'Neill and Thurnham, 1998) whereas xanthophylls are non-provitamin A carotenoids, and finally (d) the differential clearance rate of carotenoids from the plasma once absorbed. These many factors, which make it difficult to compare the actual absorption of the different carotenoids *in vivo*, can be avoided by using the *in vitro* system described here. A differential transport of carotenoids through Caco-2 cell monolayers was shown as follows: all-*trans* β-C (11%) ≈ α-C (10%) > LUT (7%) > LYC (2.5%) (Table 17.1). These *in vitro* data and several studies with animals (Bierer et al., 1995; Clark et al., 1998) and humans (Johnson et al., 1997; O'Neill and Thurnham, 1998) converge to indicate that LYC is poorly absorbed compared to other carotenoids. In addition, these data were in agreement with a human study (O'Neill and Thurnham, 1998), which showed that β-C is preferentially absorbed compared to LUT. In contrast, it was reported that plasma β-C response was lower than plasma LUT response when the two carotenoids β-C and LUT were ingested separately (Kostic et al., 1995; Gartner et al., 1996; van Vliet et al., 1999). However, in these studies, the retinyl ester fraction formed during the intestinal absorption of β-C was not analyzed, a fact that could contribute to an underestimated β-C absorption compared to LUT absorption. Interestingly, for the four individual carotenoids tested (β-C, α-C, LYC, and LUT), the extent of secretion varied over a wider range (2.5%–11%) than the extent of cellular uptake (15%–18%), indicating that the carotenoid structure might be a major determinant in its ability to be incorporated into CM.

17.2.5 Carotenoid Interaction during Intestinal Absorption

Carotenoids compete for their absorption and metabolism, but data are conflicting as indicated in a review (van den Berg, 1999). In humans, β-C reduced the apparent LUT absorption (Kostic et al., 1995; van den Berg, 1998; van den Berg and van Vliet, 1998), while LUT had either no effect (Kostic et al., 1995) or reduced the apparent β-C absorption (van den Berg, 1998; van den Berg and van Vliet, 1998). This inhibitory effect of LUT on plasma β-C response observed *in vivo* could be attributed at least partly to the fact that LUT inhibits β-C cleavage enzyme as suggested in rats (van Vliet et al., 1996), but not confirmed in humans (van den Berg, 1998; van den Berg and van Vliet, 1998). Furthermore, β-C was shown to improve the apparent LYC absorption (Johnson et al., 1997), while LYC had no effect on β-C in humans (Johnson et al., 1997; van den Berg and van Vliet, 1998). Recently, when carotenoids were provided in their natural vegetable matrices, it was reported that adding a second carotenoid to a meal that contained another carotenoid diminished the CM response of the first carotenoid (Tyssandier et al., 2002). However, in this postprandial study, it was difficult to define clearly specific interaction between two carotenoids since some of the meals contained more than two carotenoids. In addition, "pharmacological" doses of carotenoids are commonly used in these interaction studies: doses at which the efficiency of carotenoid absorption seems to decrease probably in relation to the limited capacity of micellar incorporation of carotenoids in the lumen (Olson, 1998; van den Berg, 1999; van Lieshout et al., 2003). Thus, it is difficult to interpret the results in terms of interaction at the cellular level.

Using the *in vitro* cell culture system and a range of physiological concentrations (1–5 μM), neither LUT nor β-C affected significantly the transport of each other through Caco-2 cell monolayers, while the main carotenoid interactions were observed between nonpolar carotenoids (β-C/α-C and β-C/LYC) (Figure 17.4). The discrepancy between these *in vitro* data and *in vivo* data might be due to the fact that plasma carotenoid response measured in *in vivo* studies does not reflect only intestinal absorption as mentioned earlier. Thus, the specific interactions observed in the *in vitro* study (During et al., 2002) indicate that two carotenoids exhibiting similar structural characteristics could follow a similar pathway in intestinal cells and thus compete for their cellular uptake and/or their incorporation into CM. For instance, in CM particles, carotenoids may organize themselves differently on the basis of their structural properties; the more polar carotenoids (xanthophylls) may remain at the surface and the less polar carotenoids (carotenes) in the core of CM. Finally, these mutual interactions are also consistent with the idea of a facilitated uptake process.

Thus, this *in vitro* cell culture model is useful for a better understanding of the mechanisms involved in the intestinal absorption of carotenoids at the cellular level. The concentration dependence (saturation) of β-C uptake and secretion in CM, the discrimination between β-C isomers for their cellular uptake, the differential absorption of different carotenoids as well as their interactions observed during transport through Caco-2 cells, all suggest that the intestinal transport of carotenoids might be facilitated by the participation of a specific epithelial transporter. This hypothesis was supported by the identification of a scavenger receptor with a high sequence homology to the mammalian class B scavenger receptors (SR-BI and CD36) mediating the cellular uptake of carotenoids in *drosophila* (Kiefer et al., 2002). It was demonstrated that the *in vivo* mutation of the ninaD gene encoding this epithelial receptor resulted in a defect of the cellular uptake of β-C (precursor of the visual chromophore in flies) and thus in the blindness phenotype observed in the *drosophila* mutant, ninaD (Kiefer et al., 2002). Thus, studies were conducted to ask if carotenoid uptake by intestinal cells may involve a specific epithelial transporter(s).

17.2.6 Ezetamibe Inhibits Carotenoid and Cholesterol Absorption in Caco-2 Cells but Not Retinol Absorption

The first study was conducted to determine whether carotenoids and cholesterol share common pathways (transporters) for their intestinal absorption (During et al., 2005). Differentiated Caco-2 cells on membranes were incubated (16 h) with a carotenoid (1 μmol/L) with or without ezetimibe (EZ; Zetia, an inhibitor of cholesterol transport), and with or without antibodies against the receptors, cluster determinant 36 (CD36) and scavenger receptor class B, type I (SR-BI). Carotenoid transport in Caco-2 cells (cellular uptake + secretion) was decreased by EZ (10 mg/L) as follows: β-C and α-C (50% inhibition) ≫ β-cryptoxanthin and LYC (20%) ≫ LUT:ZEA (1:1) (7%). EZ reduced cholesterol transport by 31%, but not retinol transport. β-Carotene transport was also inhibited by anti-SR-BI, but not by anti-CD36. The inhibitory effects of EZ and anti-SR-BI on β-C transport

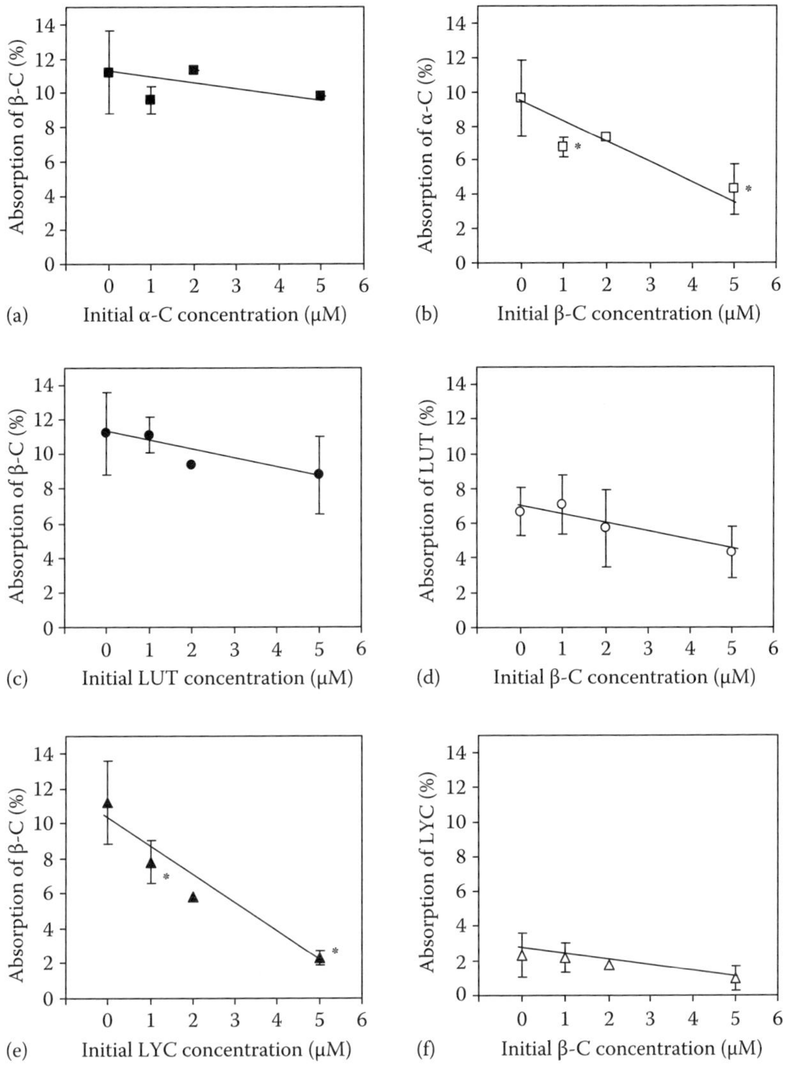

FIGURE 17.4 Interactions between carotenoids during their transport through Caco-2 cell monolayers. (a) α-C effect on β-C transport, (b) β-C effect on α-C transport, (c) LUT effect on β-C transport, (d) β-C effect on LUT transport, (e) LYC effect on β-C transport, and (f) β-C effect on LYC transport. Data with error bars are mean ± SD obtained from three or more independent experiments ($^{*}P < 0.05$ compared with the carotenoid alone). (Modified from During, A. et al., *J. Lipid Res.*, 43, 1086, 2002.)

were additive, indicating that they may have different targets. Finally, differentiated Caco-2 cells treated with EZ showed a significant decrease in mRNA expression for the surface receptors SR-BI, Niemann-Pick type C1 Like 1 protein (NPC1L1), and ATP-binding cassette transporter, subfamily A (ABCA1) and for the nuclear receptors retinoid acid receptor γ, sterol-regulatory element binding proteins 1 and 2, and liver X receptor β as assessed by real-time PCR analysis. The data indicate that (1) EZ is an inhibitor of carotenoid transport, an effect that decreases with increasing polarity of the carotenoid molecule; (2) SR-BI is involved in carotenoid transport; and (3) EZ may act, not only by interacting physically with cholesterol transporters as previously suggested, but also by downregulating expression of these proteins. The cellular uptake and efflux of carotenoids, like that of cholesterol, likely involve more than one transporter. The suggested role of SR-BI in carotenoid transport in mammalian intestine was supported by studies by Reboul et al. (2005) showing

SR-BI involvement in lutein uptake by Caco-2 cells and by the inhibition of β-C absorption in the SR-BI null mouse (van Bennekum et al., 2005).

17.2.7 Independent Pathways of Retinol and Carotenoid Absorption in Caco-2 Cells: Direct Evidence for the Participation of SR-BI in Carotenoid Absorption

As demonstrated above, the uptake of β-C at the apical membrane of differentiated Caco-2 cells occurs via a saturable, facilitated mechanism and is inhibited by Ezetimibe, a clinically used inhibitor of cholesterol absorption. Carotenoids secreted at the basolateral membrane were associated

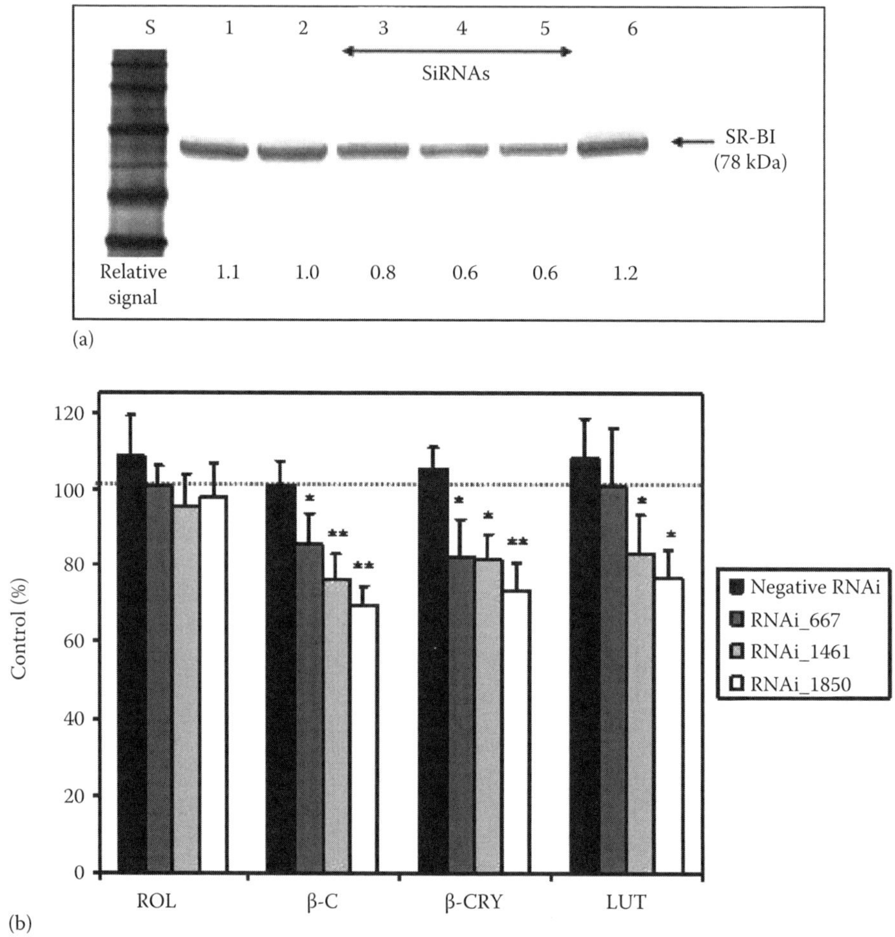

FIGURE 17.5 Effects of the small interfering RNA (siRNA or RNAi) inhibition of scavenger receptor class B type I (SR-BI) expression on the cellular uptake of ROL, β-carotene (β-C), β-cryptoxanthin (β-CRY), or lutein (LUT) in Caco-2 cells. (a) Immunoblots of SR-BI expression (using 25 μg total protein/well) in cells treated under the following conditions: lane 1, scrambled RNAi; lane 2, Lipofectamine™ 2000 (LP2000) only; lane 3, RNAi_667; lane 4, RNAi_1461; lane 5, RNAi_1850; and lane 6, no treatment. Lane S represents protein standards (MagicMark XP standard from 60 to 220 kDa). (b) Cellular uptake of ROL and carotenoids (expressed as the percentage of control cells treated with LP2000 only) after incubation of cells with ROL or a carotenoid at 2 μM for 1 h at 72 h after transfection with a RNAi against SR-BI. Data are mean ± SD of three to five independent experiments for each compound tested. $^{*}P < 0.05$, $^{**}P < 0.0001$ compared with the negative control. (From During, A. et al., *J. Lipid Res.*, 48, 2284, 2007. With permission.)

exclusively with chylomicrons. Recent studies were designed to compare mechanisms of retinol and carotenoid transport (During et al., 2007). When cells were incubated with retinol for varying times (1–24 h), cellular retinol reached plateau levels within 2 h, whereas retinyl ester formation increased continuously. Retinol and retinyl ester efflux into basolateral medium increased linearly with time. Free retinol was associated with the nonlipoprotein fraction and retinyl esters with chylomicrons. In contrast to carotenoids, retinol uptake at the apical membrane was directly proportional to initial retinol concentration over a wide range (0.5–110 μM). However, free retinol efflux from the basolateral membrane occurred via two processes: (a) a saturable process at low concentrations (<10 μM) and (b) a nonsaturable process at higher concentrations. When cells were loaded with retinol and then maintained on retinoid-free medium for 5d, free retinol, but not retinyl esters, was secreted into the basolateral medium. Glyburide (inhibitor of ABCA1 and other transporters) significantly reduced free retinol efflux, but not cellular retinol uptake. Inhibition of ABCA1 protein expression by siRNAs inhibited free retinol efflux but had no effect on carotenoid efflux from the basolateral membrane. SR-B1 inhibition did not affect retinol transport, but decreased cellular uptake of β-C, β-cryptoxanthin, and LUT. Importantly, the extent of inhibition of SR-BI expression correlated with the extent of inhibition of carotenoid absorption in this system (Figure 17.5). Inhibition of NPC1L1 expression by siRNA did not affect either retinol or carotenoid uptake. These data suggest that (a) free retinol enters intestinal cells by diffusion; (b) free retinol efflux is partly facilitated, probably by the basolateral transporter ABCA1; and (c) newly synthesized retinyl esters, but not preformed esters, are incorporated into chylomicrons and secreted. In contrast to vitamin A transport, carotenoid uptake is mediated by the apical transporter SR-B1 and carotenoid efflux occurs exclusively via their secretion in chylomicrons.

REFERENCES

T. L. Bierer, N. R. Merchen, and J. W. Erdman, Comparative absorption and transport of five common carotenoids in preruminant calves, *J. Nutr.* 125 (1995) 1569–1577.

G. Britton, Structure and properties of carotenoids in relation to function, *FASEB J.* 9 (1995) 1551–1558.

G. S. Bullerjahn and L. A. Sherman, Identification of a carotenoid-binding protein in the cytoplasmic membrane from the heterotrophic cyanobacterium Synechocystis sp. Strain PCC6714, *J. Bacteriol.* 167 (1986) 396–399.

R. M. Clark, L. Yao, L. She, and H. C. Furr, A comparison of lycopene and canthaxanthin absorption: Using the rat to study the absorption of non-provitamin A carotenoids, *Lipids* 33 (1998) 159–163.

A. During and E. H. Harrison, Mechanisms of provitamin A (carotenoid) and vitamin A (retinol) transport into and out of intestinal Caco-2 cell, *J. Lipid Res.* 48 (2007) 2284–2294.

A. During, M. M. Hussain, D. W. Morel, and E. H. Harrison, Carotenoid uptake and secretion by Caco-2 cells: β-Carotene isomer selectivity and carotenoid interactions, *J. Lipid Res.* 43 (2002) 1086–1095.

A. During, H. Dawson, and E. H. Harrison, Carotenoid transport is decreased and the expression of the lipid transporters SR-BI, NPC1L1, and ABCA1 is down-regulated in CACO-2 cells treated with ezetimibe, *J. Nutr.* 135 (2005) 2305–2312.

A. J. Edwards, C.-S. You, J. E. Swanson, and R. S. Parker, A novel extrinsic reference method for assessing the vitamin A value of plant foods, *Am. J. Clin. Nutr.* 74 (2001) 348–355.

M. I. El-Gorab, B. A. Underwood, and J. D. Loerch, The roles of bile salts in the uptake of β-carotene and retinol by everted gut sacs, *Biochim. Biophys. Acta* 401 (1975) 265–277.

J. W. Erdman, T. L. Bierer, and E. T. Gugger, Absorption and transport of carotenoids, *Ann. N. Y. Acad. Sci.* 691 (1993) 76–85.

H. R. Furr and R. M. Clark, Intestinal absorption and tissue distribution of carotenoids, *J. Nutr. Biochem.* 8 (1997) 364–377.

D. A. Garrett, M. L. Failla, and R. J. Sarama, Development of an in vitro digestion method to assess carotenoid bioavailability from meals, *J. Agric. Food Chem.* 47 (1999) 4301–4309.

D. A. Garrett, M. L. Failla, and R. J. Sarama, Estimation of carotenoid bioavailability from fresh stir-fried vegetables using ab in vitro digestion/Caco-2 cell culture model, *J. Nutr. Biochem.* 11 (2000) 574–580.

C. Gartner, W. Stahl, and H. Sies, Preferential increase in chylomicron levels of the xanthophylls lutein and zeaxanthin compared to beta-carotene in the human, *Int. J. Vitam. Nutr. Res.* 66 (1996) 119–125.

J. M. Gaziano, E. J. Johnson, R. M. Russell, J. E. Manson, M. J. Stampfer, P. M. Ridker, B. Frei, C. H. Hennekens, and N. I. Krinsky, Discrimination in absorption or transport of beta-carotene isomers after oral supplementation with either all-trans or 9-cis-β-carotene, *Am. J. Clin. Nutr.* 61 (1995) 1248–1252.

D. S. Goodman, R. Blomstrand, B. Werner, H. S. Huang, and T. Shiratori, The intestinal absorption and metabolism of vitamin A and beta-carotene in man, *J. Clin. Invest.* 45 (1966) 1615–1623.

E. T. Gugger and J. W. Erdman, Intracellular beta-carotene transport in bovine liver and intestine is not mediated by cytosolic proteins, *J. Nutr.* 126 (1996) 1470–1474.

D. Hollander and P. E. Ruble, Beta-carotene intestinal absorption: bile, fatty acid, pH, and flow rate effects on absorption, *Am. J. Physiol.* 235 (1978) E686–E691.

C. D. Jensen, T. W. Howes, G. A. Spiller, T. S. Pattison, J. H. Whittam, and J. Scala, Observations on the effects of ingesting cis- and trans-β-carotene isomers on human serum concentrations, *Nutr. Rep. Int.* 35 (1987) 413–422.

E. J. Johnson, J. Qin, N. I. Krinsky, and R. M. Russell, Ingestion by men of a combined dose of beta-carotene and lycopene does not affect the absorption of beta-carotene but improves that of lycopene, *J. Nutr.* 127 (1997) 1833–1837.

Z. E. Jouni and M. Wells, Purification of a carotenoid-binding protein from the midgut of the silkworm, Bombyx mori, *Ann. N. Y. Acad. Sci.* 691 (1993) 210–212.

C. Kiefer, E. Sumser, M. F. Wernet, and J. von Lintig, A class B scavenger receptor mediates the cellular uptake of carotenoids in *Drosophila, Proc. Natl. Acad. Sci. U. S. A.* 16 (2002) 10581–10586.

H. Kläui and J. C. Bauernfeind, in: J. C. Bauernfeind (Ed.), *Carotenoids as Colorants and Vitamin A Precursors: Technological and Nutritional Applications*, Academic Press, Inc., New York (1981), pp. 58–63.

D. Kostic, W. S. White, and J. A. Olson, Intestinal absorption, serum clearance, and interactions between lutein and beta-carotene when administered to human adults in separate or combined oral doses, *Am. J. Clin. Nutr.* 62 (1995) 604–610.

M. R. Lakshman and M. N. Rao, Purification and characterization of cellular carotenoid-binding protein from mammalian liver, *Methods Enzymol.* 299 (1999) 441–456.

C. M. Lee, A. C. Boileau, T. W. Boileau, A. W. Williams, K. S. Swanson, K. A. Heintz, and J. W. Erdman, Review of animal models in carotenoid research, *J. Nutr.* 129 (1999) 2271–2277.

Y. Lin, S. R. Dueker, B. J. Burri, T. R. Neidlinger, and A. J. Clifford, Variability of the conversion of beta-carotene to vitamin A in women measured by using a double-tracer design, *Am. J. Clin. Nutr.* 71 (2000) 1545–1554.

J. Luchoomun and M. M. Hussain, Assembly and secretion of chylomicrons by differentiated Caco-2 cells. Nascent triglycerides and preformed phospholipids are preferentially used for lipoprotein assembly, *J. Biol. Chem.* 274 (1999) 19565–19572.

A. R. Mangels, J. M. Holden, G. R. Beecher, M. R. Forman, and E. Lanza, Carotenoid content of fruits and vegetables: An evaluation of analytic data, *J. Am. Diet. Assoc.* 93 (1993) 284–296.

J. A. Novotny, S. R. Dueker, L. A. Zech, and A. J. Clifford, Compartmental analysis of the dynamics of beta-carotene metabolism in an adult volunteer, *J. Lipid Res.* 36 (1995) 1825–1838.

J. A. Olson, Carotenoids, in: M. E. Shils, J. A. Olson, M. Shike, A. C. Ross (Eds.), Modern *Human Nutrition in Health and Disease*, 9th ed., Lippincott Williams & Wilkins, Baltimore, MD (1998), pp. 525–541.

M. E. O'Neill and D. I. Thurnham, Intestinal absorption of beta-carotene, lycopene and lutein in men and women following a standard meal: Response curves in the triacylglycerol-rich lipoprotein fraction, *Br. J. Nutr.* 79 (1998) 149–159.

R. S. Parker, Carotenoids in human blood and tissues, *J. Nutr.* 119 (1989) 101–104.

R. S. Parker, Absorption, metabolism, and transport of carotenoids, *FASEB J.* 10 (1996) 542–551.

R. Peto, R. Doll, J. D. Buckley, and M. B. Sporn, Can dietary beta-carotene materially reduce human cancer rates? *Nature* 290 (1981) 201–208.

C. L. Poor, T. L. Bierer, N. R. Merchen, G. C. Fahey, M. R. Murphy, and J. W. Erdman, Evaluation of the preruminant calf as a model for the study of human carotenoid metabolism, *J. Nutr.* 122 (1992) 262–268.

J. Pollack, J. M. Campbell, S. M. Potter, and J. W. Erdman, Mongolian gerbils (Meriones unguiculatus) absorb beta-carotene intact from a test meal, *J. Nutr.* 124 (1994) 869–873.

J. M. Seddon, U. A. Ajani, R. D. Sperduto, R. Hiller, N. Blair, T. C. Burton, M. D. Farber, E. S. Gragoudas, J. Haller, D. T. Miller, L. A. Yanuzzi, and W. Willett, Dietary carotenoids, vitamins A, C, and E, and the advanced age-related macular degeneration. Eye disease case-control study group, *J. Am. Med. Assoc.* 272 (1994) 1413–1420.

W. G. Siems, O. Sommerburg, and F. J. van Kuijk, Lycopene and beta-carotene decompose more rapidly than lutein and zeaxanthin upon exposure to various pro-oxidants in vitro, *Biofactors* 10 (1999) 105–113.

W. Stahl, W. Schwarz, J. von Laar, and H. Sies, All-trans beta-carotene preferentially accumulates in human chylomicrons and very low density lipoproteins compared with the 9-cis geometrical isomer, *J. Nutr.* 125 (1995) 2128–2133.

G. Tang, J. Qin, G. G. Dolnikowski, and R. M. Russell, Vitamin A equivalence of beta-carotene in a women as determined by a stable isotope reference method, *Eur. J. Nutr.* 39 (2000) 7–11.

V. Tyssandier, N. Cardinault, C. Caris-Veyrat, M.-J. Amiot, P. Grolier, C. Bouteloup, V. Azais-Braesco, and P. Borel, Vegetable-borne lutein, lycopene, and beta-carotene compete for incorporation into cylomicrons, with no adverse effect on the medium term (3-wk) plasma status of carotenoids in humans, *Am. J. Clin. Nutr.* 75 (2002) 526–534.

V. Tyssandier, E. Reboul, J. F. Dumas, C. Bouteloup-Demange, M. Armand, J. Marcand, M. Sallas, and P. Borel, Processing of vegetable-borne carotenoids in the human stomach and duodenum, *Am. J. Physiol. Gastrointest. Liver Physiol.* 284 (2003) G913-G923.

A. van Bennekum, M. Werder, S. T. Thuahnai, C. H. Han, P. Duong, D. L. Williams, P. Wettstein, G. Schulthess, M. C. Phillips, and H. Hauser, Class B scavenger receptor-mediated intestinal absorption of dietary β-carotene and cholesterol, *Biochemistry* 44 (2005) 4517–4525.

H. van den Berg, Effect of lutein on beta-carotene absorption and cleavage, *Int. J. Vitam. Nutr. Res.* 68 (1998) 360–365.

H. van den Berg, Carotenoid interactions, *Nutr. Rev.* 57 (1999) 1–10.

H. van den Berg and T. van Vliet, Effect of simultaneous, single oral doses of beta-carotene with lutein or lycopene on the beta-carotene and retinyl ester responses in the triacylglycerol-rich lipoprotein fraction of men, *Am. J. Clin. Nutr.* 68 (1998) 82–89.

K. H. van het Hof, I. A. Brouwer, C. E. West, E. Haddeman, R. P. Steegers-Theunissen, M. van Dusseldorp, J. A. Weststrate, T. K. Eskes, and J. G. Hautvast, Bioavailability of lutein from vegetables is 5 times higher than that of beta-carotene, *Am. J. Clin. Nutr.* 70 (1999) 261–268.

M. van Lieshout, C. E. West, Muhilal, D. Permaesih, Y. Wang, X. Xu, R. B. van Breemen, A. F. L. Creemers, M. A. Verhoeven, and J. Lugtenburg, Bioefficacy of beta-carotene dissolved in oil studied in children in Indonesia, *Am. J. Clin. Nutr.* 73 (2001) 949–958.

M. van Lieshout, C. E. West, and R. B. van Breemen, Isotopic tracer techniques for studying the bioavailability and bioefficacy of dietary carotenoids, particularly beta-carotene, in humans: A review, *Am. J. Clin. Nutr.* 77 (2003) 12–28.

G. Van Poppel, Epidemiological evidence for beta-carotene in prevention of cancer and cardiovascular disease, *Eur. J. Clin. Nutr.* 50 (1996) 55S–57S.

T. Van Vliet, Absorption of beta-carotene and other carotenoids in humans and animal models, *Eur. J. Clin. Nutr.* 50 (1996) S32–S37.

T. van Vliet, W. H. Schreurs, and H. van den Berg, Intestinal beta-carotene absorption and cleavage in men: Response of beta-carotene and retinyl esters in the triglyceride-rich lipoprotein fraction after a single oral dose of beta-carotene, *Am. J. Clin. Nutr.* 62 (1995) 110–116.

T. van Vliet, F. van Schaik, W. H. Schreurs, and H. van den Berg, In vitro measurement of beta-carotene cleavage activity: Methodological considerations and the effect of other carotenoids on beta-carotene cleavage, *Int. J. Vitam. Nutr. Res.* 66 (1996) 77–85.

W. E. von Doering, C. Sotiriou-Leventis, and W. R. Roth, Thermal interconversions among 15-cis, 13-cis, and all-trans β-carotene: Kinetics, Arrhenius parameters, thermochemistry, and potential relevance to anticarcinogenicity of all-trans β-carotene, *J. Am. Chem. Soc.* 117 (1995) 2747–2757.

X.-D. Wang, N. I. Krinsky, R. P. Marini, G. Tang, J. Yu, R. Hurley, J. G. Fox, and R. M. Russell, Intestinal uptake and lymphatic absorption of beta-carotene in ferrets: a model for human beta-carotene metabolism, *Am. J. Physiol.* 263 (1992) G480-G486.

A. Y. Yemelyanov, N. B. Katz, and P. S. Bernstein, Ligand-binding characterization of xanthophyll carotenoids to solubilized membrane proteins derived from human retina, *Exp. Eye Res.* 72 (2001) 381–392.

K. J. Yeum and R. M. Russell, Carotenoid bioavailability and bioconversion, *Annu. Rev. Nutr.* 22 (2002) 483–504.

C.-S. You, R. S. Parker, K. J. Goodman, J. E. Swanson, and T. N. Corso, Evidence of cis-trans isomerization of 9-cis-beta-carotene during absorption in humans, *Am. J. Clin. Nutr.* 64 (1996) 177–183.

C.-S. You, R. S. Parker, and J. E. Swanson, Bioavailability and vitamin A value of carotenes from red palm oil assessed by an extrinsic isotope reference method, *Asia Pac. J. Clin. Nutr.* 11 (2002) S348-S442.

P. F. Zagalsky, E. E. Eliopoulos, and J. B. Findlay, The lobster carapace carotenoprotein, alpha-crustacyanin. A possible role for tryptophan in the bathchromic spectral shift of protein-bound astaxanthin, *J. Biochem.* 274 (1991) 79–83.

R. G. Ziegler, Vegetables, fruits, and carotenoids and the risk of cancer, *Am. J. Clin. Nutr.* 53 (1991) 251S–259S.

18 Competition Effects on Carotenoid Absorption by Caco-2 Cells

Emmanuelle Reboul and Patrick Borel

CONTENTS

18.1 THE HUMAN CACO-2 INTESTINAL CELL MODEL: A VALUABLE TOOL FOR STUDYING CAROTENOID ABSORPTION

The Caco-2 cell line was isolated from a human colon carcinoma, and has been characterized as one of the best *in vitro* models of intestinal epithelium. Indeed, in contrast to other intestinal cell lines, Caco-2 cells are able to constitute a homogenous monolayer and to spontaneously differentiate into polarized cells, highly similar to human mature enterocytes, after approximately 2 weeks of culture. Furthermore, the Caco-2 cells present microvillosities at the apical side and have a high transmembrane resistivity, which confirms the fact that the cells are confluent and link to one another via gap junctions. Finally, they can absorb different compounds, express many enzymes involved in intestinal metabolic pathways (Pinto et al. 1983, Musto et al. 1995, Salvini et al. 2002), and give reproducible *in vitro* results consistent with results obtained in *in vivo* studies (Artursson and Karlsson 1991).

Because the extent of oral drug absorption is highly correlated with the transport across Caco-2 monolayers, this cell system has been used in high throughput screens to model transport and metabolism of numerous drugs and their derivatives. Moreover, this cell system has also been validated for studying the intestinal absorption of many nutrients and micronutrients, such as fatty acids (Trotter et al. 1996), cholesterol (Hauser et al. 1998, Werder et al. 2001, Play et al. 2003), and numerous antioxidant molecules such as polyphenols (Manna et al. 1997), vitamin E (Traber et al. 1990), vitamin C (Kuo et al. 2001), and carotenoids (Garrett et al. 1999, 2000, Ferruzzi et al. 2001, Sugawara et al. 2001, During et al. 2002). In some of these studies, the TC-7 clone was chosen because it presents highly homogenous cells compared to the parental Caco-2 cells (Gres et al. 1998) and it displays a β-carotene 15,15′-monooxygenase activity (During et al. 1998), which is implicated in the cleavage of provitamin A carotenoids.

Competition resulting in a decrease of global carotenoid absorption can occur at many steps of the digestion process, including the transfer of carotenoids into mixed micelles, transport across the apical membrane of the enterocyte, and incorporation into chylomicrons. The *in vitro* digestion model (Garrett and Failla 1999) can provide interesting data on the competition that occurs between carotenoids and other compounds at the level or carotenoid incorporation into micelles during digestion. The postprandial studies comparing chylomicron carotenoid responses in humans (i.e., "relative carotenoid absorption") cannot differentiate between the above-mentioned steps. Thus, the *in vitro* Caco-2 model is particularly useful for studying the competition that occurs between carotenoids and other molecules at the intestinal cell level.

18.2 COMPETITION DURING UPTAKE BY ENTEROCYTES

18.2.1 Competition between Carotenoids

During the 1990s, it was suggested that carotenoids compete for both absorption and metabolism, although there were some discrepancies in both the magnitude and the direction of the interactions observed (van den Berg 1999). For example in rats, liver vitamin A storage (used as a measure of β-carotene absorption) was enhanced by a small dose of lutein, but reduced by a larger dose of lutein (High and Day 1951). In humans, β-carotene reduced the apparent lutein absorption (Kostic et al. 1995, van den Berg 1998, van den Berg and van Vliet 1998), while lutein had either no effect (Kostic et al. 1995) or reduced the apparent β-carotene absorption (van den Berg 1998, van den Berg and van Vliet 1998, Tyssandier et al. 2002). In the latest study, it was also found that lutein diminished lycopene absorption, and vice versa.

The first study of Caco-2 dedicated to evaluate the competition between carotenoids with regard to absorption was published in 1999 (Garrett et al. 1999). In this study, the researchers examined the effect of relatively high levels of extracellular or intracellular β-carotene on the uptake of micellar lutein. Beadlets containing water miscible β-carotene were used because they allowed the delivery of high concentration of β-carotene to cells. Nevertheless, this implies that β-carotene was not provided to cells incorporated in its physiological vehicle, that is, in mixed micelles. In the first experiment, Caco-2 cells were incubated for 8 h in micellar medium containing 2.9 μmol/L lutein either with or without 23.7 μmol/L β-carotene. Cellular lutein content was 25% higher ($P < 0.01$) in cells incubated in the medium containing a high level of β-carotene than in cells treated with lutein alone. Next, cells were incubated in control medium (no β-carotene) or medium containing 33 μmol/L β-carotene for 24 h to elevate cellular β-carotene content. Control and β-carotene-treated cells contained less than 1 and 432 ± 25 pmol β-carotene/mg cell protein, respectively, after 24 h. Fresh micellar medium containing 2.3 μmol/L lutein was then added to all the cells after removing spent media. Cellular accumulation of micellar lutein was not altered ($P > 0.01$) by high levels of intracellular β-carotene.

A second study using Caco-2 cells for investigating competition between carotenoids was performed by During et al. (2002). In this study, carotenoids were administrated to cells solubilized in Tween 40 micelles at physiological concentrations (from 1 to 5 μM). Neither lutein nor β-carotene significantly affected their mutual uptake by Caco-2 cell monolayers after 16 h incubation. The authors hypothesized that the inhibitory effect of lutein on β-carotene response observed *in vivo* (van den Berg 1998, van den Berg and van Vliet 1998, Tyssandier et al. 2002) could be attributed, at least partly, to an effect of lutein on β-carotene conversion into vitamin A (van Vliet et al. 1996). In the same study, During et al. showed a mutual negative interaction between lycopene and β-carotene, while lycopene had no significant effect on β-carotene absorption in humans (Johnson et al. 1997, van den Berg and van Vliet 1998) or ferrets (White et al. 1993). Their conclusion was that in an *in vitro* cell culture system, the main carotenoid interactions occur only between nonpolar carotenoids (β- and α-carotene and β-carotene/lycopene), suggesting that hydrocarbon carotenoids that exhibit similar structural characteristics could follow similar pathways for their cellular uptake and/or incorporation into chylomicrons. These mutual interactions were also consistent with the idea of a facilitated uptake process. Although the secretion of carotenoids in the basolateral chamber was also assessed in this work, we cannot make conclusions regarding competition between carotenoids during the incorporation into chylomicrons. Indeed, decreased secretion could also be the consequence of competition at the level of cellular uptake.

In a third study performed in our laboratory in 2005, lutein absorption was measured after lutein-rich mixed micelles were mixed either with carotenoid-free mixed micelles or with mixed micelles containing β-carotene and/or lycopene (Reboul et al. 2005). The carotenoids were provided at physiological concentrations (i.e., 0.90, 0.2, and 0.13 μM for lutein, β-carotene, and lycopene, respectively) while the mixed micelles contained lipids from the digestion process and biliary salts. Lutein absorption was significantly decreased when micellar lutein was co-incubated for 3 h with micellar β-carotene (approx. 20%) or with both micellar β-carotene and lycopene, but not with micellar lycopene alone (Figure 18.1). Although it is unclear if significant competition could have been observed at a reduced, more physiological time of incubation (i.e., 30 min), this last result is in agreement with human studies in which β-carotene significantly affected lutein absorption (Kostic et al. 1995, van den Berg 1998, van den Berg and van Vliet 1998, Tyssandier et al. 2002). In contrast to what was observed in humans (Tyssandier et al. 2002), the fact that lycopene did not significantly affect lutein absorption was explained either by the lower concentration of lycopene that could have been incorporated into micelles (0.13 μM instead of 0.2 μM for β-carotene), or by the fact that this carotenoid has no β-ionone rings (in contrast with β-carotene and lutein). Note that the different results obtained in this study and in the study from During et al. (2002) can be explained by the vehicle used to deliver the carotenoids (Tween 40 micelles vs. the physiological mixed micelles).

In summary, Caco-2 cells studies strongly suggest that carotenoids interact with each other at the level of cellular uptake by the enterocyte. This phenomenon has been explained by the fact that the uptake of several carotenoids involves, at least in part, the same intestinal membrane transporter: the scavenger receptor class B type I SR-BI (Reboul et al. 2005, van Bennekum et al. 2005, Moussa et al. 2008).

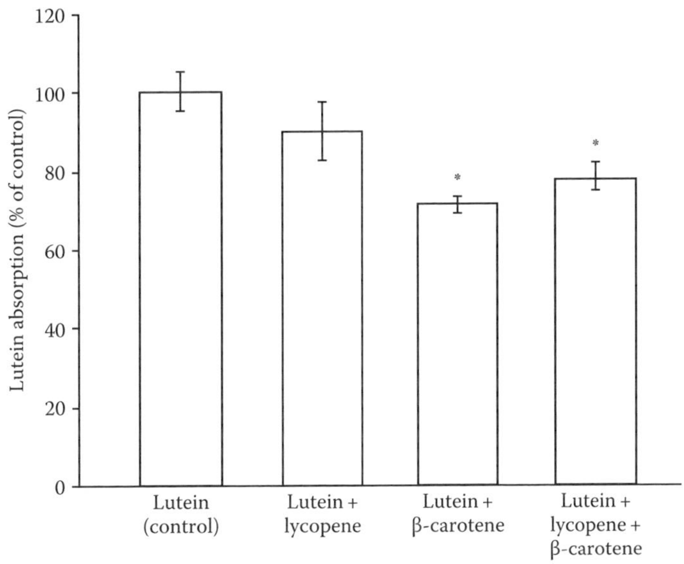

FIGURE 18.1 Competitive effects of β-carotene and lycopene on lutein absorption. The effects of β-carotene and lycopene on lutein absorption were observed in confluent Caco-2 cells. The apical side of the cells received FBS-free medium containing lutein-rich micelles (0.90 μM), or lutein-rich micelles (0.90 μM) plus lycopene-rich micelles (0.13 μM), or lutein-rich micelles (0.90 μM) plus β-carotene-rich micelles (0.20 μM), or lutein-rich micelles (0.90 μM) plus β-carotene-rich micelles (0.90 μM) plus lycopene-rich micelles (0.13 μM). The basolateral side received complete medium. Incubation time was 180 min. Data are mean ± SEM of three assays. An asterisk indicates a significant difference with control (0.90 μM lutein-rich micelles alone).

18.2.2 Competition between Carotenoids and Other Antioxidant Micronutrients

In a normal meal containing plant-derived foods, carotenoids are necessarily ingested with other dietary antioxidants, the main ones being vitamin C, E, and polyphenols. It is assumed that these antioxidants may either protect carotenoids from degradation in the gastrointestinal tract before their absorption, or compete with carotenoids for absorption. The effect of the other main dietary antioxidants on carotenoid uptake by intestinal cells has been addressed in a recent study performed in our laboratory (Reboul et al. 2007a). In this study, a full factorial design experiment was elaborated to assess the effect of vitamin C, vitamin E (an equimolar mixture of (*R,R,R*)-α-tocopherol and (*R,R,R*)-γ-tocopherol was used), and polyphenols (a mixture of gallic acid, caffeic acid, (+)-catechin and naringenin was used) on the absorption of a carotenoid model: lutein. All of the above cited antioxidants were provided to cells across a range of physiological concentrations. While the mixture of polyphenols significantly ($P < 0.05$) impaired lutein uptake, no significant effects were observed with vitamins C or E. Also, no significant degradation of lutein occurred during the duration of the experiments regardless of the presence of other antioxidant micronutrients, which indicates that the observed effects were likely due to competition and not degradation. Additionally, no interaction was observed between the different classes of micronutrients in terms of lutein uptake. The fact that the polyphenol mixture significantly impaired lutein uptake raised the question as to which specific polyphenol(s) of the mixture were responsible for this effect. Therefore, we conducted a second series of experiments to measure the individual effect of each polyphenol. These additional experiments showed that naringenin was the only polyphenol able to significantly impair lutein uptake (about 25% for 25 mM and 50% for 150 mM naringenin, respectively; $P < 0.05$). We observed that the mixture of polyphenols containing 25 mM naringenin had a similar effect on 25 mM naringenin alone, indicating that the other polyphenols tested (gallic acid, caffeic acid, and (+)-catechin) had probably no significant effect on lutein uptake. The specific effect exerted by naringenin needs to be further elucidated by additional experiments, but it was highlighted that naringenin was the most lipophilic of all the polyphenols tested ($\log P = 2.52$ vs. 0.86, 0.82, and 0.38 for gallic acid, caffeic acid, and (+)-catechin, respectively; Cooper et al. 1997). Therefore it was hypothesized that naringenin affects lutein uptake through an interaction with SR-BI, which is known to transport lipophilic molecules with low substrate specificity. A second hypothesis was that naringenin interacts with membrane lipids (Tachibana et al. 2004), thereby altering the invagination of lipid raft domains containing lutein receptors.

In the above study, vitamin E had no significant effect on lutein absorption. This was surprising since both (*R,R,R*)-α-tocopherol and (*R,R,R*)-γ-tocopherol have been shown to be transported through the SR-BI (Reboul et al. 2006), which is also involved in lutein uptake (Reboul et al. 2005). We suggest that this is likely due to the fact that vitamin E was provided over a concentration range that was close to the normal physiological concentration (maximum 5.5 μM). When tocopherols and carotenoids were incubated at higher concentrations (40 and 6 μM for α-tocopherol and lutein, respectively) (Reboul et al. 2006), lutein significantly impaired tocopherol uptake. Conversely, there was no significant effect of β-carotene or lycopene on tocopherol uptake but, as discussed above this was probably due to the low concentration of these carotenoids that could be incorporated in mixed micelles (2.8 and 0.4 μM for β-carotene and lycopene, respectively). This result is in agreement with another study in which it was shown that a mixture of carotenoids (lycopene, β-carotene, and lutein) significantly impaired α-tocopherol absorption in Caco-2 cells (Reboul et al. 2007b). The studies described above show that carotenoids likely impair tocopherol absorption. Therefore, although not yet observed in Caco-2 cell experiments, it is likely that tocopherol can impair carotenoid absorption as well. This hypothesis is supported by an *in vivo* study in rats where α-tocopherol decreased canthaxanthin absorption (Hageman et al. 1999).

18.2.3 Competition between Carotenoids and Other Fat-Soluble Compounds

Plant sterols and stanols (phytosterols) effectively diminish exogenous (dietary) and endogenous (biliary) cholesterol absorption in humans (Nguyen 1999). It is assumed that this is due to a competition between these cholesterol and phytosterols with regard to their incorporation into mixed micelles. Thus, it has been hypothesized that the absorption of carotenoids, which are also incorporated into mixed micelles during digestion, may also be impaired by phytosterols. Phytosterols might also diminish cholesterol absorption by another mechanism. Indeed it has been shown that these plant sterols may enhance cholesterol efflux back to the apical side of the cell, that is, intestinal lumen. It is tempting to suggest that such a mechanism works also for carotenoids. To support an inhibitory effect of phytosterols on carotenoid absorption it has been shown that the cellular uptake of 7.5 μM β-carotene was significantly reduced to about 50% in Caco-2 cells by the presence of 20 μM β-sitosterol in the medium (Fahy et al. 2004). Whatever the mechanism involved, that is, competition for incorporation into micelles or net uptake by the enterocyte, the inhibitory effect of phytosterols, in either free or ester form, on carotenoid bioavailability has been confirmed in a clinical study (Richelle et al. 2004).

In conclusion, the Caco-2 cell monolayer model has given original data on the competition effect of several nutrients on carotenoid uptake. Most of these data have been confirmed in several *in vivo* studies, including clinical studies, confirming that this model is a valuable tool to study competition effects on carotenoid absorption.

REFERENCES

Artursson, P. and J. Karlsson (1991). Correlation between oral drug absorption in humans and apparent drug permeability coefficients in human intestinal epithelial (Caco-2) cells. *Biochem. Biophys. Res. Commun.* **175**(3): 880–885.

During, A. et al. (1998). Characterization of beta-carotene 15,15′-dioxygenase activity in TC7 clone of human intestinal cell line Caco-2. *Biochem. Biophys. Res. Commun.* **249**(2): 467–474.

During, A. et al. (2002). Carotenoid uptake and secretion by CaCo-2 cells: Beta-carotene isomer selectivity and carotenoid interactions. *J. Lipid Res.* **43**(7): 1086–1095.

Fahy, D. M. et al. (2004). Phytosterols: Lack of cytotoxicity but interference with beta-carotene uptake in Caco-2 cells in culture. *Food Addit. Contam.* **21**(1): 42–51.

Ferruzzi, M. G. et al. (2001). Assessment of degradation and intestinal cell uptake of carotenoids and chlorophyll derivatives from spinach puree using an *in vitro* digestion and Caco-2 human cell model. *J. Agric. Food Chem.* **49**(4): 2082–2089.

Garrett, D. A. et al. (1999). Accumulation and retention of micellar beta-carotene and lutein by Caco-2 human intestinal cells. *J. Nutr. Biochem.* **10**(10): 573–581.

Garrett, D. A. et al. (2000). Estimation of carotenoid bioavailability from fresh stir-fried vegetables using an *in vitro* digestion/Caco-2 cell culture model. *J. Nutr. Biochem.* **11**(11–12): 574–580.

Gres, M. C. et al. (1998). Correlation between oral drug absorption in humans, and apparent drug permeability in TC-7 cells, a human epithelial intestinal cell line: Comparison with the parental Caco-2 cell line. *Pharm. Res.* **15**(5): 726–733.

Hageman, S. H. et al. (1999). Excess vitamin E decreases canthaxanthin absorption in the rat. *Lipids* **34**(6): 627–631.

Hauser, H. et al. (1998). Identification of a receptor mediating absorption of dietary cholesterol in the intestine. *Biochemistry* **37**(51): 17843–17850.

High, E. G. and H. G. Day (1951). Effects of different amounts of lutein, squalene, phytol and related substances on the utilization of carotene and vitamin A for storage and growth in the rat. *J. Nutr.* **43**: 245–260.

Johnson, E. J. et al. (1997). Beta-carotene isomers in human serum, breast milk and buccal mucosa cells after continuous oral doses of all-*trans* and 9-*cis* beta-carotene. *J. Nutr.* **127**(10): 1993–1999.

Kostic, D. et al. (1995). Intestinal absorption, serum clearance, and interactions between lutein and beta-carotene when administered to human adults in separate or combined oral doses. *Am. J. Clin. Nutr.* **62**: 604–610.

Kuo, S. M. et al. (2001). Dihydropyridine calcium channel blockers inhibit ascorbic acid accumulation in human intestinal Caco-2 cells. *Life Sci.* **68**(15): 1751–1760.

Manna, C. et al. (1997). The protective effect of the olive oil polyphenol (3,4-dihydroxyphenyl)-ethanol counteracts reactive oxygen metabolite-induced cytotoxicity in Caco-2 cells. *J. Nutr.* **127**(2): 286–292.
Moussa, M. et al. (2008). Lycopene absorption in human intestinal cells and in mice involves scavenger receptor class B type I but not Nienmann-Pick C1-like 1. *J. Nutr.* **138**: 1432–1436.
Musto, P. et al. (1995). All-*trans* retinoic acid for advanced multiple myeloma. *Blood* **85**: 3769–3770.
Nguyen, T. T. (1999). The cholesterol-lowering action of plant stanol esters. *J. Nutr.* **129**(12): 2109–2112.
Pinto, M. et al. (1983). Enterocyte-like differentiation an polarizationof the human colon carcinoma cell line Caco-2 in culture. *Biol. Cell* **47**: 323–330.
Play, B. et al. (2003). Glucose and galactose regulate intestinal absorption of cholesterol. *Biochem. Biophys. Res. Commun.* **310**(2): 446–451.
Reboul, E. et al. (2005). Lutein transport by Caco-2 TC-7 cells occurs partly by a facilitated process involving the scavenger receptor class B type I (SR-BI). *Biochem. J.* **387**(Pt 2): 455–461.
Reboul, E. et al. (2006). Scavenger receptor class B type I (SR-BI) is involved in vitamin E transport across the enterocyte. *J. Biol. Chem.* **281**(8): 4739–4745.
Reboul, E. et al. (2007a). Effect of the main dietary antioxidants (carotenoids, gamma-tocopherol, polyphenols, and vitamin C) on alpha-tocopherol absorption. *Eur. J. Clin. Nutr.* **61**(10): 1167–1173.
Reboul, E. et al. (2007b). Differential effect of dietary antioxidant classes (carotenoids, polyphenols, vitamins C and E) on lutein absorption. *Br. J. Nutr.* **97**(3): 440–446.
Richelle, M. et al. (2004). Both free and esterified plant sterols reduce cholesterol absorption and the bioavailability of beta-carotene and alpha-tocopherol in normocholesterolemic humans. *Am. J. Clin. Nutr.* **80**(1): 171–177.
Salvini, S. et al. (2002). Functional characterization of three clones of the human intestinal Caco-2 cell line for dietary lipid processing. *Br. J. Nutr.* **87**(3): 211–217.
Sugawara, T. et al. (2001). Lysophosphatidylcholine enhances carotenoid uptake from mixed-micelles by Caco-2 human intestinal cells. *J. Nutr.* **131**(11): 2921–2927.
Traber, M. G. et al. (1990). Vitamin E uptake by human intestinal cells during lipolysis *in vitro*. *Gastroenterology* **98**(1): 96–103.
Trotter, P. J. et al. (1996). Fatty acid uptake by Caco-2 human intestinal cells. *J. Lipid Res.* **37**(2): 336–346.
Tyssandier, V. et al. (2002). Vegetable-borne lutein, lycopene, and beta-carotene compete for incorporation into chylomicrons, with no adverse effect on the medium-term (3-wk) plasma status of carotenoids in humans. *Am. J. Clin. Nutr.* **75**(3): 526–534.
van Bennekum, A. et al. (2005). Class B scavenger receptor-mediated intestinal absorption of dietary beta-carotene and cholesterol. *Biochemistry* **44**(11): 4517–4525.
van den Berg, H. (1998). Effect of lutein on beta-carotene absorption and cleavage. *Int. J. Vitam. Nutr. Res.* **68**(6): 360–365.
van den Berg, H. (1999). Carotenoid interactions. *Nutr. Rev.* **57**(1): 1–10.
van den Berg, H. and T. van Vliet (1998). Effect of simultaneous, single oral doses of beta-carotene with lutein or lycopene on the beta-carotene and retinyl ester responses in the triacylglycerol-rich lipoprotein fraction of men. *Am. J. Clin. Nutr.* **68**(1): 82–89.
van Vliet, T. et al. (1996). *In vitro* measurement of beta-carotene cleavage activity: Methodological considerations and the effect of other carotenoids on beta-carotene cleavage. *Int. J. Vitam. Nutr. Res.* **66**: 77–85.
Werder, M. et al. (2001). Role of scavenger receptors SR-BI and CD36 in selective sterol uptake in the small intestine. *Biochemistry* **40**(38): 11643–11650.
White, W. S. et al. (1993). Interactions of oral β-carotene and canthaxanthin in ferrets. *J. Nutr.* **123**(8): 1405–1413.

Part VII

The Chemistry and Biochemistry of Carotene Oxidases, Cell Regulation, and Cancer

19 Diverse Activities of Carotenoid Cleavage Oxygenases

Erin K. Marasco and Claudia Schmidt-Dannert

CONTENTS

19.1 INTRODUCTION TO APOCAROTENOIDS

Apocarotenoids are isoprenoid compounds that contain shortened carbon backbones compared to the naturally occurring carotenoids from which they are derived by oxidative cleavage. Cleavage of the carotenoid backbone can occur through nonspecific nonenzymatic (e.g., radical formation) or enzymatic oxidation (e.g., peroxidases). However, in biological systems where apocarotenoids exhibit specific biological activities, the oxidative cleavage of the carotenoid backbone is catalyzed by a class of enzymes known as carotenoid cleavage enzymes. Unlike unspecific carotenoid cleavage, these enzymes catalyze cleavage of specific double bonds of the carotenoid backbone. Enzymes with preferences for different carotenoid substrates and activities for cleaving different sites within a specific carotenoid have been identified and are discussed in detail in this chapter. Because the mechanism by which these enzymes catalyze oxidative cleavage, either via a monooxygenase or dioxygenase mechanism, is currently controversial, we use the term carotenoid cleavage oxygenases (CCOs).

Presently more than 100 naturally occurring apocarotenoids have been described but the possible structural diversity of apocarotenoids is much larger because of the large structural diversity of carotenoids found in nature (more than 600 known structures), the many possible oxidative cleavage sites within a carotenoid molecule, and additional diversification of carotenoid cleavage products by subsequent modifications like oxidation and glycosylation reactions (Fleischmann et al. 2002, Carail and Caris-Veyrat 2006). The functions of apocarotenoids are equally diverse and we are just at the beginning of understanding their varied biological roles. Carotenoid cleavage products for example function as signaling molecules (Booker et al. 2004), hormones (e.g., abscisic acid in plants) (Tan et al. 2003) or vitamin A (e.g., retinal) precursors (Bouvier et al. 2005). Other apocarotenoids are important in microbial light-driven transport and phototaxis and phototransduction (retinal) in vertebrates (Beja et al. 2000, 2001, Spudich et al. 2000). In vertebrates, retinoids (in particular, retinoic acid [RA] and 9-*cis*-retinal) have important signaling functions and regulate metabolic and developmental processes (Lampert et al. 2003, Soprano et al. 2004, Altucci et al. 2007, Anthonise et al. 2007, Alexander et al. 2008). Many apocarotenoids are of value for food, nutraceutical, medical and agricultural applications. Carotenoid cleavage compounds are potent aroma compounds (e.g., β-ionone, pseudoionone) and are important food colorants and spices (e.g., crocin and bixin) (reviewed by Camara and Bouvier (2004) and Bouvier et al. (2005)). RA may be beneficial for the treatment and prevention of several diseases including skin and metabolic diseases, promyelocytic leukemia, and cancers (Soprano et al. 2004, Altucci et al. 2007). In plants, apocarotenoids are involved in seed development, branching, and adaptation to environmental stresses (reviewed by Booker et al. (2004), Schwartz et al. (2004), and Auldridge et al. (2006)) and therefore of agricultural interest.

In this chapter, we first describe the different types of CCOs that have been identified in plants, vertebrates and microorganisms, followed by discussions of our current understanding of their biochemistry and biological functions. A section describing current and potential commercial applications, including limitations that presently restrict biotechnological use of CCOs, concludes our review.

19.2 APOCAROTENOID FORMATION

The formation of apocarotenoids can be either through nonspecific mechanisms such as photooxidation or lipoxygenase (LOX) cooxidation (Wache et al. 2003) or through specific enzymatic activity. Nonspecific mechanisms generate a range of products in an unregulated manner (Caris-Veyrat et al. 2003, Carail and Caris-Veyrat 2006). For example, nonspecific oxidation is important for flavor development in tea fermentation and tobacco curing. Enzymatic oxidation of carotenoids frequently involves enzymes with mechanisms involving the formation of free radicals and active oxygen species such as phenoloxidase, peroxidase, LOX, and xanthine oxidase. For example, gastric mucosal homogenates with fatty acid hydroperoxides cleave β-carotene in a nonspecific manner to produce a range of products from C_{18}–C_{30} (Yeum et al. 1994). Similar metabolites are also formed by human gastric mucosal homogenates, LOX and linoleic acid hydroperoxide (Yeum et al. 1995). Plants have LOXs that catalyze the cooxidation of carotenoids in the presence of polyunsaturated fatty acids. A fatty acylperoxy radical is generated by the LOX and attacks the polyene chain randomly (Wu et al. 1999). Microorganisms are known to secrete peroxidases that break down carotenoids for utilization as carbon source (Zorn et al. 2003a,b).

In the last decade, a new class of enzymes responsible for the specific oxidative cleavage of carotenoids has been identified in plants, animals, and more recently, in microorganisms. Collectively, these enzymes are known as CCOs, although they are subdivided into different classes depending on their cleavage activities (Figure 19.1) (reviewed by Bouvier et al. [2005]). These enzymes are distinct from the oxidases described above because of the high specificity they exhibit both for a specific carotenoid and the site of cleavage. Further, carotenoid cleavage by

CCO enzymes is frequently a regulated process with additional biochemical steps that create the biologically active apocarotenoid (Iuchi et al. 2000, Simkin et al. 2004a,b, Rodrigo et al. 2006). In the following section we will discuss the different types of CCOs that have been isolated and characterized over the last decade.

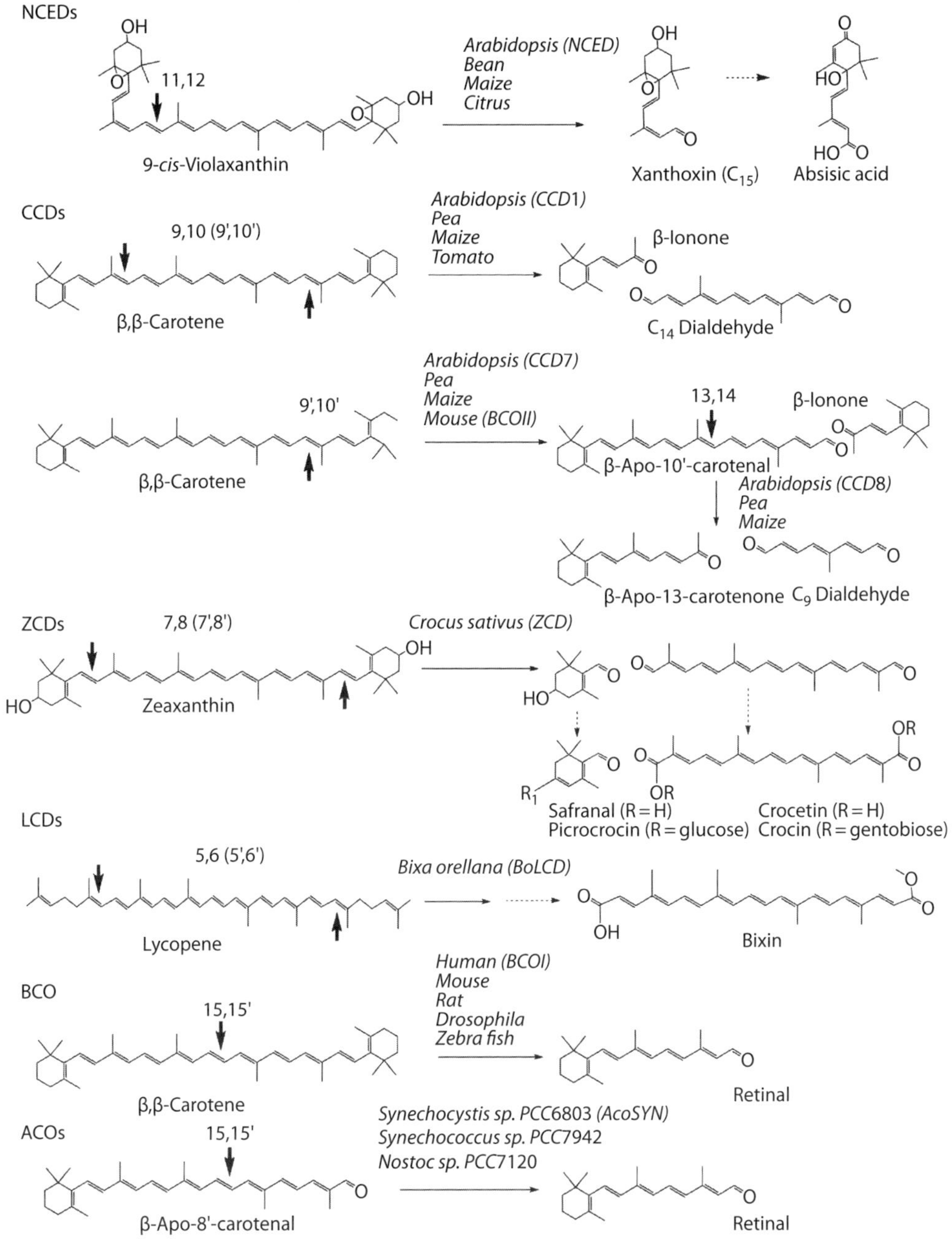

FIGURE 19.1 Representative cleavage patterns observed by assorted CCO enzymes. Carotenoid cleavage enzymes can be designated into several classes based on substrate specificity and cleavage site regioselectivity. NCEDs cleave the 11,12-double bond; CCDs cleave the 9,10-(9′,10′-) position excluding a few exceptions (LCD and ZCD); BCO1s are mammalian enzymes that cleave the 15,15′-position; ACOs are microbial enzymes that cleave apocarotenoids at the 15,15′-double bond.

19.3 CAROTENOID CLEAVAGE OXYGENASES

There has been evidence of specific carotenoid cleavage for approximately 50 years arising from research on vitamin A production. Carotenoid cleavage has also been long known to be an important rate-limiting step in the formation of the plant hormone abscisic acid (ABA). An important breakthrough came in 1997 when the first CCO was cloned and characterized from a plant. This enzyme was later found to be related to the enzyme responsible for vitamin A formation in animals. In the ensuing decade, sequence homology has led researchers to identify an ever increasing number of CCO examples. A BLAST search of genome sequences with known CCO representatives yields a large number of CCO homologs in all kingdoms (Figure 19.2). At present, over 200 examples of putative CCO homologs can be identified in sequence databases (reviewed by Kloer and Schulz (2006)). However, only a small fraction of enzymes categorized as CCOs (~30) have actually been cloned, characterized, and/or tested *in vitro* (Tables 19.1 through 19.4). Generally, all characterized CCO enzymes oxidatively cleave one or two internal carbon–carbon double bonds (reviewed by Bouvier et al. (2003), Giuliano et al. (2003), and Kloer and Schulz (2006)). While most of these enzymes likely cleave carotenoids, other family members may act on different substrates.

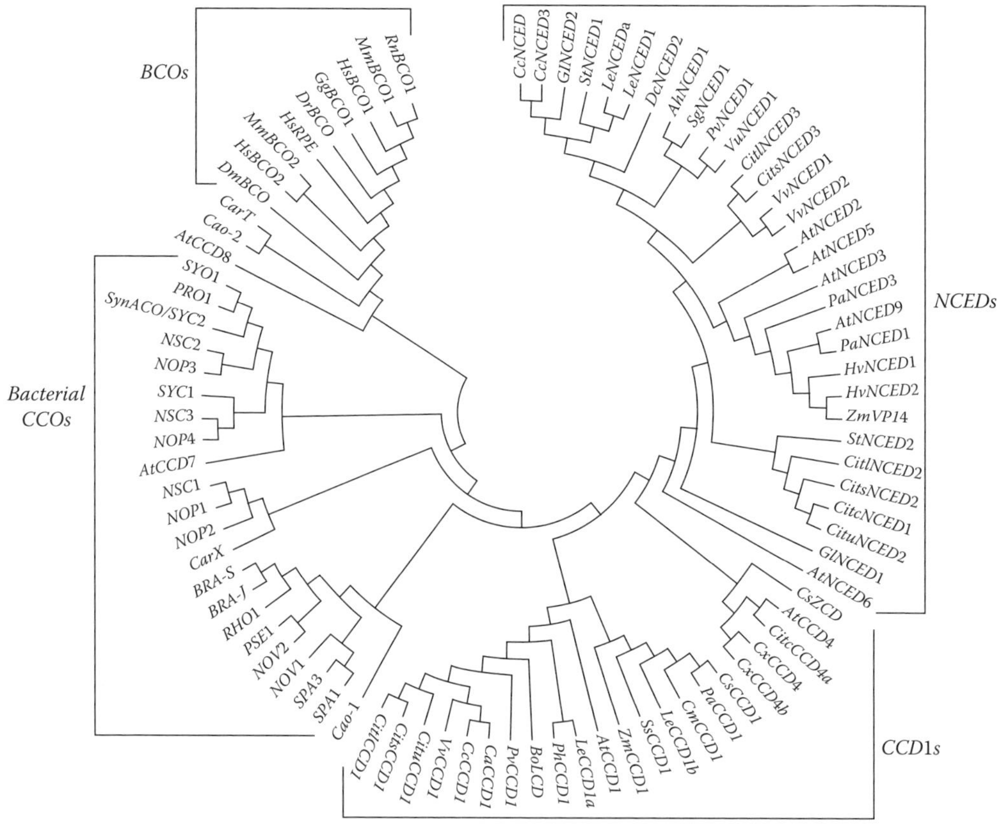

FIGURE 19.2 Radial phylogram representation of CCO homologs. Functionally characterized CCO homologs listed in Tables 19.1 through 19.4 were aligned with ClustalW (EMBL). The evolutionary history was inferred using the Neighbor-Joining method (Saitou and Nei 1987). Phylogenetic analyses were conducted in MEGA4 (Tamura et al. 2007). There is clear clustering of NCED, CCD1, BCO and of microbial CCO representatives. NCEDs cleave the 11,12-double bond; CCDs cleave the 9,10-(9′,10′-) position excluding a few exceptions (LCD, ZCD, CCD7, CCD8); BCO1s are vertebrate enzymes that cleave the 15,15′-position; BCO2s are vertebrate homologs that cleave the 9,10-double bond. Microbial CCOs are more disparate and function cannot be deduced from sequence identity. Accession numbers and enzyme names can be found in Tables 19.1 through 19.4.

TABLE 19.1
Functionally Characterized NCEDs

Name	Organism	Accession No.	References	Location[a]
AtNCED2	*Arabidopsis thaliana*	NP_193569	Tan et al. (2003)	Root
AtNCED3	*A. thaliana*	NP_188062	Tan et al. (2003)	Root, leaf, stem
AtNCED5	*A. thaliana*	NP_174302	Tan et al. (2003)	Seed
AtNCED6	*A. thaliana*	NP_189064	Tan et al. (2003)	Seed, endosperm
AtNCED9	*A. thaliana*	NP_177960	Tan et al. (2003); Lefebvre et al. (2006)	Seed, endosperm
AhNCED1	*Arachis hypogaea*	CAE00459	Wan and Li (2005)	
CitcNCED1	*Citrus clementina*	ABC26010	Agusti et al. (2007)	
CitlNCED2	*Citrus limon*	BAE92962	Kato et al. (2006)	
CitlNCED3	*C. limon*	BAE92965	Kato et al. (2006)	
CitsNCED2	*Citrus sinensis*	BAE92961	Kato et al. (2006)	
CitsNCED3	*C. sinensis*	BAE92964	Kato et al. (2006)	
CituNCED2	*Citrus unshiu*	BAE92960	Kato et al. (2006)	
CcNCED	*Coffea canephora*	ABA43901	Simkin et al. (2008)	
CcNCED3	*C. canephora*	ABD78412	Simkin et al. (2008)	
DcNCED2	*Daucus carota*	ABB52079	Soar et al. (2004)	
GlNCED1	*Gentiana lutea*	AAS47837	Zhu et al. (2007)	Flower
GlNCED2	*G. lutea*	AAS47838	Zhu et al. (2007)	
HvNCED1	*Hordeum vulgare L.*	BAF02837	Chono et al. (2006)	Root, grain
HvNCED2	*H. vulgare L.*	AB239298	Chono et al. (2006)	Root, grain, embryo
LeNCEDa	*Lycopersicon esculentum*	CAB10168	Burbidge et al. (1997)	
LeNCED1	*L. esculentum*	CAD30202	Thompson et al. (2000a,b); Thompson et al. (2004)	
PaNCED1	*Persea americana*	AAK00632	Chernys and Zeevaart (2000)	
PaNCED3	*P. americana*	AAK00623	Chernys and Zeevaart (2000)	
PvNCED1	*Phaseolus vulgaris*	Q9M6E8	Qin and Zeevaart (1999)	Fruit ripening, leaf
StNCED1	*Solanum tuberosum*	AAT75151	Destefano-Beltran et al. (2006)	
StNCED2	*S. tuberosum*	AAT75152	Destefano-Beltran et al. (2006)	
SgNCED1	*Stylosanthes guianenses*	AAY98512	Yang and Guo (2007)	
VuNCED1	*Vigna unguiculata*	BAB11932	Iuchi et al. (2000)	
VvNCED1	*Vitris vinifera*	AAR11193	Soar et al. (2004)	
VvNCED2	*V. vinifera*	AAR11194	Soar et al. (2004)	
ZmVP14	*Zea mays*	AAB62181	Schwartz et al. (1997)	Leaf, root, embryo

[a] If known.

TABLE 19.2
Functionally Characterized CCDs

Name	Organism	Accession No.	Activity	References
AtCCD1	*Arabidopsis thaliana*	CAA06712	9,10 (9′,10′)	Schwartz et al. (2001)
AtCCD4	*A. thaliana*	O49675		Iuchi et al. (2001); Naested et al. (2004); Auldridge et al. (2006)
AtCCD7	*A. thaliana*	Q7XJM2	9,10	Booker et al. (2004); Schwartz et al. (2004); Auldridge et al. (2006)
AtCCD8	*A. thaliana*	Q8VY26	Cleaves C_{27}	Schwartz et al. (2004); Bainbridge et al. (2005); Auldridge et al. (2006)
BoLCD	*Bixa orellana*	CAD71148		Bouvier et al. (2003)
CaCCD1	*Coffea Arabica*	ABA43904		Simkin et al. (2008)
CcCCD1	*Coffea canephora*	ABA43900		Simkin et al. (2008)
CitsCCD1	*Citrus sinensis*	BAE92958	9,10 (9′,10′)	Kato et al. (2006)
CitlCCD1	*Citrus limon*	BAE92959	9,10 (9′,10′)	Kato et al. (2006)
CituCCD1	*Citrus unshiu*	BAE92957	9,10 (9′,10′)	Kato et al. (2006)
CitcCCD4a	*Citrus clementina*	ABC26011		Agusti et al. (2007)
CmCCD1	*Cucumis melo*	ABB82946	9,10 (9′,10′)	Ibdah et al. (2006)
CxCCD4	*Chrysanthemum x morifolium*	BAF36654		Ohmiya et al. (2006)
CxCCD4b	*C. x morifolium*	BAF36656		Ohmiya et al. (2006)
CsCCD1	*Crocus sativus*	Q84KG5	9,10 (9′,10′)	Bouvier et al. (2003)
CsZCD	*C. sativus*	Q84K96	7,8 (7′,8′)	Bouvier et al. (2003)
LeCCD1a	*Lycopersicon esculentum*	AAT68187	9,10 (9′,10′)	Simkin et al. (2004)
LeCCD1b	*L. esculentum*	AAT68188	9,10 (9′,10′)	Simkin et al. (2004)
PaCCD1	*Persea americana*	AAK00622	9,10 (9′,10′)	Chernys and Zeevaart (2000)
PhCCD1	*Petunia hybrid*	AAT68189	9,10 (9′,10′)	Simkin et al. (2004)
PvCCD1	*Phaseolus vulgaris*	AAK38744	9,10 (9′,10′)	Schwartz et al. (2001)
SsCCD1	*Suaeda salsa*	AAY21819		Cao et al. (2005)
VvCCD1	*Vitis vinifera*	AAX48772	9,10 (9′,10′)	Mathieu et al. (2005)
ZmCCD1	*Zea mays*	AAV39613	9,10 (9′,10′)	Walter et al. (2007)

For example, stilbene oxygenases, which cleave the interphenyl α,β-double bond of stilbene derivatives, are known to be members of the CCO family of enzymes that act on substrates other than carotenoids (see Section 19.3.5).

CCOs are referred to in some literature as carotenoid cleavage dioxygenases (CCDs). But their initial classification as dioxygenases was not based on experimental evidence for such a mechanism. The term CCO will therefore be used to describe the enzymes in general. However, for the sake of clarity, names of previously published CCOs will remain as in the original literature. Characterized CCOs have been classified into several different groups (9-*cis*-epoxy carotenoid dioxygenases [NCEDs], carotenoid cleavage dioxygenase 1 [CCD1], β-carotene oxygenases [BCO], apocarotenoid cleavage oxygenases [ACO], and lignostilbene dioxygenases [LSD]) depending on their origin, sequence identity, and substrate specificity (Figures 19.1 and 9.2) (Bouvier et al. 2005). The localization and regulation of

TABLE 19.3
Functionally Characterized BCOs from Vertebrates

Name	Organism	Accession No.	Activity	Location[a]	Reference
DmBCO	*D. melanogaster*	AAF54978	15,15′	Head	von Lintig and Vogt (2000)
DrBCO	*Danio rerio*	NP_571873	15,15′		Lampert et al. (2003)
GgBCO1	*Gallus gallus*	Q9I993	15,15′		Wyss et al. (2000)
HsBCO1	*Homo sapiens*	Q9HAY6	15,15′	RPE cells, kidney, testis, liver, brain, small intestine, colon	Yan et al. (2001)
HsBCO2	*Homo sapiens*	Q9BYV7	9′,10′		Kiefer et al. (2001)
HsRPE	*Homo sapiens*	Q16518	Isomerase	RPE cells	Nicoletti et al. (1995)
MmBCO1	*Mus musculus*	Q9JJS6	15,15′	Liver, kidney, small intestines, testis	Wyss et al. (2001)
MmBCO2	*Mus musculus*	Q99NF1	9′,10′		Kiefer et al. (2001)
RnBCO1	*Rattus norvegicus*	Q91XT5	15,15′	Intestine	Takitani et al. (2006)

[a] If known.

the groups are distinct as well (see below). Sections 19.3.1 through 19.3.5 list and discuss CCOs from plants (NCEDs, CCD1s), vertebrates (BCOs), and microorganisms (CCOs, ACOs).

19.3.1 Plant CCOs

19.3.1.1 NCEDS

9-*cis*-Epoxy dioxygenases, the first CCOs described, are involved in ABA formation in plants (Schwartz et al. 1997). Xanthophylls (oxygen containing carotenoids) with a 9-*cis* conformation (i.e., violaxanthin and neoxanthin) are cleaved at the 11,12-double bond to form xanthoxin (C_{15}) which is then converted into ABA (Figure 19.1). ABA is a plant hormone that affects drought tolerance, seed development, and sugar sensing (Schwartz et al. 2003, Taylor et al. 2005). The first NCED to be cloned and described was identified in the maize (*Zea mays*) mutant, *viviparous 14* (VP14) (Schwartz et al. 1997, Tan et al. 1997). This transposon mutant had no lesion in carotenoid synthesis, but was impaired in the cleavage activity necessary to form ABA. VP14 is the prototypical NCED, but several other NCEDs from tomato (Burbidge et al. 1999), orange (Rodrigo et al. 2006), bean (Qin and Zeevaart 1999), avocado (Chernys and Zeevaart 2000), cowpea (Iuchi et al. 2000), bean (Qin and Zeevaart 1999), and *Arabidopsis* (Iuchi et al. 2001) have been identified and functionally characterized as recombinant proteins expressed in *E. coli* (Table 19.1). Known substrates of NCEDs have a 5,6-epoxy-3-hydroxy-β-ionone ring near the cleavage site (Schwartz et al. 2003). While recombinant enzymes *in vitro* act on a number of 9-*cis*-epoxy carotenoids, neoxanthin is thought to be the primary substrate in plants. For example, the K_M of recombinant maize

TABLE 19.4
Overview of Characterized CCOs from Microorganisms

Name	Organism	Accession No.	Activity	References
BRA-J	*Bradyrhizobium japonicum* USDA110	NP_772430	?	Marasco and Schmidt-Dannert (2008)
BRA-S	*Bradyrhizobium* sp. BTai	YP_001241346	?	Marasco and Schmidt-Dannert (2008)
Cao-1	*Neurospora crassa* OR74A	XP_961764	?	Saelices et al. (2007)
Cao-2	*N. crassa* OR74A	XP_958452	Tor.	Saelices et al. (2007)
CarT	*Gibberella zeae* PH-1 (*F. fujikuroi*)	XP_382801	Tor.	Prado-Cabrero et al. (2007)
CarX	*G. zeae* PH-1 (*F. fujikuroi*)	XP_383243	15,15′	Prado-Cabrero et al. (2007)
NOP1	*Nostoc punctiforme* PCC 73102	ZP_00106997	9′,10′	Marasco et al. (2006)
NOP2	*N. punctiforme* PCC 73102	ZP_00106156	?	Unpublished
NOP3	*N. punctiforme* PCC 73102	ZP_00112423	?	Marasco et al. (2006), unpublished
NOP4	*N. punctiforme* PCC 73102	ZP_00111018	15,15′	Marasco et al. (2006), unpublished
NOV1	*Novosphingobium aromaticivorans*	YP_496081	LSO	Marasco and Schmidt-Dannert (2008)
NOV2	*N. aromaticivorans*	YP_498079	LSO	Marasco and Schmidt-Dannert (2008)
NSC1	*Nostoc* sp. PCC 7120	NP_485149	9,10*	Marasco et al. (2006)
NSC2	*Nostoc* sp. PCC 7120	NP_488324	15,15′*	Marasco et al. (2006)
NSC3	*Nostoc* sp. PCC 7120	NP_488935	9′,10′	Marasco et al. (2006)
PRO1	*Prochlorococcus marinus* MIT9313	NP_895705	?	Marasco et al. (2006)
RHO1	*Rhodopseudomonas palustris* CGA0009	NP_946559	15,15′*	Unpublished
SPA1	*Sphingomonas paucimobilis*	AAC60447	LSO	Kamoda and Saburi (1993)
SPA2	*S. paucimobilis*	Heterodimer	LSO	Kamoda and Saburi (1993)
SPA3	*S. paucimobilis*	AAB35856	LSO	Kamoda and Saburi (1995)
SPA4	*S. paucimobilis*		LSO	Kamoda et al. (1997)
SYC1	*Synechocystis* PCC6803	NP_441785	?	Ruch et al. (2005); Marasco et al. (2006)
SynACO (SYC2)	*Synechocystis* PCC6803	NP_441748	15,15′	Ruch et al. (2005); Marasco et al. (2006)
SYO1	*Synechococcus elongates* PCC7942	YP_399215	15,15′*	Marasco et al. (2006), unpublished
PSE1	*Pseudomonas putida*	BAF62888	ISO	Yamada et al. (2007)

Note: Tor, torulene; LSO, lignostilbene; ISO, isoeugenol; *, full-length carotenoid cleavage; ?, activity not known.

VP14 and bean *Pv*NCED are lower for neoxanthin than 9-*cis*-violaxanthin (Qin and Zeevaart 1999, Schwartz et al. 2003).

Cleavage of xanthophylls by NCEDs occurs in the thylakoid membranes of chloroplasts. NCEDs contain signaling peptide sequences that target them to the thylakoid membranes. Transport studies with the oxygenase *Pv*NCED1 from bean showed transport into chloroplasts and association with the thylakoid membranes. Similar studies with *Vu*NCED (cowpea) also showed targeted transport to the chloroplasts (Iuchi et al. 2000). Disruptions or deletions of the VP14 N-terminus interfered with thylakoid associations (Tan et al. 2001). Following cleavage of plastidal 9-*cis*-epoxy carotenoids to xanthoxin, the C_{15} apocarotenoid is transported to the cytosol where it is oxidized and reduced to make ABA (Seo and Koshiba 2002, Auldridge et al. 2006). The transport mechanisms of xanthoxin are unknown at this time.

The NCEDs represent a multigene family with spatial and temporal regulation (Chernys and Zeevaart 2000, Iuchi et al. 2000, Lefebvre et al. 2006, Rodrigo et al. 2006, Kato et al. 2007). Many plants contain multiple NCEDs and they are thought to act during different developmental and growth phases. Of the nine CCO paralogs found in *Arabidopsis*, five are classified as NCEDs and contain signal peptides that target them to the thylakoid membranes. NCED 2 and 3 from *A. thaliana* are expressed in the roots, whereas NCEDs 6 and 9 have seed specific expression (Tan et al. 2003, Lefebvre et al. 2006). In *Phaseolus vulgaris*, NCEDs 1 and 3 are expressed during fruit ripening and NCED1 is present in the leaves (Chernys and Zeevaart 2000). Expression of NCEDs can be induced by water (Tan et al. 1997, Qin and Zeevaart 1999, Chernys and Zeevaart 2000, Tan et al. 2003, Rodrigo et al. 2006) or salt stress (Iuchi et al. 2000). Overexpression of the NCED enzymes in plants leads to an increase in ABA production and concurrent increase in drought tolerance (Thompson et al. 2000a,b, Qin and Zeevaart 2002).

19.3.1.2 CCDS

A second class of plant CCOs was discovered that had a higher affinity for carotenes than xanthophylls. These enzymes, named CCD1s, cleave cyclic carotenoids symmetrically at the 9,10- and 9′,10′-double bonds forming a C_{14} dialdehyde and two volatile C_{13} cyclohexone derivatives (e.g., β-ionone) (Figure 19.1). CCD1s are thought to play a role in carotenoid turnover; although the extent of this activity is not well characterized (Simkin et al. 2004). These enzymes have been described from a number of different sources (Table 19.2) (Schwartz et al. 2001, Simkin et al. 2004a,b, Cao et al. 2005). The β-ionone cleavage product produced by CCD1s can rearrange and be modified to a number of different molecules producing many of the aromas associated with fruits and plants (Leffingwell 2003).

The substrate specificity of these enzymes is not stringent; for example, CCD1 from tomato was also shown to cleave at the 9,10- and 9′,10′-positions of β-carotene, zeaxanthin, lutein, violaxanthin and neoxanthin all of which have different ionone ring modifications. Unlike NCEDs, CCD1 enzymes have no plastid-targeting sequences and are localized in the cytosol. It is postulated that they access the carotenoids in the plastids through a monotopic membrane association (Kloer et al. 2005). Similar to the NCEDs, the CCD1s have differential expression under different environmental conditions. The transcripts of *Ph*CCD1 (Table 19.2) in leaves oscillate according to light levels and circadian mechanisms (Simkin et al. 2004). Light exposure increases the transcript levels in leaves and transcription is diurnally regulated directly by light. In corollas, the oscillation of transcript levels is circadian in nature.

In addition to CCD1, there are three more CCO paralogs from *Arabidopsis* that preferentially cleave carotenes over 9-*cis*-epoxycarotenoids. These enzymes are targeted to the plastid and have been named CCD 4, 7, and 8. CCD7 is 9, 10 specific and like CCD1 accepts a variety of carotenoid substrates. However, unlike CCD1, this enzyme cleaves carotenoids asymmetrically: β-carotene is cleaved by CCD7 at the 9,10-position to yield β-ionone and 10′-apocarotenal. The C_{27} product is then cleaved by another CCD1 paralog, CCD8, into the C_{18} compound 13-apo-β-carotenone and a C_9 dialdehyde (Booker et al. 2004, Schwartz et al. 2004, Auldridge et al. 2006). The sequential activities of CCD7 and CCD8 synthesize a plant hormone that regulates apical dominance. The bioactive dialdehyde product has not been identified or characterized, but it is known to promote shoot branching because a loss of CCD7 or CCD8 results in a bushy growth phenotype (Auldridge et al. 2006). CCD7 and CCD8 orthologs have been found in many plant genomes suggesting the hormone has widespread importance in plant development. Another CCO paralog in the *Arabidopsis* genome is named CCD4, which cleaves carotenoids when expressed in *E. coli*, but it is not well characterized at this time (see Table 19.2 for a list of identified CCDs).

19.3.1.3 ZCD and LCD

In addition to NCEDs and CCD1s, another group of plant CCOs has been described (Table 19.2) that produces industrially important apocarotenoids such as the pigment bixin (derived from the seeds of the lipstick tree, *Bixa orellana*) and the spice saffron (derived from the stigmata of *Crocus sativus*). In *Bixa orellana,* a lycopene-specific 5,6 (5′,6′)-cleavage dioxygenase (*Bo*LCD) is responsible for the formation of bixin dialdehyde and a C_7 cleavage product MHO (6-methyl-5-hepten-2-one) (Figure 19.1). Unlike the previously discussed enzymes, this CCO was shown to exclusively cleave acyclic carotenoids. Expression of this lycopene specific CCO and two additional enzymes, bixin aldehyde dehydrogenase and norbixin carboxyl methyltransferase, in lycopene accumulating recombinant *E. coli* resulted in bixin biosynthesis (Bouvier et al. 2003). A closely related enzyme is the zeaxanthin-specific 7,8 (7′,8′)-cleavage dioxygenase (*Cs*ZCD) isolated from *Crocus sativus*. This enzyme was characterized and found to form crocetin dialdehyde and 3-hydroxy-β-cyclocitral *in vitro* (Bouvier et al. 2003). The two C_{10} cleavage products and the C_{20} product are modified to form safranal, picrocrocin and crocin, respectively (Figure 19.1). Crocins are responsible for the pigmentation of saffron, while safranal is responsible for the aroma of this spice and picrocrocins are the bitter tasting glucosides of safranal. One potential problem with the studies on *Bo*LCD and *Cs*ZCD, however, is the extremely high nucleotide sequence identity (97%) between the two reported sequences, suggesting that one of the reported sequences is probably incorrect.

19.3.2 Vertebrate CCOs

Over 40 years ago carotenoid cleavage was postulated to be the route to Vitamin A formation in vertebrates (Olson 1961, Olson and Hayaishi 1965). Although first shown with isolated protein (Lakshmanan et al. 1972), the discovery of VP14 in plants led to the cloning of vertebrate genes and the characterization of oxygenases from fruit fly (von Lintig and Vogt 2000), chicken (Wyss et al. 2000), mouse (Redmond et al. 2001), rat (Zaripheh et al. 2006), and human (Yan et al. 2001), which confirmed the reaction (Table 19.3 and Figure 19.1). Symmetric and asymmetric cleavage of C_{40} carotenoids are the only known routes to retinoids (Figure 19.3) (the generic term retinoids refers to retinal derivatives such as retinol [vitamin A] and 9-*cis*-RA). The symmetric route, cleavage at the 15,15′-double bond by the BCO1 class of enzymes, was the first to be discovered (Lakshmanan et al. 1972). BCO1 is responsible for the generation of two molecules of retinal per carotenoid cleaved. This reaction is highly specific for substrates with one nonsubstituted β-ionone ring (i.e.,

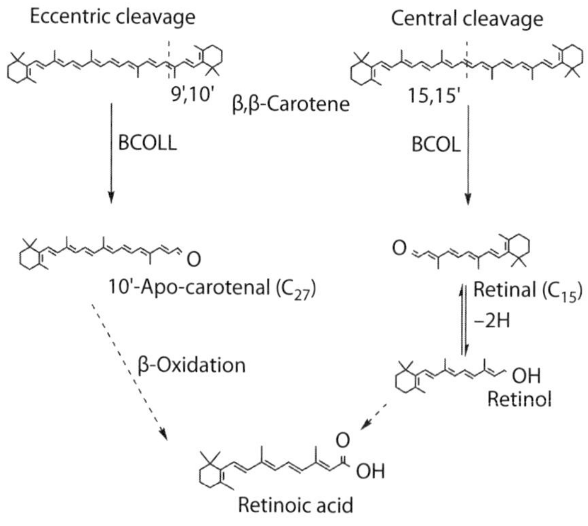

FIGURE 19.3 Pro-vitamin A cleavage and formation of retinoids via eccentric and central cleavage of β,β-carotene.

β-carotene, canthaxanthin) (Lindqvist and Andersson 2002). Human BCO1 was found specific for the pro-vitamin A carotenoids, β,β-carotene and β-cryptoxanthin (Lindqvist and Andersson 2002). Mouse BCO1, however, cleaves both lycopene and β,β-carotene in recombinant *E. coli*, but *in vitro* cleavage of lycopene required levels threefold higher than β,β-carotene (Redmond et al. 2001).

A second enzyme, BCO2, was identified that cleaves carotenoids asymmetrically at the 9,10-double bond to produce the 10-apocarotenal (C_{27}) and β-ionone (C_{13}), in a reaction similar to the *Arabidopsis* CCD7. Examples of BCO2 have been cloned from mouse, zebra fish, ferret, and human (Kiefer et al. 2001, von Lintig et al. 2005, Hu et al. 2006). Substrate studies with different BCO2s showed that these enzymes prefer acyclic carotenoids such as lycopene over cyclic carotenoids (Kiefer et al. 2001, von Lintig et al. 2005, Hu et al. 2006). These enzymes also seem to be selective for different carotenoid isomers. BCO2 from ferret for example cleaves *cis*-isomers of lycopene but not all-*trans*-lycopene (Hu et al. 2006).

19.3.3 Fungal CCOs

Some of the most recent CCO research has been the discovery of CCO homologs in fungi. Evidence from *Fusarium* carotenoid accumulation mutants suggested that CCO enzymes are responsible for the formation of some of the pigments seen in fungi. For example, the ascomycete fungus, *Fusarium fujikuroi,* synthesizes the acid apocarotenoid neurosporaxanthin by converting torulene (C_{40}) into neurosporaxanthin (C_{35}) (Prado-Cabrero et al. 2007). Another eukaryotic model organism, *Neurospora crassa*, contains an orthologous pathway to produce neurosporaxanthin (Saelices et al. 2007). Two light-induced enzymes, CarT (AM418467) from *Fusarium* and CAO-2 (XP_958452) from *Neurospora,* have been identified and found to be highly specific for cleavage of the torulene 4′,5′-double bond to produce the corresponding β-apo-4′-carotenal. Acyclic carotenoids such as lycopene were cleaved by CarT, but the bicyclic substrate β-carotene was not a substrate (Prado-Cabrero et al. 2007). No activity was observed with substrates such as γ-torulene lacking the 4′,5′-double bond (Saelices et al. 2007). Tests with monocyclic synthetic substrates showed that, in contrast to the apocarotenoid cleavage enzyme from *Synechocystis* sp. PCC6803 (SynACO) (see below), the cleavage site specificity is determined by the acyclic end of the substrate: β-apo-8′-carotenal and β-apo-10′-carotenal were converted into the 13′,14′- and 15,15′-cleavage products, respectively (Prado-Cabrero et al. 2007). *In vitro* assays with 8′-lycopenal also showed that the CarT enzyme prefers to cleave the nonoxygenated end of an acyclic carotenoid (Prado-Cabrero et al. 2007). Other CCO paralogs in *Fusarium* and *Neurospora* (CarX and CAO-1 in *F. fujikuroi* and *N. crassa,* respectively) do not perform this cleavage reaction and have a different function (see below).

Filamentous fungi contain opsin proteins that utilize light-mediated isomerization of a retinal chromophore for ion pumping or light-sensing (Prado et al. 2004). Heterologously expressed *N. crassa* opsin1 binds all-*trans* retinal to form a green-light absorbing pigment with the characteristics of archaeal sensory rhodopsin II, suggesting that the *N. crassa* opsin1 function may be important for sensory perception in *N. crassa* (Bieszke et al. 1999). Opsin1 has a role in transcriptional regulation during conidial development and carotenogenesis and plays an auxiliary role during late-stage conidial events and/or stress responses (Bieszke et al. 2007). The retinal chromophore is thought to be generated in *N. crassa* by the putative CCO enzyme CAO-1 (XP_961764.1). In *Fusarium,* the opsin protein is clustered with genes involved in carotenoid biosynthesis (carRA and carB). CarO, the opsin protein, and carX, the CCO, exhibit the same transcriptional pattern (low expression in dark and induction triggered by light illumination that peaks after 1 h of exposure). The deletion of the opsin-like protein CarO in the *car* gene cluster did not create a detectable phenotype (Prado et al. 2004), but disruption of carX (the oxygenase) resulted in induction of carotenoid biosynthesis (Prado-Cabrero et al. 2007). CarX cleaves β-carotene, γ-carotene, and torulene, but not linear carotenoids such as lycopene or neurosporaxanthin *in vitro*; unlike CarT it is not involved in neurosporaxanthin formation. The 15,15′-double bond is the target of the CarX cleavage reaction regardless of the substrate chain length. CAO-1 is predicted to have a similar function in *N. crassa*.

19.3.4 Cyanobacterial CCOs

Environmental sampling and observations identified apocarotenoid products from microbial sources including cyanobacteria. Cyanobacteria have been found to excrete apocarotenoids such as methyl ketones, *trans*-geranyl acetone, and ionone derivatives (Enzell 1985). C_{13} and C_9 volatiles, β-ionone and β-cyclocitral, respectively, were isolated from *Microcystis aeruginosa* and *Anabaena cylindrical* (Juttner 1976). In follow-up studies, an enzyme from *Microcystis* PCC7806 was found that specifically cleaves β-carotene and zeaxanthin asymmetrically at the 7,8 (7′,8′)-bonds to form volatile aroma compounds (β-cyclocitral, hydroxyl-β-cyclocitral) and crocetindial (8,8′-diapocarotene-8,8′dial). The enzyme did not cleave the major cyanobacterial carotenoids echinenone or myxoxanthophyll. In the native host, this enzyme was associated with the membrane, required iron, and is sensitive to sulfhydryl reagents, antioxidants, and chelating agents (Juttner 1985). Labeling studies confirmed that at least one of the aldehyde oxygens of the cleavage products is derived from dioxygen. This enzyme is the first microbial CCO enzyme described, but it has not yet been cloned.

The proliferation of microbial genome sequencing projects has made the identification of CCO homologs in microorganisms possible. Genome mining based on protein sequence homology has resulted in the cloning of several microbial CCOs (Marasco et al. 2006) (Figure 19.2 and Table 19.4). The first report of a cloned bacterial CCO was in 2005 for an apocarotenoid cleavage enzyme from *Synechocystis* PCC 6803 (SynACO) (Ruch et al. 2005). This enzyme was crystallized and the structure solved at 2.4 Å resolution (PDB code 2BIX; see below) (Kloer et al. 2005). *In vitro* assays testing apocarotenoids of various chain lengths, β-ring substitutions, and alcohol derivatives showed cleavage of the 15,15′ bond of apocarotenoids. β-apo-8′carotenal (C_{30}) was converted into retinal (C_{20}) (Figure 19.1) and related apocarotenoid compounds 10′ apocarotenal (C_{27}) and 3-hydroxy-β-apo-12′carotenal were also substrates. Despite reacting with a variety of carotenoid chain lengths and both aldehydes and alcohols, SynACO does not cleave C_{25}, C_{35} or full-length carotenoids. Regardless of the substrate, the 15,15′-bond was always the target suggesting the β-ionone ring is the determining factor in cleavage site specificity (Ruch et al. 2005). Cleavage specificity by a related enzyme from *Nostoc* sp. *PCC* 7120 (NosACO or NSC2; 53% identity) showed similar specificity (Scherzinger et al. 2006). Kinetic analysis showed higher binding affinity for aldehydes, but alcohols and unsubstituted rings were converted more quickly by NSC2 (Scherzinger et al. 2006). Based on modeling NSC2 on the SynACO scaffold, Scherzinger et al. (2006) predicted that the enzyme will not cleave full-length carotenoids (see below). However, β-carotene cleavage was observed by Marasco et al. (2006). The slow turnover of full-length carotenoids by the microbial CCO enzymes may account for this contradiction.

The filamentous, diazotrophic *Nostoc* sp. PCC7120 genome contains three open reading frames with homology to CCOs. Cloning and partial characterization of these enzymes identified three distinct cleavage activities (Marasco et al. 2006). NSC1 (*all1106*) cleaves full-length carotenoids and β-apo-8′apocarotenal in the 9,10-position resulting in β-ionone. NSC3 (*all4895*) was found to cleave the 9,10-bond of apocarotenoids only. As previously mentioned, NSC2 (*all4284*) cleaved the 15,15′-bond to form retinal. Included in the study was a survey of four other cyanobacterial genomes and related open reading frames (*Nostoc punctiforme*, *Synechocystis* sp. PCC6803, *Synechococcus elongatus* PCC7942, *Prochlorococcus marinus* MIT9313). The unicellular *Synechococcus elongatus* PCC7942 enzyme (SYO) bleached carotenoid containing *E. coli* in a similar manner to NSC 1 and 2 (Marasco et al. 2006). Further studies have indicated cleavage of the 15,15′-bond of β-carotene, zeaxanthin, and β-apo-8′carotenal (Marasco and Schmidt-Dannert, unpublished). The other filamentous cyanobacteria, *Nostoc punctiforme,* contained 4 CCO paralogs, two of which were found to cleave apocarotenoids and the other two were inactive against (apo)carotenoid substrates tested. NOP1 (Npun1384) cleaved 9,10-position of β-apo-8′carotenal and NOP4 (Npun5487) cleaved the 15,15′-bond to form retinal; NOP2 (Npun0528) and NOP3 (Npun6913) were inactive (Marasco and Schmidt-Dannert, unpublished). The putative CCO homolog from *Prochlorococcus marinus* MIT9313 was inactive under conditions tested as was the second CCO paralog present in *Synechocystis* sp. PCC6803 (SYC1).

Four different species of cyanobacteria (*Nostoc* sp., *Nostoc punctiforme*, *Synechocystis* PCC6803, *Synechococcus elongatus*) have been shown to cleave (apo)carotenoids into retinal which is the chromophore for retinylidene proteins (see Table 19.4 for an overview of cleavage activities). A sensory rhodopsin protein (ASR) was recently identified in *Nostoc* sp. PCC 7120 and shown to have a unique photocycle and exhibit light-induced reversible interconversion between 13-*cis*- and all-*trans*-retinal (Jung et al. 2003, Sineshchekov and Spudich 2004, Vogeley et al. 2004, Sineshchevkov et al. 2005). ASR may act as a sensor for light-regulated processes such as chromatic adaptation in *Nostoc* sp. PCC 7120. Intriguingly the other three strains that also generate retinal by specific carotenoid cleavage enzymes do not have ASR homologs in their genomes. The role of retinal in these organisms is a major unanswered question given retinal's prevalence in nature as a signaling molecule (see Section 19.5). Another interesting observation that may have biological significance is the presence of multiple cleavage enzymes with different functions in single cyanobacterial genomes. Filamentous cyanobacteria have several CCO paralogs in their genomes compared to unicellular cyanobacteria that have only one or two CCO paralogs, although carotenoid biosynthetic pathways are not significantly duplicated or more complicated in filamentous *Nostoc* species (Liang C 2006). Multiple CCO enzymes in these strains may act sequentially on cleavage products to synthesize molecules similar to those produced by the CCO activities of CCD7 and CCD8 from *Arabidopsis*. This is an area of research that merits further investigation.

19.3.5 Bacterial CCOs

The cleavage of the interphenyl α,β-double bond of lignostilbenes by molecular oxygen to the corresponding aldehydes is a reaction analogous to the carotenoid cleavage reaction catalyzed by NCED in the biosynthesis of the plant hormone ABA (Figure 19.4) (Kamoda and Saburi 1993a,b, Han et al. 2002, Schwartz et al. 2003). Enzymes (four isoenzymes corresponding to two genes) from *Sphingomonas paucimobilis* TMY1009 have been shown to cleave stilbene-type intermediates which can arise from the degradation of dimeric lignin compounds (Kamoda and Saburi 1995). Protein sequences of NCEDs and LSD isoenzymes are similar and consequently many members of the bacterial CCO family are annotated in databases as lignostilbene-α,β-dioxygenases (LSD, EC 1.13.11.43). The enzymatic reaction was shown to require molecular oxygen and ferrous iron as do other CCO catalyzed reactions. LSDs are proposed to be involved in degradation of the cell wall constituent lignin, although this has never been shown with natural substrates. Lignin degradation products include metabolites belonging to the flavonoid/stilbene class of compounds.

Descriptions of the cloning, enzyme purification, and inhibitor studies on four LSD isoforms are described in a series of papers (Kamoda and Saburi 1993a,b, 1995, Kamoda et al. 1997, 2003, 2005). The four isoforms have different substrate specificities for substituted stilbenoids (Kamoda and Saburi 1993, Kamoda et al. 2003). *In vitro* assays with stilbene derivatives containing various functional groups showed that a 4′-hydroxyl group was essential for cleavage (Kamoda et al. 2003). Inhibitor studies suggest that the binding of biphenyl substrates in LSDs is different from carotenoid binding in NCEDs as LSD inhibitors did not inhibit NCED enzymes (Han et al. 2002).

Recently, two enzymes NOV1 (YP_496081) and NOV2 (YP_498079) from *Novosphingobium aromaticivorans* DSM12444 have been identified that also cleave stilbene compounds (Figure 19.4) (Marasco and Schmidt-Dannert 2008). Both enzymes cleave stilbenes with a range of functional group substitutions and required a 4′-oxygen functional group. Unlike the *Sphingomonas* enzymes, the NOV enzymes had similar substrate specificities (Marasco and Schmidt-Dannert 2008). Both mono- and dioxygenase mechanisms have been attributed to CCO enzymes, but labeling studies with the NOV enzymes favors a monooxygenase mechanism (Marasco and Schmidt-Dannert 2008) (see Section 19.4).

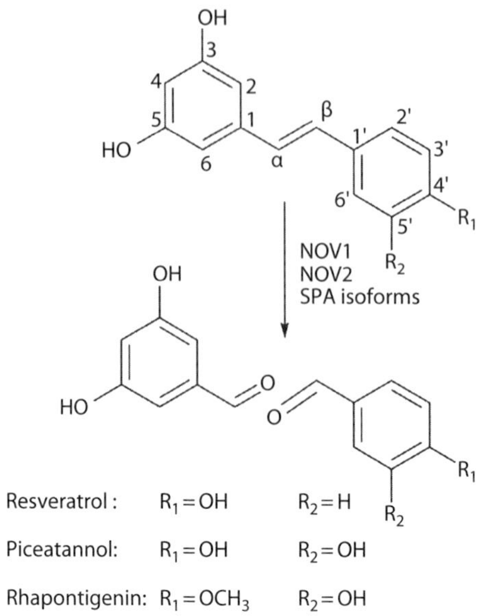

FIGURE 19.4 Cleavage of the α,β-double bond of stilbene substrates catalyzed by CCO homologs known as LSDs. The four isoforms from *Sphingomonas paucimobilis* and two enzymes from *Novophingobium aromaticivorans* cleave stilbene substrates containing a 4′-oxygen functional group at one aromatic ring.

19.4 STRUCTURE AND MECHANISM OF CCOS

CCO enzymes are mononuclear, nonheme Fe^{2+} enzymes. Enzymes with mononuclear, nonheme Fe^{2+} centers in their active site are well-known for their catalytic versatility (reviewed by Straganz and Nidetzky (2006)). Common reactions of these types of enzymes include carbon-oxygen bond formation and oxygenative bond cleavage. The predominant motif is a facial triad of two histidine ligands and one carboxylate (reviewed by Hegg and Que [1997]). This motif paradigm can be found in many mononuclear, nonheme Fe^{2+} proteins with different structural folds (i.e., catechol 2,3-dioxygenase, Rieske dioxygenase, pterin dioxygenase) (reviewed by Que and Ho [1996] and Costas et al. [2004}). Alternative active site structures for mononuclear, nonheme Fe^{2+} oxygenases exist. The majority of these proteins possess the cupin superfamily fold (i.e., quercetin dioxygenase, cysteine dioxygenase). The CCO family is unique among the mononuclear, nonheme Fe^{2+} type oxygenases because of its seven-bladed, β-propeller fold (Figure 19.5) (Kloer et al. 2005). The mononuclear iron in the CCO active site also has a rare coordination geometry not found in other structurally characterized dioxygenase enzymes. The iron is complexed by four histidines in an octahedral geometry with two missing ligands, one of the coordination positions is occupied by water while the other position is empty. The four metal-ligating histidine side chains are located at the propeller-axis with three of the histidines held in place by hydrogen bonding to glutamate residues. All four histidines are perfectly conserved among CCO family members. The rigid active site does not collapse or change shape when iron is removed. Iron-free mutants are inactive and apoenzymes produced by chelation of the iron cannot be reconstituted through the incorporation of other divalent metals (Poliakov et al. 2005, Takahashi et al. 2005, Marasco et al. 2006).

So far, only one structure has been solved for a member of the CCO family. The structure of the apocarotenoid cleavage enzyme SynACO from the cyanobacteria *Synechocystis* sp. PCC6803 was solved to 2.4 Å (Kloer et al. 2005). The structure has a rare fold, consisting of a rigid, seven-bladed propeller. Five of the blades are comprised of four antiparallel β-strands, and two blades have five strands (Figure 19.5). The regions of highest conservation amongst the carotenoid oxygenases are the β-strands of the propellers. The bottoms of the propellers are flat with the β-strands connected

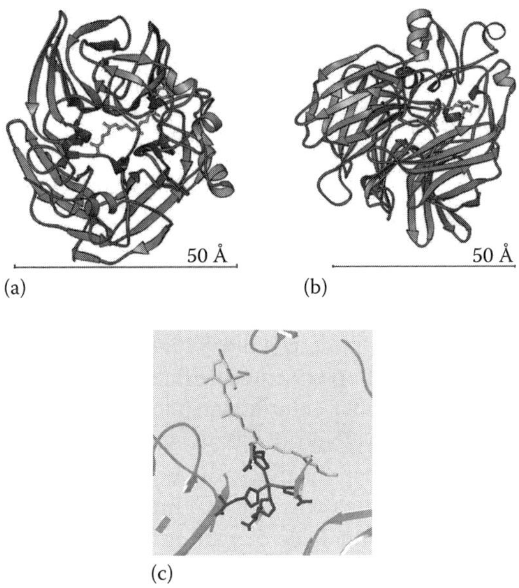

FIGURE 19.5 (See color insert following page 336.) Structure of SynACO (PDB ID 2BIW). (a) SynACO has a rare seven-bladed propeller structure with the substrate tunnel perpendicular to the propellers. The Fe^{2+} metal center (shown in orange) is coordinated by four histidines. The substrate β-apo-8′-carotenal is shown in gray. (b) Side view showing the α-helical region of the entrance loops (shown in blue). A hydrophobic patch lies over the substrate tunnel. (c) Slab view of the active site histidine residues (HIS183, HIS238, HIS304, HIS404) and the apocarotenoid substrate. The substrate is isomerized from an all-*trans* configuration to a *cis* configuration at the 13,14- and 13′,14′-double bonds. Cleavage occurs at the 15,15′-double bond.

by short loops, whereas the top forms a large dome structure because the strands are connected by extended loops with short helices. These loops define a tunnel passing through the iron active center that runs perpendicular to the propeller axis. The tunnel entrance is located near the hydrophobic patch and continues to the iron center. The exit is on the far side of the patch, following a turn in the active site. It is hypothesized that the length and sequence of the connecting loops are the parameters that allow family members to define substrate specificity. The oxidative cleavage of carotenoids by carotenoid oxygenases has been shown to be a regulated process that occurs with high regio- and stereospecificity. Electron density along the polyene chromophore and presence of functional groups may influence which cleavage site is selected (Woodall et al. 1997). Structural evidence suggests that the length and sequence of the entrance tunnel to the active site may determine cleavage site specificity, but there is little experimental evidence to support this claim (Kloer et al. 2005).

The substrate does not coordinate directly with the active site iron, and it is thought that two sites on the metal are filled with water (Kloer et al. 2005). In the substrate soaked crystal, structural features suggest that the 8-apocarotenal substrate entered the active site with the linear end, leaving the ionone-ring at the tunnel entrance. Electron density fitting suggests that the substrate has an interesting configuration near the active site metal. The double bonds on either side of the cleavage site (15,15′-double bond) are isomerized from *trans* to *cis* configuration (Figure 19.5).

Structural information about the oxygenases provided limited insight into the mechanism (Schmidt et al. 2006). The crystallized enzyme from *Synechocystis* sp. PCC6803 is membrane associated and the interaction with the membrane is believed to be mediated by a nonpolar patch on the surface of the enzyme. This hydrophobic patch is thought to provide the necessary access of the protein to the membrane-bound carotenoids. Following withdrawal from the membrane, the substrate moves through the hydrophobic tunnel toward the metal center. The substrate orients the reactive double bond near the Fe^{2+} and dioxygen coordinates in a side-on fashion and displaces both water molecules to form a ternary complex (Kloer et al. 2005). At this point, the reaction can proceed via two possible intermediates (either a dioxetane or epoxide intermediate). Both intermediates would be expected to decompose into the same cleavage product (Kloer and Schulz 2006).

The enzyme mechanism is known to consume dioxygen, but whether this enzyme catalyzes oxidative cleavage via a mono- or dioxygenase mechanism cannot be deduced from the structure. Both dioxygenase and monooxygenase mechanisms have been hypothesized based on conflicting ^{18}O labeling studies (Figure 19.6). Incorporation of both oxygen atoms of O_2 are supported by studies from *Arabidopsis thaliana* producing β-ionone (Schmidt et al. 2006) and plants producing ABA (Zeevaart et al. 1989). In contrast, the formation of retinal has been suggested at different times with different enzyme examples both as a dioxygenase mechanism and a monooxygenase-like mechanism through a postulated epoxy intermediate (Leuenberger et al. 2001). Labeling studies with *Microcystis* utilizing whole cells under an $^{18}O_2$ atmosphere showed an ^{18}O label on β-cyclocitral (86%), hydroxyl-β-cyclocitral (20.5% labeled), while the dialdehyde cleavage product crocetindial (8,8′-diapocarotene-8,8′dial) was unlabeled. The authors suggest the lack of labeling on the linear cleavage product was due to a high exchange rate of the labeled oxygen with water. The exchange rate of the aldehyde oxygen in this study was indeed high (33% after 20 h). When the cells were exposed to $H_2{}^{18}O$ in the converse labeling experiment, the β-cyclocitral was labeled at 17%. The authors suggest a dioxygenase mechanism from this evidence. Studies with endogenous 15,15′ CCO enzymes from chicken mucosa using both ^{17}O and $H_2{}^{18}O$ showed incorporation of one oxygen from O_2 and one from water through a proposed epoxide intermediate (Leuenberger et al. 2001). In this study, an equal enrichment of ^{17}O and ^{18}O (52%:41%) was observed. This study has been criticized, however, for the long incubation time and the coupling of the enzyme assay with the horse liver alcohol dehydrogenase enzyme to reduce the aldehyde to an alcohol (Schmidt et al. 2006). Very recently, labeling studies using $H_2{}^{18}O$ and $^{18}O_2$ with the stilbene cleaving enzymes NOV1 show that at least these enzymes cleave the interphenyl double bond of stilbenes with a monooxygenase mechanism (Marasco and Schmidt-Dannert 2008). Unlike carotenoid cleavage by CCOs, stilbene cleavage catalyzed by NOV2 is relatively fast, stilbenes are readily solubilized in the assays and cleavage products can be rapidly isolated and detected by GC-MS, which reduces the extent of unspecific label exchange which is a problem in studies with these enzymes.

The current controversy over the oxygenase mechanism of this family of nonheme iron enzymes stems from contradictory findings from labeling studies and a lack of rigorous biophysical studies (Leuenberger et al. 2001, Schmidt et al. 2006). The poor activities of recombinant CCOs in *in vitro* assays and cleavage of water insoluble substrates may largely be responsible for the lack of rigorous mechanistic studies of this class of nonheme iron oxygenases. To date there has been no cofactor identified that is associated with the cleavage activity of the CCOs. Given the poor reactivity *in vitro*, it is plausible that there is a nontraditional cofactor associated with the enzyme (Paik et al. 2001). Another possibility is that the membrane association of the enzymes limits their activities unless when incorporated in liposomes. As more enzyme examples are discovered and better reaction conditions developed, more careful biochemical characterization may be possible.

19.5 BIOLOGICAL FUNCTIONS OF APOCAROTENOIDS

The structural variety of apocarotenoids results in divergent biological functions. To date, the known biological activities of apocarotenoids generated by plant and microbial CCOs are more diverse than those produced in animals. Pigmentation is the most obvious function of apocarotenoids in plants; some cleavage products maintain an extended conjugated system and serve as pigments (e.g., saffron). When acting as pigments in plant tissues, they are often found in specialized plastids known as chromoplasts. In thylakoid membranes, apocarotenoids act as accessory pigments and are involved in photoprotection; smaller cleavage compounds protect against UVB by absorbing between 280 and 320 nm.

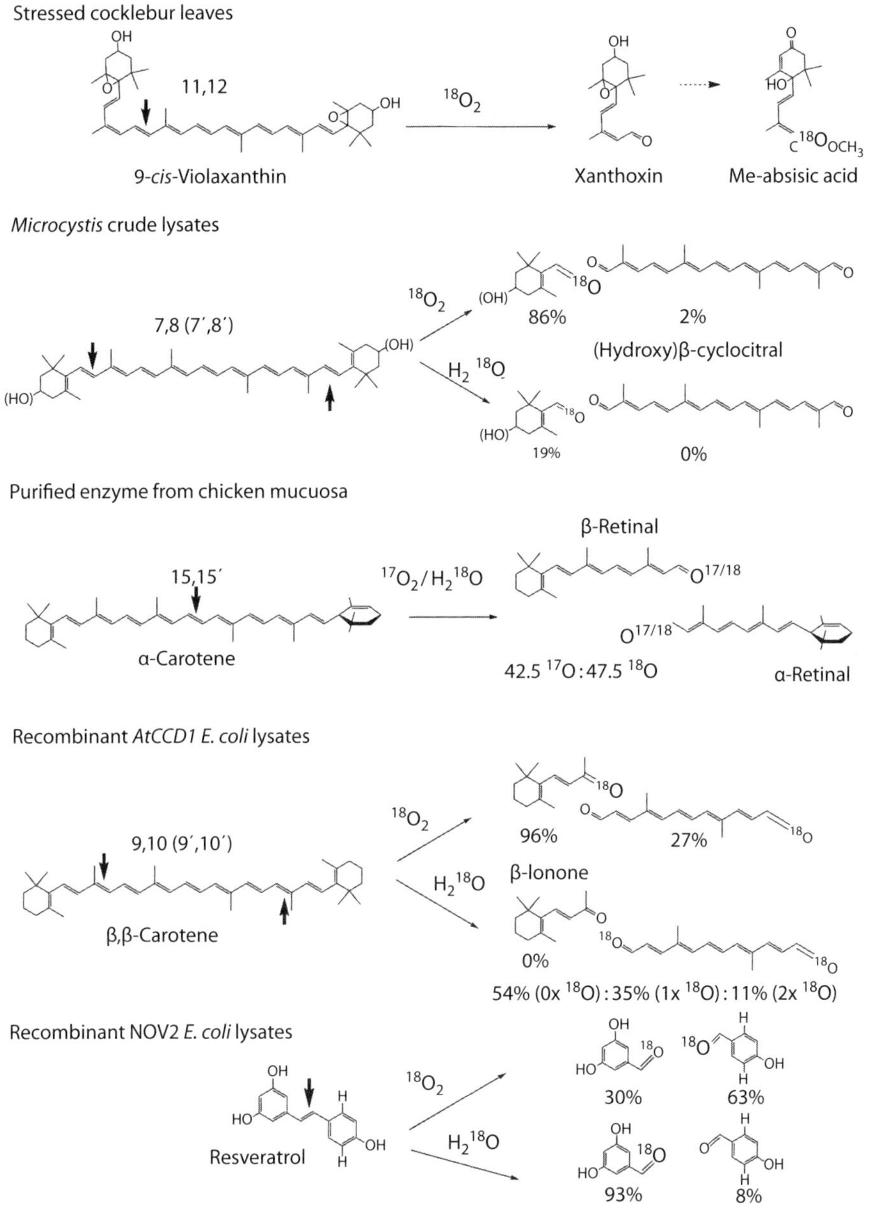

FIGURE 19.6 Summary of oxygen labeling studies data used to determine the mechanism of CCO enzymes. The reported isotopic labeling patterns observed with different CCO homologs listed in chronological order of discovery. Studies on ABA only examined one product (Zeevaart et al. 1989) and the mechanism was described as a dioxygenase mechanism; the *Microcystis* reaction was described as a dioxygenase mechanism (Juttner 1988); the coupled assay with the enzyme from chicken mucosa was described as a monooxygenase mechanism (Leuenberger et al. 2001); the reactions catalyzed by *At*CCD1 were characterized as dioxygenase (Schmidt et al. 2006); the cleavage of the interphenyl double bond of stilbenes by a *Novosphingobium* CCO homolog NOV2 was described as monooxygenase mechanism (Marasco and Schmidt-Dannert 2008). Heavy oxygen labels are shown in bold and the size of the oxygen label is reflective of the percentage labeled.

Apocarotenoids also act as chemoattractants, repellants, and growth effectors in plants and cyanobacteria. They attract pollinators to plants through the use of color similar to full-length carotenoids. Their aromas are thought to be attractants for animals and insects to facilitate in seed dispersal and pollination. Small volatile apocarotenoids lure pollinators and levels of apocarotenoids often increase in ripening fruits (Simkin et al. 2004a,b). Volatiles also function as repellants in some plants and insects. For example, grasshopper ketone (C_{13}) from *Romalea microptera* acts as an ant repellent (Meinwald and Eisner. 1968) and sunflowers produce apocarotenoids that have allelopathic activities (Macias et al. 2002). In cyanobacteria, volatile apocarotenoids such as β-ionone and β-cyclocitral are also used as allelopathic bioregulators to provide a competitive advantage (Juttner 1979, 1995, Walsh et al. 1998).

The role of apocarotenoids in arbuscular mycorrhizal (AM) associations is an exciting story that is still developing. Over 80% of land plants establish mutualistic symbiotic relationships with arbuscular mycorrhiza fungi (*Glomeromycota*). The obligate soil-borne symbionts form associations with plant roots to facilitate the uptake of mineral nutrients and exchange carbohydrates. The arbuscule structures help plants grow, support stress tolerance, and alter root secondary metabolism. Three signaling molecules (trigolactones, mycorradicin, and blumenin) (Figure 19.7) associated with arbuscular mycorrhiza formation have been linked to apocarotenoid formation (reviewed by Akiyama (2007) and Walter et al. (2007)) and 9,10-bond cleavage is responsible for the compounds that accumulate in roots of plants associated with AM fungi (Walter et al. 2000). Strigolactones are tricyclic sesquiterpene lactones that were first isolated from *Lotus japonicus* (Akiyama et al. 2005). These host-derived signaling molecules induce hyphal branching and stimulate seed germination (Akiyama et al. 2005). Strigolactones are proposed to be derived from C_{40} carotenoid cleavage catalyzed by a NCED (Matusova et al. 2005). Cleavage of the 11,12- (11′,12′-) bond of 9-*cis*-β-carotene is proposed to form a C_{14} apocarotenoid that can be converted through a series of unknown steps to the strigolactone 5-deoxystrigol. In the presence of the CCO inhibitor naproxen, wild-type maize exhibited lowered levels of seed germination comparable to the VP14 NCED mutant suggesting NCED activity was responsible (Matusova et al. 2005). The formation of the other two AM associated apocarotenoids are thought to occur via cleavage of the 9,10- (9′,10′-) double bond of still unknown carotenoid(s) yielding mycorradicin (C_{14}) and cyclohexenone (C_{13}) products (Fester et al. 2002). The C_{14} mycorradicin metabolite (10,10′-diapocarotene-10,10′-dioic acid) is a yellow pigment that forms upon fungal association with plants (Klingner et al. 1995, Walter et al. 2000). The cyclohexenone derivative was identified as a glycoside that was named blumenin (Maier et al. 1995) (Figure 19.7). Blumenin levels are AM specific and increase under colonization conditions (Maier et al. 1997). Diverse AM-specific cyclohexenone apocarotenoids have been isolated that vary in their ring substitution and the number and nature of glycosylations. For example, blumenol (C9-*O*-(2′-*O*-β-glucuronosyl)-β-glucoside) is a glycosylated cyclohexenone in cereal mycorrhizal roots

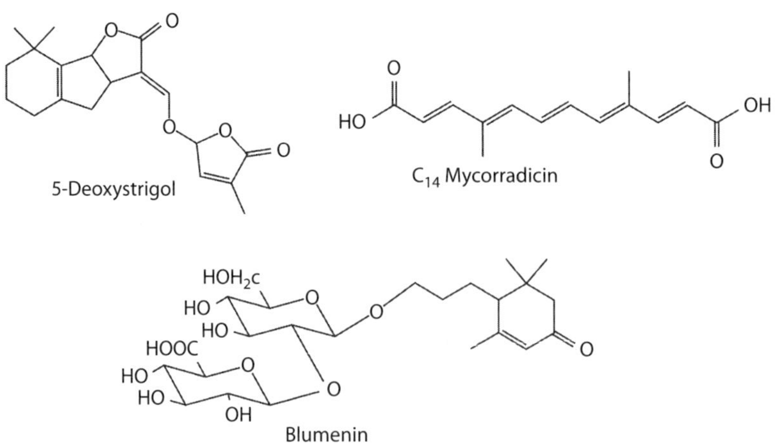

FIGURE 19.7 Representatives of apocarotenoid derived signaling molecules associated with arbuscular mycorrhiza formation.

(Maier et al. 1995, Fester et al. 1999, Vierheilig et al. 2000). Isotope trace experiments determined that the cyclohexenones are derived from the 2-C-methylerythritol phosphate or 1-deoxy-D-xylulose 5-phosphate (DXP) isoprenoid precursor pathway, which supports the observation that colonized roots exhibit elevated transcript levels of two rate-limiting DXP biosynthetic genes (*dxs* and *dxr*) (Maier et al. 1998, Walter et al. 2000, Strack et al. 2003). The CCO enzyme responsible for the formation of mycorradicin and blumenin has not been identified, but a CCO paralog in *Medicago truncululata* was found to be upregulated in mycorrhizal roots (Lohse et al. 2005). The details of apocarotenoid formation and function in AM symbiosis are still being worked out.

The signaling aspects of apocarotenoids are the least well understood and potentially the most promising areas of new research. In plants, ABA has an immensely important role in drought tolerance and seed development. The discovery of the indirect route to ABA formation was a major milestone in plant research. Discoveries such as the role CCD7 and CCD8 play in regulating lateral branching are at the forefront of current CCO research (see above). The uncharacterized phytohormone generated by CCD7 and CCD8 holds potential for new roles of apocarotenoids in signaling. The biological activities of apocarotenoids in microorganisms are currently largely unknown, but may be involved in previously undetected signaling pathways.

In animals, retinal's involvement with the visual cycle is well established, but the signaling functions of other retinoids are not as well known. Retinal (15-apo-β-carotenal; C_{20}) is the chromophore of rhodopsin in the vertebrate visual cycle (Spudich et al. 2000). Retinal can also be converted to other retinoids with potent biological activities in metazoans through oxidation to RA and then isomerization to 9-*cis*-RA. 9-*cis*-RA has been identified as an important signaling molecule in the immune system (Szondy et al. 1998, Wang et al. 2007), in development (Lampert et al. 2003), and in cancer prevention (Altucci et al. 2007). The RA derivatives signal by binding to nuclear retinoic acid receptor (RAR) and retinoid X receptor (RXR) (Altucci et al. 2007). RAs have also been examined as potential cancer therapies (Patel et al. 2007) and vitamin A as well as 9-*cis*-RA are thought to be involved in transcriptional regulation (Bachmann et al. 2002).

The recent construction of a knockout BCO1 mouse should provide more insight into the role of retinoids in metabolism and the specific role that carotenoid cleavage enzymes play in signaling in animals (Hessel et al. 2007). BCO1 deficient mice showed serious impairments in β,β-carotene metabolism suggesting that BCO1 is the key enzyme involved in vitamin A production (Hessel et al. 2007). BCO1 knockout mice accumulated β-carotene in adipose tissues and exhibited increased lipid accumulation in the liver suggesting that RA influences liver fatty acid metabolism (Hessel et al. 2007). Involvement of BCO1 in lipid metabolism is also supported by its transcriptional regulation by the peroxisome proliferator-activated receptor γ (PPAR-γ) which dimerizes with RXR to control BCO1 gene expression (Boulanger et al. 2003).

Significantly less is known about the function of the second CCO (BCO2) present in animals. Studies in rat found that BCO1 and BCO2 were differentially expressed and that lycopene, the substrate of BCO2, may modulate β-carotene or lipid metabolism (Zaripheh et al. 2006). BCO1 knockout mice showed increased hepatic BCO2 mRNA levels (fourfold) and lowered (fivefold) lycopene levels probably as the result of increased lycopene cleavage by BCO2 (Lindshield et al. 2007). Animals that have been fed lycopene accumulate lycopene cleavage products (referred to as lycopenoids) such as apo-8′-lycopenal (found at levels of ~600 pmol g^{-1} in rat liver), apo-10′-lycopenal (found at levels of ~8 pmol g^{-1} in ferret lung), apo-12′-lycopenal, and other polar products (Gajic et al. 2006, Hu et al. 2006). Lycopenoid levels in tissues are equivalent or greater than RA levels in similar tissues, and it is hypothesized that they may be agonists or antagonists for nuclear receptors such as RARs/RXRs/PPARs that are known to interact directly or indirectly with retinoids (Lindshield et al. 2007). It is known that lycopene metabolites transactivate antioxidant response element genes (Ben-Dor et al. 2005) and enhance gap junction communication in rat liver cells (Aust et al. 2003). However, further studies are needed to understand the effects of both retinoids and lycopenoids in animals and to understand their bioactivity and unravel their complex signaling activities.

19.6 COMMERCIAL RELEVANCE OF APOCAROTENOIDS

Apocarotenoids play important roles in our daily lives as aroma compounds and pigments. Production of aroma compounds and pigments and the development of new aroma compounds for mostly food applications provide opportunities for the use of enzymatic processes (either *in vivo* or *in vitro*) for their synthesis. Cooxidation systems employing LOXs, peroxidases, or xanthin oxidases are the current processes for enzymatic carotenoid cleavage to yield aroma compounds. The mechanism behind these reactions is the generation of a free radical species from a cofactor which then cleaves the carotenoid substrate. The free radical generation, however, results in a mixture of cleavage products because it lacks specificity. CCOs in general have broad substrate specificities, but cleave with high regioselectivity, making them appealing enzymes for the production of flavor and fragrance compounds (Winterhalter and Rouseff 2002). However, because of poor *in vitro* activities of most CCOs described today, biosynthesis of apocarotenoids in engineered hosts expressing recombinant CCOs appears presently to be the most feasible biotechnological application of this class of enzymes. The utilization of CCOs in *in vitro* transformation reactions first requires a better understanding of their enzymatic properties.

19.6.1 APOCAROTENOID BIOSYNTHESIS IN RECOMBINANT HOSTS

The low threshold values, characteristic aroma notes, and potency of aroma volatiles derived from carotenoids (C_{40}) has led to the isolation and structural elucidation of a wide variety of carotenoid aroma compounds (mainly C_9–C_{13}) from plant extracts (reviewed by Winterhalter and Rouseff 2002). Volatile 9–13 carbon carotenoid cleavage compounds such as β-ionone, α-ionone, dihydroactinidiolide, gerinol, damascenol, and eugenol serve in plants as pollinator attractants, antifungals, or to deter pests (Pichersky and Gershenzon 2002). The C_{13} ionones are found in many fruit flavors (raspberry, blackberry, blackcurrant, peach, apricot, melon, tomato), plant odors (violet, black tea, tobacco, carrot, vanilla), and mushrooms (Chanterelle). The structurally diverse aroma chemicals derived from oxidative cleavage of carotenoid compounds result from the tremendous structural diversity of carotenoid precursors (more than 800 known), cleavage site variations, and subsequent oxidative modifications and glycosylations of the cleavage products (Enzell 1985). The production of volatile aroma and flavor compounds from recombinant carotenoid cleavage reactions is one potential industrial application of CCOs (Marasco and Schmidt-Dannert 2003). Functional expression of CCO from all kingdoms has been achieved in recombinant *E. coli* coexpressing carotenoid biosynthetic genes and found to exhibit carotenoid cleavage activity *in vivo* (von Lintig and Vogt 2000, Kiefer et al. 2001, Schwartz et al. 2001). Thus, coexpression of many of the available carotenoid biosynthetic pathways together with different types of carotenoid cleavage enzymes opens new avenues for the production of structurally diverse carotenoid aroma compounds in engineered microbial or plant systems.

Two examples of important industrial apocarotenoid products include the pigment, bixin, and the spice, saffron. Saffron, the most expensive spice in the world (\$1000–2000 kg^{-1}), is comprised of water soluble apocarotenoid glycosides and found in the dry stigma of *Crocus sativus*. The majority of the color derives from crocetin esters which are created by the cleavage of zeaxanthin (Tarantilis et al. 1995). This cleavage reaction is catalyzed by a zeaxanthin-specific 7,8 (7′,8′) cleavage dioxygenase (*Cs*ZCD) (see Section 19.3). The heterologous expression of both the *Cs*ZCD and a suitable glucosyl transferase together with a carotenoid biosynthesis pathway would lead to a competitive alternative to natural crocin production (which is extremely tedious as it requires manual picking of *Crocus* stigmata) by production in hosts such as *E. coli* or yeast. Similar strategies using recombinant enzymes to produce bixin (an important colorant for dairy products) in a heterologous host has industrial implications. Unlike crocetin ester biosynthesis, all genes necessary for bixin biosynthesis have recently been cloned and expressed in *E. coli* making it feasible to develop a biotechnological production process (Bouvier et al. 2003).

Aroma compounds generated by enzymatic activities or fermentation (including recombinant production systems) are considered "natural" according to the U.S. Food and Drug Administration Code of Federal Regulations (CFR 1999) and European EC 1998 legislation. Biotechnological production of aroma compounds has therefore enormous economic implications because the price of "natural" compounds (as the result of consumer demand) far exceeds that of chemically synthesized compounds. For example "natural" vanillin costs \$1000–4000 kg^{-1} while the price of chemically synthesized vanillin is two orders of magnitude less (\$10–15 kg^{-1}). A promising route to natural vanillin would be the application of stilbene degrading oxygenases in engineered microbial cells that produce stilbene compounds (Watts et al. 2006).

As the signaling mechanisms of apocarotenoids in plants are elucidated, the opportunity for the development of genetically modified plants with improved drought resistance or stress tolerance can be developed. ABA mutant plants exhibited increased transpiration efficiency and root conductivity (Thompson et al. 2007). The finding of strigolactones as branching factors and the mycorrhizin and cyclohexenones as AM signals influences crop production strategies (Akiyama 2007). Clearer understanding of their interactions with plants may also improve plant growth and development.

19.6.2 Improving CCO *In Vitro* Activity

The poor solubility of the CCO enzymes is one impediment to their use in biotransformation reactions. CCOs have a tendency to form insoluble aggregates in recombinant hosts. Another challenge lies in delivering the extremely hydrophobic substrates to the enzymes *in vitro*. Because detergent-based aqueous micellar systems make enzymology difficult, CCOs are only superficially characterized *in vitro*. Comprehensive kinetic studies with purified CCO enzymes have not been published to date. Recently, several studies have investigated assay methods to address these impediments with some success (see below). However, although some improvements in *in vitro* activity have been achieved, the catalytic activities of CCOs *in vitro* are still extremely low suggesting that perhaps a critical cofactor (small molecule and/or protein) may still be missing. As discussed in Section 19.4, these enzymes represent a novel class of oxygenases with a yet to be clearly defined mechanism. It is possible that cofactors (i.e., those involved in electron transfer) are required for this class of enzymes to function properly; although commonly with oxygenases associated cofactors have no or little effect on the catalytic activity of this class of enzymes (Marasco et al. 2006).

Two distinct approaches for improved solubility of the CCO enzymes have been taken. The first relies on coexpression of molecular chaperones with CCO enzymes; expression of GroEL and GroES with microbial CCOs resulted in improved solubility in *E. coli* (Marasco et al. 2006). Other common chaperones such as *tig*, *dnaK*, *dnaJ*, and *grpE* did not improve solubility (Marasco et al. 2006). The second approach utilizes solubility-enhancing fusion proteins such as glutathione-S-transferase (GST) and the *E. coli* transcription-termination anti-termination factor NusA (Schilling et al. 2007). The specific activity of GST-AtCCD1 in cellular extracts increased throughout the stationary phase and there was a twofold increase in maximum specific activity for the GST-AtCCD1 compared to the His_{6X}-tag version (K_m = 1.81 compared to 0.90 mU mg^{-1} of total protein when a unit is described as cleavage of 1 μmol of substrate per minute). NusA fusions resulted in reduced activity (0.35–0.90 mU mg^{-1}). Turnover numbers of the purified fusion proteins (GST and NusA) were reduced compared to the His_{6X}-tagged version (0.040, 0.051–0.132). Schilling et al. (2007) suggest that weak promoters may be beneficial for CCO expression (Schilling et al. 2007). Findings by Mathieu et al. (2007) echo these reports by describing the benefits of slow, low temperature growth, harvesting cells at the end of the stationary phase, and the use of a detergent in the lysis buffer (Mathieu et al. 2007). The addition of detergents to lysis buffers aided in the extraction of soluble protein (0.08%–0.2% Triton X-100). An earlier examination of detergents in assays with BCOI found octylglucosylpyranoside to be the most beneficial detergent (Lindqvist and Andersson 2002).

CCO activation by organic solvents is another aspect of *in vitro* activity to be optimized. It was observed that the addition of a small amount of organic solvent 1%–15% improved the activity of

*At*CCD7 (Schwartz et al. 2004). Organic solvent addition (dioxane, DMSO, methanol or acetone) improved activity under low concentrations (Mathieu et al. 2007). Short chain aliphatic alcohols activated the enzymes although the reason for this activation is unclear (probably due to influences on substrate accessibility or micellar structure). An increase in activity was observed for all aliphatic alcohols tested, although the optimal concentration lessened with increasing log *P* values (Schilling et al. 2007).

19.7 CONCLUSIONS AND OUTLOOK

To date, the majority of the literature on CCO enzymes has been devoted to descriptions of the homologs from different organisms. However, despite the large number of CCOs described so far, relatively little is known about the mechanism of this novel class of oxygenases. The low *in vitro* activities obtained with recombinant enzymes is a major impediment to biochemical studies. The necessity of carotenoid cleaving enzymes to interact with membranes to access their hydrophobic carotenoid substrates creates difficulties in creating optimal assay conditions. With only one solved structure of a microbial apocarotenoid cleaving CCO, structural analysis of this class of enzymes severely lags behind other oxygenase family members.

Despite these difficulties, research on CCO enzymes represents one of the most exciting fields of carotenoid research. As more examples are discovered in different organisms, the evolutionary history and current biological functions appear even more complex. The three major areas of research being pursued that will provide the most exciting results include (1) elucidation of the signaling pathways through identification of downstream targets for the cleavage products and identification of the signals themselves (in the case of CCD7 and CCD8), (2) characterization of the enzymes' catalytic mechanism and mechanisms of substrate and cleavage site selection, and (3) a greater understanding of evolution of the nonheme iron oxygenases family in nature.

REFERENCES

Agusti, J., M. Zapater et al. (2007). Differential expression of putative 9-*cis*-epoxycarotenoid dioxygenases and abscisic acid accumulation in water stressed vegetative and reproductive tissues of citrus. *Plant Sci.* **172**(1): 85–94.

Akiyama, K. (2007). Chemical identification and functional analysis of apocarotenoids involved in the development of arbuscular mycorrhizal symbiosis. *Biosci. Biotechnol. Biochem.* **71**(6): 1405–1414.

Akiyama, K., K. Matsuzaki et al. (2005). Plant sesquiterpenes induce hyphal branching in arbuscular mycorrhizal fungi. *Nature* **435**(7043): 824–827.

Alexander, S. P. H., A. Mathie et al. (2008). Retinoic acid and retinoid X. *Br. J. Pharmacol.* 153(Suppl 2): S131–S131.

Altucci, L., M. D. Leibowitz et al. (2007). RAR and RXR modulation in cancer and metabolic disease. *Nat. Rev. Drug Discov.* **6**(10): 793–810.

Anthonise, L. F., D. R. Soprano et al. (2007). Retinoids in biological control and cancer. *J. Cell. Biochem.* **102**(4): 886–898.

Auldridge, M. E., A. Block et al. (2006). Characterization of three members of the arabidopsis carotenoid cleavage dioxygenase family demonstrates the divergent roles of this multifunctional enzyme family. *Plant J.* **45**(6): 982–993.

Auldridge, M. E., D. R. McCarty et al. (2006). Plant carotenoid cleavage oxygenases and their apocarotenoid products. *Curr. Opin. Plant Biol.* **9**(3): 315–321.

Aust, O., N. Ale-Agha et al. (2003). Lycopene oxidation product enhances gap junctional communication. *Food Chem. Toxicol.* **41**(10): 1399–1407.

Bachmann, H., A. Desbarats et al. (2002). Feedback regulation of beta,beta-carotene 15,15′-monooxygenase by retinoic acid in rats and chickens. *J. Nutr.* **132**(12): 3616–3622.

Bainbridge, K., K. Sorefan et al. (2005). Hormonally controlled expression of the arabidopsis MAX4 shoot branching regulatory gene. *Plant J.* **44**(4): 569–580.

Beja, O., L. Aravind et al. (2000). Bacterial rhodopsin: Evidence for a new type of phototrophy in the sea. *Science* **289**(5486): 1902–1906.

Beja, O., E. N. Spudich et al. (2001). Proteorhodopsin phototrophy in the ocean. *Nature* **411**(6839): 786–789.

Ben-Dor, A., M. Steiner et al. (2005). Carotenoids activate the antioxidant response element transcription system. *Mol. Cancer Ther.* **4**(1): 177–186.

Bieszke, J. A., E. L. Braun et al. (1999). The nop-1 gene of *Neurospora crassa* encodes a seven transmembrane helix retinal-binding protein homologous to archaeal rhodopsins. *Proc. Natl. Acad. Sci.* **96**(14): 8034–8039.

Bieszke, J., L. Li et al. (2007). The fungal opsin gene nop-1 is negatively-regulated by a component of the blue light sensing pathway and influences conidiation-specific gene expression in *Neurospora crassa. Curr. Genet.* **52**(3): 149–157.

Booker, J., M. Auldridge et al. (2004). MAX3/CCD7 is a carotenoid cleavage dioxygenase required for the synthesis of a novel plant signaling molecule. *Curr. Biol.* **14**(14): 1232–1238.

Boulanger, A., P. McLemore et al. (2003). Identification of beta-carotene 15,15′-monooxygenase as a peroxisome proliferator-activated receptor target gene. *FASEB J.* **17**(10): 1304–1306.

Bouvier, F., O. Dogbo et al. (2003). Biosynthesis of the food and cosmetic plant pigment bixin (annatto). *Science* **300**(5628): 2089–2091.

Bouvier, F., J. C. Isner et al. (2005). Oxidative tailoring of carotenoids: A prospect towards novel functions in plants. *Trends Plant Sci.* **10**(4): 187–194.

Bouvier, F., C. Suire et al. (2003). Oxidative remodeling of chromoplast carotenoids: Identification of the carotenoid dioxygenase CsCCD and CsZCD genes involved in crocus secondary metabolite biogenesis. *Plant Cell* **15**(1): 47–62.

Burbidge, A., T. Grieve et al. (1997). Structure and expression of a cDNA encoding a putative neoxanthin cleavage enzyme (NCE), isolated from a wilt-related tomato (*Lycopersicon esculentum* Mill.) library. *J. Exp. Bot.* **48**(317): 2111–2112.

Burbidge, A., T. M. Grieve et al. (1999). Characterization of the ABA-deficient tomato mutant *notabilis* and its relationship with maize Vp14. *Plant J.* **17**(4): 427–431.

Camara, B. and F. Bouvier (2004). Oxidative remodeling of plastid carotenoids. *Arch. Biochem. Biophys.* **430**(1): 16–21.

Cao, Y. R., X. L. Guo et al. (2005). Isolation and characterization of carotenoid cleavage dioxygenase genein halophyte *Suaeda salsa. Plant Growth Regul.* **46**(1): 61–67.

Carail, M. and C. Caris-Veyrat (2006). Carotenoid oxidation products: From villain to saviour? *Pure Appl. Chem.* **78**(8): 1493–1503.

Caris-Veyrat, C., A. Schmid et al. (2003). Cleavage products of lycopene produced by *in vitro* oxidations: Characterization and mechanisms of formation. *J. Agr. Food Chem.* **51**(25): 7318–7325.

Chernys, J. T. and J. A. D. Zeevaart (2000). Characterization of the 9-*cis*-epoxycarotenoid dioxygenase gene family and the regulation of abscisic acid biosynthesis in avocado. *Plant Physiol.* **124**(1): 343–353.

Chono, M., I. Honda et al. (2006). Field studies on the regulation of abscisic acid content and germinability during grain development of barley: Molecular and chemical analysis of pre-harvest sprouting. *J. Exp. Bot.* **57**(10): 2421–2434.

Costas, M., M. P. Mehn et al. (2004). Dioxygen activation at mononuclear nonheme iron active sites: Enzymes, models, and intermediates. *Chem. Rev.* **104**(2): 939–986.

Destefano-Beltran, L., D. Knauber et al. (2006). Effects of postharvest storage and dormancy status on ABA content, metabolism, and expression of genes involved in ABA biosynthesis and metabolism in potato tuber tissues. *Plant Mol. Biol.* **61**(4–5): 687–697.

Enzell, C. (1985). Biodegradation of carotenoids: An important route to aroma compounds. *Pure Appl. Chem.* **57**: 693–700.

Fester, T., B. Hause et al. (2002). Occurrence and localization of apocarotenoids in arbuscular mycorrhizal plant roots. *Plant Cell Physiol.* **43**(3): 256–265.

Fester, T., W. Maier et al. (1999). Accumulation of secondary compounds in barley and wheat roots in response to inoculation with an arbuscular mycorrhizal fungus and co-inoculation with rhizosphere bacteria. *Mycorrhiza* **8**(5): 241–246.

Fleischmann, P., K. Studer et al. (2002). Partial purification and kinetic characterization of a carotenoid cleavage enzyme from quince fruit (*Cydonia oblonga*). *J. Agric. Food Chem.* **50**(6): 1677–1680.

Gajic, M., S. Zaripheh et al. (2006). Apo-8′-lycopenal and apo-12′-lycopenal are metabolic products of lycopene in rat liver. *J. Nutr.* **136**(6): 1552–1557.

Giuliano, G., S. Al-Babili et al. (2003). Carotenoid oxygenases: cleave it or leave it. *Trends Plant Sci.* **8**(4): 145–149.

Han, S. Y., H. Inoue et al. (2002). Design and synthesis of lignostilbene-alpha,beta-dioxygenase inhibitors. *Bioorg. Med. Chem. Lett.* **12**(8): 1139–1142.

Hegg, E. L. and L. Que (1997). The 2-His-1-carboxylate facial triad–An emerging structural motif in mononuclear non-heme iron(II) enzymes. *Eur. J. Biochem.* **250**(3): 625–629.

Hessel, S., A. Eichinger et al. (2007). CMO1-deficiency abolishes vitamin a production from β-carotene and alters lipid metabolism in mice. *J. Biol. Chem.*: M706763200 **282**: 33553–33561.

Hu, K.-Q., C. Liu et al. (2006). The biochemical characterization of ferret carotene-9′,10′-monooxygenase catalyzing cleavage of carotenoids *in vitro* and *in vivo*. *J. Biol. Chem.* **281**(28): 19327–19338.

Ibdah, M., Y. Azulay et al. (2006). Functional characterization of CmCCD1, a carotenoid cleavage dioxygenase from melon. *Phytochemistry* **67**(15): 1579–1589.

Iuchi, S., M. Kobayashi et al. (2000). A stress-inducible gene for 9-*cis*-epoxycarotenoid dioxygenase involved in abscisic acid biosynthesis under water stress in drought-tolerant cowpea. *Plant Physiol.* **123**(2): 553–562.

Iuchi, S., M. Kobayashi et al. (2001). Regulation of drought tolerance by gene manipulation of 9-*cis*-epoxycarotenoid dioxygenase, a key enzyme in abscisic acid biosynthesis in Arabidopsis. *Plant J.* **27**(4): 325–333.

Jung, K. H., V. D. Trivedi et al. (2003). Demonstration of a sensory rhodopsin in eubacteria. *Mol. Microbiol.* **47**(6): 1513–1522.

Juttner, F. (1976). Beta-cyclocitral and alkanes in microcystis (cyanophyceae). *Zeitschrift Naturforschung C* **31**(9–10): 491–495.

Juttner, F. (1979). Algal excretion product, geranylacetone–potent inhibitor of carotene biosynthesis in synechococcus. *Zeitschrift Naturforschung C* **34**(11): 957–960.

Juttner, F. (1988). Carotene oxygenase in microcystis. *Meth. Enzymol.* **167**: 336–341.

Juttner, F. (1995). Physiology and biochemistry of odorous compounds from fresh-water cyanobacteria and algae. *Water Sci. Technol.* **31**(11): 69–78.

Juttner, F. and Hoflacher, B. (1985). Evidence of β-carotene 7,8(7′,8′) oxygenase (β-cyclocitral, crocetindial generating) in *Microcystis*. *Arch. Microbiol.* **141**(4): 337.

Kamoda, S. and Y. Saburi (1993a). Cloning, expression, and sequence-analysis of a lignostilbene-alpha,beta-dioxygenase gene from *Pseudomonas paucimobilis* TMY1009. *Biosci. Biotechnol. Biochem.* **57**(6): 926–930.

Kamoda, S. and Y. Saburi (1993b). Structural and enzymatic comparison of lignostilbene-alpha,beta-dioxygenase isozymes, I, II, and III, from *Pseudomonas paucimobilis* TMY1009. *Biosci. Biotechnol. Biochem.* **57**(6): 931–934.

Kamoda, S. and Y. Saburi (1995). Cloning of a lignostillbene-alpha,beta-dioxygenase isozyme gene from *Pseudomonas paucimobilis* TMY1009. *Biosci. Biotechnol. Biochem.* **59**(10): 1866–1868.

Kamoda, S., T. Terada et al. (1997). Purification and some properties of lignostilbene-alpha,beta-dioxygenase isozyme IV from *Pseudomonas paucimobilis* TMY1009. *Biosci. Biotechnol. Biochem.* **61**(9): 1575–1576.

Kamoda, S., T. Terada et al. (2003). A common structure of substrate shared by lignostilbenedioxygenase isozymes from sphingomonas paucimobilis TMY1009. *Biosci. Biotechnol. Biochem.* **67**(6): 1394–1396.

Kamoda, S., T. Terada et al. (2005). Production of heterogeneous dimer lignostilbenedioxygenase II from LSDA and LSDB in *Escherichia coli* cells. *Biosci. Biotechnol. Biochem.* **69**(3): 635–637.

Kato, M., H. Matsumoto et al. (2006). The role of carotenoid cleavage dioxygenases in the regulation of carotenoid profiles during maturation in citrus fruit. *J. Exp. Bot.* **57**(10): 2153–2164.

Kato, M., H. Matsumoto et al. (2007). Accumulation of carotenoids and expression of carotenoid biosynthetic genes and carotenoid cleavage dioxygenase genes during fruit maturation in the juice sacs of 'tamami,' 'kiyomi' tangor, and 'wilking' mandarin. *J. Jpn. Soc. Hort. Sci.* **76**(2): 103–111.

Kiefer, C., S. Hessel et al. (2001). Identification and characterization of a mammalian enzyme catalyzing the asymmetric oxidative cleavage of provitamin A. *J. Biol. Chem.* **276**(17): 14110–14116.

Klingner, A., H. Bothe et al. (1995). Identification of a yellow pigment formed in maize roots upon mycorrhizal colonization. *Phytochemistry* **38**(1): 53–55.

Kloer, D. P., S. Ruch et al. (2005). The structure of a retinal-forming carotenoid oxygenase. *Science* **308**(5719): 267–269.

Kloer, D. P. and G. E. Schulz (2006). Structural and biological aspects of carotenoid cleavage. *Cell. Mol. Life Sci.* **63**(19–20): 2291–2303.

Lakshmanan, M. R., H. Chansang et al. (1972). Purification and properties of carotene 15,15′-dioxygenase of rabbit intestine. *J. Lipid Res.* **13**(4): 477–482.

Lampert, J. M., J. Holzschuh et al. (2003). Provitamin A conversion to retinal via the beta,beta-carotene-15,15′-oxygenase (bcox) is essential for pattern formation and differentiation during zebrafish embryogenesis. *Development* **130**(10): 2173–2186.

Lefebvre, V., H. North et al. (2006). Functional analysis of *Arabidopsis* NCED6 and NCED9 genes indicates that ABA synthesized in the endosperm is involved in the induction of seed dormancy. *Plant J.* **45**(3): 309–319.

Leffingwell, J. (2003). Saffron. Retrieved January 8th, 2006, from http://www.leffingwell.com/saffron.htm.

Leuenberger, M. G., C. Engeloch-Jarret et al. (2001). The reaction mechanism of the enzyme-catalyzed central cleavage of beta-carotene to retinal. *Angew. Chem. Int. Ed. Engl.* **40**(14): 2614–2617.

Liang C, Z. F., W. Wei, Z. Wen, and S. Qin (2006). Carotenoid biosynthesis in cyanobacteria: Structural and evolutionary scenarios based on comparative genomics. *Int. J. Biol. Sci.* **2**(4): 197–207.

Lindqvist, A. and S. Andersson (2002). Biochemical properties of purified recombinant human beta-carotene 15,15′-monooxygenase. *J. Biol. Chem.* **277**(26): 23942–23948.

Lindshield, B. L., K. Canene-Adams et al. (2007). Lycopenoids: Are lycopene metabolites bioactive? *Arch. Biochem. Biophys.* **458**(2): 136–140.

Lohse, S., W. Schliemann et al. (2005). Organization and metabolism of plastids and mitochondria in arbuscular mycorrhizal roots of *Medicago truncatula. Plant Physiol.* **139**(1): 329–340.

Macias, F. A., A. Torres et al. (2002). Bioactive terpenoids from sunflower leaves cv. Peredovick (R). *Phytochemistry* **61**(6): 687–692.

Maier, W., K. Hammer et al. (1997). Accumulation of sesquiterpenoid cyclohexenone derivatives induced by an arbuscular mycorrhizal fungus in members of the *Poaceae. Planta* **202**(1): 36–42.

Maier, W., H. Peipp et al. (1995). Levels of a terpenoid glycoside (Blumenin) and cell wall-bound phenolics in some cereal mycorrhizas. *Plant Physiol.* **109**(2): 465–470.

Maier, W., B. Schneider et al. (1998). Biosynthesis of sesquiterpenoid cyclohexenone derivatives in mycorrhizal barley roots proceeds via the glyceraldehyde 3-phosphate/pyruvate pathway. *Tetrahedron Lett.* **39**(7): 521–524.

Marasco, E. and C. Schmidt-Dannert (2003). Towards the biotechnological production of aroma and flavor compounds in engineered microorganisms. *Appl. Biotechnol. Food Sci. Pol.* **1**(3): 145–157.

Marasco, E. and C. Schmidt-Dannert (2008). Identification of bacterial carotenoid cleavage dioxygenase homologs that cleave the interphenyl α,β double bond of stilbene derivatives via a monooxygenase reaction. *Chembiochem.* **9**(9): 1450–1461.

Marasco, E. K., K.-l. Vay et al. (2006). Identification of carotenoid cleavage dioxygenases from *Nostoc* sp. PCC 7120 with different cleavage activities. *J. Biol. Chem.* **281**(41): 31583–31593.

Mathieu, S., F. Bigey et al. (2007). Production of a recombinant carotenoid cleavage dioxygenase from grape and enzyme assay in water-miscible organic solvents. *Biotechnol. Lett.* **29**(5): 837–841.

Mathieu, S., N. Terrier et al. (2005). A carotenoid cleavage dioxygenase from *Vitis vinifera* L.: Functional characterization and expression during grape berry development in relation to C13-norisoprenoid accumulation. *J. Exp. Bot.* **56**(420): 2721–2731.

Matusova, R., K. Rani et al. (2005). The strigolactone germination stimulants of the plant-parasitic *Striga* and *Orobanche* spp. are derived from the carotenoid pathway. *Plant Physiol.* **139**(2): 920–934.

Meinwald, J., K. Erickson, M. Hartshorn, Y. C. Meinwald, and T. Eisner (1968). Defensive mechanisms of arthropods. XXIII. Anallenic sesquiterpenoid from the grasshopper *Romalea microptera. Tetrahedron Lett.* **25** 2959–2962.

Naested, H., A. Holm et al. (2004). *Arabidopsis* VARIEGATED 3 encodes a chloroplast-targeted, zinc-finger protein required for chloroplast and palisade cell development. *J. Cell. Sci.* **117**(Pt 20): 4807–4818.

Nicoletti, A., D. J. Wong et al. (1995). Molecular characterization of the human gene encoding an abundant 61 kDa protein specific to the retinal pigment epithelium. *Hum. Mol. Genet.* **4**(4): 641–649.

Ohmiya, A., S. Kishimoto et al. (2006). Carotenoid cleavage dioxygenase (CmCCD4a) contributes to white color formation in chrysanthemum petals. *Plant Physiol.* **142**(3): 1193–1201.

Olson, J. A. and O. Hayaishi (1965). The enzymatic cleavage of beta-carotene into Vitamin A by soluble enzymes of rat liver and intestine. *Proc. Natl. Acad. Sci. U. S. A.* **54**(5): 1364–1370.

Olson, J. A. (1961). The conversion of radioactive beta, beta-carotene into Vitamin A by the rat intestine *in vivo. J. Biol. Chem.* **236**(2): 349–356.

Paik, J., A. During et al. (2001). Expression and characterization of a murine enzyme able to cleave beta-carotene–The formation of retinoids. *J. Biol. Chem.* **276**(34): 32160–32168.

Patel, J. B., J. Mehta et al. (2007). Novel retinoic acid metabolism blocking agents have potent inhibitory activities on human breast cancer cells and tumour growth. *Br. J. Cancer* **96**(8): 1204–1215.

Pichersky E. and J. Gershenzon (2002). The formation and function of plant volatiles: Perfumes for pollinator attraction and defense. *Curr. Opin. Plant Biol.* **53**: 237–243.

Poliakov, E., S. Gentleman et al. (2005). Key role of conserved histidines in recombinant mouse beta-carotene 15,15′-monooxygenase-1 activity. *J. Biol. Chem.* **280**(32): 29217–29223.

Prado-Cabrero, A., A. F. Estrada et al. (2007). Identification and biochemical characterization of a novel carotenoid oxygenase: Elucidation of the cleavage step in the *Fusarium* carotenoid pathway. *Mol. Microbiol.* **64**(2): 448–460.

Prado-Cabrero, A., D. Scherzinger et al. (2007). Retinal biosynthesis in fungi: Characterization of the carotenoid oxygenase CarX from *Fusarium fujikuroi. Eukaryotic Cell* **6**(4): 650–657.

Prado, M. M., A. Prado-Cabrero et al. (2004). A gene of the opsin family in the carotenoid gene cluster of *Fusarium fujikuroi. Curr. Genet.* **46**(1): 47–58.

Qin, X. Q. and J. A. D. Zeevaart (1999). The 9-*cis*-epoxycarotenoid cleavage reaction is the key regulatory step of abscisic acid biosynthesis in water-stressed bean. *Proc. Natl. Acad. Sci. U. S. A.* **96**(26): 15354–15361.

Qin, X. Q. and J. A. D. Zeevaart (2002). Overexpression of a 9-*cis*-epoxycarotenoid dioxygenase gene in *Nicotiana plumbaginifolia* increases abscisic acid and phaseic acid levels and enhances drought tolerance. *Plant Physiol.* **128**(2): 544–551.

Que, L. and R. Y. N. Ho (1996). Dioxygen activation by enzymes with mononuclear non-heme iron active sites. *Chem. Rev.* **96**(7): 2607–2624.

Redmond, T. M., S. Gentleman et al. (2001). Identification, expression, and substrate specificity of a mammalian beta-carotene 15,15′-dioxygenase. *J. Biol. Chem.* **276**(9): 6560–6565.

Rodrigo, M. J., B. Alquezar et al. (2006). Cloning and characterization of two 9-*cis*-epoxycarotenoid dioxygenase genes, differentially regulated during fruit maturation and under stress conditions, from orange (*Citrus sinensis* L. Osbeck). *J. Exp. Bot.* **57**(3): 633–643.

Ruch, S., P. Beyer et al. (2005). Retinal biosynthesis in eubacteria: *In vitro* characterization of a novel carotenoid oxygenase from *Synechocystis* sp. PCC 6803. *Mol. Microbiol.* **55**(4): 1015–1024.

Saelices, L., L. Youssar et al. (2007). Identification of the gene responsible for torulene cleavage in the *Neurospora* carotenoid pathway. *Mol. Gen. Genomics* **278**(5): 527–537.

Saitou, N. and M. Nei (1987). The neighbor-joining method: A new method for reconstructing phylogenetic trees. *Mol. Biol. Evol.* **4**: 406–425.

Scherzinger, D., S. Ruch et al. (2006). Retinal is formed from apo-carotenoids in *Nostoc* sp PCC7120: *in vitro* characterization of an apo-carotenoid oxygenase. *Biochem. J.* **398**: 361–369.

Schilling, M., F. Patett et al. (2007). Influence of solubility-enhancing fusion proteins and organic solvents on the *in vitro* biocatalytic performance of the carotenoid cleavage dioxygenase AtCCD1 in a micellar reaction system. *Appl. Microbiol. Biotechnol.* **75**(4): 829–836.

Schmidt, H., R. Kurtzer et al. (2006). The carotenase AtCCD1 from *Arabidopsis thaliana* is a dioxygenase. *J. Biol. Chem.* **281**(15): 9845–9851.

Schwartz, S. H., X. Q. Qin et al. (2004). The biochemical characterization of two carotenoid cleavage enzymes from *Arabidopsis* indicates that a carotenoid-derived compound inhibits lateral branching. *J. Biol. Chem.* **279**(45): 46940–46945.

Schwartz, S. H., X. Q. Qin et al. (2001). Characterization of a novel carotenoid cleavage dioxygenase from plants. *J. Biol. Chem.* **276**(27): 25208–25211.

Schwartz, S. H., X. Q. Qin et al. (2003). Elucidation of the indirect pathway of abscisic acid biosynthesis by mutants, genes, and enzymes. *Plant Physiol.* **131**(4): 1591–1601.

Schwartz, S. H., X. Q. Qin et al. (2004). The biochemical characterization of two carotenoid cleavage enzymes from *Arabidopsis* indicates that a carotenoid-derived compound inhibits lateral branching. *J. Biol. Chem.* **279**(45): 46940–46945.

Schwartz, S. H., B. C. Tan et al. (1997). Specific oxidative cleavage of carotenoids by VP14 of maize. *Science* **276**(5320): 1872–1874.

Seo, M. and T. Koshiba (2002). Complex regulation of ABA biosynthesis in plants. *Trends Plant Sci.* **7**(1): 41–48.

Simkin, A. J., H. Moreau et al. (2008) An investigation of carotenoid biosynthesis in *Coffea canephora* and *Coffea arabica. J. Plant Physiol.* **165**(10):1087–1106.

Simkin, A. J., S. H. Schwartz et al. (2004a). The tomato carotenoid cleavage dioxygenase 1 genes contribute to the formation of the flavor volatiles beta-ionone, pseudoionone, and geranylacetone. *Plant J.* **40**(6): 882–892.

Simkin, A. J., B. A. Underwood et al. (2004b). Circadian regulation of the PhCCD1 carotenoid cleavage dioxygenase controls emission of beta-ionone, a fragrance volatile of petunia flowers. *Plant Physiol.* **136**(3): 3504–3514.

Sineshchekov, O. A. and J. L. Spudich (2004). Light-induced intramolecular charge movements in microbial rhodopsins in intact *E. coli* cells. *Photochem. Photobiol. Sci.* **3**(6): 548–554.

Sineshchekov, O. A., V. D. Trivedi et al. (2005). Photochromicity of *Anabaena* sensory rhodopsin, an atypical microbial receptor with a *cis*-retinal light-adapted form. *J. Biol. Chem.* **280**(15): 14663–14668.

Soar, C. J., J. Speirs et al. (2004). Gradients in stomatal conductance, xylem sap ABA and bulk leaf ABA along canes of *Vitis vinifera* cv. Shiraz: Molecular and physiological studies investigating their source. *Funct. Plant Biol.* **31**(6): 659–669.

Soprano, D. R., Q. Pu et al. (2004). Retinoic acid receptors and cancers. *Ann. Rev. Nutr.* **24**(1): 201–221.

Spudich, J. L., C. S. Yang et al. (2000). Retinylidene proteins: Structures and functions from archaea to humans. *Annu. Rev. Cell Dev. Biol.* **16**: 365–392.

Strack, D., T. Fester et al. (2003). Arbuscular mycorrhiza: Biological, chemical, and molecular aspects. *J. Chem. Ecol.* **29**(9): 1955–1979.

Straganz, G. D. and B. Nidetzky (2006). Variations of the 2-His-1-carboxylate theme in mononudear non-heme Fe-III oxygenases. *Chembiochem* **7**(10): 1536–1548.

Szondy, Z., U. Reichert et al. (1998). Retinoic acids regulate apoptosis of T lymphocytes through an interplay between RAR and RXR receptors. *Cell Death Differ.* **5**(1): 4–10.

Takahashi, Y., G. Moiseyev et al. (2005). Identification of conserved histidines and glutamic acid as key residues for isomerohydrolase activity of RPE65, an enzyme of the visual cycle in the retinal pigment epithelium. *FEBS Lett.* **579**(24): 5414–5418.

Takitani, K., C. L. Zhu et al. (2006). Molecular cloning of the rat beta-carotene 15,15′-monooxygenase gene and its regulation by retinoic acid. *Eur. J. Nutr.* **45**(6): 320–326.

Tamura, K., J. Dudley et al. (2007). MEGA4: Molecular evolutionary genetics analysis (MEGA) software version 4.0. *Mol. Biol. Evol.* **24**: 1596–1599.

Tan, B. C., K. Cline et al. (2001). Localization and targeting of the VP14 epoxy-carotenoid dioxygenase to chloroplast membranes. *Plant J.* **27**(5): 373–382.

Tan, B. C., L. M. Joseph et al. (2003). Molecular characterization of the *Arabidopsis* 9-*cis*-epoxycarotenoid dioxygenase gene family. *Plant J.* **35**(1): 44–56.

Tan, B. C., S. H. Schwartz et al. (1997). Genetic control of abscisic acid biosynthesis in maize. *Proc. Natl. Acad. Sci. U. S. A.* **94**(22): 12235–12240.

Tarantilis, P. A., G. Tsoupras et al. (1995). Determination of saffron (*Crocus sativa* L.) components in crude plant extracts using hihg-performance liquid chromatography-UV-visible photodiode array detection-mass spectrometry. *J. Chromat. A* **699**: 107–118.

Taylor, I. B., T. Sonneveld et al. (2005). Regulation and manipulation of the biosynthesis of abscisic acid, including the supply of xanthophyll precursors. *J. Plant Growth Reg.* **24**(4): 253–273.

Thompson, A. J., J. Andrews et al. (2007). Overproduction of abscisic acid in tomato increases transpiration efficiency and root hydraulic conductivity and influences leaf expansion. *Plant Physiol.* **143**(4): 1905–1917.

Thompson, A. J., A. C. Jackson et al. (2000a). Abscisic acid biosynthesis in tomato: Regulation of zeaxanthin epoxidase and 9-*cis*-epoxycarotenoid dioxygenase mRNAs by light/dark cycles, water stress and abscisic acid. *Plant Mol. Biol.* **42**(6): 833–845.

Thompson, A. J., A. C. Jackson et al. (2000b). Ectopic expression of a tomato 9-*cis*-epoxycarotenoid dioxygenase gene causes over-production of abscisic acid. *Plant J.* **23**(3): 363–374.

Thompson, A. J., E. T. Thorne et al. (2004). Complementation of notabilis, an abscisic acid-deficient mutant of tomato: Importance of sequence context and utility of partial complementation. *Plant Cell Environ.* **27**(4): 459–471.

Vierheilig, H., H. Gagnon et al. (2000). Accumulation of cyclohexenone derivatives in barley, wheat and maize roots in response to inoculation with different arbuscular mycorrhizal fungi. *Mycorrhiza* **9**(5): 291–293.

Vogeley, L., O. A. Sineshchekov et al. (2004). *Anabaena* sensory rhodopsin: A photochromic color sensor at 2.0 angstrom. *Science* **306**(5700): 1390–1393.

von Lintig, J., S. Hessel et al. (2005). Towards a better understanding of carotenoid metabolism in animals. *Biochim. Biophys. Acta* **1740**(2): 122–131.

von Lintig, J. and K. Vogt (2000). Filling the gap in vitamin A research–Molecular identification of an enzyme cleaving beta-carotene to retinal. *J. Biol. Chem.* **275**(16): 11915–11920.

Wache, Y., A. Bosser-DeRatuld et al. (2003). Effect of *cis/trans* isomerism of beta-carotene on the ratios of volatile compounds produced during oxidative degradation. *J. Agric. Food Chem.* **51**(7): 1984–1987.

Walsh, K., G. J. Jones et al. (1998). Effects of high irradiance and iron concentration on pigment and fatty acid composition in the cyanobacterium *Microcystis aeruginosa. Mar. Freshwater Res.* **49**(5): 399–407.

Walter, M. H., T. Fester et al. (2000). Arbuscular mycorrhizal fungi induce the non-mevalonate methylerythritol phosphate pathway of isoprenoid biosynthesis correlated with accumulation of the 'yellow pigment' and other apocarotenoids. *Plant J.* **21**(6): 571–578.

Walter, M. H., D. S. Flo et al. (2007). Apocarotenoid biosynthesis in arbuscular mycorrhizal roots: Contributions from methylerythritol phosphate pathway isogenes and tools for its manipulation. *Phytochemistry* **68**(1): 130–138.

Wan, X. and L. Li (2005). Molecular cloning and characterization of a dehydration-inducible cDNA encoding a putative 9-*cis*-epoxycarotenoid dioxygenase in *Arachis hygogaea* L. *DNA Seq.* **16**(3): 217–223.

Wang, X. C., C. Allen et al. (2007). Retinoic acid enhances the production of IL-10 while reducing the synthesis of IL-12 and TNF-alpha from LPS-stimulated monocytes/macrophages. *J. Clin. Immunol.* **27**(2): 193–200.

Watts, K. T., P. C. Lee et al. (2006). Biosynthesis of plant-specific stilbene polyketides in metabolically engineered *Escherichia coli. BMC Biotechnol.* **6**: 22.

Winterhalter B. and R. L. Rouseff (2002). *Carotenoid-Derived Aroma Compounds: An Introduction.* Washington DC: Amercian Chemical Society.

Woodall, A. A., S. W. M. Lee et al. (1997). Oxidation of carotenoids by free radicals: Relationship between structure and reactivity. *Biochim. Biophys. Acta* **1336**(1): 33–42.

Wu, Z., D. S. Robinson et al. (1999). Co-oxidation of beta-carotene catalyzed by soybean and recombinant pea lipoxygenases. *J. Agric. Food Chem.* **47**(12): 4899–4906.

Wyss, A., G. Wirtz et al. (2000). Cloning and expression of beta,beta-carotene 15,15′-dioxygenase. *Biochem. Biophys. Res. Commun.* **271**(2): 334–336.

Wyss, A., G. M. Wirtz et al. (2001). Expression pattern and localization of beta,beta-carotene 15,15′-dioxygenase in different tissues. *Biochem. J.* **354**: 521–529.

Yamada, M., Y. Okada et al. (2007). Purification, characterization and gene cloning of isoeugenol-degrading enzyme from *Pseudomonas putida* IE27. *Arch. Microbiol.* **187**(6): 511–517.

Yan, W., G.-F. Jang et al. (2001). Cloning and characterization of a human beta,beta-carotene-15,15′-dioxygenase that is highly expressed in the retinal pigment epithelium. *Genomics* **72**(2): 193–202.

Yang, J. and Z. Guo (2007). Cloning of a 9-*cis*-epoxycarotenoid dioxygenase gene (SgNCED1) from *Stylosanthes guianensis* and its expression in response to abiotic stresses. *Plant Cell Rep.* **26**(8): 1383–1390.

Yeum, K. J., Y. C. Leekim et al. (1994). *In vitro* metabolism of beta-carotene by lipoxygenase and human stomach mucosal homogenates. *FASEB J.* **8**(4): A192–A192.

Yeum, K. J., Y. C. Leekim et al. (1995). Similar metabolites formed from beta-carotene by human gastric-mucosal homogenates, lipoxygenase, or linoleic acid hydroperoxide. *Arch. Biochem. Biophys.* **321**(1): 167–174.

Zaripheh, S., T. Y. Nara et al. (2006). Dietary lycopene downregulates carotenoid 15,15′-monooxygenase and PPAR-γ in selected rat tissues. *J. Nutr.* **136**(4): 932–938.

Zeevaart, J. A. D., T. G. Heath et al. (1989). Evidence for a universal pathway of abscisic acid biosynthesis in higher plants from O^{18} incorporation patterns. *Plant Physiol.* **91**(4): 1594–1601.

Zhu, C., F. Kauder et al. (2007). Cloning of two individual cDNAS encoding 9-*cis*-epoxycarotenoid dioxygenase from *Gentiana lutea*, their tissue-specific expression and physiological effect in transgenic tobacco. *J. Plant Physiol.* **164**(2): 195–204.

Zorn, H., S. Langhoff et al. (2003a). Cleavage of beta,beta-carotene to flavor compounds by fungi. *Appl. Microbiol. Biotechnol.* **62**(4): 331–336.

Zorn, H., S. Langhoff et al. (2003b). A peroxidase from *Lepista irina* cleaves beta,beta-carotene to flavor compounds. *Biol. Chem.* **384**(7): 1049–1056.

20 Oxidative Metabolites of Lycopene and Their Biological Functions

Jonathan R. Mein and Xiang-Dong Wang

CONTENTS

20.1 INTRODUCTION

Considerable interest and research efforts have been expended in an effort to uncover the potential roles of carotenoids in human health and disease. While early studies focused on provitamin A carotenoids, more recent research efforts have focused on the potential roles of the non-provitamin A carotenoids (e.g., lycopene) in the health and the disease. Lycopene has been implicated as having a potential beneficial impact in a number of chronic diseases including cancer. Although evidence from epidemiological and animal studies supports a potential chemopreventive role of lycopene (Boileau et al. 2003, Canene-Adams et al. 2007, Giovannucci 1999b, Giovannucci and Clinton 1998, Siler et al. 2004), the biochemical mechanisms behind such beneficial effects have, as of yet, not been well-defined. Several reports have demonstrated the potential beneficial effects of lycopene especially in respect to antioxidant function, enhanced cellular gap junction communication, the induction of phase II enzymes through the activation of the antioxidant response element (ARE) transcription system, the suppression of insulin-like growth factor (IGF)-1 stimulated cell proliferation by induced insulin-like growth factor binding protein (IGFBP), and the inhibition of cell proliferation and the induction of apoptosis. With the cloning and the characterization of two distinct carotenoid cleaving enzymes, recent research has focused on the metabolic fate of lycopene and the subsequent metabolites created. Several reports, including our own, suggest that the biological activities of lycopene may be mediated, in part, by lycopene metabolites. Lycopene metabolites, and carotenoid metabolites in general, can possess either more or less activity than the parent compound or can have an entirely independent function. The chemical and biological metabolisms of lycopene and the potential actions of lycopene and its metabolites on chemoprevention will be highlighted in this chapter.

20.2 FORMATION OF LYCOPENE METABOLITES *IN VITRO*

Carotenoids are a class of lipophilic compounds with a polyisoprenoid structure. Most carotenoids contain a series of conjugated double bonds, which are sensitive to oxidative modification and *cis–trans* isomerization. There are six major carotenoids (β-carotene, α-carotene, lycopene, β-cryptoxanthin, lutein, and zeaxanthin) that can be routinely found in human plasma and tissues. Among them, β-carotene has been the most extensively studied. More recently, lycopene has attracted considerable attention due to its association with a decreased risk of certain chronic diseases, including cancers. Considerable efforts have been expended in order to identify its biological and physiochemical properties. Relative to β-carotene, lycopene has the same molecular mass and chemical formula, yet lycopene is an open-polyene chain lacking the β-ionone ring structure. While the metabolism of β-carotene has been extensively studied, the metabolism of lycopene remains poorly understood.

20.2.1 Chemical Oxidation of Lycopene

There have been a number of reports studying the formation of lycopene metabolites and oxidation products *in vitro*. Many of these studies have utilized various oxidizing systems and have identified several unique metabolites. Using three separate solubilization schemes (toluene, aqueous Tween 40, and liposomal suspension), Kim and colleagues (2001) identified several oxidative products after the incubation of lycopene under atmospheric oxygen, including 3,7,11-trimethyl-2,4,6,10-dodecatetraen-1-al; 6,10,14-trimethyl-3,5,7,9,13-pentadecapentaen-2-one; *acyclo*-retinal, apo-14′, and 12′-, 10′-, 8′-, and 6′-lycopenal. It was subsequently demonstrated that *acyclo*-retinal could be oxidized to the corresponding *acyclo*-retinoic acid (ACR) when incubated with pig liver homogenate (Kim et al. 2001) indicating potential *in vivo* formation. The incubation of deuterated lycopene with rat intestinal post-mitochondrial fractions and soy lipoxygenase led to the identification of several additional lycopene metabolites and oxidative products, such as 3-keto-apo-13-lycopenone; 3,4-dehydro-5,6-dihydro-15,15′-apo-lycopenal; 2-apo-5,8-lycopenal-furanoxide; lycopene-5,6,5′,6′-diepoxide; lycopene-5,8-furanoxide; and 3-keto-lycopene-5′,8′-furanoxide (Ferreira et al. 2003). Zhang et al. exposed lycopene to atmospheric oxygen and a perfusion of ozone and identified (*E*, *E*, *E*)-4-methyl-8-oxo-2,4,6-nonatrienal as a lycopene oxidative product, which induced apoptosis in HL-60 cells (Zhang et al. 2003). Using a combination of hydrogen peroxide and osmium tetroxide (Aust et al. 2003), an oxidized lycopene mixture was separated and, using an increase in gap junction communication (GJC) as a marker for bioactivity, a new metabolite was tentatively identified as 2,7,11-trimethyl-tetradecahexaene-1,14-dial (Aust et al. 2003). Using a separate oxidizing method, Caris-Veyrat et al. identified an extensive number of lycopene oxidative products (Caris-Veyrat et al. 2003). The oxidation of lycopene by potassium permanganate produced eight apo-lycopenals, three apo-lycopenones, and six apo-carotendials as detected by HPLC–DAD–MS. Taken together, the results from these studies suggest that the susceptibility of carbonyl compounds to cleavage by autooxidation, radical-mediated oxidation, and singlet oxygen occurs in carotenoids with a long-chain of conjugated double bonds. Although the significance of such oxidative metabolites remains poorly understood, these products may be produced *in vivo* (if the tissues are exposed to oxidative stress, such as smoking and drinking) and have certain biological activities.

20.2.2 Enzymatic Cleavage of Lycopene

20.2.2.1 Carotene-15,15′-Oxygenase and Lycopene

For provitamin A carotenoids, such as β-carotene, α-carotene, and β-cryptoxanthin, central cleavage is a major pathway leading to the formation of vitamin A (Goodman and Huang 1965, Olson and Hayaishi 1965). The carotene 15,15′-monooxygenase (CMO1) gene, which is responsible for central cleavage at the 15,15′ double bond (von Lintig and Vogt 2000, Wyss et al. 2000), has been cloned and characterized in a number of species including the human and the mouse (Lindqvist and Andersson 2002, Paik et al. 2001, Redmond et al. 2001, Wyss et al. 2001, Yan et al. 2001). With the molecular characterization of CMO1 (von Lintig and Vogt 2000, Wyss et al. 2000), it has been definitively shown that CMO1 catalyzes the central cleavage of β-carotene to yield two molecules of retinal, thus, contributing to vitamin A stores (Hessel et al. 2007). While many reports have focused exclusively on the ability of CMO1 to cleave β-carotene, few have explored other carotenoids, such as lycopene, as potential substrates, and those that have, have had mixed results. When lycopene was incubated with the *Drosophila* homologue of CMO1 (von Lintig and Vogt 2000) or the crude preparations of rat liver and intestine (Nagao and Olson 1994), as well as human retinal pigment epithelium CMO1 (Yan et al. 2001), no lycopene cleavage products were detected. Using lycopene-accumulating *Escherichia coli* that expresses mouse CMO1 (Redmond et al. 2001), Redmond and colleagues observed a distinct bleaching of color from red to white suggesting the cleavage of lycopene (Redmond et al. 2001). In addition, purified recombinant mouse CMO1 displayed *in vitro* cleavage activity toward lycopene. However, *acyclo*-retinal, the central cleavage product of lycopene, was only detected when the lycopene concentrations used were 2.5–3 times higher than the observed K_m for β-carotene ($K_m = 6\,\mu M$). In contrast, Lindqvist and Andersson, using a purified recombinant CMO1 isolated from a human liver cDNA library, demonstrated cleavage activity toward both β-carotene and β-crytoxanthin but no activity toward lycopene or zeaxanthin (Lindqvist and Andersson 2002). Although CMO1 was shown to cleave β-cryptoxanthin, the analysis of the apparent K_m revealed an approximate fourfold lower affinity toward β-cryptoxanthin ($K_m = 30.0 \pm 3.8\,\mu M$) than toward β-carotene ($K_m = 7.1 \pm 1.8\,\mu M$) (Lindqvist and Andersson 2002). These authors concluded that the presence of at least one unsubstituted β-ionone ring appears to be sufficient for the catalytic cleavage of the central carbon 15,15′ double bond. Taken together, these studies suggest that lycopene is a poor substrate for CMO1. However, the recent observation that *cis*-lycopene isomers were superior substrates for carotene-9,10′-oxygenase (CMO2) compared to all-*trans* lycopene (Hu et al. 2006) brings about an important question. Does CMO1 cleave *cis*-lycopene isomers to *acyclo*-retinoids? Previous *in vivo* reports using all-*trans* lycopene as a supplement observed dramatic increases in *cis*-isomers of lycopene, including 5-*cis*, 13-*cis*, and 9-*cis* isomers (Boileau et al. 1999, Liu et al. 2003, 2006, Wu et al. 2003). From the previous *in vitro* kinetic analyses (Lindqvist and Andersson 2002, Redmond et al. 2001, Yan et al. 2001), it is unclear if all-*trans* lycopene was used as the substrate in the determination of CMO1 activity toward lycopene. The chemical structures of *cis*-lycopene isomers could mimic the unsubstituted β-ionone ring structures of other carotenoid molecules and fit into the enzyme–substrate pocket enabling central cleavage (Figure 20.1). More understanding of the significance of the metabolism of *cis*-lycopene will help to further elucidate the biological functions of lycopene.

20.2.2.2 Carotene-9′,10′-Oxygenase and Lycopene

In addition to the central cleavage pathway, an alternative pathway for carotenoid metabolism in mammals, termed the excentric cleavage pathway, remained a controversial issue for several decades. The controversy centered on the existence of a dedicated enzyme responsible for excentric carotenoid metabolism. We had previously demonstrated the random cleavage of β-carotene and identified a series of homologous carbonyl cleavage products, including β-apo-14′-, 12′-, 10′-, and 8′-carotenals, β-apo-13-carotenone and retinoic acid in the tissue homogenates of humans, ferrets, and rats (Wang et al. 1991, 1996, Tang et al. 1991). This controversy was put to rest with the

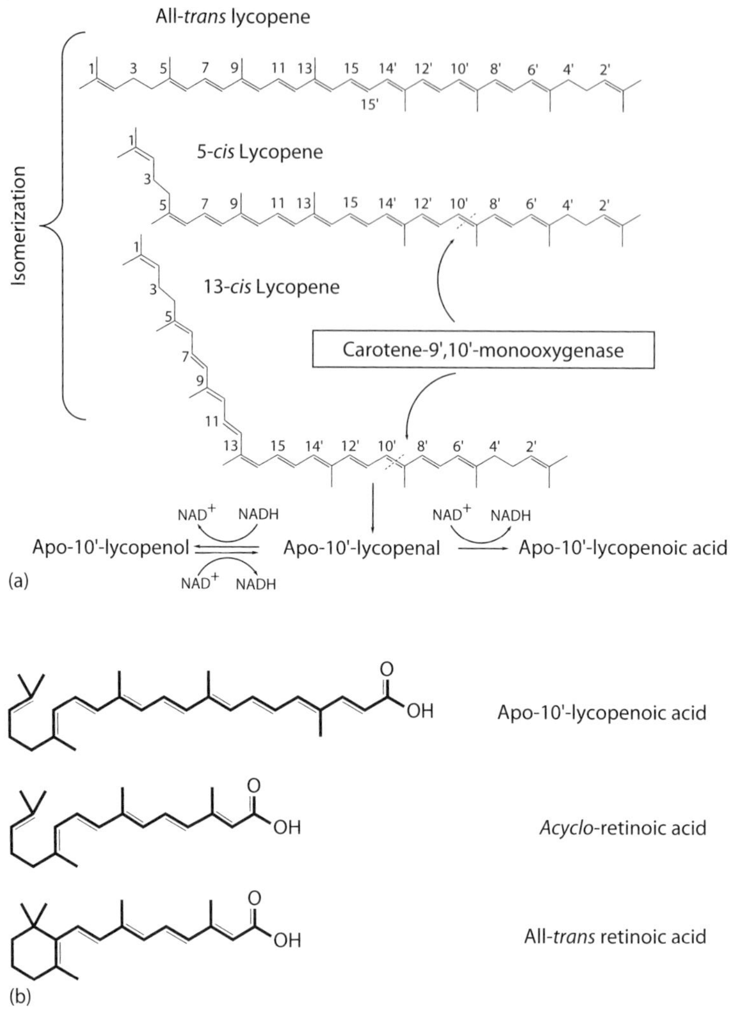

FIGURE 20.1 Schematic illustration of lycopene metabolic pathway by CMO2. (a) 5-*cis* Lycopene and 13-*cis* lycopene are preferentially cleaved by CMO2 at 9′,10′-double bond. The cleavage product, apo-10′-lycopenal, can be further oxidized to apo-10′-lycopenol or reduced to apo-10′-lycopenoic acid, depending on the presence of NAD^+ or NADH. (b) Chemical structures of apo-10′-lycopenoic acid, *acyclo*-retinoic acid, and all-*trans* retinoic acid. (Adapted from Hu, K.Q. et al., *J. Biol. Chem.*, 281, 19327, 2006. With permission.)

cloning and the characterization of the murine CMO2 by Kiefer and colleagues, thus, confirming the existence of the asymmetric cleavage pathway of carotenoids (Kiefer et al. 2001). The cleavage of β-carotene into apo-10′-carotenal was demonstrated using β-carotene synthesizing and accumulating *Escherichia coli* strains that express the mouse CMO2. When CMO2 was induced in a similar *Escherichia coli* model, which synthesizes and accumulates lycopene, a distinct color shift from red to white occurred, indicating the cleavage of lycopene. This important observation raised the question of whether CMO2 can catalyze the excentric cleavage of lycopene at the 9′,10′ double bond, forming apo-10′-lycopenal.

Because ferrets (*Mustela putorius furo*) and humans are similar in terms of carotenoid absorption, tissue distribution and concentrations, and metabolism (Wang 2005, Wang et al. 1992), we cloned and characterized the ferret CMO2 gene (Hu et al. 2006). Using the reported cDNA sequence for a carotene excentric cleavage enzyme from humans, we cloned a full-length carotene-9′,10′-oxygenase in ferrets that encodes a protein of 540 amino acids and has a 82% identity with human carotene, 9′,10′-oxygenase. Further analysis revealed that the enzyme is expressed in the testis, the liver, the lung, the prostate, the intestine, the stomach, and the kidneys of ferrets, similar to the expression pattern of human CMO2 (Lindqvist et al. 2005). Using the recombinant ferret CMO2 expressed in *Spodoptera frugiperda* (Sf9) insect cells for kinetic analysis, we found that the cleavage of carotenoids by the ferret CMO2 occurs in a pH-, incubation time-, protein dose-, and substrate dose-dependant manner (Hu et al. 2006). Notably, the optimum pH for CMO2 is 8.5, which differs from the optimum pH (7.7) for the activity of CMO1, the central cleavage enzyme for carotenoids (Lindqvist and Andersson 2002). The difference in optimum pH between these two carotenoid cleavage enzymes may indicate different roles of the two pathways in carotenoid metabolism or different functions in various pathophysiological conditions, which need further investigation. Nonetheless, similar to the CMO1, we found that the cleavage activity of ferret CMO2 for both β-carotene and lycopene was iron-dependent, indicating that iron is an essential cofactor for the enzymatic cleavage activity of carotenoids. This is supported by the existence of four conserved histidines residues in the ferret CMO2 (Hu et al. 2006). These data are in agreement with previous observations demonstrating that these conserved histidines act as putative iron-binding residues for iron coordination in apocarotenoid 15,15′-oxygenase (Kloer et al. 2005) and CMO1 (Poliakov et al. 2005) supporting the notion that the entire superfamily of oxygenases shares a common structure (Poliakov et al. 2005).

Interestingly, we demonstrated that the recombinant ferret CMO2 catalyzes the excentric cleavage of all-*trans* β-carotene and *cis*-lycopene isomers effectively but not all-*trans* lycopene at the 9′,10′ double bond (Hu et al. 2006). While we estimated a K_m of 3.5 μM for all-*trans* β-carotene based on the CMO2 expressed in SF9 cells, we could not calculate the kinetic constants of CMO2 for lycopene due to difficulty in controlling auto-isomerization, thus, necessitating the use of mixed isomers of lycopene as the substrates for kinetic analysis. Since the lycopene substrate mixture contains only ~20% as *cis* isomers and considering that the ferret CMO2 would not cleave all-*trans* lycopene, we speculate that the K_m for *cis*-lycopene is actually much lower than that of the lycopene isomer mixture. This indicates that *cis*-lycopene may act as a better substrate than all-*trans* β-carotene for the ferret CMO2. The mechanism whereby ferret CMO2 preferentially cleaves the 5-*cis* and 13-*cis*-isomers of lycopene into apo-10′-lycopenal but not all-*trans* lycopene is currently unknown. One possible explanation is that the chemical structure of *cis* isomers of lycopene could mimic the ring structure of the β-carotene molecule and fit into the substrate–enzyme binding pocket (Figure 20.1). Although this hypothesis warrants further investigation, the observation that the supplementation of all-*trans* lycopene results in a significant increase in *cis*-lycopene tissue concentration in ferrets underlies the significance of this observation (Boileau et al. 1999, Liu et al. 2003, 2006).

20.2.2.3 Regulation of Carotene Oxidases

A number of animal studies have demonstrated that CMO1 activity is affected by nutritional status, such as vitamin A status (Parvin and Sivakumar 2000, van Vliet et al. 1996). Other studies have indicated that the expression of CMO1 may be regulated at the transcriptional level through feedback regulatory mechanisms via interactions between retinoic acid and its nuclear receptors (Bachmann et al. 2002, Chichili et al. 2005). Recent molecular studies of the mouse and the human CMO1 promoters demonstrated the presence of a peroxisome proliferator response element (PPRE) (Boulanger et al. 2003, Gong et al. 2006). PPARγ (peroxisome proliferators activated receptor-γ) and RXRα (retinoid X repetor-α) agonists were shown to transactivate the CMO1 promoter–reporter when cotransfected with the corresponding nuclear receptor (Boulanger et al. 2003). The analysis of the human CMO1 promoter identified an additional enhancer element. A myocyte enhancer factor-2 (MEF2) binding site was identified and when mutated reduced luciferase activity by ~30% (Gong et al. 2006). The *in vivo* importance of the MEF2 binding site is not fully understood. Nonetheless, the regulation by PPAR and RXR indicates a regulatory link between carotenoid and lipid metabolisms. Two recent reports provided some supportive data for this relationship. In F344 rats supplemented with lycopene, CMO1 expression was significantly decreased in the adrenal gland and kidney (Zaripheh et al. 2006). Interestingly, fatty acid binding protein-3 (FABP-3), a PPARγ target gene, was downregulated in parallel with CMO1. While the production of vitamin A from β-carotene was abolished in CMO1-knock out (KO) mice, lipid metabolism was significantly altered (Hessel et al. 2007). There was a significant increase in several fatty acid metabolism–genes, including CD36 and FABP-4 expression, both PPARγ target genes in visceral adipose tissue. Additionally, there were significant increases in serum free fatty acids and total lipids resulting in hepatic steatosis. It has been suggested that the cleavage products of β-carotene may fine-tune the cross talk between the nuclear receptors that regulate lipid metabolism (Ziouzenkova and Plutzky 2008, Ziouzenkova et al. 2007a,b). The relationship between carotenoid and lipid metabolism deserves further inquiry.

Regulation of CMO2 is much less understood. A recent analysis failed to identify a PPRE within the mouse CMO2 promoter (Zaripheh et al. 2006). Work in our laboratory also failed to identify any potential nuclear receptor response elements or any other enhancer sequences within the ferret CMO2 promoter (unpublished data). In ferrets supplemented with low- and high-dose β-carotene and exposed to cigarette smoke for six weeks, we observed no change in lung and liver CMO2 expressions (Mein et al. 2006). Similar findings were observed using CMO1-KO mice. CMO1-KO mice supplemented with β-carotene accumulated significant amounts of β-carotene in various tissues (Hessel et al. 2007). However, there were no changes in CMO2 expression in the tissues analyzed. While we observed as approximately fourfold increase in CMO2 expression in the lungs of ferrets after nine weeks of lycopene supplementation (Hu et al. 2006), there was no significant effect of lycopene on CMO2 expression in several tissues of F344 rats, including lungs (Zaripheh et al. 2006). Clearly, more research is needed in order to gain a better understanding of the transcriptional regulatory mechanisms of CMO2.

20.3 FORMATION OF LYCOPENE METABOLITES *IN VIVO*

While *in vitro* oxidation studies have yielded a large number of oxidative metabolites, *in vivo* studies have yielded a much smaller catalogue of lycopene metabolites (Khachik et al. 2002, Lindshield et al. 2007). Khachik and colleagues first identified 5,6-dihydroxy-5,6-dihydrolycopene in human serum (Khachik et al. 1992a,b, 1995). It was proposed that the formation of this metabolite results from the oxidation of lycopene-forming 5,6-epoxide, which is then reduced to the 5,6-dihydroxy-5,6-dihydrolycopene metabolite. The same group also identified epimeric 2,6-cyclolycopene-1,5-diols in human milk and serum (Khachik et al. 1997). Based on the studies of lycopene oxidation with *m*-chloroperbenzoic acid (Khachik 1998a,b), it was proposed that lycopene is first oxidized at 1,2- and 5,6-positions to form lycopene 1,2-epoxide and lycopene 5,6-epoxide. Due to the instability of the epoxide rearrangement products, cyclization occurs and results in the formation of corresponding diols. Interestingly, epimeric 2,6-cyclolycopene-1,5-diols are also found in low concentrations in tomato-based products (Khachik 1998a). Whether the presence of these metabolites results from the consumption of tomato-based products or from *in vivo* oxidation or both is not fully understood.

The use of animal models has proven to be useful in the identification of *in vivo* lycopene metabolites. A study in preruminant cattle identified 5,6-dihydrolycopene and 5,6-dihydro-5-*cis* lycopene in serum after two weeks of lycopene supplementation (Sicilia et al. 2005). Using ^{14}C-labeled lycopene, Gajic et al. detected both apo-8′-lycopenal and apo-12′-lycopenal in rat liver 24 h post dosing (Gajic et al. 2006). In addition, a large quantity of very polar, unidentified short-chain compounds was detected. We have recently identified apo-10′-lycopenol in ferret lungs, which is the predicted cleavage product of CMO2, and certain unidentified compounds appearing between the retention times of 10 and 13 min in the HPLC profiles of ferret lungs (Hu et al. 2006) after lycopene supplementation for 9 weeks. Since we did not detect apo-10′-lycopenal or apo-10′-lycopenoic acid in the lung tissues of lycopene-supplemented ferrets, it is likely that apo-10′-lycopenal is a short-lived intermediate compound and can be reduced to its alcohol form *in vivo*, and apo-10′-lycopenoic acid may be present at too low a concentration to be detected in our HPLC system. This was supported by our subsequent demonstrations that the incubation of apo-10′-lycopenal with the post-nuclear fraction of hepatic tissues of ferrets resulted in both apo-10′-carotenol and apo-10′-lycopenoic acid depending upon the presence of either NAD^+ or NADH. In the presence of NADH, apo-10′-lycopenal was converted to both the alcohol and acid forms (Figure 20.1). The latter could be due to the consumption of NADH for the reduction reaction, which makes NAD^+ available for the oxidation of apo-10′-lycopenal. Nonetheless, the presence of specific metabolites has not been consistent across different animal models. Whether these differences are due to different methodological approaches, species used, or tissues analyzed should be ascertained.

20.4 BIOLOGICAL ACTIVITY OF LYCOPENE METABOLITES

The identification of lycopene metabolites *in vitro* and *in vivo* raises the question as to whether lycopene metabolites, similar to other carotenoid metabolites, which can possess either more or less activity than the parent compound or have entirely different functions (Wang 2004), may contribute, at least in part, to the biological functions ascribed to lycopene.

20.4.1 Antioxidant Properties

Much of the biological activity ascribed to lycopene has been attributed to its antioxidant capabilities. Indeed, antioxidant properties of many carotenoids have been long believed to play critical roles in their anticarcinogenic actions (Krinsky and Johnson 2005). Among naturally occurring carotenoids, lycopene has shown the strongest ability to scavenge free radicals (Miller et al. 1996) and chemically quench singlet oxygen (Conn et al. 1991, Di Mascio et al. 1989), having shown to be 2- and 10-fold more effective at quenching singlet oxygen than β-carotene and α-tocopherol, respectively (Di Mascio et al. 1989). Accordingly, several epidemiological studies have evaluated the role of lycopene as a potential *in vivo* antioxidant. Using tomatoes or tomato products, numerous studies have demonstrated decreased DNA damage (Bowen et al. 2002, Chen et al. 2001), decreased susceptibility to oxidative stress in lymphocytes (Porrini and Riso 2000, Riso et al. 1999), and decreased LDL oxidation (Agarwal and Rao 1998) or lipid peroxidation (Agarwal and Rao 1998, Bub et al. 2000). However, most *in vivo* studies have used tomato products, which also contain various micronutrients and phytochemicals, including other carotenoid, polyphenol, vitamin C, and vitamin E. Caution must be taken when attributing the beneficial effects of tomatoes and tomato products solely to lycopene.

While evidence suggests that intact lycopene functions as an antioxidant, especially *in vitro*, there is little evidence to support an antioxidant role of lycopene metabolites. We have recently provided the evidence of a possible antioxidant effect of the lycopene metabolite apo-10′lycopenoic acid in immortalized lung cells (BEAS-2B). After 24 h of treatment with apo-10′-lycopenoic acid (3–10 μM), we observed a dose-dependent decrease in endogenous reactive oxygen species (ROS) production (Lian and Wang 2008). This decrease in ROS was comparable to control cells treated with *tert*-butylhydroquinone. We next determined if apo-10′-lycopenoic acid had any effect on H_2O_2-induced oxidative damage, as measured by lactate dehydrogenase release (LDH). Pretreating BEAS-2B cells with of apo-10′-lycopenoic acid (3–10 μM) for 24 h resulted in a dose-dependent inhibition of LDH release. These results were comparable to control cells pretreated with tHBQ (Lian and Wang 2008). Taken together, our data suggests that lycopene metabolites in general, and apo-10′-lycopenoic acid in particular, may possess antioxidant functions. Further research is clearly needed in this area.

20.4.2 Gap Junction Communication

Gap junctions are cell-to-cell channels that enable connected cells to exchange nutrients, waste products, and information. Each gap junction is derived from six connexin proteins from each adjacent cell for a total of 12 connexin proteins. The connexin family has >20 connexins that are expressed in mammals with both cell and developmental specificities of expression (Bertram 2004). Although there are >20 connexins, connexin 43 (Cx43) is the most widely expressed connexin. More interestingly, Cx43 is the connexin most often induced by retinoids and carotenoids. GJC has been implicated in the control of cell growth via adaptive responses: differentiation, proliferation, and apoptosis (Trosko et al. 1998). A large body of evidence now indicates the loss of gap junctional communication as a hallmark of carcinogenesis (King and Bertram 2005).

Targeting connexins as a possible strategy for chemoprevention has been suggested (King and Bertram 2005). Retinoids and carotenoids increase GJC between normal and transformed cells (Hossain et al. 1989, Zhang et al. 1991). It was demonstrated that both provitamin A and non-provitamin A carotenoids inhibited carcinogen-induced neoplastic transformation (Bertram et al. 1991) and upregulated Cx43 mRNA expression (Hossain et al. 1989, Zhang et al. 1991). Furthermore, whereas treatment with retinoic acid increased Cx43 expression within 6 h, carotenoid treatment required approximately three times longer and produced the same response (Rogers et al. 1990, Zhang et al. 1992). This lag in activity is often attributed to the formation of active metabolites.

Several lines of *in vitro* evidence indicate that carotenoid oxidative products/metabolites may be responsible for increased GJC, especially in the case of lycopene. After the complete oxidation of lycopene with hydrogen peroxide and osmium tetroxide, Aust et al. (2003) isolated an oxidative metabolite that effectively increased gap junction communication. The compound, identified as 2,7,11-trimethyl-tetradecahexaene-1,14-dial, induced GJC to a level comparable to retinoic acid. The oxidative metabolite lycopene-5,6-epoxide, which is found in tomatoes (Khachik et al. 1995) was shown to increase Cx43 expression in human keratinocytes (Khachik et al. 1995). Stahl et al. demonstrated that the central cleavage product of lycopene, ACR, could increase GJC (Stahl et al. 2000). However, an effect of ACR on GJC was only achieved at high concentrations indicating that the contribution of ACR to the activity of lycopene on GJC appears to be minimal. More recently, we have demonstrated the cleavage of lycopene to apo-10′-lycopenal by ferret CMO2 (Hu et al. 2006). While the Cx43 promoter does not contain an retinoic acid response element (RARE), it has been reported that retonic acid receptor (RAR) antagonists inhibited upregulation by retinoids and had no effect on the effect of carotenoids (Hix et al. 2005). This is interesting due to the effect of both oxidative metabolites and enzymatic cleavage metabolites of lycopene on modulating GJC, which could provide two separate pathways of increasing GJC. Considering the bioconversion of lycopene into apo-10′-lycopenoids, whether apo-10-lycopenoids contribute to lycopene activity on GJC warrants further study.

20.4.3 Retinoid Activity

β-Carotene and its excentric cleavage metabolites can serve as direct precursors for all-*trans*-retinoic acid and 9-*cis*-retinoic acid (Napoli and Race 1988, Wang et al. 1994, 1996), which are ligands for both RARs and RXRs. Retinoid receptors function as ligand-dependent transcription factors and regulate gene expression by binding as dimeric complexes to the RARE and the retinoid X response element, which are located in the 5′ promoter region of responsive genes. RXR can not only form dimeric complexes with RAR but can also dimerize with other members of the nuclear hormone receptor superfamily, such as thyroid hormone receptors, the vitamin D receptors, PPARs, and possibly other receptors with unknown ligands designated or orphan receptors.

Upregulation of retinoid receptor expression and function by provitamin A carotenoids may play a role in mediating the growth inhibitory effects of retinoids in cancer cells (Lian et al. 2006, Prakash et al. 2004). However, it is unclear if non-provitamin A carotenoids and their metabolites may act as ligands for nuclear receptors. Using a RARE-reporter gene, Ben-Dor et al. (2001) demonstrated that ACR transactivates the RARE-reporter gene through an interaction with RARα. However, the potency of activation was approximately 100-fold lower than retinoic acid. Binding affinity studies indicated that ACR bound to RXRα with no appreciable affinity whereas ACR bound to RARα with an equilibrium dissociation constant in the range of 50–150 nM; two orders of magnitude lower than all-*trans* retinoic acid. Intact lycopene displayed a weak transactivation of the RARE-reporter gene (Ben-Dor et al. 2001). Stahl et al. demonstrated similar findings with the RAR-β2 promoter. Only when ACR was provided at concentrations 500-fold higher than retinoic acid was an effect on luciferase activity and β-galactosidase reporter activity observed (Stahl et al. 2000). There was no effect of intact lycopene on reporter transactivation at the concentrations used in this study. ACR was also found to have no significant effect on the transactivation of RAR and RXR reporter systems in a separate study (Araki et al. 1995). We recently demonstrated an increase in RARβ mRNA expression after treatment with apo-10′-lycopenoic acid in NHBE, BEAS-2B, and A549 cell lines (Lian et al. 2007). Because of the similarity in chemical structures among apo-10′-lycopenoic acid, ACR, and all-*trans* retinoic acid (Figure 20.1), we investigated whether the increased RARβ mRNA expression by apo-10′-lycopenoic acid was due to an increased transactivation of the RARβ promoter region. We found that the deletion of the promoter region between −1500 and −124 base pairs did not affect the transactivation activity of apo-10′-lycopenoic acid. However, the mutation of the RAR binding site, located between −53 and −37 base pairs, completely abolished the induction of promoter activity by both retinoic acid and apo-10′-lycopenoic acid (Figure 20.2) (Lian et al. 2007). These results suggest that the induction of RARβ by apo-10′-lycopenoic acid may be mediated through retinoid signaling.

20.4.4 Induction of Phase II Enzymes

The phase I and II enzymes respond to a variety of compounds, including drugs, environmental compounds, pollutants, carcinogens, and dietary and endogenous compounds. Phase I enzymes, such as cytochrome P450, catalyze the addition of oxygen to carcinogens thereby increasing the reactivity of carcinogens and the formation of DNA adducts known as bioactivation (Mandlekar et al. 2006). In general, the action of phase II enzymes increases the hydrophilicity of carcinogens and enhances their detoxification and excretion (Xu et al. 2005). The induction of phase II enzymes is mediated through *cis*-regulatory DNA sequences located in the promoter or enhancer regions, which are known as AREs (Talalay et al. 2003). The major ARE transcription factor, Nrf2 (nuclear factor E_2-related factor 2) is a primary factor involved in the induction of antioxidant and detoxifying enzymes (Giudice and Montella 2006) and is essential for the induction of several phase II enzymes, including glutathione *S*-transferases (GSTs) and NAD(P)H:quinone oxidoreductases (NQO1s) (Ramos-Gomez et al. 2001).

Many carotenoids including lycopene have been shown to induce several phase I and II enzymes both *in vivo* and *in vitro* (Ben-Dor et al. 2005, Breinholt et al. 2000, Gradelet et al. 1996, Zaripheh et al. 2006). Gradelet et al. observed an induction in the phase II enzymes, *p*-nitrophenol-UDP-glucuronosyltransferase and NQO1, in rats fed with various carotenoids. (Gradelet et al. 1996). Evidence in recent years has begun to accumulate indicating that the beneficial effect of lycopene may be due to the induction of phase II detoxification enzymes (Talalay 2000). For example, Breinholt et al. (2000) demonstrated a dose-dependent induction of several phase I and II enzymes in female Wistar rats supplemented with lycopene at doses ranging from 0.001 to 0.1 g/kg body weight for two weeks. Hepatic ethoxyresorufiin *O*-dealkylase and benzyloxyresorufin *O*-dealkylase increased approximately twofold and 50%, respectively, suggesting the activation of cytochrome P450 1A enzymes. In addition, several liver and red blood cell phase II enzyme activities, such as GST, glutathione reductase (GR), and quinone reductase, were significantly increased by lycopene feeding. The induction of phase II enzymes by lycopene has been reported in other animal studies (Bhuvaneswari et al. 2001, Zaripheh et al. 2006). However, it is

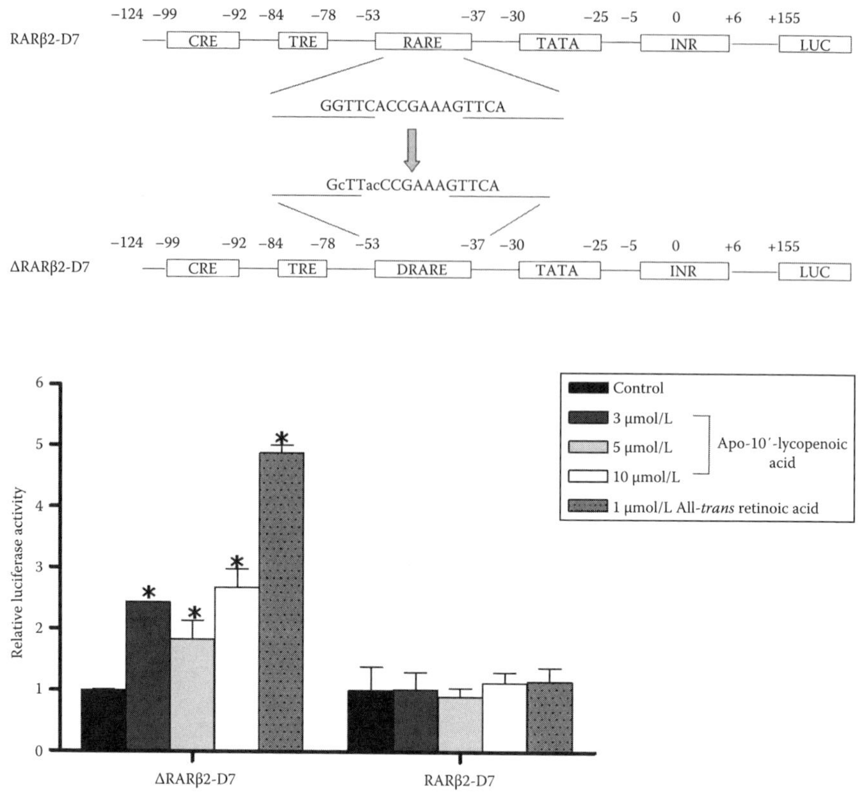

FIGURE 20.2 The involvement of RARE on apo-10′-lycopenoic acid-transactivated RARβ expression. Upper panel: Diagram of the RARβ reporter vector with wild type and mutated RAREs. Lower panel: HeLa cells transfected with the RARβ reporter vector and an internal control vector were treated with 5 μmol/L of apo-10′-lycopenoic acid or 1 μmol/L of *all-trans* retinoic acid for 24 h. Luciferase activities were measured by dual-luciferase reporter system. Values are means of ± SEM of three replicate assays. *, statistically significantly different, as compared with control in the same group, $P < 0.05$. (Adapted from Lian, F. et al., *Carcinogenesis*, 28, 1567, 2007. With permission.)

unclear if the induction of xenobiotic metabolizing enzymes is due to intact lycopene or lycopene metabolites. Ben-Dor et al. showed that lycopene induced phase II enzymes by activating Nrf2 transcription factor. More interestingly, an ethanolic extract of lycopene containing unidentified hydrophilic derivatives activated ARE-driven reporter gene at a potency similar to lycopene alone (Ben-Dor et al. 2005). Although the identity of the lycopene oxidative derivatives is unknown, this study suggested that lycopene oxidative metabolites might be responsible for the induction of phase II enzymes through ARE-induced expression. Using immortalized BEAS-2B human bronchial epithelial cells, we demonstrated a dose- and a time-dependent increase in nuclear Nrf2 protein accumulation with apo-10′-lycopenoic acid treatment (Lian and Wang 2008). In addition, apo-10′-lycopenoic acid significantly induced the mRNA expression of several phase II enzymes, including NQO1, GST, GR, heme-oxygenase-1 (HO-1), glutamate-cysteine ligase (catalytic unit and modifier unit), microsomal epoxide hydrolase 1, and UDP glucuronosyltransferase 1 family, polypeptide A6, as compared to THF alone (Lian and Wang 2008). Additionally, we observed that all three lycopenoids, including apo-10′-lycopenal, apo-10′-lycopenol, and apo-10′-lycopenoic acid, can induce HO-1 mRNA expression in BEAS-2B cells. Although the mechanisms behind the Nrf2-dependent phase II enzyme induction by these three lycopenoids remain unknown, these results suggest that the induction in phase II enzyme observed in previous studies may be a result of lycopene metabolites. Further investigation is clearly needed.

20.4.5 Interference with Growth Factors

IGFs (IGF-1 and IGF-2) are mitogens that play a central role in the regulation of cellular proliferation, differentiation, and apoptosis (Yu and Rohan 2000). By binding to membrane IGF-1 receptors, IGFs activate intracellular phosphatidylinositol 3′-kinase/Akt/protein kinase B and Ras/Raf/MAPK pathways, which regulate various biological processes, such as cell cycle progression, survival, and transformation (Clemmons et al. 1995). IGFs are sequestered in circulation by a family of binding proteins (IGFBP1–IGFBP6), which regulate the availability of IGFs to bind with IGF receptors (Clemmons et al. 1995). The disruption of normal IGF signaling, leading to hyperproliferation and survival signal expression, has been implicated in the development of several tumor types (Jerome et al. 2003). Recent epidemiological studies provide supportive evidence that lycopene may have a chemopreventive effect against a broad range of epithelial cancers, particularly prostate, breast, colorectal, and lung cancers (Arab et al. 2001, Clinton et al. 1996, Giovannucci 1999a,b, 2002). Sharoni and colleagues provided a potential mechanism whereby lycopene interfered with IGF-I stimulated cell growth (Karas et al. 1997, 2000, Levy et al. 1995). They showed that IGF-I stimulated cell growth, as well as DNA binding activity of the AP-1 transcription factor, were reduced by the physiological concentrations of lycopene in endometrial, mammary (MCF-7), and lung (NCI-H226) cancer cell lines. Lycopene has been shown to inhibit IGF-1 stimulated insulin receptor substrate 1 phosphorylation and cyclin D1 expression, block IGF-1 stimulated cell cycle progression (Karas et al. 2000, Nahum et al. 2006), and increase membrane-associated IGFBPs (Karas et al. 1997, 2000). Consistent with previous *in vitro* findings, recent epidemiological studies demonstrated that the higher dietary intake of lycopene has been associated with the lower circulating levels of IGF-1 (Mucci et al. 2001) and the higher levels of IGFBPs (Holmes et al. 2002, Vrieling et al. 2007).

We have examined the effect of lycopene on the prevention of IGF signaling in cigarette smoke–related lung carcinogenesis in the ferret model (Liu et al. 2003). We found that plasma IGF-1 levels were not affected by cigarette smoke exposure or lycopene supplementation. However, IGFBP-3 levels were increased by lycopene supplementation and decreased by smoke exposure. Interestingly, lycopene increased plasma IGFBP-3 regardless of smoke exposure status. Increased plasma IGFBP-3 was associated with the inhibition of cigarette smoke-induced lung squamous metaplasia, decreased proliferating cell nuclear antigen, phosphorylated BAD levels, and cleaved caspase 3 suggesting the inhibition of cell proliferation and the induction of apoptosis (Liu et al. 2003). These results, along with others, suggest that the interference of IGF-1 signaling may be an important mechanism by which lycopene exerts its anticancer activity. However, whether intact lycopene or its metabolites are responsible for the observed effects on IGF-1 signaling remains unknown. We have recently provided evidence that lycopene metabolites may be partly responsible. Treatment with apo-10′-lycopenoic acid (5–20 μM) resulted in a dose-dependent increase in IGFBP-3 mRNA levels in THLE-2 human liver cells. Similar concentrations of retinoic acid, lycopene, and ACR showed no significant effect on the induction of IGFBP-3 mRNA levels (unpublished results). Research into this area is going on in this laboratory.

20.4.6 Cell Proliferation and Apoptosis

The growth inhibitory effect of lycopene was first demonstrated by Levy et al., who showed that lycopene is a stronger cell growth inhibitor than β-carotene (Levy et al. 1995). This growth inhibitory effect was further observed in several cell lines, including breast cancer (Levy et al. 1995, Nahum et al. 2001), prostate cancer (Levy et al. 1995), lung cancer (Levy et al. 1995), colon cancer (Salman et al. 2007), and oral cavity cancer cells (Livny et al. 2002), as well as normal prostate epithelial cells (Obermuller-Jevic et al. 2003). The growth inhibition of lycopene on MCF-7 breast cancer cells was associated with a decreased G_1-S cell cycle progression, a decreased cyclin D1 expression, and the stabilization of p27 in the cyclin E-CDK complex (Nahum et al. 2001, 2006). In addition to cell-proliferation inhibition, the growth inhibitory effect of lycopene may also be attributed to the induction of apoptosis. Hwang et al. observed that 1 μM water-soluble lycopene inhibited the growth of LNCaP prostate cancer cells, while 5 μM lycopene blocked cells in G2/M phase and induced apoptosis (Hwang and Bowen 2004). In another study, the physiological concentrations of lycopene (0.3–3 μM) did not affect the proliferation of LNCaP cells but rather affected mitochondrial function and induced apoptosis (Hantz et al. 2005). Palozza et al. reported that lycopene (0.5–2 μM) inhibited the growth of cigarette smoke condensate-exposed immortalized RAT-1 fibroblast cells by arresting cell cycle progression and inducing apoptosis (Palozza et al. 2005).

Among identified lycopene metabolites, the central cleavage product ACR, an analog of retinoic acid, is the best studied. It has been shown to inhibit cell proliferation (Ben-Dor et al. 2001, Kotake-Nara et al. 2001, Nara et al. 2001) and induce apoptosis (Kotake-Nara et al. 2002) in a variety of cell lines. Using lycopene, ACR, and retinoic acid, Ben-Dor et al. observed a decrease in cell growth and a decreased rate of cell cycle progression, especially G_1 to S transition, in MCF-7 mammary cancer cells (Ben-Dor et al. 2001). Both ACR and retinoic acid inhibited cell growth with a similar potency (IC_{50} ~ 1–2 μM) to lycopene. Moreover, both ACR and retinoic acid decreased serum-stimulated cyclin D_1 protein expression, a finding also observed with intact lycopene (Nahum et al. 2001). In addition to ACR, the activity of other lycopene metabolites has been investigated. Zhang et al. identified an oxidative product of lycopene, (*E*,*E*,*E*)-4-methyl-8-oxo-2,4,6-nonatrienal, that induced apoptosis in HL-60 cells (Zhang et al. 2003). A dose-dependent reduction of cell viability with a concomitant increase in chromatin condensation and nuclear fragmentation, characteristic of apoptosis, were observed. Further analysis revealed an increased ratio of sub-G_1 cells, and an increase in Caspase-8 and Caspase-9 activities. These apoptotic changes were accompanied by a decrease in Bcl2 and Bcl-X_L protein expression but no changes in Bax expression. In spite of the above evidence, the physiological role of these lycopene products remains unknown since none of these metabolites have been detected in biological systems.

Recently, we have investigated the activity of the apo-10′-lycopenoic acid on cell proliferation in three cell lines: NHBE, a normal human bronchial epithelial cell line; BEAS-2B, an immortalized human bronchial epithelial cell line; and A549 cells, a non-small cell lung cancer cell, which represent the different stages of lung carcinogenesis (Lian et al. 2007). We showed that apo-10′-lycopenoic acid treatment inhibited cell growth in all three cell lines, albeit with different sensitivities. The growth inhibitory action of apo-10′-lycopenoic acid was largely due to decreased cell proliferation, as we did not observe any induction of apoptosis. Treatment with apo-10′-lycopenoic acid for 48 h significantly decreased A549 cells in S-phase from 31% in THF alone treated cells to 24% and 21% in cells treated with 3 and 5 μM apo-10′-lycopenoic acid. Accordingly, there was a concomitant increase in the number of cells in G_1/G_0 phase. These results suggested the effect of apo-10′-lycopenoic acid on cell proliferation was due to effects on cell cycle regulators, thus, we investigated potential cell cycle regulators to identify potential targets of apo-10-lycopenoic acid. The treatment of A549 cells resulted in a dose-dependent decrease in the mRNA and the protein levels of cyclin E but not cyclin D. The analysis of p21 and p27 mRNA levels revealed no significant effects of apo-10′-lycopenoic acid on transcription. However, there was a significant increase in p21 and p27 protein levels. Similar results were observed in BEAS-2B cells (Lian et al. 2007). We have also observed similar findings in human liver cells. Apo-10′lycopenoic acid, in a dose-dependent manner, inhibited cell growth and induced apoptosis in THLE-2 liver cells by stimulating the cyclin-dependent kinase inhibitor p21, and by reducing the activation of Jun N-terminal kinase and cyclin D1 gene expression (Hu et al. 2008). In order to support our *in vitro* findings, an *in vivo* study was performed to evaluate the effect of apo-10′-lycopenoic acid on tumor development in the A/J mouse model for lung cancer. A/J mice were preloaded with control diet or diet containing 10, 40, or 120 mg/kg diet of apo-10′-lycopenoic acid for two weeks before lung tumors were induced by the injection of 4-(*N*-methyl-*N*-nitrosamino)-1-(3-pyridal)-1-butanone (NNK). After 14 weeks on experimental diets, a significant decrease in tumor number but not tumor incidence was observed in treated animals (Lian et al. 2007). Interestingly, the plasma level of apo-10′-lycopenoic acid, which demonstrated the protective effect against lung tumor formation in mice, was much lower than reported for plasma lycopene concentrations in humans, suggesting that apo-10′-lycopenoic acid may, at least partially, mediate the chemopreventive activity of lycopene.

It should be pointed out that the potential use of lycopene metabolites, such as apo-10′-lycopenoic acid, as chemopreventive agents against cancers demands careful investigation. The *in vivo* metabolism of lycopene is complicated, and may be affected by a number of environmental factors, such as oxidative stress induced by cigarette smoking and alcohol consumption. For example, we have shown that high doses of β-carotene in an oxidative environment (such as, the lungs of smokers) may result in higher levels of polar metabolites, which can promote carcinogenesis; whereas lower doses of β-carotene have been shown to be protective (Liu et al. 2000, Wang et al. 1999). In a very recent study, we observed that high dose lycopene supplementation in the presence of alcohol ingestion increased hepatic inflammation and TNF-α expression (Veeramachaneni et al. 2008). While no apparent adverse effects, such as a decrease in body weight or tissue damage, were observed in our recent study of apo-10′-lycopenoic acid-supplemented, NNK-treated A/J mice (Lian et al. 2007), an earlier study has shown an enhancement of benzo[a]pyrene-induced mutagenesis in mouse lung and colon tissues after lycopene supplementation (Guttenplan et al. 2001). These results suggest that lycopene or lycopene metabolites may, such as β-carotene and its metabolites, enhance carcinogenesis. Further investigation into the dose effects of lycopene, especially in response to smoke-exposure and/or alcohol ingestion, as well as a further understanding of the metabolism of apo-10′-lycopenoids on carcinogenesis is needed.

20.5 SUMMARY

To gain a better understanding of the beneficial biological activities of lycopene, a greater knowledge of the metabolism of lycopene is needed. In particular, the identification of lycopene metabolites and oxidation products *in vivo*, the importance of tissue-specific lycopene cleavage by CMO1/CMO2, and the potential interaction between lycopene dose, and smoking and alcohol ingestion remains a vital step toward a better understanding of lycopene metabolism. An important question that remains unanswered is whether the effect of lycopene on various cellular functions and signaling pathways is a result of the direct actions of intact lycopene or its derivatives. While evidence is presented in this chapter to support the latter, more research is clearly needed to identify and characterize additional lycopene metabolites and their biological activities, which will potentially provide invaluable insights into the mechanisms underlying the beneficial effects of lycopene to humans.

ACKNOWLEDGMENTS

This material is based upon work supported by NIH Grant R01CA104932 and the U.S. Department of Agriculture, Agricultural Research Service, under agreement No. 58-1950-7-707. Any opinions, findings, conclusions, or recommendations expressed in this publication are those of the authors and do not necessarily reflect the views of the NIH or the U.S. Department of Agriculture.

ABBREVIATIONS

ACR	*acyclo*-retinoic acid
ARE	antioxidant response element
CMO1	β-carotene-15,15′-oxygenase
CMO2	carotene-9′,10′-oxygenase
Cx43	connexin 43
GJC	gap junction communication
IGF	insulin-like growth factor
IGFBP	insulin-like growth factor binding protein

RAR retinoic acid receptor
RARE retinoic acid response element
ROS reactive oxygen species
RXR retinoid X receptor

REFERENCES

Agarwal, S. and A. V. Rao. 1998. Tomato lycopene and low density lipoprotein oxidation: A human dietary intervention study. *Lipids* 33(10):981–984.

Arab, L., S. Steck-Scott, and P. Bowen. 2001. Participation of lycopene and beta-carotene in carcinogenesis: Defenders, aggressors, or passive bystanders? *Epidemiol Rev* 23(2):211–230.

Araki, H., Y. Shidoji, Y. Yamada, H. Moriwaki, and Y. Muto. 1995. Retinoid agonist activities of synthetic geranyl geranoic acid derivatives. *Biochem Biophys Res Commun* 209(1):66–72.

Aust, O., N. Ale-Agha, L. Zhang et al. 2003. Lycopene oxidation product enhances gap junctional communication. *Food Chem Toxicol* 41(10):1399–1407.

Bachmann, H., A. Desbarats, P. Pattison et al. 2002. Feedback regulation of beta,beta-carotene-15, 15′-monooxygenase by retinoic acid in rats and chickens. *J Nutr* 132(12):3616–3622.

Ben-Dor, A., A. Nahum, M. Danilenko et al. 2001. Effects of *acyclo*-retinoic acid and lycopene on activation of the retinoic acid receptor and proliferation of mammary cancer cells. *Arch Biochem Biophys* 391(2):295–302.

Ben-Dor, A., M. Steiner, L. Gheber et al. 2005. Carotenoids activate the antioxidant response element transcription system. *Mol Cancer Ther* 4(1):177–186.

Bertram, J. S. 2004. Induction of connexin 43 by carotenoids: Functional consequences. *Arch Biochem Biophys* 430(1):120–126.

Bertram, J. S., A. Pung, M. Churley et al. 1991. Diverse carotenoids protect against chemically induced neoplastic transformation. *Carcinogenesis* 12(4):671–678.

Bhuvaneswari, V., B. Velmurugan, S. Balasenthil, C. R. Ramachandran, and S. Nagini. 2001. Chemopreventive efficacy of lycopene on 7,12-dimethylbenz[a]anthracene-induced hamster buccal pouch carcinogenesis. *Fitoterapia* 72(8):865–874.

Boileau, A. C., N. R. Merchen, K. Wasson, C. A. Atkinson, and J. W. Erdman, Jr. 1999. Cis-lycopene is more bioavailable than trans-lycopene in vitro and in vivo in lymph-cannulated ferrets. *J Nutr* 129(6):1176–1181.

Boileau, T. W., Z. Liao, S. Kim et al. 2003. Prostate carcinogenesis in N-methyl-N-nitrosourea (NMU)-testosterone-treated rats fed tomato powder, lycopene, or energy-restricted diets. *J Natl Cancer Inst* 95(21):1578–1586.

Boulanger, A., P. McLemore, N. G. Copeland et al. 2003. Identification of beta-carotene 15, 15′-monooxygenase as a peroxisome proliferator-activated receptor target gene. *Faseb J* 17(10):1304–1306.

Bowen, P., L. Chen, M. Stacewicz-Sapuntzakis et al. 2002. Tomato sauce supplementation and prostate cancer: Lycopene accumulation and modulation of biomarkers of carcinogenesis. *Exp Biol Med (Maywood)* 227(10):886–893.

Breinholt, V., S. T. Lauridsen, B. Daneshvar, and J. Jakobsen. 2000. Dose-response effects of lycopene on selected drug-metabolizing and antioxidant enzymes in the rat. *Cancer Lett* 154(2):201–210.

Bub, A., B. Watzl, L. Abrahamse et al. 2000. Moderate intervention with carotenoid-rich vegetable products reduces lipid peroxidation in men. *J Nutr* 130(9):2200–2206.

Canene-Adams, K., B. L. Lindshield, S. Wang et al. 2007. Combinations of tomato and broccoli enhance antitumor activity in dunning r3327-h prostate adenocarcinomas. *Cancer Res* 67(2):836–843.

Caris-Veyrat, C., A. Schmid, M. Carail, and V. Bohm. 2003. Cleavage products of lycopene produced by in vitro oxidations: characterization and mechanisms of formation. *J Agric Food Chem* 51(25):7318–7325.

Chen, L., M. Stacewicz-Sapuntzakis, C. Duncan et al. 2001. Oxidative DNA damage in prostate cancer patients consuming tomato sauce-based entrees as a whole-food intervention. *J Natl Cancer Inst* 93(24):1872–1879.

Chichili, G. R., D. Nohr, M. Schaffer, J. von Lintig, and H. K. Biesalski. 2005. beta-Carotene conversion into vitamin A in human retinal pigment epithelial cells. *Invest Ophthalmol Vis Sci* 46(10):3562–3569.

Clemmons, D. R., W. H. Busby, T. Arai et al. 1995. Role of insulin-like growth factor binding proteins in the control of IGF actions. *Prog Growth Factor Res* 6(2–4):357–366.

Clinton, S. K., C. Emenhiser, S. J. Schwartz et al. 1996. cis-trans lycopene isomers, carotenoids, and retinol in the human prostate. *Cancer Epidemiol Biomarkers Prev* 5(10):823–833.

Conn, P. F., W. Schalch, and T. G. Truscott. 1991. The singlet oxygen and carotenoid interaction. *J Photochem Photobiol B* 11(1):41–47.

Di Mascio, P., S. Kaiser, and H. Sies. 1989. Lycopene as the most efficient biological carotenoid singlet oxygen quencher. *Arch Biochem Biophys* 274(2):532–538.

Ferreira, A. L., K. J. Yeum, R. M. Russell, N. I. Krinsky, and G. Tang. 2003. Enzymatic and oxidative metabolites of lycopene. *J Nutr Biochem* 14(9):531–540.

Gajic, M., S. Zaripheh, F. Sun, and J. W. Erdman, Jr. 2006. Apo-8′-lycopenal and apo-12′-lycopenal are metabolic products of lycopene in rat liver. *J Nutr* 136(6):1552–1557.

Giovannucci, E. 1999a. Insulin-like growth factor-I and binding protein-3 and risk of cancer. *Horm Res* 51 Suppl 3:34–41.

Giovannucci, E. 1999b. Tomatoes, tomato-based products, lycopene, and cancer: review of the epidemiologic literature. *J Natl Cancer Inst* 91(4):317–331.

Giovannucci, E. 2002. A review of epidemiologic studies of tomatoes, lycopene, and prostate cancer. *Exp Biol Med (Maywood)* 227(10):852–859.

Giovannucci, E. and S. K. Clinton. 1998. Tomatoes, lycopene, and prostate cancer. *Proc Soc Exp Biol Med* 218(2):129–139.

Giudice, A. and M. Montella. 2006. Activation of the Nrf2-ARE signaling pathway: a promising strategy in cancer prevention. *Bioessays* 28(2):169–181.

Gong, X., S. W. Tsai, B. Yan, and L. P. Rubin. 2006. Cooperation between MEF2 and PPARgamma in human intestinal beta,beta-carotene 15,15′-monooxygenase gene expression. *BMC Mol Biol* 7:7.

Goodman, D. S. and H. S. Huang. 1965. Biosynthesis of vitamin A with rat intestinal enzymes. *Science* 149:879–880.

Gradelet, S., P. Astorg, J. Leclerc et al. 1996. Effects of canthaxanthin, astaxanthin, lycopene and lutein on liver xenobiotic-metabolizing enzymes in the rat. *Xenobiotica* 26(1):49–63.

Guttenplan, J. B., M. Chen, W. Kosinska et al. 2001. Effects of a lycopene-rich diet on spontaneous and benzo[a] pyrene-induced mutagenesis in prostate, colon and lungs of the lacZ mouse. *Cancer Lett* 164(1):1–6.

Hantz, H. L., L. F. Young, and K. R. Martin. 2005. Physiologically attainable concentrations of lycopene induce mitochondrial apoptosis in LNCaP human prostate cancer cells. *Exp Biol Med (Maywood)* 230(3):171–179.

Hessel, S., A. Eichinger, A. Isken et al. 2007. CMO1 deficiency abolishes vitamin A production from beta-carotene and alters lipid metabolism in mice. *J Biol Chem* 282(46):33553–335561.

Hix, L. M., A. L. Vine, S. F. Lockwood, and J. S. Bertram. 2005. Retinoids and carotenoids as cancer chemopreventive agents: Role of upregulated gap junctional communication. In *Carotenoids and Retinoids: Molecular Aspects and Health Issues*, edited by L. Packer, Obermueller-Jevic, U., Kraemer, K., and Sies, H. AOCS Press, Champaign, IL.

Holmes, M. D., M. N. Pollak, W. C. Willett, and S. E. Hankinson. 2002. Dietary correlates of plasma insulin-like growth factor I and insulin-like growth factor binding protein 3 concentrations. *Cancer Epidemiol Biomarkers Prev* 11(9):852–861.

Hossain, M. Z., L. R. Wilkens, P. P. Mehta, W. Loewenstein, and J. S. Bertram. 1989. Enhancement of gap junctional communication by retinoids correlates with their ability to inhibit neoplastic transformation. *Carcinogenesis* 10(9):1743–1748.

Hu, K. Q., Y. Wang, R. M. Russell, and X. D. Wang. 2008. Apo-10′-lycopenoic acid functions as a peroxisome proliferator-activated receptor gamma activator and inhibits cell growth both *in vitro* and in ob/ob mice treated with diethylnitrosamine. In *American Association of Cancer Research Annual Meeting*. San Diego, CA.

Hu, K. Q., C. Liu, H. Ernst et al. 2006. The biochemical characterization of ferret carotene-9′,10′-monooxygenase catalyzing cleavage of carotenoids in vitro and in vivo. *J Biol Chem* 281(28):19327–19338.

Hwang, E. S. and P. E. Bowen. 2004. Cell cycle arrest and induction of apoptosis by lycopene in LNCaP human prostate cancer cells. *J Med Food* 7(3):284–289.

Jerome, L., L. Shiry, and B. Leyland-Jones. 2003. Deregulation of the IGF axis in cancer: Epidemiological evidence and potential therapeutic interventions. *Endocr Relat Cancer* 10(4):561–578.

Karas, M., H. Amir, D. Fishman et al. 2000. Lycopene interferes with cell cycle progression and insulin-like growth factor I signaling in mammary cancer cells. *Nutr Cancer* 36(1):101–111.

Karas, M., M. Danilenko, D. Fishman et al. 1997. Membrane-associated insulin-like growth factor-binding protein-3 inhibits insulin-like growth factor-I-induced insulin-like growth factor-I receptor signaling in ishikawa endometrial cancer cells. *J Biol Chem* 272(26):16514–16520.

Khachik, F., G. R. Beecher, M. B. Goli, W. R. Lusby, and C. E. Daitch. 1992a. Separation and quantification of carotenoids in human plasma. *Methods Enzymol* 213:205–219.

Khachik, F., G. R. Beecher, M. B. Goli, W. R. Lusby, and J. C. Smith, Jr. 1992b. Separation and identification of carotenoids and their oxidation products in the extracts of human plasma. *Anal Chem* 64(18):2111–2122.

Khachik, F., G. R. Beecher, and J. C. Smith, Jr. 1995. Lutein, lycopene, and their oxidative metabolites in chemoprevention of cancer. *J Cell Biochem Suppl* 22:236–246.

Khachik, F., L. Carvalho, P. S. Bernstein et al. 2002. Chemistry, distribution, and metabolism of tomato carotenoids and their impact on human health. *Exp Biol Med (Maywood)* 227(10):845–851.

Khachik, F., H. Pfander, and B. Traber. 1998. Proposed mechanisms for the formation of the synthetic and naturally occuring metabolites of lycopene in tomato products and human serum. *J Agric Food Chem* 46:4885–4890.

Khachik, F., C. J. Spangler, J. C. Smith, Jr. et al. 1997. Identification, quantification, and relative concentrations of carotenoids and their metabolites in human milk and serum. *Anal Chem* 69(10):1873–1881.

Khachik, F., A. Steck, U.A. Niggli, and H. Pfander. 1998. Partial synthesis and structural elucidation of the oxidative metabolites of lycopene identified in tomato paste, tomato juice and human serum. *J Agric Food Chem* 46:4885–4890.

Kiefer, C., S. Hessel, J. M. Lampert et al. 2001. Identification and characterization of a mammalian enzyme catalyzing the asymmetric oxidative cleavage of provitamin A. *J Biol Chem* 276(17):14110–14116.

Kim, S. J., E. Nara, H. Kobayashi, J. Terao, and A. Nagao. 2001. Formation of cleavage products by autoxidation of lycopene. *Lipids* 36(2):191–199.

King, T. J. and J. S. Bertram. 2005. Connexins as targets for cancer chemoprevention and chemotherapy. *Biochim Biophys Acta* 1719(1–2):146–160.

Kloer, D. P., S. Ruch, S. Al-Babili, P. Beyer, and G. E. Schulz. 2005. The structure of a retinal-forming carotenoid oxygenase. *Science* 308(5719):267–269.

Kotake-Nara, E., S. J. Kim, M. Kobori, K. Miyashita, and A. Nagao. 2002. *Acyclo*-retinoic acid induces apoptosis in human prostate cancer cells. *Anticancer Res* 22(2A):689–695.

Kotake-Nara, E., M. Kushiro, H. Zhang et al. 2001. Carotenoids affect proliferation of human prostate cancer cells. *J Nutr* 131(12):3303–3306.

Krinsky, N. I. and E. J. Johnson. 2005. Carotenoid actions and their relation to health and disease. *Mol Aspects Med* 26(6):459–516.

Levy, J., E. Bosin, B. Feldman et al. 1995. Lycopene is a more potent inhibitor of human cancer cell proliferation than either alpha-carotene or beta-carotene. *Nutr Cancer* 24(3):257–266.

Lian, F., K. Q. Hu, R. M. Russell, and X. D. Wang. 2006. Beta-cryptoxanthin suppresses the growth of immortalized human bronchial epithelial cells and non-small-cell lung cancer cells and up-regulates retinoic acid receptor beta expression. *Int J Cancer* 119(9):2084–2089.

Lian, F., D. E. Smith, H. Ernst, R. M. Russell, and X. D. Wang. 2007. Apo-10′-lycopenoic acid inhibits lung cancer cell growth *in vitro*, and suppresses lung tumorigenesis in the A/J mouse model *in vivo*. *Carcinogenesis* 28(7):1567–1574.

Lian, F. and X. D. Wang. 2008. Apo-10′-lycopenoic acid, an enzymatic metabolite of lycopene, induces Nrf2-mediated expression of phase II detoxifying/antioxidant enzymes in human bronchial epithelial cells. *Int J Cancer* (in press).

Lindqvist, A. and S. Andersson. 2002. Biochemical properties of purified recombinant human beta-carotene 15,15′-monooxygenase. *J Biol Chem* 277(26):23942–23948.

Lindqvist, A., Y. G. He, and S. Andersson. 2005. Cell type-specific expression of beta-carotene 9′, 10′-monooxygenase in human tissues. *J Histochem Cytochem* 53(11):1403–1412.

Lindshield, B. L., K. Canene-Adams, and J. W. Erdman, Jr. 2007. Lycopenoids: are lycopene metabolites bioactive? *Arch Biochem Biophys* 458(2):136–140.

Liu, C., F. Lian, D. E. Smith, R. M. Russell, and X. D. Wang. 2003. Lycopene supplementation inhibits lung squamous metaplasia and induces apoptosis via up-regulating insulin-like growth factor-binding protein 3 in cigarette smoke-exposed ferrets. *Cancer Res* 63(12):3138–3144.

Liu, C., R. M. Russell, and X. D. Wang. 2006. Lycopene supplementation prevents smoke-induced changes in p53, p53 phosphorylation, cell proliferation, and apoptosis in the gastric mucosa of ferrets. *J Nutr* 136(1):106–111.

Liu, C., X. D. Wang, R. T. Bronson et al. 2000. Effects of physiological versus pharmacological beta-carotene supplementation on cell proliferation and histopathological changes in the lungs of cigarette smoke-exposed ferrets. *Carcinogenesis* 21(12):2245–2253.

Livny, O., I. Kaplan, R. Reifen et al. 2002. Lycopene inhibits proliferation and enhances gap-junction communication of KB-1 human oral tumor cells. *J Nutr* 132(12):3754–3759.

Mandlekar, S., J. L. Hong, and A. N. Kong. 2006. Modulation of metabolic enzymes by dietary phytochemicals: a review of mechanisms underlying beneficial versus unfavorable effects. *Curr Drug Metab* 7(6):661–675.

Mein, J. R., N. Chongvirihaphan, and X. D. Wang. 2006. The effect of combined antioxidant supplementation (beta-carotene, ascorbic acid and alpha-tocopherol) on the expression of CMO1 and CMO2 in smoke-exposed ferrets. In *American Association of Cancer Research: Frontiers in Cancer Prevention Research Meeting*. Boston, MA.

Miller, N. J., J. Sampson, L. P. Candeias, P. M. Bramley, and C. A. Rice-Evans. 1996. Antioxidant activities of carotenes and xanthophylls. *FEBS Lett* 384(3):240–242.

Mucci, L. A., R. Tamimi, P. Lagiou et al. 2001. Are dietary influences on the risk of prostate cancer mediated through the insulin-like growth factor system? *BJU Int* 87(9):814–820.

Nagao, A. and J. A. Olson. 1994. Enzymatic formation of 9-*cis*, 13-*cis*, and all-*trans* retinals from isomers of beta-carotene. *Faseb J* 8(12):968–973.

Nahum, A., K. Hirsch, M. Danilenko et al. 2001. Lycopene inhibition of cell cycle progression in breast and endometrial cancer cells is associated with reduction in cyclin D levels and retention of p27(Kip1) in the cyclin E-cdk2 complexes. *Oncogene* 20(26):3428–3436.

Nahum, A., L. Zeller, M. Danilenko et al. 2006. Lycopene inhibition of IGF-induced cancer cell growth depends on the level of cyclin D1. *Eur J Nutr* 45(5):275–282.

Napoli, J. L. and K. R. Race. 1988. Biogenesis of retinoic acid from beta-carotene. Differences between the metabolism of beta-carotene and retinal. *J Biol Chem* 263(33):17372–17377.

Nara, E., H. Hayashi, M. Kotake, K. Miyashita, and A. Nagao. 2001. Acyclic carotenoids and their oxidation mixtures inhibit the growth of HL-60 human promyelocytic leukemia cells. *Nutr Cancer* 39(2): 273–283.

Obermuller-Jevic, U. C., E. Olano-Martin, A. M. Corbacho et al. 2003. Lycopene inhibits the growth of normal human prostate epithelial cells in vitro. *J Nutr* 133(11):3356–3360.

Olson, J. A. and O. Hayaishi. 1965. The enzymatic cleavage of beta-carotene into vitamin A by soluble enzymes of rat liver and intestine. *Proc Natl Acad Sci U S A* 54(5):1364–1370.

Paik, J., A. During, E. H. Harrison et al. 2001. Expression and characterization of a murine enzyme able to cleave beta-carotene. The formation of retinoids. *J Biol Chem* 276(34):32160–32168.

Palozza, P., A. Sheriff, S. Serini et al. 2005. Lycopene induces apoptosis in immortalized fibroblasts exposed to tobacco smoke condensate through arresting cell cycle and down-regulating cyclin D1, pAKT and pBad. *Apoptosis* 10(6):1445–1456.

Parvin, S. G. and B. Sivakumar. 2000. Nutritional status affects intestinal carotene cleavage activity and carotene conversion to vitamin A in rats. *J Nutr* 130(3):573–577.

Poliakov, E., S. Gentleman, F. X. Cunningham, Jr., N. J. Miller-Ihli, and T. M. Redmond. 2005. Key role of conserved histidines in recombinant mouse beta-carotene 15,15′-monooxygenase-1 activity. *J Biol Chem* 280(32):29217–29223.

Porrini, M. and P. Riso. 2000. Lymphocyte lycopene concentration and DNA protection from oxidative damage is increased in women after a short period of tomato consumption. *J Nutr* 130(2):189–192.

Prakash, P., C. Liu, K. Q. Hu et al. 2004. Beta-carotene and beta-apo-14′-carotenoic acid prevent the reduction of retinoic acid receptor beta in benzo[a]pyrene-treated normal human bronchial epithelial cells. *J Nutr* 134(3):667–673.

Ramos-Gomez, M., M. K. Kwak, P. M. Dolan et al. 2001. Sensitivity to carcinogenesis is increased and chemoprotective efficacy of enzyme inducers is lost in nrf2 transcription factor-deficient mice. *Proc Natl Acad Sci U S A* 98(6):3410–3415.

Redmond, T. M., S. Gentleman, T. Duncan et al. 2001. Identification, expression, and substrate specificity of a mammalian beta-carotene 15,15′-dioxygenase. *J Biol Chem* 276(9):6560–6565.

Riso, P., A. Pinder, A. Santangelo, and M. Porrini. 1999. Does tomato consumption effectively increase the resistance of lymphocyte DNA to oxidative damage? *Am J Clin Nutr* 69(4):712–718.

Rogers, M., J. M. Berestecky, M. Z. Hossain et al. 1990. Retinoid-enhanced gap junctional communication is achieved by increased levels of connexin 43 mRNA and protein. *Mol Carcinog* 3(6):335–343.

Salman, H., M. Bergman, M. Djaldetti, and H. Bessler. 2007. Lycopene affects proliferation and apoptosis of four malignant cell lines. *Biomed Pharmacother* 61(6):366–369.

Sicilia, T., A. Bub, G. Rechkemmer et al. 2005. Novel lycopene metabolites are detectable in plasma of preruminant calves after lycopene supplementation. *J Nutr* 135(11):2616–2621.

Siler, U., L. Barella, V. Spitzer et al. 2004. Lycopene and vitamin E interfere with autocrine/paracrine loops in the Dunning prostate cancer model. *Faseb J* 18(9):1019–1021.

Stahl, W., J. von Laar, H. D. Martin, T. Emmerich, and H. Sies. 2000. Stimulation of gap junctional communication: comparison of *acyclo*-retinoic acid and lycopene. *Arch Biochem Biophys* 373(1):271–274.
Talalay, P. 2000. Chemoprotection against cancer by induction of phase 2 enzymes. *Biofactors* 12(1–4):5–11.
Talalay, P., A. T. Dinkova-Kostova, and W. D. Holtzclaw. 2003. Importance of phase 2 gene regulation in protection against electrophile and reactive oxygen toxicity and carcinogenesis. *Adv Enzyme Regul* 43:121–134.
Tang, G. W., X. D. Wang, R. M. Russell, and N. I. Krinsky. 1991. Characterization of beta-apo-13-carotenone and beta-apo-14′-carotenal as enzymatic products of the excentric cleavage of beta-carotene. *Biochemistry* 30(41):9829–9834.
Trosko, J. E., C. C. Chang, B. Upham, and M. Wilson. 1998. Epigenetic toxicology as toxicant-induced changes in intracellular signalling leading to altered gap junctional intercellular communication. *Toxicol Lett* 102–103:71–78.
van Vliet, T., M. F. van Vlissingen, F. van Schaik, and H. van den Berg. 1996. beta-Carotene absorption and cleavage in rats is affected by the vitamin A concentration of the diet. *J Nutr* 126(2):499–508.
Veeramachaneni, S., L. M. Ausman, S. W. Choi, R. M. Russell, and X. D. Wang. 2008. High dose lycopene supplementation increases hepatic CYP2E1 protein and inflammation in alcohol-fed rats. *J Nutr* (submitted).
von Lintig, J. and K. Vogt. 2000. Filling the gap in vitamin A research. Molecular identification of an enzyme cleaving beta-carotene to retinal. *J Biol Chem* 275(16):11915–11920.
Vrieling, A., D. W. Voskuil, J. M. Bonfrer et al. 2007. Lycopene supplementation elevates circulating insulin-like growth factor binding protein-1 and -2 concentrations in persons at greater risk of colorectal cancer. *Am J Clin Nutr* 86(5):1456–1462.
Wang, X. D. 2004. Carotenoid oxidative/degradative products and their biological activities. In *Carotenoids in Health and Disease*, edited by N. I. Krinsky, Mayne, S.T., and Sies, H. Marcel Dekker, New York.
Wang, X. D. 2005. Can smoke-exposed ferrets be utilized to unravel the mechanisms of action of lycopene? *J Nutr* 135(8):2053S-2056S.
Wang, X. D., N. I. Krinsky, P. N. Benotti, and R. M. Russell. 1994. Biosynthesis of 9-cis-retinoic acid from 9-cis-beta-carotene in human intestinal mucosa in vitro. *Arch Biochem Biophys* 313(1):150–155.
Wang, X. D., N. I. Krinsky, R. P. Marini et al. 1992. Intestinal uptake and lymphatic absorption of beta-carotene in ferrets: A model for human beta-carotene metabolism. *Am J Physiol* 263(4 Pt 1):G480–G486.
Wang, X. D., C. Liu, R. T. Bronson et al. 1999. Retinoid signaling and activator protein-1 expression in ferrets given beta-carotene supplements and exposed to tobacco smoke. *J Natl Cancer Inst* 91(1):60–66.
Wang, X. D., R. M. Russell, C. Liu et al. 1996. Beta-oxidation in rabbit liver *in vitro* and in the perfused ferret liver contributes to retinoic acid biosynthesis from beta-apocarotenoic acids. *J Biol Chem* 271(43):26490–26498.
Wang, X. D., G. W. Tang, J. G. Fox, N. I. Krinsky, and R. M. Russell. 1991. Enzymatic conversion of beta-carotene into beta-apo-carotenals and retinoids by human, monkey, ferret, and rat tissues. *Arch Biochem Biophys* 285(1):8–16.
Wu, K., S. J. Schwartz, E. A. Platz et al. 2003. Variations in plasma lycopene and specific isomers over time in a cohort of U.S. men. *J Nutr* 133(6):1930–1936.
Wyss, A., G. M. Wirtz, W. D. Woggon et al. 2001. Expression pattern and localization of beta,beta-carotene-15,15′-dioxygenase in different tissues. *Biochem J* 354 (Pt 3):521–529.
Wyss, A., G. Wirtz, W. Woggon et al. 2000. Cloning and expression of beta,beta-carotene-15,15′-dioxygenase. *Biochem Biophys Res Commun* 271(2):334–336.
Xu, C., C. Y. Li, and A. N. Kong. 2005. Induction of phase I, II and III drug metabolism/transport by xenobiotics. *Arch Pharm Res* 28(3):249–268.
Yan, W., G. F. Jang, F. Haeseleer et al. 2001. Cloning and characterization of a human beta,beta-carotene-15, 15′-dioxygenase that is highly expressed in the retinal pigment epithelium. *Genomics* 72(2):193–202.
Yu, H., and T. Rohan. 2000. Role of the insulin-like growth factor family in cancer development and progression. *J Natl Cancer Inst* 92(18):1472–1489.
Zaripheh, S., T. Y. Nara, M. T. Nakamura, and J. W. Erdman, Jr. 2006. Dietary lycopene downregulates carotenoid-15,15′-monooxygenase and PPAR-gamma in selected rat tissues. *J Nutr* 136(4):932–938.
Zhang, H., E. Kotake-Nara, H. Ono, and A. Nagao. 2003. A novel cleavage product formed by autoxidation of lycopene induces apoptosis in HL-60 cells. *Free Radic Biol Med* 35(12):1653–1663.
Zhang, L. X., R. V. Cooney, and J. S. Bertram. 1991. Carotenoids enhance gap junctional communication and inhibit lipid peroxidation in C3H/10T1/2 cells: Relationship to their cancer chemopreventive action. *Carcinogenesis* 12(11):2109–2114.

Zhang, L. X., R. V. Cooney, and J. S. Bertram. 1992. Carotenoids up-regulate connexin43 gene expression independent of their provitamin A or antioxidant properties. *Cancer Res* 52(20):5707–5712.

Ziouzenkova, O., G. Orasanu, M. Sharlach et al. 2007a. Retinaldehyde represses adipogenesis and diet-induced obesity. *Nat Med* 13(6):695–702.

Ziouzenkova, O., G. Orasanu, G. Sukhova et al. 2007b. Asymmetric cleavage of beta-carotene yields a transcriptional repressor of retinoid X receptor and peroxisome proliferator-activated receptor responses. *Mol Endocrinol* 21(1):77–88.

Ziouzenkova, O. and J. Plutzky. 2008. Retinoid metabolism and nuclear receptor responses: New insights into coordinated regulation of the PPAR-RXR complex. *FEBS Lett* 582(1):32–38.

21 Lycopene Oxidation, Uptake, and Activity in Human Prostate Cell Cultures

Phyllis E. Bowen

CONTENTS

21.1 INTRODUCTION

Population studies associate tomato consumption with reduced risk to prostate cancer. The most positive associations have come from cohort studies performed before the prostate-specific antigen (PSA)-screening era, and these studies have suggested that the tomato/lycopene effect was the strongest for clinically relevant prostate cancers (Giovannucci 2007). Small human studies have shown *in vivo* antioxidant effects for tomato products but evidence for lycopene alone is weak (Chen et al. 2001, Porrini and Riso 2000, Riso et al. 2004, Zhao et al. 2006). Animal and tissue culture studies have been useful in differentiating the effects of lycopene versus the mixture of biologically active compounds in tomatoes as well as the exploration of plausible mechanisms of action. Cell culture studies have the advantage of exploring the modulation of cellular processes in single cell types using known concentrations of lycopene and can be used to evaluate possible synergies between other tomato constituents, such as polyphenolic compounds, other carotenoids, and vitamin E. In order to fully appreciate the results of cell culture studies using lycopene alone or lycopene in combination with other biologically active compounds, it is important to understand (1) prostate biology, (2) the role of the various cells in prostate function, (3) which cells are the most vulnerable to the carcinogenic process and how that process proceeds, and (4) the origin of each of the prostatic cell lines that has been used and its characteristics.

In cell culture, lycopene is a highly oxidizable nonpolar hydrocarbon supplied in an aqueous medium and is incubated at body temperature for 12–72 h. The amount of intact lycopene or its oxidation products delivered to and absorbed by various cell types is an important factor to keep in mind when evaluating the effects of lycopene on various cellular processes. Before reviewing cell culture studies designed to characterize the effects of lycopene on prostate cell biology, the characteristics of prominent prostate cell lines, and the stability and uptake of lycopene by various prostate cell lines are reviewed.

21.2 PROSTATE CELL BIOLOGY AND CARCINOGENESIS

21.2.1 THE NORMAL PROSTATE

It is important to understand the architecture of the normal prostate and the complicated cross talk between the heterogeneous cell types involved in the carcinogenic process in order to interpret the effect of lycopene on various prostate cell lines in tissue culture. The prostate is mainly a secretory organ that supplies 10%–30% of the seminal fluid for ejaculation. Figure 21.1 shows a micrograph of normal prostate tissue. The peripheral zone (Sampson et al. 2007), where 70% of tumors can be found, is marked by acinar (glands) that collect the fluid secreted by the surrounding secretory epithelial cells. The secreted fluid is alkaline and is a complex mixture containing PSA, prostate acid phosphatase (PAP), and prostasomes (Sampson et al. 2007) among other constituents. Prostasomes are exocytosed from the acinar epithelial cells as small vesicles (40–500 nm) surrounded by a cholesterol/sphingomyelin-rich membrane and contain numerous enzymes, immunosuppressants, zinc, calcium, selenium, ATP, and neuroendocrine markers, such as neuropeptide Y. They promote sperm viability and mobility (Kravets et al. 2000). Beneath the epithelial cells surrounding the acinar is a layer of basal cells backed by stromal fibroblasts and smooth muscle cells, which compose most of the prostate structure. The basal cells do not secrete PSA or PAP, have few androgen receptors, but develop these as they differentiate and move to the surface of the acinar (Miki and Rhim 2007). In summary, the prostate is composed of many different cell types: cells found in the epithelium (stem cells, transit-amplifying cells, basal cells, secretory cells, and neuroendocrine cells) and cells found in

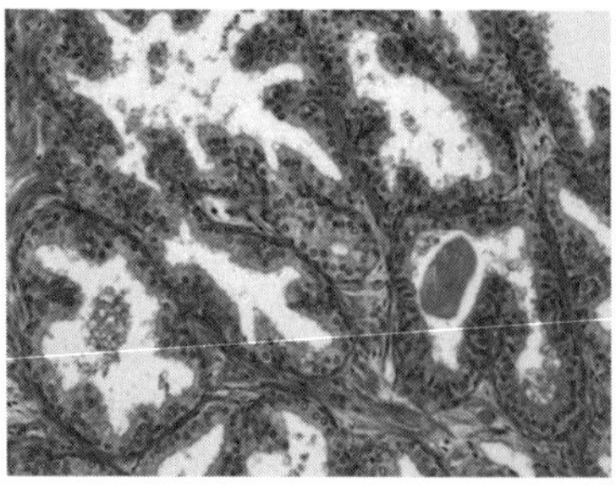

FIGURE 21.1 (See color insert following page 336.) Normal tissue from human prostate showing secretory section—hematoxilin and eosin staining showing epithelial cells lining secretory ducts backed by basal and stromal cells. (Courtesy of A. Brollo, Wikimedia Commons, 2005.)

the stromal areas (smooth muscle cells, fibroblasts, myofibroblasts, endothelial cells, and extracellular matrix) (Sampson et al. 2007). Both the epithelial and stromal compartments' turn over is relatively slow through balanced proliferation and apoptosis. Most of these cell types, and even the extracellular matrix, have been implicated in the prostatic carcinogenic process.

21.2.2 Prostate Carcinogenesis

The focus of the carcinogenic process is on the acinar epithelial cell layer and the underlying less differentiated basal cell epithelium and the basement membrane. Throughout this structure are dispersed small number of neuroendocrine cells that may also have a role. Most widely available cell lines are thought to be of acinar epithelial cell origin. Anatomically observable changes may start with the development of inflammation and the invasion of inflammatory cells resulting in atrophic epithelium or focal atrophic epithelium, which is highly prevalent in the prostate peripheral zone in older men. These areas are often associated with high-grade prostatic intraepithelial neoplasia (HGPIN), which is widely regarded as a precursor of prostate cancer (De Marzo et al. 2007). HGPIN is characterized by an increased layering of epithelial cells, enlarged nuclei, and a thinning of the basal cell layer but is distinguished from adenocarcinoma where the basement membrane has disappeared. Prostate cancer is multifocal because within one prostate, and even within one biopsy, one can find normal tissue, HGPIN, and various areas of adenocarcinoma exhibiting different levels of morphological disorganization, and at the molecular level, different levels of genetic alteration (Schulz et al. 2003). Figure 21.2 marks the progression from HGPIN to increasing grades of cellular disorganization that is the basis of the gleason scoring system. The implication of this variability is that the lycopene response of cells harvested and cultured from particular foci may not be generalizable to other cancer foci in the prostate or all prostate adenocarcinomas.

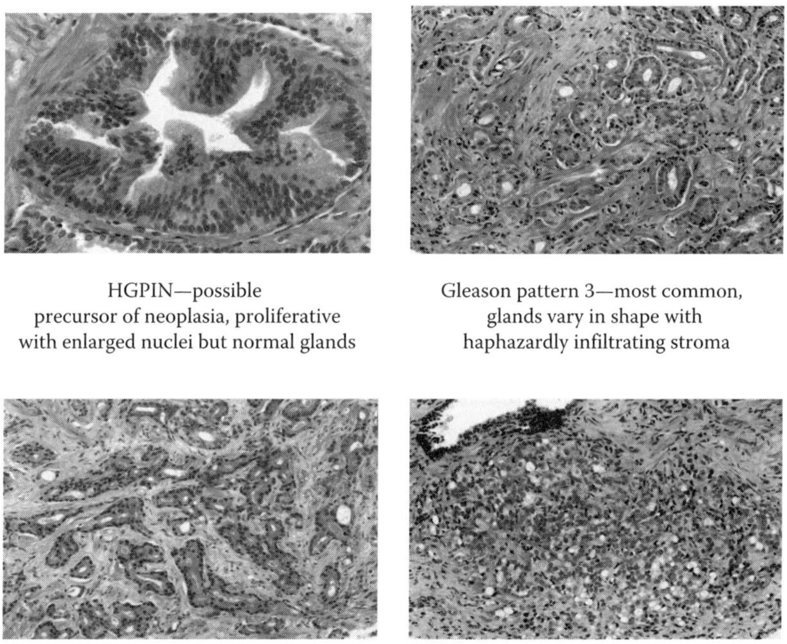

HGPIN—possible precursor of neoplasia, proliferative with enlarged nuclei but normal glands

Gleason pattern 3—most common, glands vary in shape with haphazardly infiltrating stroma

Gleason pattern 4—glands have fused

Gleason pattern 5—loss of glandular structure

FIGURE 21.2 (See color insert following page 336.) Human prostate showing progression of cancer grading from benign to severe—hemotoxilin and eosin staining showing changes in the organization of epithelial cells forming ducts and surrounding stromal cells.

Androgen withdrawal has long been a component of prostate cancer treatment because most adenocarcinomas regress without androgen stimulation. However, this is a temporary measure since continued androgen ablation leads to the development of androgen-insensitive cancer, which has a poor prognosis (Balakumaran and Febbo 2006). Therefore, one of the major distinctions of various prostate cancer cell lines is their dependence or independence from the growth-stimulating effects of androgens. Androgen receptors (AR) in cell cytoplasm bind to dihydrotestosterone (DHT) (converted within the prostate from testosterone to DHT by the enzyme 5α-hydroxylase), which stimulates AR translocation to the nucleus where its dimers activate the transcription of pathways that promote prostate cell growth and development (Comstock and Knudsen 2007, Tomlins et al. 2006). Androgen independence does not mean that these transformed cells do not express AR. In fact, AR expression is found to be increased in many hormone-refractory adenocarcinomas (Tomlins et al. 2006), which may allow for very low androgen levels or weakly androgenic compounds to inappropriately stimulate proliferation (Sampson et al. 2007) under the right circumstances. Furthermore, AR is expressed in both epithelial and stromal cells. Thus, changes in the prostate androgen–estrogen equilibrium may produce a "reactive stroma" that, in turn, stimulates epithelial cell proliferation, extracellular matrix deposition, and fibroblast transdifferentiation to myofibroblasts, all of which can be observed in both benign hyperplasia and prostate cancer (Sampson et al. 2007). Current opinion postulates a role for stem cells that survive and adapt to androgen ablation as the main culprits in the development of hormone-refractory cancer (Miki and Rhim 2007).

21.3 CHARACTERISTICS OF PROSTATE CELL LINES USED IN LYCOPENE STUDIES

Table 21.1 describes prostate cell lines that have been used for lycopene research and their characteristics, partly derived from the excellent compendium by Sobel and Sadar (2005) and Webber et al. (1997). The LNCaP cell line has been the most commonly used one for lycopene investigations and other prostate cancer-related researches because it represents the least transformed of the commercially available immortalized cells lines. It remains sensitive to androgen stimulation and continues to secrete PSA and PAP, a major function of fully differentiated prostate epithelial cells. Cells from distant metastases are recognized as epithelial in origin by the presence of the intermediate filament proteins, cytokeratin 8, and cytokeratin 18, but the cells are negative for desmin and factor VIII, which are associated with muscle and endothelial cells, respectively (Sobel and Sadar 2005). Even though DU-145 and PC-3 cells are of epithelial origin, they do not provide the platform for investigating prostate cancer prevention at the earlier stages of neoplastic events, but are of interest for later stages where androgen sensitivity is lost. Single cell line studies do not model our current notions of prostate tissue carcinogenesis, which involve considerable cellular cross talk between various cell types. Coculturing two or more cell types may better model the dynamics found in the intact tissue and represent one of the newer approaches to the study of promising agents, such as lycopene. Primary cell cultures, which are directly established from human tissues, may be a promising approach to sort out the issues of mechanism in normal, BPH, and neoplastic cells at the various stages of carcinogenesis. They are limited to about 30 population doublings that may vary with the age of the donor but with correct technique, it is possible to harvest cells with about 90% purity and 90% *in vitro* survival (Miki and Rhim 2007, Peehl 2004).

21.4 LYCOPENE STABILITY AND UPTAKE BY CULTURED PROSTATE CELLS

21.4.1 Vehicles for Lycopene Delivery

Keeping lycopene soluble and unoxidized in the warm, aqueous tissue culture medium over the incubation periods of 12–72 h is problematic. Furthermore, depending on the lycopene preparation, variable amounts remain on the filters used to prevent microbial contamination and hence the

TABLE 21.1 Characteristics of Prostate Cell Lines Used for Lycopene Studies

Cell Line	Cell Type	Source	AR Receptor	Androgen Sensitive	PSA Secreting	Tumor Forming	Metastasizing	Growth Factor Receptors	Growth Factor Secretion
PrEC (Fry et al. 2000)	Basal cell epithelia grown in PrEGM medium, Bioclonics	Primary cells Clone 6448 at third passage obtained from two different young men with BPH from Bioclonics	Not detectable	Unknown	Weak	No	No		
LNCaP (60–72 h doubling time) *63 + sublines*	Epithelial	Lymph node metastasis Moderately differentiated PCa WM treated with estrogens, orchectomy, cisplatin	Yes—7877A mutation making it more hormone sensitive	Yes—grows well on small amounts of androgen in growth serum; growth stimulated by estradiol and progesterone	Yes—and also PAP	Yes—but dependent on Matrigel in both genders	Only sub-cell lines: LNCaP-abl LNCaP–04–2 LNCap-0–1 *Lymph, lung, bone, liver*	EFG/TGFα-R + FGF-R + IFG-1R + TGFβ-R –	EGF + + TGFα + + bFGE – IGF-1 + + TGFβ –
PC-3 (33 h doubling time) *11 + sublines*	Epithelial	Lumbar vertebra metastasis Poorly differentiated adenocarcinoma WM treated with DES, orchectomy	No—weak AR receptor staining	No	No—but weakly stains for PSA	Yes	Only sub-cell lines: PC3/M PC-3/M –LN4	EFG/TGFα-R + FGF-R + + + IFG-1R + + TGFβ-R + + Akt + + + PTEN–NEP –	EGF + TGFα + bFGE + IGF-1 + + + TGFβ + + PTEN–NEP–
DU-145 (34 h doubling time) *1 + sublines*	Epithelial	Brain metastasis Moderately differentiated adenocarcinoma WM with lymphocytic leukemia and PCa treated with DES and bilateral orchectomy	No	No	No	Yes	Yes—in SCID mice *Liver, kidney, lung, spleen, adrenal, lymph nodes, diaphragm*	EFG/TGFα-R + + + FGF-R + + + IFG-1R + + + TGFβ-R + +	EGF + + TGFα + + + bFGE + + IGF-1 + + TGFβ + +

Sources: Data from Sobel, R.E. and Sadar, M.D., *J. Urol.*, 173, 342, 2005; Webber, M.M. et al., *Prostate*, 30, 58, 1997.
Note: PCa, prostate cancer; WM, white male; DES, diethylstilbestrol; AR, androgen receptor; PSA, prostate specific antigen; and PAP, prostatic acid phosphatase.

lycopene concentration of the stock "solution" must be measured and adjusted after filtration. A number of solubilization methods have been used over the years to assure the availability of lycopene to the cells in culture, such as solvents tetrahydrofuran (THF), dimethylsulfoxide (DMSO), water miscible gelatin beadlets, micelles, and prostasomes. The first successful effort was the use of THF, which tends to form a molecular cage around the lycopene molecules (Zhang et al. 1991). Great care must be taken to prevent the formation of THF peroxides. Although the use of THF/lycopene delivery with other cell lines has been successful, apparent lycopene uptake was lower than with beadlets (Shahrzad et al. 2002), and the LNCaP cells were not viable when exposed to the small amount of THF required for lycopene solubilization (Xu et al. 1999) in the hands of some investigators. Also, lycopene dissolved in THF and LNCaP culture medium, and incubated in glass under standard culture conditions was rapidly degraded with a half-life of approximately 2h (Xu et al. 1999). DMSO has also been used as a solubilizing agent for lycopene, especially in combination with lycopene-containing water dispersible beadlets (Borthakur et al. 2005, Offord et al. 2002) and does not harm LNCaP cell viability. In our laboratory, the addition of DMSO to help solubilize lycopene from beadlets showed no special advantage. Micelles containing 263 μM lycopene (maximal concentration that can be incorporated into micelles) have been used to overcome these problems and found to be stable in cell culture medium under standard incubation conditions for 50h with only slight losses at 100h. LNCaP growth over a 4 day period was slightly inhibited by 5 and 10 μM lycopene-containing micelles (Xu et al. 1999). Micelle preparations are difficult to handle since filtration to eliminate microbes in the preparation of the culture medium leaves many of the micelles on the filter. The availability of water-dispersible beadlets containing lycopene and control beadlets containing no lycopene (DSM Nutritional Products, Inc, Persippany, NJ or BASF Nutrition, Mt. Olive, NJ) have been used in recent studies. A careful evaluation of lycopene stability, and lycopene uptake in cell culture medium with and without LNCaP cells showed that 76% of the lycopene was recovered from the medium after 72h using 10% lycopene DSM beadlets that were supplied at concentrations ranging from 0.17 to 22.3 μM. The presence of LNCaP cells showed a slightly greater loss of lycopene from the medium, which would be expected if there was an uptake by prostate cells (Liu et al. 2006). Another comparison of fetal bovine serum (FBS) (with endogenous lipoproteins), micelles, or THF or THF/BHT showed an 80% loss of lycopene (2–5h) in cell free medium with the THF systems with no loss with FBS or micelles during the same time frame. At 24h, 60% of the lycopene was lost in the FBS system but the micelle formulation had only a 20% loss. Surprisingly the micelle preparation (without lycopene) was extremely cytotoxic to DU 145 cells. Lycopene uptake by DU 145 and PC-3 cells was greater with the FBS system compared to the THF system (Lin et al. 2007). Prostasomes have also been explored as a vehicle for lycopene delivery. At high (4.38 μM) and physiological (0.876 μM) lycopene concentrations, seminal prostasomes took up the lycopene in a dose-dependent manner and lycopene stability in prostasomes (normalized to prostasome protein) suspended in PBS over a 24h period was excellent (Goyal et al. 2006a). Also, lycopene supplementation of PNT2 and PC-3 cells resulted in the incorporation of lycopene into the prostasomes secreted by these cells at concentrations of 3.71 and 1.41 mg/mg of exosomal protein, respectively (Goyal et al. 2006b).

21.4.2 Presence of Lycopene Isomers and Oxidation Products in Culture Media

All-*trans* lycopene is rapidly isomerized to an equilibrium mixture with its *cis* isomers both in cell culture medium (Liu et al. 2006) and *in vivo* in prostate tissue (Clinton et al. 1996, van Breemen et al. 2002). The *cis* isomers of lycopene are absorbed better than the all-*trans* isomers when fed to humans (Unlu et al. 2007). The 5′-*cis* isomer predominates in plasma (Gustin et al. 2004). Since lycopene absorption by prostate cells might be due to facilitated diffusion (Liu et al. 2006), it is likely that the *cis* isomers of lycopene form a significant proportion of intracellular lycopene in the experiments that are reviewed later.

Lycopene loss, however small, during incubation raises the question of the nature of the lycopene oxidation products, and whether these products are taken up by cells and are more bioactive than lycopene. A number of investigators have characterized lycopene oxidation products generated *in vitro* under fairly harsh conditions (Caris-Veyrat et al. 2003, Ferreira et al. 2004, Kim et al. 2001, Nara et al. 2001, Woodall et al. 1997). Among the oxidation products identified were several lycopenals, monoxides, diepoxides, furanoxides, and acycloretinal, which could be converted to acycloretinoic acid. Acycloretinal was formed by lycopene autooxidation but at a much slower rate than other products (Kim et al. 2001). Lycopene oxidation mixtures had greater antiproliferative activity than unoxidized lycopene (Nara et al. 2001, Woodall et al. 1997, Yeh and Hu 2001) and enhanced gap junctional communication in non-prostate cell lines (Aust et al. 2003). Acycloretinoic acid reduced the viability of PC-3 and DU-145 cells but not LNCaP cells and the aforementioned phenomena occurred to a greater extent than all-*trans*-retinoic acid, geranylgeranoic acid, and 9-*cis*-retinoic acid (Kotake-Nara et al. 2002). Another isolated lycopene oxidation product, 4-methyl-8-oxo-2,4,6-nonatrienoyl, caused a reduced Bcl-2 and Bcl-XL protein expressions (anti-apoptotic proteins), an increased caspase 8 and 9 (markers of apoptosis) activities, and nuclear fragmentation in HL-60 human leukemia cells, whereas lycopene in the same high concentration (10 μM) did not (Zhang et al. 2003) bring about the same effects. In summary, lycopene oxidation products formed in the cell culture medium in small quantities could be more bioactive than lycopene (Lindshield et al. 2007). Some investigators have noted that they have changed their lycopene-containing media daily to minimize this problem but most investigators make no mention of it.

21.4.3 Lycopene Uptake by Cultured Prostate Cells

An elegant study of lycopene uptake in LNCaP, PC-3, and DU-145 cells using beadlet-delivered 1.48 μM all-*trans* lycopene (a maximal level in human plasma) found that all three cell lines rapidly took up lycopene during the first 10 h of incubation. Cells continued to accrue lycopene, but more slowly, over the next 48 h. The uptake by the LNCaP cells was 2.5-fold higher than PC-3 cells and 4.5-fold higher than DU-145 cells at 24 h of incubation but lycopene showed no affinity for the AR receptor, which is expressed in the LNCaP cells (Liu et al. 2006). LNCaP uptake followed Michaelis–Menten kinetics with a V_{max} of 66.3 pmol/10^6 cells/h and a K_m of 7.72 μM lycopene. Because of the sensitivity of their LC-MS-MS lycopene assay, Liu et al. were also able to investigate the subcellular lycopene distribution. The nuclear membrane contained 55%, the nuclear matrix 26%, and the microsomal fraction 19% of the intracellular fraction. The cytosol contained no lycopene(Liu et al. 2006).

21.5 OXIDANT AND ANTIOXIDANT EFFECTS OF LYCOPENE IN PROSTATE CELL LINES

21.5.1 Redox Characteristics of Lycopene

The antioxidant and pro-oxidant nature of lycopene chemistry has been hypothesized by some to be the principle bioactivity that drives its cellular consequences. It is important to describe these reactions and the variety of circumstances that cause lycopene to act as a pro-oxidant or an antioxidant because the cell culture environment might be quite different from what is generally encountered by prostate cells within the intact healthy prostate or tumor areas.

Truscott and his colleagues have performed a series of experiments with artificial and physiologically relevant systems to explore the redox chemistry of lycopene. Lymphoid cells were harvested from subjects who had consumed 500 mL/d of tomato juice for 2 weeks achieving plasma lycopene concentrations of 0.15–1.5 μM. The cells were subjected, *ex vivo*, to a singlet oxygen generator (rose bengal + D_2O) or nitrogen dioxide radical generator (nitronaphthalene triplet with pulsed laser excitation) (Bohm et al. 2001) Immediate cell membrane destruction was monitored in lycopene supplemented and unsupplemented cells by dye exclusion staining. The *in vivo* protection factors for $NO_2^{\cdot}$ and $^1O_2$ were 17.6 and 6.3, respectively, which compared to *in vitro* protection factors for lycopene supplementation of only 8.2 and 3.1, respectively. These investigators suggested that the difference might be due to lower lycopene uptake or lycopene molecule aggregation when lycopene was added directly to cells (*in vitro*), but it could have been due to an uptake of phenolic compounds and vitamin C by harvested lymphocytes exposed to the plasma of the tomato juice drinkers. Singlet oxygen was quenched by energy transfer to the lycopene triplet state (highly subject to irreversible loss through oxidation) that reverts to the ground state, while the nitrogen dioxide radical was quenched by electron transfer producing a lycopene radical cation. Unless terminated by ascorbate or other antioxidant this radical was highly oxidizing to amino acids, such as tyrosine and cysteine, unlike the *in vivo* situation, where adequate antioxidants are available in the cell. Lycopene radical quenching may not be as available under cell culture situations and may not mimic the *in vivo* prostate environment.

It has long been known that whether β-carotene acts as a pro-oxidant or an antioxidant, depends upon its concentration and oxygen partial pressure (Burton and Ingold 1984). Cell culture experiments are usually performed at atmospheric oxygen partial pressures of 160 mmHg (21% O_2) whereas *in vivo*, tissues are dependent on oxygen diffusion rates and may be as low as 30–70 mmHg (4%–10% O_2) (Crawford and Blankenhorn 1991). At lower partial pressures, β-carotene and lycopene tend to act as antioxidants in both organic and aqueous phase systems but some reports indicate that at atmospheric partial pressures (encountered in cell culture) they may act as pro-oxidants (Edge and Truscott 1997). Lycopene concentration, cell type, the presence of other antioxidants, and incubation environment probably all play a role in the pro-oxidative or antioxidative nature of lycopene and must be kept in mind as we review lycopene effects on cells in culture.

21.5.2 Lycopene as a Pro-Oxidant or Antioxidant in Cell Cultures

Whether lycopene is acting as a pro-oxidant or an antioxidant and under what circumstances is a continuing issue. Investigators have measured the biomarkers of oxidative stress to explore this question. The hexane extracts of tomato paste or lycopene decreased malondialdehyde (MDA) adduct formation (a measure of lipid peroxide formation) at physiological lycopene concentrations (0.1–1 μM) in LNCaP cells incubated for 24 and 48 h. But MDA levels were increased in cells incubated in 5 and 10 μM lycopene (Hwang and Bowen 2005a). DNA damage, measured as 8-OH deoxyguanosine/guanosine ratio (8OHdG/dG), was also increased with 5 μM lycopene and physiologic concentrations were not protective of DNA except 1 μM lycopene at 48 h. However, LNCaP cell growth inhibition of 55% was seen at 1 μM lycopene (a concentration associated with lower MDA and no change in DNA damage) indicating that the lycopene radical may not have been the direct cause (Hwang and Bowen 2005a). Nevertheless, lycopene concentration and incubation time are important variables in the ensuing discussion of lycopene effects on prostate cell function.

The pro-oxidative and antioxidative effects of lycopene have been explored in non-prostatic cell lines. Yeh and Hu used foreskin fibroblasts (Hs68 cells) with two different oxidation generators and found that 20 μM lycopene acted as a lipid antioxidant in one system and a pro-oxidant in another system, while DNA damage was not affected by the presence of either lycopene or β-carotene (Yeh and Hu 2000). When they incubated cells with 20 and 40 μM oxidized lycopene or β-carotene (60°F for 1 h resulting in 80% and 35% loss of color, respectively) decreased cell viability was evident by 4 h and progressed dose-dependently, which was attributed to both apoptosis and cell lysis. Unoxidized lycopene and β-carotene had no effect. Oxidized lycopene increased 8OHdG levels in calf thymus DNA, and both 8OHdG and DNA chain breaks were increased in incubated fibroblast cells whereas the unoxidized carotenoids had no effect (Yeh and Hu 2001). A dose effect was also noted by Lowe et al. using HT29 cells (colon carcinoma) where 1–3 μM lycopene suppressed DNA damage induced by xanthine/xanthine oxidase but 4–10 μM lycopene offered no protection and increased damage over baseline levels (Lowe et al. 1999). Skin fibroblasts were incubated with lycopene, β-carotene, or lutein in liposomes and then exposed to UVB light for 20 min, followed by 1 h of incubation to allow lipid peroxides to develop. Carotenoid concentrations were measured in the harvested fibroblasts and lipid peroxides were measured by thiobarbituric acid reactive substances (TBARS). Lycopene afforded the greatest protection from the generation of lipid peroxides at 0.05 versus 0.40 and 0.30 nmol/mg protein for lycopene, β-carotene, and lutein, respectively (Eichler et al. 2002). However, using UVA irradiation of human skin fibroblasts, Offord et al. found that 0.5–1.0 μM lycopene or β-carotene afforded no protection, as measured by the induction of MMP-1 mRNA (matrix metallopeptidase 1, an interstitial collagenase), without added vitamins C and E, indicating the necessity to protect the carotenoids from oxidation (Offord et al. 2002). Finally, CVI-P monkey cell oxidation, generated by a Fe-NTA/ascorbate system, reduced TBARS by 86% and 8OHdG levels by 77% at 20 pmol lycopene/10^8 cells (Matos et al. 2000).

In summary, unoxidized lycopene can act as a lipid and a DNA antioxidant at physiological concentrations but oxidized lycopene or high concentrations of lycopene, and depending upon the oxidizing conditions, may increase lipid peroxidation and oxidative DNA damage. Furthermore, the pro-oxidant effects may result in an increased apoptosis and a decreased cell viability, which should be kept in mind as studies on proliferation and apoptosis are reviewed.

21.6 LYCOPENE EFFECTS ON PROLIFERATION, CELL CYCLE, AND APOPTOSIS

Table 21.2 summarizes studies that have evaluated some aspect of the effects of lycopene on proliferation, cell cycle, or apoptosis. Most of the studies have used the three major malignant epithelial cell lines, LNCaP, PC-3, and DU-145. Only one study used a non-neoplastic epithelial cell line but the results with lycopene were consistent with the common cancer cell lines (Obermuller-Jevic et al. 2003). Overall, lycopene can reduce prostate cancer cell proliferation and cause cell cycle arrest at the G0/G1 phase with a slight depression of cyclin D, the earliest stimulator of DNA synthesis necessary for cell production (Hwang and Bowen 2004, Ivonov et al. 2007, Obermuller-Jevic et al. 2003, Tang et al. 2005). Some of these effects could be seen at physiologically feasible concentrations in the hands of some investigators (Forbes et al. 2003, Ivonov et al. 2007, Kim et al. 2002, Obermuller-Jevic et al. 2003). More consistent was the finding of a loss of cancer cell viability with lycopene treatment and this was due to apoptosis rather than cell necrosis. Some investigators found an increased apoptosis at physiologically feasible lycopene concentrations (Hantz et al. 2005, Hwang and Bowen 2004, 2005b). Since lycopene oxidation products were not measured in cells or culture media, it is not clear whether lycopene or lycopene oxidation products were responsible and what the mechanism of action might be. However, studies in breast and endometrial cancer cell lines have identified that the lycopene-induced cell cycle arrest was associated with a reduction in cyclin D concentrations, an increase in the CDK inhibitor protein p21, and the retention of the proliferative protein p27 in the cyclin E-cdk2 complex leading to the inhibition of G_1 phase of CDK activities (Nahum et al. 2001). The apoptotic effects of lycopene appear to be via the mitochondrial pathway since the MTT assay for cell viability (used by several investigators) depends on functioning mitochondria. Hantz et al. found that mitochondrial oxidation was decreased by 20% at 1 μM lycopene concentrations with an increased release of cytochrome *c*. These investigators noted that these apoptotic phenomena occur at lycopene levels that maintain plasma membrane integrity (Hantz et al. 2005). Since lycopene accumulates in cell membranes, is it possible that lycopene could also accumulate in mitochondrial membranes and alter membrane function? Alternatively, could lycopene oxidation products induce mitochondrial membrane lipid peroxidation and perforation with the subsequent exit of cytochrome *c* and the loss of mitochondrial efficiency?

TABLE 21.2 Lycopene Effects on Prostate Cell Proliferation, Cell Cycle, and Apoptosis

Reference	Cell Type	Lycopene Treatment	Solvent	Incubation	Measurement and Methods	Outcome
Barber et al. (2006)	Primary benign epithelial cells (PECs) isolated at the time of surgery ($n = 6$) *PSA-secreting* LNCaP used as comparison	1–20 μM lycopene (Sigma)	THF	48 h	Proliferation measured as BrdU incorporation into DNA	Both normal and cancer cells showed reduced BrdU incorporation (49%–23%) as the dose of lycopene was increased from 1 to 15 μM
Obermuller-Jevic et al. (2003)	PrEC-non-neoplastic cells	All-*trans*/5% *cis* isomers: 0.5–5.0 μM	THF affected cell cycle, corrected for THF alone values	48 h medium changed every 24 h	Lycopene uptake	0.885 and 0.704 nmol/g protein at 24 and 72 h, respectively
					^{3}H thymidine incorporation for proliferation	30% inhibition at 0.5 μM 60% inhibition at 1.0 μM 80% inhibition at 2.0 μM
					Cell cycle by flow cytometry and cyclins D and E	Arrest in GO/G1 stage only at 5 μM Cyclin D_1 slightly depressed at 0.5 μM, completely suppressed at 5 μM cyclin E not affected
Kim et al. (2002)	LNCaP	Lyco-Red 10^{-9}–10^{-4} M	Water-dispersible	24, 48, 72 h	Direct cell counts	Cell numbers 20% lower at 0.1 μM 24% lower at 1.0 μM at 24 h 25.6% lower at 10 μM at 48 h
Hwang and Bowen (2004)	LNCaP	Roche 10% water miscible beadlets 0.1–5.0 μM	DMSO in H_2O	6, 24, 48 h	Lycopene uptake	50 ng/μg protein at 1 μM at 24 h 62 ng/μg protein at 5 μM at 24 h
					Cell viability by Trypan blue exclusion	No change at 6 h 25% inhibition at 0.1 μM at 24 and 48 h 40% inhibition at 1.0 μM at 24 h 65% inhibition at 5.0 μM at 24 h
					Cell cycle by flow cytometry	Cell cycle arrest only at 5 μM—S phase change from 45% to 29%
Hwang and Bowen (2005b)		Tomato paste hexane extract			Apoptosis by Annexin V binding	Increased 1.5- and 2.5-fold at 1 μM at 24 and 48 h Increased 2.7-, 8.2-, and 6.6-fold at 5 μM at 6, 24, and 48 h, respectively; lycopene alone slightly less effective

Kotake-Nara et al. (2001)	LNCaP PC-3 DU-145	Lyco-Mato 6% α-, and β-carotene fucoxanthin neoxanthin capxanthin zeaxanthin lutein cryptoxanthin 5, 10, and 20 μM	THF	72 h medium changed daily	Cell viability by MTT	80%–90% inhibition at 20 μM for fucoxanthin and neoxanthin all cell types 25% inhibition at 5 μM for lycopene for LNCaP, 30% for DU-145, and 30% PC-3 No inhibition for other carotenoids
					Apoptosis by TUNEL	Apoptosis increased for fucoxanthin and neoxanthin
Tang et al. (2005)	LNCaP DU-145 PC-3	95% pure natural lycopene 6% oleoresin 10–50 μM	THF	24 h	Cell viability by MTT	No inhibition at 24 h in any cell type Negligible at 48 h in any cell type 25% inhibition at 72 h in DU-145 and PC-3 at 20 μM No inhibition in LNCaP until 50 μM
					Cell cycle by flow cytometry	Increased cells in G_0/G_1 at 32 μM at 72 h
					Apoptosis by flow cytometry	Doubling of apoptotic cells at 8 μM Doubled again at 16 μM, 32 μM increased 20% more
Hantz et al. (2005)	LNCaP	Lycopene (Sigma) 0.3–3.0 μM	THF	20 h 72 h for DNA banding	Lycopene uptake	5.5–36.7 pmol/10^6 cells with 3 μM the highest
					Proliferation by BrdU	No difference at any concentration
					Trypan Blue exclusion and viable cell counts	No difference in cell number or viability
					Cell viability by MTT	Decreased linearly 61%–83% with increasing lycopene concentration
					Necrosis by LDH	No necrosis
					Mitochondrial transmembrane potential by DiOC6 and JC-1 fluorescence	Increase membrane dysfunction of 20% and 13% at 1 μM and 3 μM for DiOC6. Fivefold increase in apoptotic cells using JC-1 at 3 μM
					Apoptosis by cytochrome *c* release Annexin V binding DNA fragmentation by agarose gel electrophoresis	Increased with 0.3–3.0 μM 2.7-, 1.6-, and 3.3-fold increase with 0.3–3.0 μM None even at 72 h

(*continued*)

TABLE 21.2 (continued) Lycopene Effects on Prostate Cell Proliferation, Cell Cycle, and Apoptosis

Reference	Cell Type	Lycopene Treatment	Solvent	Incubation	Measurement and Methods	Outcome
Ivonov et al. (2007)	LNCaP PC-3 PTEN deficient cell lines	Lycopen (37.8% lycopene, water miscible) LycoTrue (92.4% lycopene in THF) 0.1–10μM	H_2O THF/ BHT	Various times 24, 48, and 72h	Lycopene stability	LycoTrue more stable in culture than LycoPen
					Cell proliferation: Crystal violet	Decreased by 15% at 1μM in LNCaP for Lycopen; Decreased by 60% at 1μM in LNCaP, PC-3 for LycoTrue at 48h but not at 24h; incorporation at 0.4, 0.8μM in LNCaP but modest reduction for PC-3 cells
					Proliferation by MTS staining	Decreased proliferation 30% at 0.5μM, 90% at 2.5μM of LycoTrue in LNCaP
					Cell cycle by flow cytometry + BrdU to determine whether cells were senescent or apoptotic	Increased G0/G1:S + G2/M ratio twofold at 0.08μM in LNCaP; 38%, 69% increase in ratio at 0.4, 0.8μM in PC-3 cells; Drastic increase in % cells below G0/G1 phase only in LNCaP cells
					Phosphorylated cell signaling/ regulatory proteins (50) by immunoblotting	Direct interference with proteins connected to G1/S cell cycle transition; IGF-IR suppressed with 10-fold increase in $IGFBP_2$ in the presence and the absence of androgen
Kotake-Nara et al. (2002)	LNCaP PC-3 DU-145	All-*trans* RA 9-*cis* RA 4 OH retinamide Acyclo RA-lycopene oxidation product 5,10, and 20μM	THF	72h	Cell viability by MTT	Acyclo RA no effect on LNCaP but 4HPR reduced viability Acyclo RA 50% reduced in PC-3 at 20μM at 25% at 10μM; 35% in DU145 at 20μM No effect of all- *trans* RA or 9-*cis* RA
					Apoptosis by laddering in phase-contrast microscope + TUNEL	Acylo RA induced apoptosis in PC-3 and DU145

Forbes et al. (2003)	PC-3 PC-3MM$_2$—very invasive	Sigma lycopene solubilized in 1:2000 dilution of chloroform 1 μM	Chloroform	8 days media refreshed every 4 days	Proliferation by counts	Growth reduction in PC-3 but not PC-3MM$_2$
					Connexin 43	No detectable levels in PC-3MM$_2$ but slight up-regulation in PC-3
					In vitro invasion through trans-well polycarbonate membranes 8.0 μm pores	100% increase in invasiveness in PC-3MM$_2$
					UPAR expression	Increased in PC-3MM$_2$
					UPA expression	No difference
					PA-1 expression	No difference
						No effect on PC-3 cells
Pastori et al. (1998)	PC-3 DU-145	1–5 μM lycopene + 50 μM α-tocopherol	THF	Unknown	Proliferation by cell count and thymidine labeling	Lycopene alone had modest effect (48% in DU 145) but 1 μM lycopene + 50 μM α-tocopherol inhibited cell growth by 40% in PC-3 and 88% in DU 145 cells. Degree of synergy was > 50% for 0.2–1.5 μM lycopene but dropped dramatically at ≥ 2 μM lycopene
		50 μM β-tocopherol, probucol, ascorbic acid				No inhibition synergy with these antioxidants

Note: THF = tetrahydrofuran; BrdU = bromodeoxyuridine; Lyco-Red = 10% lycopene water miscible beadlet containing small amounts of phytoeine, phytofluen, tocopherol from LycoRed Corp, Orange, NJ; DMSO = dimethysulfoxide; Roche beadlets = 10% lycopene water miscible beadlet now from DSM Nutritional Products, Inc., Persippany, NY; TUNEL = terminal deoxynucleotidyl transferase dUTP nick end labeling; RA = retinoic acid; uPAR = urokinase plasminogen activator receptor; uPA = urokinase; PA-1 = plasminogen activator 1; LNCaP, PC-3, DU-145 = neoplastic prostate epithelial cells (see Table 21.1).

There are several alternative pathways associated with the balance between proliferation and apoptosis that are affected by lycopene treatment, especially the insulin-like growth factor (IGF) signaling pathway. Another is the possibility that lycopene or one of its breakdown products has retinoid activity. Kotake-Nara et al. compared acyclo-retinoic acid, an *in vitro* oxidation product of lycopene, to four actively researched anticarcinogenic retinoids. Acycloretinoic acid was found to more actively reduce PC-3 and DU-145 cell viabilities (but not LNCaP) through apoptosis in a medium already containing small amounts of natural retinoids. But study concentrations were 20 μM, far above physiologically relevant lycopene concentrations, let alone the smaller concentration of one of its breakdown products. Acycloretinoic acid had a very low affinity for the retinoid X receptors (RXR) and retinoic acid receptors (RAR) receptors (Kotake-Nara et al. 2002).

21.7 LYCOPENE AND THE INSULIN-LIKE GROWTH FACTOR SIGNALING PATHWAY

The mechanism for lycopene's possible interference with IGF signaling has recently become a focus for cell culture studies since lycopene has been shown to interfere with both IGF signaling and androgen pathways in the normal and cancerous rat prostate glands (Herzog et al. 2005, Siler et al. 2004). Furthermore, elevated plasma concentrations of IGF-1 and the ratio IGF-1/IGFBP-3 (insulin-like growth factor binding protein 3) are risk factors for prostate cancer (Li et al. 2003, Miyata et al. 2003, Mucci et al. 2001, Oliver et al. 2004) and there is an inverse association between plasma IGF-1 concentration and tomato intake (Gunnell et al. 2003, Mucci et al. 2001). It appears that lycopene supplementation may increase IGF binding proteins but less so circulating levels of IGF-1 (Graydon et al. 2007, Riso et al. 2006, Voskuil et al. 2008, Vrieling et al. 2007). However, these circulating levels might be of liver origin and may not reflect what is happening in various types of prostate cells.

The IGF pathway has been the focus of research for several different cancers and is associated with mitogenesis and the down-regulation of apoptosis. It is composed of a number of components: IGFs (secreted by not only liver cells but also by a number of other tissues), IGF binding proteins (most of plasma IGF-1 is bound to IGFBP-3), and IGF receptors. In the prostate, stromal cells are the principal secretors of IGF-1 with small amounts detectable in LNCaP and DU-145 cells. When stromal PrSC cells are cocultured with either LNCaP or DU-145 cell, these cancer epithelial cells are more proliferative, both *in vitro* and *in vivo*, and when PrSC cell IGF-1 secretion is blocked or the IGF-1 receptor (IGF-IR) is blocked, cancer cell proliferation is decreased (Kawada et al. 2006). Interestingly, a study of IGF-IR staining in normal and prostate cancer specimens showed that IGF-IR was expressed in normal and prostate cancer epithelial cells but rarely in adjacent stromal cells (Ryan et al. 2007). Therefore, IGF-1 is an important mediator of the stromal cell support of epithelial cell carcinogenesis.

Table 21.3 summarizes the tissue culture studies that have explored the action of lycopene on the IGF axis using prostate cells. Both Ivonov et al. (2007) and Kanagaraj et al. (2007) found detectable IGF-1 in cell culture medium, presumably produced from their PC-3 epithelial cells, but lycopene treatment (40–60 μM) had no effect on IGF-1 concentrations. Both groups of investigators found 1.5- to 2-fold increases in IGFBP-3 in cells and the culture medium that corroborates the human studies, even though the lycopene doses used were much higher than could be achieved physiologically. Both investigative groups also reported that when PC-3 cells were stimulated to increased growth by IGF-1 exposure the lycopene decreased the IGF-1 stimulated IGF-IR expression (Table 21.3). Liu et al. (2008) continued to explore the stromal, epithelial cell interactions of IGF signaling using DHT (stimulator) and lycopene (inhibitor) as probes. They used the PrSC stromal cell lines or stromal fibroblasts (6S) (derived from a prostate cancer patient), which were cocultured with primary epithelial cells (NPE) without the androgen receptor. Lycopene, at physiological concentrations, decreased by 50% the sixfold induction of IGF-1 mRNA caused by DHT in PrSC stromal cells, but had no effect on non-DHT-stimulated cells. The lycopene effect may be due to a reduction (60%–70%) in

TABLE 21.3 Lycopene Effects on Prostate Cell IGF Signaling

Reference	Cell Type	Lycopene Treatment	Solvent	Incubation	Measurement and Methods	Outcome
Ivonov et al. (2007)	PC-3	LycoPen = 37.8% lycopene, water miscible LycoTrue 92.4% in THF 0.1–10 μM	H_2O THF/ BHT	48 h	IGF -1 and IGFBP-3 in culture media by IRMA kit	No difference in IGF-1 secretion by cells treated with 20–60 μM lycopene; IGFBP-3 secretion increased by 1.5- to 2-fold
					Cellular IGF-1 and IGF-IR by specific antibody reactions with SDS-PAGE isolated protein	Increased cellular IGFBP-3 with 60 μM lycopene independent of IGF-1 stimulation; Lycopene down-regulated IGF-1 stimulated IGF-IR production
Kanagaraj et al. (2007)	PC-3	97% pure lycopene from tomatoes 20, 40, 60 μM IGF-1 at 50 ng/mL	THF	24, 48, 72, 96 h	Cell viability by Trypan Blue	Increased dead cell number; lycopene + IGF-1 increased dead cell number further
					Cell viability by MTT	25%–30% reduced mitochondria-mediated viable cells
					Proliferation by thymidine labeling	Reduced IGF-1 stimulated proliferation by 50% at 40 μM
					IGF-1 expression by Western blot; secretion by IRA	No difference in IGF-1 in culture media
					IGFBP-3 expression by Western blot; secretion by IRA	Increased by 1.5- to 2-fold at 40 μM in cells and culture media; IGF-1 stimulation + lycopene increased IGFBP-3 levels while IGF-1 alone decreased BP by twofold
					IGF-IR expression by Western blot	No difference; IGF-1 stimulated IGF-IR expression but added lycopene decreased its expression
					Apoptosis by flow cytometry	22% early, 13% late apoptosis at 20 μM for 24 h
					DNA fragmentation by agarose gel electrophoresis	Observed after 48 h

(continued)

TABLE 21.3 Lycopene Effects on Prostate Cell IGF Signaling

Reference	Cell Type	Lycopene Treatment	Solvent	Incubation	Measurement and Methods	Outcome
Liu et al. (2008)	(1)Normal prostate stromal cells (PrSC); (2) 6S stromal fibroblasts prostate, cancer-derived; (3) NPE (PREC) primary prostate epithelial cells lacking androgen receptor: Also as cocultures	0.3–1 μM lycopene	THF	72 h followed by 24 h of CM Cocultures of (1) NPE + PRSC or (2) NPE + 6s cells + camptothecin (CM) IGF-1(0.15 or 1 ng/mL) stimulated NPE-6s cocultures w/wo 1 μM lycopene	DNA fragmentation	DHT(dihydrotestosterone) inhibited DNA fragmentation of NPE cells induced by CM when cocultured with 6s cells, but not when cocultured with PrSC cells. Lycopene at 1 μM in the NPE-6s cocultures treated with CM or CM + DHT overcame the antiapoptotic effect of DHT. DHT increases IGF-1 mRNA sixfold and lycopene decreased this DHT induction by 50% but had no effect on non-induced IGF-1 mRNA; IGF-1 not induced by DHT in PRSC cells
					Androgen receptor (AR) and β-catenin expression	24 h lycopene incubation reduced DHT-induced AR expression by 70% and AR nuclear localization by 60% in 6s cells
					Cell viability by MTT using 0.15 ng/mL IGF-1 (amount secreted by 6s cell with DHT induction	NPE or PrEC monocultures: IGF-1 increased, CM reduced (NPE), lycopene reduced viability with IGF-1 stimulation but had no effect without it.
					1 μM lycopene 16 h pretreatment with immunoblots for Akt and GSK3 phosphorylation at 0 and 4 h after IGF-1 stimulation	Lycopene decreased Akt phosphorylation 70% and completely prevented IGF-1 induced increase in GSK3 phosphorylation. IGF-1 decreased but lycopene + IGF-1 rectified tyrosine phosphorylation in GSK3.

Note: THF = tetrahydrofuran; BHT = butylated hydroxytoluene, an antioxidant; Camptothecin (CM) = causes inhibition of the DNA enzyme topoisomerase (Top 1) which induces DNA damage and apoptosis; DHT = dihydrotestosterone; PrEC = normal prostate stromal cells; LNCaP, PC-3, DU-145 = neoplastic prostate epithelial cells (See Table 21.1).

observed AR expression and nuclear localization in 6s cells. These authors suggested that the ability of lycopene to inhibit the anti-apoptotic effect of DHT treatment is due to its down-regulation of IGF-1 production in the stromal cells and attenuating its signal reception in the epithelial cells. IGF-1 stimulates the Akt/Gsk3β pathway toward cell growth, but lycopene attenuates the serine phosphorylation of Akt and the tyrosine phosphorylation of GSK3 thus limiting cell growth stimulated by either IGF-1 (coming from normal stromal cells) or androgen (Liu et al. 2008).

21.8 OTHER LYCOPENE ACTIVITIES

Lycopene treatment appears to have other bioactivities that do not fall into the broad category of the modulation of cell proliferation or apoptosis. Only those that have been explored in prostate cell lines are discussed below.

21.8.1 Enhancement of Gap-Junction Communication by Connexin 43 Up-Regulation

Connexin 43 is one of the cell membrane gap junction proteins that allow for cell to cell communication. Cancer cell transformation is correlated with the loss of Connexin 43 expression (Hussain et al. 1989, Zhang et al. 1992). Lycopene increased Connexin 43 expression in fetal skin fibroblasts (Stahl et al. 2000), oral cancer KB-1 cells grown in organotypic rafts (Livny et al. 2003) and breast cancer cell lines (Chalabi et al. 2007).

Gitenay et al. used the prostate PC-3AR (expresses the androgen receptor) to explore the effect of lycopene, its possible biological metabolites, and other bioactive compounds found in tomatoes, on cell growth, and Connexin 43 expression. They did this in a unique way. Rats were fed with BASF lycopene beadlets, or lyophilized red or yellow tomatoes for 6 weeks, then sacrificed to obtain their respective sera for the cell culture studies. Although they did not measure harvested rat sera for lycopene, its metabolites, or tomato phenolics, they did find that PC-3AR cells cultured with lycopene or the serum of rats fed with red tomatoes took up equivalent amounts of lycopene whereas cells exposed to control sera or the sera of rats fed with yellow tomatoes failed to take up lycopene. Connexin 43 expression and protein amounts were higher for all treatments compared to cells exposed to control sera but lycopene-only sera did not increase Connexin43 protein as much as red or yellow tomato exposed sera and the increase was not statistically significant. Sera from red or yellow tomato-fed rats decreased cell numbers but lycopene-only treatment did not. These observations led Gitenay et al. to conclude that other tomato ingredients or their metabolites found circulating in the sera of tomato-fed rats were responsible for the up-regulation of Connexin 43 in the prostate cell line. They also recognized that rat is a very poor carotenoid absorber and so they suggested that the tomato phenolic compounds in their sera may have played a more prominent role compared to the tomato carotenoids in this experiment (Gitenay et al. 2007). Quercetin, the predominant phenolic compound in tomatoes, was found to repress the expression of the androgen receptor in the LNCaP cell line (Xing et al. 2001) but quercetin mainly circulates as multiple glucuronides in the rat (Graf et al. 2006) and the human (Moon et al. 2008). Unlike quercetin, its glucuronide metabolites do not act as pro-oxidants (Lodi et al. 2008).

Forbes et al. saw a slight up-regulation of Connexin 43 in PC-3 cells (androgen unresponsive) with pure lycopene treatment but no Connexin 43 expression (with or without lycopene) in the highly metastatic PC-3MM2 variant (Table 21.2) indicating that with increasing metastatic conversion either the ability to turn on Connexin 43 is lost and/or the receptors responsive to lycopene are lost (Forbes et al. 2003).

21.8.2 Metastatic Invasiveness

Another focus has been on the invasiveness of prostate cancer cells since prostate cancer lethality is tied to metastatic processes. Metastatic cells have acquired the ability to invade basement membranes and even extracellular matrix by the up-regulation of a system of MMPs via the urokinase plasminogen activator system (activator + receptor + inhibitor). The up-regulation of the receptor, urokinase plasminogen activated receptor (uPAR), has been shown to enhance invasion and prostate cancer cell growth (Festuccia et al. 1998). Forbes et al. explored the effect of lycopene to prevent invasion of a multipass bone metastatic prostate cell line (PC-3MM2) through a transwell polycarbonate membrane coated with a Matrigel basement membrane matrix. Lycopene treatment at the physiological dose of 1.0 μM caused a doubling of the cells that invaded the matrix and an increased expression of the uPAR without increasing the activator (uPA) and the inhibitor (PAI-1). Lycopene had no effect on uPAR, uPA, or PAI-1 in regular PC-3 cells (Forbes et al. 2003) indicating that as transformation proceeds the mechanisms affected by lycopene may be lost, while other processes may come to the fore that may be enhanced by the presence of lycopene. Since this paper has appeared, clinicians have been concerned over the risk of lycopene supplementation for patients with advanced metastatic prostate cancer (Ablin 2005). However, Huang et al. used highly invasive liver SK-Hep-1 cells in the polycarbonate transwell matrigel system and found that lycopene suppressed invasion, but this suppression was sensitive to lycopene concentration. Five μM of lycopene suppressed invasion by 81%–91% but higher concentrations were less suppressive. β-carotene also suppressed cell migration, but not invasion, and was far less effective than lycopene. The protein nm23-H1 has been identified as an important suppressor of functions that are necessary for metastasis. Lycopene treatment at 2.5 and 5 μM concentrations increased nm23-H1 protein level by 220% and its mRNA by 153%. Higher levels of lycopene reduced its expression (Huang et al. 2005). Kozuki et al. also found that lycopene, as well as other carotenoids, inhibited hepatoma AH109A from invading mesothelial cell membranes in coculture, and from the figures presented in their paper, it appeared that lycopene suppressed the invasion at a lower concentration (2.5 μM) than other carotenoids (Kozuki et al. 2000).

In summary, lycopene must have some specific effect on unknown cellular processes that control the modulation of multiple pathways. General properties, such as antioxidation or pro-oxidation, are unlikely to explain these effects. Since the activation, silencing or loss of pathway control is different for each cell type and its degree of transformation, we do not have enough information to predict whether lycopene may be beneficial or detrimental under different circumstances in various prostate cell lines and in the different stages of prostate cancer.

21.9 IS THERE A CENTRAL MECHANISM FOR LYCOPENE ACTION?

The promise of cell culture studies is the elucidation of the mechanism(s) by which lycopene might act to prevent the initiation, the promotion, or the progression of prostate cancer. Can the studies performed, so far, point to these mechanisms? Wertz et al. suggested six possible modes of action for lycopene in prostate health promotion as (1) antioxidant function, (2) the direct inhibition of cell cycle progression, (3) the direct initiation of apoptosis, possibly by lycopene oxidation products, (4) the inhibition of IGF-1 signal transduction, (5) the suppression of inflammation, especially modulated by IL-6, and (6) the inhibition of androgen activation and signaling (Wertz et al. 2004). The cell culture studies have corroborated animal and human circumstantial evidences for most of these functions. Lycopene, often at physiologically relevant concentrations, interferes with cell cycle and proliferation via cyclin D_1 down-regulation. But is the concomitant increase in apoptosis, often seen in the same studies, an independent effect? Several studies have found an up-regulation of IGFBP-3 and a smaller decrease in IGF-1 output, especially from the stromal cells underlying the epithelial cells. Epithelial cells seem to undergo the most radical neoplastic transformations. Interference suggested in androgen signaling by lycopene may not be directly testable in cell culture and aside from determining lycopene effects on androgen dependent- and independent-prostate cancer cell lines or observing lycopene effects with and without DHT. The improvement of gap-junction communication via the increased synthesis of Connexin 43 in prostate cancer cells by lycopene appears to be a completely different mode of action and may or may not be connected with pathways that control cell invasion.

The fact that lycopene appears to have so many different bioactivities in prostate cancer cell lines points to a common mechanism that might explain a variety of its effects. What are the candidates for a common modality of action and is it a chemically feasible mechanism for lycopene action? There are four modalities that could be explored through cell culture studies: (1) the effects of lycopene on shifting the aberrant DNA methylation pattern that is seen as prostate cells undergo carcinogenic transformation, (2) the modulation of retinoid receptor signaling, (3) the modulation of redox controlled cell signaling mechanisms, and (4) selective binding to catalytic and signaling proteins with common structural motifs.

21.9.1 Lycopene and Gene Methylation

Cytosine is methylated to form 5-methylcytosine, which combines with guanine to form cytosine–guanine dinucleotides (CpGs) that cluster in key regulatory regions called CpG islands. CpG methylation is a key epigenetic regulatory mechanism. Hypermethylation tends to inactivate genes, and promoter hypermethylation and the silencing of associated genes are widespread in prostate cancer (Murphy et al. 2008). The identification of a signature gene methylation pattern that would differentiate clinically significant prostate cancer from indolent cancer is a current research focus. For example, the gene for glutathione-S-transferase pi (GSTP1) has been found to be methylated in >75%–90% of prostate cancers but not methylated in normal epithelium (Lin et al. 2001, Woodson et al. 2004). The gene for IGFBP-3 has also been found to be methylated in prostate cancer (Perry et al. 2007) as is RARβ, the retinoid receptor (Bastian et al. 2007, Woodson et al. 2004). Increased DNA methylation of the connexin promoter region and the down-regulation of Connexin 43 expression are features of human lung tumors (Chen et al. 2003).

King-Batoon et al. have found that the physiological concentrations of lycopene (2 μM) partially demethylated the promoter for GSTP1 and restored its expression in the breast cancer cell line MDA-MB-468, but RARβ was not demethylated. However, lycopene did induce the demethylation of RARβ in MCF10A fibrocystic cells (King-Batoon et al. 2008). Genistein, in this study, was less active compared to lycopene.

Is it chemically feasible to suppose that lycopene could directly modulate methylation/demethylation? A family of DNA methyltransferases is responsible for methylation, and it has been recently proposed that the DNA methyltransferases DNMT3A and DNMT3B are also responsible for active demethylation through a very complicated mechanism that cycles every 100 min during transcriptional cycling (Kangaspeska et al. 2008, Metivier et al. 2008). So how could lycopene intervene in this process? Interestingly, Kangaspeska et al. (2008) used doxorubicin to reduce methylation with the hypomethylation of the proximal pS2 promotor occurring about 45 min after its introduction into MDA-MB-231, estrogen receptor-negative breast cancer cells. Doxorubicin is thought to act by intercalating within the DNA structure and this interferes with the unwinding of DNA for replication (Formari et al. 1994). It is possible that lycopene could act in the same manner since all-*trans* lycopene has a flat, rod-like structure and 5′ *cis* lycopene is almost as flat, and 55% of the lycopene is found in the nuclear membrane and 26% in the nucleus matrix of prostate cancer cells (Liu et al. 2006). Bathaie et al. found that the saffron carotenoids, crocetin, dimethylcrocetin, and crocin bind to calf thymus DNA in the outside groove-binding pattern (Bathaie et al. 2007). However, both the doxorubicin and the saffron carotenoids are highly oxygenated and water-soluble, whereas lycopene is the most hydrophobic of the carotenoids. Furthermore, how could lycopene, concentrated in the nuclear bi-membrane be in proximity to DNA that is about to undergo replication? Of interest is the observation that silenced genes, methylated and/or tightly wrapped around histones, are more likely to be found peripherally, nearest to the nuclear membrane (Shaklai et al. 2007). Alternatively, lycopene or a lycopene oxidation product could be acting as a weak androgen or estrogen antagonist or even agonist (Hirsch et al. 2007, Wertz et al. 2004). Estrogen stimulates cell division and thus methylation/demethylation cycling.

21.9.2 Lycopene Modulation of Retinoid Receptor Signaling

Retinoic acid is the natural ligand for the RAR and the heterodimer RAR/RXR DNA receptors that activate the response elements, RAR response element (RARE) and RXR response element (RXRE) which, in turn, regulate the expression of a wide range of target genes engaged in cellular differentiation and the local modulation of chromatin structure. RARE and RXRE activations are generally considered antiproliferative. Retinoids inhibit the growth of several prostate cancer cell lines and suppress the development of prostate carcinogenesis in some animal models but seem to be more effective when combined with other chemopreventive agents. Surprisingly, antagonists to RARβ and RARγ were even more effective (Pasquali et al. 2006). The structural analysis of the ligand-binding pockets of both RAR and RXR retinoid receptors indicates the need for a negatively charged ligand, which is attracted to a cluster of positively charged amino acids (two arginines and three lysines). The pocket for RAR appears to be larger than that of RXR, which may explain why both retinoic acid and 9-*cis*-retinoic acid activate RAR but only 9-*cis*-retinoic acid activates RXR. Interestingly, some of the synthetic retinoids appear to have no resemblance to linear retinoic acid but all have a hydrophobic moiety that mimicks the β-ionone ring of the two natural ligands. A good ligand fit changes the conformation of these receptors such that they are able to recruit crucial enzymes and factors that regulate DNA transcription (Klaholz and Moras 1998). Differences in the conformation of the response elements on different genes explain the wide variation in physiological responses to various retinoid-like ligands.

Investigators have focused on the more polar metabolites of lycopene as possible RAR ligands. Eighty percent of radio-labeled lycopene fed to rats accumulated in the prostate as polar products and the level of lycopenoids, such as 10′-lycopenal, was comparable to retinoic acid concentrations in the ferret lung (Lindshield et al. 2007). Acycloretinal (20′ lycopenal) is a common *in vitro* product formed with the cleavage at the central double bond of lycopene. Acycloretinal has been shown to be converted to acycloretinoic acid spontaneously *in vitro* and by rat liver homogenates (Nagao 2004). Stahl and coworkers found that acycloretinoic acid was a much weaker ligand (50 μM) compared to retinoic acid (0.1 μM) for the transactivation of the RARβ2 promoter and while acycloretinoic acid stimulated gap -junction communication in fetal skin fibroblasts and stabilized Connexin 43 mRNA, it was 10-fold less effective than retinoic acid (Stahl et al. 2000). Vine et al. found that RARα is the responsive nuclear receptor for the retinoid up-regulation of Connexin 43 expression but lycopene and other non-provitamin A carotenoids act in a manner independent of RARs on the RARE. They did not directly evaluate acycloretinoic acid (Vine et al. 2005). Ben-Dor et al. used a transient transfection system to explore retinoic acid versus acycloretinoic acid as ligands for RAR and RXR, and the extent of transactivation of reporter genes containing three different RAREs. Although acycloretinoic acid and retinoic acid (5 μM, respectively) were equally potent in producing G0/G1 cell cycle arrest in breast cancer MCF-7 cells, acycloretinoic acid was about 100-fold less potent as a ligand for RAR and its transactivation, compared to retinoic acid. It was not a ligand for RXR. Since acycloretinoic acid had similar activity to lycopene in producing cell cycle arrest, these investigators concluded that acycloretinoic acid was likely not an active metabolite of lycopene (Ben-Dor et al. 2001). The minimal direct transcriptional activation seen with RAR still may be due to polar lycopene oxidation products. These products are almost impossible to eliminate during *in vitro* experiments, which may explain why lycopene is sometimes reported as directly involved in RARE activation. However, the 100-fold weaker effect of acycloretinoic acid in RAR-mediated actions discourages the pursuit of retinoid-like effects of lycopene or its oxidative metabolites.

21.9.3 Modulation of Redox-Controlled Signaling Pathways

Although the cellular concentrations of lycopene or its oxidation products may be too low to have a general antioxidant or pro-oxidant effect on cells, there is sufficient evidence of its *in vivo* effect on the classical measures of oxidative stress to indicate that its participation in the redox state of redox sensitive pathways, that control cell proliferation and apoptosis, may be in play. Nuclear factor kappa B (NF-κB) has been postulated to play a role in the initiation and the progression of various cancers, particularly, in prostate cancer (Suh and Rabson 2004). The NF-κB family is composed normally of inactive transcription factors, which can be uninhibited by a number of pathways that appear to be redox sensitive, since any number of antioxidant compounds can block NF-κB stimulation by TNFα, IL-1, lipopolysaccharide (LPS), or H_2O_2 and its translocation to the nucleus (Mercurio and Manning 1999). NF-κB, once attached to various DNA response elements, coordinates immune and inflammatory responses as well as cell proliferation and survival especially at low levels of oxidative stress (Trachootham et al. 2008). Therefore its quiescence would be consistent with the effect of lycopene on cell cycle and apoptosis. NF-κB is over-expressed in the androgen-insensitive prostate cell lines, PC-3 and DU-145 and prostate carcinoma xenographs, whereas its expression is low in the hormone-responsive LNCaP cells (Paule et al. 2007, Suh and Rabson 2004). Kim et al. found that lycopene (5–10 μM) inhibited the maturation of dendritic cells, which are responsible for antigen-presentation in the stimulation of naive T lymphocytes. This suppression of the immune response was associated with an inhibition of LPS-induced up-regulation of p-ERK, p-p38, and p-JNK, all part of the redox-sensitive mitogen-activated protein kinase (MAPK) signaling pathway that can stimulate the cytosolic liberation of NF-κB. Indeed, these investigators found that 10 μM lycopene suppressed the NF-κB p65 nuclear translocation usually seen with LPS stimulation (Kim et al. 2004). Dendritic cells are known to sustain some of the chronic inflammatory diseases (Poulter and Janossy 1985) and since inflammation may be a component of prostate carcinogenesis it may have a role in the whole prostate. Can lycopene or one of its oxidation products act by changing the redox microenvironment? The only evidence that a change in redox balance may be its mechanism of action comes from the studies of β-carotene. Palozza et al. found that β-carotene (10–30 μM) reduced the growth of HL-60 leukemic cells and was positively correlated with the ROS content of the cells. This effect could be prevented by the addition of 5 μM α-tocopherol pointing to a change in redox balance toward oxidation by some oxidation product of β-carotene. DNA binding of NF-κB was seen in cells treated with 10 μM β-carotene with only 3 h of incubation and this effect was greatly attenuated with α-tocopherol treatment (Palozza et al. 2003). These studies used pharmacologic doses of β-carotene and since apoptosis rather than cell preservation was associated with NF-κB binding in these experiments, the dual nature of NF-κB as a transcription factor that can also stimulate cell death under severe oxidative stress was probably in play. We would expect lycopene to have the same effect, at these high concentrations, since it has an even greater propensity to oxidize in cell culture compared to β-carotene. However, Huang et al. observed that 1–10 μM lycopene induced the inhibition of direct binding of NF-κB to binding sites in the MMP-9 promotor (MMP-9 is a matrix metalloproteinase responsible for tumor invasion and angiogenesis) in SK-Hep-1 human hepatoma cells and was not redox dependent. The evidence for some other mechanism that shifts the redox balance in these experiments was (1) coincubation with H_2O_2 did not limit lycopene's inhibitory effect at physiologic concentrations even though it abolished lycopene's antioxidant activity and (2) incubation with β-carotene had the same antioxidant effect as lycopene but had no effect on the inhibition of cell invasion. They concluded that lycopene action was likely associated with effects on the IGF signaling pathway (Huang et al. 2007).

Nuclear factor-E2 related factor 2 (Nrf2) is another nuclear transcription factor that can be found in its inactivated state in the cytoplasm. Oxidative stress and electrophiles are the major activators of Nrf2, which translocates to the nucleus and heteromerizes with small Maf proteins that then bind to an antioxidant response element (ARE) for over 200 genes involved in the synthesis of proteins that act as antioxidants, phase II detoxification enzymes, proteosomes, heat-shock proteins, and glutathione-synthesis proteins (Trachootham et al. 2008). Ben-Dor et al. explored the effect of tomato carotenoids on this system in breast cancer MCF-7 and hepatic cancer $HepG_2$ cells. Lycopene (6 μM), more so than phytoene, astaxanthin, or tert-butylhydroquinone (tBHQ), a well-known antioxidant and ARE activator, produced a three- to fourfold activation of the reporter gene for γ-glutamylcysteine synthase and NAD(P)H:quinone oxidoreductase, both Phase II enzymes turned on by Nrf2 in breast and liver cancer cells. The mRNA and protein levels of these enzymes were also induced to a much larger extent by lycopene compared to the other carotenoids and there was a corresponding increase in glutathione synthesis. The functional Nrf2 protein and the ARE transcription system were essential for the induction of these Phase II enzymes. These investigators, using an ethanolic extract of pure lycopene containing no lycopene but probably its oxidation products, found that it was as potent as lycopene itself in showing the same effects. Lycopene, as well as β-carotene and tBHQ, caused the migration of Nrf2 to the nuclei of the $HepG_2$ cells that colocalized with PML nuclear bodies. Surprisingly, tBHQ, lycopene, phytoene, and astaxanthin but not β-carotene lowered intracellular ROS in both cell types even though they had greatly different ARE activation concentrations (Ben-Dor et al. 2005). These authors concluded that the modulation of ROS does not explain the variable effects of these antioxidants on Nrf2 migration and ARE activation, and the equivalent activity of the putative lycopene oxidation products points to the generation of an electrophilic α,β-unsaturated carbonyl compound as the possible ARE activation moiety during the course of the lycopene incubation.

In summary, the apparent redox modulation of lycopene certainly affects two important redox sensitive transcription factors at higher concentrations of lycopene. However, electrophilic lycopene oxidation products cannot be ruled out as the major activators and the activation may be due to specific molecular interactions.

21.9.4 Selective Binding to Catalytic and Signaling Proteins

Except for changing the redox environment of redox sensitive signaling proteins, other actions must be mediated by some sort of lycopene–protein interaction. Carotenoids are commonly found to be bound to proteins in nature. The biosynthesis of the most common plant cyclic carotenoids, such as β-carotene, lutein, and neoxanthin, is tightly coordinated with the biogenesis and the assembly of protein-rich photosynthetic structures. The inhibition of their syntheses limits the synthesis, the assembly, and the stability of carotenoid–chlorophyll binding proteins as well as the nuclear genes that encode these proteins (Cunningham et al. 1996). So there are chemically feasible precedents in nature for protein–carotenoid interactions. Lycopene does not have rings or oxygen atoms for stable binding with proteins and less is known concerning specific lycopene–protein interactions. Various lycopene oxidation products could also act as ligands for particular proteins.

There have been only a few studies that have directly studied lycopene binding to particular proteins. Lycopene (2 μM) was shown to directly bind with platelet-derived growth factor (PDGF-BB), but not vascular endothelial growth factor (VEGF) or MMP, inhibiting its ability to stimulate EGF-induced ERK1/2 phosphorylation. Also, lycopene inhibited PDGF-BB-induced mouse smooth muscle cell and human fibroblast invasion. Furthermore, lycopene was found to bind to PDGF-BB in human plasma. The binding was specific for lycopene because β-carotene showed negligible binding (Chiang et al. 2007, Lo et al. 2007). Hazai et al. used molecular modeling to explore protein binding to 5-lipoxygenase by all-*trans*-, 5 *cis*-, and dihydroxy-lycopene and two lycopenals. The three forms of lycopene, C40 rodlike structures, were found to bind with a high affinity into the long, linear groove at the interface of the N- and C-terminal domains (β-barrel and catalytic domain, respectively) of 5-LOX, which would alter the activity of this enzyme by allosteric means. This interface is called the cleavage site because when the two domains are split the enzyme loses its activity. The association and the movement of these two domains regulate enzyme activity. The two smaller lycopenals fit (with high affinity) into interior pockets next to the catalytic site and would be excellent competitive inhibitors of 5-LOX activity, perhaps through radical scavenging (Hazai et al. 2006). This study, although a computer modeling exercise, points to the feasibility of lycopene–protein binding of a myriad of enzymes or signaling molecules with common structural motifs. Such lycopene–protein binding may explain the wide array of inhibitory activities and even up-regulatory activities (in the case of Connexin 43) that are observed in cell culture as well as *in vivo* studies of lycopene and its oxidation products. Since lycopene isomers and their oxidation products could be existing simultaneously within prostate cells and in culture media, one could even postulate the array of activities, ascribed to lycopene in this review, may be due to a variety of attachments to bioactive proteins.

21.10 CONCLUSIONS

There is no doubt that lycopene and/or its oxidation products exhibit many important bioactivities in prostate cancer cell lines and the consequences of these are correlated with measured responses in animal and human studies. The "elephant in the room" is the existence of unidentified lycopene oxidation products in the cell culture media and the prostate cells themselves in almost all studies that have been reviewed herein. These are difficult to avoid and the complexity of these products may render them almost impossible to completely characterize in every experiment. The working out of chemically feasible mechanisms of action for lycopene is obviously dependent upon clean experiments with pure molecules of known structure. Several solvent delivery systems do much better than others in preserving the integrity of the lycopene molecule and the addition of a protective antioxidant, such as tocopherol, to the incubation medium may be of advantage. There are commercially available incubation chambers where the partial pressure of oxygen can be controlled to provide redox environments that are similar to the one experienced by prostate cells in a living prostate gland. The major physiologically relevant lycopene oxidation products should be identified and compared with lycopene action in future studies.

The complexity of prostate cell cross talk that may be partially assessed by prostate cell cocultures should add to our understanding of how lycopene or its oxidation products participate. However, of utmost importance is the characterization of lycopene or lycopene oxidation product binding to particular proteins that shift their function and therefore the pathways in which they act. Such characterization is foundational to understanding the mechanism of action of lycopenoids. Simpler model systems where even the whole cell is too complex may be useful in working out these mechanisms of action.

The use of cell cultures to work out the modes of action of lycopene either in the prevention or the modulation of prostate cancer progression is important. The proper design of clinical trials is severely hampered if the mechanism of action is poorly understood. There are many questions of clinical importance that can be addressed as a first step in finding answers through the use of cell culture techniques. These include (1) the interaction of lycopene with common forms of prostate cancer therapy, such as radiation or androgen ablation, (2) synergy with other bioactive compounds, such as selenium, vitamin E, and various phytochemicals, (3) the specific modes of action in increasingly transformed and metastasized cells, and (4) lycopene effects on intercellular communication. Lycopene is an intriguing compound and there are many tantalizing questions that are worth answering. Cell culture studies will continue to play an important role in our understanding of prostate carcinogenesis and whether lycopene or its oxidation products have a place in prostate cancer prevention or therapy.

REFERENCES

Ablin, AJ. 2005. Lycopene: A word of caution. *Am J Health Syst Pharm* 62:899.

Aust, O, N Ale-Agha, LX Zhang, H Wollersen, H Sies, and W Stahl. 2003. Lycopene oxidation product enhances gap junctional communication. *Food Chem Toxicol* 41(10):1399–1407.

Balakumaran BS and PG Febbo. 2006. New insights into prostate cancer biology. *Hematol Oncol Clin North Am* 20:773–796.

Barber, NJ, X Zhang, G Zhu et al. 2006. Lycopene inhibits DNA synthesis in primary prostate epithelial cells in vitro and its administration is associated with a reduced prostate-specific antigen velocity in a phase II clinical study. *Prostate Cancer Prostatic Dis* 9:407–413.

Bastian, PJ, J Ellinger, LC Heukamp, P Kahl, SC Muller, and A von Rucker. 2007. Prognostic value of CpG island methylation of PTGS2, RAR-beta, EDNRB, and other gene loci in patients undergoing radical prostatectomy. *Eur Urol* 51(3):665–674.

Bathaie, SK, A Bolhasan, R Hoshyar, B Ranjbar, F Sabouni, and A-A Moosavi-Movahedi. 2007. Interaction of saffron carotenoids as anticancer compounds with ctDNA, Oligo (dG.dC)15, and Oligo (dA.dT)15. *DNA Cell Biol* 26(8):533–540.

Ben-Dor, A, A Naham, M Danilenko et al. 2001. Effect of *acyclo*-retinoic acid and lycopene on activation of the retinoic acid receptor and proliferation of mammary cancer cells. *Arch Biochem Biophys* 391(2):295–302.

Ben-Dor, A, M Steiner, L Gheber et al. 2005. Carotenoids activate the antioxidant response element transcription system. *Mol Cancer Ther* 4(1):177–186.

Bohm, F, R Edge, M Burke, and TG Truscott. 2001. Dietary uptake of lycopene protects human cells from singlet oxygen and ntirogen dioxide—ROS components from cigarette smoke. *J Photochem Photobiol B* 64:176–178.

Borthakur G, K Sprandel, ES Hwang et al. 2005. Effect of chronic dosing of lycopene as a tomato-based beverage on oxidative stress biomarkers. *FASEB J* 19(4):A472.

Burton, GW and KU Ingold. 1984. Beta-carotene: An unusual type of lipid antioxidant. *Science* 224:569–573.

Caris-Veyrat, C, A Schmid, M Carail, and V Bohm. 2003. Cleavage products of lycopene produced by in vitro oxidations: Characterization and mechanisms of formation. *J Agric Food Chem* 51(25):7318–7325.

Chalabi, N, L Delort, S Satih, P Dechelotte, YJ Bignon, and DJ Bernard-Gallon. 2007. Immunohistochemical expression of RARalpha, RARbeta, and Cx43 in breast tumor cell lines after treatment with lycopene and correlation with RT-QPCR. *J Histochem Cytochem* 55(9):877–883.

Chen, JT, YW Cheng, MC Chou et al. 2003. The correlation between aberrant connexin 43 mRNA expression induced by promoter methylation and nodal micrometastasis in non-small cell lung cancer. *Clin Cancer Res* 9:4200–4204.

Chen, L, M Stacewicz-Sapuntzakis, C Duncan et al. 2001. Oxidative DNA damage in prostate cancer patients consuming tomato sauce-based entrees as a whole food intervention. *J Natl Cancer Inst* 93:1872–1879.

Chiang, H-S, W-B Wu, J-Y Fang et al. 2007. Lycopene inhibits PDGF-BB-induced signaling and migration of human dermal fibroblasts through interaction with PDGF-BB. *Life Sci* 81:1509–1517.

Clinton, SK, C Emenhiser, SJ Schwartz et al. 1996. *Cis-trans* lycopene isomers, carotenoids, and retinol in human prostate. *Cancer Epidemiol Biomarkers Prev* 5(10):823–833.

Comstock, C and KE Knudsen. 2007. The complex role of AR signaling after cytotoxic insult. *Cell Cycle* 6(11):1307–1313.

Crawford, BH and DH Blankenhorn. 1991. Arterial wall oxygenation, oxyradicals, and atherosclerosis. *Atherosclerosis* 89:97–108.

Cunningham, FX, B Pogson, Z Sun, KA McDonald, D Della Penna, and E Gantt. 1996. Functional analysis of the beta and epsilon lycopene cyclase enzymes of arabidopsis reveals a mechanism for control of cyclic carotenoid formation. *Plant Cell* 8:1613–1626.

De Marzo, A, N Yasutomo, and WG Nelson. 2007. Inflammation, atrophy, and prostate carcinogenesis. *Urol Oncol* 25:398–400.

Edge, R and TG Truscott. 1997. Prooxidant and antioxidant reaction mechanisms of carotene and radical interactions with vitamins E and C. *Nutrition* 13(11/12):992–994.

Eichler, O, H Sies, and W Stahl. 2002. Divergent optimum levels of lycopene, beta-carotene and lutein protecting against UVB irradiation in human fibroblasts. *Photochem Photobiol* 75(5):503–506.

Ferreira, ALA, KJ Yeum, RM Russell, NI Krinsky, and G Tang. 2004. Enzymatic and oxidative metabolites of lycopene. *J Nutr Biochem* 15:493–502.

Festuccia, C, V Dolo, F Guerra et al. 1998. Plasminogen activator system modulated invasive capacity and proliferation in prostatic tumor cells. *Clin Exp Metastasis* 16:513–528.

Forbes, K, K Gillette, and I Sehgal. 2003. Lycopene increases urokinase receptor and fails to inhibit growth or connexin expression in a metastatically passaged prostate cancer cell line: a brief communication. *Exp Biol Med* 228:967–971.

Formari, FA, JK Randolph, JC Yalowich, MK Ritke, and DA Gewirtz. 1994. Interference by Doxorubicin with DNA unwinding in MCF-7 breast tumor cells. *Mol Pharmacol* 45:649–656.

Fry, PM, DL Hudson, MH O'Hare, and JRW Masters. 2000. Comparison of marker protein expression in benign prostatic hyperplasia *in vivo* and *in vitro*. *BJU Int* 85:504–513.

Giovannucci, E. 2007. Does prostate-specific antigen screening influence the results of studies of tomatoes, lycopene, and prostate cancer risk? *J Natl Cancer Inst* 99(14):1060–1062.

Gitenay, D, B Lyan, J Talvas, A Mazur, S George, C Caris-Veyrat, and E Rock. 2007. Serum from rats fed red or yellow tomatoes induced Connexin-43 expression independently from lycopene in a prostate cancer cell line. *Biochem Biophys Res Commun* 364:578–582.

Goyal, A, GH Delves, M Chopra, BA Lwaleed, and AJ Cooper. 2006a. Can lycopene be delivered into semen via prostasomes? *In vitro* incorporation and retention studies. *Int J Androl* 29:528–533.

Goyal A, GH Delves GH, M Chopra , BA Lwaleed , and AJ Cooper. 2006b. Prostate cells exposed to lycopene *in vitro* liberate lycopene-enriched exosomes. *BJU Int* 98(4):907–911.

Graf, BA, C Ameho, GG Doinikowski, PE Mibury, CY Chen, and JB Blumberg. 2006. Rat gestrointestinal tissues metabolize quercetin. *J Nutr* 136(1):39–44.

Graydon, R, SE Gilchrist, IS Young, U Obermuller-Jevic, O Hasselwander, and JV Woodside. 2007. Effect of lycopene supplementation on insulin-like growth factor-1 and insulin-like growth factor binding protein-3: a double-blind, placebo-controlled trial. *Eur J Clin Nutr* 61(10):1196–2000.

Gunnell, D, SE Oliver, TJ Peters et al. 2003. Are diet-prostate cancer associations mediated by the IGF axis? A cross-sectional analysis of diet, IGF-1 and IGFBP-3 in healthy middle-aged men. *Br J Cancer* 88(11):1682–1686.

Gustin, DM, Rodvold KA, Sosman JA et al. 2004. Single-does pharmakinetic study of lycopene delivered in a well-defined food-based lycopene delivery system (tomato pasted-oil mixture) in healthy adult male subjects. *Cancer Epidemiol Biomarkers Prev* 13(5):850–860.

Hantz, H, LF Young, and KR Martin. 2005. Physiologically attainable concentrations of lycopene induce mitochondrial apoptosis in LNCaP human prostate cancer cells. *Exp Biol Med* 230:171–179.

Hazai, E, Z Bikadi, F Zsila, and SF. Lockwood. 2006. Molecular modeling of the non-covalent binding of the dietary tomato carotenoids lycopene and lycophill, and selected oxidative metabolites with 5-lipoxygenase. *Bio Med Chem* 14:6859–6867.

Herzog, A, U Siler, V Spitzer et al. 2005. Lycopene reduced gene expression of steroid targets and inflammatory markers in normal rat prostate. *FASEB J* 19:272–274.

Hirsch, K, A Atzmon, M Danilenko, J Levy, and Y Sharoni. 2007. Lycopene and other carotenoids inhibit estrogenic activity of 17 beta-estradiol and genistein in cancer cells. *Breast Cancer Res Treat* 104:221–230.

Huang, C-S, Y-E Fan, C-Y Lin, and M-L Hu. 2007. Lycopene inhbitis matrix metalloproteinase-9 expression and down-regulates the binding activity of nuclear facter-kappa B and stimulatory protein -1. *J Nutr Biochem* 18:449–456.

Huang, C-S, S Ming-Kuei, C-H Chuang, and M-L Hu. 2005. Lycopene inhibits cell migration and invasion and upregulates Nm23-H1 in a highly invasive hepatocarinoma, SK-Hep-1 cells. *J Nutr* 135:2119–2123.

Hussain, MZ, LR Wilkens, PP Mehta, WR Lowenstein, and JS Bertram. 1989. Enhancement of gap junctional communication by retinoids correlates with their ability to inhibit neoplastic transformation. *Carcinogenesis* 10:1743–1748.

Hwang, E-S and PE Bowen. 2004. Cell cycle arrest and induction of apoptosis by lycopene in LNCaP human prostate cancer cells. *J Med Food* 2004(3):284–289.

Hwang, E-S and PE Bowen. 2005a. Effects of lycopene and tomato paste extracts on DNA and lipid oxidation in LNCaP human prostate cells. *BioFactors* 23:97–105.

Hwang, E-S and PE Bowen. 2005b. Effects of tomato paste extracts on cell proliferation, cell-cycle arrest and apoptosis in LNCaP human prostate cancer cells. *Biofactors* 23:75–84.

Ivonov, NI, SP Cowell, P Brown, PS Rennie, ES Guns, and ME Cox. 2007. Lycopene differentially induces quiescence and apoptosis in adrogen-responsive and -independent prostate cell lines. *Clin Nutr* 26:252–263.

Kanagaraj, P, MR Vijayababu, B Ravisankar, J Anbalagan, MM Aruldhas, and J Aranakaran. 2007. Effect of lycopene on insulin-like growth factor-1, IGF binding protein-3, and IGF type-1 receptor in prostate cancer cells. *J Cancer Res Clin Oncol* 133:351–359.

Kangaspeska, S, B Stride, R Metivier et al. 2008. Transient cyclical methylation of promoter DNA. *Nature* 452:112–115.

Kawada, M, H Inoue, T Masuda, and D Ikeda. 2006. Insulin-like growth factor 1 secreted from prostate stromal cells mediates tumor-stromal cell interactions of prostate cancer. *Cancer Res* 66(8):4419–4425.

Kim, G-Y, KimJ-H, AhnS-C et al. 2004. Lycopene suppresses the lipopolysaccharide-induced phenotypic and functional maturation of murine dendritic cells through inhibition of mitogen-activated protein kinases and nuclear factor NF-kappa B. *Immunology* 113:203–211.

Kim, JS, E Nara, H Kobayashi, J Terao, and A Nagao. 2001. Formation of cleavage products by autooxidation of lycopene. *Lipids* 36(2):191–199.

Kim, L, AV Rao, and LG Rao. 2002. Effect of lycopene on prostate LNCaP cancer cells in culture. *J Med Food* 5(4):181–187.

King-Batoon, A, JM Leszczynska, and CB Klein. 2008. Modulation of gene methylation by genistein or lycopene in breast cancer cells. *Environ Mol Mutagen* 46:36–45.

Klaholz, BP and D Moras. 1998. A structural view of ligand binding to the retinoid receptors. *Pure Appl Chem* 70(1):41–47.

Kotake-Nara, E, JS Kim, M Kobori, K Miyashita, and A Nagao. 2002. *Acylo*-retinoic acid induces apoptosis in human prostate cancer cells. *Anticancer Res* 22(2A):689–95.

Kotake-Nara, E, M Kushiro, Zhang H, T Sugawara, K Miyashita, and A Nagao. 2001. Carotenoids affect proliferation of human prostate cancer cells. *J Nutr* 131:3303–3306.

Kozuki, Y, Y Miura, and K Yagasaki. 2000. Inhibitory effects of carotenoids on the invasion of rat ascites hepatoma cells in culture. *Cancer Lett* 151:111–115.

Kravets, FG, J Lee, B Singh, A Trocchia, SN Pentyala, and S Khan. 2000. Prostasomes: Current concepts. *Prostate* 43:169–174.

Li, L, H Ya, F Schumacher, G Casey, and JS Witte. 2003. Relations of serum insulin-like growth factor-1-(IGF-1) and IGF binding protein-3 to risk of prostate cancer(United States). *Cancer Causes Control* 14:721–726.

Lin C-Y, Huang C-S, and Hu M-L. 2007. The use of fetal bovine serum as delivery vehicle to improve the uptake and stability of lycopene in cell culture studies.*Br J Nutr* 98:226–232.

Lin, X, M Tascilar, WH Lee et al. 2001. GSTP1 CpG island hypermethylation is responsible for the absence of GSTP1 expression in human prostate cancer cells. *Am J Pathol* 159:1815–1826.

Lindshield, BL, K Canene-Adams, and JW Erdman. 2007. Lycopenoids: Are lycopene metabolites bioactive? *Arch Biochem Biophys* 458:136–140.

Liu, A, N Pajkovic, Y Pang, D Zhu, B Calamini, AL Mesecar, and RB van Breemen. 2006. Absorption and subcellular localization of lycopene in human prostate cancer cells. *Mol Cancer Ther* 5(11):2879–2885.

Liu, X, JD Allen, JT Arnold, and MR Blackman. 2008. Lycopene inhibits IGF-1 signal transduction and growth of normal prostate epithelial cells by decreasing DHT-modulated IGF-1 production in cocultured reactive stromal cells. *Carcinogenesis* 29(4):816–823.

Livny, O, I Kaplan, R Reifen, S Polak-Charcon, Z Madar, and B Schwartz. 2003. Oral cancer cells differ from normal oral epithelial cells in tissuelike organization and in response to lycopene treatment: An organotypic cell culture study. *Nutr Cancer* 47(2):195–209.

Lo, H-M, C-F Hung, Y-L Tseng, B-H Chen, J-S Jian, and W-B Wu. 2007. Lycopene binds PDGF-Bb and inhibits PDGD-BB-induced intracellular signaling transduction pathway in rat smooth muscle cells. *Biochem Pharmacol* 74:54–63.

Lodi, F, R Jimenez, C Menendez, PW Needs, J Buaret, and E Perez-Vizcaino. 2008. Glucuronidated metabolites of the flavonoid of quercetin do not auto-oxidize, do not generate free radicals and do not decrease nitric oxide bioavailability. *Planta Med* 74(7):741–746.

Lowe, GM, LA Booth, AJ Young, and R Bilton. 1999. Lycopene and beta-carotene protect against oxidative damage in HT29 cells at low concentrations but rapidly lose this capacity at higher doses. *Free Radic Res* 30:141–151.

Matos, HR, P Di Mascio, and MHG Medieros. 2000. Protective effect of lycopene on lipid peroxidation and oxidative DNA damage in cell culture. *Arch Biochem Biophys* 383(1):56–59.

Mercurio, F and AM Manning. 1999. NF-kappa B as a primary regulator of the stress response. *Oncogene* 18:6163–6171.

Metivier, R, R Gallais, C Tiffoche et al. 2008. Cyclical DNA methylation of a transcriptionally active promoter. *Nature* 452:45–50.

Miki, J and JS Rhim. 2007. Prostate cell cultures as *in vitro* models for the study of normal stem cells and cancer stem cells. *Prostate Cancer Prostatic Dis* 11(1):32–39.

Miyata, Y, H Sakai, T Hayashi, and H Kanetake. 2003. Serum insulin-like growth factor binding protein-3/prostate-specific antigen ratio is a useful predictive marker in patients with advanced prostate cancer. *Prostate* 54:125–132.

Moon, YJ, L Wang, R DiCenzo, and ME Morris. 2008. Quercetin pharmacokinetics in humans. *Biopharm Drug Dispos* 29:205–217.

Mucci, LA, R Tamimi, P Lagiou et al. 2001. Are dietary influences on the risk of prostate cancer mediated through the insulin-like growth factor system. *BJU Int* 87:814–820.

Murphy, TM, AS Perry, and M Lawler. 2008. The emergence of DNA methylation as a key modulator of aberrant cell death in prostate cancer. *Endocr Relat Cancer* 15:11–25.

Nagao A 2004. Oxidative conversion of carotenoids to retinoids and other products. *J Nutr* 143:237s-240s.

Nahum A, Kirsch K, Danilenko M et al. 2001. Lycopene inhibition of cell cycle progression in breast and endometrial cancer cells is associated with reduction in cyclin D levels and retention of p27-Kip1 in the cyclin E-cdk2 complexes. *Oncogene* 20:3428–3436.

Nara, E, H Hayashi, M Kotake, K Miyashita, and A Nagao. 2001. Acyclic carotenoids and their oxidation mixtures inhibit the growth of HL-60 human promyelocytic leukemia cells. *Nutr Cancer* 39(2):273–83.

Obermuller-Jevic, UC, E Olano-Martin, AM Carbach et al. 2003. Lycopene inhibits the growth of normal human prostate epithelial cells in vitro. *J Nutr* 133:3356–3360.

Offord, EA, JC Gautier, O Avanti et al. 2002. Photoprotective potential of lycopene, beta-carotene, vitamin E, vitamin C and carnosic acid in UVA-irradiated human skin fibroblasts. *Free Radic Biol Med* 32(12):1293–1303.

Oliver, SE, D Gunnell, J Donovan et al. 2004. Screen-detected prostate cancer and the insulin-like growth factor axis: results of a population-based case-control study (United States). *Int J Cancer* 108(6):887–892.

Palozza, P, S Serini, A Torsello et al. 2003. Beta-carotene regulates NF-kappa B DNA-binding activity by a redox mechanism in human leukemia and colon adenocarcinoma cells. *J Nutr* 133:381–388.

Pasquali, D, V Rossi, G Bellastella, A Ballastella, and AA Sinisi. 2006. Natural and synthetic retinoids in prostate cancer. *Curr Pharm Des* 12:1923–1929.

Pastori, M, H Pfander, D Boscoboinik, and A Azzi. 1998. Lycopene in association with alpha-tocopherol inhibits at physiological concentrations proliferation of prostate carcinoma cells. *Biochem Biophys Res Commun* 250:582–585.

Paule, B, S Terry, L Kheuang, P Soyeux, F Vacherot, and A de la Taille. 2007. The NF-kappa B/IL-6 pathway in metastatic androgen-independent prostate cancer: new therapeutic approaches. *World J Urol* 25:477–489.

Peehl, DM. 2004. Are primary cultures realistic models for prostate cancer? *J Cell Biochem* 91:185–195.

Perry, AS, B Loftus, R Moroose et al. 2007. In silico mining identifies IGFBP-3 as a novel target of methylation in prostate cancer. *Br J Cancer* 96:1587–1594.

Porrini, M and P Riso. 2000. Lymphocyte lycopene concentration and DNA protection from oxidative damage is increased in women after a short period of tomato consumption. *J Nutr* 130:189–192.

Poulter LW and Janossy G. 1985. The involvement of dendritic cells in chronic inflammatory disease. *Scand J Immunol* 21:401–411.

Riso, P, A Brusamolino, A Martinetti, and M Porrini. 2006. Effect of a tomoto drink intervention on insulin-like growth factor (IGF)-1 serum levels in health subjects. *Nutr Cancer* 55(2):157–162.

Riso, P, F Visioli, D Erba, G Testolin, and M Porrini. 2004. Lycopene and vitamin C concentrations increase in plasma and lymphocytes after tomato intake. Effects on cellular antioxidant protection. *Eur J Clin Nutr* 58:1350–1358.

Ryan, CJ, CM Haqq, J Simko et al. 2007. Expression of insulin-like growth factor-1 receptor in local and metastatic prostate cancer. *Urol Oncol* 2007(2):134–140.

Sampson, N, G Untergasser, E Plas, and P Berger. 2007. The ageing male reproductive tract. *J Pathol* 211:206–218.

Schulz, WA, Burchardt M, and Cronauer MV. 2003. Molecular biology of prostate cancer. *Mol Hum Reprod* 9(8):437–448.

Shahrzad, S, E Cardenas, A Sevanian, and L Packer. 2002. Impact of water-dispersible beadlets as a vehicle for the delivery of carotenoids to cultured cells. *Biofactors* 16:83–91.

Shaklai, S, N Amariglio, G Rechavi, and AJ Simon. 2007. Gene silencing at the nuclear periphery. *FEBS J* 274:1383–1392.

Siler U, Barella L, Spitzer V et al. 2004. Lycopene and vitamin E interfere with autocrine/paracrine loops in the Dunning prostate cancer model. *FASEB J* 18:1019–1021.

Sobel, RE and MD Sadar. 2005. Cell lines used in prostate cancer research: A compendium of old and new lines- Part 1. *J Urol* 173:342–359.

Stahl, W, J von Laar, H-D Martin, T Emmerich, and H Sies. 2000. Stimulation of gap junctional communication: comparison of *acyclo*-retinoic acid and lycopene. *Arch Biochem Biophys* 373(1):271–274.

Suh, J and AB Rabson. 2004. NF-kappa B activation in human prostate cancer: important mediator or epiphenomenon. *J Cell Biochem* 91:100–112.

Tang L, Jin T, Zeng X, Wang J-S. 2005. Lycopene inhibits the growth of human androgen-independent prostate cancer cells *in vitro* and in BALB/c nude mice. *J Nutr* 135:287–290.

Tomlins, SA, MA Rubin, and AM Chinnaiyan. 2006. Integrative biology of prostate cancer progression. *Annu Rev Pathol Mech Dis* 1:243–271.

Trachootham, D, W Lu, MA Ogasawara, N Rivera-Del Valle, and P Huang. 2008. Redox regulation of cell survival. *Antioxid Redox Signal* 10(8):1342–1374.

Unlu, NZ, T Bohn, DM Francis, HN Nagaraja, SK Clinton, and SJ Schwartz. 2007. Lycopene from heat-treated *cis*-isomer-rich tomato sauce is more bioavailable than from all-*trans*-rich tomato sauce in human subjects. *Br J Nutr* 98(1):140–146.

van Breemen, RB, X Xu, MA Viana et al. 2002. Liquid chromatography-mass spectrometry of *cis*- and *trans*-lycopene in human serum and prostate tissue after dietary supplementation with tomato sauce. *J Agric Food Chem* 50(8):2214–2219.

Vine, AL, YM Leung, and JS Bertram. 2005. Transcriptional regulation of connexin 43 expression by retinoids and carotenoids: similarities and differences. *Mol Carcinog* 43:75–85.

Voskuil, DW, A Vrieling, CM Korse et al. 2008. Effects of lycopene on the insulin-like growth factor (IGF) system in premenopausal breast cancer surviviors and women at high familial breast cancer risk. *Nutr Cancer* 60(3):342–353.

Vrieling, A, DW Voskull, JM Bonfrer et al. 2007. Lycopene supplementation elevates circulatng insulin-like growth factor binding protein-1 and -2 concentrations in persons at great risk of colorectal cancer. *Am J Clin Nutr* 86(5):1456–1462.

Webber, MM, D Bello, and S Quader. 1997. Immortalized and tumorgenic adult human prostatic epithelial cell lines: characteristics and applications. Part 2: Tumorogenic cell lines. *Prostate* 30:58–64.

Wertz, K, U Siler, and R Goralczyk. 2004. Lycopene: Modes of action to promote prostate health. *Arch Biochem Biophys* 430:127–134.

Woodall, AA, SW Lee, RJ Weesie, MJ Jackson, and G Britton. 1997. Oxidation of carotenoids by free radicals: Relationship between structure and reactivity. *Biochim Biophys Act* 1336(1):33–42.

Woodson, K, J Gillespie, J Hanson, M Emmert-Buck, JM Phillips et al. 2004. Heterogeneous gene methylation patterns among pre-invasive and cancerous lesions of the prostate: a histopathologic study of whole mount prostate specimens. *Prostate* 60(1):25–31.

Xing, N, Y Chen, SH Mitchell, and CYF Young. 2001. Quercetin inhibits the expression and function of the androgen receptor in LNCaP prostate cancer cells. *Carcinogenesis* 22(3):409–414.

Xu, X, Yan Wang, AI Constantinou, M Stacewicz-Sapuntzakis, PE Bowen, and RB van Breemen. 1999. Solubilization and stabilization of carotenoids using micelles: Delivery of lycopene in cells in culture. *Lipids* 34:1031–1036.

Yeh, SL, and ML Hu. 2000. Antioxidant and pro-oxidant effects of lycopene in comparison with beta-carotene on oxidant-induced damage in Hs68 cells. *J Nutr Biochem* 11:548–554.

Yeh, SL and ML Hu. 2001. Induction of oxidative DNA damage in human foreskin fibroblast HS68 cells by oxidized beta-carotene and lycopene. *Free Radic Res* 35(2):203–213.

Zhang, H, E Kotake-Nara, H Ono, and A Nagao. 2003. A novel cleavage product formed by autooxidation of lycopene induces apoptosis in HL-60 cells. *Free Radic Biol Med* 35(12):1653–1663.

Zhang, LX, RV Cooney, and JS Bertram. 1991. Carotenoids enhance gap junctional communication and inhibit lipid peroxidation in C3H/10T1/2 cells: Relationship to their cancer prevention action. *Carcinogenesis* 12(11):2109–2114.

Zhang, LX, RV Cooney, and JS Bertram. 1992. Carotenoids up-regulate Connexin 43 expression independent of their provitamin A or antioxidant properties. *Cancer Res* 52:5707–5712.

Zhao, X, G Aldini, E Johnson, H Rasmussen, K Kraemer, H Woolf, N Musaeus, NI Krinsky, RM Russell, and K Yeum. 2006. Modification of lymphocyte DNA damage by carotenoid supplementation in postmenopausal women. *Am J Clin Nutr* 83:163–169.

22 Carotenoids as Modulators of Molecular Pathways Involved in Cell Proliferation and Apoptosis

Paola Palozza, Assunta Catalano, and Rossella Simone

CONTENTS

22.1 INTRODUCTION

Extensive research in the last few years has revealed that the regular consumption of certain fruits containing carotenoids, an important group of phytochemicals derived from such fruits and vegetables, is involved in cancer prevention. Both prospective and retrospective epidemiological studies have consistently and clearly shown that an increased intake of fruits and vegetables rich in carotenoids is associated with a decreased risk of cancer (Mayne, 1996; Peto et al., 1981; Ziegler et al., 1996). Accordingly, a number of *in vitro* and *in vivo* studies reported that β-carotene and other carotenoids inhibit the growth of cancer cells (Gerster, 1995, Palozza, 2005; Palozza et al., 2004b, 2006a). In particular, carotenoids have been reported to attenuate or delay chemical- or ultraviolet (UV)-induced carcinogenesis in animal models. They have been also found to act as potent growth-inhibitory agents in several tumor cells, including colon, melanoma, prostate, oral, lung, and breast cancer cells and to enhance the cytotoxicity of well-known chemotherapeutics (Palozza et al., 2006a). Anticarcinogenic activities have been demonstrated for both provitamin A and nonprovitamin A carotenoids (Krinsky, 1993). In human mammary epithelial cells, morphologic changes suggesting differentiation were observed to accompany a reduced proliferative capacity in response to both β-carotene and canthaxanthin (Rock et al., 1995). Recently, carotenoids have been found to modulate several pathways of the apoptotic process (Palozza et al., 2004b). In contrast, results from human intervention studies, the β-Carotene and Retinol Efficacy Trial (CARET) (Omenn et al., 1996) and the α-Tocopherol, β-Carotene Cancer Prevention Study (ATBC) (the α-tocopherol, Beta-Carotene Cancer Prevention Study Group, 1994), indicated that the exposure of subjects taking supplemental β-carotene to cigarette smoke increased lung cancer incidence. The Australian Polyp Study also showed that β-carotene supplementation was associated with higher risk of recurrence of large colorectal adenomas (Wahlqvist et al., 1994). In view of these contradictory findings, there has been considerable interest in elucidating the mechanism(s) by which carotenoids affect cell growth.

One of the most attractive hypotheses to explain the modulatory effects of carotenoids on cell growth is that these molecules may act as modulators of redox status and intracellular reactive oxygen species (ROS) production (Palozza, 2005). The ROS have been reported to play a major physiological role in several aspects of intracellular signaling and regulation (Palmer and Paulson, 1997). It has been clearly demonstrated that ROS interfere with the expression of a number of genes and signal transduction pathways (Thannickal and Fanburg, 2000). Because ROS are oxidants by nature, they influence the redox status and may, according to their concentration, cause either a positive (cell proliferation) or a negative cell response (growth arrest or cell death).

However, recently, several other "non-redox" mechanisms have been implicated in the modulation of cell growth by carotenoids, which include the direct modulation of the expression of proteins and transcription factors involved in cell proliferation, differentiation and apoptosis.

This review reports the more recent evidence for the ability of β-carotene and other carotenoids to modulate cell signaling related to cell growth and implicated in a lot of pathological events, including cancer, inflammation, and atherosclerosis by both redox and non-redox mechanisms.

22.2 MODULATION OF NUCLEAR FACTOR-KAPPA B (NF-κB) ACTIVATION PATHWAY

Research over the past decade has revealed that NF-κB is an inducible transcription factor for genes involved in cell survival, differentiation, and growth. In most resting cells, NF-κB is sequestered in the cytoplasm by binding to the inhibitory IκB proteins that blocks the nuclear localization sequences of NF-κB. NF-κB is activated by a variety of stimuli such as carcinogens, inflammatory agents, tumor promoters including cigarette smoke, phorbol esters, okadaic acid, H_2O_2, and tumor necrosis factor α (TNFα). These stimuli promote dissociation of IκBα through phosphorylation, ubiquitination and its ultimate degradation in the proteasomes. This process unmasks the nuclear localization sequence of NF-κB, facilitating its nuclear entry, binding to κB regulatory elements and activating transcription of target genes. Many of the target genes that are activated are critical to the establishment of early and late stages of aggressive cancers such as expression of cyclin D1, apoptosis suppressor proteins such as Bcl-2 and Bcl-xl, and those required for metastasis and angiogenesis such as matrix metalloproteases (MMP) and vascular endothelial growth factor (VEGF). Several laboratories have demonstrated that treatment of cells with H_2O_2 can activate the NF-κB pathway (La Rosa et al., 1994). The observation that inducers of NF-κB activity, such as TNFα, interleukin (IL)-1, lypopolysaccharide (LPS), phorbol myristate acetate, UV and ionizing radiation, generated elevated levels of ROS, prompted speculation that ROS may function as common mediators of NF-κB activation.

We recently reported that β-carotene induced a significant increase in ROS production and/or in oxidized glutathione content in HL-60 cells (Palozza et al., 2002b) as well as in colon adenocarcinoma cells (Palozza et al., 2001a). These effects were always accompanied by a sustained elevation of NF-κB and by a significant inhibition of cell growth (Palozza et al., 2003b). Interestingly, in all cell lines studied, α-tocopherol and *N*-acetylcysteine inhibited the effects of β-carotene on cell growth and apoptosis, and normalized the increased expression of c-myc induced by the carotenoid, suggesting that a redox regulation of NF-κB induced by β-carotene may be involved in the growth-inhibitory and pro-apoptotic effects of the carotenoid in tumor cells (Palozza et al., 2003b).

In addition, it has been recently reported that lycopene significantly inhibited MMP-9 levels and the binding abilities of NF-κB and stimulatory protein 1 (Spl) in the human hepatoma cell line SK-Hep-1, suppressing the invasive ability of these cells. Such an effect was also accompanied by a decrease in the expression of the insulin-like growth factor-1 receptor (IGF-1R) and in the intracellular level of ROS (Huang et al., 2007). Moreover, it has been also reported that combinations of vitamin D3 and dietary antioxidants, including β-carotene, may be useful in overcoming the differentiation block present in acute promyelocytic HL-60 leukemia cells through a mechanism involving a marked reduction in the nuclear content of NF-κB (Sokoloski et al., 1997). Recent data suggest that carotenoid molecules may represent nontoxic agents for the control of pro-inflammatory genes through a mechanism involving NF-κB. In fact, lycopene prevented macrophage activation induced by gliadin and IF-γ through an inhibition of the activation of NF-κB, interferon regulatory factor-1 and signal transducer and activator of transcription-1α and lowered the levels of both nitric oxide synthase and cyclooxygenase-2 (COX-2) (De Stefano et al., 2007). Similarly, astaxanthin (Lee et al., 2003) and β-carotene (Bai et al., 2005) inhibited the production of inflammatory mediators by blocking NF-κB activation in both LPS-stimulated RAW264.7 cells and primary macrophages. A mechanism of inhibition of NF-κB by lycopene seems to be also involved in the ability of the carotenoid to suppress the LPS-induced maturation of dendritic cells (Kim et al., 2004).

22.3 MODULATION OF ACTIVATED PROTEIN-1 ACTIVATION PATHWAY

Activated protein-1 (AP-1) is another transcription factor that regulates the expression of several genes that are involved in cell differentiation and proliferation. The functional activation of the AP-1 transcription complex is implicated in tumor promotion as well as in malignant transformation. This complex consists of either homo- or heterodimers of the members of the jun and fos family of proteins (Eferl and Wagner., 2003). This AP-1 mediated transcription of several target genes can also be activated by a complex network of signaling pathways that involves external signals such as growth factors, mitogen-activated protein kinases (MAPK), extracellular-signal regulated protein kinases (ERK), and c-Jun N-terminal kinases (JNKs). Some of the target genes that are activated by AP-1 transcription complex mirror those activated by NF-κB and include cyclin D1, Bcl-2, Bcl-xl, VEGF, MMP, and urokinase-type plasminogen activator (uPA). Expression of genes such as MMP and uPA especially promotes angiogenesis and invasive growth of cancer cells. Most importantly, AP-1 can also promote the transition of tumor cells from an epithelial to mesenchymal morphology, which is one of the early steps in tumor metastasis. These oncogenic properties of AP-1 are primarily dictated by the dimer composition of the AP-1 family proteins, and their posttranscriptional and translational modifications.

Recent observations suggest that carotenoids may modulate the AP-1 activation process. It has been recently reported in mammary tumor cell lines that β-carotene and its cleavage products were able to decrease the activation of AP-1 (Tibaduiza et al., 2002). Moreover, lycopene was also shown to downregulate AP-1 in mammary cancer cells (Karas et al., 2000). In addition, a pharmacological dose, but not a physiological dose of β-carotene (Liu et al., 2000) given to ferrets exposed to tobacco smoke caused elevated expression of the AP-1 proteins c-jun and c-fos (Wang et al., 1999). The authors suggested that the activation of AP-1 observed at high doses of the carotenoid may partially explain the increased risk of lung cancer among smokers and asbestos workers, observed in some β-carotene clinical trials (the α-tocopherol, Beta-Carotene Cancer Prevention Study Group, 1994; Omenn et al., 1996).

22.4 MODULATION OF RETINOID RECEPTORS

Several studies indicate that most, if not all, action of retinoic acid (RA) is due to its ability to alter gene transcription through changes in retinoid acid nuclear receptors, namely retinoic acid receptors (RARs) and retinoid X receptors (RXRs). Increasing evidence seems to suggest the possibility that carotenoids can regulate several cell functions through the modulation of these transcription factors. In particular, β-carotene has been reported to work as a chemopreventive agent by upregulating the expression of retinoid receptors in mouse skin (Ponnamperuma et al., 2000). RA concentration and RARβ gene expression, but not expression of RARα and RARγ, has been reported to be reduced by smoke and high concentrations of β-carotene in lung tissue of ferrets (Wang et al., 1999). Interestingly, such an effect was not observed when the carotenoid was given at low doses (Liu et al., 2000) suggesting that only a pharmacological but not a physiological dose of the carotenoid in association with smoke was responsible for changes in the expression of retinoid receptors. Interestingly, in these studies, the down-regulation of RARβ expression by an association of β-carotene at high dose and smoke was associated with an increased cell proliferation (Wang et al., 1999). Possible changes of RARβ isoforms in an AJ-mouse model exposed to the tobacco smoke carcinogen 4-(*N*-methyl-*N*-nitrosamino)-1-(3-pyridyl)-1-butanone (NNK) were also studied (Goralczyk et al., 2005). While NNK reduced the expression of all isoforms, β-carotene alone in non-initiated mice tended to increase RARβ expression, especially RARβ2 and RARβ4. However, in the groups initiated with NNK and supplemented with the carotenoid, the suppressing effect of NNK dominated and β-carotene was not able to restore RARβ expression. In addition, the authors show that the modulation of RA responsive gene expression by NNK and/or β-carotene was not predictive for later tumor development (Goralczyk et al., 2005). It has been hypothesized that lycopene derivatives may also act as ligands for a nuclear receptor in analogy to RA. Consistent with this, synthesized acyclo-retinoids were able to cause transactivation of an RAR reporter gene, comparable to that obtained by RA (Araki et al., 1995). In a study from Ben-Dor et al. (2001) the inhibition of human mammary MCF-7 cancer cell growth and the transactivation of the RAR reporter gene by synthetic acyclo-RA, the open chain analog of RA, was compared to the effects of lycopene and RA in the same systems. Acyclo-RA was remarkably less potent in activating the RA response element than RA (Ben-Dor et al., 2001). Lycopene exhibited only very modest activity in this system. In contrast to the results from the transactivation studies, acyclo-RA, RA, and lycopene inhibited cell growth with a similar potency, suggesting that the effects of acyclo-RAs are not entirely mediated by the RAR. Similar results were obtained by Stahl et al. (2000). These authors demonstrated that RA is much more potent than acyclo-RA in the transactivation of the RA responsive promoters of RAR-β2.

22.5 MODULATION OF PEROXISOME-PROLIFERATOR ACTIVATED RECEPTORS

Peroxisome-proliferator activated receptors (PPARs) are lipid-activated transcription factors exerting several functions in development and metabolism. PPARα is implicated in the regulation of lipid metabolism, lipoprotein synthesis, and inflammatory response in liver and other tissues.

PPARγ plays an important role in the regulation of proliferation and differentiation in several cell types. Until recently, the physiological functions of PPARδ remained elusive. It has been shown that treatment of obese animals by specific PPARδ agonists resulted in normalization of metabolic parameters and reduction of adiposity. The presence of PPARγ receptors in various cancer cells and their activation by fatty acids, prostaglandins, and related hydrophobic agents make these ligand-dependent transcription factors an interesting target for carotenoid derivatives. Moreover, it should be pointed out that in the nucleus, PPARγ is always found as a dimer with RXR. Sharoni et al. (2003) recently studied the efficacy of several carotenoids in transactivation of PPAR response element (PPARE). The results indicate that lycopene, phytoene, phytofluene and β-carotene are able to transactivate PPARE in MCF-7 cells co-transfected with PPARγ. Recently, it has been reported that fucoxanthin, the major carotenoid found in edible seaweed, such as *Undaria pinnatifida* and *Hijikia fusiformis*, enhanced the antiproliferative effect of a PPARγ ligand, troglitazone in CaCo-2 colon cancer cells (Hosokawa et al., 2004).

Moreover, fucoxanthin and fucoxanthinol inhibited the adipocyte differentiation of 3T3-L1 cells through down-regulation of PPARγ (Maeda et al., 2006). Recently, it has been demonstrated that increased PPARγ mRNA and protein levels were implicated, in association with an increased ROS production, in the apoptotic effects of β-carotene in MCF-7 cancer cells. In this cell line, the carotenoid also increased the cyclin-dependent kinase inhibitor p21(WAF1/CIP1) expression and decreased the prostanoid synthesis rate-limiting enzyme COX-2 expression. The authors clearly demonstrated that the addition of 2-chloro-5-nitro-*N*-phenylbenzamide (GW9662), an irreversible PPARγ antagonist, partly attenuated the cell death caused by the carotenoid (Cui et al., 2007). β-Carotene and its metabolites exert a broad range of effects, in part by regulating transcriptional responses through specific nuclear receptor activation. The symmetric cleavage of β-carotene can yield 9-*cis* retinoic acid (9-*cis* RA), the natural ligand for the nuclear receptor RXR, the obligate heterodimeric partner for numerous nuclear receptor family members. A significant portion of β-carotene can also undergo asymmetric cleavage to yield apocarotenals, a series of poorly understood naturally occurring molecules whose biologic role, including their transcriptional effects, remains essentially unknown. Recently, it has been shown that β-apo-14′-carotenal (apo14), but not other structurally related apocarotenals, repressed PPARγ and PPARα responses (Ziouzenkova and Plutzky, 2008). These results also suggest a novel model of molecular endocrinology in which metabolism of a parent compound, β-carotene, may alternatively activate 9-*cis* RA or inhibit apo14 specific nuclear receptor responses.

It has been recently published that lycopene significantly decreased the expression of PPARγ and its target gene, FABP3, in the adrenal glands and kidney, suggesting that this carotenoid may act as a downregulator of PPARγ expression and may play an important role in the modulation of retinoid, and/or lipid metabolism (Zaripheh et al., 2006). Similarly, all-*trans*-RA, a β-carotene metabolite, has been reported to inhibit the expression of PPARγ-2 and that of CCAAT/enhancer binding protein-a in adipose tissue (Yun et al., 2002). Such a repression has been suggested to be mediated by an induction of DEC1/Stra13 (Yun et al., 2002).

It has been shown that the expression of connexin 43 (Cx43) is upregulated by cancer-preventive retinoids and carotenoids, which correlate with the suppression of carcinogen-induced transformation in 10T1/2 cells. Recently, it has been reported that Cx43 induction by astaxanthin, but not by a RAR-specific retinoid, was inhibited by GW9662, a PPARγ antagonist (Bertram et al., 2005).

22.6 MODULATION OF THE ANTIOXIDANT RESPONSE ELEMENT

Induction of phase II enzymes, which conjugate reactive electrophiles and act as indirect antioxidants, appears to be the means for achieving protection against a variety of carcinogens in animals and humans. Transcriptional control of the expression of these enzymes is mediated, at least in part, through antioxidant response element (ARE) found in the regulatory regions of their genes. The transcription factors NrF2, which bind to ARE, seem to be essential for the induction of phase II enzymes, such as glutathione S-transferases, NAD(P)H quinone oxidoreductase (NQO1), as well as heme oxygenase-1 (HO-1) and the thiol containing reducing factor thioredoxin. Carotenoids have been shown to modulate tumor growth acting as potent inducers of these enzymes (Ben-Dor et al., 2005). β-Carotene has been found to modulate the expression of HO-1, decreasing it, as observed in cultured FEK4 cells (Trekli et al., 2003) or fibroblasts (Offord et al., 2002) exposed to UVA or increasing it, as observed in human skin fibroblasts enriched with the carotenoid and exposed to UV-light (Obermuller-Jevic et al., 2001). In this study, the prooxidant effects of β-carotene were totally suppressed by vitamin E, but only moderately by vitamin C (Obermuller-Jevic et al., 2001). The modulation of this enzyme may occur through an activation of MAPK leading to induction of ARE, as suggested for other dietary chemopreventive compounds (Owuor et al., 2002). An alternative mechanism to explain the regulation of HO-1 expression by β-carotene has been recently suggested (Palozza et al., 2005b). In this study, the carotenoid controlled HO-1 expression through the induction of Bach1, known to act as a HO-1 repressor gene, in fibroblasts exposed to cigarette smoke condensate (Palozza et al., 2006b).

Several lines of evidence suggest that lycopene also acts as an inducer of the activity and/or of the expression of phase II enzymes in healthy animals (Breinholt et al., 2000) as well as in animals bearing tumors, including gastric (Velmurugan et al., 2002) and DMBA-induced hamster buccal pouch (Bhuvaneswari et al., 2002) tumors. At the same time, enzymes of oxidative defense were induced and lipid peroxidation was reduced (Bhuvaneswari et al., 2002, Velmurugan et al., 2002) by the carotenoid.

It has been recently reported that in transiently transfected cancer cells lycopene transactivated the expression of reporter genes fused with ARE sequences. Other carotenoids such as phytoene, phytofluene, β-carotene, and astaxanthin had a much smaller effect. An increase in protein as well as mRNA levels of the phase II enzymes NQO1 and γ-glutamylcysteine synthetase was observed in nontransfected cells after carotenoid treatment. The potency of the carotenoids in ARE activation did not correlate with their effect on intracellular reactive oxygen species and reduced glutathione level, which may indicate that ARE activation is not solely related to their antioxidant activity. The increase in phase II enzymes was abolished by a dominant-negative Nrf2, suggesting that carotenoid induction of these proteins depends on a functional Nrf2 and the ARE transcription system (Ben-Dor et al., 2005).

22.7 MODULATION OF XENOBIOTIC AND OTHER ORPHAN NUCLEAR RECEPTORS

Orphan receptors are structurally related to nuclear hormone receptors but lack known physiological ligands. Xenobiotic receptors represent a family of orphan receptors and make up part of the defense mechanism against foreign lipophilic chemicals (xenobiotics). They include the steroid and xenobiotic receptor/pregnane X receptor (PXR), constitutive androstane receptor, and the aryl hydrocarbon receptor. These receptors respond to a wide variety of drugs, environmental pollutants, carcinogens, dietary and endogenous compounds and regulate the expression of cytochrome P450 (CYP) enzymes, conjugating enzymes and transporters involved in the metabolism and elimination of xenobiotics. It has been reported that carotenoids may modulate the expression of detoxification enzymes (Paolini et al., 1999, 2001; Perocco et al., 1999). Recently, it has been shown that β-carotene can act as an inducer of several carcinogen-metabolising enzymes in the lung of Sprague-Dawley rats (Paolini et al., 2001). They include CYP1A1/2, CYP3A, CYP2B1, and CYP2A. Such inductions have been associated to an overgeneration of reactive oxygen centered radicals (Paolini et al., 2001). In addition, many tobacco smoke pro-carcinogens are themselves CYP inducers and could act in a synergistic way with β-carotene or with some of its oxidation products, such as β-apo-8′-carotenal, further contributing to the overall carcinogenic risk (Wang et al., 1999). Moreover, induction of transformation by benzo[a]pyrene and cigarette smoke condensate in BALB/c 3T3 cells was markedly enhanced by the presence of β-carotene in either acute or chronic treatment (Perocco et al., 1999). Such an enhancement has been related to the boosting effect of the carotenoid on P450 apparatus (Perocco et al., 1999). β-Carotene has been also reported to enhance ethanol-hepatotoxicity by an induction of CYP2E1 and CYP4A1 in both rodents and nonhuman primates (Kessova et al., 2001).

Other carotenoids, such as canthaxanthin and astaxanthin have been recognized as potent inducers of CYP1A1 and 1A2 in rat liver (Gradelet et al., 1998). The administration of lycopene to rats was shown to induce liver CYP types 1A1/2, 2B1/2, and 3A in a dose-dependent manner (Breinholt et al., 2000). The observation that these enzymatic activities were induced at very low lycopene plasma levels suggests that modulation of drug metabolising enzymes by carotenoids may be relevant to humans (Breinholt et al., 2000).

Recently, β-carotene has been shown to act as an activator of phase I enzymes in the human liver via PXR-mediated mechanism (Ruhl, 2005).

22.8 MODULATION OF P53

p53 is a tumor-suppressor and transcription factor. It is a critical regulator in many cellular processes including cell signal transduction, cellular response to DNA-damage, genomic stability, cell cycle control, and apoptosis. The protein activates the transcription of downstream genes such as p21WAF1 and Bax to induce the apoptotic process, inhibiting the growth of cells with damaged DNA or cancer cells (el-Deiry et al., 1993; Vogelstein and Kinzler, 1992). Mutant p53 loses its ability to bind DNA effectively, and as a consequence the p21 protein is not made available to regulate cell division. Thus, cells divide uncontrollably and form tumors.

At the moment, the role of p53 and its related genes in the regulation of cell growth signaling by carotenoids is not well understood and the results appear controversial. It is possible that several factors may influence the modulatory effects of β-carotene and other carotenoids on p53 levels, including the concomitant presence of smoke, the type and the concentration of the carotenoid, the association with other antioxidants as well as the biological cellular environment.

Liu et al. (2004) found that the combined presence of smoke and β-carotene can modify the levels as well as the phosphorylation of p53, JNK, and p38 MAP kinase in lungs of ferrets depending on the dose of the carotenoid. While high concentrations were responsible for an increase of these parameters, low concentrations decreased them (Liu et al., 2004). In different cell models, including RAT-1 immortalized fibroblasts, Mv1Lu lung, MCF-7 mammary, Hep-2 larynx and LS-174 colon cancer cells, exposed to an association of cigarette smoke condensate (TAR) and β-carotene, the concomitant presence of the carotenoid and TAR increased the levels of 8-OHdG, a biomarker of oxidative DNA damage and an index of mutagenesis and carcinogenesis. Such an effect was strictly accompanied by an increased number of proliferating cells, due to a deregulation of p53 expression, affecting the levels of p21WAF1 and cyclin D1 (Palozza et al., 2004a). Interestingly, in contrast to β-carotene, the arrest of cell cycle progression by lycopene in RAT-1 fibroblasts exposed to TAR did not involve p53 pathway (Palozza et al., 2005b). In such cells lycopene counteracted the effects of TAR on p53 expression by significantly decreasing it. Such a finding is not surprising in view of the fact that p53 responds to stress signals that can cause oncogenic alterations, such as DNA damage. Cells increased their content of 8-OHdG as a consequence of smoke exposure and this results in an increased expression of p53. Given its antioxidant function, lycopene may protect cells against TAR-induced DNA oxidation. Similarly, smoke-elevated total p53 and phosphorylated p53 in gastric mucosa of ferrets were markedly attenuated by lycopene, administered at both low and high doses (Liu et al., 2006).

Interestingly, the combined supplementation of β-carotene, α-tocopherol, and ascorbic acid has been reported to be protective against 4-(methylnitrosamino)-1-(3-pyridyl)-1-butanone (NNK)-induced lung carcinogenesis in smoke-exposed ferrets through maintaining normal tissue levels of RA, inhibiting the activation of MAPK pathways, cell proliferation, and phosphorylation of p53 (Kim et al., 2006).

Another factor responsible for regulating the levels of p53 by β-carotene could be the dose employed. At high carotenoid concentrations, an increase in p53 expression was observed in SCC cells (Schwartz, 1993) and in HL-60 cells (Palozza et al., 2002b). In HL-60 cells, the treatment with the carotenoid induced a remarkable increase in ROS production, accompanied by an enhanced expression of p21WAF1 and by a concomitant arrest of cell cycle at the G0/G1 phase (Palozza et al., 2002b). An arrest of cell cycle, accompanied by apoptosis induction, was also observed following dietary supplementation with lutein (Chew et al., 2003). The inhibition of mouse mammary tumor growth by lutein was also supported by the observed increase in the expression of p53 and Bax induced by the carotenoid (Chew et al., 2003).

Interestingly, while it has been reported that the inhibition of cell growth by carotenoids in colon (Palozza et al., 2001b, 2007a) as well as in prostate (Williams et al., 2000) adenocarcinoma cancer cells was independent of p53 and p21 status, HL-60 cells increased their p21 expression as a consequence of the treatment with β-carotene (Palozza et al., 2002b). In addition, the antiproliferative effects of β-carotene required p21 expression in human fibroblasts (Stivala et al., 2000). In contrast, mammary and endometrial cancer cells decreased p21 levels, following lycopene treatment (Nahum et al., 2001).

22.9 MODULATION OF MITOGEN-ACTIVATED PROTEIN KINASES

In addition to NF-κB and Akt pathways, MAPK pathway has received increasing attention as a target molecule for cancer prevention and therapy. The MAPK cascades include ERKs, JNKs/stress-activated protein kinases (SAPKs), and p38 kinases. ERKs are believed to be strongly activated and to play a critical role in transmitting signals initiated by growth-inducing tumor promoters, including 12-*O*-tetradecanoyl-phorbol-13-acetate, epidermal growth factor (EGF), and platelet-derived growth factor (PDGF) (Cowley et al., 1994; Minden et al., 1994). On the other hand, stress-related tumor promoters, such as UV irradiation and arsenic, potently activate JNKs/SAPKs and p38 kinases (Bode and Dong, 2000, 2003, Kallunki et al., 1994). The MAPK pathway consists of a cascade in which a MAP3K activates a MAP2K that activates a MAPK (ERK, JNK, and p38), resulting in the activation of NF-κB, cell growth, and cell survival (Seger and Krebs, 1995).

A recent observation shows that β-carotene was able to counteract the dangerous effect of 7-ketocholesterol in human macrophages by limiting the apoptotic processes; reducing the intracellular ROS production; and inhibiting the phosphorylation of p38, JNK, and ERK1/2 induced by the oxysterol (Palozza et al., 2007b).

22.10 MODULATION OF CELL-CYCLE-RELATED PROTEINS

Several proteins are known to regulate the timing of the events in the cell cycle. The loss of this regulation is the hallmark of cancer. Major control switches of the cell cycle are the cyclins and the cyclin-dependent kinases. Cyclin D1, a component subunit of cyclin-dependent kinase (CDK)-4 and CDK6, is a rate-limiting factor in progression of cells through the first gap (G1) phase of the cell cycle (Baldin et al., 1993). Dysregulation of the cell cycle check points and overexpression of growth-promoting cell cycle factors such as cyclin D1 and CDKs are associated with tumorigenesis (Diehl et al., 2002).

Increasing evidence demonstrates that carotenoids are able to regulate cell growth by controlling cell cycle progression, and by interfering with these cell cycle regulatory pathways. It has been reported that lycopene may act as a potent modulator of cyclin D1 in both normal and tumor cultured cells. Moreover, a modulation of cyclin D1 by lycopene in cells exposed *in vitro* and *in vivo* to cigarette smoke has been recently demonstrated. Lycopene was able to inhibit the growth of mammary (MCF-7 and T-47D) and endometrial (ECC-1) cancer cells through down-regulation of cyclin D1 and cyclin D3. The reduction in cyclin D1 levels by lycopene has been suggested to have two consequences. The first one is a direct effect causing reduction in cyclin D-CDK4 complexes resulting in a decrease of both CDK4 and CDK2 kinase activity and in a reduction of the hypophosphorylation of pRb. The second one is a retention of p27 in cyclin E-CDK2 complexes, an indirect effect that leads to the inhibition of CDK2 activity (Nahum et al., 2001).

In a recent study, LNCaP and PC3 prostate cancer cells treated with lycopene-based agents have been reported to undergo mitotic arrest. Lycopene's antiproliferative effects were likely achieved through a block in G1/S transition mediated by decreased levels of cyclins D1 and E and cyclin-dependent kinase4 and suppressed retinoblastoma phosphorylation (Ivanov et al., 2007).

We recently reported that tomato added to cultured colon (HT-29 and HCT-116) cancer cells by an *in vitro* digestion procedure was able to induce an arrest of cell cycle progression at the G0/G1 phase (Palozza et al., 2007a). Such an effect was accompanied by a dose-dependent decrease in the expression of cyclin D1. Although tomato digestates contain a complex mix of compounds besides lycopene, including a large variety of micronutrients and microcostituents, such as polyphenols and other non provitamin A carotenoids, this observation seems to support the notion that lycopene may be a molecule that is extremely important in the regulation of intracellular levels of cyclin D.

Similarly, lycopene was also able to inhibit cell cycle progression at the G0/G1 phase and to reduce cell proliferation by a mechanism involving cyclin D1 in normal cells. It has been reported that, after the stimulation of synchronized human normal prostate epithelial cells with growth factors, cyclin D1 protein expression increases in lycopene-untreated cells. Such an increase was lower or even absent following treatment with lycopene at the concentration of 0.5 mmol/L and 5.0 mmol/L, respectively. Interestingly, it was specific for cyclin D1, since cyclin E levels remained constant and were unaffected by lycopene treatment (Obermüller-Jevic et al., 2003).

Moreover, we recently reported that lycopene was able to enhance the arrest of cell cycle progression induced by TAR in RAT-1 immortalized fibroblasts. TAR-exposed cells treated with lycopene showed a delay in cell cycle at the G0/G1 phase and a concomitant reduction in S phase. Such effects were accompanied by a dose-dependent decrease in cyclin D1 levels. On the other hand, fibroblasts treated with lycopene alone showed the same effects, although to a lower extent. The down-regulation of cyclin D1 observed in this study was dose-dependent and occurred at lycopene concentration achievable *in vivo* after carotenoid supplementation (Palozza et al., 2005b).

In accord with these *in vitro* studies, treatment with lycopene *in vivo* has been also reported to induce modulatory effects on cyclin D1 expression. It has been reported that smoke exposure substantially decreased the levels of $p21^{Waf1/Cip1}$and increased those of cyclin D1 and proliferating cellular nuclear antigen (PCNA) in gastric mucosa from ferrets. Supplementation of ferrets with either low or high doses of lycopene prevented the changes in $p21^{Waf1/Cip1}$, cyclin D1, and PCNA caused by smoke exposure in a dose-dependent fashion (Liu et al., 2006).

Although further studies are needed to clarify the mechanism(s) of lycopene interference with cell signaling leading to down-regulation of cyclin D1 and ultimately to cell cycle arrest in both normal and tumor cells, these reports suggest that the reduction in cyclin D1 by lycopene treatment may be a key event in the ability of the carotenoid to arrest cell cycle progression.

The regulation of cell cycle-related proteins by other carotenoids is less investigated.

In a recent study, the antiproliferative effect of different carotenoids, including β-carotene, lycopene and lutein, on PCNA and cyclin D1 expression in human KB cells have been studied. The results indicate that carotenoids suppressed cell growth by acting as inhibitors of the expressions of PCNA and cyclin D1, although in a different extent (Cheng et al., 2007). On the other hand, β-carotene was able to induce a cell cycle delay in G2/M phase by decreasing the expression of cyclin A in human colon adenocarcinoma cells (Palozza et al., 2002a).

Excentric cleavage products of β-carotene inhibited the growth of estrogen receptor positive and negative breast cancer cells through the down-regulation of cell cycle regulatory proteins, such as E2F1 and Rb and through the inhibition of AP-1 transcriptional activity (Tibaduiza et al., 2002).

Recently, fucoxanthin has been shown to inhibit the proliferation of human colon cancer cells by a mechanism involving an up-regulation of p21WAF1/Cipl (Das et al., 2005).

22.11 MODULATION OF APOPTOSIS-RELATED PROTEINS

Apoptosis helps to establish a natural balance between cell death and cell renewal in mature animals by destroying the excess, damaged, or abnormal cells. However, the balance between survival and apoptosis often tips toward the former in cancer cells. In fact, the intricate network of apoptotic signaling is often dysregulated in cancer cells. One example of this is the up-regulation of the anti-apoptotic protein Bcl-2 in human colon cancer cells. The Bcl-2 protein family is important for the regulation of apoptosis and it includes anti-apoptotic members such as Bcl-2 and Bcl-xl, as well as the pro-apoptotic members such as Bax, Bak, Bad (Bcl-2/Bcl_{XL}-antagonist, causing cell death), and Bid. Their main site of action is believed to be the mitochondria, where they facilitate or impede the release of cytochrome c.

Several studies show that carotenoids are able to act as apoptosis inducers by modulating different molecular pathways involved in the apoptotic process, as recently reviewed (Palozza et al., 2004a).

22.11.1 Bcl-2 Family Proteins

It has been demonstrated that β-carotene is able to decrease the expression of Bcl-2 and Bcl_{XL} in colon cancer cells (Palozza et al., 2001b) and to diminish that of Bcl-2 in HL-60 cells (Palozza et al., 2002b). Such effects were strictly related to apoptosis induction and to ROS production by the carotenoid. This finding is particularly interesting in the light of the data supporting a role for Bcl-2 in an antioxidant pathway, whereby this protein prevents programmed cell death by decreasing formation of reactive oxygen species and lipid peroxidation products (Kane et al., 1993).

In addition, recent data suggest that carotenoids modulate Bid (Palozza et al., 2006a; Prasad et al., 2006), Bad (Liu et al., 2003), Bcl-xl (Prasad et al., 2006), and Bax (Palozza et al., 2004a) expression in different experimental models. In particular, it has been recently shown that the increase in Bax expression and in apoptosis induction caused by cigarette TAR was prevented by the addition of β-carotene in RAT-1 fibroblasts as well as in different tumor cell lines (Palozza et al., 2004a). In addition, recently, it has been reported that lycopene can induce apoptosis of PC-3 cells, downregulating the expression of cyclin D1 and Bcl-2 and upregulating that of Bax (Wang et al., 2007).

22.11.2 Caspase Cascade

The mechanism(s) of caspase cascade activation during β-carotene-induced apoptosis has been recently investigated in tumor cells (Palozza et al., 2003a). β-carotene-induced apoptotic pathway requires activation of caspase-3, which has been defined as a key player in apoptosis induced by many stimuli and is also necessary for the nuclear changes associated with apoptosis, such as chromatin condensation. The carotenoid can initiate caspase-3 cascade mainly by interacting with a signal complex on cell membrane, which induces caspase-8 activation, and then by operating through a non-receptor signaling pathway within cytoplasm, which induces caspase-9 activation (Palozza et al., 2003a). How caspase-8 can be activated by β-carotene is not clear. The authors suggest different hypotheses. It is possible that β-carotene activates TNF/Fas ligand system/receptor. Moreover, it is known that ROS may be implicated in the activation of caspase-8 and the carotenoid has been reported to act as a redox agent. Finally, the carotenoid may induce conformational changes of some membrane-associated "death" receptors, resulting in the activation of caspase-8. A direct activation of caspase-9 via caspase-8 by β-carotene has also been reported in tumor cells. The carotenoid increased the expression of the truncated form of Bid, which translocates to the mitochondria and acts as a potent inducer of apoptosis, by releasing cytochrome c and activating caspase-9 (Palozza et al., 2003a).

Moreover, a recent study also revealed that ROS generation led to the activation of caspase-2 during β-carotene-induced apoptosis in the human leukemic T cell line Molt 4. The apoptosis progressed by simultaneous activation of caspase-8 and caspase-9, and a cross talk between these initiator caspases was mediated by the pro-apoptotic protein Bid. Inhibition of caspases 2, 8, 9, and 3 independently suppressed the caspase cascade. The cleavage of the anti-apoptotic protein BclXL was found to be another important event during β-carotene-induced apoptosis, suggesting the presence of an extensive feedback amplification loop in β-carotene-induced apoptosis (Prasad et al., 2006).

22.11.3 Mitochondrial Proteins

The involvement of mitochondria in the pro-apoptotic effects of carotenoids has been clearly demonstrated by the fact that β-carotene induces the release of cytochrome c from mitochondria and alters the mitochondrial membrane potential ($\Delta\psi m$) in different tumor cells (Palozza et al., 2003a). Moreover, the highly polar xanthophyll neoxanthin has been reported to induce apoptosis in colon cancer cells by a mechanism that involves its accumulation into the mitochondria and a consequent loss of mitochondrial transmembrane potential and releas of cytochrome c and apoptosis-inducing factor (Terasaki et al., 2007).

22.11.4 Cyclooxygenase

Numerous preclinical studies point out the importance of regulation of cyclooxygenase-2 (COX-2) expression in the prevention and, most importantly, in the treatment of several malignancies. This enzyme is overexpressed in practically every premalignant and malignant conditions involving colon, liver, pancreas, breast, lung, bladder, skin, stomach, head and neck and esophagus cancers (Eberhart et al., 1994). Overexpression of this enzyme is a consequence of a deregulation of transcriptional and/or posttranscriptional control. Several growth factors, cytokines, oncogenes, tumor promoters stimulate COX-2 transcription. Expression of COX-2 is increased in HER2/neu expressing breast carcinomas owing to enhanced *Ras* signaling. Depending upon the stimulus and the cell type, different transcription factors, including AP-1, NF-IL-6, NF-κB, can stimulate COX-2 transcription. One of the possible mechanisms by which COX-2 can induce tumorigenesis is through its ability to act as an anti-apoptotic gene. A recent study in our laboratory shows that β-carotene is able to downregulate the expression of COX-2 in colon cancer cells and such an effect was accompanied by apoptosis induction (Palozza et al., 2005a). This observation is particularly interesting in view of the fact that COX-2 expression is regulated by PPARγ and PPAR receptors have been suggested to be modulated by carotenoids (Sharoni et al., 2002).

22.12 MODULATION OF DIFFERENTIATION-RELATED PROTEINS

The induction of differentiation may be an effective mechanism for chemoprevention of chronic diseases. It has been reported that lycopene alone induced differentiation of HL-60 promyelocytic leukemia cells (Amir et al., 1999). A similar effect was also observed for other carotenoids, including β-carotene and lutein (Liu et al., 1997; Sokoloski et al.,1997).

The differentiation effect of lycopene was associated with elevated expression of several differentiation-related proteins, such as cell surface antigen (CD14), oxygen burst oxidase and chemotactic peptide receptors (Amir et al., 1999). Recently, it has also been reported that lycopene was also able to stimulate the differentiation marker alkaline phosphatase activity in

SaOS-2 osteoblasts (Kim et al., 2003). Such an effect was shown to be dependent on the stage of cell differentiation. The mechanism of the differentiating activity of lycopene is still unclear. One of the most reliable hypothesis is that the carotenoid may activate the expression of nuclear hormone receptors, such as RAR and RXR (Sharoni et al., 2002).

22.13 MODULATION OF GROWTH FACTORS

Growth factors are proteins that bind to receptors on the cell surface, with the primary result of activating cellular proliferation and/or differentiation. Some of the growth factors implicated in carcinogenesis are EGF, PDGF, fibroblast growth factors, transforming growth factors TGF-α, TGF-β, erythropoietin, IGF, IL-1, IL-2, IL-6, IL-8, TNF, interferon-g, and colony-stimulating factors (CSFs). The potent cell proliferation signals generated by various growth factor receptors such as the EGF receptor, IGF-1R, and VEGF-receptor networks constitute the basis for receptor-driven tumorigenicity in the progression of several cancers (Choi et al., 2001).

Abnormal growth factor signaling pathways lead to increased cell proliferation, suppression of apoptotic signals, and invasion contributing to metastasis.

22.13.1 IGF and PI3K/Akt Pathways

The IGFs play a pivotal role in regulating mitogenic and apoptotic pathways (Yu and Rohan , 2000). Several lines of evidence implicate IGF-1 and its receptor, IGF-IR, in lung cancer and other malignancies (Giovannucci, 1999; Yu and Rohan, 2000). IGF has been reported to be one of the most important factors that activates PI3K/Akt pathway. The over-expression of Akt isoforms has been described in breast, colon, ovarian, pancreatic, prostate, and bile duct cancers (Altomare et al., 2003; Kandel and Hay, 1999; Roy et al., 2002; Tanno et al., 2004; Vivanco and Sawyers, 2002). Moreover, Akt activity promotes resistance to chemo- and radiotherapy (Brognard et al., 2001; Clark et al., 2002; Tanno et al., 2004). The mechanism of Akt activation remains a target for researchers in the hunt for suitable options for cancer therapy.

Available data suggest that lycopene and tomato products are able to modulate IGF-1 pathway *in vitro* as well as *in vivo* models. Clinical data shows that higher intake of cooked tomatoes or lycopene is significantly associated with lower circulating levels of IGF-1 (Mucci et al., 2001) and higher levels of insulin-like growth factor–binding protein-3 (IGFBP-3) (Holmes et al., 2002). Moreover, it has been recently demonstrated that, in the MatLyLu Dunning prostate cancer model, IGF-1 expression was decreased locally in prostate tumors by lycopene supplementation (Siler et al., 2004). In addition, lycopene treatment has been reported to strongly reduce the IGF-1 stimulation of activator protein 1 binding in MCF-7 mammary cancer cells and such an effect was associated with a delayed Gl-S cell cycle progression (Karas et al., 2000). The attenuation of cyclin Dl levels by lycopene seems to be an important mechanism for the reduction of the mitogenic action of IGF-I1 (Nahum et al., 2006).

Lycopene has been also reported to significantly increase the levels of IGFBP-3 in PC-3 prostate cancer cells. Such an inhibition was accompanied by a decrease in cell proliferation and by an increase in apoptosis induction (Kanagaraj et al., 2007). Interestingly, in a recent study, the arrest of cell cycle and the decrease in cyclins D1 and E induced by lycopene in androgen-responsive LNCaP and androgen-independent PC3 prostate cancer cells correlated with decreased IGF-1R expression and activation, increased IGF binding protein-2 expression and decreased Akt activation, further confirming that lycopene can suppress PI3K-dependent proliferative and survival signaling (Du et al., 1998).

Two recent studies seem to demonstrate that lycopene is able to counteract the dangerous effects of smoke by acting through a mechanism involving IGF-1 and/or Akt pathways. In the lung of ferrets, it has been shown that cigarette smoke-induced lesions (e.g., squamous metaplasia,

PCNA over-expression, and diminished apoptosis) were associated with reduced plasma IGFBP-3 concentrations and increased IGF-1/IGFBP-3 ratios. Such changes significantly affected the status of cell proliferation and apoptosis in the lung of ferrets. Smoke exposure significantly decreased cleaved caspase-3 protein and increased PCNA. Furthermore, smoke exposure suppressed Bad-mediated apoptosis by inducing the phoshorylation of Bad at both Ser136 and Ser112. These smoke-induced changes were prevented by lycopene supplementation in a dose-dependent manner. The carotenoid was able to increase IGFBP-3 levels and decrease IGF-1/IGFBP-3 ratio. Moreover, it decreased Bad phosphorylation at both Ser136 and Ser112 and increased cleaved caspase-3, preventing cigarette smoke-induced squamous metaplasia and the increase in PCNA (Liu et al., 2003).

A recent *in vitro* study also suggests that the modulation of Akt pathway may have a key role in the pro-apoptotic effects of lycopene under smoke conditions (Palozza et al., 2005b). In fact, while RAT-1 fibroblasts exposed to cigarette smoke condensate (TAR) exhibited high levels of phosphorylated Akt, cells exposed to a combination of TAR and lycopene strongly decreased them. Moreover, the exposition of RAT-1 fibroblasts to TAR alone suppressed Bad-mediated apoptosis by inducing the phosphorylation of Bad at Ser136. Conversely, lycopene was able to completely prevent the phosphorylation of Bad induced by TAR, confirming *in vitro* the results obtained *in vivo* by Liu et al. (2003). In our laboratory, similar results have been recently found in the human prostate DU-145 cancer cells exposed to lycopene in association with TAR. The carotenoid was able to prevent TAR-induced Akt and Bad phosphorylation. Moreover, in the same study, the expression of the heat shock protein, Hsp90, was increased following TAR exposure (Palozza et al., 2005a). Such an increase was counteracted by lycopene. This finding is particularly interesting in view of a previous report showing that Hsp90 maintains Akt activity by binding to Akt and by preventing PP2A-dependent dephosphorylation of Akt (Sato et al., 2000). Moreover, Hsp90 has been reported to prevent proteasome-dependent degradation of PDK1, which is known to activate Akt (Fujita et al., 2002). On the other hand, the finding that lycopene is able to counteract the effect of TAR on Hsp90 is not surprising in view of the fact that heat shock proteins increase as a consequence of oxidative stress, including smoke (Pinot et al., 1997) and that lycopene acts as a potent antioxidant (Conn et al., 1991; Di Mascio et al., 1989). The modulation of Hsp90 by lycopene under smoke conditions could be a further suggestive intracellular mechanism to explain the modulatory activity of lycopene on Bad.

22.13.2 Platelet-Derived Growth Factor-BB

In a recent study, it has been found that lycopene inhibited PDGF-BB–induced signaling and cell migration in human cultured skin fibroblasts through a novel mechanism of action, i.e., direct binding to PDGF-BB. The trapping of PDGF by lycopene also compromised melanoma-induced fibroblast migration and attenuated signaling transduction in fibroblasts simulated by melanoma-derived conditioned medium, suggesting that lycopene may interfere with tumor–stroma interactions. The trapping activity of lycopene on PDGF suggests that it may act as an inhibitor on stromal cells, tumor cells and their interactions, which may contribute to its antitumor activity (Wu et al., 2007).

22.14 MODULATION OF HORMONES

Androgens are implicated in prostatic neoplasia, including benign prostatic hyperplasia and prostate cancer. Studies in prostate suggest that 5-α-dihydrotestosterone is the principal androgen responsible for both normal and hyperplastic growth of the prostate gland. 5-α-dihydrotestosterone is produced from testosterone by steroid 5-α-reductase. It has been recently reported that lycopene reduced the expression of 5-α-reductase I in prostate tumors in the rat MatLyLu Dunning prostate cancer model (Siler et al., 2004). As a consequence of this down-regulation, several androgen target genes were drastically downregulated, including cystatin-related protein 1 and 2, prostatic spermine-binding protein, prostatic steroid-binding protein C1, C2, and C3 chain, and probasin (Siler et al., 2004). Moreover, the carotenoid was found to affect androgen signaling in normal prostatic tissues from young rats (Herzog et al., 2005).

In a recent study aimed to determine whether carotenoids are able to inhibit the signaling of steroidal estrogen and phytoestrogen in breast (T47D and MCF-7) and in endometrial (ECC-1) cancer cells, lycopene, phytoene, and phytofluene have been found to inhibit estrogen-induced transactivation of ERE that was mediated by both estrogen receptors (ERs) ERα and ERβ. These data suggest that these compounds may be possible candidates to inhibit the deleterious effect of both 17β-estradiol and genistein in hormone-dependent mammary and endometrial malignancies (Hirsch et al., 2007).

22.15 MODULATION OF GAP JUNCTION COMMUNICATION PROTEINS

Gap junction proteins play a key role in the maintenance of tissue homeostasis and the loss of gap-junctional communication (GJC) may be important for malignant transformation and its restoration may reverse the malignant process (Hotz-Wagenblatt and Shalloway, 1993). One of the first observations regarding the effects of carotenoids on the modulation of protein expression was made by Zhang et al. (1991, 1992) with the finding that carotenoids were able to increase GJC and induce the synthesis of Cx43, a component of the gap junction structure. The expression of Cx43 has been reported to be upregulated by both retinoids and carotenoids and it was found to correlate with the suppression of carcinogen-induced transformation in 10T1/2 cells. However, in this cell model, the molecular mechanisms for this up-regulation seem to be different between provitamin A carotenoids and non-provitamin A carotenoids, as recently discussed (Vine and Bertram, 2005). In fact, the RAR antagonist Ro 41-5253 suppressed retinoid-induced connexin-43 protein expression but did not suppress protein expression induced by the non-provitamin A carotenoid, astaxanthin. On the other hand, connexin-43 induction by astaxanthin, but not by a RAR-specific retinoid, was inhibited by GW9662, a PPARγ antagonist (Vine and Bertram, 2005).

Although the influence of lycopene and proliferation of carcinoma cells appears not limited to its ability to modulate Cx43 expression, lycopene as well as its oxidation products have been reported to enhance GJC in cultured cells (Livny et al., 2002; Stahl et al., 2000). Recent data indicate that lycopene may indeed increase connexin-43 expression in human prostate (Kucuk et al., 2001).

22.16 CONCLUSIONS

This chapter has summarized some of the recent observations on the modulatory effects of β-carotene and other carotenoids on molecular pathways involved in cell proliferation, differentiation, and apoptosis in animal models and in cultured cells. The studies provide evidence that carotenoids influence several cellular and molecular processes that can be implicated in the effects of these molecules in human health. However, it is difficult at the moment to directly relate the available experimental data to human pathophysiology. This is due to the lack of adequate *in vitro* methods of delivering carotenoids to cells, high carotenoid concentrations used in some studies, which are not achievable *in vivo* in human plasma, as well as due to the difficulty in determining sensitive *in vivo* markers of long-term health effects at an early stage. Moreover, synergistic interactions may occur *in vivo*, which may be related to the different dietary compounds to modulate a network of proteins and transcription factors. In addition, several changes on cell signaling may be mediated *in vivo* by carotenoid derivatives rather than to the intact carotenoid molecules themselves. Finally, inter-individual polymorphisms can mask the response to carotenoids and thereby complicate this undertaking to an even greater extent. Nevertheless, understanding the effects of carotenoids on cell signaling is fundamental to improve the knowledge of strategies in the prevention of chronic diseases.

REFERENCES

Altomare, D.A., Tanno, S., De Rienzo, A. et al. 2003. Frequent activation of AKT2 kinase in human pancreatic carcinomas. *J. Cell. Biochem.* 88:470–476.

Amir, H., Karas, M., Giat, J. et al. 1999. Lycopene and 1,25-dihydroxyvitamin D3 cooperate in the inhibition of cell cycle progression and induction of differentiation in HL-60 leukemic cells. *Nutr Cancer* 33:105–112.

Araki, H., Shidoji, Y., Yamada, Y., Moriwaki, H., and Muto, Y. 1995. Retinoid agonist activities of synthetic geranyl geranoic acid derivatives. *Biochem Biophys Res Commun* 209:66–72.

Bai, S.K., Lee, S.J., Na, H.J. et al. 2005. Beta-carotene inhibits inflammatory gene expression in lipopolysaccharide-stimulated macrophages by suppressing redox-based NF-kappaB activation. *Exp Mol Med.* 37:323–334.

Baldin, V., Lukas, J., Marcote, M.J., Pagano, M., and Draetta, G. 1993. Cyclin D1 is a nuclear protein required for cell cycle progression in G1. *Genes Dev* 7:812–821.

Ben-Dor, A., Nahum, A., Danilenko, M. et al. 2001. Effects of acyclo-retinoic acid and lycopene on activation of the retinoic acid receptor and proliferation of mammary cancer cells. *Arch Biochem Biophys* 391:295–302.

Ben-Dor, A., Steiner, M., Gheber, L. et al. 2005. Carotenoids activate the antioxidant response element transcription system. *Mol Cancer Ther* 4:177–186.

Bertram, J.S. and Vine, A.L. 2005. Prevention by retinoids and carotenoids: Independent action on a common target. *Biochim Biophys Acta* 1740:170–178.

Beta-Carotene Cancer Prevention Study Group. The Alpha-tocopherol. 1994. The effect of vitamin E and beta carotene on the incidence of lung cancer and other cancers in male smokers. *N Engl J Med* 330:1029–35.

Bhuvaneswari, V., Velmurugan, B., and Nagini, S. 2002. Induction of glutathione-dependent hepatic biotransformation enzymes by lycopene in the hamster cheek pouch carcinogenesis model. *J Biochem Mol Biol Biophys* 6:257–260.

Bode, A.M. and Dong, Z. 2000. Signal transduction pathways: targets for chemoprevention of skin cancer. *Lancet Oncol* 1:181–188. Review.

Bode, A.M. and Dong, Z. 2003. Mitogen-activated protein kinase activation in UV-induced signal transduction. *Sci STKE* (167):RE2.

Breinholt, V., Lauridsen, S.T., Daneshvar, B., and Jakobsen, J. 2000. Dose-response effects of lycopene on selected drug-metabolizing and antioxidant enzymes in the rat. *Cancer Lett* 154:201–210.

Brognard, J., Clark, A.S., Ni, Y., and Dennis. P.A. 2001. Akt/protein kinase b is constitutively active in non-small cell lung cancer cells and promotes cellular survival and resistance to chemotherapy and radiation. *Cancer Res* 61:3986–3997.

Cheng, H.C., Chien, H., Liao, C.H., Yang, Y.Y., and Huang, S.Y. 2007. Carotenoids suppress proliferating cell nuclear antigen and cyclin D1 expression in oral carcinogenic models. *J Nutr Biochem* 18:667–675.

Chew, B.P., Brown, C.M., Park, J.S., Mixter, P.F. 2003. Dietary lutein inhibits mouse mammary tumor growth by regulating angiogenesis and apoptosis. *Anticancer Res* 23:3333–3339.

Choi, J.H., Kim, C., Lim, H.Y. et al. 2001. Vascular endothelial growth factor in the serum of patients with non-small cell lung cancer: Correlation with platelet and leukocyte counts. *Lung Cancer* 33:171–179.

Clark, A.S., West, K., Streicher, S., and Dennis, P.A. 2002. Constitutive and inducible Akt activity promotes resistance to chemotherapy, trastuzumab, or tamoxifen in breast cancer cells *Mol Cancer Ther* 1:707–717.

Conn, P.F., Schalch, W., and Truscott, T.G. 1991. The singlet oxygen and carotenoid interaction. *Photochem Photobiol* 11B:41–47.

Cowley, S., Paterson, H., Kemp, P., and Marshall, C.J. 1994. Activation of MAP kinase is necessary and sufficient for PC12 differentiation and for transformation of NIH 3T3 cells. *Cell* 77:841–52.

Cui, Y., Lu, Z., Bai, L., Shi, Z., Zhao, W.E., and Zhao, B. 2007. Beta-carotene induces apoptosis and up-regulates peroxisome proliferator-activated receptor gamma expression and reactive oxygen species production in MCF-7 cancer cells. *Eur J Cancer* 43:2590–2601.

Das, S.K., Hashimoto, T., Shimizu, K., and Kanazawa, K. 2005. Fucoxanthin induces cell cycle arrest at G0/G1 phase in human colon carcinoma cells through up-regulation of p21WAF1/Cip1. *Biochim Biophys Acta* 30:328–335.

De Stefano, D., Maturi, M.C., Simeon, V. et al. 2007. Lycopene, quercetin and tyrosol prevent macrophage activation induced by gliadin and IFN-gamma. *Eur J Pharmacol* 566:192–199.

Di Mascio, P., Kaiser, S., and Sies. H. 1989. Lycopene as the most efficient biological carotenoid singlet oxygen quencher. *Arch Biochem Biophys* 274:532–538.

Diehl, J.A. 2002. Cycling to cancer with cyclin D1. *Cancer Biol Ther* 1:226–231.

Du, K. and Montminy. M. 1998. CREB is a regulatory target for the protein kinase Akt/PKB. *J Bio Chem* 273:32377–32379.

Eferl, R. and Wagner, E.F. 2003. AP-1: A double-edged sword in tumorigenesis. *Nat Rev Cancer* 3:859–868.

el-Deiry, W.S., Tokino, T. Velculescu, V.E. et al. 1993. WAF1, a potential mediator of p53 tumor suppression. *Cell* 1 75:817–825.

Fujita, N., Sato, S., Ishida, A., and Tsuruo. T. 2002. Akt-dependent phosphorylation of p27Kip1 promotes binding to 14-3-3 and cytoplasmic localization. *J Biol Chem* 277:10346–53.

Gerster, H. 1995. Carotene, vitamin E and vitamin C in different stages of experimental carcinogenesis. *Eur J Clin Nutr* 49:155–168.

Giovannucci, E. 1999. Insulin-like growth factor-I and binding protein-3 and risk of cancer. *Horm Res* 51(Suppl. 3):34–41.

Goralczyk, R., Wertz, K., Lenz, B. et al. 2005. Beta-carotene interaction with NNK in the AJ-mouse model: Effects on cell proliferation, tumor formation and retinoic acid responsive genes. *Biochim Biophys Acta* 1740:179–188.

Gradelet, S., Le Bon, A.M., Berges, R., Suschetet, M., and Astorg, P. 1998. Dietary carotenoids inhibit aflatoxin B1-induced liver preneoplastic foci and DNA damage in the rat: Role of the modulation of aflatoxin B1 metabolism. *Carcinogenesis* 19:403–411.

Herzog, A., Siler, U., Spitzer, V. et al. 2005. Lycopene reduced gene expression of steroid targets and inflammatory markers in normal rat prostate. *FASEB J* 19: 272–274.

Hirsch, K., Atzmon, A., Danilenko, M., Levy, J., and Sharoni, Y. 2007. Lycopene and other carotenoids inhibit estrogenic activity of 17beta-estradiol and genistein in cancer cells. *Breast Cancer Res Treat* 104:221–230.

Holmes, M.D., Pollak, M.N., Willett, W.C., and Hankinson, S.E. 2002. Dietary correlates of plasma insulin-like growth factor I and insulin-like growth factor binding protein 3 concentrations. *Cancer Epidemiol Biomarkers Prev* 11:852–861.

Hosokawa, M., Kudo, M., Maeda, H., Kohno, H., Tanaka, T., and Miyashita, K. 2004. Fucoxanthin induces apoptosis and enhances the antiproliferative effect of the PPARgamma ligand, troglitazone, on colon cancer cells. *Biochim Biophys Acta* 1675:113–119.

Hotz-Wagenblatt, A. and Shalloway, D. 1993. Gap junctional communication and neoplastic transformation. *Crit Rev Oncog* 4:541–558.

Huang, C.S., Fan, Y.E., Lin, C.Y., and Hu, M.L. 2007. Lycopene inhibits matrix metalloproteinase-9 expression and down-regulates the binding activity of nuclear factor-kappa B and stimulatory protein. *J Nutr Biochem* 18:449–456.

Ivanov, N.I., Cowell, S.P., Brown, P. Rennie, P.S. Guns, E.S., and Cox. M.E. 2007. Lycopene differentially induces quiescence and apoptosis in androgen-responsive and -independent prostate cancer cell lines. *Clin Nutr* 26:252–263.

Kallunki, T., Su, B., Tsigelny, I. et al. 1994. JNK2 contains a specificity-determining region responsible for efficient c-Jun binding and phosphorylation. *Genes Dev* 8:2996–3007.

Kanagaraj, P., Vijayababu, M.R., Ravisankar, B. et al. 2007. Effect of lycopene on insulin-like growth factor-I, IGF binding protein-3 and IGF type-I receptor in prostate cancer cells. *J Cancer Res Clin Oncol* 133:351–359.

Kandel, E.S. and Hay. N. 1999. The regulation and activities of the multifunctional serine/threonine kinase Akt/PKB. *Exp Cell Res* 253:210–229.

Kane, D.J., Sarfian, T.A., Anton, R., Hahn, H., Gralla E.B., and Valentie, J.S. 1993. Bcl-2 inhibition of neural death: decreased generation of reactive oxygen species. *Science* 262:1274–1277.

Karas, M., Amir, H., Fishman, D. et al. 2000. Lycopene interferes with cell cycle progression and insulin-like growth factor I signaling in mammary cancer cells. *Nutr Cancer* 36:101–11.

Kessova, I.G., Leo, M.A., and Lieber, C.S. 2001. Effect of beta-carotene on hepatic cytochrome P-450 in ethanol-fed rats. *Alcohol Clin Exp Res* 25:1368–1372.

Kim, G.Y., Kim, J.H., Ahn, S.C. et al. 2004. Lycopene suppresses the lipopolysaccharide-induced phenotypic and functional maturation of murine dendritic cells through inhibition of mitogen-activated protein kinases and nuclear factor-kappaB. *Immunology* 113:203–211.

Kim, L., Rao, A.V., and Rao, L.G. 2003. Lycopene II–effect on osteoblasts: The carotenoid lycopene stimulates cell proliferation and alkaline phosphatase activity of SaOS-2 cells. *J Med Food* 6:79–86.

Kim, Y., Chongviriyaphan, N., Liu, C., Russell, R.M., and Wang, X.D. 2006. Combined antioxidant (beta-carotene, alpha-tocopherol and ascorbic acid) supplementation increases the levels of lung retinoic acid and inhibits the activation of mitogen-activated protein kinase in the ferret lung cancer model. *Carcinogenesis* 27:1410–1419.

Krinsky, N. 1993. Micronutrients and their influence on mutagenicity and malignant transformation. *Ann N Y Acad Sci* 28:229–242. Review.

Kucuk, O., Sarkar, F.H., Sakr, W. et al. 2001. Phase II randomized clinical trial of lycopene supplementation before radical prostatectomy. *Cancer Epidemiol Biomarkers Prev* 10:861–868.

La Rosa, F.A., Pierce, J.W., and Sonenshein, G.E. 1994. Differential regulation of the c-myc oncogene promoter by the NF-kB Rel family of transcription factor. *Mol Cell Biol* 14:1039–1044.

Lee, S.J,. Bai, S.K., Lee, K.S. et al. 2003. Astaxanthin inhibits nitric oxide production and inflammatory gene expression by suppressing I(kappa)B kinase-dependent NF-kappaB activation. *Mol Cells* 16:97–105.

Liu, C., Lian, F., Smith, D.E., Russell, R.M., and Wang, X.D. 2003. Lycopene supplementation inhibits lung squamous metaplasia and induces apoptosis via up-regulating insulin-like growth factor-binding protein 3 in cigarette smoke-exposed ferrets. *Cancer Res* 63:3138–3144.

Liu, C., Russell, R.M., and Wang, XD. 2004. Low dose beta-carotene supplementation of ferrets attenuates smoke-induced lung phosphorylation of JNK, p38 MAPK, and p53 proteins. *J Nutr* 134:2705–2710.

Liu, C., Russell, R.M., and Wang, X.D. 2006. Lycopene supplementation prevents smoke-induced changes in p53, p53 phosphorylation, cell proliferation, and apoptosis in the gastric mucosa of ferrets. *J Nutr* 136:106–111.

Liu, C., Wang, X.D., Bronson, R.T., Smith, D.E., Krinsky, N.I., and Russell, R.M. 2000. Effects of physiological versus pharmacological beta-carotene supplementation on cell proliferation and histopathological changes in the lungs of cigarette smoke-exposed ferrets. *Carcinogenesis* 21:2245–2253.

Liu, Y., Chang, R.L., Cui, X.X., Newmark, H.L. and Conney, A.H. 1997. Synergistic effects of curcumin on all-trans retinoic acid- and 1alpha,25-dihydroxyvitamin D3-induced differentiation in human promyelocytic leukemia HL-60 cells. *Oncol Res* 9:19–29.

Livny, O., Kaplan, I., Reifen, R., Polak-Charcon, S., Madar, Z., and Schwartz, B. 2002. Lycopene inhibits proliferation and enhances gap-junction communication of KB-1 human oral tumor cells. *J Nutr* 132:3754–3759.

Maeda, H., Hosokawa, M., Sashima, T., Takahashi, N., Kawada, T., and Miyashita, K. 2006. Fucoxanthin and its metabolite, fucoxanthinol, suppress adipocyte differentiation in 3T3-L1 cells. *Int J Mol Med* 18:147–152.

Mayne, S.T. 1996. Beta-carotene, carotenoids, and disease prevention in humans. *Faseb J* 10:690–701.

Minden, A., Lin, A., McMahon, M. et al. 1994. Differential activation of ERK and JNK mitogen-activated protein kinases by Raf-1 and MEKK. *Science* 266:1719–1723.

Mucci, L.A., Tamimi, R., Lagiou, P. et al. 2001. Are dietary influences on the risk of prostate cancer mediated through the insulin-like growth factor system?. *BJU Int* 87:814–820.

Nahum, A., Hirsch, K., Danilenko, M. et al. 2001. Lycopene inhibition of cell cycle progression in breast and endometrial cancer cells is associated with reduction in cyclin D levels and retention of p27(Kip1) in the cyclin E-cdk2 complexes. *Oncogene* 20:3428–3436.

Nahum, A., Zeller, L., Danilenko, M. et al. 2006. Lycopene inhibition of IGF-induced cancer cell growth depends on the level of cyclin D1. *Eur J Nutr* 45: 275–282.

Obermüller-Jevic, U.C., Olano-Martin, E., Corbacho, A.M. et al. 2003. Lycopene inhibits the growth of normal human prostate ephitelial cells in vitro. *J Nutr* 133:3356–3360.

Obermuller-Jevic, UC. Schlegel, B. Flaccus, A., and Biesalski, HK. 2001. The effect of beta-carotene on the expression of interleukin-6 and heme oxygenase-1 in UV-irradiated human skin fibroblasts in vitro. *FEBS Lett* 509:186–190.

Offord, EA., Gautier, J.C., Avanti, O. et al. 2002. Photoprotective potential of lycopene, beta-carotene, vitamin E, vitamin C and carnosic acid in UVA-irradiated human skin fibroblasts. *Free Radic Biol Med* 32:1293–1303.

Omenn, G.S., Goodman, G.E., Thornquist, M.D. et al. 1996. Effect of combination of beta carotene and vitamin A on lung cancer disease. *N Engl J Med* 334:1150–1155.

Owuor, ED. and Kong, AN. 2002. Antioxidants and oxidants regulated signal transduction pathways. *Biochem Pharmacol* 64:765–770.

Palmer, H.J. and Paulson, K.E. 1997. Reactive oxygen species and antioxidants in signal transduction and gene expression. *Nutr Rev* 55:353–361.

Palozza, P. 2005. Can beta-carotene regulate cell growth by a redox mechanism? An answer from cultured cells. *Biochim Biophys Acta* 1740:215–221.

Palozza, P., Calviello, G. Serini S. et al. 2001a. Beta-carotene at high concentrations induces apoptosis by enhancing oxy-radical production in human adenocarcinoma cells. *Free Radic Biol Med* 30:1000–1007.

Palozza, P., Serini, S., Di Nicuolo, F. et al. 2001b. Mitogenic and apoptotic signaling by carotenoids: Involvement of a redox mechanism. *IUBMB Life* 52:77–81.

Palozza, P., Serini, S., Maggiano, N. et al. 2002a. Induction of cell cycle arrest and apoptosis in human colon adenocarcinoma cell lines by beta-carotene through down-regulation of cyclin A and Bcl-2 family proteins. *Carcinogenesis* 23:11–18.

Palozza, P., Serini, S., Torsello, A. et al. 2002b. Regulation of cell cycle progression and apoptosis by beta-carotene in undifferentiated and differentiated HL-60 leukemia cells: possible involvement of a redox mechanism. *Int J Cancer* 97:593–600.

Palozza, P., Serini, S., Torsello, A. et al. 2003a. Mechanism of activation of caspase cascade during beta-carotene-induced apoptosis in human tumor cells. *Nutr Cancer* 47:76–87.

Palozza, P., Serini, S., Torsello, A. et al. 2003b. Beta-carotene regulates NF-kappaB DNA-binding activity by a redox mechanism in human leukemia and colon adenocarcinoma cells. *J Nutr* 133:381–388.

Palozza, P. Serini, S. Di Nicuolo, F. et al. 2004a. Beta-carotene exacerbates DNA oxidative damage and modifies p53-related pathways of cell proliferation and apoptosis in cultured cells exposed to tobacco smoke condensate. *Carcinogenesis* 25:1315–1325.

Palozza, P., Serini, S., Di Nicuolo, F., and Calviello, G. 2004b. Modulation of apoptotic signalling by carotenoids in cancer cells. *Arch Biochem Biophys* 430:104–109.

Palozza, P., Serini, S., Maggiano, N. et al. 2005a. Beta-carotene downregulates the steady-state and heregulin-alpha-induced COX-2 pathways in colon cancer cells. *J Nutr* 135:129–136.

Palozza, P., Sheriff, A., Serini, S. et al. 2005b. Lycopene induces apoptosis in immortalized fibroblasts exposed to tobacco smoke condensate through arresting cell cycle and down-regulating cyclin D1, pAKT and pBad. *Apoptosis*10:1445–1456.

Palozza, P., Serini, S., and Calviello, G. 2006a. Carotenoids as modulators of signaling pathways. *Curr Signal Transduc Ther* 1:325–335.

Palozza, P., Serini, S., Currò, D., Calviello, G., Igarashi, K., and Mancuso. C. 2006b. β-carotene and cigarette smoke condensate regulate heme oxygenase-1 and its repressor factor Bach1: Relationship with cell growth. *Antiox Redox* Signal 8:1069–1080.

Palozza, P., Serini, S., Boninsegna, A. et al. 2007a. The growth-inhibitory effects of tomatoes digested in vitro in colon adenocarcinoma cells occur through down regulation of cyclin D1, Bcl-2 and Bcl-xl. *Br J Nutr* 98:789–795.

Palozza, P., Serini, S., Verdecchia, S. et al. 2007b. Redox regulation of 7-ketocholesterol-induced apoptosis by beta-carotene in human macrophages. *Free Radic Biol Med* 42:1579–1590.

Paolini, M., Antelli, A., Pozzetti, L. et al. 2001. Induction of cytochrome P450 enzymes and over-generation of oxygen radicals in beta-carotene supplemented rats. *Carcinogenesis* 22:1483–1495.

Paolini, M., Cantelli-Forti, G., Perocco, P., Peduli, G.F., Abdel-Rahman, S.Z., and Legator, M.S. 1999. Co-carcinogenic effect of beta-carotene. *Nature* 398:760–761.

Perocco, P., Paolini, M. Mazzullo, M. Biagi, GL., and Cantelli-Forti, G. 1999. Beta-carotene as enhancer of cell transforming activity of powerful carcinogens and cigarette-smoke condensate on BALB/c 3T3 cells in vitro. *Mutat Res* 440:83–90.

Peto, R., Doll, R., Buckley, J.D., and Sporn, M.B. 1981. Can dietary beta-carotene materially reduce human cancer rates? *Nature* 290:201–208.

Pinot, F., el Yaagoubi, A., Christie, P., Dinh-Xuan, A.T., and Polla, B.S. 1997. Induction of stress proteins by tobacco smoke in human monocytes: Modulation by antioxidants. *Cell Stress Chaperones* 2:156–161.

Ponnamperuma, R.M., Shimizu, Y., Kirchhof, S.M., and De Luca, L.M. 2000. Beta-carotene fails to act as a tumor promoter, induces RAR expression, and prevents carcinoma formation in a two-stage model of skin carcinogenesis in male Sencar mice. *Nutr Cancer* 37:82–88.

Prasad, V., Chandele, A., Jagtap, J.C., Sudheer, K.P., and Shastry P. 2006. ROS-triggered caspase 2 activation and feedback amplification loop in beta-carotene-induced apoptosis. *Free Radic Biol Med* 41:431–442.

Rock, C.L., Kusluski, R.A., Galvez, M.M., Ethier, S.P. 1995. Carotenoids induce morphological changes in human mammary epithelial cell cultures. *Nutr Cancer* 23:319–333.

Roy, H.K., Olusola, B.F., Clemens, D.L. et al. 2002. AKT proto-oncogene overexpression is an early event during sporadic colon carcinogenesis. *Carcinogenesis* 23:201–305.

Ruhl, R. 2005. Induction of PXR-mediated metabolism by beta-carotene. *Biochim Biophys Acta* 1740: 162–169.

Sato, S., Fujita, N., and Tsuruo, T. 2000. Modulation of Akt kinase activity by binding to Hsp. *Proc Natl Acad Sci* 97:10832–10837.

Schwartz, J.L. 1993. In vitro biological methods for determination of carotenoid activity. *Methods Enzymol* 214:226–256.

Seger, R. and Krebs, E.G. 1995. The MAPK signaling cascade. *FASEB J* 9:726–735.

Sharoni, Y., Agraria, R., Amir, H. et al. 2003. Modulation of transcriptional activity by antioxidant carotenoids. *Mol Aspects Med* 24:371–384.

Sharoni, Y., Danilenko, M., Walfisch, S. et al. 2002. Role of gene regulation in the anticancer activity of carotenoids. *Pure Appl Chem* 74:1469–1477.

Siler, U., Barella, L., Spitzer, V., Schnorr, J. et al. 2004. Lycopene and vitamin E interfere with autocrine/paracrine loops in the Dunning prostate cancer model. *FASEB J* 18:1019–1021.

Sokoloski, J.A., Hodnick, W.F., Mayne, S.T., Cinquina, C., Kim, C.S., and Sartorelli, A.C. 1997. Induction of the differentiation of HL-60 promyelocytic leukemia cells by vitamin E and other antioxidants in combination with low levels of vitamin D3: Possible relationship to NF-kappaB. *Leukemia* 11:1546–1553.

Stahl, W., von Laar, J., Martin, H.D., Emmerich, T., and Sies, H. 2000. Stimulation of gap junctional communication: comparison of acyclo-retinoic acid and lycopene. *Arch Biochem Biophys* 373:271–274.

Stivala, L.A., Savio, M., Quarta, S. et al. 2000. The antiproliferative effect of beta-carotene requires p21waf1/cip1 in normal human fibroblasts. *Eur J Biochem* 267:2290–2296.

Tanno, S., Yanagawa, N A., Habiro, K. et al. 2004. Serine/threonine kinase AKT is frequently activated in human bile duct cancer and is associated with increased radioresistance. *Cancer Res* 64:3486–3490.

Thannickal, V. J. and Fanburg, B.L. 2000. Reactive oxygen species in cell signaling *Am J Physiol Lung Cell Mol Physiol* 279:L1005–L1028.

Terasaki , M., Asai, A., Zhang, H. and Nagao, A. 2007. A highly polar xanthophyll of 9′-cis-neoxanthin induces apoptosis in HCT116 human colon cancer cells through mitochondrial dysfunction. *Mol Cell Biochem* 300:227–237.

Tibaduiza, E.C., Fleet, J.C., Russell, R.M., and Krinsky, N.I. 2002. Excentric cleavage products of beta-carotene inhibit estrogen receptor positive and negative breast tumor cell growth in vitro and inhibit activator protein-1-mediated transcriptional activation. *J Nutr* 132:1368–1375.

Trekli, M.C., Riss, G., Goralczyk, R., and Tyrrell, R.M. 2003. Beta-carotene suppresses UVA-induced HO-1 gene expression in cultured FEK4. *Free Radic Biol Med* 34:456–464.

Velmurugan, B., Bhuvaneswari, V., Burra, U.K., and Nagini, S. 2002. Prevention of N-methyl-N′-nitro-N-nitrosoguanidine and saturated sodium chloride-induced gastric carcinogenesis in Wistar rats by lycopene. *Eur J Cancer Prev* 11:19–26.

Vine, A.L. and Bertram, J.S. 2005. Up-regulation of connexin 43 by retinoids but not by non-provitamin A carotenoids requires RARs. *Nutr Cancer* 52:105–113.

Vivanco, I. and Sawyers, C.L. 2002. The phosphatidylinositol 3-Kinase AKT pathway in human cancer. *Nat Rev Cancer* 2:489–501.

Vogelstein, B. and Kinzler, K.W. 1992. p53 function and dysfunction. *Cell* 70:523–526.

Wang, A. and Zhang L. 2007. Effect of lycopene on proliferation and cell cycle of hormone refractory prostate cancer PC-3 cell line. *Wei Sheng Yan Jiu* 36:575–578.

Wang, X.D., Liu, C., Bronson, R.T., Smith, D.E., Krinsky, N.I., and Russell, M. 1999. Retinoid signaling and activator protein-1 expression in ferrets given beta-carotene supplements and exposed to tobacco smoke. *J Natl Cancer Inst* 91:60–66.

Williams, A.W., Boileau, T.W., Zhou, J.R., Clinton, S.K., and Erdman, J.W. Jr. 2000 Beta-carotene modulates human prostate cancer cell growth and may undergo intracellular metabolism to retinol. *J Nutr* 130:728–732.

Wu, W.B., Chiang, H.S., Fang, J.Y., and Hung, C.F. 2007. Inhibitory effect of lycopene on PDGF-BB-induced signalling and migration in human dermal fibroblasts: A possible target for cancer. *Biochem Soc Trans* 35:1377–1378.

Yu, H. and Rohan. T. 2000. Role of the insuline-like growth factor family in cancer development and progression. *J Natl Cancer Inst* 92: 1472–1489.

Yun, Z., Maecker, H.L., Johnson, R.S., and Giaccia, A.J. 2002. Inhibition of PPAR gamma 2 gene expression by the HIF-1-regulated gene DEC1/Stra13: A mechanism for regulation of adipogenesis by hypoxia. *Dev Cell* 2:331–341.

Zaripheh, S., Nara, T.Y., Nakamura, M.T., and Erdman, J.W.Jr. 2006. Dietary lycopene downregulates carotenoid 15,15′-monooxygenase and PPAR-gamma in selected rat tissues. *J Nutr* 136:932–938.

Zhang, L.X., Cooney, R.V., and Bertram, J.S. 1991. Carotenoids enhance gap junctional communication and inhibit lipid peroxidation in C3H/10T1/2 cells: Relationship to their cancer chemopreventive action. *Carcinogenesis* 12:2109–2114.

Zhang, L.X., Cooney, R.V., and Bertram, J.S. 1992. Carotenoids up-regulate connexin43 gene expression independent of their provitamin A or antioxidant properties. *Cancer Res* 52:5707–5712.

Ziegler, R.G., Mayne, S.T., and Swanson, C.A. 1996. Nutrition and lung cancer. *Cancer Causes Control* 7:157–177.

Ziouzenkova, O. and Plutzky, J. 2008. Retinoid metabolism and nuclear receptor responses: New insights into coordinated regulation of the PPAR-RXR complex. *FEBS Lett* 582:32–38.

Part VIII

Carotenoids and Carotenoid Biochemistry in Animal Systems

23 Control and Function of Carotenoid Coloration in Birds: Selected Case Studies

Kevin J. McGraw and Jonathan D. Blount

CONTENTS

23.1 INTRODUCTION

Many animals deposit carotenoids into external body tissues, such as skin, feathers, or other keratinized structures like the beak, where they impart rich red, orange, or yellow colors (e.g., in flamingos and guppies) or even purple and green hues when in combination with other color-generating mechanisms (e.g., melanin pigments, structural coloration) (McGraw 2006). The carotenoid basis for such colors has been known for 75 years (Volker 1938), and with that has come widespread interest in the biological causes and consequences of carotenoid-based color displays. Special interest has been shown in species where the sexes differ in coloration (e.g., Badyaev and Hill 2000, Gray 1996); in most cases, males display larger areas of or more intense carotenoid coloration and use such colors as a means of signaling their worth as a mate to females of their species or of signaling their competitive advantages to rival males. Sexual selection is recognized as a powerful and important evolutionary force, which has shaped variation in behavior, physiology, and morphology, not least variation in carotenoid coloration within and among species (Andersson 1994). However, sometimes adult females or even young animals can display this form of coloration, and investigations into the nature and role of carotenoids in these instances have proven to be excellent tests of the limits and generalities of theories on animal signal use (e.g., Jawor et al. (2004) and Tschirren et al. (2005)).

Carotenoid colors in birds have also served as ideal models for investigating the factors that keep animal signals "honest." Many forms of signaling in animals, such as songs or dances or large body size, have evolved as reliable modes of communication because they are challenging or costly to produce or maintain, and thus only the highest quality individuals are able to produce them to the fullest extent (Grafen 1990, Johnstone 1995, Zahavi 1975). Carotenoid colors serve as tractable systems for evaluating such costs because their molecular currency can be quantified directly and traced during the process of signal formation, from food to integument. Contributions of carotenoid chemistry to this research avenue have been immeasurable (Brush and Power 1976, Stradi et al. 1995), and careful tracking of carotenoid-related processes have revealed important dietary and health challenges associated with producing vibrant carotenoid pigmentation (see more below).

It is our goal here to provide a detailed synthesis of the most up-to-date research on the control and function of carotenoid coloration in birds, the best studied of all animal taxa in this respect. We recognize that many similar reviews have been published over the past decade (Blount and McGraw 2008, Hill 1999, 2002, 2006, McGraw 2006, Møller et al. 2000, Olson and Owens 1998). These have appeared in various journals, both theory-based and taxon-specific, and books, both single-authored and edited, which has allowed this information to be disseminated to a variety of specialists in both chemistry and biology. Charged with writing another review on this topic, but to a different carotenoid-centric audience, we chose to adopt a different format for this synthesis. We tackle core principles of carotenoid color production and use by reviewing several of the best case studies in the field, each of which brings with it a unique set of research questions, tools, and outcomes. This species-level approach should allow nonspecialists a unique look at the advantages and disadvantages of studying certain carotenoid-related phenomena (i.e., diet, behavior) in particular animals (i.e., captive vs. wild, young vs. old, males vs. females). Our selection of diverse topics and systems should also demonstrate how rewarding and far-reaching this line of avian carotenoid research has been, touching on key scientific topics that range from antioxidant biology to environmental contamination and conservation.

23.2 "YOU ARE WHAT YOU EAT": DIETARY CONTROL OF CAROTENOID COLORATION IN THE HOUSE FINCH

Among the earliest studies of animal carotenoids that were performed in the context of animal communication were those on house finches (*Carpodacus mexicanus*), which not coincidentally is the free-ranging bird species for which we probably know the most about carotenoid accumulation and use (McGraw et al. 2006b). Male house finches (members of the songbird family *Fringillidae*, which includes old and new world crossbills, siskins, canaries, and goldfinches) display unparalleled intraspecific variability in the intensity of carotenoid pigmentation in feathers from the crown, breast, and rump, ranging from brilliant red through orange to drab yellow. Several lines of field and lab evidence indicate that males with redder plumage are preferred as mates and have higher fitness (Hill 2002). Brush and Power (1976) were the first researchers to formally consider the carotenoid basis of coloration in this wild bird species (see Table 1 in their paper for a nice review of earlier articles on carotenoids in the feathers of other bird species). They aimed to build on a long history of research on the role of diet in controlling the red/orange/yellow coloration of pet, zoo, and domesticated birds (i.e., chickens, canaries; reviewed in McGraw (2006) and Hill (2002)); they were particularly attracted to studying this species not only because of their natural color variations but also because prior observations among aviculturalists indicated that male house finches uniformly grew pale yellow feathers when caged during the molt period. Brush and Power (1976) delivered supplemental canthaxanthin and β-carotene to molting male house finches fed a typical seed diet (e.g., commercially available parakeet or finch mix) in captivity and found that canthaxanthin-fed birds grew red feathers and that β-carotene-fed males grew reddish-orange or orange feathers. This was an important first step in documenting how high supplies of certain dietary carotenoids can influence plumage colors, but, because color was not quantified objectively (instead assessed by eye) and because the authors recognized that one of the carotenoids they used (canthaxanthin) is likely not a part of the natural diet of house finches, additional work was clearly needed. In interpreting their results, Brush and Power (1976) gave equal attention to carotenoid metabolism in this process and surmised that red pigments in house finch feathers are synthesized from important dietary precursors that are lacking in the typical seed diet (see more below).

Hill (1992, 1993) then performed numerous, extensive follow-up studies on house finch diet and coloration, concurrent to his work on the mate-choice function of this plumage trait (Hill 1990, 1991, 1994). Male house finches exhibit extensive variation in the size and color intensity of carotenoid-based plumage patches among populations, and one of his new aims was to understand the basis for such geographic variability. Interestingly, in captive diet experiments using canthaxanthin supplementation, males from various populations (e.g., Michigan, New York, California, Hawaii) converged on a similar color appearance (based on subjective quantitative assessments of color when matched to standardized color chips) when fed the same diet, suggesting that variations in dietary carotenoid intake among populations explains their different color expressions (Hill 1993). The size of colorful patches in some populations, however, remained unchanged from their previous condition, which is consistent with the idea that the area of coloration is not as environmentally sensitive and instead under stronger genetic control. Hill (1992) also examined the degree to which factors like carotenoid storage or age were related to color expression. Regardless of whether birds were captured from the wild just prior to molt or were fed a carotenoid-deficient (plain seed) diet for months prior to molt, male house finches consistently grew drab plumage when fed a carotenoid-deficient diet at the time they were growing their feathers (Hill 1992); this stands as one of the better experimental tests of the idea that carotenoid storage in internal tissues, such as liver or adipose, contributes minimally to proximate color acquisition (but see McGraw et al. (2006b) and more below). Hill (1992) also observed that first-year males in the wild tended to have less red plumage than older adults, and this may be associated with a poor diet among young birds, though in this correlational study other factors like health (see more below in this section) could not be ruled out.

With one or two exceptions (e.g., Slagsvold and Lifjeld 1985), this compilation of excellent studies stood for several years as our only means of understanding carotenoid color variation in nondomestic birds (until work by Bortolotti and colleagues on American kestrels in the mid-1990s; Bortolotti et al. 1996), but even so it still lacked a clear naturalistic context. Dietary carotenoids in wild birds had never been studied, until Hill and colleagues undertook such an investigation of male house finches from California and Mexico (Hill et al. 2002). Naturally foraging male house finches that were in the process of growing their colorful feathers were captured, euthanized, and their gut contents extracted and analyzed for total carotenoid content (HPLC was not used to identify individual compounds). Hill et al. (2002) found that, regardless of age, there was a significant positive correlation between the plumage redness of males from California and gut carotenoid concentration (Figure 23.1). This relationship between diet and color had been previously established in guppies (*Poecilia reticulata*; Grether et al. 1999), but it was an essential validation of theories of dietary control for coloration in a wild bird.

One additional piece of the puzzle that was of interest to Hill and colleagues studying house finches was whether particular carotenoid types in the diets of house finches were valuable for acquiring red coloration. Inouye et al. (2001) used HPLC to describe the diversity of types of carotenoids responsible for yellow and red coloration in house finch feathers, and used likely biochemical product-precursor relationships to hypothesize that a special pathway for acquiring red coloration involved the metabolism of dietary β-cryptoxanthin (thought to be a rare dietary compound) into red feather ketocarotenoids like 3-hydroxy-echinenone. For the first time in a lab experiment on house finches, Hill (2000) administered supplemental dietary β-cryptoxanthin (in the form of tangerine juice) during molt and found that males occasionally developed red feathers. This hinted at a very specific, natural dietary control agent for coloration in this species. Since then, field studies

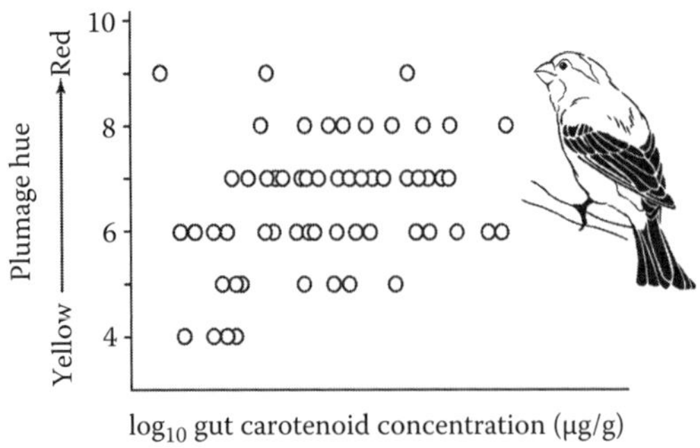

FIGURE 23.1 Relationship between the redness of growing carotenoid-containing feathers in wild, adult, male house finches (*Carpodacus mexicanus*) captured in San Jose, CA and the concentration of carotenoids measured from their gut contents (e.g., crop, proventriculus, gizzard). Plumage coloration was scored by visual comparison to color chips in the *Methuen Handbook of Colour*. Diet samples were ground in the presence of organic solvent and carotenoid concentration was determined with visible-light spectrophotometry.

of carotenoids accumulated by house finches have confirmed the strong predictive power of β-cryptoxanthin in the development of red feathers in males (McGraw et al. 2006b). Males with more β-cryptoxanthin in blood circulation and in liver tissue during the molt period were more likely to grow red, ketocarotenoid-containing feathers (McGraw et al. 2006b); it was not apparent from this correlational study, however, if liver carotenoids played a key coloring role or were just correlated with other factors (like plasma carotenoids) to determine color. Additional challenges that lie ahead in this line of work include understanding the natural sources of this pigment in house finch food during molt as well as testing whether house finches adopt a specialized foraging strategy for acquiring this pigment at this time of year.

It is noteworthy that there has been an extensive parallel line of inquiry on the effects of parasites and health on house finch coloration (reviewed in Hill (2002), also see Hill et al. (2004)). Some studies have gone as far as saying that one disease—the avian pox—may have been the initial driving force behind the unusual range of intrapopulational color variability seen among males in this species (Zahn and Rothstein 1999; but see Hill (2001) for a critique). This well-studied system of mechanisms is a perfect example of the complexities of carotenoid intake and use, especially among wild animals, and the difficulties in isolating the relative importance of competing control mechanisms without detailed, comparative, experimental approaches within a naturalistic context.

23.3 PHYSIOLOGICAL AND GENETIC CONTROL OF CAROTENOID COLORATION IN A DOMESTIC SONGBIRD MODEL, THE ZEBRA FINCH

Even before Hill's extensive work on house finch carotenoids had been published, there was clear evidence in one other avian species—the zebra finch (*Taeniopygia guttata*)—that females used variation in an elaborate carotenoid-based color signal to make mate selection decisions. Male zebra finches, an Australian grassland passerine species from an entirely different family of finches (*Estrildidae*) than house finches, display an intense red beak that derives its color from carotenoids (McGraw et al. 2002). These birds are ideal for controlled studies because they have been domesticated for centuries and freely mate and breed under laboratory conditions. Burley and Coopersmith (1987) gave females the choice of assessing and spending time with males that varied in beak coloration and showed that females significantly preferred to associate with red males; this pattern has recently been corroborated by Blount et al. (2003), though there have been some objections to the strength of mate selection on this trait under certain conditions or relative to other traits (Collins et al. 1994).

Much like the parallel work on house finches, researchers began to investigate mechanisms underlying color variation among males (i.e., what factors make some males more attractive and worth mating with more than others?). Because these animals were housed under identical conditions, given identical foods, and yet displayed variable beak colors, initial emphases were placed on how physiological and genetic parameters should impact beak color expression. Burley et al. (1992) provided an excellent first description of the degree to which bill color changed over varying environmental contexts; intense reproduction and social crowding, for example, appeared to have strongest effects on color, in these cases fading the beak from red to orange. Birkhead et al. (1998) showed that birds with redder beaks were in better body condition. Parasite loads also were oddly positively correlated with bill redness (Burley et al. 1991), however, unlike the traditional prediction that parasites should debilitate birds and fade beak color. Nonetheless, these initial studies were suggestive of physiological control agents of coloration, whereby the hormones and energetic investments associated with breeding and competing in groups, not diet, are key in shaping carotenoid accumulation and coloration.

Little was done to test these energy/hormone hypotheses for many years, however, and instead attention was devoted next to genetic mechanisms. For sexual traits like coloration to persist over evolutionary time, they must have heritable bases; however, virtually no studies, aside from those in chickens and aquacultured fish (where the sexual selection role in maintaining coloration is unclear; e.g., McGraw and Klasing [2006]), had investigated the genetic architecture underlying carotenoid pigmentation. Still some of the best work on this topic in birds was done by Price and colleagues using domesticated zebra finches. In their controlled breeding and selection experiments, they found strong heritability of male beak redness (and also orange coloration of the female beak; Price 1996, Price and Burley 1993) as well as significantly positive selection differentials and gradient coefficients, such that redder males had higher reproductive rates (Price and Burley 1994). Very recently, this issue has been revisited in the context of the heritability of other information that beak redness might reveal, including health and condition (Birkhead et al. 2006; also see more below in this section). These researchers found strong additive genetic variation in and genetic correlations among male beak redness, condition, and health (as measured by response to a tetanus immunization), such that females gain good genes by mating with males in good condition and a redder beak. Despite consistent support for this genetic model of color control, it is important to point out that in none of these studies were differential maternal effects—specifically, varying amounts of yolk carotenoids contributed by mothers—carefully controlled for; in the study by Birkhead et al. (2006), females were paired twice, each time with a different male, to increase chances of detecting maternal effects, but it is possible that many of the effects seen here are linked to carotenoid accumulation patterns, with more carotenoid-replete birds depositing greater amounts of carotenoids in yolk and thus having offspring with more carotenoids and redder beaks, independent of their genes. In fact, in zebra finches, McGraw et al. (2005) found such a positive correlation between maternal beak redness, yolk carotenoid investment, and the redness of the beak developed by sons when sexually mature.

This trans-generational view of carotenoids raises the prospects of the direct physiological actions that carotenoids have on animals, as it relates to their color development. In another subsection, we detail the antioxidant role of carotenoids in another avian species, but zebra finches have been among the best-studied birds in the context of the immunoenhancing roles of carotenoids. The humoral and cell-mediated aspects of immunocompetence have been shown to increase in response to increased dietary carotenoid intake (Blount et al. 2003, McGraw and Ardia 2003). Alonso-Alvarez et al. (2004) also found that circulating carotenoid levels in zebra finches changed in parallel with a measure of antioxidant defense (resistance of red blood cells to free radical attack). Taken together, these studies reveal key health roles of carotenoids, consistent with the view that there are strong physiological inputs into the carotenoid color signaling system of this species. But might diet also play a role in coloration then? Clearly in many studies (the three listed above) providing experimental supplements with carotenoids can enhance beak coloration, but Birkhead et al. (1999) first showed experimentally that raising nestling zebra finches on a poor-quality diet did not affect adult beak redness. McGraw et al. (2003) then performed a detailed analysis of natural diet (food intake) as a function of carotenoid accumulation and coloration and found that food consumption was unrelated to beak coloration, and instead that physiological accumulation of carotenoids (in blood) was a strong positive predictor of bill redness.

Somewhat derived from these diet-related findings, four other attributes of zebra finch physiology have been considered within the last few years in the context of color expression. If circulation of carotenoids predicts beak redness (McGraw et al. 2003), it stood to reason that factors that affect lipid circulation might physiologically limit color expression. Lipoproteins are the blood carrier molecules that deliver nutrients like carotenoids to peripheral tissues (Trams 1969). In a correlational and experimental study, a component of lipoproteins (cholesterol) was measured as well as manipulated, both via a dietary increase and a drug-facilitated (statin) decrease, in order to test relationships with carotenoid circulation and bill coloration. McGraw and Parker (2006) found that cholesterol was a strong predictor and determinant of carotenoid status; birds fed supplemental cholesterol had more carotenoids in circulation and in the beak, while statin administration decreased systemic cholesterol along with carotenoid levels in the blood and bill (Figure 23.2b). Because lipid circulation has been linked to steroid hormone production, especially sex steroids like testosterone, which may also be associated with the expression of sexual signals (Kimball 2006), the link between lipoproteins, testosterone, and carotenoids was subsequently tested. Testosterone levels were correlated with cholesterol and carotenoid levels as well as beak color in adult male zebra finches; moreover, testosterone manipulations had strong impacts on cholesterol, carotenoids, and color (Figure 23.2) (McGraw et al. 2006a). In a precocial gamebird, the red-legged partridge (*Alectoria rufa*), Blas et al. (2006) also found that testosterone increased carotenoid accumulation and coloration. These complex physiological (e.g., hormone, health, nutrition) links to color are yet to be tested in other species, especially wild birds, but based on the commonalities of carotenoid limitations and value across species there is good reason to think they are broadly applicable (Peters 2007).

In addition to lipoproteins and hormones, noncarotenoid antioxidants and cold ambient temperatures have been uniquely identified as factors affecting carotenoid physiology and coloration in male zebra finches. Melatonin is a hormone antioxidant protector of DNA and, when supplemented experimentally in the diet, was found to enhance bill redness (Bertrand et al. 2006). Beak color faded in zebra finches that were exposed to cold ambient temperatures (6°C) for one month (Eraud et al. 2007). These new avenues for research further deepen our understanding of the intrinsic and extrinsic forces that shape how carotenoids are utilized and allocated to attractiveness at the expense of other physiological functions. They now should be integrated with the other nutritional, endocrinological, immunological, and physiological parameters that have been studied to develop a comprehensive framework for the requirements and trade-offs of carotenoid assimilation in this model species.

23.4 THE ANTIOXIDANT ROLE OF CAROTENOIDS IN A COLORFUL RAPTOR

In 1999, von Schantz and coworkers published an influential paper in which they suggested that carotenoid coloration may be a signal of an individual's ability to resist oxidative stress (von Schantz et al. 1999). Because carotenoids have multiple functions, including as colorants and antioxidants, they suggested that under conditions of oxidative stress, individuals may face a trade-off in carotenoid allocation, where carotenoids are required to be utilized as antioxidants, resulting in reduced carotenoid availability for sexual signal expression (von Schantz et al. 1999). Because oxidative stress has been implicated in the etiology of many diseases and is thought to be an important agent underlying cellular and organismal ageing (Kohen and Nyska 2002; see Chapter 10), it should benefit females to choose a male that has a lower susceptibility to oxidative stress (von Schantz et al. 1999).

This is a relatively hard theory to test. It requires an experimental manipulation of susceptibility to oxidative stress in males, preferably both upward and downward, coupled with measurements

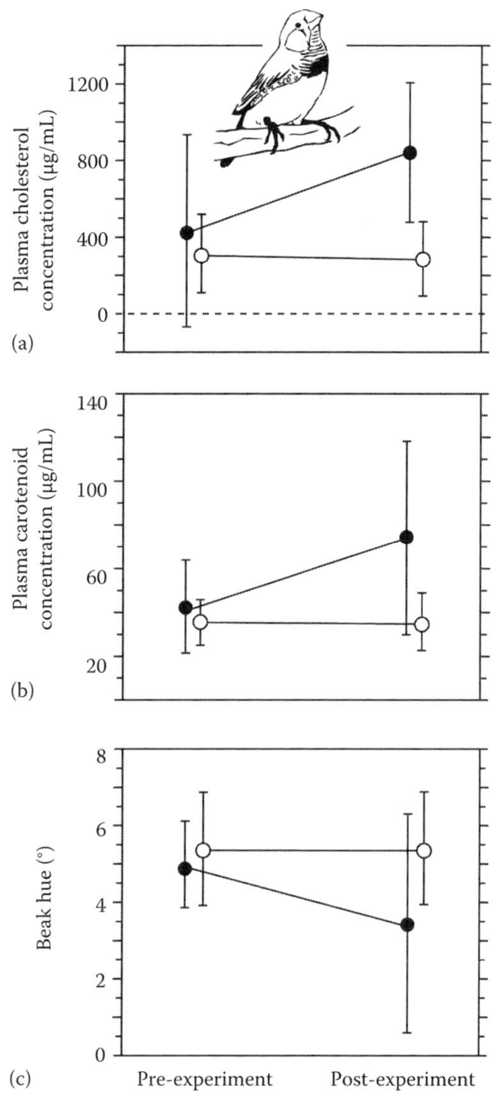

FIGURE 23.2 Effect of testosterone administration on (a) cholesterol circulation through blood, (b) carotenoid circulation through blood, and (c) carotenoid-based beak coloration in captive male zebra finches (*Taeniopygia guttata*). Testosterone was delivered via subcutaneous Silastic capsule implants for an eight week period. Blood was drawn and beak hue was scored (using a handheld visible light reflectance spectrophotometer) both before and after the eight week study. Plasma cholesterol was determined using a colorimetric and spectrophotometric commercial assay, and plasma carotenoid content was assessed using high-performance liquid chromatography. Means ± SD are shown.

of corresponding changes in carotenoid coloration and female mate choice. Such an experiment is yet to be carried out in any species. However, important advances in our understanding of the relationship between oxidative stress and sexual signaling have been made by David Costantini and coworkers, using Eurasian kestrels (*Falco tinnunculus*). Kestrels utilize two carotenoids, lutein and zeaxanthin (~9:1), which they deposit in the bare parts of their integument, i.e., the skin of the cere (base of the bill), lores (around the eyes) and tarsi (legs), giving a yellow-orange color (Casagrande et al. 2006). In a Mediterranean study population, kestrels obtain these carotenoids by feeding on lizards, small birds, and insects (Costantini et al. 2005). Carotenoid coloration has been shown to be affected by dietary intake of carotenoids in juvenile, wild kestrels, suggesting that carotenoids are environmentally and physiologically limiting (Casagrande et al. 2007). This begs the question of whether carotenoid availability may be limiting for protection against oxidative stress, and whether the susceptibility of individuals to oxidative stress is signaled by carotenoid coloration. It is indeed apparent that the stimulation of the T-cell mediated arm of the immune system poses an oxidative challenge in kestrels; injection of phytohaemagglutinin (PHA) into the wing-web caused increased levels of reactive oxygen metabolites (ROMs), decreased total antioxidant capacity (OXY) against *in vitro* hypochlorite-induced oxidation, and increased levels of carotenoids in plasma (Costantini and Dell'Omo 2006). This suggests that the immune challenge invoked oxidative stress, and carotenoids were mobilized from storage organs to counter this and/or to bolster the immune response (Costantini and Dell'Omo 2006).

Other studies of the relationship between carotenoids and oxidative stress in kestrels have yielded contradictory results. In rehabilitated, captive adults, dietary carotenoid supplementation has been shown to result in elevated blood levels of carotenoids, but no effect on OXY, whilst there were increases in levels of ROMs and oxidative stress (ratio between ROMs and OXY) and loss of body mass (Costantini et al. 2007a). This suggests that above a certain threshold, carotenoids may actually have detrimental effects (Costantini et al. 2007a). In similar work using nestlings, dietary carotenoid supplementation had no effect on OXY, ROMs or the ROM:OXY ratio, or the body mass of individuals (Costantini et al. 2007b). The strongest predictor of ROMs and OXY was age, with younger individuals having higher levels (Figure 23.3). Similarly, brood size has been found to be an important determinant of oxidative stress in nestling kestrels, with larger broods being more susceptible, suggesting an effect of intrabrood competition in determining oxidative stress (Costantini et al. 2006). However, levels of ROMS, OXY and the ROM:OXY ratio were unrelated to levels of carotenoids in circulation (Costantini et al. 2006). This suggests that, in kestrels, other types of antioxidants are likely to be more important than carotenoids in mitigating oxidative stress (Costantini et al. 2006). Indeed, it has recently been suggested that carotenoids are minor antioxidants for bird species in general (Costantini and Møller 2008; also see Isaksson and Andersson (2008)). In a review of 20 published studies of 6 species, Costantini and Møller (2008) concluded that there was highly significant heterogeneity in effect sizes among studies, with little evidence that carotenoids are important antioxidants in birds. Lutein is the most abundant carotenoid in circulation in all bird species studied to date, although in many species additional carotenoids are present, and these may vary greatly in antioxidant activity as affected by the polarities of functional groups and the number of conjugated double bonds (Miller et al. 1996). It would therefore be interesting to study relationships between levels of individual carotenoids and antioxidant activity, in a greater range of species (Costantini and Møller 2008). If carotenoids are indeed unimportant antioxidants in birds, carotenoid coloration might still reveal an individual's OXY by reflecting concentrations of other (colorless) antioxidants such as vitamins C and E (Hartley and Kennedy 2004) and melatonin (Bertrand et al. 2006).

23.5 DEVELOPMENTAL PREDICTORS OF COLORATION IN ALTRICIAL BIRDS: STUDIES OF BLUE TITS AND GREAT TITS

Honest signaling based on carotenoid coloration requires that carotenoids are limiting, such that not all individuals can afford to produce a maximal display. This raises the question of what limits genetic and environmental factors impose on carotenoid coloration. Studies of blue tits (*Cyanistes caeruleus*) and great tits (*Parus major*) during the nestling developmental period have provided insights into the importance of genetic and environmental determinants of carotenoid coloration. These tit species are convenient models because of their wide distribution and abundance in the Western Palearctic, the fact that they readily accept bird boxes, and are reasonably tolerant of human disturbance during breeding (Gosler 1993). Both species have carotenoid-based ventral plumage

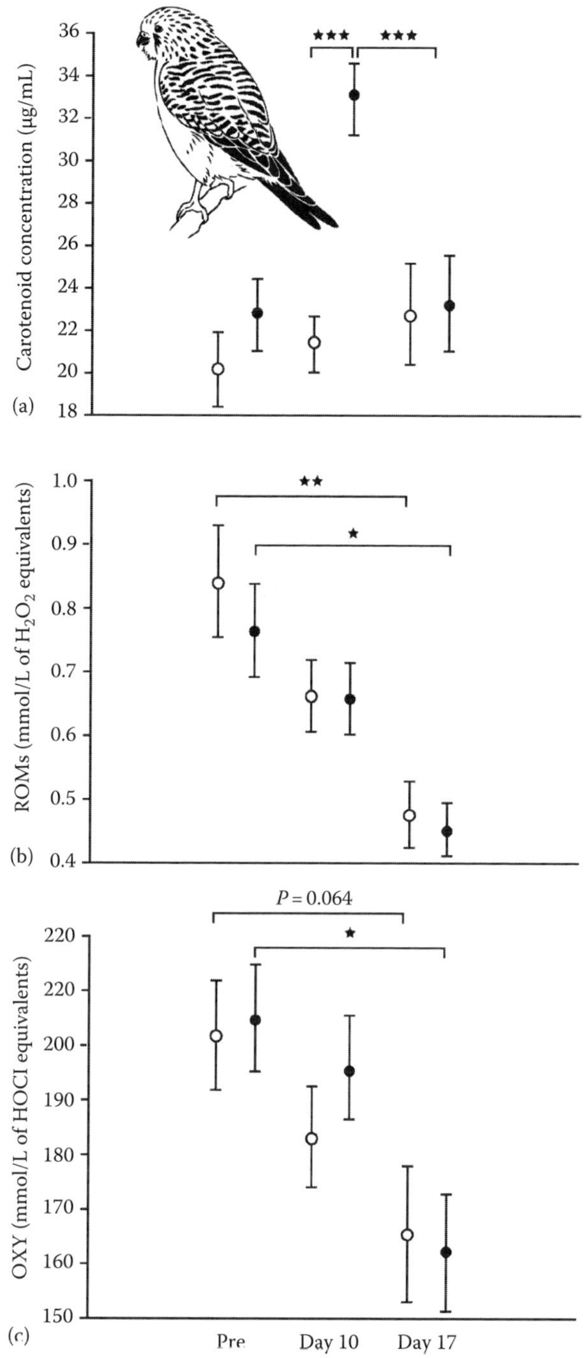

FIGURE 23.3 Effects of dietary supplementation with carotenoids on (a) serum carotenoid concentrations, (b) serum concentrations of ROMs, and (c) serum antioxidant capacity (OXY), in nestling European kestrels (*Falco tinnunculus*). Open circles are nonsupplemented controls, and closed circles are carotenoid-supplemented birds. Means ± SE are shown. Means that are significantly different are denoted by asterisks ($^{*}P<0.05$; $^{**}P<0.01$; $^{***}P<0.001$).

coloration produced by lutein and zeaxanthin (*P. major*), or by lutein and zeaxanthin plus traces of β-carotene (*C. caeruleus*) (Partali et al. 1987). Tits are also well-suited to a variety of experimental manipulations, including dietary carotenoid supplementation and cross-fostering, where broods are partially mixed between the nests of different parents to enable partitioning of the variance in carotenoid coloration due to genetic and environmental factors.

Using such approaches, it has been shown that carotenoid coloration in nestling great tits is significantly influenced by their nest of origin (i.e., it has a partially genetic basis), but is most strongly influenced by environmental factors, becoming elevated in carotenoid-supplemented nestlings (Hadfield and Owens 2006, Tschirren et al. 2003) and in experimentally reduced broods (Tschirren et al. 2003). Similarly, Fitze et al. (2003a) demonstrated that there was a positive correlation between the color saturation of nestlings' plumage and that of their foster father, whereas no such correlation existed with the coloration of their true father (Fitze et al. 2003a). Taken together, these studies indicate that environmental factors operating during the rearing period most strongly determine variation in carotenoid coloration (Fitze et al. 2003a, Hadfield and Owens 2006, Tschirren et al. 2003). Indeed, the development of carotenoid coloration is most sensitive early in the nestling period: dietary carotenoid supplementation before six days of age had a greater influence on coloration than supplementation later in the nestling period (Fitz et al. 2003b). Variation in carotenoid coloration, however, is largely influenced by post-ingestion processes rather than by variation in carotenoid intake *per se*: Hadfield and Owens (2006) found that dietary carotenoid supplementation did not reduce variation in plumage coloration in blue tit nestlings, suggesting that natural variation in carotenoid intake may play a minor role in maintaining variation in carotenoid coloration in this species.

What post-ingestion processes, therefore, may maintain variation in carotenoid coloration? Individual nestlings must partition their carotenoids between plumage coloration and alternative physiological functions such as antioxidant and immune defenses, which could give rise to carotenoid allocation trade-offs (Biard et al. 2006). However, studies of great tits and blue tits have found no effects of carotenoid supplementation on immune defenses, suggesting these are not carotenoid-limited (Biard et al. 2006). Whether carotenoid availability may be limiting for antioxidant protection during nestling development has not been investigated, although a recent study has shown that blood levels of carotenoids and antioxidant activity were not correlated (Isaksson et al. 2007). Alternatively, individuals may vary in their capacities to absorb and transport carotenoids as affected by the sizes of lipoprotein populations, which themselves may be influenced by lipid and protein intake in the diet (McGraw and Parker 2006). Lipids and proteins play various important roles, e.g., in immune function and metabolic energy production, so trade-offs in the allocation of such macronutrients may actually underpin the signal honesty of carotenoid coloration (McGraw and Parker 2006). This idea has received support in a recent study of nestling Great Tits, where it was shown that experimental immune activation (injection of PHA into the wing-web) caused reduced carotenoid coloration of plumage; however, dietary supplementation with the specific carotenoids responsible for feather coloration (lutein and zeaxanthin) did not boost the immune response. Instead, dietary supplementation with β-carotene, which occurs naturally in the diet, but is not itself found in the feathers of this species, resulted in enhanced immune responses (Figure 23.4). These results indicate that the kinds of carotenoids used for coloration differ from the kind(s) used for running the immune system in this species, and there is no trade-off in carotenoid allocation between these two functions. Instead the honesty of carotenoid coloration appears to be enforced by an individual's physiological limitation to absorb and/or transport carotenoids, which is sensitive to nutritional condition more generally (Fitze et al. 2007). Preferences for more intense carotenoid coloration in sexual or parent-offspring signaling contexts may therefore select for individuals that are generally in better nutritional condition and more competitive, rather than specifically for immunocompetence (Fitze et al. 2007, McGraw and Parker 2006).

Interestingly, the signaling function of carotenoid coloration in nestling tits remains obscure. The absence of a correlation between the color of nestlings and their true fathers suggests that carotenoid coloration in nestlings is not a sexually selected trait. Similarly, nestling plumage color

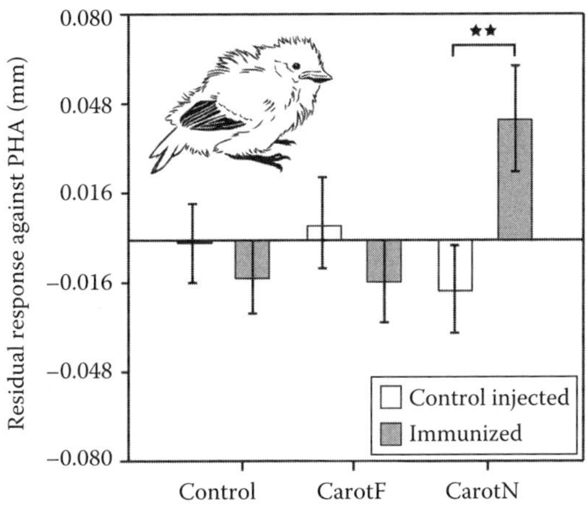

FIGURE 23.4 Effects of dietary supplementation with carotenoids on cell-mediated immune responses to intradermal injection with phytohaemagglutinin in the wing-web in nestling great tits (*P. major*). Values are the residuals (means ± SE) from an analysis of variance including nest as a random factor, therefore statistically controlling for the nonindependence of related individuals within nests. Means that are significantly different are denoted by asterisks ($^{**}P<0.01$). CarotF = feather carotenoids; CarotN = nonfeather carotenoids.

is not correlated with coloration as a breeding adult the following year, suggesting that these color traits have different signal functions (Fitze et al. 2003a). One possibility is that nestling coloration functions as a signal in parent–offspring communication, where parents decide how to allocate care among their chicks based on their coloration. However, no effect of experimentally manipulated offspring color on parental preference has been found (Tschirren et al. 2005). Moreover, chick coloration is unrelated to survival post-fledging, suggesting that this plumage trait is not under natural selection (Fitze and Tschirren 2006). Clearly, more work is needed in this area.

23.6 DEVELOPMENTAL PREDICTORS OF ADULT COLORATION IN PRECOCIAL CAPTIVE BIRD MODELS

While numerous investigations have been performed on the ontogenetic conditions that influence carotenoid accumulation and coloration, even in wild birds, they still have made limited contributions to our understanding of carotenoid sexual signaling because the color is developed in sexually immature animals and has no known visual communication function (see Section 23.5). In contrast, there are several insightful studies on domesticated birds, especially gamebirds (e.g., pheasant, chicken), where the effects of developmental stressors were examined in relation to later-life color expression, including some adult sexual ornaments. This mechanistic approach to "organizational control" of sexual traits provides an excellent supplement to the dominant lines of research on "activational control" of ornamental coloration in adults (see Sections 23.2 and 23.4). Until very recently, studies of organizational effects on animal signals were largely limited to the neurobiological, endocrinological, and nutritional investigations of bird song learning and development (e.g., Nowicki et al. (1998)). From this framework emerge several key principles in nutrition, immunology, and development that make carotenoid color systems excellent models for studying the ontogeny of signal production.

Domestic chicken (*Gallus gallus domesticus*) has long been an ideal species for studying carotenoids, given the industrial production needs for high carotenoid accumulation in legs and yolk and the various problems that can prevent this from happening (e.g., pale bird syndrome, due to parasites and poor diets; Bauernfeind 1981). Within the last five years, however, a new line of work has emerged in chickens, with the goal of understanding how early-life access that animals receive to carotenoids affects their carotenoid accumulation abilities and skin coloration later in life. Leg color is sexually dichromatic in this species, though it is thought to be a product of artificial selection more than adaptive trait expression favored originally by natural or sexual selection (McGraw and Klasing 2006). Koutsos and colleagues have conducted several excellent nutritional, immunological, and ontogenetic studies of carotenoids and uncovered several important relationships that factor into adult coloration. Because carotenoids at a very early age can come either from mother (via egg yolk) or from the neonate diet, Koutsos et al. (2003) manipulated levels of carotenoids from both sources and examined carotenoid accumulation into body tissues four weeks later (admittedly still months before full sexual maturity). They found that yolk carotenoids were especially key for carotenoid accumulation later in life; when birds hatched from carotenoid-free eggs, even when they were fed high supplies of dietary carotenoids post-hatch, they were never able to accumulate as much carotenoids as those who received maternal yolk carotenoids. We can posit then that, if this constraint continued into adulthood, the ability of a bird to become sexually colorful can be permanently impaired by the diet it received from its mother even prior to hatching. Blount et al. (2003) similarly demonstrated in altricial zebra finches (see Section 23.3) that brief exposure to a low-quality nestling diet reduces antioxidant (including carotenoid) circulation in adults, although an irreversible effect on adult male coloration or attractiveness was not apparent in this study.

Carotenoid intake/exposure during development may have long-term effects on avian immunocompetence also. In the above-mentioned study, deposition of carotenoids (e.g., lutein) into immune tissues (e.g., thymus, bursa) was sensitive to the same embryonic carotenoid mechanism (Koutsos et al. 2003), as was lutein accumulation in monocytes from one month-old chickens (Selvaraj et al. 2006). Koutsos et al. (2006) went on to directly measure immune system performance, by assessing the systemic inflammatory response to lipopolysaccharide injection (which simulates an infection), and showed important *in ovo* carotenoid effects on immunity. Saino et al. (2003) found a similar result in free-ranging barn swallows (*Hirundo rustica*), using injections with carotenoids directly into egg yolk as opposed to manipulating the maternal diet (and thus perhaps affecting many other maternal processes and products, not just carotenoids, that change the composition of the egg and yolk). These studies bring into question the actual mechanism for carotenoid "organization" of health (and perhaps coloration); as antioxidants, carotenoids may protect maturing immune cells from damage, hence permitting optimal formation of the immune system and thus optimal performance later in life; this would leave fewer carotenoids required to maintain good health and thus more for color development. Alternatively, early carotenoid exposure may affect carotenoid accumulation mechanisms (e.g., gut lining, lipoproteins) and permanently increase assimilation efficiency, allowing optimal antioxidant, immunoenhancing, and coloring actions of carotenoid supplies later in life. Careful physiological and immunological manipulations and measurements will be required to disentangle these alternative hypotheses.

Even more relevant to sexual signal developments have been the developmental nutrition studies conducted on ring-necked pheasants (*Phasianus colchicus*) in Europe. Male pheasants are elaborately adorned with many colorful plumage features, as well as rich red facial wattles that contain carotenoid pigments like astaxanthin (Brockmann and Volker 1934). Wattles are enlarged during sexual encounters, and females prefer to mate with males that have redder wattles (Hillgarth and Wingfield 1997); wattle size is also correlated with dominance in juvenile males (Papeschi et al. 2003). Pheasants are also ideal study subjects for this line of work because, as a precocial species, nutrition can be manipulated independent of any parental involvement. Ohlsson and colleagues have conducted intricate experiments to investigate the role of early nutritional conditions on wattle coloration and size as well as on adult immune system performance. In their first study (Ohlsson et al. 2002), dietary protein was manipulated (either 27% as a high dose or 20.5% as a normal growth amount) in a factorial design at two developmental stages (0–3 weeks = very early; 4–8 weeks = early), wattles were scored at weeks 20 and 40, and immunocompetence (measured as antibody response to the human diphtheria-tetanus vaccine and as wing-web swelling in response to local

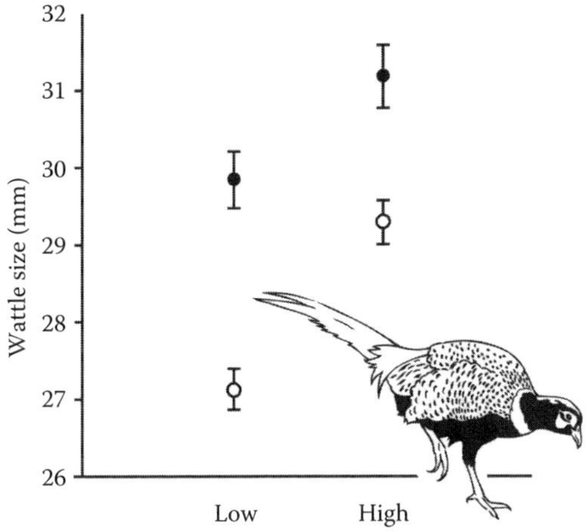

FIGURE 23.5 Effect of feeding captive male ring-necked pheasant (*Ph. colchicus*) young a high- or low-protein feed for the first three weeks of life on the expression of wattle coloration (mean ± SE) at 20 (open circles) and 40 (filled circles) weeks of age. Coloration was determined using a principal components analysis (PCA) of tristimulus scores (hue, saturation, and brightness) obtained with a Colortron II reflectance spectrophotometer.

mitogen injection) was determined at week 20. Antibody production and swelling response were not associated with early-life nutrition, but Ohlsson et al. (2002) found that higher protein intake increased wattle size and redness; the very-early-life manipulation was especially strong for wattle size (Figure 23.5). Though it is not yet apparent *how* protein intake should mechanistically impact processes associated with carotenoid accumulation and deposition in skin, this study stands as one of the best demonstrations of an organizational effect on carotenoid coloration in birds. A different organizational mechanism involving yolk testosterone was experimentally tested in a very recent study, but had no effect on wattle characteristics (Rubolini et al. 2006).

In follow-up work, emphasis has been placed on factors in adult pheasants that affect their current levels of wattle ornamentation. In fact, until this point, the relative roles of early- versus late-life effects on carotenoid colors have not been wholly apparent in any system yet (i.e., no factor had been uniformly studied in both developmental and adult life stages). Ohlsson et al. (2003) conducted the same protein manipulation in adult pheasants and found similar effects of high protein intake on wattle color (though no effect on wattle size was evident). With respect to wattle color, the question now becomes which life phase is more crucial for exaggerated expression of wattle color in adult male pheasants. Smith et al. (2007) went on to study another adult condition—carotenoid intake—as it related to wattle coloration and found comparatively stronger effects of carotenoid consumption, compared to protein intake, on adult wattle coloration. Taken together, these studies paint a clear picture of nutritional control of fleshy coloration in a precocial bird, both organizationally and activationally, but, to fully understand the differential roles of different nutrients in this system, comprehensive, lifetime experimental manipulations are needed, including more naturalistic carotenoids (e.g., canthaxanthin was used), to isolate the strongest limitations for becoming sexually attractive (see Hill (2006) for another detailed discussion of the multivariate forces that shape carotenoid ornamentation in birds).

23.7 CAROTENOID SIGNALING OF SOCIAL STATUS IN WIDOWBIRDS

Signals that have evolved through sexual selection are either designed to attract the opposite sex, or to intimidate rivals of the same sex. Males often compete with each other for access to limited or valuable territories, food or mates, and have been shown to use signals that communicate their readiness and ability to fight, which ensures that not all contests escalate to the point of injury or death (Maynard-Smith and Harper 2003). The dominance status signaling hypothesis was originally devised for male birds that live in nonbreeding foraging flocks, where individuals regularly encounter new, unfamiliar foes and could benefit from using signals that reveal their aggressive nature (Rohwer 1975). We now know that carotenoid coloration, among other sexually selected traits, can serve as a signal of social status in various species of birds and fish (reviewed by Blount and McGraw (2008)).

One group of birds has been especially well studied in the field of carotenoid coloration as a social status signal—the widowbirds (*Euplectes* spp). Widowbirds exhibit a long tail, and, in many species, also striking yellow or red color patches on their body or shoulders (epaulettes). Long tails and carotenoid coloration are two examples of sexually selected plumage traits, prompting the question of what selection pressures have maintained the elaboration of two, distinct and costly ornaments (Andersson et al. 2002). In his classic sexual selection experiment, Malte Andersson demonstrated that female long-tailed widowbirds (*E. progne*) preferred to mate with males that had had their tails experimentally elongated (rather than males where the tail was left unmanipulated or was of reduced length) (Andersson 1982). This female preference for longer-tailed males has also been shown in Jackson's widowbirds (*E. jacksoni*), red-collared widowbirds (*E. ardens*), and red-shouldered widowbirds (*E. axillaries*) (Andersson 1992, Pryke and Andersson 2002, 2005, Pryke et al. 2001a), indicating a generalized female bias for long tails in widowbirds (Pryke and Andersson 2002).

Carotenoid coloration in males of these widowbird species, however, plays no role in female mate choice (Pryke and Andersson 2003a, Pryke et al. 2001a). In a series of elegant studies using experimental removal and replacement of territory holding males in the wild, and experimental manipulations of the size and redness of color patches, coupled with captive studies of aggressive interactions, Sarah Pryke, Staffan Andersson and coworkers have shown that carotenoid coloration functions in male–male competition, with males that have larger and/or redder epaulettes outcompeting males with smaller and/or less red signals (Pryke and Andersson 2003a,b, Pryke et al. 2001b, 2002; Figure 23.6). Territory acquisition and defense is a prerequisite for reproductive success in males, because territories are in limited supply and are used to display to females, which base their

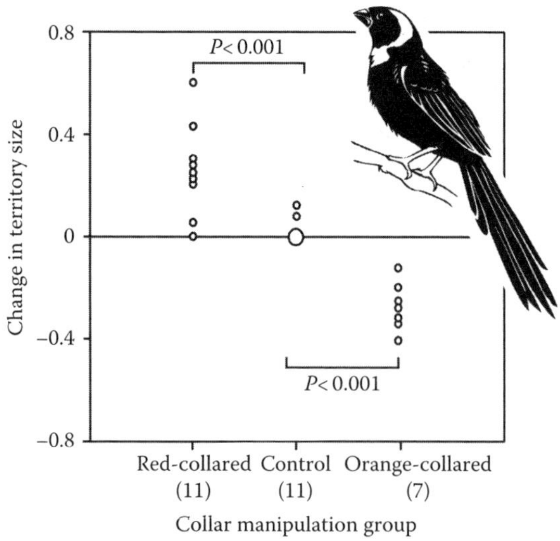

FIGURE 23.6 Effects of experimentally manipulated collar color on changes in territory size of red-collared widowbirds (*Euplectes ardens*), i.e., territory size before collar treatment subtracted from territory size after treatment. *Red collar* males had their collar feathers bleached before being repainted with average-sized red collars; *control collar* males had their feathers bleached before being repainted to their original size and color; *orange collar* males had their feathers bleached before being repainted with average-sized orange collars.

mate choice decisions on male tail length (see above). Male tail length itself does not affect the outcome of contest competition: individuals with experimentally shortened or control tails have been found to be equally successful in acquiring and holding territories as longer-tailed males (Andersson 1982, Pryke and Andersson 2005).

Do tail length and carotenoid coloration correlate in males? Interestingly, the expression of these two traits is inversely related in widowbirds, with this negative correlation being steeper in territory-holding males compared to nonresident “floaters.” This suggests that males face a physiological trade-off in the elaboration of tails and coloration, respectively, which is amplified by costs associated with territory defense (Andersson et al. 2002). This comprehensive body of work using widowbirds therefore provides an important lesson in the interpretation of multiple sexual ornaments: carotenoid coloration in males, whilst elaborate, is not necessarily aimed to impress prospective female mates. In the case of widowbirds, females ultimately base their mate choice decisions on the resource-holding potential of males, as signaled by tail length, whereas carotenoid coloration functions as a signal in male–male competition.

23.8 FEMALE COLORATION AND MUTUAL SEXUAL SIGNALING IN NORTHERN CARDINALS

Species in which both sexes exhibit some form of sexual display are ideal for testing evolutionary models of communication and sexual selection (Kraaijeveld et al. 2007). In birds, where females do a majority of nest construction and egg incubating, it has long been thought that selection should favor females with cryptic plumage, so that predators have a difficult time locating the key nest-visiting parent as well as eggs and nestlings (Wallace 1889). When some hint of the male condition is seen in females, however, it is not clear whether this limited trait expression is a function of the genetic correlation of the phenotype that must be expressed in some form in both sexes or represents adaptive signal use (Amundsen 2000).

The northern cardinal (*Cardinalis cardinalis*) is a North American passerine from the new world family Cardinalidae (including saltators and cardinals) in which males display striking red carotenoid-based plumage coloration across the body (McGraw et al. 2001) and females exhibit splashes of red plumage on the crest, wings, tail, and underwing. Both sexes have deep orange, carotenoid-based bills as well. The signaling role and value of carotenoid coloration in males attracted the initial attention of behavioral ecologists, who found that males with redder plumage defended higher-quality territories, raised more offspring, and tended to be naturally better competitors in the winter (Wolfenbarger 1999a,b), though plumage manipulation experiments in the lab did not support a “status signaling” role or female mate preferences for this trait (Wolfenbarger 1999c). This may be because cardinals show signs of exceptional behavioral and physiological stress when held under captive conditions (personal observation) and thus may not have behaved as they would in a natural setting. Redder males also are better fathers, at least in terms of the proportion of feedings to their nestlings that they perform (Linville et al. 1998; but see Jawor and Breitwisch (2006) for the lack of a relationship between male color and mate feedings).

In more recent work, researchers have sought to identify the signaling role of *female* coloration in northern cardinals. A first indication of a quality-signaling aspect to female plumage redness would be positive assortative mating by coloration, with the idea that higher-quality females would associate only with higher-quality males, either via active mate choice (in either sex) or intrasexual competition. Jawor et al. (2003), but not Linville et al. (1998), detected positive assortative mating for plumage redness in cardinals (Figure 23.7). These researchers went on to examine whether plumage and beak color intensity in females is associated with body size, body condition, and indices of reproductive success. They found that females with deeper bill coloration and redder underwings were larger and in better condition, and that redder underwing color was positively correlated with the timing of reproduction and the number of offspring produced in a year (Jawor et al. 2004).

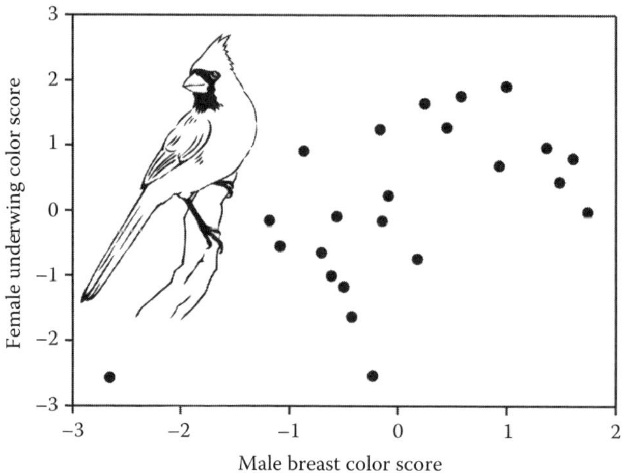

FIGURE 23.7 Correlational evidence for assortative mating by carotenoid-based plumage coloration in pairs of wild northern cardinals (*Cardinalis cardinalis*). Tristimulus color scores were obtained using a digital swatchbook reflectance spectrophotometer and entered into a PCA to acquire PC1 for statistical analysis.

All of this work, even if correlational, is consistent with the idea that female coloration in northern cardinals serves as a valuable signal of mate quality and/or competitive behavior. But if the signal is so useful, why is it not more extensively exaggerated and displayed across the body? Jawor and Breitwisch (2003) have shown that female cardinals have a discrete "hidden," underwing flashing display behavior (i.e., "jacket-opening") that they reserve only for males in close proximity. This suggests that females would pay a price to otherwise having these flashy signals permanently viewable, and this idea of a "coverable badge" has been advocated both in the context of predator avoidance and mate competition (i.e., flash only when needed, in the presence of the signal receiver). Interestingly, in light of this, northern cardinals are one of the few species for which the nest-predation benefits of reduced ornamentation have been tested. Miller (1999) creatively tested this concept in a field experiment in which he placed quail eggs inside of old cardinal nests and set either a brown piece of cardboard or a red piece of cardboard atop each nest, simulating the presence of a drab and bright female cardinal, respectively. He failed to find an effect of cardboard type on nest predation though, which either indicates that the stimuli were insufficiently representative of the presence of female cardinals or that predation does not exert a strong pressure on color reduction, at least in this habitat (southeastern United States). Clearly, more carefully designed experiments are now needed to identify signal receivers (whether they be predators, males, or other females) and signaling contexts.

In closing, it is noteworthy that in this species ornament associations with behavior, especially parenting and assortative mating, varied from study to study. To our knowledge, no clear variables have been isolated as of yet to try to explain these discrepancies; in the case of parental investment, not all paternal behaviors are necessarily correlated (Jawor et al. 2004), and since two different metrics were used (nesting feedings vs. mate feedings) it is conceivable that higher-quality red males place a higher priority on feeding more needy offspring compared to their mates. It is also possible that behavioral plasticity interplays, such that shifting annual environmental circumstances (e.g., food shortages, cold snaps) create differences in the strength of associations between quality metrics like color and behavior. For example, in a year when temperatures were extremely cold and fruit crop production was poor, plumage color in male northern cardinals was less red at the population level (Linville and Breitwisch 1997), which may have had cascading effects on the value and information of plumage as a signal that year. As studies are now more often conducted at an annual or single-time-period scale, we must continue to value these multi-year lines of research and the deeper insights into patterns and strengths of selection pressures that they may provide.

23.9 CAROTENOID COLORATION AND ENVIRONMENTAL CONTAMINATION: GREAT TITS AS BIOINDICATORS

The idea that signal expression in animals provides a reliable indicator of an individual's condition, social status or health underpins our understanding of the evolution of carotenoid-based coloration (Olson and Owens 1998). Such animal signals have evolved to convey information to prospective mates and competitors. However, carotenoid coloration in wild birds has also been identified as a useful bioindicator of exposure to pollution (Eeva et al. 1998, Hõrak et al. 2000, 2001). The great tit (*P. major*) has become a classic model in this area of study, being a widely distributed and common small passerine species of birds found in most parts of the Western Palearctic (Gosler 1993), and having carotenoid-based yellow ventral plumage color produced by a combination of lutein and zeaxanthin (Partali et al. 1987). Great tits readily accept bird boxes and are reasonably tolerant of human disturbance and experimental manipulations, making them a particularly tractable and therefore well-studied species in behavioral ecology (Gosler 1993). Effects of exposure to pollution on toxin accumulation, coloration, health, and productivity in Great tits have been extensively studied by several research groups in different parts of Europe (e.g., Finland, Belgium, Sweden), using two main empirical approaches: comparisons of birds living in rural and urban (polluted) habitats, respectively; and observations of birds living along pollution gradients extending from point source pollution centers such as smelters.

It is well-established that a diverse range of pollutants may bioaccumulate in great gits, including heavy metals such as copper, lead, cadmium, arsenic, and zinc in feathers and other body tissues (Dauwe et al. 2000, 2004, Eens et al. 1999, Llacuna et al. 1995), and certain organochlorine compounds in body fat and muscle (Dauwe et al. 2003). Exposure to such pollutants has been associated with reduced carotenoid coloration. Great tits have relatively pale carotenoid-based plumage coloration in urban areas compared to rural ones, and in areas closer to a point source of heavy metal pollution along a pollution gradient (Eeva et al. 1998, Hõrak et al. 2000, 2001, Isaksson et al. 2005). Furthermore it appears that the effects of being produced in a polluted environment on carotenoid coloration can manifest early in life; in a cross-fostering experiment, nestlings from an urban population did not increase in carotenoid coloration when switched to be reared in a rural population (Hõrak et al. 2000). Potentially, this might point to the importance of egg carotenoids in determining nestling coloration; it is known that egg levels of carotenoids are lower in urban nests compared to rural nests (Hõrak et al. 2002).

Carotenoid coloration in Great Tits is, therefore, a useful biomarker of pollution exposure. But is the link between pollution exposure and carotenoid coloration indirect (i.e., mediated through food availability and quality), or direct (i.e., mediated through oxidative stress)? Great tits are heavily dependent on Lepidopteran larvae for successful breeding (e.g., Gosler (1993) and Perrins (1991)), these being an important source of carotenoids (Partali et al. 1987). In a classic study of 500 nesting boxes used by great tits within 6 km of a copper smelter in southwest Finland, Eeva et al. (1998) showed that both the abundance of green caterpillars and the intensity of plumage color in nestlings decreased with increasing proximity to the pollution source. They therefore argued that paler carotenoid coloration in pollution-exposed birds was a consequence of reduced food (caterpillar) availability. It has also been shown that the carotenoid-content of caterpillars, i.e., food quality, is lower in urban compared to rural environments (although paradoxically, in this study caterpillar abundance was actually higher at urban compared to rural sites, which may have offset the reduced quality of food in urban areas) (Isaksson and Andersson 2007).

It has been suggested that there could also be direct, toxic effects of pollutants on great tits, which causes reduced carotenoid coloration (Hõrak et al. 2000, 2001, Isaksson et al. 2005). Indeed birds may be exposed to pollutants through the air that they breathe, but also in their diet; body levels of metal contaminants in caterpillars have been found to correlate positively with proximity to a pollution source, which is reflected in the excreta of great tit nestlings (Dauwe et al. 2004). A wide range of chemical pollutants has the potential to cause oxidative stress in vertebrates, and this may be predicted to place an increased demand on body levels of carotenoids to function as antioxidants, resulting in reduced availability of carotenoids for pigmenting feathers. This hypothesis has been investigated by Isaksson and coworkers, who have found that great tits living in an urban environment had higher levels of oxidized to reduced glutathione, compared to birds in a rural environment, indicative of increased oxidative stress in more pollution-exposed birds (Isaksson et al. 2005). At the population level, urban birds also had paler carotenoid coloration on average than urban birds, but at the individual level there were no significant correlations between carotenoid coloration and oxidative stress (Figure 23.8). Therefore, whilst living in an urban environment is associated with increased oxidative stress, and also with reduced carotenoid coloration, it is unclear whether these effects are directly and causally linked (Isaksson et al. 2005). The idea that carotenoids could play an important antioxidant role in protecting against pollution induced oxidative stress has been rendered more unlikely by the finding that plasma antioxidant activity was not related to carotenoid

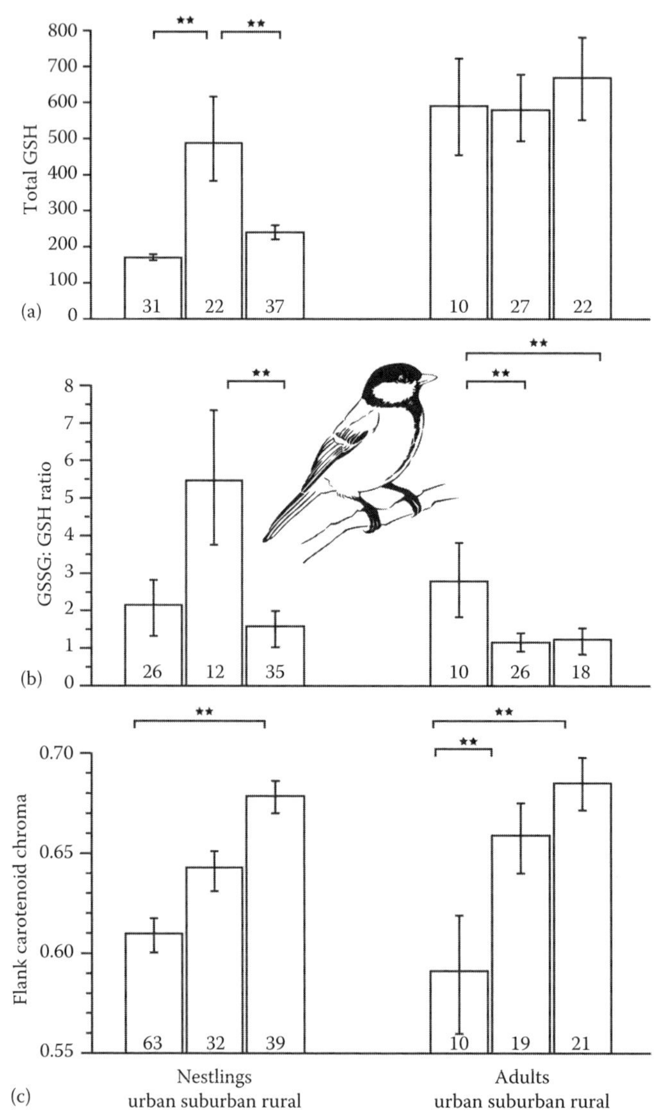

FIGURE 23.8 Average levels in great tits of (a) total glutathione (tGSH, nmol/ml), (b) oxidative stress (ratio of oxidized to reduced glutathione, GSSG:GSH), and (c) carotenoid coloration ("carotenoid chroma," calculated using data obtained by reflectance spectroradiometry). Means ± SE are shown. Means that are significantly different are denoted by asterisks ($^{**}P<0.01$).

coloration or to plasma concentrations of carotenoids in great tits (Isaksson et al. 2007). It remains possible that carotenoid coloration could be an indicator of the body's enzymatic antioxidant defenses against pollutants; this has not yet been investigated.

ACKNOWLEDGMENTS

During manuscript preparation, Jon Blount was supported by a University Research Fellowship from The Royal Society. Kevin McGraw was supported by the National Science Foundation (grant # IOS-074636) and by the School of Life Sciences at Arizona State University.

REFERENCES

Alonso-Alvarez, C., Bertrand, S., Devevey, G., Gaillard, M., Prost, J., Faivre, B., and Sorci, G. 2004. An experimental test of the dose-dependent effect of carotenoids and immune activation on sexual signals and antioxidant activity. *Am. Nat.* 164: 651–659.

Amundsen, T. 2000. Why are female birds ornamented? *Trends Ecol. Evol.* 15: 149–155.

Andersson, M. 1982. Female choice selects for extreme tail length in a widowbird. *Nature* 299: 818–820.

Andersson, M. 1994. *Sexual Selection*. Princeton University Press.

Andersson, S. 1992. Female preference for long tails in lekking Jackson's widowbirds: Experimental evidence. *Anim. Behav.* 43: 379–388.

Andersson, S., Pryke, S. R., Örnborg, J., Lawes, M. J., and Andersson, M. 2002. Multiple receivers, multiple ornaments, and a trade-off between agonistic and epigamic signaling in a widowbird. *Am. Nat.* 160: 683–691.

Badyaev, A. V. and Hill, G. E. 2000. Evolution of sexual dichromatism: Contribution of carotenoid- versus melanin-based coloration. *Biol. J. Linn. Soc.* 69: 153–172.

Bauernfeind, J. C. 1981. *Carotenoids as Colorants and Vitamin A Precursors*. Academic Press, New York.

Bertrand, S., Faivre, B., and Sorci, G. 2006. Do carotenoid-based sexual traits signal the availability of non-pigmentary antioxidants? *J. Exp. Biol.* 209: 4414–4419.

Biard, C., Surai, P. F., and Moller, A. P. 2006. Carotenoid availability in diet and phenotype of blue and great tit nestlings. *J. Exp. Biol.* 209: 1004–1015.

Birkhead, T. R., Fletcher, F., and Pellatt, E. J. 1998. Sexual selection in the zebra finch *Taeniopygia guttata*: Condition, sex traits and immune capacity. *Behav. Ecol. Sociobiol.* 44: 179–191.

Birkhead, T. R., Fletcher, F., and Pellatt, E. J. 1999. Nestling diet, secondary sexual traits and fitness in the zebra finch. *Proc. R. Soc. Lond. B* 266: 385–390.

Birkhead, T. R., Pellatt, E. J., Matthews, I. M., Roddis, N. J., Hunter, F. M., McPhie, F., and Castillo-Juarez, H. 2006. Genic capture and the genetic basis of sexually selected traits in the zebra finch. *Evolution* 60: 2389–2398.

Blas, J., Perez-Rodriguez, L., Bortolotti, G. R., Vinuela, J., and Marchant, T. A. 2006. Testosterone increases bioavailability of carotenoids: Insights into the honesty of sexual signaling. *Proc. Natl. Acad. Sci. U. S. A.* 103: 18633–18637.

Blount, J. D. and McGraw, K. J. 2008. Signal functions of carotenoid colouration. In: *Carotenoids. Volume 4: Natural Functions* (G. Britton, S. Liaaen-Jensen, and H. Pfander, eds.). Birkhauser, Basel, Switzerland, pp. 213–236.

Blount, J. D., Metcalfe, N. B., Birkhead, T. R., and Surai, P. F. 2003. Carotenoid modulation of immune function and sexual attractiveness in zebra finches. *Science* 300: 125–127.

Bortolotti, G. R., Negro, J. J., Tella, J. L., Marchant, T. A., and Bird, D. M. 1996. Sexual dichromatism in birds independent of diet, parasites and androgens. *Proc. R. Soc. Lond. B* 263: 1171–1176.

Brockmann, H. and Volker, O. 1934. Der gelbe Federfarbstoff des Kanarienvogels (*Serinus canaria canaria*) und das Vorkommen von Carotinoiden bei Vogeln. *Hoppe-Seyl. Z.* 224: 193–215.

Brush, A. H. and Power, D. M. 1976. House finch pigmentation: Carotenoid metabolism and the effect of diet. *Auk* 93: 725–739.

Burley, N. and Coopersmith, C. B. 1987. Bill color preferences in zebra finches. *Ethology* 76: 133–151.

Burley, N., Tidemann, S. C., and Halupka, K. 1991. Bill colour and parasite levels of zebra finches. In *Bird-Parasite Interactions: Ecology, Evolution and Behaviour* (Loye, J. E. and Zuk, M., eds.). Oxford University Press, Oxford, pp. 359–376.

Burley, N. T., Price, D. K., and Zann, R. A. 1992. Bill color, reproduction and condition effects in wild and domesticated zebra finches. *Auk* 109: 13–23.

Casagrande, S., Costantini, D., Fanfani, A., Tagliavini, J., and Dell'Omo, G. 2007. Patterns of serum carotenoid accumulation and skin colour variation in kestrel nestlings in relation to breeding conditions and different terms of carotenoid supplementation. *J. Comp. Physiol. B* 177: 237–245.

Casagrande, S., Csermely, D., Pini, E., Bertacche, V., and Tagliavini, J. 2006. Skin carotenoid concentration correlates with male hunting skill and territory quality in the kestrel (*Falco tinnunculus*). *J. Avian Biol.* 37: 190–196.

Collins, S. A., Hubbard, C., and Houtman, A. M. 1994. Female mate choice in the zebra finch—The effect of male beak colour and male song. *Behav. Ecol. Sociobiol.* 35: 21–25.

Costantini, D. and Dell'Omo, G. 2006. Effects of T-cell-mediated immune response on avian oxidative stress. *Comp. Biochem. Physiol. A* 145: 137–142.

Costantini, D. and Møller, A. P. 2008. Carotenoids are minor antioxidants for birds. *Funct. Ecol.* 22: 367–370.

Costantini, D., Casagrande, S., De Filippis, S., Brambilla, G., Fanfani, A., Tagliavini, J., and Dell'Omo, G. 2006. Correlates of oxidative stress in wild kestrel nestlings (*Falco tinnunculus*). *J. Comp. Physiol. B* 176: 329–337.

Costantini, D., Casagrande, S., Di Lieto, G., Fanfani, A., and Dell'Omo, G. 2005. Consistent differences in feeding habits between neighbouring breeding kestrels. *Behaviour* 142: 1409–1421.

Costantini, D., Coluzza, C., Fanfani, A., and Dell'Omo, G. 2007a. Effects of carotenoid supplementation on colour expression, oxidative stress, and body mass in rehabilitated captive adult kestrels (*Falco tinnunculus*). *J. Comp. Physiol. B* 177: 723–731.

Costantini, D., Fanfani, A., and Dell'Omo, G. 2007b. Carotenoid availability does not limit the capability of nestling kestrels (*Falco tinnunculus*) to cope with oxidative stress. *J. Exp. Biol.* 210: 1238–1244.

Dauwe, T., Chu, S. G., Covaci, A., Schepens, P., and Eens, M. 2003. Great tit (Parus major) nestlings as biomonitors of organochlorine pollution. *Archiv. Environ. Contamin. Toxicol.* 44: 89–96.

Dauwe, T., Bervoets, L., Blust, R., Pinxten, R., and Eens, M. 2000. Can excrement and feathers of nestling songbirds be used as biomonitors for heavy metal pollution? *Archiv. Environ. Contamin. Toxicol.* 39: 541–546.

Dauwe, T., Janssens, E., Bervoets, L., Blust, R., and Eens, M. 2004. Relationships between metal concentrations in great tit nestlings and their environment and food. *Environ. Pollut.* 131: 373–380.

Eens, M., Pinxten, R., Verheyen, R. F., Blust, R., and Bervoets, L. 1999. Great and blue tits as indicators of heavy metal contamination in terrestrial ecosystems. *Ecotoxicol. Environ. Saf.* 44: 81–85.

Eeva, T., Lehikoinen, E., and Rönkä, M. 1998. Air pollution fades the plumage of the great tit. *Funct. Ecol.* 12: 607–612.

Eraud, C., Devevey, G., Gaillard, M., Prost, J., Sorci, G., and Faivre, B. 2007. Environmental stress affects the expression of a carotenoid-based sexual trait in male zebra finches. *J. Exp. Biol.* 210: 3571–3578.

Fitze, P. S., Kolliker, M., and Richner, H. 2003a. Effects of common origin and common environment on nesting plumage coloration in the great tit (*Parus major*). *Evol.* 57: 144–150.

Fitze, P. S., and Tschirren, B. 2006. No evidence for survival selection on carotenoid-based nesting coloration in great tits (*Parus major*). *J. Evol. Biol.* 19: 618–624.

Fitze, P. S., Tschirren, B., and Richner, H. 2003b. Carotenoid-based colour expression is determined early in nestling life. *Oecologia* 137: 148–152.

Fitze, P. S., Tschirren, B., Gasparini, J., and Richner, J. 2007. Carotenoid-based plumage colors and immune function: Is there a trade-off for rare carotenoids? *Am. Nat.* 169: S137–S144.

Gosler, A. G. 1993. *The Great Tit*. Hamlyn, London.

Grafen, A. 1990. Biological signals as handicaps. *J. Theor. Biol.* 144: 517–546.

Gray, D. A. 1996. Carotenoids and sexual dichromatism in North American passerines. *Am. Nat.* 148: 453–480.

Grether, G. F., Hudon, J., and Millie, D. F. 1999. Carotenoid limitation of sexual coloration along an environmental gradient in guppies. *Proc. R. Soc. Lond. B* 266: 1317–1322.

Hadfield, J. D. and Owens, I. P. F. 2006. Strong environmental determination of a carotenoid-based plumage trait is not mediated by carotenoid-availability. *J. Evol. Biol.* 19: 1104–1114.

Hartley, R. C. and Kennedy, M. W. 2004. Are carotenoids a red herring in sexual display? *Trends Ecol. Evol.* 19: 353–354.

Hill, G. E. 1990. Female house finches prefer colorful males: Sexual selection for a condition-dependent trait. *Anim. Behav.* 40: 563–572.

Hill, G. E. 1991. Plumage coloration is a sexually selected indicator of male quality. *Nature* 350: 337–339.

Hill, G. E. 1992. The proximate basis of variation in carotenoid pigmentation in male house finches. *Auk* 109: 1–12.
Hill, G. E. 1993. Geographic variation in the carotenoid plumage pigmentation of male house finches (*Carpodacus mexicanus*). *Biol. J. Linn. Soc.* 49: 63–86.
Hill, G. E. 1994. Geographic variation in male ornamentation and female mate preference in the house finch: A comparative test of models of sexual selection. *Behav. Ecol.* 5: 64–73.
Hill, G. E. 1999. Mate choice, male quality, and carotenoid-based plumage coloration. *Proc. Int. Ornithol. Congr.* 22: 1654–1668.
Hill, G. E. 2000. Energetic constraints on expression of carotenoid-based plumage coloration. *J. Avian Biol.* 31: 559–566.
Hill, G. E. 2001. Pox and plumage coloration in the house finch: A critique of Zahn and Rothstein. *Auk* 118: 256–260.
Hill, G. E. 2002. *A Red Bird in a Brown Bag: The Function and Evolution of Colorful Plumage in the House Finch*. Oxford University Press, New York.
Hill, G. E. 2006. Environmental regulation of ornamental coloration. In *Bird Coloration. Volume I. Mechanisms and Measurements* (Hill, G. E. and McGraw, K. J., eds.). Harvard University Press, Cambridge, MA, pp. 507–560.
Hill, G. E., Farmer, K. L., and Beck, M. L. 2004. The effect of mycoplasmosis on carotenoid plumage coloration in male house finches. *J. Exp. Biol.* 207: 2095–2099.
Hill, G. E., Inouye, C. Y., and Montgomerie, R. 2002. Dietary carotenoids predict plumage coloration in wild house finches. *Proc. R. Soc. Lond. B* 269: 1119–1124.
Hillgarth, N. and Wingfield, J. C. 1997 Parasite-mediated sexual selection: endocrine aspects. In *Host–Parasite Evolution. General Principles and Avian Models* (Clayton, D. H. and Moore, J., eds.). Oxford University Press, Oxford, pp. 78–104.
Hõrak, P. 1993. Low fledging success of urban great tits. *Ornis Fennica* 70: 168–172.
Hõrak, P. and Lebreton, J.-D. 1998. Survival of adult great tits in relation to sex and habitat: A comparison of urban and rural populations. *Ibis* 140: 205–209.
Hõrak, P., Ots, I., Vellau, H., Spottiswoode, C., and Møller, A. P. 2001. Carotenoid-based plumage coloration reflects hemoparasite infection and local survival in breeding great tits. *Oecologia* 126: 166–173.
Hõrak, P., Surai, P. F., and Møller, A. P. 2002. Fat-soluble antioxidants in the eggs of great tits *Parus major* in relation to breeding habitat and laying sequence. *Avian Sci.* 2: 123–130.
Hõrak, P., Vellau, H., Ots, I., and Møller, A. P. 2000. Growth conditions affect carotenoid-based plumage coloration of great tit nestlings. *Naturwissenschaften* 87: 460–464.
Inouye, C. Y., Hill, G. E., Stradi, R. D., and Montgomerie, R. 2001. Carotenoids pigment in male house finch plumage in relation to age, subspecies, and ornamental coloration. *Auk* 118: 900–915.
Isaksson, C. and Andersson, S. 2007. Carotenoid diet and nestling provisioning in urban and rural great tits *Parus major. J. Avian Biol.* 38: 564–572.
Isaksson, C. and Andersson, S. 2008. Oxidative stress does not influence carotenoid mobilization and plumage pigmentation. *Proc. R. Soc. Lond. B* 275: 309–314.
Isaksson, C., McLaughlin, P., Monaghan, P., and Andersson, S. 2007. Carotenoid pigmentation does not reflect total non-enzymatic antioxidant activity in plasma of adult and nestling great tits, *Parus major. Funct. Ecol.* 21: 1123–1129.
Isaksson, C., Örnborg, J., Stephensen, E., and Andersson, S. 2005. Plasma glutathione and carotenoid coloration as potential biomarkers of environmental stress in great tits. *EcoHealth* 2: 138–146.
Jawor, J. M. and Breitwisch, R. 2003. A unique female ornament display in northern cardinals. *Wilson Bull.* 115: 464–467.
Jawor, J. M. and Breitwisch, R. 2006. Is mate provisioning predicted by ornamentation? A test with northern cardinals (*Cardinalis cardinalis*). *Ethology* 112: 888–895.
Jawor, J. M., Gray, N., Beall, S. M., and Breitwisch, R. 2004. Multiple ornaments as indicators of individual quality in female northern cardinals (*Cardinalis cardinalis*). *Anim. Behav.* 67: 875–882.
Jawor, J. M., Linville, S. U., Beall, S. M., and Breitwisch, R. 2003. Assortative mating by multiple ornaments in northern cardinals (*Cardinalis cardinalis*). *Behav. Ecol.* 14: 515–520.
Johnstone, R. A. 1995. Sexual selection, honest advertisement and the handicap principle: Reviewing the evidence. *Biol. Rev.* 70: 1–65.
Kimball, R. T. 2006. Hormonal control of coloration. In *Bird Coloration, Volume I, Mechanisms and Measurements* (Hill, G. E. and McGraw, K. J., eds.). Harvard University Press, Cambridge, MA, pp. 431–468.

Kohen, R. and Nyska, A. 2002. Oxidation of biological systems: Oxidative stress phenomena, antioxidants, redox reactions, and methods for their quantification. *Toxicol. Pathol.* 30: 620–650.

Koutsos, E. A., Clifford, A. J., Calvert, C. C., and Klasing, K. C. 2003. Maternal carotenoid status modifies the incorporation of dietary carotenoids into immune tissues of growing chickens (*Gallus gallus domesticus*). *J. Nutr.* 133: 1132–1138.

Koutsos, E. A., Garcia Lopez, J. C., and Klasing, K. C. 2006. Carotenoids from in ovo or dietary sources blunt systemic indices of the inflammatory response in growing chicks (*Gallus gallus domesticus*). *J. Nutr.* 136: 1027–1031.

Kraaijeveld, K., Kraaijeveld-Smit, F. J. L., and Komdeur, J. 2007. The evolution of mutual ornamentation. *Anim. Behav.* 74: 657–677.

Linville, S. U. and Breitwisch, R. 1997. Carotenoid availability and plumage coloration in a wild population of northern cardinals. *Auk* 114: 796–800.

Linville, S. U., Breitwisch, R., and Schilling, A. J. 1998. Plumage brightness as an indicator of parental care in northern cardinals. *Anim. Behav.* 55: 119–127.

Llacuna, S., Gorriz, A., Sanpera, C., and Nadal, J. 1995. Metal accumulation in three species of passerine birds (*Emberiza cia, Parus major,* and *Turdus merula*) subjected to air-pollution from a coal-fired power plant. *Arch. Environ. Contam. Toxicol.* 28: 298–303.

Maynard-Smith, J. and Harper, D. 2003. *Animal Signals*. Oxford University Press, Oxford.

McGraw, K. J. 2006. Mechanics of carotenoid-based coloration. In *Bird Coloration, Volume I, Mechanisms and Measurements* (Hill, G. E. and McGraw, K. J, eds.). Harvard University Press, Cambridge, MA, pp. 177–242.

McGraw, K. J. and Ardia, D. R. 2003. Carotenoids, immunocompetence, and the information content of sexual colors: An experimental test. *Am. Nat.* 162: 704–712.

McGraw, K. J. and Klasing, K. C. 2006. Carotenoids, immunity, and integumentary coloration in red junglefowl (*Gallus gallus*). *Auk* 123: 1161–1171.

McGraw, K. J. and Parker, R. S. 2006. A novel lipoprotein-mediated mechanism controlling sexual attractiveness in a colourful songbird. *Physiol. Behav.* 87: 103–108.

McGraw, K. J., Adkins-Regan, E., and Parker, R. S. 2002. Anhydrolutein in the zebra finch: A new, metabolically derived carotenoid in birds. *Comp. Biochem. Physiol. B* 132: 811–818.

McGraw, K. J., Adkins-Regan, E., and Parker, R. S. 2005. Maternally derived carotenoid pigments affect offspring survival, sex ratio, and sexual attractiveness in a colourful songbird. *Naturwissenschaften* 92: 375–380.

McGraw, K. J., Correa, S. M., and Adkins-Regan, E. 2006a. Testosterone upregulates lipoprotein status to control sexual attractiveness in a colorful songbird. *Behav. Ecol. Sociobiol.* 60: 117–122.

McGraw, K. J., Gregory, A. J., Parker, R. S., and Adkins-Regan, E. 2003. Diet, plasma carotenoids, and sexual coloration in the zebra finch (*Taeniopygia guttata*). *Auk* 120: 400–410.

McGraw, K. J., Hill, G. E., Stradi, R., and Parker, R. S. 2001. The influence of carotenoid acquisition and utilization on the maintenance of species-typical plumage pigmentation in male American goldfinches (*Carduelis tristis*) and northern cardinals (*Cardinalis cardinalis*). *Physiol. Biochem. Zool.* 74: 843–852.

McGraw, K. J., Nolan, P. M., and Crino, O. L. 2006b. Carotenoid accumulation strategies for becoming a colorful house finch: Analyses of plasma and liver pigments in wild molting birds. *Funct. Ecol.* 20: 678–688.

Miller, M. W. 1999. Relative effects of plumage coloration and vegetation density on nest success. *Condor* 101: 255–261.

Miller, N. J., Sampson, J., Candeias, L. P., Bramley, P. M., and Rice-Evans, C. A. 1996. Antioxidant activities of carotenes and xanthophylls. *FEBS Lett.* 384: 240–242.

Møller, A. P., Biard, C., Blount, J. D., Houston, D. C., Ninni, P., Saino, N., and Surai, P. F. 2000. Carotenoid-dependent signals: Indicators of foraging efficiency, immunocompetence, or detoxification ability? *Avian Poult. Biol. Rev.* 11: 137–159.

Nowicki, S., Peters, S., and Podos, J. 1998. Song learning, early nutrition and sexual selection in songbirds. *Am. Zool.* 38: 179–190.

Ohlsson, T., Smith, H. G., Raberg, L., and Hasselquist, D. 2002. Pheasant sexual ornaments reflect nutritional conditions during early growth. *Proc. R. Soc. Lond. B* 269: 21–27.

Ohlsson, T., Smith, H. G., Raberg, L., and Hasselquist, D. 2003. Effects of nutrition on sexual ornaments and humoral immune responsiveness in adult male pheasants. *Ethol. Ecol. Evol.* 15: 31–42.

Olson, V. A. and Owens, I. P. F. 1998. Costly sexual signals: Are carotenoids rare, risky or required? *Trends Ecol. Evol.* 13: 510–514.

Papeschi, A., Carroll, J. P., and Dessi-Fulgheri, F. 2003. Wattle size is correlated with male territorial rank in juvenile ring-necked pheasants. *Condor* 105: 362–366.

Partali, V., Liaaen-Jensen, S., Slagsvold, T., and Lifjeld, J. T. 1987. Carotenoids in food chain studies. II. The food chain of *Parus* spp. monitored by carotenoid analysis. *Comp. Biochem. Physiol. B* 87: 885–888.

Perrins, C. M. 1991. Tits and their caterpillar food supply. *Ibis* 133: 49–54.

Peters, A. 2007. Testosterone and carotenoids: An integrated view of trade-offs between immunity and sexual signaling. *BioEssays* 29: 427–430.

Price, D. K. 1996. Sexual selection, selection load and quantitative genetics of zebra finch bill colour. *Proc. R. Soc. Lond. B* 263: 217–221.

Price, D. K. and Burley, N. T. 1993. Constraints on the evolution of attractive traits: Genetic (co)variance of zebra finch bill colour. *Heredity* 71: 405–412.

Price, D. K. and Burley, N. T. 1994. Constraints on the evolution of attractive traits: Selection in male and female zebra finches. *Am. Nat.* 144: 908–934.

Pryke, S. R. and Andersson, S. 2002. A generalized female bias for long tails in a short-tailed widowbird. *Proc. R. Soc. Lond. B* 269: 2141–2146.

Pryke, S. R. and Andersson, S. 2003a. Carotenoid-based epaulettes reveal male competitive ability: Experiments with resident and floater red-shouldered widowbirds. *Anim. Behav.* 66: 217–224.

Pryke, S. R. and Andersson, S. 2003b. Carotenoid-based status signaling in red-shouldered widowbirds (*Euplectes axillaris*): Epaulet size and redness affect captive territorial competition. *Behav. Ecol. Sociobiol.* 53: 393–401.

Pryke, S. R. and Andersson, S. 2005. Experimental evidence for female choice and energetic costs of male tail elongation in red-collared widowbirds. *Biol. J. Linn. Soc.* 86: 35–43.

Pryke, S. R., Andersson, S., and Lawes, M. J. 2001a. Sexual selection of multiple handicaps in the red-collared widowbird: Female choice of tail length but not carotenoid display. *Evolution* 55: 1452–1463.

Pryke, S. R., Andersson, S., Lawes, M. J., and Piper, S. E. 2002. Carotenoid status signaling in captive and wild red-collared widowbirds: Independent effects of badge size and color. *Behav. Ecol.* 13: 622–631.

Pryke, S. R., Lawes, M. J., and Andersson, S. 2001b. Agonistic carotenoid signaling in male red-collared widowbirds: Aggression related to the colour signal of both the territory owner and model intruder. *Anim. Behav.* 62: 695–704.

Rohwer, S. A. 1975. The social significance of avian winter plumage variability. *Evolution* 29: 593–610.

Rubolini, D., Romano, M., Martinelli, R., Leoni, B., and Saino, N. 2006. Effects of prenatal yolk androgens on armaments and ornaments of the ring-necked pheasant. *Behav. Ecol. Sociobiol.* 59: 549–560.

Saino, N., Ferrari, R., Romano, M., Martinelli, R., and Møller, A. P. 2003. Experimental manipulation of egg carotenoids affects immunity of barn swallow nestlings. *Proc. R. Soc. Lond. B* 270: 2485–2489.

Selvaraj, R. K., Koutsos, E. A., Calvert, C. C., and Klasing, K. C. 2006. Dietary lutein and fat interact to modify macrophage properties in chicks hatched from carotenoid deplete or replete eggs. *J. Anim. Physiol. Anim. Nutr.* 90: 70–80.

Slagsvold, T. and Lifjeld, J. T. 1985. Variation in plumage colour of the great tit *Parus major* in relation to habitat, season and food. *J. Zool.* 206: 321–328.

Smith, H. G., Raberg, L., Ohlsson, T., Granbom, M., and Hasselquist, D. 2007. Carotenoid and protein supplementation have differential effects on pheasant ornamentation and immunity. *J. Evol. Biol.* 20: 310–319.

Stradi, R., Celentano, G., and Nava, D. 1995. Separation and identification of carotenoids in bird's plumage by high-performance liquid chromatography-diode-array detection. *J. Chromatogr. B* 670: 337–348.

Trams, E. G. 1969. Carotenoid transport in the plasma of the scarlet ibis (*Eudocimus ruber*). *Comp. Biochem. Physiol.* 28: 1177–1184.

Tschirren, B., Fitze, P. S., and Richner, H. 2003. Proximate mechanisms of variation in the carotenoid-based plumage coloration of nestling great tits (*Parus major* L.). *J. Evol. Biol.* 16: 91–100.

Tschirren, B., Fitze, P. S., and Richner, H. 2005. Carotenoid-based nestling coloration and parental favouritism in the great tit (*Parus major*). *Oecologia* 143: 477–482.

Volker, O. 1938. The dependence of lipochrome-formation in birds on plant carotenoids. *Proc. Int. Ornithol. Congr.* 8: 425–426.

von Schantz, T., Bensch, S., Grahn, M., Hasselquist, D., and Wittzell, H. 1999. Good genes, oxidative stress and condition-dependent sexual signals. *Proc. R. soc. Lond. B* 266: 1–12.

Wallace, A. R. 1889. *Darwinism: An Exposition of the Theory of Natural Selection with Some of Its Applications.* Macmillan, London.

Wolfenbarger, L. L. 1999a. Red coloration of male northern cardinals correlates with mate quality and territory quality. *Behav. Ecol.* 10: 80–90.
Wolfenbarger, L. L. 1999b. Is red coloration of male northern cardinals beneficial during the nonbreeding season?: A test of status signaling. *Condor* 101: 655–663.
Wolfenbarger, L. L. 1999c. Female mate choice in northern cardinals: Is there a preference for redder males? *Wilson Bull.* 111: 76–83.
Zahavi, A. 1975. Mate selection—A selection for a handicap. *J. Theor. Biol.* 53: 205–214.
Zahn, S. N., and Rothstein, S. I. 1999. Recent increase in male house finch plumage variation and its possible relationship to avian pox disease. *Auk* 116: 35–44.

24 Transport of Carotenoids by a Carotenoid-Binding Protein in the Silkworm

Takashi Sakudoh and Kozo Tsuchida

CONTENTS

24.1 INTRODUCTION

The concentrations and types of carotenoids in animals differ substantially among tissues. They occasionally undergo significant chemical transformations, and they can be a critical factor for survival. For example, the green larvae of the swallowtail butterfly, *Papilio xuthus*, become green pupae when the larvae pupate on green twigs of plants. The green pupal cuticles contain β-carotene and lutein (Ohnishi 1959). However, if pupation takes place on dead branches, brown pupae are usually produced, likely in response to odorant stimuli (Hidaka 1961). Neither β-carotene nor lutein have been detected in the brown pupal cuticle, but papilioerythrin and an astaxanthin-like carotenoid, presumably canthaxanthin, were found (Ohnishi 1959, Harashima et al. 1972). The concentrations and types of carotenoids in the pupae of *P. xuthus* vary significantly among hemolymph, fat bodies, and cuticles (Harashima et al. 1972). A protective role of the color of the cuticle has been proposed based on an experiment with fowl as the predator (Hidaka et al. 1959), which suggested that the carotenoid composition is a critical factor for survival. Carotenoid composition is also significant in human medicine. For example, high macular levels of lutein and zeaxanthin are associated with a decreased risk of age-related macular degeneration (Loane et al. 2008), the main cause of blindness in the developed world.

Animals cannot synthesize carotenoids *de novo*. To deposit carotenoids in the proper tissues in the proper amounts, they must acquire carotenoids from dietary sources and transport them to target sites. Knowledge of the molecular mechanisms of carotenoid transport, however, is still

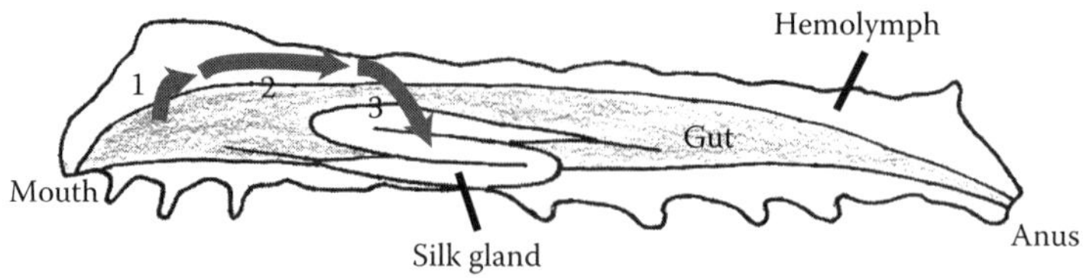

FIGURE 24.1 The pathway of carotenoid transport in the silkworm. Carotenoids are absorbed from dietary mulberry leaves into the intestinal mucosa, transferred to the hemolymph (1), transported in the hemolymph by plasma lipoproteins (2), and accumulated in the silk gland (3).

poor. With rare exceptions, we do not know what genes control the concentrations and forms of carotenoid in each tissue.

How can the molecular mechanism of the transport system of carotenoids be studied? A hint comes from the recognition that carotenoids are generally hydrophobic. In order to move carotenoids through the aqueous biological environment, carotenoid-binding proteins (CBPs) that cover the hydrophobic surface of carotenoids are thought to be needed. In particular, selective uptake of certain types of carotenoids likely demands highly specific CBPs. Thus, identification and analysis of CBPs are expected to provide new insights into the transport system of carotenoids (Bhosale and Bernstein 2007).

The silkworm, *Bombyx mori*, is a good model organism for studying CBPs for the following reasons. Wild-type silkworm larvae feed on carotenoid-rich mulberry leaves, where carotenoids are absorbed into the intestinal mucosa and transferred to the hemolymphal lipoprotein called lipophorin (Chino 1985). Here, the carotenoids are accumulated in the silk gland, the largest tissue in the late stage of the last instar and the site of silk protein synthesis (Figure 24.1), resulting in the formation of a yellow cocoon. In addition, the silkworm is relatively large and easily reared in large numbers, allowing us to obtain substantial amounts of carotenoid-rich materials for purification of CBPs. Furthermore, during the long history of sericulture, several mutants that produce white cocoons due to defects in the carotenoid transport system have been found and maintained as genetic resources (Tazima 1964, Banno et al. 2005, Fujii and Banno 2005). We can investigate the relationship between CBPs and the mutants, which will enable us to dissect the transport system genetically.

The hemolymphal transport of carotenoids by lipophorin has been elucidated and documented (Law and Wells 1989, Tsuchida et al. 1998, Arrese et al. 2001, Canavoso et al. 2001), as has plasma transport by mammalian lipoproteins (Paker 1996, Yeum and Russell 2002). Lipophorin serves as a shuttle that moves carotenoids from one tissue to another without itself entering the cells, in stark contrast to the vertebrate low-density lipoprotein (LDL) (Brown and Goldstein 1986), which is endocytosed and metabolized in the cell. Here, we focus on the recent biochemical and genetic studies of the intracellular CBP of the silkworm, which mainly transports lutein. We hope this review provides insights into the studies of CBPs in other organisms.

24.2 IDENTIFICATION OF THE CAROTENOID-BINDING PROTEIN IN THE SILK GLAND

24.2.1 Purification and Cloning of CBP

The purification strategy for CBPs is conceptually simple. Proteins from carotenoid-rich tissues are separated under nondenaturing and relatively aqueous conditions where carotenoids are expected to remain bound to the CBPs. CBPs are then detected by the color of the carotenoids.

Several researchers have tried to isolate cellular CBPs from the silkworm. In Nakajima's study (1963), the whole midgut mucosa was homogenized and the proteins separated with a gel-filtration chromatography column. Carotenoids were found in certain fractions containing proteins, suggesting the existence of CBPs in the midgut. Jouni and Wells purified a 35 kDa protein containing lutein from the larval midgut, which they named lutein-binding protein (LBP) (Jouni and Wells 1996). Immunoblotting analysis indicated that LBP is distributed in the midgut, silk gland, fat body, ovary, and testis. The sequences of these proteins, however, have not been reported to date.

Purification and cloning of the silkworm CBP from the silk gland were first reported in 2002 (Tabunoki et al. 2002). First, the larval silk glands were dissected out with ice-cold phosphate-buffered saline and centrifuged. The supernatant was purified by a combination of ammonium sulfate fractionation and four chromatographic procedures: anion exchange chromatography, gel filtration, chromatofocusing, and hydroxyapatite fractionation. Three yellow fractions were obtained after the anion exchange chromatography, and the flow-through fraction, which showed the highest carotenoid:protein ratio along with a characteristic carotenoid spectrum, was used in the subsequent purification. The purified protein, named CBP, is a 33 kDa protein. The bound carotenoids were extracted from CBP and identified as lutein (88%), β-carotene (9%), and α-carotene (3%), which is consistent with the carotenoid composition of lipophorin (Tsuchida et al. 2004b). The carotenoid: protein ratio is calculated as approximately 1:1, which is the typical ratio of general intracellular lipid-transfer proteins (Holthuis and Levine 2005). The absorption spectrum of CBP is characterized by three absorbance maxima in the visible region at 436, 461, and 493 nm. These absorbance maxima represent a significant red shift of 22 nm compared with the lutein spectrum in hexane, and can be attributed to a bathochromic shift upon binding to CBP.

To obtain the sequence for CBP, a cDNA library from the silk gland was prepared and screened using a anti-CBP antibody. One positive clone was identified from 200,000 plaques, and rapid amplification of 5′ complementary DNA ends (5′-RACE) was performed to obtain the full length of the CBP cDNA (Figure 24.2a). The predicted sequence encodes a 297-residue polypeptide of

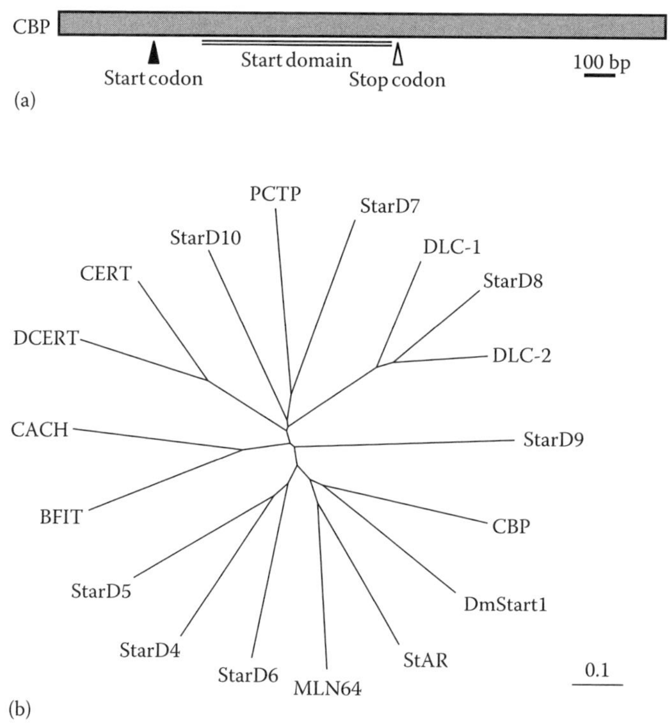

FIGURE 24.2 CBP is a member of a START domain–containing gene family. (a) Schematic structure of the CBP cDNA sequence. (b) Phylogenetic tree of CBP, DmStart1, and DCERT of the fruit fly *Drosophila melanogaster*, and 15 mammalian START domain–containing genes. The tree was constructed using ClustalW base on an amino acid sequences of their START domains.

molecular mass of 33.6 kDa, consistent with the mass of the purified CBP. The deduced amino acid sequence agreed with the three polypeptide sequences acquired from the digestion of the purified CBP. Recombinant CBP produced with *Escherichia coli* was recognized by the anti-CBP antibody, providing further support that the obtained clone encodes CBP.

24.2.2 Characterization of the CBP Sequence and Its Mammalian Homologs

CBP is predicted by PSORT analysis to be a cytoplasmic protein (Horton et al. 2007) and to contain a steroidogenic acute regulatory protein (StAR)–related lipid transfer (START) domain of 200 amino acids at its C-terminus (Figure 24.2a).

The START domain–containing gene family is conserved in plants and animals and is thought to serve as a versatile binding interface for lipids and to function in intracellular lipid transport (Soccio and Breslow 2003, Schrick et al. 2004, Alpy and Tomasetto 2005). The properties and function of the START domain have been most intensively investigated in mammalian START domain–containing genes (Figure 24.2b). The x-ray crystal structures of four mammalian START domains have been reported: apo-forms of metastatic lymph node 64 (MLN64) (Tsujishita and Hurley 2000), StarD4 (Romanowski et al. 2002), and CERT (Kudo et al. 2008), and lipid-bound forms of phosphatidylcholine transfer protein (PCTP) (Roderick et al. 2002) and CERT (Kudo et al. 2008). All structures share the same fold with nine β-strands, four α-helices, and an internal cavity to accommodate lipid binding. By producing mutants of StAR and CERT (including DCERT, the ortholog of CERT in the fruit fly *Drosophila melanogaster*), researchers have shown these proteins to be necessary for the transport of cholesterol and ceramide, respectively (Caron et al. 1997, Hanada et al. 2003, Rao et al. 2007). StAR mediates the transport of cholesterol from the outer to the inner mitochondrial membrane, which is the rate-limiting step in mammalian steroidogenesis (Stocco 2001, Miller 2007). CERT extracts ceramide from the membrane of the endoplasmic reticulum and transports it to the membrane of the Golgi apparatus in a nonvesicle manner (Hanada et al. 2007). The START domain–containing gene family is also significant from a medical viewpoint: StAR is a gene responsible for congenital lipoid adrenal hyperplasia (Lin et al. 1995), and CERT is a key regulator of the response of cancer cells to taxanes, important anticancer agents (Swanton et al. 2007). To date, no START domain–containing gene other than CBP has been reported to bind or transport carotenoids.

A homology search and phylogenetic analysis of the mammalian START domain–containing gene family revealed that the START domain of CBP is most homologous to those of StAR (~26% identity at the amino acid level) and MLN64 (~29% identity) (Figure 24.2b). MLN64 contains four transmembrane (TM) helices called a MENTAL domain in the N-terminus and a START domain in the C-terminus (Alpy and Tomasetto 2006). The MENTAL domain localizes MLN64 to the late endosome membrane with the START domain facing the cytosol (Alpy et al. 2001). The START domain of the MLN64 protein was shown to bind cholesterol (Tsujishita and Hurley 2000, Holtta-Vuori et al. 2005). MLN64 is implicated in the cellular transport of cholesterol by *in vitro* studies utilizing cultured cells (Alpy and Tomasetto 2006); however, mice with a targeted mutation in the MLN64 START domain were fertile and exhibited only modest alterations in cellular sterol metabolism (Kishida et al. 2004).

24.2.3 Distribution of CBP

Western analysis revealed that CBP is expressed not only in the silk gland but also in several tissues with yellow color: the midgut, testis, and ovary (Figure 24.3a) (Tabunoki et al. 2002). However, expression of CBP was not observed in tissues lacking carotenoids, such as the fat body, the malpighian tubule, and the integument. Expression of CBP in the silk gland is clearly limited to the middle division (Tabunoki et al. 2002, Tsuchida et al. 2004a). In the silk gland, the portion where carotenoid transport occurs is similarly limited to the middle division (Nakajima 1963). Studies

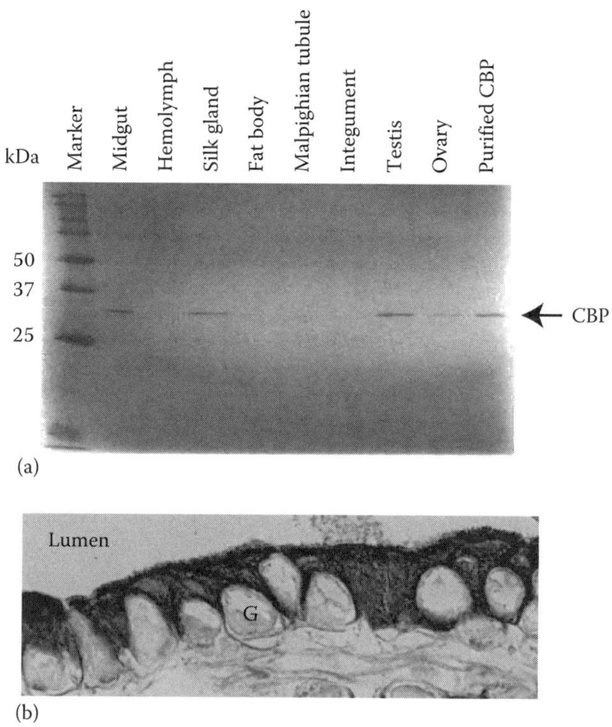

FIGURE 24.3 Distribution of CBP. (a) Analysis of tissue distribution by Western blotting. (b) Immunohistochemistry of the midgut epithelium. CBP was not detected in the goblet cells (G) of the midgut epithelium, which are thought to be involved in ion transport.

of the developmental profile of CBP expression in the midgut revealed that CBP reached the highest concentration on days 4 and 5 of the fifth larval instar, which coincides with the highest food consumption (Tsuchida et al. 2004a). Immunohistochemistry demonstrated that CBP was located in the epithelial cells of the midgut and silk gland that are in contact with the intestinal lumen and hemolymph, respectively (Figure 24.3b) (Tsuchida et al. 2004a). In the midgut, CBP was uniformly expressed along the brush border of the columnar cells, providing a large surface for carotenoid absorption. These data on the distribution of CBP are consistent with the view that CBP is involved in the cellular uptake of carotenoids from the diet or lipophorin at the apical membrane and that CBP functions as a key determinant of the concentration of carotenoids in each tissue.

24.3 LINK BETWEEN CBP AND A GENETIC LOCUS RESPONSIBLE FOR CAROTENOID TRANSPORT

24.3.1 Silkworm Genes Responsible for Carotenoid Transport

The expression profile of CBP suggests its involvement in carotenoid transport, which led to the investigation of the relationship between CBP and the genetic loci for carotenoid transport. The genetic loci for carotenoid transport are summarized in Figure 24.4a. Among them, we describe the *Y* (*yellow blood*), *I* (*yellow inhibitor*), and *C* (*yellow cocoon*) genes. The functions of these three loci are illustrated diagrammatically in Figure 24.4b. The *Y* gene (Toyama 1906, 1912) controls the uptake of carotenoids from the midgut lumen into the intestinal mucosal cells and from the hemolymph into the cells in the middle division of the silk gland (Nakajima 1963). Larvae of mutants with the $+^Y/+^Y$ genotype ($+^Y$ indicates a recessive allele of the *Y* gene) inadequately absorb dietary

Symbol of gene	Locus	Phenotype
Acp	?	Light yellow hemolymph, even though individuals have *Y* and *I*
C	12–7.2	Making yellow cocoon in combination with *Y* and $+^{I}$
F	6–34.7	Making yellowish flesh cocoon in combination with *Y* and $+^{I}$
I	9–16.2	Colorless hemolymph, even though individuals have *Y*
Pk	2–?	Making pink cocoon in combination with *Y*, $+^{I}$ and *F*
Y	2–28.6	Yellow hemolymph
Ymc	?	Making light yellow cocoon in combination with Y^{A}, an allele of *Y*

(a)

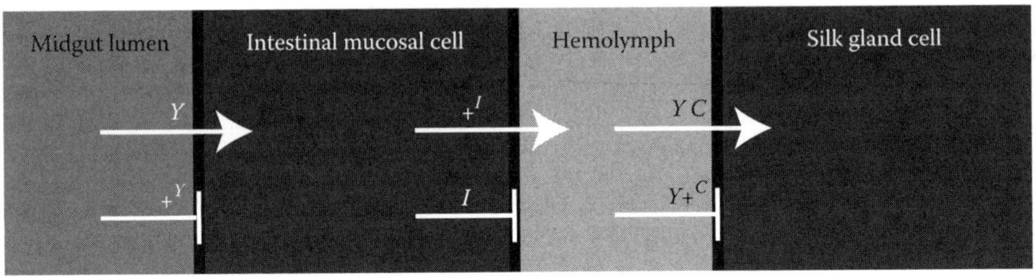

(b)

FIGURE 24.4 Silkworm genetic loci responsible for carotenoid transport. (a) List of the genetic loci. $+^{I}$ indicates a recessive allele of *I*. (b) Schematic illustration of the function of the *Y*, *I*, and *C* genes. Only larvae with the genotype [*Y* $+^{I}$ *C*] transport carotenoids into the silk gland and create yellow cocoons.

carotenoids, resulting in colorless (carotenoid-deficient) hemolymph. The *I* gene (Toyama 1912) controls the transfer of carotenoids from the intestinal mucosal cells into the hemolymph (Nakajima 1963). Larvae of mutants with the dominant *I* allele are incompetent at transferring carotenoids from intestinal mucosal cells, resulting in the accumulation of carotenoids in the intestinal mucosa. A larval hemolymph of the *I* allele strain is therefore colorless even if the strain bears the *Y* allele. The *C* gene (Uda 1919) controls the uptake of carotenoids, especially xanthophylls, from the hemolymph into the cells in the middle part of the middle division of the silk gland. In contrast to the *Y* gene, the *C* gene does not affect the uptake of carotenoids from the midgut lumen into the midgut epithelium (Nakajima 1963). Only larvae with the genotype [*Y* $+^{I}$ *C*] create yellow cocoons. All other combinations make white cocoons.

24.3.2 CBP Is a Product of the *Yellow Blood* Gene

Western blotting and immunohistochemistry clearly demonstrated that CBP is expressed in the larvae of the *Y* allele strain and is not expressed in the larvae of the $+^{Y}$ allele strain (Figure 24.5a) (Tabunoki et al. 2002, Tsuchida et al. 2004a). The genotype with respect to *I* and *C* genes had no relationship with the level of the expression of CBP (Tabunoki et al. 2002, Tsuchida et al. 2004a). Restriction fragment length polymorphism (RFLP) mapping with a cDNA probe of the CBP gene revealed that the CBP gene was on the second chromosome, which is the same as the *Y* gene and different from the *I* (the 9th chromosome) and the *C* (the 12th chromosome) genes (Hara et al. 2007). As indicated above, CBP is expressed in both the midgut and silk gland, where the *Y* gene functions. These observations suggested that the *Y* gene corresponds directly to the CBP gene.

Next, the difference in the genomic and cDNA sequences of CBP between the *Y* and $+^{Y}$ alleles were determined (Figure 24.5b) (Sakudoh et al. 2005, 2007). The *Y* allele strain had at least two copies of the CBP gene, all of which could be classified into one of two types: a *Y*-a sequence

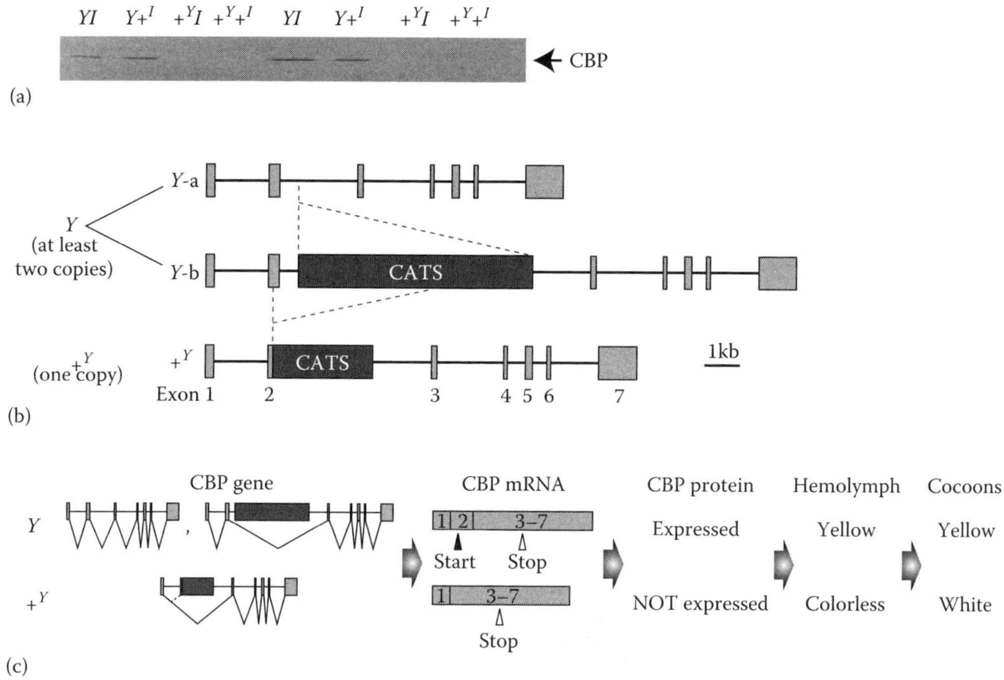

FIGURE 24.5 Link between CBP and the *Y* gene. (a) Expression analysis of CBP in *Y* and *I* mutants by Western blotting. CBP is expressed only in the *Y* allele strain. (b) Schematic structure of the CBP genomic sequence in the *Y* and $+^Y$ allele strains. Note that the copy number differs between strains. CATS is a non-LTR retrotransposon. CATS was named after CBP-associated TRAS-like sequence because its sequence has homology with TRAS, a site-specific retrotransposon that targets telomeric repeats of silkworm, while the insertion site of CATS does not contain telomeric repeats (Sakudoh et al. 2005). (c) Model of the mechanism by which the difference in the CBP gene affects CBP protein expression and the resulting phenotype. In the $+^Y$ allele strain, the splicing out of exon 2, likely associated with the CATS insertion and subsequent genomic deletion, generates a nonfunctional mRNA devoid of the true start methionine. This results in an inability to produce CBP protein and the formation of colorless hemolymph and white cocoons.

comprised of seven exons, or a *Y*-b sequence containing a non-LTR retrotransposon named CATS between exon 2 and exon 3 of the *Y*-a sequence. The $+^Y$ allele strain contained a single copy of the $+^Y$ sequence, which contains a truncated CATS coupled with a 169 bp deletion at the 3′ terminus of exon 2 corresponding to the open reading frame of the CBP gene. Deletion of the 4.5 kb region of the *Y*-b sequence, including the 3′ terminus of exon 2 and the 5′ terminus of CATS, is supposed to have produced the $+^Y$ sequence. Northern blotting and reverse transcription-PCR analysis revealed that exon 2 was absent in the CBP cDNA of the $+^Y$ allele strain (Sakudoh et al. 2007). The lack of exon 2 containing the start methionine results in the absence of expression of CBP, leading to colorless hemolymph and white cocoons in the $+^Y$ allele strain (Figure 24.5c). Thus, the CBP gene of the $+^Y$ allele strain is considered a null allele.

In the silkworm, a stable germline transformation has been developed (Tamura et al. 2000), and tissue-specific expression with the binary GAL4/upstream activating sequence system can be induced (Imamura et al. 2003, Uchino et al. 2006). The recovery of carotenoid uptake by the expression of CBP into the colorless hemolymph strain in this transgenic system was examined (Sakudoh et al. 2007). CBP was successfully expressed in the midgut and silk gland and the yellow color of the hemolymph and cocoon was restored, verifying the function of CBP as the *Y* gene product. A hypomorphic phenotype of cocoon decoloration by the injection of a double-stranded RNA probe of the CBP gene is also consistent with the function of the *Y* gene (Tabunoki et al. 2004).

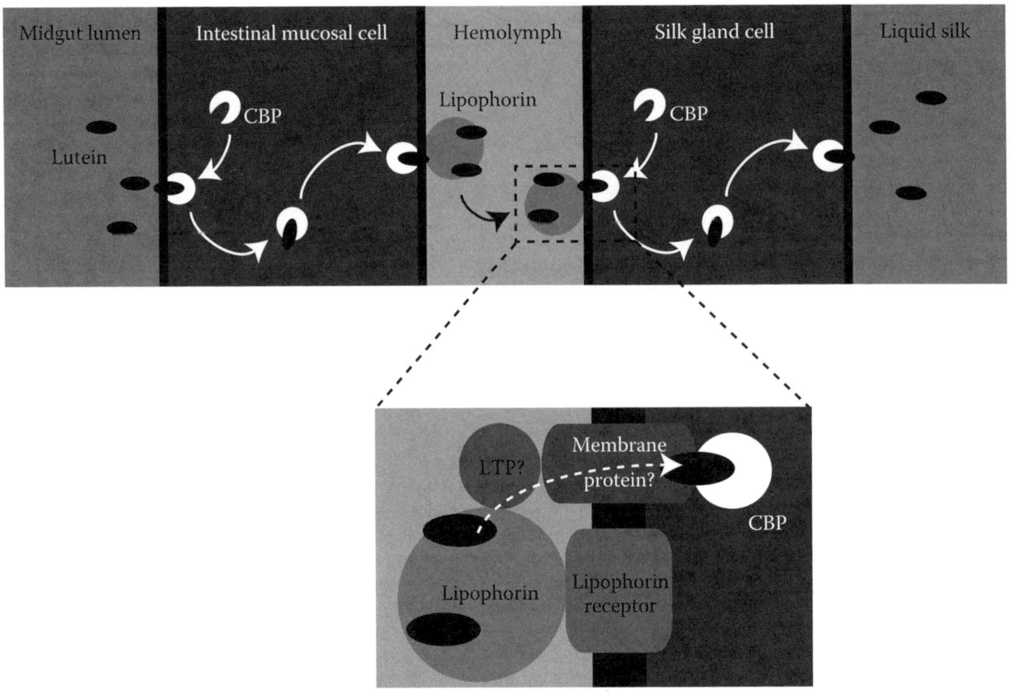

FIGURE 24.6 Model of the molecular function of CBP. CBP moves in the cytosol and relays carotenoid in combination with the lipophorin in the hemolymph. At the membrane, other factors might be involved in the carotenoid transport (magnification, see the text for explanation).

Although precise mapping is needed to draw a strict conclusion, all of the data strongly suggest that the *Y* gene corresponds directly to the CBP gene. Furthermore, CBP is the first intracellular molecule involved in carotenoid transport to be genetically characterized. Figure 24.6 presents a "gondola" model for CBP function in which CBP picks up a load (carotenoid) on one bank (apical membrane), moves it into the lake (cytosol) in a boatlike manner, and sets it down on the other bank (basal membrane). Immunohistochemical and confocal microscopic analysis (Alpy et al. 2001), or live imaging of mutants tagged with fluorescent protein (Zhang et al. 2002) might provide insights into the validity of this model.

Analysis of lipid composition differences in the hemolymph, midgut, fat body, and silk gland between the *Y* and $+^Y$ allele strains suggested that the *Y* gene does not affect the metabolism of general lipids, hydrocarbon, triacylglycerol, diacylglycerol, cholesterol, or phospholipid (Tsuchida et al. 2004b). This specificity is expected to be determined by the ligand specificity of CBP for carotenoids.

24.4 MEMBRANE-SPANNING ISOFORM OF CBP

Considering the function of its homologous genes, CBP can be expected to have a function in cholesterol transport: StAR and MLN64, the closest mammalian homologs of CBP (Figure 24.2b), are known to bind and transport cholesterol. Furthermore, DmStart1, the closest homolog of CBP, StAR, and MLN64 in *D. melanogaster*, is highly expressed in the prothoracic gland cells where ecdysone, the molting hormone that plays a pivotal role in the control of the insect developmental schedule, is synthesized from cholesterol (Roth et al. 2004). DmStart1 contains a MENTAL domain in its N-terminus and a START domain in its C-terminus, as does MLN64. Its developmental expression pattern in the larval stage correlates with the humoral ecdysone level. No other homologs of StAR,

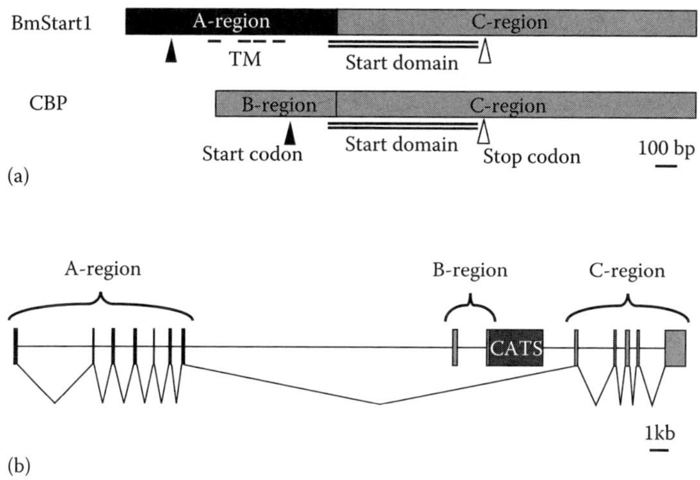

FIGURE 24.7 BmStart1, an alternative splicing isoform of CBP. (a) Schematic structure of BmStart1 and CBP cDNA sequences. BmStart1 consists of the BmStart1-specific A-region and the common C-region, while CBP consists of the CBP-specific B-region and the C-region. A-region codes the four putative TM helices. (b) Schematic structure of the BmStart1/CBP genomic sequence in the $+^Y$ allele strain. Connected lines indicate the structure of the BmStart1 cDNA. A-region and C-region are not disrupted by CATS, allowing the expression of BmStart1 in the $+^Y$ allele strain.

MLN64, and DmStart1 other than CBP could be found in the sequence database of *B. mori*. Does CBP play a role in ecdysteroidogenesis? What is the counterpart of StAR/MLN64 in the $+^Y$ allele strain, where CBP is absent?

It may be noteworthy that CBP has an alternative splicing isoform, BmStart1 (Figure 24.7a) (Sakudoh et al. 2005). BmStart1 is comprised of 12 exons, five of which are identical to the exon of the 3′ terminus of CBP (Figure 24.7b). BmStart1 shares the START domain in the C-terminus with CBP, and contains a MENTAL domain in its N-terminus. The exons of BmStart1 are not disrupted in the $+^Y$ allele strain, and BmStart1 was expressed in various tissues including ecdysteroidogenic tissues, prothoracic gland, testis, and ovary in both the *Y* and $+^Y$ allele strains. BmStart1 mRNA abundance in the prothoracic gland, the main ecdysteroidogenic tissue in *B. mori*, was relatively elevated in the latter larval stage, when ecdysteroid synthesis is highly active. Thus, BmStart1 could be a candidate for the counterpart of StAR/MLN64 in the silkworm, though it could also be a membrane-tethered CBP. If BmStart1 transports cholesterol and not carotenoid, an alternative splicing could produce unique protein isoforms whose endogenous ligands are structurally and functionally different, perhaps sterol or carotenoid. Elucidation of the function of BmStart1 in lipid transport will provide insights into the functional evolution of the genes homologous to CBP.

24.5 CONCLUDING REMARKS AND FUTURE PERSPECTIVES

CBP has been purified from silk gland and has been successfully shown, by utilizing the genetic resources and technologies developed for sericulture, to have functional significance in carotenoid transport. The study of CBP demonstrates that identification of CBPs indeed will be a powerful way to dissect the transport system for carotenoids in each organism. To date, several CBPs have been identified from various species and tissues (Bhosale and Bernstein 2007). Genetic analysis of these proteins might also establish their functional significance in each transport system of carotenoids.

We consider that there must be CBPs other than CBP in the silkworm. Lipophorin binds carotenoids (Tsuchida et al. 1998). Fujii et al. isolated a CBP of 60 kDa from the larval hemolymph, and using immunoblotting analysis showed that this protein is expressed in both hemolymph and midgut (Fujii et al. 1988a,b). The lack of cross-reaction of their antibodies and the difference in tissue distribution revealed that LBP and CBP definitely are different molecules (Jouni and Wells 1996, Tabunoki et al. 2002). Furthermore, two independent yellow fractions were omitted in the process of purification of CBP (Tabunoki et al. 2002). Mobilizing hydrophobic carotenoids in the inter- and intracellular aqueous environment may be a substantial task for an organism, and many CBPs may therefore work in a coordinated manner. Further identification of CBPs from the silkworm will be an effective way to clarify the transport system of carotenoids.

Lipophorin acts as a reusable shuttle between the membrane-bound lipophorin receptors in tissues (Tsuchida and Wells 1990, Gopalapillai et al. 2006) and is not generally endocytosed in the cells (Law and Wells 1989, Arrese et al. 2001, Canavoso et al. 2001). Thus, the intracellular CBP alone seems not to be able to pick up carotenoid from the lipophorin that resides outside of the cell. Cell surface components are thought to be necessary to allow intracellular delivery of carotenoids (Figure 24.6, magnification) (Arrese et al. 2001). The lipid transfer particle (LTP) (Blacklock and Ryan 1994, Tsuchida et al. 1997) on the outer surface of membranes and unknown membrane-spanning factors that specifically transfer carotenoids might be candidates.

Although there are several mutations that affect carotenoid transport, no gene except for the *Y* gene has been identified. In recent years, the whole genome sequence database (Mita et al. 2004, Xia et al. 2004), EST database (Mita et al. 2003), BAC library (Suetsugu et al. 2007), microsatellite markers (Miao et al. 2005), and SNP markers (Yamamoto et al. 2006) of the silkworm have been developed, which will enable us to identify the gene by a forward genetic approach. Positional cloning of the genes responsible for carotenoid transport has been achieved in *D. melanogaster*, a well-established model organism for molecular genetics (von Lintig et al. 2005). A blindness mutant of *D. melanogaster*, NinaD, which has a defect in the cellular uptake of carotenoids, was shown to encode a class B scavenger receptor (Kiefer et al. 2002). A paralog of NinaD, SANTA MARIA (Wang et al. 2007), and its mammalian homologs, SR-BI and CD36 (Reboul et al. 2005, van Bennekum et al. 2005, During and Harrison. 2007), were subsequently shown to be involved in carotenoid uptake. Identification of other silkworm genes by the forward genetic approach will also enhance our understanding of the molecular system of carotenoid transport in animals.

Comparison of the genomic sequence of CBP between the *Y* and $+^Y$ allele strains (Figure 24.5b) not only reveals the function of the CBP gene but also enables us to guess the process by which the cocoon color has changed through the history of sericulture. The deletion in the CBP gene of the $+^Y$ allele strain suggests that the *Y* allele, rather than $+^Y$, would be the ancient form of *B. mori*, and yellow cocoons rather than white are believed to be the original form of *B. mori*. Humans might have found and selected one or more white cocoon(s) from among the mass of yellow cocoons and spread the allele by selective breeding. This hypothesis is supported by the yellowish color of the *B. mandarina*, a putative ancestral species of *B. mori*. Analysis of more strains and more genes will provide the details of the history of cocoon color.

Besides its biological significance, the silkworm has economic value. Silk has been a major natural fiber used in textile production for millennia. By utilizing CBP, coloration of a natural fiber by transport of a natural pigment based on molecular genetic engineering has been achieved (Sakudoh et al. 2007). Determination of other genes for cocoon color may lead to the ability to produce custom-colored silks, which may have an impact on the textile industry.

ACKNOWLEDGMENTS

The authors thank Drs. T. Tamura, K. Yamamoto, and Y. Banno for helpful suggestions and technical assistance. This work was supported by grants from Teimei Empress Memorial Foundation, the Futaba Electronics Memorial Foundation, and Grant-in-Aid for Scientific Research of Japan Society for the Promotion of Science.

REFERENCES

Alpy, F. and Tomasetto, C. 2005. Give lipids a START: The StAR-related lipid transfer (START) domain in mammals. *J. Cell Sci.*, 118(Pt 13):2791–2801.

Alpy, F. and Tomasetto, C. 2006. MLN64 and MENTHO, two mediators of endosomal cholesterol transport. *Biochem. Soc. Trans.*, 34(Pt 3):343–345.

Alpy, F., Stoeckel, M. E., Dierich, A. et al. 2001. The steroidogenic acute regulatory protein homolog MLN64, a late endosomal cholesterol-binding protein. *J. Biol. Chem.*, 276(6):4261–4269.

Arrese, E. L., Canavoso, L. E., Jouni, Z. E., Pennington, J. E., Tsuchida, K., and Wells, M. A. 2001. Lipid storage and mobilization in insects: Current status and future directions. *Insect Biochem. Mol. Biol.*, 31(1):7–17.

Banno, Y., Fujii, H., Kawaguchi, Y. et al. 2005. *A Guide to the Silkworm Mutants 2005*. Fukuoka, Japan: Silkwork Genetics Division, Institute of Genetic Resources, Kyushu University.

Bhosale, P. and Bernstein, P. S. 2007. Vertebrate and invertebrate carotenoid-binding proteins. *Arch. Biochem. Biophys.*, 458(2):121–127.

Blacklock, B. J. and Ryan, R. O. 1994. Hemolymph lipid transport. *Insect Biochem. Mol. Biol.*, 24(9):855–873.

Brown, M. S. and Goldstein, J. L. 1986. A receptor-mediated pathway for cholesterol homeostasis. *Science*, 232(4746):34–47.

Canavoso, L. E., Jouni, Z. E., Karnas, K. J., Pennington, J. E., and Wells, M. A. 2001. Fat metabolism in insects. *Annu. Rev. Nutr.*, 21:23–46.

Caron, K. M., Soo, S. C., Wetsel, W. C., Stocco, D. M., Clark, B. J., and Parker, K. L. 1997. Targeted disruption of the mouse gene encoding steroidogenic acute regulatory protein provides insights into congenital lipoid adrenal hyperplasia. *Proc. Natl. Acad. Sci. U.S.A.*, 94(21):11540–11545.

Chino, H. 1985. Lipid transport: Biochemistry of hemolymph lipophorin. In *Comprehensive Insect Physiology, Biochemistry, and Pharmacology*, eds. Kerkut G. A. and Gilbert L. I., Oxford: Pergamon Press, 10:115–134.

During, A. and Harrison, E. H. 2007. Mechanisms of provitamin A (carotenoid) and vitamin A (retinol) transport into and out of intestinal Caco-2 cells. *J. Lipid Res.*, 48(10):2283–2294.

Fujii, H., Morooka, J., Tochihara, S., Kawaguchi, Y., and Sakaguchi, B. 1988a. Existence of carotenoids binding protein in larval hemolymph of the yellow blood strain of *Bombyx mori*. *J. Seric. Sci. Jpn.*, 57(2):94–99.

Fujii, H., Matsui, T., Tochihara, S., and Kawaguchi, Y. 1988b. Purification of carotenoids binding protein from larval hemolymph of the yellow blood strain of *Bombyx mori*. *J. Seric. Sci. Jpn.*, 57(5):398–404.

Fujii, H. and Banno, Y. 2005. Silkworm base. http://www.shigen.nig.ac.jp/silkwormbase/about_kaiko.jsp (accessed January 15, 2008).

Gopalapillai, R., Kadono-Okuda, K., Tsuchida, K. et al. 2006. Lipophorin receptor of *Bombyx mori*: cDNA cloning, genomic structure, alternative splicing, and isolation of a new isoform. *J. Lipid Res.*, 47(5):1005–1013.

Hanada, K., Kumagai, K., Yasuda, S. et al. 2003. Molecular machinery for non-vesicular trafficking of ceramide. *Nature*, 426(6968):803–809.

Hanada, K., Kumagai, K., Tomishige, N., and Kawano, M. 2007. CERT and intracellular trafficking of ceramide. *Biochim. Biophys. Acta*, 1771(6):644–653.

Hara, W., Sosnicki, S., and Banno, Y. et al. 2007. Mapping analysis of carotenoid-binding protein of *Bombyx mori* by restriction fragment length polymorphism. *J. Insect Biotechnol. Sericol.*, 76(3):149–154.

Harashima, K., Ohno, T., Sawachika, T., Hidaka, T., and Ohnishi, E. 1972. Carotenoids in orange pupae of the swallowtail, *Papilio xuthus*. *Insect Biochem.*, 2:29–48.

Hidaka, T. 1961. Recherches sur le mécanisme endocrine de l'adaptation chromatique morphologique chez les nymphes de *Papilio xuthus*. *J. Fac. Sci. Tokyo Univ.*, sec IV, 9:223–261.

Hidaka, T., Kimura, T., and Onosaka, M. 1959. Experiments on the protective coloration of pupae of the swallowtail, *Papilio xuthus* L. *Doubutsugaku zasshi*, 68(6):222–226.

Holthuis, J. C. and Levine, T. P. 2005. Lipid traffic: Floppy drives and a superhighway. *Nat. Rev. Mol. Cell. Biol.*, 6(3):209–220.

Hölttä-Vuori, M., Alpy, F., Tanhuanpää, K., Jokitalo, E., Mutka, A. L., and Ikonen, E. 2005. MLN64 is involved in actin-mediated dynamics of late endocytic organelles. *Mol. Biol. Cell.*, 16(8):3873–3886.

Horton, P., Park, K. J., Obayashi, T. et al. 2007. WoLF PSORT: Protein localization predictor, *Nucleic Acids Res.*, 35(Web Server issue):W585–W587.

Imamura, M., Nakai, J., Inoue, S., Quan, G. X., Kanda, T., and Tamura, T. 2003. Targeted gene expression using the GAL4/UAS system in the silkworm *Bombyx mori*. *Genetics*, 165(3):1329–1340.

Jouni, Z. E. and Wells, M. A. 1996. Purification and partial characterization of a lutein-binding protein from the midgut of the silkworm *Bombyx mori*. *J. Biol. Chem.*, 271(25):14722–14726.

Kiefer, C., Sumser, E., Wernet, M. F., and von Lintig, J. 2002. A class B scavenger receptor mediates the cellular uptake of carotenoids in Drosophila. *Proc. Natl. Acad. Sci. U.S.A.*, 99(16):10581–10586.

Kishida, T., Kostetskii, I., Zhang, Z. et al. 2004. Targeted mutation of the MLN64 START domain causes only modest alterations in cellular sterol metabolism. *J. Biol. Chem.*, 279(18):19276–19285.

Kudo, N., Kumagai, K., Tomishige, N. et al. 2008. Structural basis for specific lipid recognition by CERT responsible for nonvesicular trafficking of ceramide. *Proc. Natl. Acad. Sci. U.S.A.*, 105(2):488–493.

Law, J. H. and Wells, M. A. 1989. Insects as biochemical models. *J. Biol. Chem.*, 264(28):16335–16338.

Lin, D., Sugawara, T., Strauss, J. F. 3rd et al. 1995. Role of steroidogenic acute regulatory protein in adrenal and gonadal steroidogenesis. *Science*, 267(5205):1828–1831.

Loane, E., Nolan, J. M., O'Donovan, O., Bhosale, P., Bernstein, P. S., and Beatty, S. 2008. Transport and retinal capture of lutein and zeaxanthin with reference to age-related macular degeneration. *Surv. Ophthalmol.*, 53(1):68–81.

Miao, X. X., Xub, S. J., Li, M. H. et al. 2005. Simple sequence repeat-based consensus linkage map of *Bombyx mori*. *Proc. Natl. Acad. Sci. U.S.A.*, 102(45):16303–16308.

Miller, W. L. 2007. Steroidogenic acute regulatory protein (StAR), a novel mitochondrial cholesterol transporter. *Biochim. Biophys. Acta*, 1771(6):663–676.

Mita, K., Kasahara, M., Sasaki, S. et al. 2004. The genome sequence of silkworm, *Bombyx mori*. *DNA Res.*, 11(1):27–35.

Mita, K., Morimyo, M., Okano, K. et al. 2003. The construction of an EST database for *Bombyx mori* and its application. *Proc. Natl. Acad. Sci. U.S.A.*, 100(24):14121–14126.

Nakajima, M. 1963. Physiological studies on the function of genes concerning carotenoid permeability in the silkworm. *Bull. Fac. Agric. Tokyo Univ. Agric. Technol.*, 8:1–80.

Ohnishi, E. 1959. Pigment composition in the pupal cuticles of two colour types of the swallowtails, *Papilio xuthus* L. and *P. protenor demetrius* Cramer. *J. Insect Physiol.*, 3:132–145.

Parker, R. S. 1996. Absorption, metabolism, and transport of carotenoids. *FASEB J.*, 10(5):542–551.

Rao, R. P., Yuan, C., Allegood, J. C. et al. 2007. Ceramide transfer protein function is essential for normal oxidative stress response and lifespan. *Proc. Natl. Acad. Sci. U.S.A.*, 104(27):11364–11369.

Reboul, E., Abou, L., Mikail, C. et al. 2005. Lutein transport by Caco-2 TC-7 cells occurs partly by a facilitated process involving the scavenger receptor class B type I (SR-BI). *Biochem J.*, 387:455–461.

Roderick, S. L., Chan, W. W., Agate, D. S. et al. 2002. Structure of human phosphatidylcholine transfer protein in complex with its ligand. *Nat. Struct. Biol.*, 9(7):507–511.

Romanowski, M. J., Soccio, R. E., Breslow, J. L., and Burley, S. K. 2002. Crystal structure of the *Mus musculus* cholesterol-regulated START protein 4 (StarD4) containing a StAR-related lipid transfer domain. *Proc. Natl. Acad. Sci. U.S.A.*, 99(10):6949–6954.

Roth, G. E., Gierl, M. S., Vollborn, L., Meise, M., Lintermann, R., and Korge, G. 2004. The *Drosophila* gene *Start1*: A putative cholesterol transporter and key regulator of ecdysteroid synthesis. *Proc. Natl. Acad. Sci. U.S.A.*, 101(6):1601–1606.

Sakudoh, T., Sezutsu, H., Nakashima, T. et al. 2007. Carotenoid silk coloration is controlled by a carotenoid-binding protein, a product of the *Yellow blood* gene. *Proc. Natl. Acad. Sci. U.S.A.*, 104(21):8941–8946.

Sakudoh, T., Tsuchida, K., and Kataoka, H. 2005. BmStart1, a novel carotenoid-binding protein isoform from *Bombyx mori*, is orthologous to MLN64, a mammalian cholesterol transporter. *Biochem. Biophys. Res. Commun.*, 336(4):1125–1135.

Schrick, K., Nguyen, D., Karlowski, W. M., and Mayer, K. F. 2004. START lipid/sterol-binding domains are amplified in plants and are predominantly associated with homeodomain transcription factors. *Genome Biol.*, 5(6):R41.

Soccio, R. E. and Breslow, J. L. 2003. StAR-related lipid transfer (START) proteins: Mediators of intracellular lipid metabolism. *J. Biol. Chem.*, 278(25):22183–22186.

Stocco, D. M. 2001. StAR protein and the regulation of steroid hormone biosynthesis. *Annu. Rev. Physiol.*, 63:193–213.

Suetsugu, Y., Minami, H., Shimomura, M. et al. 2007. End-sequencing and characterization of silkworm (*Bombyx mori*) bacterial artificial chromosome libraries. *BMC Genomics*, 8:314.

Swanton, C., Marani, M., Pardo, O. et al. 2007. Regulators of mitotic arrest and ceramide metabolism are determinants of sensitivity to paclitaxel and other chemotherapeutic drugs. *Cancer Cell*, 11(6):498–512.

Tabunoki, H., Sugiyama, H., Tanaka, Y. et al. 2002. Isolation, characterization, and cDNA sequence of a carotenoid binding protein from the silk gland of *Bombyx mori* larvae. *J. Biol. Chem.*, 277(35):32133–32140.

Tabunoki, H., Higurashi, S., Ninagi, O. et al. 2004. A carotenoid-binding protein (CBP) plays a crucial role in cocoon pigmentation of silkworm (*Bombyx mori*) larvae. *FEBS Lett.*, 567(2–3):175–178.

Tamura, T., Thibert, C., Royer, C. et al. 2000. Germline transformation of the silkworm *Bombyx mori* L. using a *piggyBac* transposon-derived vector. *Nat. Biotechnol.*, 18(1):81–84.

Tazima, Y. 1964. *The Genetics of the Silkworm.* London: Logos Press.

Toyama, K. 1906. Studies on the hybridology of insects. I. On some silkworm crosses, with special reference to Mendel's law of heredity. *Bull. Coll. Agr. Tokyo Imp. Univ.*, 7:259–393.

Toyama, K. 1912. On the varying dominance of certain white breeds of the silk-worm, *Bombyx mori*, L. *Mol. Genet. Genomics*, 7(1):252–288.

Tsuchida, K. and Wells, M. A. 1990. Isolation and characterization of a lipoprotein receptor from the fat body of an insect, *Manduca sexta. J. Biol. Chem.*, 265(10):5761–5767.

Tsuchida, K., Soulages, J. L., Moribayashi, A., Suzuki, K., Maekawa, H., and Wells, M. A. 1997. Purification and properties of a lipid transfer particle from *Bombyx mori*: Comparison to the lipid transfer particle from *Manduca sexta. Biochim. Biophys. Acta*, 1337(1):57–65.

Tsuchida, K., Arai, M., Tanaka, Y. et al. 1998. Lipid transfer particle catalyzes transfer of carotenoids between lipophorins of *Bombyx mori. Insect Biochem. Mol. Biol.*, 28(12):927–934.

Tsuchida, K., Jouni, Z. E., Gardetto, J. et al. 2004a. Characterization of the carotenoid-binding protein of the *Y*-gene dominant mutants of *Bombyx mori. J. Insect Physiol.*, 50(4):363–372.

Tsuchida, K., Katagiri, C., Tanaka, Y. et al. 2004b. The basis for colorless hemolymph and cocoons in the *Y*-gene recessive *Bombyx mori* mutants: A defect in the cellular uptake of carotenoids. *J. Insect Physiol.*, 50(10):975–983.

Tsujishita, Y. and Hurley, J. H. 2000. Structure and lipid transport mechanism of a StAR-related domain. *Nat. Struct. Biol.*, 7(5):408–414.

Uchino, K., Imamura, M., Sezutsu, H. et al. 2006. Evaluating promoter sequences for trapping an enhancer activity in the silkworm *Bombyx mori. J. Insect Biotechnol. Sericology*, 75:89–97.

Uda, H. 1919. On the relations between blood color and cocoon color in silkworms, with special reference to Mendel's law of heredity. *Genetics*, 4(5):395–416.

van Bennekum, A., Werder, M., Thuahnai, S. T. et al. 2005. Class B scavenger receptor-mediated intestinal absorption of dietary beta-carotene and cholesterol. *Biochemistry*, 44(11):4517–4525.

von Lintig, J., Hessel, S., Isken, A. et al. 2005. Towards a better understanding of carotenoid metabolism in animals. *Biochim. Biophys. Acta*, 1740(2):122–131.

Wang, T., Jiao, Y., and Montell, C. 2007. Dissection of the pathway required for generation of vitamin A and for *Drosophila* phototransduction. *J. Cell Biol.*, 177(2):305–316.

Xia, Q., Zhou, Z., Lu, C. et al. 2004. A draft sequence for the genome of the domesticated silkworm (*Bombyx mori*). *Science*, 306(5703):1937–1940.

Yamamoto, K., Narukawa, J., Kadono-Okuda, K. et al. 2006. Construction of a single nucleotide polymorphism linkage map for the silkworm, *Bombyx mori*, based on bacterial artificial chromosome end sequences. *Genetics*, 173(1):151–161.

Yeum, K. J. and Russell, R. M. 2002. Carotenoid bioavailability and bioconversion. *Annu. Rev. Nutr.*, 22:483–504.

Zhang, M., Liu, P., Dwyer, N. K. et al. 2002. MLN64 mediates mobilization of lysosomal cholesterol to steroidogenic mitochondria. *J. Biol. Chem.*, 277(36):33300–33310.

25 Specific Accumulation of Lutein within the Epidermis of Butterfly Larvae

John T. Landrum, Derick Callejas, and Francesca Alvarez-Calderon

CONTENTS

25.1 INTRODUCTION

Carotenoids are abundant phytopigments and are essential to the coloration in many birds, fishes, and insects (Weedon 1971, Kayser 1982). There is growing evidence that non-provitamin A carotenoids are also significant phytonutrients in humans (Krinsky et al. 2005). It has been demonstrated that lutein and zeaxanthin are specifically accumulated in the human retina and function there to protect the retina from oxidative damage (see Chapter 13; Landrum and Bone 2001). Carotenoids present in insect species affect their coloration but evidence also supports the conclusion that they function physiologically as antioxidants and photoprotectants in a manner similar to that ascribed to carotenoids in humans (Felton and Summers 1995, Jenkins et al. 1999, Heller et al. 2000, Carroll and Berenbaum 2002). The coloration patterns of butterflies and moths are often striking in both adult and larval stages and the importance of coloration to the success in butterfly and moth populations is widely recognized (Carroll and Berenbaum 2002). The bright coloration of larval butterflies may also be a factor in predator avoidance; it is known that these organisms can accumulate large quantities of toxic cardenolide glycosides (Rothschild and Mummery 1986, Mebsa et al. 2005).

In adult butterflies, the bright coloration of the wings results from the presence of crystalline pterin pigments within the scales (Britton 1983). In the larval stages, patterns, including yellow coloration such as that found in the banding of the Monarch butterfly larvae (*Danaus plexippus*), are not uncommon, Figure 25.1. Surprisingly, little has been reported about the components

FIGURE 25.1 (See color insert following page 336.) Larval butterflies have distinctive coloration patterns: (a) Monarch (*Danaus plexippus*; yellow, white, and black), (b) Swallowtail (*Papilio polyxenes asterius*; yellow, black, and green), (c) Queen (*Danaus gillipus*; black, white, and yellow, some red coloration is also observable in this specimen), and (d) Atala (*Eurnaceus atala florida*; red with yellow spots).

responsible for coloration patterns in larval butterflies. Many studies of these organisms date from the 1970s or earlier and few have utilized the modern analytical methodology essential for accurate identification and quantitative measurement of carotenoids. In silk worm larvae, it has been reported that the specific uptake of the carotenoid, lutein, is responsible for the characteristic orange coloration observed in the silk (Jouni and Wells 1996, Tabunoki et al. 2002). Carotenoids have also been detected in the epidermis of the parsnip web worm where they may function to protect the organism from exposure to UV light that is capable inducing the formation of toxic $^1O_2$ and/or oxidizing, free radicals (Carroll and Berenbaum 2002). The role of lutein in the pigmentation of the pupal epidermis has been reported for Monarch, Swallowtail, and Peacock butterflies (Ohnishi 1959, Oldroyd 1971, Harashima et al. 1972, 1976, Rothschild et al. 1978, Starnecker 1997, Yamanaka et al. 2004). In Peacock pupae, lutein has been shown to accumulate in the pupal epidermis under the influence of the pupal melanization reducing factor (Starnecker 1997). Similarly, lutein accumulation is reported to be responsible for the shiny "*golden glance*" that is the hallmark characteristic of the *Monarch chrysalis* (Rothschild et al. 1978). This information and the abundant availability of xanthophylls in the diet of the Monarch larvae and those of related species argue in support of the hypothesis that the yellow coloration present in the patterns observed in the epidermis of these larvae results from the accumulation of the carotenoid, lutein. Local, selective accumulation of carotenoids in coloration patterns would be an evidence that these organisms possess a well-developed and carefully regulated mechanism for carotenoid transport and binding, a characteristic shared with the human retina (Landrum and Bone 2004). Consequently, insect carotenoid metabolism may be an important model for the study of the specific accumulation and transport of xanthophylls and may provide an opportunity to use a comparative physiological approach to developing our insight into the carotenoid metabolism within the human retina. In this chapter, we describe the application of modern HPLC, UV-visible, and mass spectrometric analysis of pigmentation in four species of South Florida butterfly larvae with particular emphasis on the abundant Monarch butterfly (*Danaus plexippu*).

25.2 EXTRACTION OF CAROTENOIDS

Butterfly larvae (Monarch, *Danaus plexippus*; Queen, *Danaus gillipus*; Eastern Black Swallowtail, *Papilio polyxenes asterius*; and Atala, *Eurnaceus atala florida*) were collected in South Florida approximately seven to eight days after hatching. The larvae were carefully dissected to remove the gut to prevent the contamination of the epidermis with the intestinal contents. The epidermis was

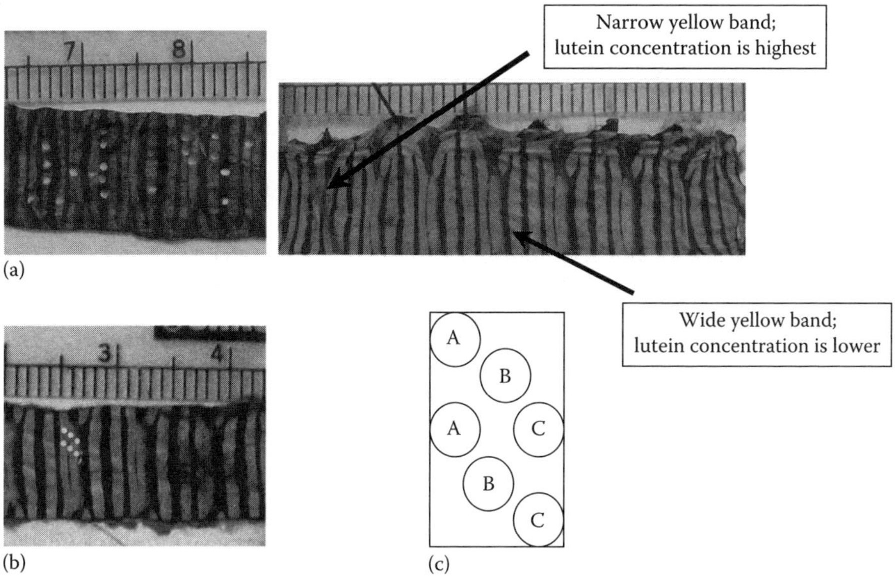

FIGURE 25.2 (See color insert following page 336.) The sampling of colored bands in the epidermis of the Monarch was accomplished using the trephines of different sizes suited for providing samples of separate regions. (a) The dissected epidermis and cuticle showing the complete banding pattern of the Monarch and the positions where punches were obtained from one specimen. (b) Sampling of wider yellow bands using a smaller (0.39 mm) diameter trephine is shown. (c) Drawing showing how the sampling positions using the small trephine from across the wide yellow band were combined.

rinsed in normal saline and fat bodies were removed from the internal surface. Punches taken from the different regions of coloration using trephines made from stainless steel tubing (diameters, ranging from 0.390 to 1.05 mm) enabled the accurate sampling of separate regions. Samples of the fresh, food-plant foliage were also collected, extracted for analysis, and compared to epidermis extracts.

Punches were taken from the three different colored bands (yellow, white, and black) of Monarch larvae, see Figure 25.2, and similar methods were applied to all species studied. The trephine size was chosen to enable the exclusive sampling of the desired region without contamination from neighboring bands. Typically, 2–8 punches with a combined area ranging from 0.36 to 1.73 mm^2 were pooled, combined with a known amount of an internal standard (monopropyl lutein), and homogenized with acetone in a tissue homogenizer. The resulting solution was filtered through a 0.2 μm nylon filter, reduced in volume using a stream of nitrogen gas, dissolved in 40 μL of ethanol, and analyzed by HPLC and LC–MS.

The foliage of the food plants was ground and the pigments were extracted into warm methanol and saponified in 4% sodium hydroxide. The carotenoids were extracted into dichloromethane, dried, and redissolved in ethanol prior to an analysis by HPLC.

25.3 ANALYSIS OF CAROTENOID EXTRACTS

25.3.1 HPLC

A Waters 2690 Alliance HPLC equipped with a 996 photodiode array and a 896 UV/Vis detector was used for carotenoid analysis. The column (Phenomenex, Torrance, CA) was a 250×4.6 mm Ultracarb 3 μm C-18 stationary phase and elution was carried out isocratically at a flowrate of 1.0 mL/min with 85:15 (v:v) acetonitrile:methanol (HPLC grade) containing 0.1% triethyl amine to prevent on-column carotenoid decomposition.

25.3.2 Mass Spectrometry

LC–MS was carried out using a ThermoSeparations HPLC composed of the Spectra System P4000 pump and the Spectra System AS3000 auto-injector coupled to a ThermoQuest Finnigan navigator mass spectrometer, and was run in the positive atmospheric pressure chemical ionization mode (APCI+). The cone voltage was set to 5 V and the data were collected in both the full-scan mode and single-ion monitoring mode observing the lutein parent $M+1$ (569 *e/z*) and principle $M+1-H_2O$ (551 *e/z*) ions and the parent $M+1$ (391 *e/z*) ion for 3-hydroxy-10′-apo-β-carotenal. The chromatography was carried out using a 150×4.6 mm Luna 3 μm C-18 column (Phenomenex) under elution conditions identical to those described above but without the triethyl amine. (Triethyl amine interferes the carotenoid ionization in the mass sensitive detector.)

25.3.3 Carotenoid Identification

Tissue samples obtained from the different colored regions of the larvae were separately analyzed by HPLC. The white, black, and yellow bands of Monarchs all contained a single, major carotenoid component, lutein (all *E*-3,3′-dihydroxy-β,ε-carotene), Figure 25.3a. The amount of lutein present in the black and white bands was markedly lower (~15x) than that in the yellow bands, see below. A small quantity of 13-*cis*-lutein and zeaxanthin were observed to elute immediately following lutein in the chromatogram and the lutein peak was preceded by a unique metabolite that is formed by the cleavage of lutein, see Section 25.4.

The HPLC analysis of milkweed, the food-plant source for Monarch butterflies, demonstrates that it contains a complex mixture of carotenoids including lutein, several other xanthophylls, xanthophyll epoxides, and β-carotene, Figure 25.3b. There is a component in the leaf extract that is observed to elute near 8 min, which has a typical carotenoid spectrum but is not identical to that of the lutein metabolite observed at near the same retention time in the extracts from larval tissue.

The identity of lutein extracted from samples was confirmed by its retention time, by co-injection with an authentic lutein standard, by the measurement of the characteristic UV-visible absorption spectrum, and mass spectra. Figure 25.4a shows the photodiode array spectrum of the lutein peak, which possesses a principal maximum at 447 nm and the distinctive clear separation of the central, principal maximum from the long wavelength maximum at 475 nm. The ratio of these two peaks as measured from valley to peak is equal to 0.6, consistent with the presence of the ε-ionone ring. The mass spectrum of the lutein peak, Figure 25.4b, shows the 569 *e/z* parent $M+1$ ion and the principle 551 *e/z* ion resulting from the loss of water from the parent ion. In addition to lutein, an unidentified component was detected that eluted prior to lutein in the extracts from larvae. Its spectrum exhibits a single maximum at 435 nm in acetonitrile/methanol, see Figure 25.5a. The UV/visible spectrum, retention time, and mass spectrum of this component are consistent with the structure, 3-hydroxy-10′-apo-β-carotenal, a product of the cleavage of lutein at the 9′-10′ carbon–carbon double bond. The mass spectrum of this component (see Figure 25.5b) has a strong parent ion at 391 *e/z*, consistent with the assignment. This component was not detected in the leaves of the food plant.

25.3.4 Comparison of Carotenoid Content in Different Colored Regions of Larvae

25.3.4.1 Monarchs

The different colored bands within individual animals were found to contain significantly different amounts of lutein. Punches of the black, white, and yellow bands were analyzed from each of nine different animals. Because of the challenge in handling and analyzing the small punches of tissue, three to eight punches were collected, pooled, and analyzed as a single sample. This procedure

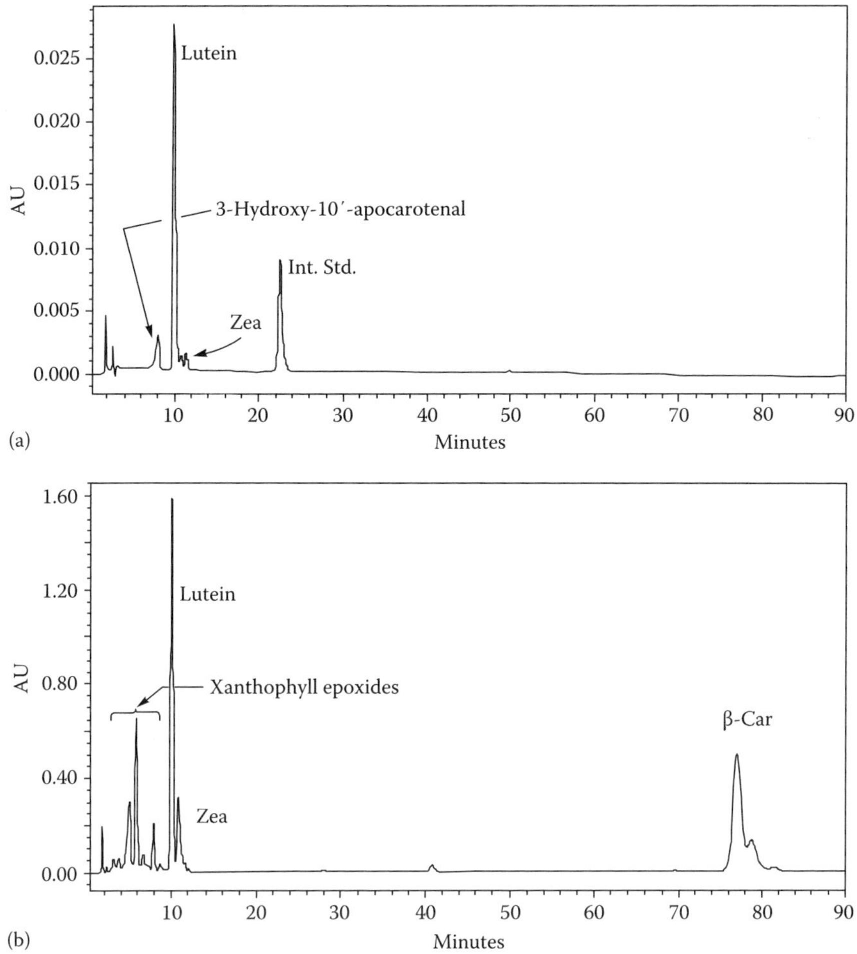

FIGURE 25.3 (a) The HPLC chromatogram of the extract obtained from a yellow-pigmented sample of Monarch epidermis. Peaks seen at 8.7, 10, and 82 min are 3-hydroxy-10′-apo-β-carotenal, lutein, zeaxanthin, and β-carotene, respectively. The peak seen eluting at 22 min is the internal standard, monopropyl lutein ether. (b) The chromatogram obtained from an extract of the leaves of the milkweed plant. Peaks eluting prior to lutein are xanthophylls and epoxy xanthophylls, identified components include lutein, zeaxanthin, β-carotene, and its *cis*-isomer, eluting at 10, 11, 41, 77, and 79 min, respectively.

also ensured that we obtained an average concentration value for each band type in each animal. In only two cases did the yellow bands have less than 10 times the level of lutein found in the other bands, see Figure 25.6. The average concentration of lutein in the yellow bands was 15.4 ± 7.2 pmole/mm^2. The corresponding average concentrations in the black and white regions were, 1.0 ± 0.5 and 1.3 ± 0.7 pmole/mm^2, respectively. Ratios of the carotenoid concentrations present in the separate bands for each animal are a clear evidence that lutein is locally concentrated in yellow versus black or white bands. The average yellow/black and yellow/white ratios are 14.6 ± 7.4 and 15.4 ± 11.2, respectively. The average white/black ratio is near unity, 1.2 ± 0.5.

Some of the yellow bands are appreciably wider than others, ~2 versus 0.5 mm, see Figure 25.2a and b. A gradient in the coloration of the larger band was observable. Through the use of small diameter trephines, 0.39 mm, three punches could be obtained at three different positions across the band as seen in Figure 25.2c. A comparison of these punches, taken from four different animals, showed that the concentration of lutein varies consistently from the front to the back of the band,

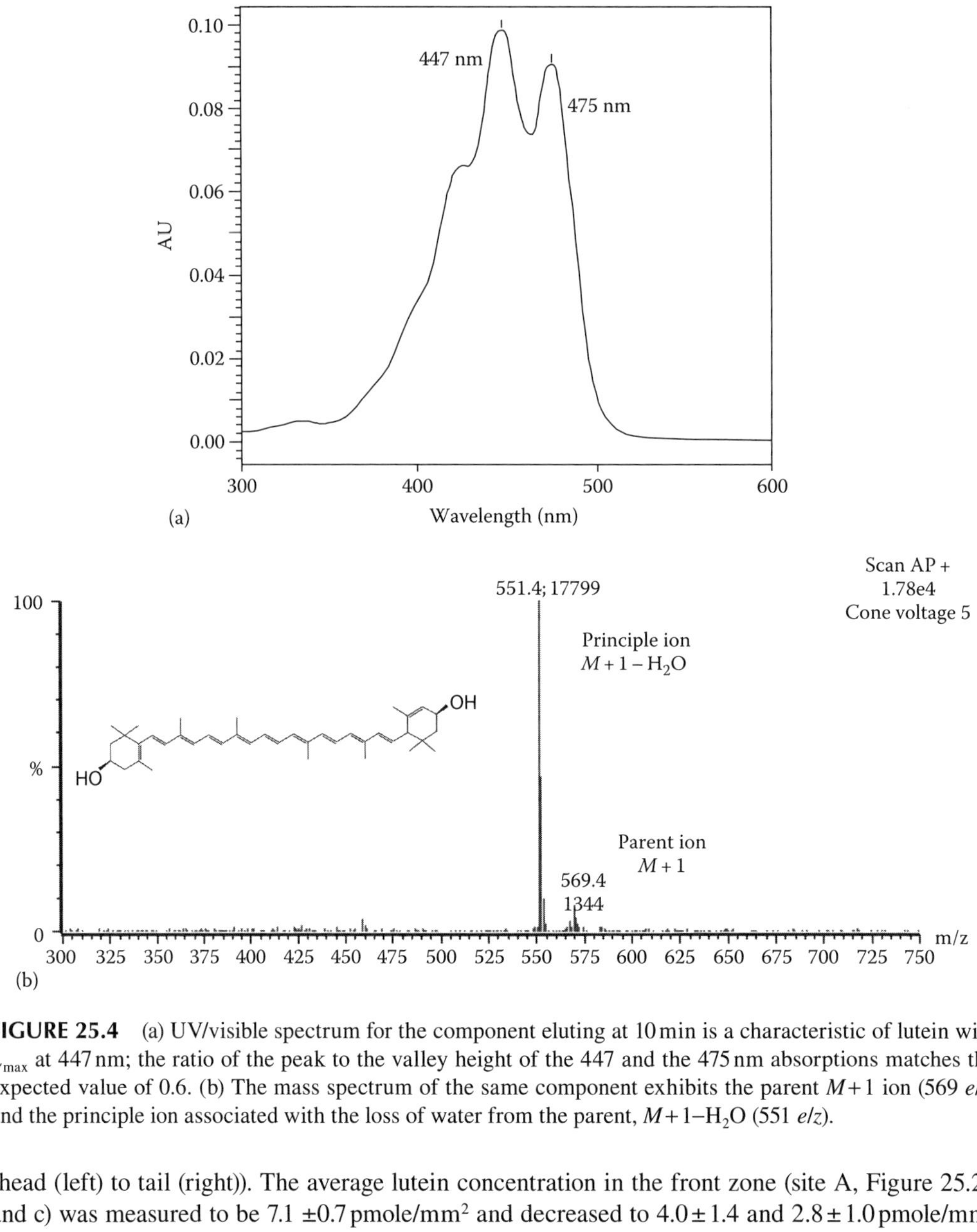

FIGURE 25.4 (a) UV/visible spectrum for the component eluting at 10 min is a characteristic of lutein with λ_{max} at 447 nm; the ratio of the peak to the valley height of the 447 and the 475 nm absorptions matches the expected value of 0.6. (b) The mass spectrum of the same component exhibits the parent $M+1$ ion (569 *e*/*z*) and the principle ion associated with the loss of water from the parent, $M+1–H_2O$ (551 *e*/*z*).

(head (left) to tail (right)). The average lutein concentration in the front zone (site A, Figure 25.2b and c) was measured to be 7.1 ±0.7 pmole/mm^2 and decreased to 4.0 ± 1.4 and 2.8 ± 1.0 pmole/mm^2 at sites B and C, respectively (Figure 25.7). The somewhat lower concentration of lutein observed in these wider bands distinguishes them from that observed from the sampling of narrower yellow bands, see Figures 25.2 and 25.6.

25.3.4.2 Queen, Eastern Black Swallowtail, and Atala Butterflies

Queen butterfly larva (*Danaus gillipus*) belongs to the same genus as the Monarch and shares a similar coloration pattern consisting of yellow, black, and white markings although the shapes differ and there is also a reddish purple pigmentation in the darker pigmented regions, see Figure 25.1c. The Eastern Black Swallowtail (*Papilio polyxenes asterius*) is characterized by green, black, and yellow markings, Figure 25.1b. The small (~1 cm) Atala larvae (*Eurnaceus atala florida*) are uniquely red with seven pairs of intense yellow spots (~1 mm), Figure 25.1d. Figure 25.8 shows a comparison of the amount of lutein found in the various colored regions of these organisms to that

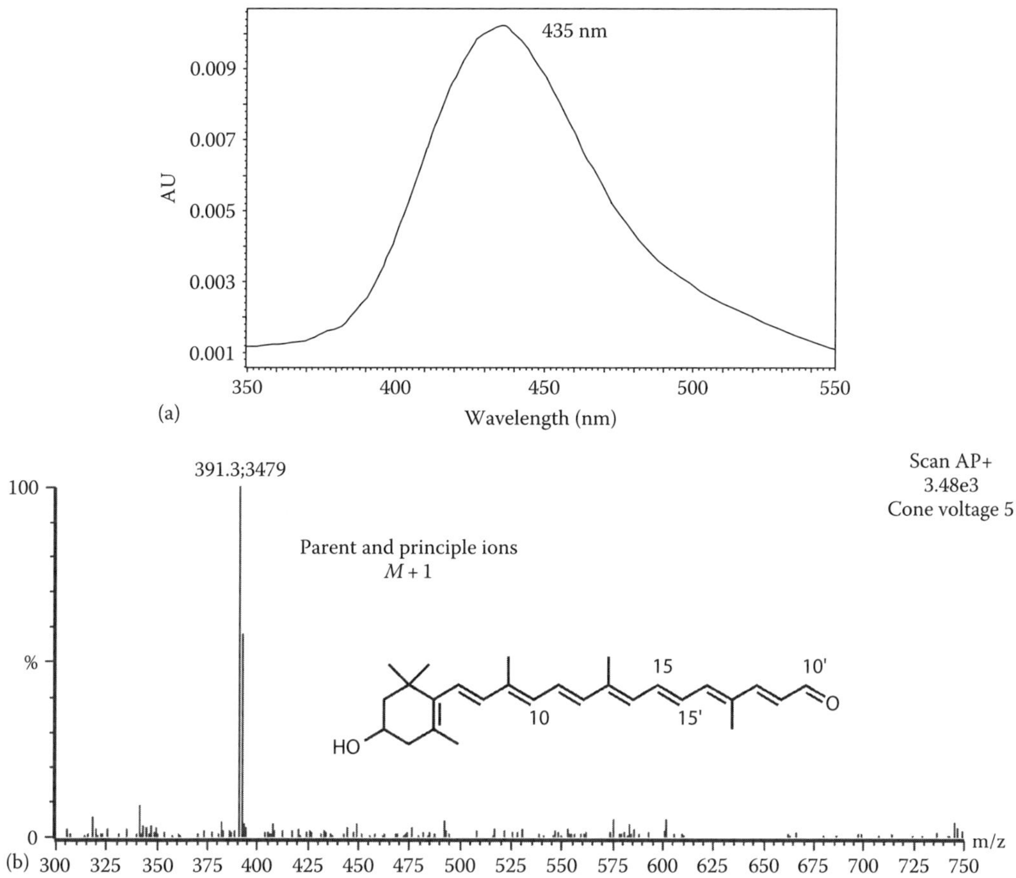

FIGURE 25.5 (a) The UV/visible spectrum of 3-hydroxy-10′-apo-β-carotenal obtained during elution; and (b) the mass spectrum shows a single major peak, the principle ion ($M+1$, 391 *e/z*) matching that of the expected $C_{27}H_{35}O_2$ chemical formula.

found in those of the Monarch. Remarkably, the Atala has much higher concentrations of lutein in all regions than the other species investigated in this study.

25.4 DISCUSSION

The presence of high concentrations of lutein and the striking absence of measurable quantities of other carotenoids in the yellow bands of the Monarch larvae as well as those of other species confirm that the coloration in these yellow pigmented regions results from a localized, specific lutein accumulation. Analysis of food-plant extracts showed that numerous carotenoids, in addition to lutein, are abundant in the diet of these larvae, including zeaxanthin, xanthophyll epoxides, and β-carotene, Figure 25.3b. A comparison of the chromatogram of the leaf extracts with that obtained from the larvae shows that lutein in the epidermis extracts is dramatically enriched relative to zeaxanthin and that the other carotenoids are undetectable. The presence of small but readily measured levels of lutein (but not other carotenoids) in other colored regions suggests that its presence, albeit at low levels, in these regions could serve a photoprotective function throughout the epidermis. Notably, the accumulation of carotenoids by parsnip webworms serves to protect them from photosensitized damage during the exposure to UV light (Carroll and Berenbaum 2002). Considerable data demonstrate that even low levels of carotenoids are effective at protecting the epidermis from light-induced

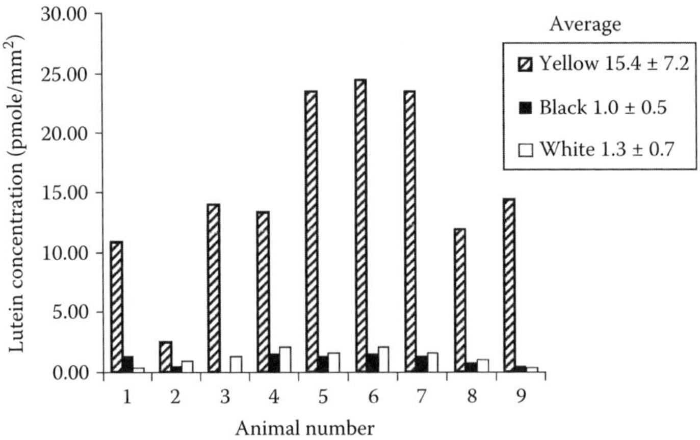

FIGURE 25.6 Amounts of lutein present (pmole/mm^2) in yellow, black, and white epidermal samples from nine individual animals. Each value is the combined result of analysis of between 3 and 8 punches from a single animal providing samples ranging from 0.39 to 1.7 mm^2. The absolute detection limit of the HPLC for carotenoids was 0.1 pmole. The black sample for animal 3 was lost during analysis.

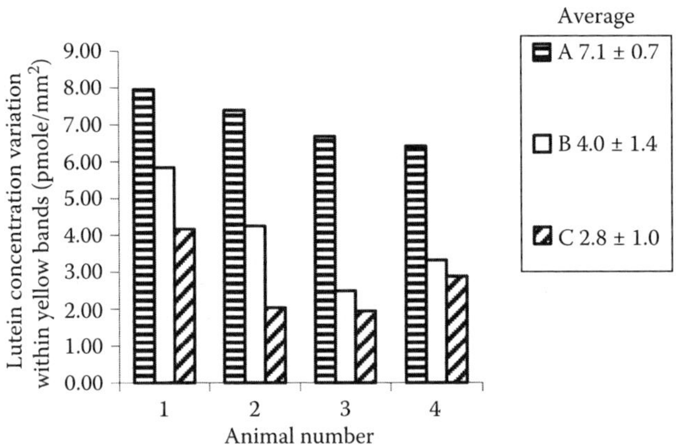

FIGURE 25.7 Analysis of sites within a single band for four separate animals shows a clear gradient in the concentration of lutein from head to tail. The average ratio of head position to middle position (*A*/*B*) is 1.9 ± 0.6 and that of the head to tail position (*A*/*C*) is 2.8 ± 0.9.

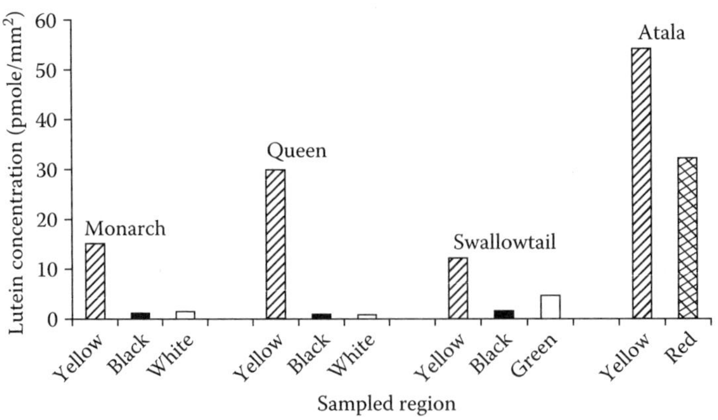

FIGURE 25.8 The relative variation in the concentration of lutein among the colored regions of the four species of butterfly larvae.

damage both in the mouse and human (Mathews-Roth and Krinsky 1970, 1985, Mathews-Roth 1986; Stahl and Sies 2004). The specific, local accumulation of lutein in Monarch and other butterfly larvae demonstrates that a mechanism exists for the active uptake and the transport of the lutein into the yellow regions within these organisms. In total, an average of ~4 mg of lutein is present in the epidermis of a typical Monarch larva. Even higher levels were found in other species, Figure 25.8. The existence and the identity of a lutein-specific binding protein responsible for the epidermal accumulation remains a significant and unanswered question.

The concentration of lutein in the yellow regions in all of these larvae is exceptionally high. By way of comparison, the total amount of carotenoids present in the human epidermis rarely reaches visibly detectable levels, although consumption of high doses of β-carotene or canthaxanthin are known to result in visible skin coloration (Mathews-Roth and Krinsky 1984, 1985, 1987, Prince and Frisoli 1993, Gonzalez et al. 2003, Stahl and Sies 2004). We estimate that the lutein concentration in the yellow bands of Monarch larvae is in the mM range, and thus even in the black and white regions where the concentration is lower (~15x) it is still quite significant. Carotenoids are known to act as the quenchers of singlet oxygen and free radicals at μM levels (Di Mascio et al. 1992). Hence, it is clear that the lutein concentration throughout the epidermis of Monarch larva is more than sufficient to provide an effective protection from photoinduced oxidative damage. A similar conclusion is appropriate for the other species.

The variation of the lutein concentration across the wide yellow bands of Monarch may be related to growth within this tissue. Growth occurs at a remarkable rate for these organisms, and it is probable that the carotenoid concentrations within the rapidly growing regions may lag behind slowly growing regions. Alternatively, the tissue thickness and microstructure, which were not measured, may vary and contribute to the observation.

The chromatograms from the yellow bands of Monarch show the presence of the apo-carotenoid, 3-hydroxy-10′-apo-β-carotenal, in significant, although variable quantities. In some instances, the amount of this component was >20% of the total carotenoid, as judged from the relative peak height. The observation of a single oxidation product corresponding to cleavage at a specific carbon–carbon double bond site is surprising. Multiple products would be expected to result from nonspecific oxidative cleavage of the double bonds present in the polyene chain (see Chapter 11 and Caris-Veyrat et al. 2003). The presence of a single product is suggestive of an enzyme capable of specifically cleaving lutein. It is known that β-carotene mono-oxygenase can cleave lycopene eccentrically at the C10–C9 double bond (Kiefer et al. 2001). A mono-oxygenase capable of cleaving lutein asymmetrically has not been previously reported but would be consistent with our observations. It is well known that the visual system of insects utilizes a 3-hydroxy-retinal whereas the mammalian vitamin A lacks the 3-hydroxyl group. While there may exist a functional role for 3-hydroxy-10′-apo-β-carotenal it is also possible that it is a marker for other metabolic activities. Further work to quantify the amounts, locations, and conditions that correspond to the observation of higher levels of 3-hydroxy-10′-apo-β-carotenal might provide an insight on this question.

The yellow coloration in the Monarch as well as the larva of three other species of butterfly from South Florida is exclusively due to the specific accumulation of exceptionally high levels of lutein producing a pigmented epidermis. This active accumulation, reminiscent of the specific accumulation that occurs in the primate macula, indicates that butterfly larva is an excellent animal model for the study of carotenoid transport and binding. As such, elucidation of the mechanism of transport and binding of lutein in the epidermis and other tissues of these butterfly larvae may provide insight into xanthophyll uptake within the human eye (Bhosale et al. 2004).

ACKNOWLEDGMENTS

The authors wish to acknowledge the FIU College of Arts and Sciences for their partial support of this project, and Myron Georgiardis from the Department of Chemistry and Biochemistry Mass Spectrometry Facility for his assistance.

REFERENCES

Bhosale P., Larson A. J., Frederick J. M., Southwick K., Thulin C. D., and Bernstein P. S. (2004). Identification and characterization of a Pi isoform of glutathione S-transferase (GSTP1) as a zeaxanthin-binding protein in the macula of the human eye. *J Biol Chem* **47**: 49447–49454.

Britton G. (1983). *The Biochemistry of Natural Pigments*. Cambridge, U.K.: Cambridge University Press.

Caris-Veyrat C., Schmid A., Carail M., and Bohm V. (2003). Cleavage products of lycopene produced by *in vitro* oxidations: characterization and mechanisms of formation. *J Agric Food Chem* **51**(25): 7318–7325.

Carroll M. J. and Berenbaum M. R. (2002). Behavioral responses of the parsnip webworm to host plant volatiles. *J Chem Ecol* **28**: 2191–2201.

Di Mascio P., Sandquist A., Devasagayam T., and Sies H. (1992). Assay of lycopene and other carotenoids as singlet oxygen quenchers. *Methods Enzymol* **213**: 429–438.

Felton G. W. and Summers C. B. (1995). Antioxidant systems in insects. *Arch Insect Biochem Physiol* **29**: 187–197.

Gonzalez S., Astner S., An W., Goukassian D., and Pathak M. A. (2003). Dietary lutein/zeaxanthin decreases ultraviolet B-induced epidermal hyperproliferation and acute inflammation in hairless mice. *J Invest Dermatol* **121**: 399–405.

Harashima K., Ohno T., Sawachika T., Hidaka T., and Ohnishi E. (1972). Carotenoids in orange pupae of the swallowtail, *Papilio xuthus. Insect Biochem* **2**: 29–48.

Harashima K., Nakahara J.-I., and Kato G. (1976). Papilioerthrinone: A new ketocarotenoid in integuments of orange pupae of a swallowtail, *Papilio xuthus*, and carapaces of a crab, *Paralithodes brevipes* (*Hanasakigani* in Japanese). *Agr Biol Chem* **40**: 711–717.

Heller K.-G., Fleischmann P., and Lutz-Roder A. (2000). Carotenoids in the spermatophores of bushcrickets (Orthoptera: *Ephippigerinae*). *Proc R Soc Lond B* **267**: 1905–1908.

Jenkins R. L., Loxdale H. D., Brookes C. P., and Dixon A. F. G. (1999). The major carotenoid pigments of the grain aphid, *Sitobion avenae* (F.) (Hemiptera: *Aphididae*). *Physiol Entomol* **24**: 171–178.

Jouni Z. and Wells M. A. (1996). Purification and partial characterization of a lutein-binding protein from the midgut of the silkworm *Bombyx mori. J Biol Chem* **271**: 14722–14726.

Kayser H., Ed. (1982). Carotenoids in insects. In *Carotenoid Chemistry and Biochemistry*. New York: Pergamon Press.

Kiefer C., Hessel S., Lampert J. M., Vogt K., Lederer M. O., Breithaupt D. E., and von Lintig J. (2001). Identification and characterization of a mammalian enzyme catalyzing the asymmetric oxidative cleavage of provitamin A. *J Biol Chem* **276**(17): 14110–14116.

Krinsky N. I., Mayne S. T., and Sies H. (2005). Carotenoids: Looking forward. In *Carotenoids in Health and Disease*. N. I. Krinsky, S. T. Mayne, and H. Sies (Eds.). New York: Marcel Dekker, pp. 547–551.

Landrum J. T. and Bone R. A. (2001). Lutein, zeaxanthin, and the macular pigment. *Arch Biochem Biophys* **385**: 28–40.

Landrum J. T. and Bone R. A., Eds. (2004). Mechanistic evidence for eye diseases and carotenoids. In *Carotenoids in Health and Disease*. New York: Marcel Dekker.

Mathews-Roth M. M. (1986). Carotenoids quench evolution of excited species in epidermis exposed to UV-B (290–320 nm) light. *Photochem Photobiol* **43**: 91–93.

Mathews-Roth M. M. and Krinsky N. I. (1970). Studies on the protective function of the carotenoid pigments of Sarcina lutea. *Photochem Photobiol* **11**(6): 419–428.

Mathews-Roth M. M. and Krinsky N. I. (1984). Effect of dietary fat level on UV-B induced skin tumors, and anti-tumor action of beta-carotene. *Photochem Photobiol* **40**(5): 671–673.

Mathews-Roth M. M. and Krinsky N. I. (1985). Carotenoid dose level and protection against UV-B induced skin tumors. *Photochem Photobiol* **42**(1): 35–38.

Mathews-Roth M. M. and Krinsky N. I. (1987). Carotenoids affect development of UV-B induced skin cancer. *Photochem Photobiol* **46**(4): 507–509.

Mebsa D., Reussa E., and Schneider M. (2005). Studies on the cardenolide sequestration in African milkweed butterflies (Danaidae). *Toxicon* **45**: 581–584.

Ohnishi E. (1959). Pigment composition in the pupal cuticles of two colour types of the swallowtails, *Papilio xuthus* L. and *P. protenor demetrius* Cramer. *J Inst Physiol* **3**: 132–145.

Oldroyd S. M. (1971). Biochemical investigations on various forms of some *Papilio* species. *Entomologist* **104**: 111–123.

Prince M. R. and Frisoli J. K. (1993). Beta-carotene accumulation in serum and skin. *Am J Clin Nutr* **57**: 175–181.

Rothschild M. and Mummery R. (1986). Carotenoids of butterfly models and their mimics (*Lep: Papilionidae* and *Nymphalidae*). *Biol J Linn Soc* **28**: 359–372.

Rothschild M., Gardiner B., and Mummery R. (1978). The role of carotenoids in the "golden glance" of *danaid* pupae. *J Zool Lond* **186**: 351–358.

Stahl W. and Sies H. (2004). Carotenoids in systematic protection against sunburn. In *Carotenoids in Health and Disease*. N. I. Krinsky, S. T. Mayne and H. Sies (Eds.). New York: Marcel Dekker, pp. 491–502.

Starnecker G. (1997). Hormonal control of lutein incorporation into pupal cuticle of the butterfly *Inachis io* and the pupal melanization reducing factor. *Physiol Entomol* **22**: 65–72.

Tabunoki H., Sugiyama H., Tanaka Y., Fujii H., Banno Y., Jouni Z. E., Kobayashi M., Sato R., Maekawa H., and Tsuchida K. (2002). Isolation, characterization, and cDNA sequence of a carotenoid binding protein from the silk gland of *Bombyx mori* larvae. *J Biol Chem* **277**: 32133–32140.

Weedon B. C. L. (1971). Occurrence. In *Carotenoids*. O. Isler (Ed.). Basel, Switzerland: Birkhauser-Verlag, pp. 29–59.

Yamanaka A., Imai H., Adachi M., Komatsu M., and Islam A. T. M. F. (2004). Hormonal control of orange coloration of diapause pupae in the swallowtail butterfly, *Papilio xuthus* L. (Lepidoptera: *Papilionidae*). *Zool Sci* **21**: 1049–1055.

Index

F

G

H

I

L

O

P

R

Z